• THE ENCYCLOPEDIA OF •
ANIMALS
MAMMALS • BIRDS • REPTILES • AMPHIBIANS

CONSULTANT EDITORS

Dr Harold G. Cogger
Deputy Director, Australian Museum,
Sydney, Australia

Dr Edwin Gould
Curator of Mammals, National Zoological Park,
Smithsonian Institution, Washington DC, USA

Joseph Forshaw
Research Associate, Department of Ornithology,
Australian Museum, Sydney, Australia

Dr George McKay
School of Biological Sciences
Macquarie University, Sydney, Australia

Dr Richard G. Zweifel
Curator Emeritus, Department of Herpetology and Ichthyology,
American Museum of Natural History, New York, USA

ILLUSTRATIONS BY

Dr David Kirshner

· THE ENCYCLOPEDIA OF ·
ANIMALS
MAMMALS · BIRDS · REPTILES · AMPHIBIANS

FOG CITY PRESS

Published by Fog City Press
814 Montgomery Street
San Francisco, CA 94133 USA

Chief Executive Officer: John Owen
President: Terry Newell
Publisher: Lynn Humphries
Project Coordinator: Helen Cooney
Copy Editors: Lesley Dow, Maureen Colman, Beverly Barnes
Assistant Editor: Veronica Hilton
Picture Research: Annette Crueger, Grant Young,
Terence Lindsey, Esther Beaton
Captions: Ronald Strahan, Tom Grant, Colin Groves,
Terence Lindsey, David Kirshner
Index: Dianne Regtop
Diagrams: David Kirshner, Alistair Barnard
Series Design: Anita Rowney
Design and Art Direction: Denese Cunningham
Cover Design: Heather Menzies, John Witzig
Production Manager: Caroline Webber
Production Coordinator: James Blackman
Sales Manager: Emily Jahn
Vice President International Sales: Stuart Laurence
European Sales Director: Vanessa Mori

Printed by Kyodo Printing Co. (S'pore) Pte Ltd
Printed in Singapore

A WELDON OWEN PRODUCTION

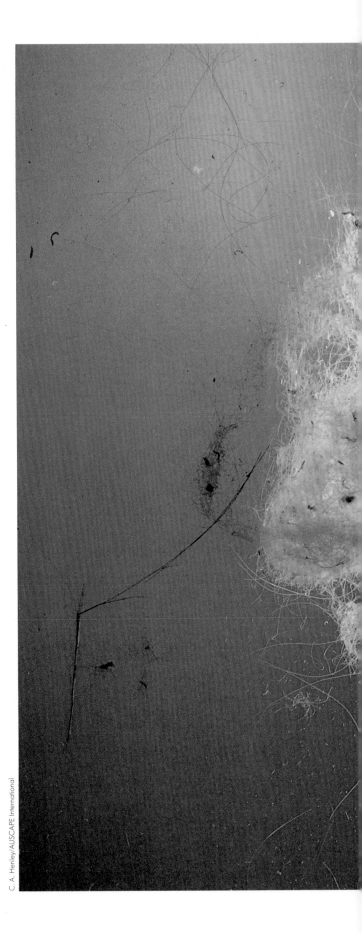

C. A. Henley/AUSCAPE International

Page 1: An ermine or stoat in its summer coat of chestnut with a white underside.
In winter, the entire coat turns white.
Photo by Hans Reinhard/Bruce Coleman Limited
Pages 2–3: Emperor penguins make their way across the frozen landscape of Antarctica.
Photo by Graham Robertson/AUSCAPE International
Pages 4-5: A female spotted grass frog Limnodynastes tasmaniensis with her egg mass.
Pages 12–13: Wild boar forage for food in family parties.
Pages 14–15: Meerkats warm themselves in the morning sun.
Photo by David Macdonald/Oxford Scientific Films
Pages 234–235: A flock of lesser flamingos photographed at Lake Nakuru, Kenya.
Photo by D. Parer & E. Parer-Cook/AUSCAPE International
Pages 454–455: A marine iguana Amblyrhynchus cristatus returns to the shore after grazing on algae.
Photo by Mark Jones/AUSCAPE International

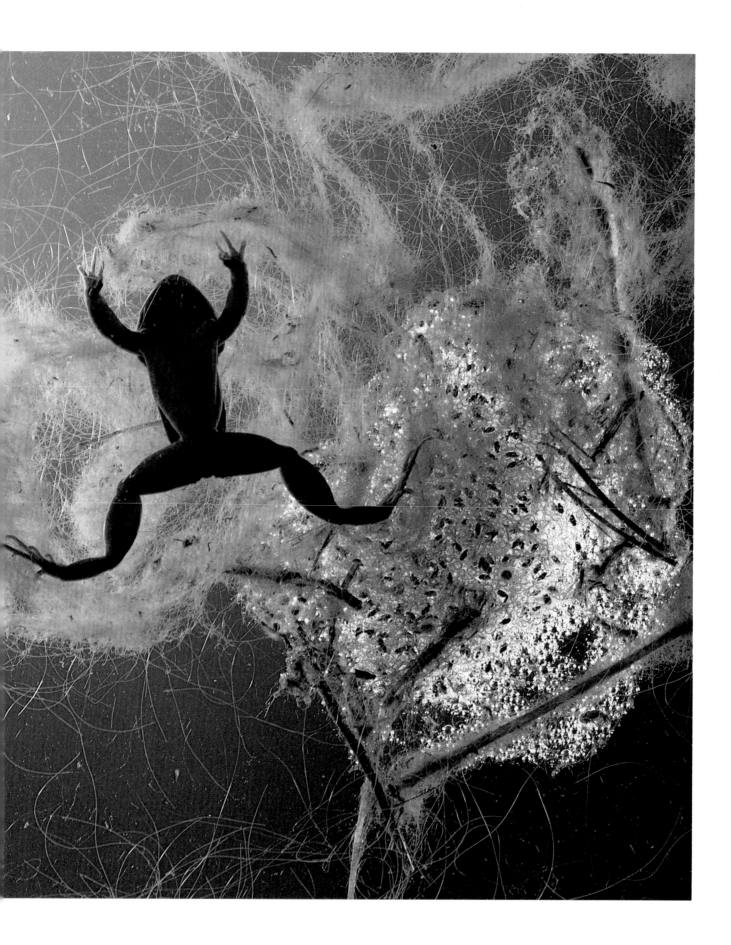

CONSULTANT EDITORS

MAMMALS

Dr Edwin Gould
Curator of Mammals, National Zoological Park,
Smithsonian Institution, Washington D.C., USA

Dr George McKay
School of Biological Sciences,
Macquarie University,
Australia

BIRDS

Joseph Forshaw
Research Assosicate, Department of
Ornithology, Australian Museum, Sydney,
Australia

REPTILES & AMPHIBIANS

Dr Harold G. Cogger
Deputy Director, Australian Museum, Sydney,
Australia

Dr Richard G. Zweifel
Department of Herpetology and Ichthyology,
American Museum of Natural History,
New York, USA

CONTRIBUTORS

Dr George W. Archibald
International Crane Foundation, Wisconsin,
USA

Dr Luis F. Baptista
California Academy of Sciences, San Francisco,
USA

Dr Aaron M. Bauer,
Villanova University, Pennsylvania, USA

Dr Anthony H. Bledsoe
University of Pittsburgh, USA

Dr Walter J. Bock
Columbia University, New York, USA

Dr M.M. Bryden
University of Sydney, Australia

Dr Charles C. Carpenter
University of Oklahoma and Oklahoma
Museum of Natural History, USA

Dr P.A. Clancey
Durban Natural Science Museum, South Africa

Dr Charles T. Collins
California State University, Long Beach, USA

Dr Joel Cracraft
University of Illinois at Chicago, USA

Dr Francis H.J. Crome
Commonwealth Scientific and Industrial
Research Organization, Australia

Dr John P. Croxall
British Antarctic Survey, Cambridge, England

G.R. Cunningham-Van Someren
formerly National Museums of Kenya,
Nairobi

Dr S.J.J.F. Davies
Curtin University of Technology,
Perth, Australia

Dr M.J. Delany
University of Bradford, England

Dr William E. Duellman
Museum of Natural History and University of
Kansas, USA

Dr Jon Fjeldså
Zoological Museum, University of
Copenhagen, Denmark

Dr Hugh A. Ford
University of New England, Armidale,
Australia

Clifford B. Frith
Queensland Museum, Australia

Dr Carl Gans
University of Michigan, USA

Dr Stephen Garnett
Biologist and Scientific Editor, Australia

Dr Valerius Geist
University of Calgary, Canada

Dr Tom Grant
University of New South Wales,
Australia

Dr Brian Groombridge
World Conservation Monitoring Centre,
Cambridge, England

Dr Colin Groves
Australian National University

Dr Collin J.O. Harrison
formerly British Museum (Natural History),
London, England

Dr Harold Heatwole
North Carolina State University, USA

J.E. Hill
formerly British Museum (Natural History),
London, England

Dr Alan Kemp
Transvaal Museum, Pretoria, South Africa

Dr Tom Kemp
University Museum, Oxford, England

Judith E. King
formerly British Museum (Natural History),
London, England

Dr Gordon L. Kirkland, Jr
Shippensburg University,
Pennsylvania, USA

Dr Anne LaBastille
Lecturer, Photographer, and Author, USA

Dr Scott M. Lanyon
Field Museum of Natural History, Chicago, USA

Dr Benedetto Lanza
University of Florence, Italy

Terence Lindsey
Associate of the Australian Museum,
Sydney, Australia

Dr Kim W. Lowe
Arthur Rylah Institute for Environmental
Research, Melbourne, Australia

Dr William E. Magnusson
Instituto Nacional de Pesquisas da Amazonia,
Brazil

S. Marchant
formerly Exploration Manager, Woodside
Petroleum Company, Australia

Dr Helene Marsh
James Cook University, Townsville, Australia

Dr H. Elliott McClure
formerly Walter Reed Army Institute of
Research, Washington, USA

Dr Donald G. Newman
Department of Conservation, Wellington,
New Zealand

Dr Annamaria Nistri
University of Florence, Italy

Dr Ronald A. Nussbaum
Museum of Zoology, University of Michigan,
USA

Dr Fritz Jürgen Obst
State Museum of Zoology, Dresden, Germany

Penny Olsen
Commonwealth Scientific and Industrial
Research Organization, Australia

Dr Norman Owen-Smith
University of Witwatersrand, South Africa

Dr Kenneth C. Parkes
Carnegie Museum of Natural History,
Pittsburgh, USA

Dr Robert B. Payne
Museum of Zoology, University of Michigan,
USA

Dr Christopher Perrins
Edward Grey Institute of Field Ornithology,
Oxford University, England

Dr Michael R.W. Rands
International Council for Bird Preservation,
Cambridge, England

Dr Olivier C. Rieppel
Field Museum of Natural History, Chicago, USA

Ian Rowley
Commonwealth Scientific and Industrial
Research Organization, Australia

Dr Jay M. Savage
University of Miami, Coral Gables,
Florida, USA

E.A. Schreiber
Los Angeles County Natural History Museum,
USA

Dr Richard Shine
University of Sydney, Australia

Dr Lester L. Short
City University of New York, USA

Dr Jeheskel (Hezy) Shoshani
Cranbrook Institute of Science, USA

Professor Peter J.B. Slater
University of St. Andrews, Scotland

Dr G.T. Smith
Commonwealth Scientific and Industrial
Research Organization, Australia

Alison Stattersfield
International Council for Bird Preservation,
Cambridge, England

Dr D. Michael Stoddart
University of Tasmania, Australia

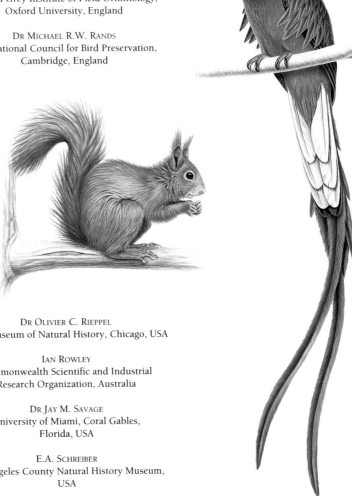

Dr R. David Stone
World Wide Fund for Nature (WWF), Gland,
Switzerland

Ronald Strahan
National Photographic Index of Australian
Wildlife, The Australian Museum, Sydney

Dr Frank Sturtevant Todd
Ecocepts International, San Diego, USA

Dr Stefano Vanni
University of Florence, Italy

Dr Edwin O. Willis
Universidade Estadual Paulista, São Paulo, Brazil

Dr W. Chris Wozencraft
Smithsonian Institution, Washington D.C.,
USA

CONTENTS

PART TWO
BIRDS

THE WORLD OF BIRDS

KINDS OF BIRDS

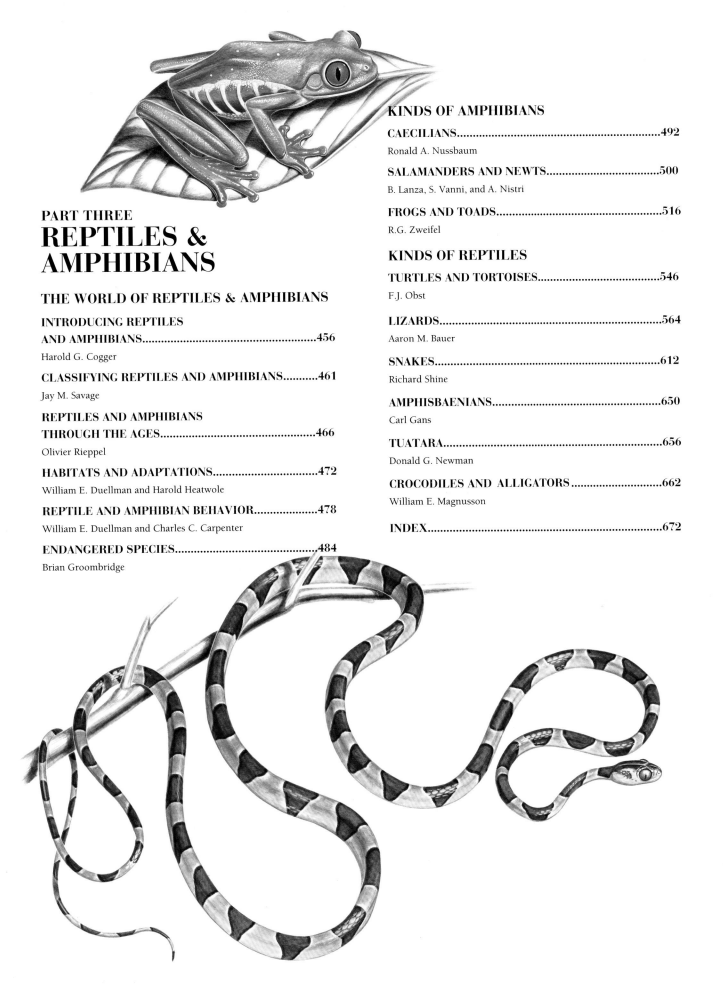

INTRODUCTION

The sheer variety of living animal species is staggering. So too is the diversity of form, behavior, and adaptation to environmental conditions they display. Even in seemingly inhospitable environments—the frozen vastness of the Arctic and Antarctic, the arid heat of deserts, and the rocky bleakness of some alpine regions—countless species of animals have evolved characteristics that enable them not only to survive, but also positively to flourish.

This encyclopedia presents a wide-ranging survey of three major groups of living things. Mammals, which evolved from reptiles more than 200 million years ago, first appeared as mouse-sized insect-eaters, eventually replacing the dinosaurs as the Earth's dominant land-dwellers. They now occupy niches in all environments—land, sea, and air. Birds, characterized by their covering of feathers and, in most species, their capacity for flight, are also descendants of reptiles, with a lineage that stretches back 160 million years. Modern reptiles and amphibians have a history longer than that of either mammals or birds. Their primitive forebears populated the land more than 300 million years ago. Considering the constraints imposed by their limited body forms and their cold-bloodedness, they display an amazing array of life styles and survival strategies.

Common as some of the animals described and illustrated in this encyclopedia are, many of them are vulnerable to extinction. While extinctions are a necessary part of the process of evolutionary change, the rise of that most successful of all predators—human beings—means that species are now disappearing at a rate that far exceeds the evolution of new ones. The destruction of habitats as a result of ever-burgeoning human populations has made life impossible for many kinds of animals. It is a sad, but real, possibility that a number of the animals encountered in these pages may well have disappeared by the end of our century.

It is to be hoped that by reading through this volume readers will gain a greater understanding of the structural diversity and the behavioral richness of animal life. Each chapter opens with a side panel summarizing essential information about the order or group, including number of species, size range, and conservation status, as well as a map outlining their world distribution. The authoritative text is written by a team of acknowledged international experts in the zoological sciences, and is carefully constructed to be read and enjoyed by everyone with an interest in natural history. It is also hoped that readers will gain a heightened appreciation of the urgent need for more effective conservation measures in order to protect these magnificent creatures that are an essential part of the world's environment.

Hans Reinhard/Bruce Coleman Ltd

PART ONE
MAMMALS

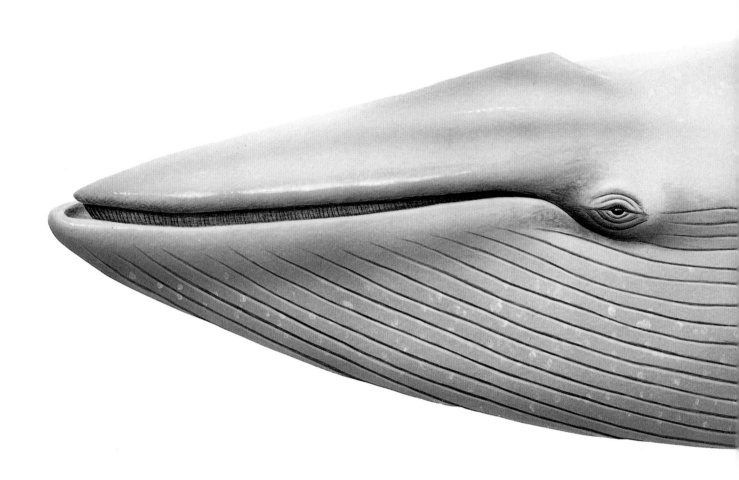

INTRODUCING MAMMALS

G. McKAY

Scientists divide the animal kingdom into several major groups for classification purposes. By far the largest group is the invertebrates: it contains about 95 percent of the millions of known species of animals, including sponges, mollusks, arthropods, and insects. Groups of vertebrates, or animals with backbones, contain the other 5 percent of known species. They can be divided roughly into fishes, amphibians, reptiles, birds, and mammals. The class Mammalia consists of fewer than 5,000 species, but the sheer diversity of mammals is astonishing. From tiny field-mice to the mighty blue whale, from a hippopotamus to a bat, from armadillos to gorillas, the class Mammalia encompasses some of the best-known and most-studied, as well as some of the least-known, members of the animal kingdom. It also includes human beings. We are mammals, classed along with monkeys, lemurs, and apes in the order Primates.

FEATURES THAT LINK MAMMALS

Living mammals are warm-blooded animals that suckle their young on milk and have a body covering of hair or fur, prominent external ears, and a mouth armed with teeth. These features—which, incidentally, are not even shared by all mammals—serve to differentiate mammals from all other living vertebrates. For scientists, however, the single feature that defines mammals is the method by which the dentary bone of the lower jaw—the one that houses the teeth—articulates directly with the skull.

Mammals evolved from a group of carnivorous reptiles, and the most primitive mammals were, like their ancestors, small flesh-eaters. As they evolved, mammals spread and adapted to many different habitats and now occupy a wide variety of niches, from small flying insect-eaters to large terrestrial grazers, and from tiny burrowing flesh-eaters to the largest of living animals, the plankton-feeding whales.

SKULLS, JAWBONES, AND TEETH

The mammal-like reptiles from which mammals evolved are known as the therapsids. Many changes in the skull occurred during the evolution of the therapsids, but the most important concerned the jaws. Early during therapsid evolution the upper jaw became firmly attached to the rest of the skull and lost the independent mobility that can still be seen in such reptiles as

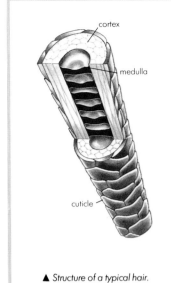

cortex

medulla

cuticle

▲ Structure of a typical hair.

A TYPICAL MAMMAL'S HAIR

Each hair consists of three concentric layers of cells, which develop in the follicle and then die as they are pushed outward. The innermost layer, called the medulla, is made up of the dead remnants of cells that contain air. This increases the insulating ability of the fur and in some cases provides buoyancy when the animal is swimming. Very thin hairs often lack a medulla, as do the wool fibres of sheep.

Surrounding the medulla are the cells of the cortex, which become invaded with fibrils of keratin and which may also contain pigment granules. The outer layer is the cuticle, the individual cells of which resemble microscopic scales. The cuticle cells and medulla cells show a wide variety of patterns which frequently differ between hair types and between species, so a species can often be identified from the structure of its hair.

▼ Like most mammals living in cold climates, the mink has an insulating coat of soft, close-packed hair. For thousands of years, we have used the fur of such animals as insulation for our own almost naked bodies.

snakes. Later, there were significant changes to the composition of the lower jaw: the bones housing the teeth became greatly enlarged and the other bones became smaller, until the dentary became the only bone in the lower jaw. In the earliest mammals, the dentary bones achieved direct contact with the squamosal bone of the skull.

One of the important developments in the mammal-like reptiles was the reduction in the number of bones in the skull and the strengthening of those that remained. The bones around the eye became less important while those surrounding the brain enlarged to accommodate an increasingly large nervous system.

As teeth are the major food-gathering structures of most mammals, they have been subject to a great deal of evolutionary change and have developed a number of distinctive features. They have, for example, different shapes in different parts of the jaw; they are replaced only once, if at all, and not many times as in fish and reptiles; and they are anchored in sockets in the bone.

SKIN AND FUR

The skin of mammals shows a number of characteristics not found in other backboned animals. These include the continuous growth and replacement of the epidermis, the outermost of the three layers of skin; the presence of hairs; and a number of new types of skin glands. The waterproof epidermis houses the hairs and skin glands, but contains no other structures. The second layer, or dermis, contains supporting collagen fibres, blood vessels, a variety of sensory nerve endings, and muscles that can cause the hairs to become erect. Attached to the inner boundary of the dermis is the sub-dermal fatty layer, which provides insulation.

Hairs, which vary greatly in shape and size, grow from the base of follicles and have a complex structure. Many mammals have at least two main hair types in their body covering: long guard hairs, and shorter hairs or underfur. The combination of these two hair types provides efficient thermal insulation by trapping a layer of still air close to the skin. In some large mammals, hair is sparse or absent, and these species rely on the insulating ability of the skin. Each individual hair is subject to wear and has a limited usable life. Replacement of hairs, or molting, may occur continuously or, in colder climates, at specific times of the year. Some species, such as the ermine, have separate summer and winter coats.

Specialized hairs, or "vibrissae", occur at various points on the body, usually on the head but also on the forearms and feet. Vibrissae have specific sensory cells associated with them and provide important tactile information.

Of the different types of glands occurring in the skin, the most important are the mammary glands. These secrete the milk that nourishes the young. Mammary glands may occur in both sexes in placental mammals but are functional only in the female. Females of all mammals possess mammary glands, but projecting teats or nipples occur only in the marsupials and placentals; in the monotremes the glands open in an areolar area which the young can nuzzle with their short beaks.

Sebaceous glands are present in most mammals, in close association with the hair follicles. They produce an oily secretion that serves to lubricate and protect the hairs. Many mammals have sweat glands, which provide evaporative cooling and elimination of metabolic wastes. Sweat glands may be widely distributed over the body, as in humans,

S. Roberts/Ardea London

or restricted to particular areas. They may even be completely lacking. Scent glands are highly specialized structures that produce volatile odorous secretions. These are important as a means of communicating information; in some species, they are important as a means of self-defense.

BACKBONES AND LIMBS

Living mammals have a double contact between the skull and the atlas—the first cervical (neck) vertebra. Most living mammals have seven cervical vertebrae, although this number does vary among sloths and is reduced to six in manatees. In most mammals, ribs are associated only with the chest vertebrae; the exceptions are the monotremes, which have cervical ribs.

The limbs of mammals have retained the basic vertebrate five-fingered plan with few modifications, even in such diverse groups as bats, whose hands have become wings, and whales, whose forelimbs are oar-like. In the hoofed mammals, there has been a progressive reduction in the number of toes from five to two, or even one. The growth of limb bones in mammals differs from that of reptiles in that growth is not indefinite and is confined to two growing points near either end of each long bone.

Monotremes, like a number of early mammal groups, have retained a primitive shoulder girdle with a number of separate bones, but the other living mammals have a shoulder girdle which consists of a scapula (shoulder-blade) and a clavicle (collar-bone). The clavicle is frequently lost in such groups as hoofed mammals, which have evolved extreme flexibility of the limbs to increase their running speed, but is retained and indeed strengthened in those primates that rely on their strong arms for climbing.

The pelvic girdles of mammals consist basically of three fused bones, as in most reptiles, although those of the monotremes and marsupials have an additional bone which provides for the insertion of muscles supporting the ventral body wall.

BRAIN AND SENSES

Mammals have much larger brains, relative to body size, than other vertebrates. The increased brain size is due to an expansion of the cerebral hemispheres. A new component of the cerebrum, the neopallium, which occurs as a very small region in some reptiles, has expanded to provide a covering of gray cells over the ancestral reptilian brain. While all mammals possess a neopallium, the placentals also have a connection between the two cerebral hemispheres: the corpus callosum.

The sense of smell is highly developed in most mammals, although it is less important in many primates and is greatly reduced or absent in cetaceans. The sense of taste, however, is relatively unspecialized. Touch receptors occur widely on the body but are especially important in association with the specialized hairs called the vibrissae.

The mammalian eye is well developed and is basically similar in structure and function to the reptilian eye. Color vision has arisen independently in several mammals, including primates and some rodents, and binocular vision, which permits efficient estimation of distance, is particularly well developed in primates. Many nocturnal mammals have a reflective layer at the back of the eye which increases visual acuity by reflecting weak incoming light back onto the retina.

Hearing is important to most mammals and a number of specializations have evolved, particularly the amplifying function of the middle ear. Most mammals have an external ear that collects sound and concentrates it on the opening to the middle ear. Many mammals can hear sounds of very high pitch, and this has been exploited—by bats, for instance—for echolocation, by which the animal detects obstacles by listening to the reflections of its own sound pulses.

HOW MAMMALS HEAR

The mammalian middle ear is a complex series of membranes and bones that transmit and amplify sound from the outside to the auditory cells in the inner ear. Sound waves impinge on the tympanic membrane, supported by the tympanic bone, and are transmitted via the malleus, incus, and stapes to the oval window of the inner ear. Here they set up an enhanced series of compression waves in the fluid of the inner ear, which in turn excite the sensory cells of the auditory epithelium. An interesting evolutionary phenomenon is the transfer of function of two former articulating bones (the malleus and the incus) from controlling the

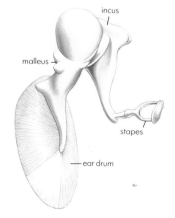

lower jaw movement to assisting and amplifying hearing.

In the more primitive mammals this middle ear structure lies outside the skull, but in more specialized forms it is protected by a bony covering, or bulla, to which the tympanic bone becomes fused.

◄ *The three bones in a mammal's ear: the malleus, incus, and stapes. Two of these, the malleus and incus, form part of the jaw hinge in reptiles.*

▲ *A female meerkat suckling young. Mammals are named after the mammary glands, found only in this group of animals. Milk is a complete food, providing all the nutrients the young need until they are ready to eat by themselves.*

HOW MAMMALS REPRODUCE

Mammalian reproduction is characterized by internal fertilization which results in an amniote egg. The female reproductive system consists of a pair of ovaries and their associated ducts. Under the influence of hormones produced by the pituitary gland, mature ova are released into the fallopian tubes. The control of ovulation may be environmental, as in many highly seasonal breeders, or it may be induced by copulation, as in cats. Males produce sperm in the testes, which are located within the body cavity or outside in a scrotum, where the sperm can develop at a temperature lower than normal body temperature. After copulation and ejaculation, the sperm migrate rapidly to the upper portion of the female fallopian tubes, where they fertilize the ova.

In monotremes the fertilized egg is surrounded in the uterus by a shell membrane and a leathery shell. It is then incubated in the uterus for about two weeks before being laid and further incubated externally. In marsupials a shell membrane, but no shell, is deposited. In placental mammals shell membranes are absent.

The young of marsupials are born in a very undeveloped state after a very short gestation. They attach firmly to the teat of a mammary gland which may or may not be located in a pouch on the mother's belly. In placental mammals gestation is relatively longer and the newborn young are in a more advanced state of development, although in some species they are still naked and helpless. A major trend in the evolution of placental mammals has been the lengthening of the gestation period and the birth of more highly developed young. This is usually associated with the production of fewer young in each litter.

OTHER ORGAN SYSTEMS

In mammals, a muscular diaphragm separates the chest and abdominal cavities. Contraction of the diaphragm, together with the action of the muscles connecting the ribs, draws air into the lungs. The upper respiratory system consists of a trachea which branches out to form the bronchi, leading to the lungs. At the upper end of the trachea is the larynx, where sound is produced.

The circulatory system is dominated by a four-chambered heart, which has a different arrangement from the four-chambered hearts of crocodiles and birds. Separation of pulmonary and systemic circulations is complete in the adult. The red blood cells bind and transport oxygen and carbon dioxide from and to the lungs. A variety of white blood cells perform the housekeeping functions of the body's immune system. All mammals can maintain their body temperature above that of their surroundings, and many are also capable of lowering their body temperature—either on a daily basis or for a prolonged period of torpor or hibernation.

Mammalian kidneys are "kidney-shaped" rather than elongated structures as in other vertebrates. Like many fish, amphibians, and turtles, mammals excrete nitrogenous wastes in the form of urea rather than as uric acid, as most reptiles and all birds do.

The digestive system shows a number of specializations in mammals, particularly among those groups that ferment plant material either in a highly modified stomach or in a cecum, which is a sac occurring at the junction of the small intestine and the proximal colon.

CLASSIFYING MAMMALS

J.E. HILL

There are several million different kinds of plants and animals. The purpose of classification is to provide each of these with a unique name by which it may be known; to describe it so that it may be recognized; and to place it in a formal arrangement or hierarchy that will express its relationship to others. The science of arranging plants and animals in this way is called taxonomy or sometimes systematics.

VERNACULAR NAMES

Many mammals, and indeed other animals and plants, have vernacular or common names, although many do not. These are not used in classification because they may differ from country to country, or from one language to another, and can be ambiguous.

SCIENTIFIC NAMES

Scientific names are always in Latin, partly because this was the common language of the early naturalists, but chiefly because, although now archaic, Latin is universally accepted by scientists as a means of avoiding the problems of translation and ambiguity.

The scientific name of any kind or species of plant or animal always consists of two words: the first or generic name indicates the relationship of the species to others, the second or specific name signifies the particular species concerned. A third or subspecific name may be used to indicate small variants within the species.

Scientific names are often descriptive, drawing attention to some significant feature of their subject, or they may refer to the country or place whence it originated. Sometimes names commemorate a particular person, and some are classical in origin, but they can also be quite arbitrary.

SPECIES

The species is the basic category in classification. About 4,250 species of modern mammals are recognized and although some, such as the giant panda, the lion, or the tiger, are easily seen to be distinct, the majority are much less obviously so and are defined by differences in external form and structure, in the structure of the skull and teeth, or even in such attributes as chromosome pattern and structure. The interpretation of similarities and differences is often subjective, and for this reason the total number of mammalian species varies slightly from one authority to another. Also, from time to time new species are discovered and described, even today.

GENERA

Species with a number of features in common are grouped together in genera, although occasionally a very strongly characterized species may be the sole member of its own genus. Thus the genus *Vulpes* includes a number of species of fox besides the red or common fox *Vulpes vulpes*, but the Arctic fox is considered sufficiently distinct from these and other foxes to be placed by itself in a separate genus as *Alopex lagopus*.

Genera are based on more far-reaching characters than species and are sometimes divided into subgenera to emphasize the features of a particular species or group of species. Again, the interpretation of their characters is subjective and the number of modern genera of mammals is not irrevocably fixed: currently about 1,050 are listed.

FAMILIES

In the same way that genera signify groups of species, or sometimes one strongly characteristic species, so families comprise groups of genera that share common characteristics or similarities: a family may include only one very characteristic genus, on occasion with only a single species.

A family may be divided into two or more subfamilies when its included genera are sufficiently diverse. The New World bat family Phyllostomidae, for instance, encompasses a wide range of genera that fall into several distinct groups, each recognized as a subfamily. There are 132 families of modern mammals.

ORDERS

Families that share major characteristics are grouped into orders. As a rule these are quite distinctive and separated from others by obvious features, but sometimes the distinction may be less immediately apparent. For example, elephants occupy an order of their own, the Proboscidea, which can be easily recognized, but some Australian mammals that belong to the order Marsupialia are very similar in some respects to members of other orders which have, however, a very different reproductive system.

THE RED FOX: AN EXAMPLE OF MAMMAL CLASSIFICATION

The basis of mammalian classification can be illustrated by taking as an example the red or common fox *Vulpes vulpes* of northern Europe, Asia, and northern America. The genus *Vulpes* contains besides this species several other species of fox, and is grouped with other genera of foxes, and with dogs, wolves, and jackals in the family Canidae. Together with a number of other families of carnivorous mammals, such as the Ursidae (bears and pandas), Procyonidae (raccoons), Hyaenidae (hyenas), and Felidae (cats), the Canidae are part of the order Carnivora, which in turn belongs to the infraclass Eutheria, and the subclass Theria.

Class: Mammalia
Infraclass: Eutheria
Subclass: Theria
Order: Carnivora
Family: Canidae
Genus: *Vulpes*
Species: *Vulpes vulpes*

▼ *Red fox cubs of the species Vulpes vulpes. Classification enables us to catalogue animals conveniently and serves to indicate the degree of evolutionary relationship between different species.*

Hans Reinhard/Bruce Coleman Ltd

Most orders include several or many families, but a number besides the Proboscidea contain but one, and one order, the Tubulidentata, has only a single genus and species, the African aardvark or ant bear *Orycteropus afer*. Orders may be subdivided into suborders to emphasize differences among their component families. There are 20 orders of modern mammals, the Rodentia or rodents being the largest, with about 40 percent of known mammal species, followed by the Chiroptera or bats, which have about a further 23 percent.

HIGHER CLASSIFICATION

Mammals belong to the class Mammalia, in which two major divisions are recognized. One, the Prototheria, includes only the spiny anteaters of Australia and New Guinea and the Australian duck-billed platypus, which are unique among mammals in that they reproduce by laying eggs. The other, the Theria, includes all other modern mammals, but itself has two divisions. One, the Metatheria, is reserved for the order Marsupialia, or marsupial mammals, limited today to Central and South America and to Australasia, whose young are born in a very undeveloped state and complete their development in a pouch or fold on the mother's abdomen. The remainder of modern mammals belong to the second group, the Eutheria, whose young develop to a relatively advanced stage within the body of the mother. The Prototheria and Theria are considered to be subclasses of the class Mammalia; the Metatheria and Eutheria are often thought to be infraclasses within the Theria, but their classification is subject to some divergence of opinion.

FOSSIL MAMMALS

Modern mammals constitute a very small part of the total kinds that have existed since mammals first evolved 200 million years ago. The most recent fossils may represent genera or even species known today, and many fossil mammals belong to the same orders and families as their modern counterparts. Many fossils, however, represent totally different mammals that have evolved, thrived, and then become extinct during the long period since mammals first appeared.

CHANGES IN CLASSIFICATION

New ways of assessing the similarities and differences between mammals, or of extending the range of characters upon which classification depends beyond those used traditionally, lead from time to time to changes in classification, especially among species and genera. Also, the interpretation and meaning of their features may differ from one authority to another. For these reasons there is no immutable classification, and although the families and orders of modern mammals are firmly established, in some instances there is no complete agreement as to their relationships.

ORDERS AND FAMILIES OF MAMMALS

The following list is based on *A World List of Mammalian Species* by G.B. Corbet and J.E. Hill, 2nd edition, 1986, Facts on File Publications, New York/British Museum (Natural History), London. Some adjustments have been made within orders in accordance with each author's preferences.

CLASS MAMMALIA

Subclass Prototheria

ORDER
MONOTREMATA — **MONOTREMES**
Tachyglossidae — Spiny anteaters
Ornithorhynchidae — Duck-billed platypus

Subclass Theria

Infraclass Metatheria

ORDER MARSUPIALIA — **MARSUPIALS**
Didelphidae — American opossums
Microbiotheriidae — Colocolos
Caenolestidae — Shrew-opossums
Dasyuridae — Marsupial mice, etc.
Myrmecobiidae — Numbat
Thylacinidae — Thylacine
Notoryctidae — Marsupial mole
Peramelidae — Bandicoots
Thylacomyidae — Rabbit-bandicoots
Vombatidae — Wombats
Phascolarctidae — Koala
Phalangeridae — Phalangers
Petauridae — Gliding phalangers
Burramyidae — Pygmy possums
Tarsipedidae — Honey possum
Macropodidae — Kangaroos, wallabies

Infraclass Eutheria

ORDER EDENTATA — **EDENTATES**
Myrmecophagidae — American anteaters
Bradypodidae — Three-toed sloths
Megalonychidae — Two-toed sloths
Dasypodidae — Armadillos

ORDER INSECTIVORA — **INSECTIVORES**
Solenodontidae — Solenodons
Tenrecidae — Tenrecs, otter shrews
Chrysochloridae — Golden moles
Erinaceidae — Hedgehogs, moonrats
Soricidae — Shrews
Talpidae — Moles, desmans

ORDER SCANDENTIA — **TREE SHREWS**
Tupaiidae — Tree shrews

ORDER DERMOPTERA — **FLYING LEMURS, COLUGOS**
Cynocephalidae — Flying lemurs, colugos

ORDER CHIROPTERA — **BATS**
Pteropodidae — Old World fruit bats
Rhinopomatidae — Mouse-tailed bats
Emballonuridae — Sheath-tailed bats
Craseonycteridae — Hog-nosed bat, bumblebee bat
Nycteridae — Slit-faced bats
Megadermatidae — False vampire bats
Rhinolophidae — Horseshoe bats
Hipposideridae — Old World leaf-nosed bats
Noctilionidae — Bulldog bats
Mormoopidae — Naked-backed bats

Phyllostomidae — New World leaf-nosed bats
Natalidae — Funnel-eared bats
Furipteridae — Smoky bats
Thyropteridae — Disc-winged bats
Myzopodidae — Old World sucker-footed bat
Vespertilionidae — Vespertilionid bats
Mystacinidae — New Zealand short-tailed bats
Molossidae — Free-tailed bats

ORDER PRIMATES — **PRIMATES**
Cheirogaleidae — Dwarf lemurs
Lemuridae — Large lemurs
Indridae — Leaping lemurs
Daubentoniidae — Aye-aye
Loridae — Lorises, galagos
Tarsiidae — Tarsiers
Callitrichidae — Marmosets, tamarins
Cebidae — New World monkeys
Cercopithecidae — Old World monkeys
Hylobatidae — Gibbons
Hominidae — Apes, man

ORDER CARNIVORA — **CARNIVORES**
Canidae — Dogs, foxes
Ursidae — Bears, pandas
Procyonidae — Raccoons, etc.
Mustelidae — Weasels, etc.
Viverridae — Civets, etc.
Herpestidae — Mongooses
Hyaenidae — Hyenas
Felidae — Cats
Otariidae — Sealions
Odobenidae — Walrus
Phocidae — Seals

ORDER CETACEA — **WHALES, DOLPHINS**
Platanistidae — River dolphins
Delphinidae — Dolphins
Phocoenidae — Porpoises
Monodontidae — Narwhal, white whale
Physeteridae — Sperm whales
Ziphiidae — Beaked whales
Eschrichtiidae — Gray whale
Balaenopteridae — Rorquals
Balaenidae — Right whales

ORDER SIRENIA — **SEA COWS**
Dugonidae — Dugong
Trichechidae — Manatees

ORDER PROBOSCIDEA — **ELEPHANTS**
Elephantidae — Elephants

ORDER PERISSODACTYLA — **ODD-TOED UNGULATES**
Equidae — Horses
Tapiridae — Tapirs
Rhinocerotidae — Rhinoceroses

ORDER HYRACOIDEA — **HYRAXES**
Procaviidae — Hyraxes

ORDER TUBULIDENTATA — **AARDVARK**
Orycteropodidae — Aardvark

ORDER ARTIODACTYLA — **EVEN-TOED UNGULATES**
Suidae — Pigs
Tayassuidae — Peccaries
Hippopotamidae — Hippopotamuses
Camelidae — Camels, llamas
Tragulidae — Mouse deer
Moschidae — Musk deer
Cervidae — Deer
Giraffidae — Giraffe, okapi
Antilocapridae — Pronghorn
Bovidae — Cattle, antelopes, etc.

ORDER PHOLIDOTA — **PANGOLINS, SCALY ANTEATERS**
Manidae — Pangolins, scaly anteater

ORDER RODENTIA — **RODENTS**
Aplodontidae — Mountain beaver
Sciuridae — Squirrels, marmots, etc.
Geomyidae — Pocket gophers
Heteromyidae — Pocket mice
Castoridae — Beavers
Anomaluridae — Scaly-tailed squirrels
Pedetidae — Spring hare
Muridae — Rats, mice, gerbils, etc.
Gliridae — Dormice
Seleviniidae — Desert dormouse
Zapodidae — Jumping mice
Dipodidae — Jerboas
Hystricidae — Old World porcupines
Erethizontidae — New World porcupines
Caviidae — Guinea pigs, etc.
Hydrochaeridae — Capybara
Dinomyidae — Pacarana
Dasyproctidae — Agoutis, pacas
Chinchillidae — Chinchillas, etc.
Capromyidae — Hutias, etc.
Myocastoridae — Coypu
Octodontidae — Degus, etc.
Ctenomyidae — Tuco-tucos
Abrocomidae — Chinchilla-rats
Echymidae — Spiny rats
Thryonomyidae — Cane rats
Petromyidae — African rock-rat
Bathyergidae — African mole-rats
Ctenodactylidae — Gundis

ORDER LAGOMORPHA — **LAGOMORPHS**
Ochotonidae — Pikas
Leporidae — Rabbits, hares

ORDER MACROSCELIDEA — **ELEPHANT SHREWS**
Macroscelididae — Elephant shrews

Alistair Barnard

▲ *One of the earliest and most primitive of the "mammal-like reptiles", Dimetrodon grew to about 3 meters in length. It inhabited parts of what is now North America 300 million years ago.*

MAMMALS THROUGH THE AGES

T.S. KEMP

The mammals of the world today are the survivors of a long history that started about 195 million years ago. In rocks of that age, the very first unmistakable mammal fossils occur, as tiny insect-eating animals that look a little like shrews. For almost the first two-thirds of their subsequent history, mammals remained small, inconspicuous animals, probably active only at night. During all this time — the Jurassic and Cretaceous periods of the geologists' time chart — they shared their habitat with the dinosaurs. But when the great extinction of the dinosaurs occurred at the end of the Cretaceous, the land appears to have been opened up for exploitation by mammals. From that moment on, throughout the following 65 million years to the present, many different kinds of mammals, large and small, carnivorous and herbivorous, terrestrial and aquatic, have evolved, flourished, and disappeared, to be replaced by yet newer kinds.

THE ORIGIN OF MAMMALS

Mammals fossilize relatively easily, because their skeletons are generally robust and withstand well the rigors of the fossilization processes. The teeth and jaws in particular are often perfectly preserved, which is doubly fortunate because the exact structure of any particular mammal's teeth tells a great deal about the evolution and biology of that mammal. Consequently the history of the mammals is better known than that of any other comparably sized group of organisms.

Three hundred million years ago the land was populated by primitive amphibians and reptiles, living in and around the extensive wet tropical swamps of the time. Among the reptile fossils, remains of a few forms that were rather larger than the rest, and which had a pair of windows, or temporal fenestrae, in the hind part of their bony skulls, have been found. These windows are still found in a modified way in mammals, indicating that these animals were the first of the "mammal-like reptiles", or synapsids as they are correctly called. The importance of this group is that it includes the animals from which mammals eventually evolved. The early members of the group were not, however, very like mammals, for they still had simple teeth with weak jaws, and clumsy, sprawling limbs; certainly they were not warm-blooded. A well-known example of the early, primitive synapsids is *Dimetrodon,* the finback reptile, which lived over a wide area of what is now North America.

In due course, more advanced kinds of mammal-like reptiles called the Therapsida evolved from *Dimetrodon*-like ancestors. Therapsids had developed more powerful jaw-closing muscles, which they could use with a variety of more elaborate dentitions such as very enlarged canine teeth, or even horny beaks. They also possessed longer, more slender limbs which must have allowed them to run more rapidly and with greater agility in pursuit of their prey, or to escape from predators. Therapsids had probably become more warm-blooded and larger brained as well, although these things are more difficult to tell from fossils.

One particular group of advanced therapsids, called the Cynodontia, had evolved several other distinctly mammalian characteristics. They had much more mammalian-like teeth, with several cusps, which worked by a shearing and crushing action between the upper and the lower teeth. Along with the modifications to the teeth themselves, very powerful yet also very accurate jaw muscles had to evolve to give these teeth greater biting power. This involved a great enlargement of the bone that carried the lower teeth, the dentary bone, so that it could also carry the attachments of the lower jaw muscles. The other bones of the lower jaw, called the postdentary bones, were correspondingly reduced and weakened. In the cynodonts, these postdentary bones had adopted a new function. Along with the old hinge bone or quadrate of the skull, they transmitted sound waves from a rudimentary ear drum, attached to the jaw, to the auditory region of the animal's braincase. In this way, the cynodonts show an approach towards the fully mammalian arrangement where only the dentary bone is left in the lower jaw, and it has

▼ Cynognathus was a carnivorous cynodont, an advanced version of the "mammal-like reptiles". The size of a badger, it had powerful jaws and mammal-like teeth.

Alistair Barnard

formed a new jaw hinge directly with the squamosal bone of the skull. The mammalian postdentary bones and quadrate have lost their contact with the jaw altogether and have become the sound-conducting ossicles of the middle ear.

Cynodonts also evolved a much more mammalian kind of skeleton, with very slender limbs held much closer to the body, which made for more agile locomotion. It is often thought, too, that the cynodonts had evolved many of the physiological characters of mammals, such as full warm-bloodedness, although not everybody accepts this. There is no good evidence either way about whether they were furry.

Sometime around the end of the Triassic period, about 195 million years ago, the first fossils that have a fully developed new jaw hinge between the dentary and the skull are found. They are generally accepted as the earliest true mammals, and undoubtedly they evolved from advanced cynodonts. However, as they possess small postdentary bones and quadrate still attached to the lower jaw, they are very primitive. The best known of these earliest of mammals are the morganucodontids, which have been found in Europe, South Africa, North America, and China, which indicates that they had a worldwide distribution. They were all very small, with a skull length of 2 to 3 centimeters (just under to just over an inch) and an overall body length of around

▲ Morganucodontids, among the earliest of the true mammals, evolved about 195 million years ago. They were very small — about the size of a shrew.

Alistair Barnard

12 centimeters (about 5 inches). Their teeth were sharp and multi-cusped, and, from the wear patterns that developed during life, it seems that they were used to capture and chew insects and perhaps other terrestrial invertebrate prey. To judge from this and from their apparently

enhanced sense of hearing and smell, morganucodontids are believed to have been adapted to a nocturnal hunting existence.

THE MESOZOIC MAMMALS

Morganucodontids and related early mammals were the start of an evolutionary radiation, through the rest of the Mesozoic era, into several different groups. However, without exception they remained very small animals. Even the largest were no larger than a domestic cat, and the great majority were very much smaller than that. Furthermore, most of them remained insectivorous, although one group, the multituberculates, adopted a herbivorous habit and looked superficially like the rodents of today.

The most important evolutionary step during the Mesozoic was the development of a more complex kind of tooth, the tribosphenic molar. These teeth have a triangle of three main cusps on both the upper and the lower teeth, while the lower also have an extended basin at the back end. They were more effective in their action, and in the Late Cretaceous, around 80 million years ago, two main groups of fossil tribosphenic mammals can already be distinguished. These are the marsupials and the placentals, which were soon to be the dominant mammals of the world. Indeed, even before the end of the Cretaceous several different kinds of placentals had already evolved, and thus the scene was set for the great post-dinosaur radiation of mammals.

Why all the Mesozoic mammals should have remained so small is a mystery. Possibly it was simply because they could not compete with the dinosaurs in the habitats appropriate for large terrestrial animals and therefore had to remain restricted to a nocturnal way of life suitable for small animals. Alternatively, the climate may have been such that the physiology of these early mammals was unsuitable for large animals.

There is also still much argument about what caused the extinction of the dinosaurs and the numerous other groups of animals, and to a lesser extent plants, that suffered at this time. But what is clear is that the mammals not only survived, but were for some reason particularly well placed to take advantage of the relatively empty new world set before them.

THE TERTIARY RADIATIONS

The evolutionary story of mammals over the last 65 million years to the present day, that is the Tertiary era, began from the few groups of small, insect-eating mammals that survived the extinction of the end of the Cretaceous. The story is very complicated, because as well as the extinction of the dinosaurs, the end of the Cretaceous also saw the beginning of the break-up of the great land mass into separate pieces that proceeded to drift apart, eventually forming the continents familiar

Alistair Barnard

▲ Uintatherium *was one of the first large herbivores, comparable in size to a present-day African rhinoceros. It had three pairs of bony protruberances on its head.*

today. For a brief time, Australia and South America were connected to one another by Antarctica, but they were shortly to become completely isolated island continents, on which quite different groups of mammals evolved. On the other hand, North America, Europe, and Asia remained effectively connected to one another. A further complicating factor was a series of climatic changes that affected mammal evolution from time to time by causing the extinction of certain groups and allowing other, usually more advanced, groups to survive and flourish. Thus there was an almost continual turnover of species and families as the millions of years passed.

In any event, within no more than 5 million years or so, several important new kinds of mammals had evolved. Looking at the major part of the world, as represented by North America and Eurasia, the early new mammals to evolve from the little insectivorous placental ancestors included the creodonts, which were the first of the large predaceous mammals. These were very like modern carnivores, with specialized shearing molar teeth and powerful sharp-clawed limbs. Several other groups of mammals had evolved into

large, browsing herbivores, with various patterns of large, flattened grinding molars, and elongated legs with little hoofs on the ends of the toes to improve their running ability. Some were quite strange looking, such as the uintatheres of North America, which had bony protuberances over the skull, and *Arsinoitherium* from Egypt with its pair of massive horns. Small herbivores, occupying the niches later adopted by the rodents, were represented, surprisingly, by a group of primitive primates called the plesiadapids.

As time passed, new and more modern groups of animals evolved, although often in rather primitive and unfamiliar form. By about 50 million years ago the true Carnivora, and the earliest of the horse family represented by *Hyracotherium,* the eohippus, had appeared, as had the first whales (though these still had serrated teeth) and other groups later to be highly successful, such as the bats, rodents, artiodactyls, elephants, and lemurs.

Around 40 million years ago, the climate deteriorated for a while, causing the extinction of about one-third of the existing mammalian families. The main sufferers were the more archaic, primitive kinds, and consequently the fauna took

Alistair Barnard

▲ Indricotherium, a hornless form of rhinoceros that lived during the Miocene, attained a height of 5.5 meters (18 feet) at the shoulder and weighed perhaps 20 tonnes (over 40,000 pounds), making it the largest land mammal ever.

on an increasingly modern appearance. This phase was followed by a long period of very favorable climate referred to as the Miocene. It was the heyday of mammalian evolution, when more groups of mammals were present than ever before or since.

One of the most significant new developments of the Miocene was the spread of grasses, to form great plains on which herds of grazing animals could thrive. Horses, and even more so artiodactyls — deers, antelopes and so on — radiated. Although the majority of groups of mammals were those familiar today, they included many bizarre species, particularly among the perissodactyls, such as the chalicotheres, which had longer front than back legs, and Indricotherium, the largest land mammal ever to have lived, which stood over 5 meters at its shoulder. Among the carnivores were a number of large sabre-toothed cats.

During the Miocene the apes also evolved from more primitive primates, and a little later, about 4 million years ago, the first members of Australopithecus had appeared in Africa, as the immediate forerunners of the humans that were to arise 2 million years later.

SOUTH AMERICAN MAMMALS

During most of the Tertiary, South America was an isolated island continent, and consequently the history of its mammal fauna differed from that of the rest of the world. Several unique orders of

herbivorous placental mammals evolved, such as the litopterns which resembled horses but were entirely unrelated to them, and other orders superficially like rhinoceroses, hippopotamuses, and so on. Another important group were the edentates, with such strange animals as the huge armored Glyptodon, and the giant ground sloth Megatherium. This order still survives today and includes some of the oddest mammals, the armadillos and South American anteaters.

The most unexpected feature of South American mammal evolution concerns carnivores, for these consisted exclusively of marsupial mammals, a group absent from the Tertiary of North America, Europe, and Asia. There were smaller forms, such as the didelphids and caenolestids, which have survived to the present. But there were also large predators, the borhyaenids, which played the same role as carnivores in other parts of the world, with forms as large as lions, and even an equivalent to the placental sabre-toothed tiger in the form of Thylacosmilus, which had huge upper canine teeth.

The only placental mammals (apart from bats, which easily migrate worldwide) that managed to invade South America during the earlier Tertiary were rodents and primates. They probably came from North America via the Caribbean Islands, and once established they evolved into the characteristic South American groups, the hystricomorph rodents and the platyrrhine or New World monkeys.

Owing to tectonic changes about 3 million years ago, South America eventually drifted northwards to connect with the North American continent. Because of the ensuing climatic changes, there followed what is often called "the Great American Interchange", which resulted in a dramatic alteration to the South American mammalian fauna. The South American herbivore groups disappeared, to be replaced by invaders from the north: representatives of such modern placental groups as horses, tapirs, llamas, and deer. Similarly, the large marsupial carnivores disappeared and in their place came dogs, cats, bears, and so on. On the other hand, many remnants of the old South American fauna do persist, such as edentates, small marsupials, rodents, and monkeys; indeed several of them invaded North America, like the didelphid opossums and the porcupine.

It is particularly clear in South America how the present mammalian fauna is a result of many processes, evolutionary, biogeographic, and climatic, that together caused a long history of change.

AUSTRALIAN MAMMALS

The history of Australian mammals is hardly known yet because there are practically no known fossil deposits containing mammals until well into the Tertiary. The main difference from elsewhere is that there is no evidence at all that placental mammals reached Australia until late in the Tertiary, and then only as a few rodent and bat species from Asia. All other placental species, such as the dingo and rabbit, are human introductions.

The egg-laying monotremes diverged quite early from the tribosphenic mammals, the marsupials and placentals. Fossil teeth of a primitive monotreme have been discovered in the Cretaceous of Australia, but they never seem to have been a particularly important group.

With these minor exceptions, all Australian mammals, past as well as present, are marsupials. They radiated into many of the niches associated with placental mammals elsewhere in the world, such as the large carnivorous thylacines, the grazing, herd-living kangaroos, and the insectivore-like opossums.

THE PLEISTOCENE EXTINCTIONS

The final drama in the history of mammals occurred only 10,000 years ago, in the late Pleistocene epoch. A phase of extinction is shown by the disappearance from the fossil record of many mammal species, but mainly the largest forms. There had been giant members of many groups throughout the world, such as the familiar mammoths, Irish elk, and the sabre-toothed tigers. Less well known but equally remarkable were such animals as giant apes and wart hogs in Africa, a giant lemur in Madagascar, a giant tapir and ground sloth in South America, and giant kangaroo, wombat, and platypus in Australia. All these became extinct at about the same time.

It is much argued whether this was due to some climatic change that made life more difficult for larger species, or whether it was connected with the rapid spread of humans throughout the world at that time. Perhaps this was the first, though by no means the last, baleful effect of human beings upon the mammals of Earth.

▼ Glyptodon, *closely related to armadillos, was roughly the size of a small motor car and was covered in a rigid bony carapace.*

HABITATS AND ADAPTATION

GORDON L. KIRKLAND JR

Mammals have evolved to successfully occupy nearly every habitat on the face of the Earth. Only the interior of the Antarctic continent remains uninhabited by mammals. To achieve their remarkable success, mammals have evolved various locomotor and feeding adaptations to permit them to survive and prosper in habitats as diverse as tropical rainforests, grasslands, tundra, and deserts. Much of the diversity in the external form of mammals reflects adaptations to specific habitats or environments.

MAMMALS OF THE FOREST

Because trees are a key habitat component of forest-dwelling mammals, particularly as sources of food, shelter, and routes of travel and escape, many mammals of the forest are excellent climbers and exhibit various degrees of specialization for moving through the trees. In more generalized species, such as many mice and rats, adaptation takes the form of long tails, which act as counterbalances as these mammals move along branches. Other more highly specialized arboreal (tree-dwelling) species have prehensile tails that serve as a "fifth hand". Prehensile tails have evolved independently in forest-dwelling representatives of several groups of mammals, including New World opossums, Australian possums and cuscuses, South American anteaters, pangolins or scaly anteaters in Africa and southern Asia, New World monkeys, South American porcupines, and some tropical rats. Opposable big toes, or in some cases thumbs, which aid in grasping branches, are also specializations for movement through trees and have evolved in several groups of primates, New World opossums, Australian possums and cuscuses, the koala, and the noolbenger or honey possum. Gliding as a means of movement in forests has evolved independently in flying squirrels of North America and Eurasia, African scaly-tailed squirrels, colugos or gliding lemurs in the jungles of Southeast Asia, and in members of the Australian gliding marsupials.

Perhaps the most specialized arboreal mammals are sloths, which spend virtually their entire lives in trees, usually coming to the ground only to

▼ Most agile of the arboreal mammals of South America, spider monkeys swing through the trees with their long arms and legs and extremely prehensile tail: they are "five-limbed". However, the even more agile gibbons of Asia have no tail.

Francois Gohier/Auscape International

defecate, on average about once every seven days. Although almost helpless on the ground, sloths are excellent swimmers, an adaptation to crossing the numerous streams and rivers that characterize their lowland forest habitats in the New World tropics. The minimal level of activity so characteristic of sloths is an adaptation to a diet of leaves containing high concentrations of toxic compounds that are manufactured by plants as a defense against herbivores. The high fiber and cellulose content of these leaves further reduces their nutritional value. By coupling minimal activity with a low metabolic rate and a dense coat for a tropical mammal, sloths substantially reduce their food requirements and consequently their intake of toxic chemicals. In addition, sloths meet some of their energy requirements by basking in the sun in the canopy of trees. The koala, which also subsists on a diet high in secondary compounds, possesses many of the same adaptations, including slow deliberate movements and relatively low metabolic rate.

Partridge Productions Ltd/Oxford Scientific Films

▲ The sloths of tropical and subtropical American forests are adapted to a slow way of life. Subsisting entirely on leaves (which are a poor source of food), they have a low metabolic rate, poor temperature control, and very slow movements. These strategies enable them to be the dominant herbivores in their environment.

◄ Many arboreal mammals leap from one tree to another with legs outstretched. It was a simple evolutionary step to change a leap into a glide by developing a web of skin between the body and the legs — as in this yellow-bellied glider, an Australian marsupial. Gliding has evolved independently in three marsupial families, two rodent families, and the flying lemurs or colugos.

Jean-Paul Ferrero/Auscape International

Richard Matthews/Planet Earth Pictures

▲ *The maned wolf has extremely long legs, employed in a bounding gait when pursuing prey — usually through tall grass. It is only distantly related to the true wolves and does not associate in packs.*

▼ *Ground squirrels are a major source of food for many carnivorous mammals, birds, and snakes. Out of their dens, they live in a constant state of nervous alertness. They benefit from living in social groups that provide multiple sense organs to detect predators.*

Francois Gohier/Auscape International

GRASSLAND MAMMALS

In contrast to forest-dwelling species, mammals living in grasslands often rely on speed, rather than concealment, as a method of escape. The long legs of hoofed mammals of the open plains function to increase their speed as a principal means of predator avoidance, as well as to permit them to efficiently travel long distances in search of food and water. One of the most striking grassland mammals is the maned wolf, a South American carnivore that has evolved extremely long legs to permit it to see over and move with ease through the tall grasses of the pampas.

The open nature of grasslands has promoted the evolution of social behavior in many species of grassland mammals, such as large ungulates (hoofed mammals), kangaroos, wallabies, ground squirrels, prairie dogs, baboons, and banded mongooses. Individuals in social groups benefit from multiple sense organs (ears, eyes, and noses) which aid in the detection of potential predators at greater distances than would be possible alone. Herding behavior in grassland ungulates is strongly reinforced by predators who selectively prey on individuals that do not remain part of the herd. Social predators such as lions and cheetahs are also characteristic of grassland habitats.

Many smaller grassland mammals are adapted for burrowing and utilize their burrow systems to escape predators and to provide relief from high temperatures during mid-summer days. Numerous members of the squirrel family, such as North American ground squirrels and prairie dogs, are semi-fossorial, meaning that they utilize extensive burrow systems for shelter and nest sites, but forage on the surface for herbaceous plants and seeds. Several groups of rodents, such as New World pocket gophers, South American tuco-tucos, and African mole-rats, have abandoned life on the surface to exploit rich subterranean plant resources in the form of roots, bulbs, and tubers. Pocket gophers and tuco-tucos utilize powerful forelimbs in burrowing, whereas African mole-rats use their large incisors to excavate burrows. These burrowing or fossorial rodents are characterized by short coats, small ears, small eyes, and short tails.

MAMMALS OF THE TUNDRA

Mammals of the Arctic tundra possess numerous adaptations to permit them to survive the physical rigors of this region. The immense shaggy coats of musk oxen mark this species as a true Ice Age survivor and one of the best adapted of Arctic mammals. One of the most familiar adaptations of Arctic mammals is the ability of some species to turn white in winter. The white winter coat of Arctic hares, Arctic foxes, and collared lemmings provides both camouflage in a white world and excellent insulation.

Locomotion during Arctic winters is often difficult, and tundra mammals possess several distinctive adaptations to increase efficiency. The surface area of the feet of Arctic hares is greatly increased by growth of dense brush-like fur on the pads. The unusually broad hooves of caribou aid locomotion in both winter snow and the marshy ground of summer. As an aid to digging through snow, ice, and frozen ground, the claws of the third and fourth toes on the forefeet of collared lemmings become greatly enlarged during winter.

Caribou or reindeer are unique among members of the deer family in that females typically possess antlers, albeit smaller than those of males. During winter both males and females use their antlers to scrape away snow and ice to expose vegetation on which they feed. Like many large mammals inhabiting grasslands, musk oxen and caribou have responded to living in an open environment by evolving herding behavior as a defense against predation. This is epitomized by the classic "circle the wagons" defense of musk oxen.

Another anti-predator adaptation in caribou is synchronized calving: 80 to 90 percent of calves in a population are born within a 10-day period in late May or early June each year. This brief calving season apparently represents a strategy designed to overwhelm and satiate predators with large numbers of young, and thereby allow some of the young to escape predation. The highly precocial newborn calves are soon able to join the herd and gain its protection. Although synchronized calving in caribou might represent a response to constraints imposed by the very short summer season of the high Arctic, synchronized calving also occurs in some grassland ungulates, such as the wildebeest in Africa, which are not under constraints of an extremely short growing season.

MAMMALS OF THE DESERT

Desert mammals must be adapted to succeed in harsh physical environments characterized by shortages of moisture and temperature extremes. Most desert mammals minimize the physical stress of these two factors by being active at night (nocturnal) and spending their days in burrows where temperatures are cooler and relative humidities are higher than on the surface. Insects and seeds are extensively utilized by desert mammals both for their nutritional value and as sources of water; the bodies of insects have a high moisture content. The moisture requirements of seed-eating rodents are met largely or entirely from water produced as a byproduct of their metabolism

summer

winter

▲ The third and fourth claws on the forefeet of the collared lemming become greatly enlarged each winter, as an aid to digging through frozen ground.

▼ A thick, shaggy coat protects the musk ox from the Subarctic cold, and adults are so large that they have no natural enemies. The young, however, are vulnerable to wolves. In the event of an attack, the adults surround the calves and form a defensive ring, with horns directed outwards: the classic "circle the wagons" defense.

Steve Kaufman/Bruce Coleman Ltd

▲ *The Arabian oryx inhabits deserts where the days are extremely hot and the nights correspondingly cold. In adaptation to this harsh environment, it feeds and moves mostly at night; its white coat reflects solar radiation; and its splayed feet are suited to walking on sand. Perhaps as a means of conserving energy, it does not maintain a constant body temperature, nor is there much aggression between members of a herd.*

of carbohydrates in seeds, termed metabolic water.

Antelope ground squirrels in the deserts of North America are active during the day and thus must directly confront the challenge of high environmental temperatures. Because lethal body temperatures are quickly reached while foraging, these squirrels periodically return to their burrows to "dump" heat in the cooler environment of the burrow. They resume their foraging after the body temperature returns to normal.

Large desert mammals, such as camels and various species of oryx, cannot escape to burrows and are exposed to daily temperature extremes. One way they survive without expending excessive amounts of water on evaporative cooling is to let their body temperature fluctuate. At night they take advantage of radiational cooling to allow the body temperature to fall below what is normal for mammals. During the day, they permit the body temperature to slowly rise as they are warmed by the sun, and only late in the day as the body temperature approaches lethal limits do they utilize evaporative cooling, thus conserving considerable water. Their pale coat helps both to reflect sunlight and to provide insulation against high environmental temperatures. Oryx also feed at night when the moisture content of plants is higher than during the day.

CONVERGENT EVOLUTION

In examining the adaptations of mammals to various habitats, it is obvious that certain adaptive types appear best suited to particular habitats. As a consequence, we find many examples of unrelated mammals that have evolved remarkably similar adaptations because they live in similar habitats and have similar ecological roles in different parts of the world. This phenomenon is termed convergent evolution.

Among desert mammals a notable example of convergent evolution involves seed-eating rodents, including kangaroo rats in North America, gerbils and jirds in Asia and Africa, jerboas in Asia and Africa, and Australian hopping mice. These rodents exhibit remarkable degrees of convergence in morphology and ecology. All have pale, sand-colored coats, enlarged hindlegs and long tails, and employ bipedal or hopping locomotion as a principal means of escape from predators.

The convergence of seed-eating desert rodents as well as other examples noted in this chapter suggests that there are optimal evolutionary answers to the challenges of adapting to exploit specific types of habitats. Because of their widespread distribution and success in adapting to so many different habitats, mammals are excellent models to illustrate the phenomenon of convergent evolution.

MAMMAL BEHAVIOR

D.M. STODDART

All animals behave—that is, they react to specific stimuli in specific ways—but mammalian behavior is characterized by the fact that mammals, with only two exceptions, give birth to living young. In addition, the young of all species are dependent upon their mothers for food, in the form of milk secreted by specialized and modified sweat glands. This necessitates a special type of social behavior that bonds the mother to its young.

DEPENDENT MAMMALS

Some types of mammals remain dependent upon their mothers for many months—even years; such is the case with primates (including humans), and with whales and elephants. Others, such as rodents, are weaned in less than three weeks. Whatever the length of this dependence, it is an example of social behavior that can be defined as an interaction between two individuals of the same species. Even the so-called "solitary" species of mammal, such as the Australian bandicoot or the orang utan, which as adults interact socially only to mate, have a close behavioral interaction with their mothers when they are young. All other behaviors shown by mammals are classified as something other than social. These include behaviors necessary for bodily maintenance, such as feeding, grooming, urinating, defecating, and huddling.

MAINTENANCE BEHAVIOR

It is difficult to dissociate maintenance behavior completely from social behaviors, since frequently the two are interrelated. For example, passing urine is a biological necessity associated with the physiological processes of nitrogen excretion. Cows, seals, whales, and female dogs and mice simply void urine whenever the distension of the bladder reaches the point where the urination center in the rear of the brain is stimulated. After urination, the stretch receptors in the bladder wall relax, the flow of messages to the brain stops, and the animal no longer feels the urge to urinate.

Adult male dogs, mice, and many other mammals, however, do not urinate in this way. In them the presence of the sex hormone testosterone—produced in the testes—influences behavior, so that urine is voided at specific points around the animal's environment where other males will encounter it. This is an example of territorial demarcation. So maintenance and social behaviors are sometimes intertwined.

The same is true for feeding behavior—many carnivorous animals hunt cooperatively in order to fill their bellies—and grooming behaviors which,

▼ Food supply is the responsibility of the female members of a pride of lions. Individual females may capture small prey, but two or more usually cooperate in bringing down antelopes, gazelles, and zebras. Even with such relatively large prey, cost-benefit considerations apply: if the animal cannot be caught after a relatively short dash, it is not worth expending more energy in a prolonged chase.

Purdy & Matthews/Planet Earth Pictures

in primates, serve an important social function. Nevertheless, in all these behaviors the underlying drive is the maintenance of the individual; the social aspect is superimposed upon it.

SOCIAL BEHAVIOR

Apart from the relatively short interactions between mothers and their young, social behavior may be divided into four main categories: play, the acquisition of social dominance, the acquisition of territory, and sexual behavior. The eventual goal of all of these behaviors is to differentiate the fitter individuals from the less fit, allowing only the fittest to mate and to pass on their genes. The notion of equality has no place in nature.

Play

Play is a special form of social behavior indulged in by the young of primates and of the carnivorous mammals, especially dogs, cats, weasels, and mongooses. It is noticeably absent in the young of the large herbivores, although occasional glimpses may be seen in lambs, foals, and calves. Play serves to provide the young animal with an opportunity to develop the skills it will later need to support itself in the wild. Young carnivores invest much time in mock fights, during which they practise attacking. Watch two puppies tumble about: they make frequent lunges to the throat and neck area, where a killing bite will be most effective. Primates are the most highly social of all mammals, and among them play serves to teach the young how to form social relationships and how to relate to other individuals. Play also teaches them how to react to signals sent out by another individual which indicate social status and mood.

▼ Young chamois at play. Play is usually restricted to the young of social species while they are under the protection of adults. It involves "sketching" or "rehearsal" of behavior — particularly food capture and combat — that will be important in later life. Play also improves an individual's agility, a matter of importance in the rock-climbing chamois.

Gunter Ziesler/Bruce Coleman Ltd

Acquiring social dominance

Many mammals do not live in one place all the time, but migrate from one place to another. Frequently they manifest a social hierarchy, in which one—or a few—individuals are dominant over all the others. In a troop of baboons, for example, the dominant males travel in the center, while younger aspiring males make up the vanguard and rearguard. The dominant males are easily distinguishable by their size and by the thickness of their manes. Some sort of adornment makes dominant members of any group distinctive: it may be visual, as in the baboon or the blackbuck, in which only the dominant buck is black (all other males are light brown in color); acoustic, as in the South American howler monkey, in which the dominant males have the loudest voices; or olfactory, as in the Australian sugar glider, in which the scent gland on the heads of the most dominant males secrete more copiously than

Gerald Cubitt/Bruce Coleman Ltd

◄ In most social mammals, one individual is the leader of a group. In some species the dominant individual is a female, but much more commonly it is a male that has attained the position after competition with rivals and monopolizes (or attempts to monopolize) breeding. This herd of blackbuck is led by the strikingly marked male; the dark individual in the background is a subadult male, too young to be a threat to the leader.

▼ Individuals (or pairs) of many species occupy well-defined territories and attack trespassers — which usually retreat. When two individuals from neighboring territories meet at a border, the tendencies of each to attack or retreat are delicately balanced, leading to a "stand off", as in these short-tailed shrews.

those of the subordinate males. In all cases, the dominants are responsible for most of the matings.

Acquiring territory

A large number of mammals are territorial, which means that the males compete for, and repel all other males from, a piece of land. The territory may be very large, as is the case with the large cats, containing enough food and other resources for rearing a family, or it may be a token patch barely big enough for the male and his consorts. The latter type of territory is seen in elephant seals and some ungulates, such as the Uganda kob antelope. Territorial fights may be aggressive, but as a rule the loser is not killed. Horns and antlers have evolved to minimize the risk of accidental death through goring. In most cases the territorial sex is the male, who may be substantially larger than the females, in order to give him the best chance of winning his encounter. Almost invariably the

Dwight R. Kuhn

males show the visual, acoustic, or olfactory adornment necessary to demarcate and defend the territory against infiltration. Among small and medium-sized mammals, scent marking of obvious points on the boundary and within the hinterland of territory is commonplace. Such scenting places, redolent of the scent signatures of their visitors, act as information exchanges for all who pass by.

Sexual behavior

Mammalian sexual behavior varies from the short and rough, as seen in bandicoots, to the highly stylized and ritualized performances of creatures such as the Uganda kob antelope. Whether mating is promiscuous, as in these two examples, or whether male and female mate for life, as in many of the small forest-dwelling antelopes, or whether it is something in between, as in most mammals, depends on many complex and intertwined environmental factors. Non-promiscuous species show a degree of courtship, during which a male and female annexe themselves from the group. In a number of species, the close proximity of the male acts to bring the female on heat, and only at this time will he attempt to mate. It is thought that pheromones, or signalling odors, in the male's urine act to stimulate the female's reproductive system. In some species—the domestic dog and the hamster are familiar examples—the female may emit a special odor which lures males from afar. She may then choose with whom she will mate. In those species in which both sexes need to help with rearing the young, for instance most primates and many carnivores, courtship will be of sufficient duration for the offspring of any previous alliance to become apparent before mating occurs. If the female is pregnant, a pair-bond may not form. In this way the male will be assured that the subsequent offspring, which he is helping to rear, is truly his own. Courtship in these species is characterized by subtle and covert cues, not intended for a general audience.

In stark contrast, those species that mate promiscuously advertise their sexual readiness in

▼ When not resting, mature male lions are occupied with defending a territory against other males and with mating. Mating involves rather fine judgement on the part of the male, since a female that is not on heat will reject his advances quite vigorously.

D. Parer & E. Parer-Cook/Auscape International

Jonathan Scott/Planet Earth Pictures

▲ *Mutual grooming in olive baboons. Reciprocal grooming is common in mammals and birds, usually serving to reinforce a pair-bond. In monkeys and apes, mutual grooming frequently extends throughout a group as an expression of group solidarity and of subtle differences in social rank.*

flamboyant behaviors. Female chimpanzees, who are strongly promiscuous, are adorned with a large patch of white sexual skin in the perivaginal area which inflates massively during the heat. Males are strongly attracted to this and line up patiently waiting their turn in the mating queue. Interestingly there is little interference and aggression seen in this whole bizarre process.

THE SURVIVAL OF THE SPECIES

Mammalian behavior is not only complex, but also highly adaptive. That is, it contributes to the continued survival of the species just as much as any other biological attribute. Research into mammalian groups that occupy a wide range of habitats, such as primates or antelopes, is revealing that the ultimate factor that governs which type of behavior evolves is the environment. Thus the strange behavior of marmosets, in which the female always has twins but is large enough to carry only one youngster, may be explained by a consideration of the nature of their territories. These are so large in relation to the tiny size of the animals that they can only be adequately defended by both the male and the female—a female on her own is unable to rear a family and protect the territory. To keep the male handy, natural selection has evolved a reproductive strategy in which, if the male does not remain with the female to help carry and protect the young—and incidentally help defend the territory—his genetic investment in the next generation will be impaired. This apparent act of genetic blackmail on the part of the female is no more than the influence of the environment on aspects of the species biology that lead to increased chances of survival.

C.A. Henley/Auscape International

Frieder Sauer/Bruce Coleman Ltd

▲ *Female brown antechinus with 8-week-old young. At birth these marsupials, not much larger than a grain of wheat, attach themselves to the mother's teats, where they remain for about five weeks. They are then left in a nest and suckled daily for another eight weeks.*

◄ *Male impalas rubbing heads. For much of the year, male and female impalas associate in separate herds. Within a male herd, individuals may signal group ties by rubbing heads and exchanging scents produced by glands on their faces. These relationships break down in the breeding season.*

ENDANGERED SPECIES

ANNE LABASTILLE

Ever since the first living forms inhabited the Earth over 3.5 thousand million years ago, species have become extinct. During that huge span of time, an estimated 250 to 500 million different forms of life have existed, and 98 percent of these have gradually died out and have been replaced with new types. The average worldwide rate of extinction of all species over that time was somewhere between a minimum of one species per thousand years and a maximum of one species per year. But extinction is speeding up. Today, everywhere on Earth, species are disappearing at an ever faster rate—be they birds, fishes, plants, mammals, invertebrates, or reptiles. And they are not being replaced.

"EXTINCTION SPASM"

The life forms existing today number between 3 and 10 million species, half of which live in tropical rainforests. No one knows exactly how many there are, because possibly millions—mainly invertebrates such as insects and lower plants—have yet to be discovered and named. Zoologists are certain, however, that approximately 4,300 kinds of mammals exist. That's all.

Between AD 1600 and 1900, scientists estimate that 75 species, mostly birds and mammals, were killed off. Another 75 disappeared between 1900 and the 1960s. Since then, the rate has soared. Already, according to the records of CITES (Convention on International Trade in Endangered Species), 500 species and subspecies of mammals are listed as being "at risk", including the big cats, primates (except human beings), and all cetaceans. Dr Norman Myers, the British ecologist, warns that during the next 25 years we can expect to lose anywhere from a minimum of 1,000 species of plants and animals per year to a maximum of 100 species of plants and animals per day.

If this downward-swirling trend continues, it means that those readers alive in AD 2015 will have witnessed the extinction of maybe a million living things. Such a violent and rapid event has never before happened in Earth's long history. It is called an "extinction spasm".

THE HUMAN FACTOR

What is causing this crisis, and how did it begin?

▼ The black-footed ferret became extinct in the wild when human activities eliminated its major food source, prairie dogs, from its habitat. Two United States government agencies have set up a captive breeding program for the last remaining black-footed ferrets in the hope of re-establishing this animal in its original habitat.

Franz J. Camenzind/Planet Earth Pictures

Gerard Laez/NHPA

The human population explosion is probably a significant factor. Human beings are exploiters. We always have been; we always will be. Humans are among the smartest and most skillful of mammals. It hasn't taken us long, in geological time, to gain control over other mammals. Ever since we emerged as intelligent hunters and food-gatherers, we have deliberately utilized other mammals for food, milk, clothing, footwear, weapons, tools, and oil. And over the last few thousand years we have successfully spread into every nook and cranny of the world. With our highly organized societies, our increasingly sophisticated forms of transportation, weaponry, and communication, and our ever more demanding types of agriculture and mariculture, we rule the planet. At the same time, human numbers have skyrocketed. We are now the major threat to most other life forms and the leading cause of biotic impoverishment.

A key factor has been the way crude weapons have evolved into efficient and fearsome killing machines. Once hunters used clubs and rocks, then sling-shots and spears. They moved on to bows and arrows, and then firearms, progressing from single-shot muskets to explosive harpoon guns to rapid-fire machine-guns. Transportation

has developed in a similar way—from using two feet, to riding beasts of burden, to piloting sailboats and wagons, to running trains, planes, ships, and cars with fossil fuels.

The dreadful impact of these developments on wildlife can be shown by case histories. For example, archeologists sleuthing through ancient "boneyards" of the late Pleistocene (100,000 to 10,000 years ago) have found proof that prehistoric people, despite their small size and puny weapons, were able to group together to kill the huge mammals, or megafauna, of that era in large numbers. Skeletons have been unearthed with stone arrowheads or spearheads in them.

Each time that humans arrived on a new land mass, a wave of extinctions appeared to follow. For instance, shortly after humans crossed the Bering Strait to North America, around 10,000 to 15,000 years ago, toward the end of the Ice Age, there was a swift disappearance of giant beavers, elephants, mastodons, camels, wooly mammoths, sabre-toothed tigers, and giant bison. At least 50 types of megafauna vanished. The same coincidental extinctions seem to have occurred in Europe, Africa, Latin America, and Australasia.

The Malagasy Republic (Madagascar) is an area

▲ Since it is impossible for humans and tigers to live in harmony, the species can only be conserved by setting aside areas of forest large enough to provide prey for these carnivores. By conserving tigers, we simultaneously conserve thousands of other species of plants and animals.

Richard Matthews/Planet Earth Pictures

▲ *The golden lion tamarin of South America. Until recently, the most vulnerable forests in South America were the hardwood forests in the coastal region of southeastern Brazil. They were wiped out save for a few patches by the early 1970s – and with them went their primates, including the golden lion tamarin. The three species were reduced to tiny remnant populations. Despite attempts to re-establish wild populations with captive-bred animals, these species remain endangered.*

that was more recently invaded by human beings. The Indonesians arrived about AD 20, and the disappearance of giant lemurs, elephant birds, and dwarf hippos followed. A manmade overkill took place. No sign of climatic change or other biological threats exists to explain the die-offs.

Neither the Ice Age nor the Madagascar extinctions, however, seem as terrible as the slaughters during historical times. After white people arrived in America, they practised roundups and slayings of millions of bison and other large mammals. One such event was the Great Pennsylvania Circle Hunt. In 1760, hunters on foot formed a 170 kilometer (100 mile) circle. Spaced just under 1 kilometer (about half a mile) apart, they marched inward, killing everything they met with guns. The total for the hunt: 41 cougars, 109 wolves, 112 foxes, 1 otter, 12

wolverines, 3 beavers, 114 bobcats, 10 black bears, 2 elk, 98 deer, 111 bison, and 3 fishers!

More advanced weaponry and transportation allow humans to hunt wolves and polar bears today in Alaska. Hunters use light aircraft to track, tire, and approach wolves on land, while polar bears are scouted on the immense ice floes by helicopter. The animals are then shot on land using high-powered rifles with scopes. Such high-technology killing denies game animals any chance of survival.

WORLDWIDE HABITAT DESTRUCTION

The main threat to the survival of mammals today, however, is modern environmental problems caused by humans. Among the worst is the destruction of their habitat—be it Arctic tundra, islands, coastal wetland, prairies, or tropical moist forest. Every ecosystem on Earth is being tampered with as people dam rivers, log woodlands, drain wetlands, build pipelines, graze domestic stock, and construct highways and suburbs. Mammals are rapidly losing their homes, food and water, and open space.

Development in the Amazon rainforest is the most frightening example. Extinction rates will be higher than anywhere else on Earth because this ecosystem supports half the Earth's living species.

The Amazon basin is so vast—6 million square kilometers (over 2 million square miles)—so seemingly fertile, so green, that it seems like an untapped paradise. Several Latin American nations, which own part of it, are trying to colonize it. Brazil is the leading country. Its government built the lengthy Transamazonica Highway, and many spur roads, in order to resettle thousands of poor peasants along its edges. These colonists have each cut and burned small tracts of virgin forest to plant rice, pepper, bananas, and cacao. When colonist cropland is hacked out of the jungle, the habitat of jaguars, howler monkeys, scarlet macaws, and other rainforest animals fragments and diminishes.

Most of the Amazon basin is, in fact, a false paradise for colonists. Soils are infertile and contain toxic aluminum, and rains are torrential. Once the green umbrella of tall trees is gone, soils rapidly compact and erode under the scorching sun and harsh rains. Eventually, they are worthless. Most farming projects have failed.

Now the Brazilian government is promoting huge cattle ranches and timber or palm plantations. Most are owned by multinational corporations and cover thousands of hectares. When the rainforest is cleared, huge fires burn wildly for weeks. In fact, the largest manmade fire ever reported (detected by satellite) was set by a multinational corporation to prepare pastures.

It is estimated that an average of 20 hectares (50 acres) of rainforest is burned or disturbed in the world *each minute*. That's equivalent to an area the size of Great Britain *every year*. Already, 37 percent of the moist tropical forest in Latin America has been converted into less productive ecosystems; 38 percent in Southeast Asia, and 52 percent in Africa. This astonishing changeover from a complex, balanced, healthy, long-lasting ecosystem with huge mature forest trees, to short-term cropland with oil palms and pepper bushes, or land for cattle grazing, is one of the most disturbing ecological and economic trade-offs happening today. If the destruction continues, scientists predict that by AD 2000 to 2010 only pockets of rainforest will remain in remote places.

Another form of habitat destruction is occurring in northern and western Africa. A region called the Sahel is experiencing swift desertification. Largely because of overgrazing by domestic animals, poor rainfall, and high population rates, the Sahara Desert is stealthily encroaching on savanna lands. An estimated 250 million humans were recently stricken by drought, along with their herds, and wild mammals such as gazelles, antelope, cheetahs, and addaxes died of thirst.

THREATS FROM ALL SIDES

Many environmental abuses indirectly affect mammals. Such abuses include air pollution, such as acid rain, fresh and salt water pollution, and soil degradation. We have no way of knowing how many mammals take sick or die from drinking filthy water, eating plants growing in contaminated earth, or living in sea water polluted with toxic and nuclear wastes. Research carried out on roe deer living in a Polish national forest downwind of a major steel manufacturing city showed deformities in antler growth and a declining birth rate. Wildlife biologists have surmised that the deer eat vegetation that has been doused with acid rain and contains traces of toxic metals from the steel mills. Reindeer of Lapland are known to have become radioactive after eating lichens that absorbed nuclear fallout from the Chernobyl disaster in the Soviet Union.

One looming environmental factor is global warming. As more and more carbon dioxide and methane are pumped into the atmosphere from tropical wildfires and combustion of fossil fuels, the Earth will heat up. A Pandora's box of climatic changes lies in store: warmer temperatures, more intense storms, rising sea levels, and torrential rains at latitudes unaccustomed to them. Some mammals will be able to migrate to find comfortable new locales, and some may adapt. Others will become too stressed and will succumb. Global warming may produce the greatest extinction spasm ever.

The introduction of exotic animals into existing ecosystems, for food or recreational hunting, is another threat to native wildlife. Foreign species can not only bring in and spread unsuspected diseases or parasites, but they are also more likely to survive than residents and often experience a population explosion. Fierce competition ensues and local fauna can be destroyed.

All kinds of human activity—including scientific research, and the testing of various cosmetics and drugs—continue to destroy wildlife. Human aggression during wars kills countless mammals. So does exploitation for the production of fashion items.

THE TALE OF FIVE SPECIES

To illustrate the dangers confronting mammals, here are five examples from the land, air, and sea. All five species are listed as endangered in the Red Data Book of the International Union for the Conservation of Nature.

Maned wolf

On the dry, grassy pampas of Argentina, Paraguay, and Brazil roams the maned wolf. It looks like a rangy, red dog on stilts, with 18 centimeter (7 inch) ears, black stockings, and a black "cape" over the shoulders. Weighing over 20 kilograms (close to 50 pounds), it is wolf-sized but is actually a fox. Maned wolves can cover over 30 kilometers (20 miles) a night on their long legs, and hunt mainly for pacas, rodents, rabbits, and birds. They are hunted by collectors for zoos and by local people for charms: in Brazil, people believe that the body parts have medicinal value and that the left eye—if plucked from a *live* animal—is lucky. Because of these problems, plus parasites and diseases, fewer than 2,000 maned wolves are left today.

▼ *Slaughtered for the supposed magical properties of its parts, the maned wolf of South America is now reduced to about 2,000 animals and is unlikely to survive in the wild.*

Francisco Erize/Bruce Coleman Ltd

▼ ▲ *The existence of the American bison was not threatened until the arrival of Europeans. Between the middle and the end of the last century, "buffalo" hunters reduced the population from about 50 million to fewer than 800. Protection and management in reserves has permitted numbers to increase to about 40,000.*

American bison

The American bison and its close relative the European bison (a subspecies) have both had near brushes with extinction in the past hundred or so years. Once, an estimated 50 to 60 million animals covered the American prairies, looking like an ocean of black dots. Migrating bison were considered the greatest natural spectacle that humans had ever seen. The Plains Indians depended heavily on bison, or buffalo as they became known, and revered them too. But it was the arrival of Europeans that almost eliminated the shaggy 1 tonne (2,200 pound) beasts.

Professional hide and meat hunters slaughtered bison by the millions. Often only the tongues, considered a delicacy, were taken and the meat left to rot. Buffalo Bill once counted 4,280 animals he alone killed in one year. Railways pushing west also played a part in bison destruction. Passengers on the early trains took pot shots at buffalo simply to while away the time. By 1884 the bison had practically disappeared from the United States, but at the last moment, reserves were established and the few remaining animals protected. There are now several thousand in Yellowstone and Wood Buffalo National Parks, the Wichita Mountains Wildlife Refuge, and in other refuges. But because the animals are confined to these isolated sanctuaries, there is little genetic exchange. Inbreeding and gene loss will possibly harm the species. The bison may be "saved", but huge herds will never again thunder across the plains.

Humpback whale

The fate of the humpback whale is a striking example of exploitation. The 19 meter (62 foot),

James Watt/Planet Earth Pictures

black-colored baleen feeders once numbered 100,000. By the 1960s, only 6,000 remained.

People have hunted whales for centuries. Norse fishermen made whale carvings 4,000 years ago and Eskimos have eaten them for at least 3,500 years. But commercial whaling, the real culprit, started as early as the tenth to twelfth centuries AD. Whales were hunted for their meat, oil, and bones. Humpbacks were one of the easiest whales to catch, so they were harpooned by the hundreds. As long as whalers used small boats and killed and butchered by hand, whales had a chance. But when huge factory ships with sonar, helicopters, explosive harpoon guns, and onboard processing appeared, whales were doomed.

In 1946, the International Whaling Commission was established to regulate whaling. Some nations paid no attention. In 1973 the United States Marine Mammal Act prohibited taking these animals and their products, except by native peoples. At the same time, the United Nations Conference on the Human Environment set a worldwide ten year moratorium on whaling. The Commission did not abide by this ruling. Finally, thanks to the concern of many scientists and the growing concern of the public, a ban on all commercial whaling took effect in 1986. Conservation groups like Greenpeace and their Save the Whales campaign have helped people to learn about and respect whales. Modern technology allows humans to hear the odd patterns of chirps, groans, and cries that male humpbacks sing in the breeding season. Female whales (and hydrophones) can hear these songs up to 160 kilometers (100 miles) away under water.

This is the first time in human history that whales have been protected. Even so, Japan, Iceland, and Norway still harvest whales under the guise of research, and 95 percent of the humpbacks are now gone.

Flying foxes

Bats make up nearly one-quarter of all mammal species. Of these, the 173 different kinds of flying fox are the most fascinating. The largest bat on Earth is the Samoan flying fox, which has a wing span of nearly 2 meters (6 feet). Since these bats feed on flowers and fruits, nectar and pollen, they use their keen eyes and noses rather than sharp hearing and echolocation to find food.

▲ The use of the harpoon gun and factory ships brought most of the large whales to the verge of extinction. Since a recent ban on commercial whaling, numbers of some species, such as the humpback whale shown here, appear to be increasing slowly.

These secretive creatures with large, luminous eyes and dog-like faces are hunted for meat in Malaysia and Indonesia, where their flesh is considered a delicacy. Bat meat is also thought to cure asthma. A hunter with a modern shotgun can exterminate a colony of a thousand flying foxes in a short time. Some bats are only wounded, however, or become orphaned youngsters. They crawl away to die and are never collected or used.

Old World tropical plants rely on fruit bats for seed dispersal and pollination. These "bat plants" give humans more than 450 useful products, such as black dye for baskets, delicious fruits, kapok fiber, charcoal, and medicines. In West Africa, the straw-colored flying fox disperses seeds of the iroko tree, which is the basis of a $100-million-a-year industry.

But flying foxes are declining, even disappearing, on many islands, including Guam, the Mariana chain, Yap, Samoa, Rodrigues, and the Seychelles. The destruction of their habitats by agriculture and fuelwood gathering is the main threat to the species. Little conservation work is done locally. But in Malaysia and Indonesia, the World Wide Fund for Nature has promoted public seminars and news coverage on the value of flying foxes and other bats.

Gorillas

Anyone who has read *National Geographic* magazine's article on gorillas, or the book *Gorillas in the Mist* by Dian Fossey, or seen the movie of the same name, knows the truth about these largest of living primates. They are, indeed, gentle giants.

Dr Fossey spent almost 20 years studying them in Africa and developed a close relationship with several groups of mountain gorillas. Despite her work and the breakthroughs she made, humans remain the gorillas' worst enemy. Even though African laws prohibit killing gorillas, enforcement is usually lacking. Poachers still sell their heads and hands as souvenirs. The lowland subspecies numbers between 5,000 and 13,000; there are as few as 350 to 500 mountain gorillas left. Even in the Virunga Volcanoes National Park, the gorilla habitat is not safe. Recently, a sizable section was gobbled up to grow pyrethrum, an insecticide, and loggers are constantly nibbling their way up the volcano slopes.

The greatest hope of saving gorillas is through tourism. Already, special groups visit Virunga to photograph the primates without harming them in any way. This leads also to local residents, rather than poachers, gaining income.

CONSERVATION MEASURES CAN STOP THE ROT

As grim as the outlook is for many mammals, conservation efforts are underway in practically every country. Hundreds of government agencies and non-government organizations exist. Some of the more notable are the International Union for the Conservation of Nature (IUCN), the World Wide Fund for Nature, and the United Nations Educational, Scientific, and Cultural Organization (UNESCO). In the United States, the National Audubon Society, the Sierra Club, the Nature Conservancy, and Planned Parenthood are also prominent.

Many laws protect wildlife and wildlands all over the world, but the problem lies in providing sufficient enforcement.

International cooperation is, however, often very successful in saving species. For example, the population of the elephant, the largest land

▼ *Flying foxes are usually numerous, but their habit of congregating in dense roosts makes them vulnerable to mass slaughter. Island species such as the Marianas fruit bat have become endangered by human predation.*

Merlin D. Tuttle/Bat Conservation International

D. Parer & E. Parer-Cook/Auscape International

mammal, has plummeted from 1.3 million to 625,000. Poachers and rebels still shoot elephants for ivory to provide income and finance civil wars, but recently ivory was placed on a list of prohibited products by CITES (the Convention on International Trade in Endangered Species). Because at least 75 nations and hundreds of non-government organizations support CITES and abide by its rules, the animals have a chance.

Captive breeding in zoos can help some endangered species. These establishments have bred 19 percent of all living mammal species over the years. A particularly successful captive breeding program has been that of the golden lion tamarin, and there has been some success with the Arabian oryx. Recently, some of the tamarins have been released back into small pockets of suitable habitat, but efforts to reintroduce species to the wild are as yet in their infancy. The main function of zoos remains the education of the public about habitat destruction and disappearing species.

Game ranching and farming is another technique for fostering rare animals. Elands, wildebeeste, wild pigs, capybaras, and pacas are some that can be raised profitably. Often, game meat contains more protein, less fat, and brings in more money than meat from domestic animals. In

Venezuela, capybara meat may be eaten during Lent whereas all other animal meat may not. Catholic monks in the sixteenth century considered them fish!

Without question, the preservation of habitat through outright purchase, lease, easements, or donations is the key to saving species. At present, national parks and equivalent reserves protect 1.6 million square kilometers (618,000 square miles)—1 percent of the world's land surface. It's not enough. Reserves large enough to protect all resident species in every ecosystem on Earth are urgently needed.

A growing interest in compensatory payment to people or nations that save animals and wildland may help. Countries can earn "conservation credits" to pay off their debts in return for setting aside parkland. And conservationists may some day be able to deduct expenses or take tax exemptions when protecting natural resources.

Ultimately, human attitudes must expand beyond utilitarian, economic, or recreational interests in animals and plants. All living things should be valued in and of themselves and have the right to exist. If humans act now to halt the slide towards extinction of the world's remaining mammals, perhaps it will not be too late.

▲ *Gorillas are threatened by reduction of habitat, poaching of young animals for some unscrupulous zoos, and slaughter for souvenirs. Controlled tourism may provide sufficient revenue to encourage local people to protect this species.*

MONOTREMES

Order Monotremata
2 families, 3 genera, 3 species

SIZE
Platypus *Ornithorhynchus anatinus*, males length 45 to 60 centimeters (18 to 24 inches); females 39 to 55 centimeters (15 to 22 inches); males weight 1 to 2.4 kilograms (2⅓ to 5 pounds); females 0.7 to 1.6 kilograms (1½ to 3½ pounds).
Echidna long-beaked *Zaglossus bruijnii*, length 45 to 90 centimeters (18 to 35 inches), weight 5 to 10 kilograms (11 to 22 pounds); short-beaked *Tachyglossus aculeatus*, length 30 to 45 centimeters (12 to 18 inches), weight 2.5 to 8 kilograms (6 to 18 pounds).

CONSERVATION WATCH
Long-beaked echidna populations are vulnerable. The other two species are common, although the platypus is considered vulnerable due to its specific habitat requirements.

▼ *The platypus propels itself through the water with its forefeet, using its hindfeet to control or change direction. The white patch of fur beneath the eye closes the eye and the ear opening when the animal is underwater.*

There are only three living species of monotreme in the world: the duck-billed platypus *Ornithorhynchus anatinus*, the long-beaked echnidna *Zaglossus bruijnii*, and the short-beaked echidna *Tachyglossus aculeatus*. The platypus, found only in the eastern states of Australia, is such an unusual animal that when the first specimen was sent back to Britain in 1798 it was widely assumed to be a fake, made by stitching together the beak of a duck and the body parts of a mammal! The two species of echidna, or spiny anteater, are found on the island of New Guinea, and the short-beaked echidna is also widely distributed in Australia where the long-beaked echidna is now unknown. The origins of this small group of mammals are obscure, as neither they nor their fossils have been found anywhere else in the world.

MAMMALS THAT LAY EGGS

It is known from fossil evidence that monotremes have been in Australia for at least 15 million years, and possibly even since the Cretaceous period (100 million years ago).

Although long-beaked echidnas are no longer found in Australia, fossils attributable to the genus *Zaglossus* indicate that several species did occur on the Australian mainland and in Tasmania until the late Pleistocene (100,000 to 12,000 years ago).

The order name Monotremata, meaning "one hole", refers to the fact that all three species have only one opening to the outside of their bodies, through which the products of the reproductive, digestive, and excretory systems are passed. This feature, however, is not exclusive to the monotremes, as marsupials also have this arrangement. While all three species have some anatomical similarities to reptiles — for example, they have ribs on the neck vertebrae and extra bones in their pectoral, or shoulder, girdles — their anatomy and physiology are largely mammalian in nature.

But the most distinctive feature of the group is that they lay eggs, instead of giving birth to live young as all other mammalian species do. The eggs are soft-shelled, measure approximately 15 x 17 millimeters (½ x ⅔ inch) and are incubated for

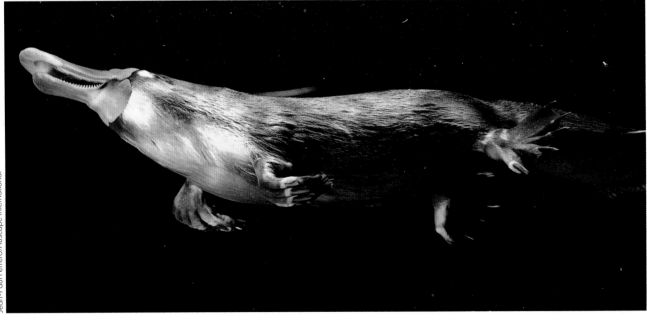

A UNIQUE BUT VENOMOUS AUSTRALIAN

The platypus is one of a very few venomous mammals. Two species of Caribbean solenodon and a few species of shrew use poisonous saliva to subdue prey, often larger than themselves, but the platypus has something completely different and of unknown function.

Rear-ankle spurs are found in all three species of monotreme. With their associated glands, they are known as the crural system. In females of all species of monotreme the spur is lost during the first year of life. Although the spurs and glands persist in male echidnas of both species, echidnas do not seem to use the system to inject venom; the platypus does.

Changes in the structure of the spur in male platypuses can be used to age animals up to 15 months after they have left the breeding burrow. Fully grown adult spurs are around 15 millimeters (½ inch) in length, can be everted away from the ankles, and can be driven into an object by the action of the muscles of the rear legs. The puncture alone is painful, but the venom injected can lead to symptoms ranging from local pain and swelling to paralysis of a whole limb in humans. When the species was hunted for its pelt, animals were stunned by a heavy caliber shot fired under them in

Curved and hollow, the platypus's spur (above) is connected by ducts to the venom glands under the thigh muscles (below).

venom gland

venom duct

spur

the water and a gun dog was sent out to retrieve the animal. There are numerous stories of dogs being killed by the platypus recovering and spurring them in the muzzle.

The nature of the venom is unknown, as is its function. An early naturalist suggested that the male used his spurs to subdue females during mating, but more recent observations indicate that the female may initiate much of the courtship and need no such subduing. The glands associated with the spur advance and recede with the testes during the breeding season, and it is assumed that their function is related to breeding. Certainly, males are more aggressive toward each other (and toward any other captor) at this time of the year. The social system in platypuses is completely unknown, but the retention of the crural system only in male platypuses seems to indicate that it may be used in establishing territories or access to mates during the breeding season. In the wild, animals presumably have the opportunity to escape from such encounters, which have been known to result in deaths in captivity. The spurs certainly represent a deterrent to predators, but their loss in females suggests that this is not their primary function.

---

around 10 days, after which the young are nourished on milk. This suckling of the young, as well as the possession of hair, a single bone making up each side of the lower jaw, and a muscular diaphragm for breathing, make the monotremes undoubtedly part of the class Mammalia.

THE PLATYPUS

This small amphibious animal, with its pliable duck-like bill, thick fur, and strongly webbed forefeet, is perhaps one of the most unusual animals alive today. The species is common over its present distribution in the streams, rivers, and lakes of much of eastern Australia, where individuals can often be seen diving for the small insect larvae and other invertebrates that make up their food supply. Between dives animals spend a few minutes chewing the food material collected in their cheek pouches, using the horny pads that replace the teeth in adults. Dives usually last for less than two minutes, and during them the groove that encloses both the ears and eyes is closed and the platypus locates its food using its bill. This organ has both sensitive touch and electrosensors, enabling the animal to locate its prey and to find its way around underwater at night, and in waters that are often quite turbid. Although individuals are most often observed around dawn and dusk, they can be active throughout the night and may, especially in winter, extend their feeding into daylight hours.

Once thought to be primitive because of its presumed poor ability to regulate its body temperature, the platypus has been found to be a competent homeotherm, maintaining a body temperature of 32°C (90°F) even when swimming for extended periods in near-freezing water. The animal raises its metabolism to generate more heat in cold conditions. This process is made more efficient by good body insulation, including a coat of fine dense fur, which retains an insulating layer of air when the animal is in water. (This fur was once prized by humans and the species was widely hunted until its protection early this century.)

When it is out of water the platypus occupies a burrow, dug into the bank of a river, where it is buffered from the extremes of environmental temperatures. These outside temperatures can vary from around –15°C (5°F) in the winter to as high as 40°C (104°F) in the summer over parts of the range of the species. There is some evidence that the platypus may hibernate during winter, although this has not been substantiated.

Not all females in a platypus population breed each year and neither males nor females appear to breed until at least their second breeding season. Females lay from one to three eggs, with two being the most common number. The eggs are thought to be incubated between the tail and body of the female as she lies curled up in a special nesting burrow of up to 30 meters (33 yards) in length. When the young hatch, they remain in this burrow

▲ The small size of the platypus can be gauged from this photograph. This is a female, which grows to about 450 millimeters (18 inches); the male can reach 600 millimeters (24 inches). When it is feeding at the surface, a platypus can submerge almost totally, with only the nostrils and the top of the head and back above the water.

for three to four months and are fed on milk by the mother throughout that time. The milk oozes out onto the fur from two nipple-like patches on the underside of the mother's body, and it is presumed that the young take the milk up from the fur. The breeding season is extended, with mating lasting from July to October. After a gestation period (thought to be around three weeks), incubation of around ten days, and the long period of suckling in the burrow, the young enter the outside environment between January and March each year.

In captivity platypuses are known to survive for up to 20 years, but the longest recorded longevity in the wild is 12 years, although the two animals recording this age were already adult when first captured and therefore could have been much older. Practically nothing is known of mortality in wild populations, although there is some predation by foxes and birds of prey. Juvenile dispersal occurs each year and it is assumed that this results in considerable mortality, especially in drier years when the amount of suitable habitat is reduced. The platypus carries a range of external and internal parasites, including its own unique species of tick, but it is not known if any of these contribute to mortality. The species has been bred only once in captivity and, because individuals are easily stressed, it is difficult to keep them successfully under captive conditions.

ECHIDNAS

Long-beaked and short-beaked echidnas are animals with a snout modified to form an elongated beak-like organ. They have no teeth, a long protrusible tongue and, in addition to normal hair, they have a number of special hairs on the sides and back which are modified to form sharp spines. The long-beaked species, at 45 to 90 centimeters (18 to 35 inches) in total length and 5 to 10 kilograms (11 to 22 pounds) in weight, is much larger than the short-beaked species, which is only 30 to 45 centimeters (11 to 18 inches) long and 2.5 to 8 kilograms (6 to 18 pounds) in weight. In the short-beaked echidna, males are larger than females. In both species only the male retains the spur on the ankle of each rear leg.

The status of the long-beaked echidna is in doubt, as the area of its distribution is poorly studied. The short-beaked echidna is distributed throughout mainland Australia and Tasmania, where its status can be regarded as common. In Papua New Guinea it is still considered to be common in lowland areas, although both species are known to be preyed upon by humans for food.

Unlike the platypus, the ears and eyes of echidnas are not housed in the same groove; the ear opening (with little visible external ear) is well behind the eye. The snout and protrusible tongue are both used in feeding. The short-beaked echidna eats mainly termites and ants although insect larvae are also taken. It procures ants and termites by excavating the mounds, galleries, and nests of these insects with the large claws on its front feet. The echida then picks up the ants or termites with its sticky tongue. It can push its elongated snout into small spaces and extend its tongue into small cavities to gain access to these insects. The generic term *Tachyglossus* actually means "swift tongue". The long-beaked echidna is chiefly a worm eater. It uses spines housed in a groove in its tongue to draw the worms into its mouth. In both species, mucous secretions make the tongue sticky and, in the absence of teeth, food material is ground between spines at the base of the tongue and at the back of the palate.

Little is known of the activities of the New Guinea echidnas, but in Australia echidnas can be active at any time of the day, although they seem to be less active and stay buried in soil or shelter under rocks or vegetation in extremes of heat or cold. They also seem to be less active during rainy weather. Like the platypus, they are unable to tolerate high temperatures and will die of heat stress if shade is not available. The burrowing ability of the short-beaked echidna is legendary, with individuals able to burrow vertically down into the earth to disappear in less than a minute.

Echidnas are endothermic and, like platypuses, can regulate their body temperatures well above that of environmental temperatures by raising their metabolism and using insulation — fur and fat in the case of the echidnas. In all three species of monotreme the temperature maintained is lower than that found in many other mammalian species, but is usually maintained within a few degrees of

▼ *The long-beaked echidna is hairier and less spiny than the short-beaked, perhaps because it lives in the cooler highland areas of New Guinea. It is a worm-eater and probes for earthworms in the ground with its long beak. It then uses its long grooved tongue to grasp the worms and pull them into its mouth. The nostrils and mouth are at the end of the beak.*

D. Parer & E. Parer-Cook/Auscape International

32°C (90°F) while the animals are active. It is now known that the short-beaked echidna sometimes hibernates for two to three weeks during winter in the Australian Alps, when body temperatures of individuals can fall to 4 to 9°C (39 to 48°F).

Little is known of the breeding cycle of the long-beaked echidna. In the short-beaked species, a pouch develops during the breeding season, into which one egg is laid. After about 10 days of incubation, the young hatches and is nourished on milk suckled from the milk patches in the pouch, the prodding of the young stimulating the milk to flow. Lactation lasts for up to six months, but once the young begins to grow spines (around nine weeks after hatching), it is left in a burrow to which the mother returns to feed it. As in the platypus, the breeding season is extended, and mating normally occurs in July and August. The length of gestation, before the female lays the egg, is not known exactly, but is thought to be about three weeks. Like platypus females, not all adult females in a short-beaked echidna population breed each year but the reasons for this are unknown.

Both species of echidna are long-lived. One short-beaked echidna in the Philadelphia Zoo lived for 49 years, and a marked individual in the wild was found to be at least 16 years of age. An individual long-beaked echidna survived for 31 to

36 years in Berlin Zoo, through both of the world wars, but nothing is known of the longevity of this species in the wild. Dingoes are known to prey on echidnas, in spite of the echidnas' ability to burrow and their armory of spines. Foxes, feral cats, and goannas take young from burrows during the suckling period, but perhaps the greatest mortality factor is the automobile. The role of parasites or diseases in mortality is largely unknown. Echidnas are readily maintained in captivity but rarely breed successfully under captive conditions.

D. Parer & E. Parer-Cook/Auscape International

▲ Echidnas dig straight down into the ground as a defense mechanism, until only the very tips of the spines are visible. This gives them some protection from predators, but dingoes can dig them out and eat them, spines and all. This photograph of a mother and her offspring is unusual as the young are normally kept in the burrow, to which the mother returns every few days.

◀ The short-beaked echidna's long sticky tongue can be pushed out of the mouth and into holes in termite mounds to lick up the occupants. Echidnas can be seen foraging on anthills, covered in ants. They seem to favor female ants, perhaps because females have more fat on them and therefore make tastier morsels.

D. Parer & E. Parer-Cook/Auscape International

MARSUPIALS

RONALD STRAHAN

Order Marsupialia
16 families, 77 genera,
c. 260 species

SIZE
Smallest Pilbara ningaui
Ningaui timealeyi, head–body
length 46 to 57 millimeters
(1⅘ to 2⅕ inches), tail length
59 to 79 millimeters (2³⁄₁₀ to
3¹⁄₁₀ inches).
Largest Red kangaroo
Macropus rufus, head–body
length 165 centimeters (65
inches), tail length 107
centimeters (42 inches),
weight 90 kilograms (198
pounds).

CONSERVATION WATCH
The following species are
listed as endangered in the
IUCN Red Data Book of
threatened mammals:
southern dibbler
Parantechinus apicalis, numbat
Myrmecobius fasciatus, greater
bilby *Macrotis lagotis*,
Leadbeater's possum
Gymnobelideus leadbeateri,
northern hairy-nosed,
wombat *Lasiorhinus krefftii*,
mountain pygmy-possum
Burramys parvus, brush-tailed
bettong *Bettongia penicillata*,
bridled nailtail wallaby
Onychogalea fraenata.
Many other species are
classified as vulnerable or of
indeterminate status.

Although marsupials (Metatheria) and non-marsupial mammals (Eutheria) have followed separate evolutionary paths for at least 100 million years, there are few fundamental differences between them. They are grouped together as members of the Theria (in contrast to the Prototheria, of which the monotremes are the only living survivors). Although anatomists can point to certain differences between all marsupials and all eutherians, the two groups are distinguished mainly by their reproduction: marsupial young are born in a much more undeveloped state than eutherian young.

WHAT MAKES MARSUPIALS DIFFERENT

Other differences are demonstrated by many, but not all, marsupials. Marsupials have a greater maximum number of incisors (five pairs in the upper jaw, four in the lower) than eutherians (never more than three pairs in each jaw). In most marsupials, the first toe of the hindfoot is opposable to the other four, as in primates, and it always lacks a claw; many terrestrial marsupials, however, have lost this toe. In general, marsupials have a somewhat smaller brain than eutherians of equivalent size. Their body temperature and rate of metabolism are slightly lower than those of eutherians; it is tempting to regard marsupials as "inferior" in this respect but, if so, eutherians would have to be regarded as inferior to birds.

▶ The woolly opossum (top) spends its life in the forest canopy, feeding mainly on fruits. The large eyes face forward, giving it a monkey-like appearance. Although most marsupials can swim, the yapok or water-opossum (bottom) is the only truly amphibious species, and it has webbed hindfeet. It feeds on invertebrate animals that live on the bottom of freshwater lakes and streams, finding them by groping with the sensitive fingers on its hands as its eyes shut tight underwater.

Classification of the marsupials is still a matter of debate but current expert opinion is that the modern fauna comprises two very distinct assemblages: the Ameridelphia, restricted to the Americas, and the Australidelphia, comprising all the species found in Australia, New Guinea, and nearby islands.

AMERICAN MARSUPIALS

There are approximately 75 species of living Ameridelphians, all of which are referred to as opossums, although they comprise three distinct families: Didelphidae (American opossums), Caenolestidae (shrew-opossums) and Microbiotheriidae (colocolo).

American opossums These "true" opossums range from the cat-sized Virginian opossum *Didelphis virginiana* to the appropriately named mouse-opossums, genus *Marmosa*. Most are omnivorous and able to climb, but many spend more time on the ground than in trees, and the tails of the more terrestrial species are shorter and less prehensile. The carnivorous short-tailed opossums, genus *Monodelphis,* are mainly terrestrial, while the woolly opossums, genus *Caluromys,* are strongly arboreal and feed mostly upon fruits and nectar. The lutrine opossum *Lutreolina crassicaudata* is a predator and enters water to swim in pursuit of prey. The yapok *Chironectes minimus* swims below

the surface to take prey on the bottom.

Shrew-opossums The seven species of shrew-opossums are survivors of a group well known from fossil remains over the past 35 million years. Now restricted to cool mist-forests of the Andes, these mouse-sized to rat-sized marsupials are characterized by a pair of long, chisel-edged incisors that project forward from the lower jaw and are used to stab the large insects and small vertebrates upon which they prey.

Colocolo The colocolo or monito del monte *Dromiciops australis,* from the cool rainforests of southern Chile, is the only surviving member of this family, which has a long fossil history. The size of a small rat, it feeds mainly upon insect larvae and pupae. Most experts on marsupial evolution believe the colocolo to be related to the American opossums but increasing fragments of evidence suggest an affinity with the Australidelphia: if so, the colocolo would constitute an important evolutionary link between the marsupials of the two continents.

AUSTRALIAN MARSUPIALS

The 200 or so living species of Australidelphians are much more diverse than the living Ameri-delphians. The 16 families fall into 4 very distinct assemblages: the carnivorous marsupials; the marsupial mole; the bandicoot group; and a large

▲ *In many respects, the Tasmanian devil is the marsupial equivalent of the hyena: it takes some live prey but is essentially a scavenger. Its powerful jaws enable it to completely consume a dead sheep, including the skull.*

A.P. Smith/Australasian Nature Transparencies

▲ *The red-tailed phascogale is one of the few dasyurid marsupials to spend much time in trees hunting prey.*

group known as diprotodonts, which includes the koala, wombats, possums, and kangaroos.

Carnivorous marsupials

These comprise three families: the Dasyuridae (dasyurids), Thylacinidae (thylacine), and Myrmecobiidae (numbat). They have three or four pairs of narrow, pointed upper incisors and three pairs of similar shape in the lower jaw: the hindfeet have four or five toes, none of which are joined.

Dasyurids Dasyurids range in size from the tiny, flat-headed planigales, genus *Planigale*, weighing about 4 grams (less than ¼ ounce) and with a head and body length less than 6 centimeters (2 ½ inches), to the terrier-sized Tasmanian devil *Sarcophilus harrisii*. Most are terrestrial but, although the tail is never prehensile, many can climb, and several, particularly the phascogales, genus *Phascogale*, are actively arboreal. There is very little variation in body shape except in the kultarr *Antechinomys laniger*, which has very long hindlegs and tail, which it employs in an unusual bounding gait.

Thylacine The thylacine or "Tasmanian tiger" *Thylacinus cynocephalus* probably became extinct in the 1940s. With a head and body length of more than 1 meter (3 feet), it was the largest carnivorous marsupial to have survived into historical times. Except for its broad-based and rather inflexible tail, it had the general conformation of a wolf, although it was striped rather like a tiger.

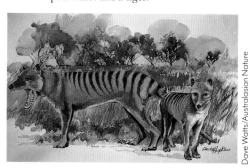

Dave Watts/Australasian Nature Transparencies

▲ *Because it sometimes attacked sheep, the thylacine was persecuted in Tasmania and the last known individual died in a zoo in 1933.*

▶ *The spotted-tailed quoll (top) is a cat-sized predator. Although an agile climber, it spends more time on the forest floor than in trees. The bilby (bottom left) lives in the Central Australian deserts. The silky fur provides thermal insulation, while the long ears probably act as radiators of excess body heat. The numbat (bottom right) is the only marsupial that feeds mainly on termites.*

Numbat Sole member of the family Myrmecobiidae, the numbat *Myrmecobius fasciatus* is the only marsupial to be habitually active during the day. With the aid of a long sticky tongue, it feeds exclusively on ground-dwelling termites; its teeth are reduced to simple pegs.

Marsupial mole

The marsupial mole *Notoryctes typhlops* spends its life moving below the desert sands, feeding on burrowing insects and reptiles. It is the only living member of the family Notoryctidae. Very little is known of its behavior and its relationship to other marsupials remains a mystery.

The bandicoot group

A dozen or so omnivorous species from Australia and New Guinea have an arrangement of teeth rather similar to that of the carnivorous marsupials, but they differ from these in many other characteristics, notably in having the second and third toes of the hindfoot fused together to form what looks like one toe with two claws.

The group is usually divided into the family Peramelidae, or "true" bandicoots, and the Thylacomyidae, with only one living species, the bilby *Macrotis lagotis*. However, it has recently been proposed that the division should lie between the predominantly New Guinean spiny bandicoots of the genera *Peroryctes* and *Echymipera*, and all the other species. All are omnivorous, digging with their strongly clawed forefeet for burrowing insects and their larvae, and for succulent underground parts of plants.

"True" bandicoots Typical members of the family Peramelidae, such as the long-nosed

M.W. Gillam/Auscape International

bandicoot *Perameles nasuta,* have short, bristly fur, short ears, and a short, sparsely furred, non-prehensile tail. The legs are powerfully muscled and rather short. Living bandicoots have a bounding gait but the pig-footed bandicoot, which became extinct in the nineteenth century, appears to have run on the tips of its toes.

Bilby The bilby or rabbit-bandicoot *Macrotis lagotis* is the only surviving member of the family Thylacomyidae. It differs from "true" bandicoots in having long, silky fur, very long ears, rather long legs, and a long, furred tail. A desert-dweller, it spends the day in a deep burrow.

▲ *Remarkably similar to the golden moles of Africa, the Australian marsupial mole is a striking example of convergent evolution. It does not make a tunnel but "swims" through the sand, which collapses behind it. Apparently it obtains oxygen from the air that exists between the grains of sand.*

Robert W.G. Jenkins/Australasian Nature Transparencies

◄ *The northern brown bandicoot uses its strongly clawed forelimbs to dig for succulent roots and underground insects. The powerful hindquarters and long hindfeet are employed in a bounding gait — a sort of "bunny hop".*

▶ "Koala" is said to be an Australian Aboriginal word meaning "doesn't drink", but it is more likely to mean "biter". Koalas are very aggressive toward each other and can defend themselves effectively with tooth and claw against dogs and humans. For its size, the koala has a remarkably small brain.

J. Cancalosi/Auscape International

▼ With a head and body length of no more than 8 centimeters (3 inches), the feathertail glider is the smallest of the volplaning mammals. The featherlike tail is used to steer it in flights of more than 20 meters (65 feet).

Jean-Paul Ferrero/Auscape International

Diprotodonts

More than half of the Australidelphian marsupials belong to this diverse group, characterized by having only one pair of well-developed incisors in the lower jaw, and by the fusion of the second and third toes of the hindfoot (as in the bandicoot group) to form a two-clawed grooming comb. Most are herbivorous, but many eat insects and some lap plant exudates.

Wombats The three living species of wombats in the family Vombatidae are stocky, grazing animals with very short tails and a somewhat bear-like appearance. They walk on the soles of their feet, which have short toes armed with strong claws. Wombats excavate long burrows in which they sleep during the day.

Koala The koala *Phascolarctos cinereus* is the only living member of the family Phascolarctidae. A long-limbed, almost tailless arboreal marsupial, it feeds on the leaves of certain eucalypts. It is related, but not closely, to the wombats.

Cuscuses and brushtail possums The family Phalangeridae includes the cuscuses, genus *Phalanger*, and others; the cuscus-like scaly-tailed possum *Wyulda squamicaudata;* and the brushtail possums, genus *Trichosurus.* Except for the brushtail possums, they have a strongly prehensile tail, much of the free end of which is bare. They are arboreal, feeding mainly upon leaves, supplemented by buds, fruits, and shoots.

Ringtail possums and greater glider All but one of the members of the family Pseudocheiridae are ringtail possums, so called in reference to the long, slender, short-furred, and strongly prehensile tail: the rock ringtail *Pseudocheirus dahli*, which makes its nest on the ground, has a shorter tail than the other, exclusively arboreal, species.

The greater glider *Petauroides volans*, largest member of the family, is exceptional in having a membrane between the elbow and ankle on each side of the body, which is employed in gliding from tree to tree: the long, well-furred tail is only weakly prehensile. All members of the family feed mainly on leaves, the greater glider being restricted to those of eucalypts.

Wrist-webbed gliders and striped possums The family Petauridae includes four gliding species in the genus *Petaurus:* they differ from the greater glider in having a membrane that extends from the wrists (not the elbows) to the ankles. Leadbeater's possum *Gymnobelideus leadbeateri* resembles the sugar glider *Petaurus breviceps*, but lacks a gliding membrane. All feed on plant exudates and insects. Four species of striped possums, genus *Dactylopsila*, are characterized by a skunk-like coloration of black and white, and by having a very long, slender fourth finger; they feed exclusively on wood-boring insects.

Pygmy-possums The five species in the family Burramyidae include four arboreal pygmy-possums, genus *Cercartetus*, weighing 7 to 30

▼ *The greater glider (below) is, at 1 meter (3 feet), the largest of the volplaning marsupials.*

grams (¼ to 1 ounce). They have long, slender, prehensile tails which are used to assist their agile climbing in shrubs and trees in search of nectar, pollen, and insects. The fifth species, the mountain pygmy-possum *Burramys parvus*, which at 40 grams (1½ ounces) is the largest member of the family, is predominantly terrestrial and feeds on insects, green plants, and seeds. It is the only Australian marsupial to live above the snowline.

Feathertails The two species in this family are characterized by a row of long, stiff hairs projecting horizontally on each side of the tail. In the tiny feathertail glider *Acrobates pygmaeus*, weighing about 12 grams (½ ounce), these hairs are so closely apposed that the tail resembles the flight feathers of a bird; the tail is employed, in combination with a membrane between the elbows and the knees, in gliding from tree to tree. In the larger feathertail possum *Distoechurus pennatus*, which lacks a gliding membrane, the lateral hairs are not closely apposed.

Honey possum The honey possum *Tarsipes rostratus* is the only member of the family Tarsipedidae. It is a small marsupial: males weigh only 9 grams (⅓ ounce), and females, 12 grams (½ ounce). The honey possum uses its long, fringed tongue to feed exclusively upon nectar.

▲ *The squirrel glider (top left) is one of several gliding marsupials in which the gliding membrane extends from the ankle to the wrist. It inhabits the canopy of open forest. The striped possum (top right) of New Guinea and the tropical rainforests of Australia digs into the tunnels of wood-boring insects with its chisel-like lower incisors and extracts the insects with its long, slender fourth finger. Most cuscuses (above, bottom) are found in New Guinea and nearby islands. With strongly grasping hands and feet and a powerfully prehensile tail, they move deliberately through the rainforest canopy. The face tends to be flat and the eyes face forward; early European explorers often mistook them for monkeys.*

John Cancalosi/Auscape International

Rat-kangaroos Members of the family Potoroidae retain some relatively unspecialized characters that have been lost in the closely related "true" kangaroos of the family Macropodidae. Most "primitive" of the group is the musky rat-kangaroo *Hypsiprymnodon moschatus*, which has a head and body length of only 23 centimeters (9 inches). It is notable for having five toes on the hindfoot (the other potoroids and "true" kangaroos have four), and for its bounding (rather than hopping) gait when it moves fast.

Other members of the group are the potoroos, genus *Potorous,* and bettongs, genus *Bettongia,* and others, most of which feed on underground tubers, bulbs, corms, and fungi, sometimes supplemented by green plant material. The tail, employed as a balance when hopping, is weakly prehensile and is used to carry nesting material.

Kangaroos and wallabies Probably the most recently evolved group of the marsupials, the Macropodidae comprises 11 genera and more than 50 species of wallabies, kangaroos, and tree-kangaroos. All, except the tree-kangaroos, genus *Dendrolagus,* share a similar body shape, with short forelimbs and large, powerful hindlimbs, long hindfeet, and a very large fourth toe. When moving fast, they hop on the hindlimbs, using the long and rather inflexible tail to provide balance. Rock-wallabies, genus *Petrogale,* which have somewhat shorter feet than the typical terrestrial kangaroos, are very agile denizens of cliffs and rock piles.

Tree-kangaroos, which evolved from terrestrial macropodids, have very long, strong forelimbs and shortened, broad hindfeet: they can walk along a horizontal branch or climb vertically by gripping a branch or stout vine with the claws of all four feet. All members of the family are herbivorous: in general, the more "primitive" species tend to be browsers, the more "advanced" to be grazers.

HOPPING, GLIDING, AND SWIMMING

The familiar hopping gait of kangaroos probably evolved from the bounding gait that has been retained by the diminutive and primitive musky

▲ One of the largest macropods, the western grey kangaroo feeds on grasses. Long after it is able to move about on its own, the young of the western grey kangaroo enters its mother's pouch to avoid danger, to sleep, or to travel.

▶ Rock-wallabies live on cliffs, boulder falls, or rockpiles. Their hindfeet have granulated ("non-slip") soles and are shorter than those of typical kangaroos. Pictured is the yellow-footed rock-wallaby.

A. Fox/Auscape International

► *Goodfellow's tree kangaroo (top right) is the brightest-colored of the tree kangaroos, which have evolved from terrestrial kangaroos. In adaptation to climbing, their forelegs became longer and stronger, while the hindfeet became shorter and broader. In many respects the musky rat-kangaroo (centre) provides evidence of the evolutionary origins of kangaroos from possums. It retains an opposable first toe; it bounds rather than hops; its tail is prehensile; and it gives birth to twins. The red kangaroo (bottom left) is the largest of the living marsupials. Old males may be 2 meters (8 feet) high when sitting, considerably more when propped on the tail. Females, and about 30 percent of males, are not red but bluish gray. The large-eyed forest wallaby of New Guinea (bottom right) inhabits the dimly lit rainforest floor. The tail, which has a scaly patch at the tip, is bent into a rightangled prop when the wallaby sits.*

Jean-Paul Ferrero/Auscape International

▲ When hopping fast, a kangaroo uses less energy than a four-legged animal of the same size, moving at the same speed. A hopping kangaroo operates like a pogo stick.

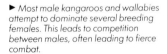

▶ Most male kangaroos and wallabies attempt to dominate several breeding females. This leads to competition between males, often leading to fierce combat.

Hans & Judy Beste/Auscape International

rat-kangaroo. The hindlimbs became increasingly larger than the forelimbs and the hindfoot became longer, providing a very effective means for fast, hopping locomotion. At equivalent speed, and making allowance for the differences in weight, a hopping kangaroo uses less energy than a running dog or horse.

But such specialization is not without cost — a typical kangaroo is unable to walk. When moving slowly, it raises the hindquarters on a tripod formed by the forelimbs and the tail pressed down to the ground, then swings both hindlimbs forward and always together. Terrestrial kangaroos cannot move each hindleg independently while supporting the body, although they can and do kick them alternately when swimming. Tree-kangaroos, which have undergone a secondary shortening of the hindfeet, can also move their hindlegs alternately when walking along a branch.

In typical kangaroos, the tail is not very flexible but is moved up and down to assist with balance during the hopping gait and acts as a fifth limb to support the body when moving slowly. In the more primitive rat-kangaroos the tail is moderately prehensile in the vertical plane, and is used to carry bundles of nesting material.

The ability to glide has arisen independently in three families of marsupials: the Pseudocheiridae, Petauridae, and Acrobatidae. In each case, this involves a membrane of skin between the forelimbs and hindlimbs, which, when the legs are extended, expands into a rectangular, kite-like airfoil. Leaping from a high tree, a glider can volplane quite long distances, steering itself by altering the tension of the membrane on either side, balancing with the outstretched tail, and finally orienting the body vertically to land on the trunk of another tree with all four feet. No other gliding mammal has anything comparable to the tail of the feathertail glider: each side bears a thin row of stiff, closely packed hairs, all of the same length, forming a structure very similar to the vane of a feather.

Although it seems that most marsupials are able to swim when necessary, only the yapok can be regarded as truly aquatic. With alternate strokes of its webbed hindfeet, this Central American species swims to the bottom of a pond or stream with its eyes closed, feeling with its long, spatulate fingers for living prey, which it grasps in its mouth and takes to the shore to be eaten. The rear-opening pouch of a female yapok is closed by a strong sphincter muscle and sealed with water-repellent secretions when the animal is swimming.

The most extreme locomotory specialization of any marsupial is seen in the small, sausage-shaped marsupial mole. It is blind, lacks external ears, and has a horny shield over the snout and surrounding the nostrils. The limbs are short, with very strong bones and powerful musculature, and the spade-like forefeet have two immense, triangular claws on the third and fourth digits, with smaller claws on the other digits. The forefoot is employed in digging away soil in front of the animal and pulling it forward. The hindfoot, which has four short claws and, uniquely among marsupials, a claw-like structure on the first toe, is used to kick sand backward from the body. The marsupial mole does not construct a tunnel but "swims" through sand, usually at a depth of 10 to 20 centimeters (4 to 8 inches), but sometimes descending to a depth of 2 meters (6½ feet) or more.

The feathertail glider, the honey possum and pygmy-possums, genus *Cercartetus*, are very small marsupials that climb by gripping with the expanded tips of their fingers and toes. Except on the conjoined second and third toes of the hindfoot, the claws are reduced and nail-like, lying above the tips of the digits. The pads on the fingers and toes of the feathertail glider are microscopically grooved, like those of geckos, enabling them to cling to a smooth surface such as a vertical sheet of glass and even to hold themselves, albeit briefly, to the underside of a horizontal sheet.

SPECIALIZED DIETS

The first marsupials probably fed mainly on insects and other small invertebrates, as do the least specialized of the living opossums, dasyurids, and possums. It was a comparatively simple step in evolution for the larger dasyurids and the thylacine to graduate to preying upon vertebrates. The greatest insect-eating specialization is found in the striped possums, which use their sharp and protruberant lower incisors to dig into branches to expose burrowing insect larvae, which are then extracted with the claw of a very long, thin, fourth finger. A lemur-like primate, the aye-aye of Madagascar, feeds in a remarkably similar manner.

No marsupial feeds entirely upon fruits, but many opossums, possums, and even some dasyurids, include these in an omnivorous diet. Gliders of the genus *Petaurus* feed partly on insects but also on the gum and sap of trees, promoting the flow of sap from eucalypts by cutting grooves into the trunk with their sharp lower incisors (a method of feeding also used by the pygmy marmoset). Bandicoots and the bilby dig in the earth for insect larvae, succulent tubers, bulbs, and corms (subterranean stems). Rat-kangaroos also make shallow excavations in search of underground fungi, but some also eat insects and the soft parts of green plants, particularly shoots.

Many diprotodont marsupials eat leaves. Tree-

Jean-Paul Ferrero/Auscape International

▲ *The tiny honey possum feeds exclusively on nectar and has a long, brush-tipped tongue.*

▼ *The numbat uses its long, sticky tongue to lick up termites.*

Dick Whitford/Australasian Nature Transparencies

kangaroos and cuscuses feed mainly on the soft, broad leaves of rainforest trees, supplemented by fruits, although cuscuses also eat large insects, eggs, and nestling birds. Brushtail and ringtail possums rely to a large extent upon the tough leaves of eucalypts, while the koala and greater glider rely exclusively on this source, which, although abundant, is not very nutritious and has a high content of toxic substances. Many wallabies browse on the leaves of shrubs and low trees but may also eat some grasses. The only marsupials to feed mainly on grasses are wombats, some species of wallaby, and *Macropus* kangaroos.

▼ *This sugar glider is about to land on the trunk of a tree, which it will grip with its hands and feet. Note the thumb-like first toe, ready to grip against the other toes.*

C. & S Pollitt/Australasian Nature Transparencies

Because mammals lack the enzymes necessary to break down plant fiber into its constituent sugars, leaves and grasses are difficult to digest, although this can be managed with the aid of microorganisms if these are given sufficient space and time to act. Most marsupials promote the digestion of plant fiber by diverting chewed food into the cecum (a capacious diverticulum of the intestine) where it remains for some time, undergoing microbial fermentation before being returned to the intestine. Others, notably the wombats, chew grass very finely and pass it very slowly through an elongated large intestine. The koala is notable for having, proportionately, the largest cecum of any mammal, as well as a very long large intestine. Ringtail possums have a large cecum, but it appears to be inadequate for complete digestion of plant material — the contents of the cecum (so-called "soft feces") are evacuated once a day and re-eaten; this material passes through the body once more, the unassimilated fraction being voided as hard fecal pellets.

Kangaroos manage the digestion of tough grasses in a quite different manner. Finely chewed food is retained for microbial digestion in a compartment of the stomach before it passes on to the rest of the alimentary canal. Although comparable in some respects with the ruminant digestion of sheep and cattle, the details of anatomy and function are quite different in kangaroos, since rumination (chewing the cud) is not an essential part of the process.

RESTING IN THE COLD

Under cold conditions, much of the energy that a very small mammal obtains from its food is expended in maintaining its body temperature. Since cold is greatest in winter, when food also tends to be scarce, such mammals face considerable difficulties. One solution to the problem is to hibernate, permitting the temperature of the body to drop almost to that of the surrounding environment and to enter a state of almost suspended animation. This reduces both the rate of heat loss and the need for food.

There is some debate about whether any marsupials truly hibernate, but the sugar glider, the honey possum, the feathertail glider, most pygmy-possums, and some small dasyurids are known to respond to cold or to a shortage of food by becoming torpid overnight or for periods of a week or so. A torpid animal "switches off" its thermal regulation and cools down to a temperature a little above that of its surroundings. Its heartbeat and metabolism slow down accordingly. Most species that become torpid also conserve body heat by huddling together in a communal nest.

DAYTIME SHELTERS

Most marsupials sleep during the day in a nest of some sort, usually of dry foliage. A few species,

Hans & Judy Beste/Auscape International

such as pygmy-possums and gliders of the genera *Petaurus* and *Acrobates*, construct compact woven nests, which are usually situated in a tree hollow but are sufficiently robust to be self-supporting if built in a forked branch. Most other marsupials make a less structured nest in a tree hole, hollow log, or other crevice; planigales and some other very small, flat-headed dasyurids make nests during the dry season in the deep cracks that develop in clay soils. Bandicoots typically make a scrape in the ground and heap vegetation over it to make a nest that appears to be without structure

but may have a definite entrance and exit. Most rat-kangaroos build nests, often against the trunk of a tree or under a bush; rock-wallabies and the wallaroo *Macropus robustus* escape the heat of the day by sheltering under rock overhangs or in caves, while other wallabies and kangaroos shelter in the shade of trees or bushes. Wombats, the bilby, the burrowing bettong, and most desert-dwelling dasyurids make nests in burrows. Tree-kangaroos, the larger cuscuses, and the koala have neither nests nor dens. Tree foliage provides shelter from sunlight, but they have no protection from rain.

▲ *Dunnarts look rather like small rodents and were once called "marsupial mice" A better comparison would be with shrews, for dunnarts are fierce little predators. This one is eating a desert grasshopper.*

Dave Watts/Australasian Nature Transparencies

▲ *Wombats produce one young, which is accommodated in a backward-opening pouch. Once able to move about, a young wombat follows its mother at heel.*

BIRTH AND ATTACHMENT

As in other mammals, the right and left ovaries of a female marsupial shed their unfertilized eggs into corresponding right and left oviducts (fallopian tubes). In other ways, however, the anatomy of the marsupial female reproductive system is quite different. Each oviduct continues into a separate uterus and thence into a lateral vagina. The two vaginas join a median tube (the urogenital sinus) that opens, together with the rectum, into a shallow cavity or cloaca. (Monotremes are usually defined as mammals that possess a cloaca, but this does not distinguish them from female marsupials, which also pass feces, urine, and newborn young out through a single cloacal opening.)

This reproductive system would appear to be a simpler, more primitive arrangement than is found in eutherian mammals, where the oviducts open into a single uterus (sometimes paired at its apex) and thence into a single vagina, which opens to the exterior. However, the marsupial condition is also more specialized — indeed unique — in having a third, median, vagina. This is a passage connecting the two uteri to the urogenital sinus and through which the young are born. In most marsupials it is a temporary structure that develops just before parturition and disappears after the birth has taken place, but in kangaroos and the honey possum it remains permanently once it has been formed for the first birth.

Male marsupials differ from male eutherians in having the scrotum in front of the penis. The penis is forked in many species, presumably to direct semen into each of the lateral vaginas. Sperms pass up the vaginas and into the oviducts, where they fertilize one or more eggs. As a fertilized egg passes down the oviduct, it becomes enclosed in a very thin shell-membrane, a vestige of the ancestral eggshell. This disintegrates in the course of gestation and the embryo becomes attached to the wall of the uterus. While enclosed in a shell-membrane, a marsupial embryo is nourished by fluids secreted into the cavity of the uterus from glands in its wall. Embryos of some species make only a tenuous connection with the uterine wall: these receive most of their nourishment from the uterine secretions. Those that develop a significant placenta are able to supplement this source with materials passing almost directly from the maternal blood into the blood of the embryo.

Gestation is short in marsupials, ranging from about 9 days in the eastern quoll *Dasyurus viverrinus* to 38 days in the eastern gray kangaroo *Macropus giganteus*. Newborn marsupials are very small. The female honey possum weighs approximately 12 grams (½ ounce) and gives birth to young weighing about 5 milligrams ($\frac{1}{5,000}$ ounce); the female red kangaroo *Macropus rufus*, weighing approximately 27 kilograms (60 pounds), produces a neonate weighing only 800 milligrams (¹⁄₃₀ ounce) (0.003 percent of the maternal weight, compared with 5 percent in humans).

All marsupial neonates are remarkably similar: the skin is bare, thin, and richly supplied with blood (possibly acting as a respiratory surface); the eyes and ears are embryonic and without function (although the ear may be sensitive to gravity); and the hindlimbs are short, five-lobed buds. However the nostrils are immense; the sense of smell is well developed; the mouth is large, with a large tongue; the lungs and alimentary canal are functional; and the forelimbs are disproportionately large,

Jean-Paul Ferrero/Auscape International

▲ *Firmly attached to one of its mother's four teats, this red kangaroo pouch embryo is more than a month old. Although still blind and hairless, it has hindlimbs and a tail.*

powerful, and equipped with needle-sharp claws.

Birth is rapid—a simple "popping out". Immediately it is free of its embryonic membranes, the neonate begins to move toward a teat, waving its head from side to side and dragging itself by alternate movements of each forelimb through the forest of hairs on its mother's belly. Once it has located a teat, the neonate takes it firmly into its mouth, the tip of the teat expanding to fill the mouth cavity. All of this is accomplished in no more than a few minutes.

A large glottis at the back of the neonate's mouth shuts off the teat from the air passage between the nostrils and the lungs, permitting it to breath while suckling, without danger of choking. Firmly attached to a teat, the neonate settles down to a period of passive growth and development, which is considerably longer than gestation. During this period it is called a "pouch embryo".

Birth and attachment to a teat have been most intensively studied in kangaroos, which have a deep, forward-opening pouch. The neonate must therefore climb upward to the lip of the pouch, then clamber (or tumble) down into it to locate a teat. But kangaroos are not typical marsupials since, apart from possums, they are the only marsupials to have a pouch that opens anteriorly. In other marsupials the pouch either opens to the rear or consists of little more than lateral folds around the mammary area; a pouch may even be lacking in some small marsupials. Obviously, a rear-opening pouch is more convenient for a newborn marsupial and this is probably the primitive arrangement.

Kangaroos are also misleading models because, although they have four teats, they normally give birth to only one young at a time. Among opossums, litters of six or more are common and the pale-bellied opossum *Marmosa robinsoni* sometimes carries a pouch young on each of its fourteen teats. From six to ten young are carried by some dasyurids; bandicoots commonly carry three or four; and pygmy-possums may have litters of from four to six.

Some marsupials give birth to more young than can be accommodated on their teats. A Virginian opossum is known to have produced an excess of at least nine, while the eastern quoll and the kowari can produce at least three extra. Little is known of the extent of such overproduction, since most supernumerary young are probably eaten or lost in the nest litter and they are unlikely to be noticed unless birth takes place on a smooth, clean surface.

In various eutherian mammals, including certain seals and bats, a fertilized egg develops into a tiny, hollow ball of cells and then becomes quiescent for a period. Known as embryonic diapause, this process also occurs in most kangaroos (although not in potoroids), some pygmy-possums and, possibly, the honey possum. In most kangaroos and wallabies the embryo from

a female's first mating passes through a normal gestation, is born, and attaches to a teat. Within a day or so of that birth, the female mates for a second time but the embryo from this mating becomes quiescent at the blastocyst stage and remains so until the first young is about to give up its attachment to its teat. The blastocyst then resumes normal development and, around the time of its birth, a third mating takes place and so the process is continued. Thereafter, throughout her reproductive career, the female will normally be carrying one blastocyst, one pouch embryo, and one young which is still suckling but able to move in and out of the pouch. The mammary gland of a teat that is supplying a pouch embryo produces milk of quite different composition from a teat that is supplying its older sibling.

Female marsupials play a passive role during the birth and teat-attachment of their young. Care of the pouch embryos is limited to cleaning the pouch or mammary area. Species that produce numerous offspring usually lack a sufficiently large pouch and, once their young have detached themselves from the teat, they are left in a nest during the night while the female is foraging. During the day she sleeps with them, suckles them, and grooms them. Young that stray a short distance from the nest may be retrieved in response to their distress calls but this behavior is not universal. When the young are fully formed and capable of independent movement, they may follow the mother while she is foraging or, more commonly, cling to her fur with their teeth and claws — creating a very considerable burden.

The larger marsupials tend to have capacious pouches and fewer (usually only one) young. In these species, the general pattern is for the young to inhabit the pouch long after they have become detached from the teat and, during the period of weaning, to use the pouch as a means of transport and a place to sleep; this is particularly the case in kangaroos, bandicoots, and wombats. In the case of the koala and the larger species of possum, partially weaned young are often carried on the mother's back, but they may be "parked" while the mother is on a long foraging excursion, and maternal care ceases at the end of weaning. Males are never involved in parental care.

Marsupials differ most notably from eutherians in the relatively small size of their newborn young — a marsupial mother's initial investment in reproduction is low. However, by the time that the young are weaned, the ratio between her weight and the weight of her offspring (whether a litter or a single individual) is similar to that of a eutherian mother and the difference is essentially a matter of the proportion of the period of development that is spent in the womb. There is no evidence that marsupial reproduction is inferior to that of eutherians; as in some other aspects of their biology, marsupials are simply different.

Female eutherian
Genital aperture separate from anus; one vagina and one uterus.

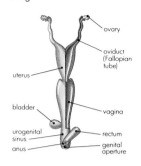

Non-pregnant female marsupial
Rectum and urogenital sinus open to outside through same (cloacal) aperture; two vaginas and two uteruses.

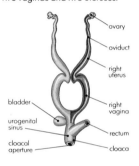

Pregnant marsupial
Median vagina develops, through which birth takes place.

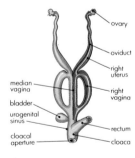

A.P. Smith/Australasian Nature Transparencies

▲ *Many male marsupials, like the sugar glider shown here, have a forked penis.*

Order Edentata
4 families, 13 genera,
29 species

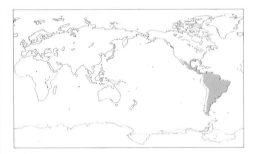

SIZE

Smallest Lesser fairy armadillo *Chlamyphorus truncatus*, head and body length 125 to 150 millimeters (5 to 6 inches); tail 25 to 30 millimeters (1 to 1⅕ inches); weight 80 to 100 grams (2⅘ to 3½ ounces).

Largest Giant anteater *Myrmecophaga tridactyla*, head and body length 100 to 120 centimeters (40 to 48 inches); tail 70 to 90 centimeters (27 to 36 inches); weight 20 to 40 kilograms (44 to 88 pounds).

CONSERVATION WATCH

The maned sloth *Bradypus torquatus* is listed as endangered in the IUCN Red Data Book of threatened mammals. Other species are classified as vulnerable or of indeterminate status.

▼ *The collared anteater or tamandua climbs in trees with the aid of a prehensile tail. The three recurved claws on each forefoot are used to open ant and termite nests.*

EDENTATES

R. D. STONE

The order Edentata comprises a bizarre group of animals that radiated in South America between the Paleocene and Pliocene epochs (65 to 2 million years ago). Four main lines evolved from this ancient stock: a lineage of armored terrestrial grazing herbivores; an arboreal (tree-dwelling) non-armored group specializing in browsing; a fossorial (burrowing) omnivore/insectivore group; and a lineage supremely adapted to feeding on ants and termites. The first lineage is represented today by the armadillos (20 species); the second, by the sloths (five species); the third is now extinct; and the fourth is represented by the American anteaters (four species). Only a few of the weird and wonderful animals of this era are still to be found, but these relicts are some of the most fascinating and, ironically, some of the least-studied living mammals.

ANIMALS WITHOUT TEETH

"Edentate" literally means without teeth, but this is a somewhat misleading term for members of this order, since only the anteaters are strictly toothless; sloths and armadillos have rootless molars, which continue to grow throughout the animal's life. However, even these vestiges are simple and lack the outer coating of enamel that protects the teeth of most mammalian species against wear.

Edentates are strictly a New World group, originating in North America. At the beginning of the Paleocene epoch, a land bridge connected the North and South American land masses, permitting the free passage of animals and plants between the two continents. In this way, South America was gradually colonized by the ancestors of many modern mammals.

When this umbilical land bridge was severed or submerged for at least 70 million years, those animals that had successfully colonized the southern "island" were able to evolve in complete isolation. This led to the evolution of many new orders, families and genera with distinctive and often bizarre features.

ANTEATERS

Anteaters were probably one of the first groups of mammals to have reached the South American continent before it became an island and would, as such, have evolved in the warm, moist seclusion of the tropical forest.

Giant anteaters

The giant anteater *Myrmecophaga tridactyla,* which ranges from 100 to 120 centimeters (40 to 48 inches) in length and 20 to 40 kilograms (44 to 88 pounds) in weight (though males are often 10 to 20 percent heavier than females), is gray with a black and white shoulder stripe. It is a familiar inhabitant of the South American savannas and open woodlands, but, on occasion, it may also venture into the tropical rainforest in search of its favorite foods — ants and termites. Endowed with only an average sense of vision, this magnificent animal possesses a highly attuned sense of smell; with its elongated, proboscis-like snout, it constantly sniffs the air and ground to locate potential prey. Scientific tests have shown that the anteater's sense of smell is at least 40 times more acute than that of humans.

Upon locating a potential feeding place, the anteater digs an entrance into the nest with rapid movements of its toughened claws. The long, worm-shaped (vermiform) tongue is protrusible and is covered with sticky saliva secreted by enlarged glands situated at the base of the neck.

Haroldo Palo Jr./NHPA

The giant anteater is the only member of the family Myrmecophagidae that is not predominantly arboreal. Although it can climb reasonably well, it is usually found on the ground moving with a characteristic rolling gait. Instead of supporting its body weight on the soles of its feet, as is the norm for most mammals, it is compelled to walk on the hard edges of each foot, turning the two largest claws (the second and the third) inward. The reason for this ungainly gait is that the lengthy, powerful claws are non-retractile and are, therefore, a hindrance to locomotion on the ground. However, the seemingly cumbersome movements of giant anteaters are deceptive, as they are able to outrun a human if necessary.

When the anteater is confronted by predators, its fearsome talons on the forelimbs serve another function apart from ripping open the concrete-like termite mounds — defense. When confronted with danger — for example, if surprised by a jaguar — the giant anteater will rear up to its full size on its hindlegs. The wielding, slashing actions of the forelimbs seriously challenge further approaches from would-be predators, which generally retreat to avoid sustaining serious injury.

The astonishing evolutionary adaptations of the giant anteater and other members of the family Myrmecophagidae to such specialized diets are doubtless the result of millions of years of gradual development by ancestral forms in conditions of comparative isolation. Although mammals that feed on ants are to be found in most other zoogeographical regions, no other single species demonstrates the same degree of specialization as the giant anteater of South America. Others, particularly the African aardvark and pangolins, furnish an interesting example of convergent evolution for, like the anteaters of the South American tropics, they all make use of similar weapons to extract and trap their prey — strong, curving claws and a sticky, protractile tongue. They also possess the same type of muscular stomach, which is capable of digesting the toughened exoskeletons of many species.

Little is known about the social behavior of the giant anteater; we can say only that its gestation period is about 190 days.

Tamanduas and silky anteaters
The other representatives of the family Myrmecophagidae are the two species of collared anteater or tamandua — *Tamandua mexicana* in the

▲ The immense claws and powerful forelegs of the giant anteater are used to dig into rock-hard termite mounds and to defend it against larger predators. It sleeps on its side, using the fan-like tail as a blanket.

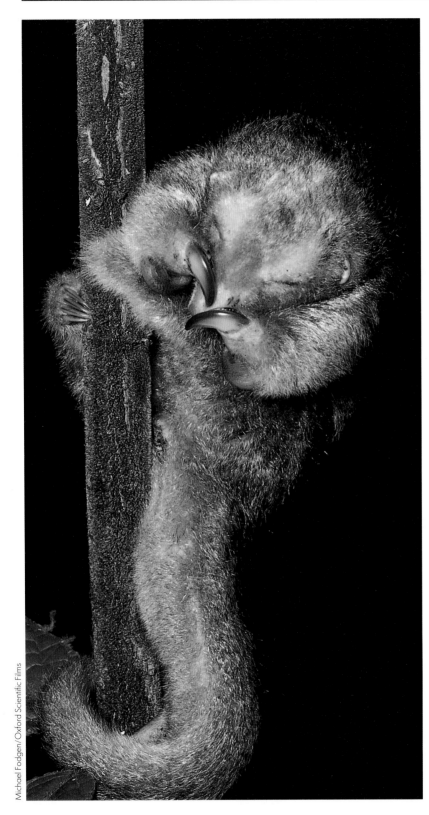

▲ The strictly arboreal silky anteater has only two functional toes on the forefeet, both armed with immense, hook-like claws. The furry tail is strongly prehensile.

north and *T. tetradactyla* in the south — as well as the two-toed or silky anteater *Cyclopes didactylus*. Unlike their much larger relative, these species prefer a forest or dense woodland habitat and, as might be expected in this habitat, are tree-dwelling animals. They seldom descend to the ground where, lacking the impressive defense of the giant anteater, they would be at the mercy of forest predators.

The two species of tamandua are about half the size of the giant anteater, with distributions that considerably overlap that of the giant anteater. Apart from color — the tamanduas' smooth fawn coats contrast strongly with the giant anteater's gray shaggy appearance — the main difference between these two genera is that tamanduas have a naked, prehensile tail which they use when climbing. Also, unlike the giant anteater, tamanduas are strictly nocturnal, using hollow trees as sanctuaries during daylight hours. But, as for the giant anteater, we know little about the tamanduas' reproduction, except that the gestation period is about 140 days.

The much smaller and even more secretive silky anteater is also nocturnal and strictly arboreal. Unlike its close relatives, the silky anteater rarely feeds on termites, preferring instead the delicate flavor of ants that inhabit the stems of lianas and tree branches. Its muzzle is proportionately shorter than that of its relatives, but its tongue is similar in structure and stickiness. Its fur is long, yellow in color, and silky in texture; the prehensile tail is longer than the body. Because of its superficial resemblance to tree-dwelling monkeys, this species is often referred to as the long-tailed tree monkey. Even though it may receive some degree of cover from its shady environment, the silky anteater is often threatened by aerial predators, such as the harpy eagle, hawk eagle, and spectacled owl. To defend itself against such attacks, it reacts like other members of its family by rearing on its hindlimbs and lashing out with its forelimbs.

SLOTHS

The sloths represent the survivors of an early mammalian radiation to South America that adapted to browsing. Strictly herbivorous, sloths are primarily leaf eaters and only rarely descend to the ground. In fact these animals are so highly modified for a specialized form of arboreal locomotion that they have almost lost the ability to move on the ground. The conversion of structural cellulose to simple sugars, as well as the detoxification of the many tannins found in leaves, was accomplished by the evolution of a chambered stomach where leaves are fermented with the aid of symbiotic protozoan and bacterial organisms.

The two families of sloths each contain a single species: Bradypodidae, the three-toed sloth, and Megalonychidae, the two-toed sloth. Both are strictly New World species and are confined to the tropical rainforests of Central and South America.

Sloths have rounded heads and flattened faces with tiny ears concealed under a dense, shaggy fur, and their hands and feet end in curved claws that are 8 to 10 centimeters (3 to 4 inches) in length. In both species the forelimbs are considerably longer than the hindlimbs, making movement on the ground awkward and ungainly. In fact, sloths' limbs are unable to support their body weight, and they move on the ground by slowly dragging the body forward with the forelimbs. They climb in an upright position by embracing a branch or by hanging upside down and moving along hand over hand. Sloths spend a considerable amount of time hanging upside down in the tree canopy — at times even sleeping in this position — where they resemble animated coat hangers.

The entire body is covered with a dense, soft fleece from which protrude larger, coarser tufts of hair. Each of these hairs is grooved, vertically and horizontally, and because of the almost constant humidity of the forest canopy, these narrow slits harbor a multitude of microscopic green algae. Consequently, although the basic color of the sloth's hair is gray or brown, the profusion of algae in the outer hairs gives the coat a distinctive greenish tinge, which effectively camouflages the animal from avian and terrestrial predators, such as the harpy eagle and jaguar.

Sloths are long-lived, solitary animals that appear to have no fixed breeding season. Following a gestation period of about one year the single young is born, high within the canopy, and initially helped to the teat by the mother, where it will feed for almost one month. The male does not participate in rearing the young. No fixed nesting site is used; instead the mother acts as a mobile nest with the infant clinging on to her. Following weaning the young remains with its mother, usually being carried, for a further six months. During this time the infant begins to feed from the same trees as its mother; adult sloths seem to have individually distinctive feeding preferences which they appear to pass onto their offspring.

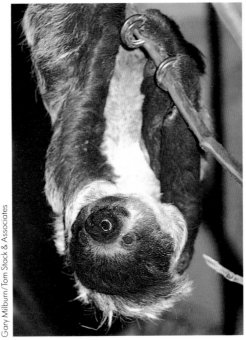

Gary Milburn/Tom Stack & Associates

◀ *The two-toed sloth lives in mountainous rainforests, where it feeds on the broad leaves of a small number of species of rainforest trees. While awake, it feeds almost continuously.*

Joe McDonald/Tom Stack & Associates

◀ *The three-toed sloth seldom descends to the ground except to defecate—about once a week. It moves very clumsily on the forest floor because the muscles of its limbs are more adapted to hanging upside down than to supporting the body.*

▶ *Sloths do not make nests or use dens. The single young must therefore be carried by its mother from the time of birth. The three-toed sloth, shown here, carries its young until it is about six months old.*

▼ *Sloths have long, shaggy hair. Each hair has a groove along its length and algae grow in this space. In wet conditions, the algae multiply and the hairs become green—perhaps contributing to camouflage.*

The vast majority of a sloth's active time is spent in feeding and, when full, an animal's stomach may account for almost one-third of its body weight, which ranges from 4 to 8 kilograms (9 to 18 pounds). In keeping with the animal's pace of life, digestion is a relatively slow process: up to a month may elapse before foodstuffs finally pass from the multichambered stomach to the small intestine.

Unlike most arboreal mammals, sloths descend from the trees to defecate on the ground, an exercise they engage in only once a week, but considering their lethargic way of life, this must consume a considerable amount of body energy. When on the ground three-toed sloths, at least, scoop out a small depression in the soil at the base of the tree before defecating. The site is apparently carefully selected, and, apart from serving as a useful olfactory signal to other sloths in the area, it has been suggested that sloths are actually fertilizing their favorite feeding trees by this action.

Such visits to the ground are also important for other species: the shaggy coat of most sloths harbors a wide range of insects, including moths, beetles, ticks, and mites. Although the exact interrelationship between these species and their "habitat" has been poorly studied, it is now known that at least one species of moth living in the fur lays its eggs on the dung that the sloth deposits at the base of the tree.

ARMADILLOS

When the first Spanish explorers set eyes on these small, scurrying creatures covered in hard, armor-like scales, they called these amazing animals "armadillios" (from the Spanish *armado*, meaning "armed creature"). As their name suggests, armadillos are armored with a flexible horny shield, backed by bone, over the shoulders, another over the hips, and a varying number of bands around the waist connecting the two. This body shell actually develops from the skin and is

Michael Fogden/Bruce Coleman Ltd

Francisco Erize/Bruce Coleman Ltd.

▲ Other armored mammals are protected by thickened skin or horny structures, but the armor of armadillos consists of plates of bone with a horny covering. The body of this Patagonian armadillo is mostly enclosed by eight hinged bands of these plates, the head and rear being covered by solid plates.

composed of strong bony plates overlaid by horn. Only the upper surface of the body and the limbs are fully armored; the ventral surface of the body is covered by a soft, hairy skin.

Less specialized feeders than the anteaters, the armadillos (family Dasypodidae) also eat their share of ants and termites from the savannas and woodlands of the South American tropics. Although most of the 20 species feed primarily on insects, they also eat a variety of invertebrates, small vertebrates, and vegetable matter. In terms of the number of species and the breadth of distribution — from Argentina, through Central America to Florida — this is the most successful edentate family.

Comparable in size to the giant anteater, the giant armadillo *Priodontes maximus* takes its food-finding task very seriously. Instead of fastidiously poking an elegant nose into a small corridor of the termite mound as, for example, an anteater would, the giant armadillo excavates a sizable tunnel directly into the mound until it reaches the heart of the colony, apparently oblivious to the bites of several thousand angry termite soldiers.

Most armadillos live in deep underground tunnels, which they excavate in sites not prone to flooding. Armadillos are generally nocturnal, often spending the day curled up with head and tail

folded neatly over the belly. Although at first appearance the toughened carapace of an armadillo would seem to provide adequate deterrent to most predators, it is, in fact, quite vulnerable to attack. The normal reaction for most species of armadillo is not to curl up immediately into a defensive position, but instead to flee, which is most effective in dense undergrowth, or to burrow, which is often accomplished at an astonishing rate. Two steppe-dwelling species, the fairy armadillo *Chlamyphorus truncatus* and the greater pichiciago *Burmeisteria retusa*, block the entrance to their burrows by obstructing the passage with their armor-plated hindquarters. Only one species, the three-banded armadillo *Tolypeutes tricinctus*, does not burrow since it is able to roll into a tight ball.

As a rule, armadillos live in semi-arid and even desert zones; it is rare for them to venture into forests. The giant armadillo, however, does have a preference for forest habitat, although it also frequents the savanna in search of food. The most widespread species is the common long-nosed or nine-banded armadillo *Dasypus novemanctus*, which is found in a range of habitats from North America to Argentina. The six species in this genus don't appear to be limited to a fixed breeding season; if environmental conditions are favorable, they will breed. The unparalleled success of members of this genus is probably due to their flexible reproductive behavior and diet, which allow them to exploit all but the most arid areas.

THREATS TO SURVIVAL

One distressing characteristic shared by the sloths, anteaters, and armadillos is that several species are severely threatened in their natural ecosystem. The magnificent giant anteater has already been exterminated from large areas of Brazil and Peru through direct hunting for trophies and also for the trade in live-animal collections. The skin of these animals is also put to some local use.

Probably the greatest single threat to the survival of these magnificent, bizarre creatures is, however, habitat destruction, which will undoubtedly have a severe impact on animals with such highly specialized, restricted feeding regimes. The vast savannas of Latin America are undergoing major habitat alterations, being used to feed the millions of cattle stocked by a number of ranches, while, elsewhere, the once extensive tracts of tropical rainforests are being severely eroded and irreparably damaged by deforestation for logging, slash and burn cultivation, and inundation as a result of hydroelectric dam construction. The

Francisco Erize/Bruce Coleman Ltd.

◄ Many armadillos curl up when asleep or disturbed but, even in this posture, some parts of the body are exposed to predators. The three-banded armadillo curls into a completely enclosed ball.

edentates need not fear the challenge from more adaptable animal competitors in future years; their present and future survival depends upon the actions of humans and their efforts to protect the remaining vestiges of their habitats. Without such efforts, these highly specialized species are certainly doomed.

▼ The armor of an armadillo has some chinks, which are found by defending soldier ants and termites when their nests are raided. This nine-banded armadillo is rubbing off ants that have penetrated its defenses.

Jeff Foott/Bruce Coleman Ltd

INSECTIVORES

**Order Insectivora
6 families, 60 genera,
c. 400 species**

SIZE
Smallest Pygmy white-
toothed shrew *Suncus
etruscus*, head to tail 35 to 48
millimeters (1³/₁₀ to 2 inches);
weight 2 grams (7/100
ounce).
Largest Greater moonrat
Echinosorex gymnurus, head
and body 26 to 45
centimeters (10 to 18 inches);
tail 20 to 21 centimeters (7⅘
to 8³/₁₀ inches); weight 1,000
to 1,400 grams (2 to 3
pounds).

CONSERVATION WATCH
The most severely threatened
genus is *Solenodon*. The giant
golden mole *Chrysospalax
trevelyani* is listed as
endangered in the IUCN Red
Data Book of threatened
mammals. Many other species
are listed as vulnerable or of
indeterminate status.

▼ *Solenodons, small animals about 300
millimeters (12 inches) long with a tail
nearly as long again, investigate their
surroundings with a long, mobile snout
armed with sensory hairs. The forelimbs
bear strong claws for digging and to
hold prey, which rapidly succumbs to a
toxic saliva injected by the solenodon's
bite.*

The order Insectivora (insect eaters) is ancient, and as a group the "true" insectivores are generally considered to be the most primitive of living placental mammals and therefore representative of the ancestral stock from which modern mammals are derived. They are a diverse and ragged assemblage of animal groups that share a tendency—if not an extreme specialization—toward eating insects. Insectivores are probably best described as being small, highly mobile animals with long, narrow, and often elaborate snouts.

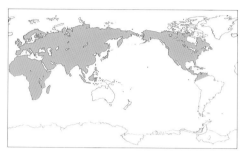

SPECIALIZED AND DIVERSE MAMMALS

With over 400 species, the insectivores are the third largest order of mammals. They are largely confined to the northern temperate zones of North America, Canada, Europe, and the Soviet Union, Africa, and southern Asia. Broadly speaking, the insectivores have primitive brains and depend more on their sense of smell than on vision; they usually have specialized dentitions; their limbs are almost always unspecialized, and they have generalized quadrupedal locomotion.

The taxonomy of this diverse group has been, and still is, the subject of much discussion and controversy, with the group often being treated as a common dumping ground for those species that do not share clear affinities with other groups or that did not, when classified, merit being placed in their own distinctive order. Probably no other group of mammals has such a clouded history as those that have developed the specialization of insectivory, or insect eating. Fossil evidence now indicates that the most primitive placental mammals were insectivores, and, since their descendants have retained dentitions that are still adapted to an insect diet, taxonomists have tended to group all types of this primitive form together with all modern descendants of its ancient lineage. Today, however, it has been recognized that the morphological differences between different forms are sufficient to warrant a revision of this classification, and three separate orders of insect-eating mammals are now recognized: the Insectivora, which include such familiar species as hedgehogs, moles, tenrecs, and shrews; the order Scandentia or tree shrews of Asia; and the order Macroscelididae or African elephant-shrews.

The order Insectivora is rich in examples of convergent evolution (where animals not closely related have adapted behaviorally or morphologically to fit the demands of a specific habitat or way of life). As an example, three diverse and geographically separated species—the Pyrenean desman *Galemys pyrenaicus,* the Mount Nimba otter shrew *Micropotamogale lamottei,* and the Madagascan aquatic tenrec *Limnogale mergulus*—have evolved in total isolation from each other, but, probably in the absence of aquatic predators and in response to the availability of an unexploited niche, each has evolved toward an aquatic way of life. The result is that they all share some physical modifications: a streamlined body shape with a reduction of external appendages, partially webbed feet, a long tail that can act as a rudder, a specialized breathing apparatus, a dense, waterproof coat, and specialized sensory mechanisms for detecting prey underwater. Other, more general, examples of convergent evolution within this order include exploitation of the fossorial (burrowing) niche by both European moles and African golden moles.

SOLENODONS

Among the largest of the living insectivores, the secretive solenodons (family Solenodontidae) are

P. Morris/Ardea London

▼▶ *Moonrats (top right) are not rodents but close relatives of hedgehogs. Their name refers to the naked, rat-like tail. Least specialized of the insectivores, they feed on a wide variety of prey. The tenrec family, largely confined to Madagascar, has undergone an amazing evolutionary radiation from shrew-like creatures into forms resembling hedgehogs, moles and otters. The skunk-like coloration of the streaked tenrec (below, left) advertises that it is unpalatable. Like the typical moles, to which they are related, desmans have small eyes and a sensitive snout. The Pyrenean desman (below, right) lives in fast mountain streams and feeds on bottom-dwelling invertebrates.*

currently confined to Cuba and Hispaniola in the Greater Antilles. Represented today by just two species—*Solenodon paradoxus* and *S. cubanus*—these relict species are now unfortunately recognized as being among the most highly endangered species on earth.

Before the arrival of humans on these islands, the solenodon was probably one of the top carnivores of the ecosystem, although it, in turn, would probably have been prey to large owls and boas. The arrival of humans, however, almost certainly spelled doom for these predominantly nocturnal creatures. Dogs and rats soon invaded the islands from fishing vessels, and, in an attempt to control the rats, which were causing widespread destruction and disease, the mongoose was introduced at a later stage. Each of these exotic introductions represented new, highly intelligent predators and competitors for the solenodons. In fact, there is evidence that the now extinct genus of shrews *Nesophontes* was once widespread throughout the West Indies, and its disappearance coincided with the arrival of the Spaniards. Today,

the few remaining solenodons appear to be destined to suffer the same fate.

As in most nocturnal insectivores, brain size is relatively small, but the sense of touch is highly developed. When feeding, the solenodon's extended and highly sensitive snout is operated like a tentacle, probing for prey in cracks and crevices while its strong claws are used to dig or break up rotting vegetation. Once the prey is located, it is quickly bitten, and a small amount of toxic saliva is injected into the victim, causing paralysis. The solenodon probably uses this technique more often when capturing prey larger than insects, such as frogs, lizards, or small birds.

The natural history of solenodons is poorly known, and the animals have not been studied in great detail in the wild or in captivity. Their life history is characterized by a relatively long life-span and low reproductive potential (only one offspring is produced each 12 to 18 months). The single young is born in a specially created nesting burrow and, once weaned, accompanies the mother on her nightly foraging trips. Solenodons

S.C. Bisserot

▲ *The resemblance of the hedgehog tenrec to "true" hedgehogs is the result of convergent evolution in two distinct families of insectivores. Tenrecs retain some primitive mammalian features: males lack a scrotum and (like monotremes and marsupials) females have a cloaca.*

are the only insectivores to practise teat transport, a technique that probably facilitates the rapid movement of young from one burrow to another if the animal is disturbed. All other insectivores carry their young in their mouths.

Solenodons appear to be strictly solitary in their habits, the mother–offspring tie being the only period of extended social behavior. Being nocturnal, solenodons rely on auditory and olfactory signals to locate their prey and also to monitor the presence of other animals.

Solenodons inhabit well-established open forests that generally have a considerable amount of scrub or ground cover. Unfortunately, almost nothing is known about their social or spatial behavior in the wild.

TENRECS AND OTTER SHREWS
Resembling hedgehogs in their general appearance and, in part, in their way of life, the tenrecs of

Madagascar and the otter shrews of Central Africa provide a striking example of convergent evolution and diversification. Fossil records indicate that members of the family Tenrecidae were already well established within the general African fauna 25 million years ago. Today, however, the only surviving members of this ancient lineage on the African mainland are the three species of otter shrew, which are confined to small rivers in the tropical forest belt of West and Central Africa.

The tenrecs were among the first mammals to arrive on Madagascar after its separation from mainland Africa around 150 million years ago. Arriving on an island with no worthy competitors, the founder species radiated into a wide variety of ecological niches. The radiation of tenrecs is one of the classic examples of radiation, with representatives of fossorial, terrestrial, semi-aquatic, and even semi-arboreal tendencies. In general, tenrecs are small animals—the largest,

Tenrec ecaudatus, weighs 1 kilogram (2 pounds), and the smallest, *Geogale* species, a termite-feeding specialist, weighs less than 10 grams (⅖ ounce). They are either crepuscular (active at twilight) or fully nocturnal in habit and most of them are solitary. An exception to this rule are members of the species *Hemicentetes semispinosus,* which appear to form colonies during the breeding season.

GOLDEN MOLES

Golden moles—so called because of the iridescent bronze glint on their fur—are strictly African in their distribution. Specialized for a fossorial life, the 18 species of the family Chrysochloridae display physical and behavioral adaptations to a wide range of terrestrial habitats.

The golden mole's streamlined body is covered in a relatively coarse, backward-facing fur, which is moisture repellant, superimposed upon a much softer, denser undercoat, which provides insulation. The eyes are minuscule and probably rudimentary, being covered with a hairy skin. The ear openings are covered in dense fur, and the nose is protected by a tough leathery pad, which may be used to move loose soil when digging.

The stocky, thickset body gives a semblance of strength and, indeed, one early naturalist remarked on how a captive golden mole could exert a force equal to 150 times the animal's own weight. The forelimbs are extremely powerful and are armed with four short digging claws. When digging, the hindfeet are firmly braced against the tunnel wall and the soil at the digging face is loosened by repeated, sharp, downward strokes of alternate forelimbs. Displaced soil is either packed against the side or roof of the tunnel using the broadened head and shoulders, to form surface ridges, or is kicked backward along the tunnel for some distance from where it is pushed to the surface by the head, to form surface molehills. The digging behavior of golden moles, which use a combined action of head, feet, and body, differs significantly from that of European and North American moles, which use only their feet.

Golden moles are solitary animals that spend more time on burrowing than on any other single activity. So great is this need that some species spend up to 75 percent of their active time engaged in burrowing. The style and size of the burrow system depends upon the species and its habitat. Mountain-dwelling and plain-dwelling species, such as the giant golden mole *Chrysospalax trevelyani* or the Hottentot golden mole *Amblysomus hottentotus,* build a semipermanent series of tunnels, which may descend to almost 1 meter (3 feet) in depth. These, and similar species, trap and eat food within the tunnels as European and North American moles do. Food is eaten immediately; it is not cached or stored. Desert-dwelling species, however, such as Grant's golden mole *Eremitalpa granti,* do not form permanent

tunnels, but instead push their way through the upper reaches of their sand-dune ecosystem almost in a swimming action, locating prey by touch and perhaps hearing. Desert moles may occasionally hunt on the surface of the sand, where they trap legless lizards or insects.

HEDGEHOGS AND MOONRATS

Probably one of the most familiar insectivores, which in several ways resembles a diminutive solenodon, is the European hedgehog *Erinaceus europaeus.* Hedgehogs belong to the family Erinaceidae, which has representative members throughout Europe, Africa, and Asia. This group has about 17 species, some of which (for example the stocky European hedgehog and the desert hedgehogs of Asia and North Africa) bear spines, while others, such as the intriguingly named moonrats or gymnures of Southeast Asia, are covered with coarse hair and often have long tails.

The most distinctive characteristic of the hedgehog is its dense coat of spines—an adult will have as many as 5,000 needle-sharp spines 2 to 3 centimeters (¾ to 1 inch) in length. The spines are actually modified hairs, each of which is filled with multiple air chambers and strengthening ridges that run down the inside wall of each tube. The spines cover the entire dorsal surface of the body, while the ventral surface is covered with a toughened skin from which protrudes a covering of coarse hair. Normally the spines are laid flat against the body but, when erect (for example, if the animal is threatened) they stick out at a variety of angles, overlapping and supporting each other to create a truly formidable defense system. As if this were not sufficient protection, hedgehogs are also able to curl into a tight ball, thereby fully concealing and protecting the softer underparts of the body. Confronted with such defense, few predators would even attempt to attack.

▼ *The golden moles of Africa look remarkably like the marsupial mole of Australia—a fascinating example of convergent evolution. The eyes are tiny and probably useless; there is no external ear; the snout is protected by a horny shield; and the forelimbs are powerful, spade-like structures.*

G.J. Broekhuysen/Ardea London

Hedgehogs are solitary animals and, depending upon the species and climate, have quite different breeding patterns. Arid-dwelling species, such as desert *Paraechinus* species and long-eared *Hemiechinus* species, breed only once a year. In temperate zones two litters are known to occur regularly. In tropical zones, where food availability and climate (two of the major determinants of the timing and duration of the breeding season) are more predictable, hedgehogs can breed all year.

The unavailability of or difficulty in locating prey is a serious consideration for all insectivores, because of their high energy demands. To combat the need to remain active when prevailing environmental conditions are not favorable, many species have developed the ability to undergo a period of dormancy (torpor or hibernation), during which the body temperature is allowed to decrease to a level close to that of the surrounding air. For example, oxygen requirements for a hedgehog may decline from an average (normal) level of 500 milliliters per kilogram per hour to about 10 milliliters per kilogram per hour. This strategy enables species that are experiencing some degree of crisis, for example food and temperature extremes, to dramatically reduce their energy expenditure, which effectively enables the animals to survive longer on fewer reserves.

Hibernation is not a species trait but is dictated by environmental conditions. Tropical hedgehogs will not normally hibernate but, if artificially exposed to low food levels or low ambient temperatures, these species also exhibit the ability to hibernate. The European hedgehog, which undergoes a seasonal hibernation in its native land, forgoes this behavior in New Zealand, where it was introduced at the beginning of this century.

Temperate-dwelling hedgehogs are commonly found in deciduous woodlands, arable land surrounded by hedgerows, and urban gardens. During the summer months, a simple nest is

▼ *Like other hedgehogs, the long-eared hedgehog has a head and upper body covered with spines. These hedgehogs are arid zone dwellers; their long ears act as heat radiators. They and the desert dwelling species breed only once a year.*

Eyal Bartov/Oxford Scientific Films

constructed from leaves and grasses, and is usually placed among the undergrowth at the base of a tree, rather than underground. Such nests are rapidly constructed and may be abandoned after just a few days in preference to another. In winter, more care is taken when choosing the nest site, since this must be well insulated against the cold.

When active, hedgehogs are constantly on the lookout for prey, and, in the course of a single night's foraging, some animals have been observed to travel over 3 kilometers (2 miles). Their favorite food is earthworms, although they will eat most ground-dwelling invertebrates, as well as seeds and fruit. In Europe, many people actually attempt to attract hedgehogs to their gardens—their appetite

Hans Reinhard/Bruce Coleman Ltd.

for slugs and chafer beetles making them the gardener's friend—by leaving out bowls of bread and milk, which the animals do appear to relish.

From what little is known about the Asian gymnures they, like hedgehogs, are predominantly solitary and do not appear to defend a fixed territory. Being nocturnal, their primary mode of communication appears to be olfaction (the sense of smell), a feature that is keenly developed in the greater moonrat *Echinosorex gymnurus* and in the lesser moonrat *Hylomys suillus,* both of which have well-developed anal scent glands that exude what is, to the human sense of smell, a foul odor. When placed in strategic parts of the animal's domain, this substance indicates that the area is already

occupied and may possibly identify the sex, age (and hence the dominance status), or stage of sexual maturity of the resident animal.

Moonrats are among the largest of the insectivores, occasionally weighing up to 2 kilograms (4½ pounds). They look quite ferocious with their coarse, shaggy coat and impressive, open-mouthed, threatening gestures. Moonrats inhabit lowland areas, including mangrove swamps, rubber plantations, and primary and secondary forests. Largely nocturnal and terrestrial, they rest during the day in hollow logs, under the roots of trees, or in empty holes. They appear to prefer wet areas and often enter water to hunt for insects, frogs, fish, crustaceans, and mollusks.

▲ Although hedgehogs are solitary animals, the partially weaned young accompany their mother to forage at night for insects and worms. These four infants appear to have become separated.

A. & E. Bomford/Ardea London

► Like most members of its family, the common shrew lives on the forest floor, actively feeding on insects, worms, and other invertebrates. Most shrews are so small that they must feed continuously to maintain body temperature. Without food, a shrew can die of starvation in as little as four hours.

▼ The Eurasian water shrew feeds mainly on bottom-living invertebrates but also takes frogs almost as large as itself. Such prey is killed by venomous saliva which is injected when the teeth of the shrew penetrate its flesh.

SHREWS

Of all insectivores, the diminutive shrews are the most successful single group. They are successful both in terms of their evolutionary radiation (almost 250 species) and also in their geographical distribution, which includes all of North and Central America, Europe, Asia, and most of Africa. Shrews are small; the smallest species, *Suncus etruscus*, weighing only 2 grams ($7/100$ ounce), is the world's smallest mammal. Shrews are secretive mammals, and are characterized by long, pointed noses, relatively large ears, tiny eyes, and dense velvety fur.

The family Soricidae, to which shrews belong, is divided into two groups: the so-called red-toothed shrews, which have pigmented teeth and are northern European in origin, and the white-toothed shrews, which originated in Asia and North Africa and have now colonized much of mainland Europe.

Shrews are very active animals and their hectic pace of life means that they must consume disproportionately large amounts of food for their body size. Recent research has shown that many species of shrew have a much higher metabolic rate than rodents of comparable size. To cope with such constant energy demands, shrews tend to live in

Dwight R. Kuhn

habitats that are highly productive. To survive, they must feed every two or three hours, and they have been recorded to consume 70 to 130 percent of their own body weight each day. Not surprisingly, therefore, shrews are generally highly opportunistic feeders, taking a wide range of invertebrate prey, supplemented by occasional carrion, fruit, seeds, and other plant material. Digestion in shrews is fairly rapid, and the gut may be emptied in under three hours. In temperate climates, at least, most species are highly territorial, aggressively defending a fixed piece of terrain against all intruding shrews, thereby ensuring that they will have undisturbed access to the maximum amount of prey.

Shrews occupy a wide range of feeding habitats, exploiting aquatic, terrestrial, and fossorial niches. Some fossorial species make use of existing tunnels dug by moles and rodents, but some red-toothed shrew species, for example the short-tailed shrew *Blarina brevicauda,* have adapted toward a truly fossorial way of life, and their body form is similar to that of the true moles. Prey, detected mainly by hearing and, to a lesser extent, by smell, is rapidly seized and killed. At least two species—the American short-tailed shrew *Blarina brevicauda* and the Eurasian water shrew *Neomys fodiens*—are known to inject their prey with venomous saliva (as solenodons do), which may be particularly useful when dealing with large prey such as frogs or fish. If there is an abundance of prey, part of a catch may be cached or stored for later use. These caches are usually defended quite vigorously against other shrews.

Shrews are essentially solitary animals and it is only at breeding times that they will approach one another in anything resembling an amicable manner. Like moles, shrews are sexually receptive for a very brief period. In some shrews this may be seasonal, but others may breed at any time of the year. In *Suncus,* the house shrew, for example, the female may not be reproductively active. A male mounts her and bites at her neck; this occurs repeatedly and eventually she becomes receptive, when she emits a characteristic chirp. He mounts and mates. The aggressive advances and odor marking of the male bring the female into heat.

The young are born after a gestation period of around three weeks and remain confined for some time in a specially prepared nest chamber, which is usually underground for protection. The mother is highly attentive and communicates with the litter of three to eight (depending on species) by making faint squeaks, which are barely audible to the human ear. If the young stray from the nest the mother will retrieve them, carrying them in her mouth back to the nest site.

At three weeks, the young are encouraged to leave the nest and accompany the mother on short foraging trips. A peculiar habit, characteristic of the white-toothed shrews, is caravanning, where the young, when foraging, actually line up and grab a tuft of hair on the rump of the preceding animal—the one at the front holding onto the mother in the same way. This peculiar behavior is first and foremost a mechanism for the young to avoid a predator.

As with moles, if the young persist in staying at the maternal nest they will be driven away by aggressive gestures. By the time the litter has dispersed, the female will probably have mated again. In this way, adult breeding females can give birth to two or even three litters a year, although the adults themselves will not survive the winter period. Compared with other similar-sized mammals, shrews have a very short life-span of 9 to 12 months on average, although all can live longer in captivity, the record being four years for a greater white-toothed shrew *Crocidura russula.*

Stephen Dalton/NHPA

▲ *The dense fur of the water shrew is strongly water-repellent. When the animal dives, its body is surrounded by air trapped between the hairs. Where streams are polluted by detergent, the fur of water shrews becomes wet and they die of cold.*

MOLES AND DESMANS

The family Talpidae includes the moles, shrew moles, and desmans, all of which are confined to the north of North America and Eurasia. These predominantly burrowing insectivores (29 species in 12 genera) are highly secretive and because of their way of life have, in general, been poorly studied. The species that has, to date, received most attention from naturalists and biologists alike is the European mole *Talpa europaea,* whose way of life and behavior are probably quite similar to many of the other species within this family.

Moles are highly specialized for a subterranean, fossorial way of life. Their broad, spade-like forelimbs, which have developed as powerful digging organs, are attached to muscular shoulders and a deep chestbone. The skin on the chest is thicker than elsewhere on the body as this region supports the bulk of the mole's weight when it digs or sleeps. Behind the enormous shoulders the body is almost cylindrical, tapering slightly to narrow hips, with short sturdy hindlimbs (which are not especially adapted for digging), and a short, club-shaped tail, which is usually carried erect. In most species, both pairs of limbs have an extra bone that increases the surface area of the paws, for extra support in the hindlimbs, and for moving earth with the forelimbs. The elongated head tapers to a hairless, fleshy, pink snout that is highly sensory. In the North American star-nosed mole *Condylura cristata,* this organ bears 22 tentacles each of which bears thousands of sensory organs.

The function of a mole's burrow is often misunderstood. Moles do not dig constantly or specifically for food. Instead the tunnel system, which is the permanent habitation of the resident

Andy Purcell/Bruce Coleman Ltd.

animal, acts as a food trap constantly collecting invertebrate prey such as earthworms and insect larvae. As they move through the soil column invertebrates fall into the animal's burrow and often do not escape before being detected by the vigilant, patrolling resident mole. Once prey is detected, it is rapidly seized and, in the case of an earthworm, decapitated. The worm is then pulled forward through the claws on the forefeet, thereby squeezing out any grit and sand from the worm's body that would otherwise cause severe tooth wear—one of the common causes of death in moles. If a mole detects a sudden abundance of prey, it will attempt to capture as many animals as possible, storing these in a centralized cache, which will usually be well defended. This cache, often located close to the mole's single nest, is packed into the soil so that the eathworms remain

alive but generally inactive for several months. Thus, if an animal experiences a period of food shortage it can easily raid this larder instead of using essential body reserves to search for scarce prey. In selecting such prey for the store, moles appear to be highly selective, generally choosing only the largest prey available.

Tunnel construction and maintenance occupy much of a mole's active time. A mole digs actively throughout the year, although once it has established its burrow system, there may be little evidence above ground of the mole's presence. Moles construct a complex system of burrows, which are usually multi-tiered. When a mole begins to excavate a tunnel system, it usually makes an initial, relatively straight, exploratory tunnel for up to 20 meters (22 yards) before adding any side branches. This is presumably an attempt to locate neighboring animals, while at the same time forming a food trap for later use. The tunnels are later lengthened and many more are formed beneath these preliminary burrows. This tiered-tunnel system can result in the burrows of one animal overlying those of its neighbors without them actually being joined together. In an established population, however, many tunnels between neighboring animals are connected.

Moles have a keen sense of orientation and often construct their tunnels in exactly the same place every year. In permanent pastures, existing tunnels may be used by many generation of moles. Some animals may be evicted from their own tunnels by the invasion of a stronger animal and, on such occasions, the loser will have to go away and establish a new tunnel system. These "master engineers" are highly familiar with each part of their own territory and are suspicious of any changes to a tunnel, which makes them difficult to capture. If, for example, the normal route to the nest or feeding area is blocked off, a mole will dig either around or under the obstacle, rejoining the original tunnel with minimum digging.

Our knowledge of the sensory world of moles is very limited. They are among the exclusively fossorial species; the eyes are small and concealed by dense fur or, as in the blind mole *Talpa caeca*, covered by skin. Shrew moles, however, forage not only in tunnels beneath the ground but also above ground among leaf litter. Although they may have a keener sense of vision than other species, they are still probably only able to perceive shadows rather than rely heavily on vision for detecting prey or for purposes of orientation. The apparent absence of ears on almost all species is due to the lack of external ear flaps and the covering of thick fur over the ear opening. It has, however, been suggested that ultrasonics may be an important means of communication among fossorial and nocturnal species. But of all the sensory means olfaction appears to be the most important medium—a fact supported by the elaborate nasal region of many

◄ *"True" moles of the family Talpidae are found over much of Eurasia and North America. Their eyes are tiny and useless for vision; there is no external ear but their hearing is acute. The long snout is extremely sensitive and the sense of smell is well developed.*

Dwight R. Kuhn

▲ *The sensory tentacles on the snout of the star-nosed mole are unique among mammals. This species burrows in swampland but also spends much time above ground and even enters the water in search of prey, paddling with its spade-like forelimbs.*

their mother for warmth. The young are fed entirely on milk for the first month, during which they rapidly gain weight. Juveniles remain in the nest until they are about five weeks old, at which time they begin to make short exploratory forays in the immediate vicinity of the nest chamber. Shortly thereafter they begin to accompany their mother on more extensive explorations of the burrow system and may disperse from there of their own accord; those that do not leave will soon be evicted by the mother.

A very different way of life is exhibited by the desmans, of which there are two species. The Pyrenean desman *Galemys pyrenaicus* is confined to permanent, fast-flowing streams of the Pyrenees mountain range and parts of northern Iberia; the Russian desman *Desmana moschata* is found only in the slower moving waters and lakes of the western and central Soviet Union. Just as moles are superbly adapted for a fossorial way of life, so too are the desmans for water. The streamlined body of the Pyrenean desman enables it to glide rapidly through the water, propelled by powerful webbed hindlimbs and steered, to some extent, by a long, broad tail. For any animal living in the snow-fed mountain streams, feeding and retaining body heat are top priorities. Unlike hedgehogs or tenrecs, desmans do not undergo periods of hibernation or torpor and must, therefore, live in optimum habitats to ensure their survival during the winter months when prey is most scarce.

Desmans feed on the larvae of aquatic insects such as the stone fly and caddis fly, as well as on small crustaceans, which they locate by probing their proboscis-like snouts beneath small rocks and by clearing away debris from the stream bed with their sharp elongated claws. Prey is consumed at the surface where, following each dive, a rigorous body grooming is carried out. This is an essential activity as it ensures that the fur is not only kept clean and in good condition but also maintains its water-repellent properties by spreading oil all over the body from sebaceous glands.

Desmans construct their nests in the banks of streams. The Russian desman actually excavates a complex burrow, which it may share with other desmans, while the smaller Pyrenean species occupies a strictly solitary nest, usually created by enlarging an already existing tunnel or crevice. Nests are composed of leaves and dried grasses and are always located above the water level.

Little is known about the breeding behavior of desmans. In the Pyrenean desman, mating takes place in spring (March to April), and, as these animals usually form a stable pair bond, competition for mates by solitary males is often quite severe. At this time of year, an interesting phenomenon occurs for each pair of animals: males become far more protective, spending most of their active time at the upper and lower reaches of their riverine territory. Energy is thus spent on

species, together with the battalion of sensory organs stored within this area.

The brief breeding season is a frantic period for moles, as females are receptive for only 24 to 48 hours. During this time males usually abandon their normal pattern of behavior and activity, spending large amounts of time and energy in locating potential mates. Mating takes place within the female's burrow system and this is the one period of non-aggressiveness between the sexes. The young, with an average of three to the litter, are born in the nest four weeks later. Weighing less than 4 grams (⅐ ounce), the pink, naked infants cannot control their body temperature and rely on

David Thompson/Oxford Scientific Films

protecting the feeding resources of that territory and, more importantly, the female. Females, in contrast, spend most of their active time feeding, surveying for a suitable nest site, and gathering nesting materials.

Young are born after a gestation period of about four weeks and are cared for solely by the female. Juveniles first leave the nest at about seven weeks, at which stage they are already proficient swimmers. Juveniles remain within the parents' territory until they are about two and a half months old, at which stage they leave in order to secure a mate and breeding territory for the coming year.

CONSERVATION

From a conservation standpoint, many species of insectivores are now severely threatened. Unlike many other mammals, this threat does not come from direct human exploitation of the animals but indirectly through such activities as deforestation, introduction of exotic species to the ecosystem, pollution, and other activities related to habitat

destruction. These activities almost invariably arise as a consequence of human exploitation of or interference with the balance of the ecosystem.

Probably one of the most severely threatened genera is *Solenodon*. No recent estimates of the number of surviving animals exist, but it is known that their habitat—a mixture of forest and rocky outcrops—is rapidly diminishing.

In fact, habitat destruction throughout the world appears to be the single greatest threat to insect-eating organisms. It is now proving to be a major threat to the giant golden mole of Africa, as the relict forests and open grasslands where this localized species occurs are being rapidly converted into second-class grazing and arable pastures. Aquatic insectivores too are facing serious threats to their survival. In Europe, the Pyrenean desman is threatened by the construction of hydroelectric dams in its mountainous refuge; the endemic otter shrews of Africa and the aquatic tenrec of Madagascar are threatened by deforestation and increased siltation of the rivers.

▲ *The European mole feeds mostly on the earthworms that enter its tunnels. When earthworms are plentiful, it bites the heads off some and stores them, still alive, in a "larder" close to its nest.*

SIZE
Smallest Northern smooth-tailed tree shrew *Dendrogale murina*, head and body length 115 millimeters (4½ inches), tail length 105 to 130 millimeters (4 to 5 inches), weight 35 to 55 grams (1¼ to 1⁹⁄₁₀ ounces).
Largest Common tree shrew *Tupaia glis*, head and body length 230 millimeters (9 inches), tail length 148 to 198 millimeters (5⅕ to 7⅘ inches), weight 85 to 185 grams (2⁹⁄₁₀ to 6½ ounces).

CONSERVATION WATCH
All species are seriously threatened by human encroachment and habitat destruction.

▼ *The common tree shrew of Malaysia and Sumatra is arboreal but frequently hunts on the ground for insects and lizards. The young are reared in a nest separate from that of the mother and are suckled every other day. Unusually among small mammals, tree shrews are active by day.*

TREE SHREWS

R. DAVID STONE

Tree shrews, of the order Scandentia, are for the most part arboreal mammals, although some species are also found living at ground level. Primarily inhabitants of tropical rainforests, these fascinating but relatively unknown animals are believed to be one of the most primitive forms of placental mammals and, as such, representative of the ancestral stock from which present-day mammals have evolved.

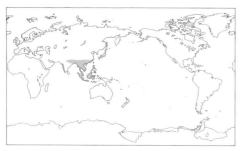

SHREWS LIKE SQUIRRELS
The 19 living species of tree shrews are all confined to eastern India and Southeast Asia. The order contains a single family, the Tupaiidae, which is further subdivided into two subfamilies, the Ptilocercinae and the Tupaiinae. Superficially, tree shrews resemble small tree squirrels—the genus name *Tupaia* is derived from the Malay *tupai*, meaning "squirrel-like animal"—but they differ in both anatomy and behavior. Tree shrews are relatively small mammals, averaging 70 to 100 grams (2½ to 3½ ounces) weight. Generally a russet brown color, they have a long, pointed muzzle with 38 sharp, pointed teeth, and 5 clawed

digits on each foot. Tail length usually exceeds body length, but the tail is not prehensile.

FEATHER-TAILED TREE SHREW
Within the subfamily Ptilocercinae is just a single species, the feather-tailed or pen-tailed tree shrew *Ptilocercus lowei*, which is nocturnal and is found only on the Malay Peninsula and Borneo. Weighing less than 80 grams (3 ounces) and measuring 30 to 35 centimeters (12 to 14 inches) from head to tail, this strange-looking gray-brown animal has large ears and lengthy facial hairs. Its tail, uniquely among tree shrews, is covered in scales, apart from the tip, which has long white hairs growing out of opposite sides, giving a feather-like appearance. The tail is quite long and serves as an organ of balance, support, and touch: when the animal is awake, the tail twitches continuously.

Feather-tailed tree shrews are arboreal and have a number of distinctive adaptations to this niche: the hands and feet are relatively larger than in other tree shrews, and the toes can be spread more widely, giving a better grip when climbing and also permitting the shrew to grasp insects with only one hand. The digits are flexed so that the claws are always in contact with the surface of the branch,

and the footpads are also larger and softer than in other species. This species appears to be sociable, living in small groups and often sharing a common sleeping nest. It feeds predominantly on a mixture of fruit and insects, particularly cockroaches, beetles, ants, and termites.

OTHER TREE SHREWS

The remaining 18 species of tree shrew are classified within the subfamily Tupaiinae and are distributed throughout eastern India, Southeast Asia, and various parts of the Malaysian archipelago. Most species are omnivorous and, depending on the species and habitat, are either arboreal or terrestrial in habit. They range from lowland forest through secondary and primary rainforests and even to montane habitats where one species, the mountain tree shrew *Tupaia montana*, is found. These diurnal tree shrews differ from their predominantly nocturnal feather-tailed relatives in their better-developed vision and greater brain development. These characteristics separate the tree shrews from other insectivores, such as hedgehogs, shrews, and moles.

Tree shrews are highly active, nervous, inquisitive, and generally aggressive animals. They are very fond of water and often bathe in water-filled hollows of trees.

The common tree shrew *Tupaia glis* lives in permanent pairs. Although the partners of each pair exhibit solitary daily ranging behaviors, they occupy the same territory, which they defend vigorously against other members of their species of the same sex. As a rule, the territories of different pairs overlap only slightly. Young animals apparently live with their parents, thus forming family groups. After attaining sexual maturity, young animals are forced to leave, and they adopt a nomadic existence for several months until they find a suitable mate and establish their own territory. Tree shrews may live for between two and three years in the wild and display a strong degree of fidelity to their mates.

LIVING IN THE TERRITORY

Scent marking appears to be a very important part of the daily routine for all species of tree shrew. Scent marks generally transmit information about the animal that deposited them and may be dispersed in either a passive manner, for example general body odor that exudes through skin pores, or in a highly selective, active fashion, whereby specific scents are deliberately deposited at a strategic part of the animal's territory, with the intention that these will be detected and interpreted by similar species. Thus, if an animal is able to detect odor and differentiate the many components of a scent mark, it may be able to gain information about the sex and social status of the depositor, and how recently the mark had been made. Tree shrews regularly distribute secretions

Rod Williams/Bruce Coleman Ltd

from specialized scent glands to new objects within the territory, and more often at specific parts of the territory, to indicate to neighboring animals that that site is already occupied and is being defended. Such boundary marking sites are often used by several neighboring animals and thus serve as communication points, providing a wealth of local information to the resident animals.

In addition to saturating the territory with odor, pair-living animals will daily cover each other with their own respective odors. Parents will also mark juveniles once they begin to leave the nest; captive females have been observed to mark their young over one hundred times in a single day.

The breeding behavior of free-living tree shrews has not been investigated in any detail, but observation of captive animals has revealed some intriguing facts. There is no indication of a fixed breeding season: breeding may occur throughout the year. A few days before the female is due to give birth, the male prepares a "maternal nest" of leaves and will then leave and not return to the nest site until the young are at least one month old. In the forest, nest sites are usually located in holes in fallen trees, hollow bamboos, or similar sites.

After a gestation period of 40 to 50 days, the naked young are born in the nest and are suckled immediately. Litter size varies according to species, but an average of three young are born. The mother assumes all responsibilities for feeding and defending the young until they have been weaned. If the young are disturbed at the nest, they make a very loud, abrupt sound and simultaneously thrust out all four legs, thus loudly rustling the leaves in the nest and creating a quite startling display that may act to discourage a predator. The major natural predators are birds of prey, small carnivores, and snakes.

▲ *In the past, tree shrews have been classified as insectivores and as primates. The reality appears to be that they are very unspecialized mammals that share some features with these two groups. The large tree shrew, shown here, is terrestrial and has a long snout and strong digging claws; the more arboreal tree shrews have shorter faces and more slender claws.*

FLYING LEMURS

J.E. HILL

Order Dermoptera
1 family, 1 genus, 2 species

SIZE
Head and body 34 to 42
centimeters (13 to 17 inches);
tail 22 to 27 centimeters (9 to
11 inches); weight 1 to 1.75
kilograms (2 to 4 pounds).

CONSERVATION WATCH
Not endangered, though they
are threatened by
deforestation and habitat
destruction; the Malayan
flying lemur appears to be
relatively abundant and
widespread.

▶ *Flying lemurs feed on leaves, shoots
and buds of rainforest trees. The
Malayan flying lemur (shown here) is a
nuisance in plantations because of its
habit of eating coconut flowers.*

▼ *Flying lemurs are the largest of the
gliding mammals. Unlike the others, they
have a membrane between the tip of
the tail and the ankles, which can be
folded into a soft pouch for carrying the
young. The young are also carried
clinging to the mother's belly.*

Flying lemurs, or colugos, belong to the very small order Dermoptera, which includes the two living species, genus *Cynocephalus*, in the family Cynocephalidae from Southeast Asia, and also the extinct family Plagiomenidae, known from a few fossil jaws and teeth from the Paleocene and Eocene of North America, and the Eocene of France. The common name "flying lemurs" is misleading since they do not fly but glide, and they are not lemurs, but the surviving representatives of an evolutionary lineage possibly related to the ancestral primates.

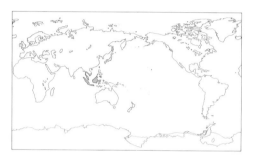

MAMMALS THAT GLIDE
Flying lemurs are cat-sized, with grayish or reddish brown backs, which are usually speckled and mottled with black and grayish white markings. The arrangement of the unusual and distinctive incisor teeth is similar to that of herbivorous mammals such as cattle or deer. The upper incisors are at the sides of the jaw, leaving a gap at the front, and do not oppose the forwardly directed lower incisors. These are broad and comb-like, with as many as 20 "tines" arising from one root; they may provide a scraping action for abrading and straining food, and grooming and cleaning the fur. The feet are furnished with strong claws.

The order name Dermoptera, meaning skin wing, aptly describes their most striking feature—the gliding membrane, or patagium, that stretches from the neck to the finger tips, along the sides of the body, between the toes, and joins the legs and tail. The gliding membrane enables flying lemurs to glide up to a recorded 136 meters (150 yards). The Philippine flying lemur *Cynocephalus volans* is slightly smaller than the Malayan flying lemur *Cynocephalus variegatus,* and the fur on its back is darker and less spotted.

Flying lemurs live in primary and secondary tropical forest, and in rubber and coconut plantations, where their spectacular gliding ability enables them to move easily from tree to tree while searching for food in the canopy. They are nocturnal, spending the day in a tree hollow or hanging beneath a branch and moving off at night through the trees to a preferred feeding area, often by a regular route. Flying lemurs climb clumsily in a lurching fashion, gripping the tree or branch with outspread limbs, and moving first the forefeet and then the hindfeet. Gliding from the upper part of one tree to a lower point on another, they climb again to make another glide to a further tree, so progressing through the forest, often along a regular route and with several animals following each other. Flying lemurs are herbivores, subsisting on leaves, shoots, buds, and perhaps flowers and fruit. They pull twigs and small branches within reach with the forefeet, stripping the leaves with the lower incisors and long, strong tongue. Their digestive system is specialized for this vegetarian diet, with an extended stomach and long, convoluted intestine.

Gestation takes 60 days. Usually a single young is born, although occasionally twins are produced. Newborn flying lemurs are relatively undeveloped (like newborn marsupials) and, until weaned, are carried clinging to the belly of the mother, who can also fold the gliding membrane near the tail into a soft, warm pouch for this purpose.

D. &W. Ward/Oxford Scientific Films

Order Chiroptera
18 families, c. 187 genera,
c. 977 species

SIZE
Smallest Hog-nosed bat
Craseonycteris thonglongyai,
forearm length 22.5 to 26
millimeters (9/10 to 1 inch),
wing span 15 centimeters (6
inches), weight 1.5 to 2 grams
(5/100 to 7/100 ounce).
Largest Large flying fox
Pteropus vampyrus, forearm
length 22 centimeters (8½
inches), wing span 2 meters
(79 inches), weight up to 1.2
kilograms (2⅗ pounds).

CONSERVATION WATCH
The following species are
listed as endangered in the
IUCN Red Data Book of
threatened mammals:
Rodrigues flying fox *Pteropus
rodricensis*, Guam flying fox
Pteropus tokudae, Singapore
roundleaf horseshoe bat
Hipposideros ridleyi, gray bat
Myotis grisescens, Virginia big-
eared bat *Pleotus townsendii
virginianus*. Many other
species are listed as
vulnerable or of
indeterminate status.

BATS

J. E. HILL

Bats occur almost worldwide, except in the Antarctic, the colder area north of the
Arctic Circle, and a few isolated oceanic islands, although more species are
concentrated in the tropics and subtropics than in the temperate zones. Some
977 species are currently recognized, second only in number to the rodents. Bats are the
only mammals capable of true flight, and their wings are their most obvious feature. To
navigate, avoid obstacles, and feed at night, when most are active, bats have developed
superb hearing and the use of high frequency echolocation.

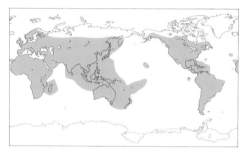

MAMMALS THAT FLY

The order Chiroptera is divided into two
suborders: the Megachiroptera, or Old World fruit
bats, that include only 1 family with about 41
genera and 163 species, and the Microchiroptera,
basically insect-eating bats, with 17 families, about
146 genera and 814 species. The terms
Megachiroptera (large bats) and Microchiroptera
(small bats) are not totally definitive—many
species in the suborder Microchiroptera are larger
than the smaller species in the suborder
Megachiroptera. The numbers of genera and species
are not immutable: they alter from time to time as
classification changes or new bats are described.
  Little is known of the origins of bats. The

earliest known fossils come from the Eocene (55
million years ago) of West Germany and North
America. These fossils are similar to modern bats
and it is suspected that bats originated much
earlier, perhaps 70 to 100 million years ago,
possibly from small, quadrupedal, arboreal, and
possibly insectivorous ancestors that developed
gliding membranes and may have been related to
ancestral primates or insectivores.
  The wings of bats are their most obvious
features. The bat wing is essentially a modified
hand — hence the Greek name Chiroptera,
meaning "hand wing". The digits, except the
thumb, are greatly elongated to support the flight
membranes with the aid of a lengthened forearm.
The thumb is usually largely free of the membrane
and has a claw, although sometimes this is small;
the second digit in most Megachiroptera also has a
claw but otherwise claws or nails are lacking on the
wings of modern bats. Feet are usually relatively
small, with five toes, each with a strong claw.
  The flight membranes, which in many bats also
join the legs and tail, are extensions of the body
integument or skin. The membranes are muscular
and tough but very flexible, with a high
concentration of blood vessels. Apparently hairless,
the membranes do, in fact, have many short

► *There are nearly 1,000 species of bats
but only three of these feed on blood:
vampire bats have given the rest of the
bats an undeservedly bad reputation.
The common vampire in this photograph
is hopping — vampires are the only bats
to move this way on the ground.*

Stephen Dalton/NHPA

transparent or translucent hairs, sometimes in bands or fringes, and the body fur may extend onto the inner part of the membrane.

Bats are the only mammals capable of true flight, an ability among modern vertebrates shared only with birds; other so-called "flying" mammals in fact glide with the aid of an outspread flexible membrane along the sides of the body and sometimes between the hindlegs and tail. Grounded bats scurry awkwardly but sometimes rapidly, although the common vampire *Desmodus rotundus* can walk, hop, and run with considerable agility using thumbs, wrists, elbows, and feet, while some free-tailed bats in the family Molossidae and the New Zealand short-tailed bat *Mystacina tuberculata* crawl and climb quite readily.

The wings are pulled downward by muscles on the chest and under the upper arm, and raised by other muscles and muscle groups on the back that act on the upper humerus and scapula (shoulder blade). This is very different from birds, where far fewer but relatively larger muscles on the chest provide the power for upward and downward movements of the wings. Further muscles along the arms in bats control the extension and

retraction of the wing, and its orientation in flight. Wing shapes vary quite widely. In general, bats that fly slowly through vegetation or other obstacles with very maneuverable flight have relatively short, broad wings, while fast-flying species that hunt in open spaces have longer, narrower wings.

Bats' ears vary widely in size and shape, some bats having exceptionally long ears. Many also have a tragus — a small projecting flap just inside the ear conch that obscures the ear opening. Contrary to popular belief, all bats have functional eyes: the eyes of fruit-eating species are large and adapted to poor light, but in most other species the eyes are relatively small and in a few they are nearly hidden in the fur. Several bat families have a noseleaf consisting of fleshy structures of the skin surrounding and surmounting the nostrils. Some bats have no tail or only the rudiment of a tail; in others the tail extends across and within the tail membrane or protrudes from its upper surface; and in some species the tail extends freely beyond the edge of the membrane. Many bats are drably colored in shades of brown or gray but some have striking color patterns.

▲ *Bats fall into two very distinct groups: the herbivorous Megachiroptera and the largely insectivorous Microchiroptera, which make up about 80 percent of bat species. The long-eared bat is a typical microchiropteran.*

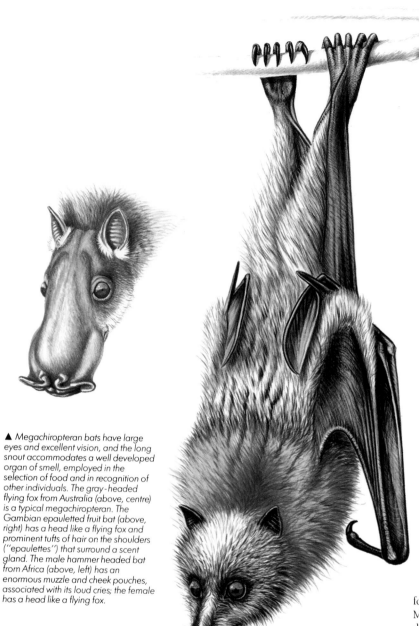

▲ *Megachiropteran bats have large eyes and excellent vision, and the long snout accommodates a well developed organ of smell, employed in the selection of food and in recognition of other individuals. The gray-headed flying fox from Australia (above, centre) is a typical megachiropteran. The Gambian epauletted fruit bat (above, right) has a head like a flying fox and prominent tufts of hair on the shoulders ("epaulettes") that surround a scent gland. The male hammer headed bat from Africa (above, left) has an enormous muzzle and cheek pouches, associated with its loud cries; the female has a head like a flying fox.*

OLD WORLD FRUIT BATS (MEGACHIROPTERA)

Old World fruit bats, flying foxes and dog-faced fruit bats belong to the one family Pteropodidae in which there are over 160 species. Found in the tropics and subtropics of the Old World (Europe, Asia, Australia, Africa), these are medium to very large bats, with a head and body length of 50 to 400 millimeters (2 to 16 inches) and a forearm of 40 to 230 millimeters (1½ to 9 inches). They have no noseleaf and a simple ear, the edge of which forms a complete ring (in contrast to Microchiroptera in which the ear edge is discontinuous). They lack a tragus (a small projecting ear flap) and the tail is very short, rudimentary, or absent, except in long-tailed fruit bats, genus *Notopteris,* which have a 60 millimeter (2⅓ inch) tail outside the narrow tail membrane.

All species in this family live on fruit, flowers, and flower products. Only the genus *Rousettus* is known for certain to utilize a simple form of echolocation. Many species are brownish or blackish, sometimes with a brighter mantle or with a gray or silvery tinge; others have speckled ears and membranes or a facial pattern of white spots or stripes, perhaps to aid concealment in foliage. Most roost in trees or in dimly lit areas of caves. The fruit-eating species usually have large, flat-crowned grinding teeth and relatively short, strong jaws; those that eat nectar and pollen have long muzzles with lightly built jaws and teeth, and a long, extensible tongue with brush-like papillae.

INSECT-EATING BATS (MICROCHIROPTERA)

Mouse-tailed bats
Mouse-tailed bats belong to the family Rhinopomatidae, in which there are three species in one genus. Found from North Africa to southern Asia, these small to medium-sized bats have a length of 53 to 90 millimeters (2 to 3½ inches) and a forearm of 45 to 75 millimeters (1¾ to 3 inches). They have no noseleaf, but the muzzle is swollen with a transverse ridge above slit-like nostrils. The ears are large, joined at the base, and have a small tragus. Mouse-tailed bats have a long, thread-like tail, which is similar in length to the forearm and projects from the edge of the narrow tail membrane. Grayish brown to dark brown in color, these bats are insectivorous and adapted for life in arid and semi-arid regions. They are gregarious and colonial, often roosting in artificial structures.

Sheath-tailed bats
There are 49 species of sheath-tailed, sac-winged, pouched, tomb, and ghost bats in the family Emballonuridae. Widely distributed in the tropics and subtropics they are small to medium in size: head and body 37 to 135 millimeters (1½ to 5¼ inches) in length, with a forearm of 32 to 95 millimeters (1¼ to 3¾ inches). They have no noseleaf. The ears vary among species but are often joined at the base, with a small or moderate tragus. The tail is partially enclosed in the tail membrane with its tip protruding from the upper surface of the membrane. In some species there is a glandular sac in the wing membrane anterior to the forearm near the elbow or a glandular pouch on the throat. Most species are brown or grayish brown in color; some species are almost black, and ghost bats, genus *Diclidurus,* are white or grayish white. All species are insectivorous and roosts vary from caves to hollow trees and foliage.

Hog-nosed bat
The single species of hog-nosed or bumblebee bat belongs to its own family, Craseonycteridae, and is found in southwest Thailand. It is a very small species, with a head and body length of only 29 to 33 millimeters (1 to 1⅓ inches) and a forearm of 23 to 26 millimeters (1 inch). The hog-nosed bat has a muzzle with a low transverse ridge above the nostrils but no noseleaf. Its ears are very large with a swollen tragus. The tail membrane is extensive, but there is no external tail. Brown to reddish gray with a paler underside, it is insectivorous and roosts in caves. The sole species, *Craseonycteris thonglongyai,* was discovered as recently as 1973.

Merlin D. Tuttle/Bat Conservation International

It is the smallest known bat and is among the smallest mammals.

Slit-faced bats
There are 14 species in one genus of slit-faced, hollow-faced, or hispid bats (family Nycteridae), which are found in Africa, and Southwest and Southeast Asia. The total head and body length is 43 to 75 centimeters (1¾ to 3 inches) with a forearm of 35 to 66 millimeters (1⅓ to 2½ inches). The slit-faced bat has a complex noseleaf, with folds and outgrowths flanking a deep longitudinal groove. The ears are large and joined at the base with a small tragus. The tail membrane is extensive and encloses the tail, which has a small T-shaped cartilaginous tip at the edge of the membrane. In color, slit-faced bats are usually brown to reddish brown. Their diet includes large arthropods, such as spiders and scorpions, although the large slit-faced bat *Nycteris grandis* also takes small vertebrates. Slit-faced bats utilize a variety of roosts from caves to tree holes and even the abandoned burrows of other mammals.

False vampire bats
False vampire bats and yellow-winged bats belong to the family Megadermatidae, which is found in Central Africa, Southeast Asia and Australia, with five species in four genera. Medium to large in size, with a head and body length of 65 to 140 millimeters (2½ to 5½ inches) and a forearm of 50

▲ *In most microchiropterans the tail helps to support a membrane between the hindlegs, but the three species of mouse-tailed bats have no such membrane and the long, dangling tail gives them a peculiar appearance.*

Stephen Dalton/NHPA

▲ The so-called false vampire bats of the Old World and the Australasian region have no resemblance to the true vampires of Central America. They are carnivores and, uniquely among the Microchiropteran bats, combine echolocation with good vision. They capture small mammals, such as mice, on the ground . . .

to 115 millimeters (2 to 4½ inches), they have a conspicuous, long, erect noseleaf. The ears are large and joined at the base with a prominent bifurcated tragus. Although the tail membrane is extensive the tail is short or absent. Color varies from blue-gray to gray-brown; the flight membranes are pinkish white in the Australian false vampire *Macroderma gigas* and yellowish in the African yellow-winged bat *Lavia frons*. Their diet includes large insects and small vertebrates such as frogs, birds, rodents, and other bats, and they hang in wait for passing prey. They roost in caves, rock crevices, hollow trees, or foliage.

Horseshoe bats
There are 63 species of horseshoe bats in the family Rhinolophidae, found in the tropics, subtropics,

and temperate zones of the Old World. They are small to medium in size, with a head and body length of 35 to 110 millimeters (1⅓ to 3 inches) and a forearm of 30 to 75 millimeters (1¼ to 3 inches). They have a distinctive horseshoe-shaped noseleaf with a strap-like sella (flat structure) above the nostrils and the central part of the noseleaf, usually with an upright, triangular, cellular, and bluntly pointed posterior projection or lancet extending the noseleaf to the rear. The ears are relatively large but have no tragus. The tail membrane encloses the tail. Most species are brown or reddish brown. Generally tropical or subtropical, but with a few temperate species that hibernate in the winter, rhinolophids forage for insects near the ground or among foliage, roosting in caves, mines, hollow trees, or buildings.

Stephen Dalton/NHPA

Old World leaf-nosed bats

The family Hipposideridae contains 66 species of Old World leaf-nosed bats and occurs in the tropics and subtropics. There is a wide variation in size among species, ranging from 28 to 110 millimeters (1 to 4 inches) in head and body length. The noseleaf is similar to that of the horseshoe bats, but lacks the sella above and behind the nostrils and the upright lancet; it is usually rounded and not triangular at the rear. The noseleaf is sometimes complex. There is no tragus and the ears are moderate-sized. Most species are some shade of brown and in all species the tail membrane encloses the tail. Old World leaf-nosed bats eat a wide variety of insects, as their size variation might suggest, and generally roost in caves where they sometimes form large colonies.

Bulldog bats

Bulldog bats, sometimes called hare-lipped, mastiff, or fisherman bats, belong to the family Noctilionidae, in which there are two species in one genus. Ranging in length from 57 to 132 millimeters (2¼ to 5 inches) with a forearm of 54 to 92 millimeters (2 to 3½ inches), bulldog bats have a pointed muzzle with a pad at the end, and lips and cheeks that form pouches. They have no noseleaf and large, slender ears with a small tragus. The moderate tail membrane encloses the tail, with its tip emerging from the upper surface. The greater bulldog bat or fisherman bat *Noctilio leporinus* catches fish up to 10 centimeters (4 inches) in length (which it detects just beneath the water surface) by seizing them with its long, sharp claws and enormous feet. Bulldog bats roost in

▲ *. . . killing them with a bite to the head or neck. They also take large insects and frogs — some even catch other bats on the wing. Some species of false vampires have a wingspan of about 30 centimeters (1 foot).*

▲ *Bulldog or fisherman bats use their long, rake-like feet to capture fishes and large insects at the surface of lakes or calm rivers. The short fur is water-repellent.*

▶ *The Californian leaf-nosed bat is a member of the New World family Phyllostomidae, which is characterized by an immense dietary range. This species eats insects and fruits.*

caves, rock crevices, or hollow trees in the tropical regions of the New World where they are found.

Naked-backed bats

Naked-backed, moustached, or ghost-faced bats belong to the family Mormoopidae found in the southern United States and Antilles to Brazil, with eight species in two genera. Small to medium in size, they have a head and body length of 40 to 77 millimeters (1½ to 3 inches) and a forearm of 35 to 65 millimeters (1⅓ to 2½ inches). They have no noseleaf, but the chin and lips have complex plates

and folds of skin. The large ears, sometimes joined at the base, have a complex tragus. There is an extensive tail membrane with about one-third of the tail protruding from the upper surface of the membrane. Brown or reddish brown in color, some wing membranes attach along the midline of the back, hence the name naked-backed. Insectivorous and often gregarious, naked-backed bats form moderate to large colonies in hot, humid caves. Most fly swiftly, often near the ground.

New World leaf-nosed bats

New World leaf-nosed bats in the family Phyllostomidae comprise 152 species in 51 genera. Bats of the New World tropics and subtropics, they vary in size from small to large: head and body measure 40 to 135 millimeters (1½ to 5 inches); forearm 25 to 110 centimeters (1 to 4 inches). Generally the muzzle has a simple, spear-shaped noseleaf, but the ears vary in size and shape, although all have a tragus. Some species have tails, and, when present, the tail is usually enclosed in the tail membrane with its tip sometimes projecting slightly beyond the edge of the membrane. Most species are reddish or reddish brown, some with white stripes on the face or back; one species — the white bat *Ectophylla alba* — is, as its name suggests, white. Few species are exclusively insectivorous. Many species are more or less omnivorous, eating insects, fruit, and in some cases they even prey on small vertebrates; others are frugivorous. Some species are nectar and pollen feeders, with long muzzles, reduced jaws and teeth, and an extensible tongue with brush-like papillae. The three species of vampire — *Desmodus rotundus*, *Diaemus youngi*, and *Diphylla ecaudata* — are specialized for an exclusive diet of blood. This family embraces almost the entire spectrum of bat food habits, except fish-eating, and its members exploit a wide variety of roosts from caves to trees and include the few genera (*Uroderma*, *Artibeus*, *Ectophylla*) that construct a rudimentary shelter.

Funnel-eared bats

The family Natalidae contains five species (in one genus) of funnel-eared or long-legged bats. Distributed from Mexico to Brazil and the Antilles, funnel-eared bats are small with a head and body length of only 35 to 55 millimeters (1⅓ to 2 inches) and a forearm 27 to 41 millimeters (1 to 1½ inches). Their nostrils are close together near the top lip and they have no noseleaf. The ears are large and funnel-shaped, with a variously distorted and thickened tragus. The extensive tail membrane encloses the tail and the hindlimbs are long and slender. Gray or yellowish to reddish brown in color, males have a curious bulbous "natalid" organ of unknown function on the forehead. All species are insectivorous and utilize caves, tunnels, or rock overhangs as roosts.

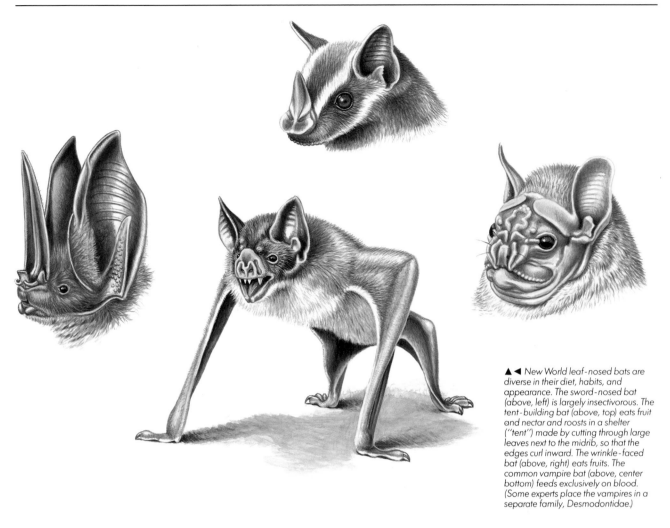

▲ ◄ *New World leaf-nosed bats are diverse in their diet, habits, and appearance. The sword-nosed bat (above, left) is largely insectivorous. The tent-building bat (above, top) eats fruit and nectar and roosts in a shelter ("tent") made by cutting through large leaves next to the midrib, so that the edges curl inward. The wrinkle-faced bat (above, right) eats fruits. The common vampire bat (above, center bottom) feeds exclusively on blood. (Some experts place the vampires in a separate family, Desmodontidae.)*

Smoky bats

Smoky bats or thumbless bats belong to the family Furipteridae found in Central America and northern and western South America. The two species in two genera are small, with a head and body length of 37 to 58 millimeters (1½ to 2¼ inches) and a forearm of 30 to 40 millimeters (1 to 1½ inches). They have large funnel-shaped ears with a small tragus, but no noseleaf. A rudimentary thumb is enclosed in the flight membrane, which extends to the base of a minute claw, and the tail is enclosed in a moderate tail membrane. Grayish or brownish gray in color, they are insectivorous and roost in caves.

Disc-winged bats

There are two species in one genus of disc-winged or New World sucker-footed bats in the family Thyropteridae, which is found from southern Mexico to Peru and Brazil. Both species are small: head and body 34 to 52 millimeters (1⅓ to 2 inches) long, and the forearm 270 to 380 millimeters (1 to 1½ inches) long. The long,

▲ *A vampire bat lapping blood from the foot of a fowl.*

Merlin D. Tuttle/Bat Conservation International

Merlin D. Tuttle/Bat Conservation International

► *Disc-winged bats have an adhesive sucker on each thumb and foot and can hang from any one sucker. They nest in naturally curling leaves, such as those of banana trees.*

slender muzzle has no noseleaf. The ears are large and funnel-shaped, with a prominent tragus. The extensive tail membrane encloses the tail, which projects slightly from the edge. There is a large sucker-shaped adhesive disc on a short stalk at the base of thumb and on the side of foot. In color they are reddish brown to light brown on the back, with a whitish or brownish underside. Disc-winged bats roost in family groups in a rolled leaf or frond in an unusual head-upward position; the adhesive discs on the thumbs and feet provide a grip, and the bats move to a new roost as the leaf unfurls.

Old World sucker-footed bat
The single species of Old World sucker-footed bat (family Myzopodidae) is found exclusively in Madagascar. It is a medium-sized bat with a head and body length of 57 millimeters (2¼ inches) and a forearm of 46 millimeters (1¾ inches). There is

no noseleaf. The ears are long and slender, with a small square tragus partially fused to the front edge of the ear and a curious mushroom-shaped process consisting of a short stalk supporting a flat expansion at the base of the back edge of each ear, which partially obscures the ear opening. The extensive tail membrane encloses the tail, which projects slightly beyond the edge. There is a sucker-shaped disc at the base of the thumb and the side of each foot. The toes are joined by webbing that extends almost to the base of the claws. Its habits are apparently similar to those of the disc-winged bats.

Vespertilionid bats
Vespertilionid bats (family Vespertilionidae), of which there are 350 species in 43 genera, are worldwide in distribution (except for polar regions and some oceanic islands). Species vary in size from very small to large, with head and body lengths of 32 to 105 millimeters (1¼ to 4 inches), and a forearm of 24 to 80 millimeters (1 to 3 inches). No species in this family has a noseleaf, but some have a transverse ridge. Ears vary in size and shape, with a short to long tragus. The tail membrane is usually extensive, enclosing the tail, which sometimes projects slightly from the edge. A few species have adhesive pads at the base of the thumb or foot. Most species are brown, grayish, or blackish, but some species are more brightly colored or have white spots or stripes. One species — the spotted bat *Euderma maculatum* — is patterned black and white, while another — the painted bat *Kerivoula picta* — has wing membranes patterned in black and orange. The members of this family occupy a wide range of habitats from semi-desert to tropical forest; species in temperate

▼ *This photograph of a long-eared bat shows that the structure of a bat's wing is fundamentally the same as that of a human arm. There is a short upper arm, a longer forearm, and a tiny wrist. The thumb is a claw and the four long fingers support the wing membrane.*

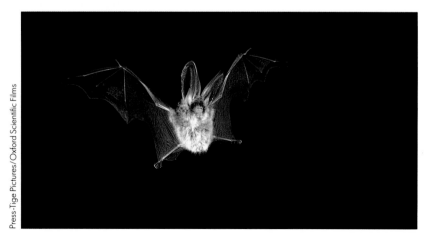

Press-Tige Pictures/Oxford Scientific Films

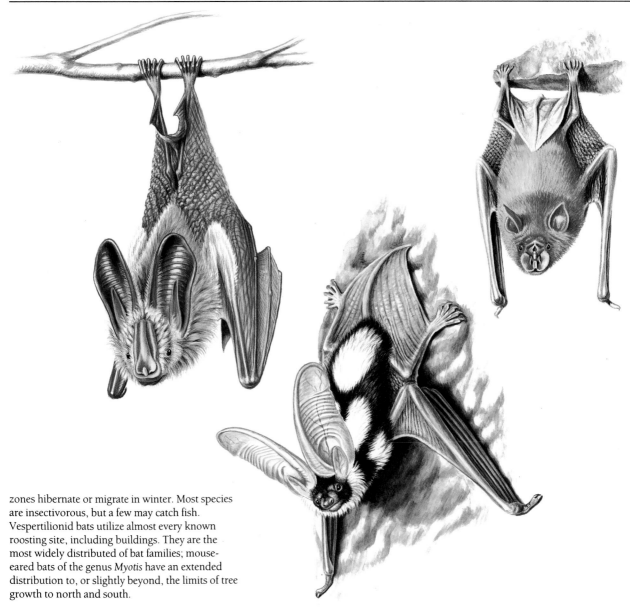

zones hibernate or migrate in winter. Most species are insectivorous, but a few may catch fish. Vespertilionid bats utilize almost every known roosting site, including buildings. They are the most widely distributed of bat families; mouse-eared bats of the genus *Myotis* have an extended distribution to, or slightly beyond, the limits of tree growth to north and south.

New Zealand short-tailed bats
As their name suggests, the two species of New Zealand short-tailed bat in the family Mystacinidae are found exclusively in New Zealand. (One species is thought to have become extinct recently.) They are small to medium in size with a head and body length of 60 millimeters (2⅓ inches) and a forearm of 40 to 48 millimeters (1½ to 1¾ inches). The front of the muzzle is obliquely truncate and there is no noseleaf. The ears are large and pointed, with a long narrow tragus. The wings are thick and leathery. The tail membrane is moderately long and thin, enclosing the short tail with its tip protruding from the upper surface of the membrane. The legs are short; the claws have a small additional talon; and the feet are wrinkled. These bats are grayish brown, brown, or blackish

brown in color; the fur is soft and velvety. Well adapted for moving about on the ground or in trees and chiefly insectivorous, New Zealand short-tailed bats possibly also eat fruit or flower products. They roost in caves, crevices, and hollow trees but apparently do not undergo prolonged hibernation. Conventionally thought to be related to vespertilionid bats and the free-tailed bats, some evidence suggests a closer relationship to the New World bulldog bats, naked-backed bats, and New World leaf-nosed bats.

Free-tailed bats
The family Molossidae comprises approximately 89 species in 13 genera and is found in both Old and New World tropics and subtropics, with some species of free-tailed bats extending into the

▲ *Most bats have dull brown to gray fur but some species are strikingly colored, as in the yellow-winged false vampire bat (above, left), the vespertilionid spotted bat (above, center), and the orange horseshoe bat (above, right). Orange-colored individuals are not uncommon in otherwise drab-colored species of horseshoe bats.*

▲ *Many bats roost communally in caves, often in large numbers. It has been estimated that communities of the Mexican free-tailed bat, here seen emerging from a maternity cave, number more than 40 million. Some flying fox roosts shelter more than a million bats.*

temperate zones. There is great size diversity within this family, with some very small species and some large; head and body lengths vary from 40 to 130 millimeters (1½ to 5 inches), and forearms 27 to 85 millimeters (1 to 3⅓ inches). The muzzle is broad and, as in the New Zealand short-tailed bats, is obliquely truncate. The lips of many species are wrinkled, and there is no noseleaf. Ears are usually moderate in length, but thick and often joined at the base, with a small tragus. The flight membranes in some species are rather tough and leathery while the tail membrane is relatively narrow, enclosing the stout tail that projects considerably beyond its edge. The legs are short and strong. The body fur is usually brown or blackish brown, but wing membranes are sometimes whitish. One species — the hairless bat, genus *Cheiromeles* — is, as its name implies, effectively devoid of fur. The Mexican free-tailed bat *Tadarida brasiliensis* at least is migratory and a few other species enter short periods of winter inactivity in the more temperate parts of the range. Free-tailed bats are strong, fast flyers, catching insects on the wing. They roost in caverns, tunnels, hollow trees, foliage, under rocks or bark, or sometimes in buildings. Most species are highly gregarious and some form very large colonies.

A VARIED DIET

Approximately 70 percent of bat species feed on insects and other small arthropods such as spiders or scorpions; the size of their aerial prey ranges from gnats to large moths. Insects are captured in flight using the mouth; the tail membrane may be curled into a scoop from which the insect may be retrieved, or the insect may be deflected towards the mouth with the wing. Some bats glean insects from foliage, take them from the ground, or even skim them from the surface of water. Small insects can be consumed in flight, but larger items are carried to a nearby perch to be eaten. Large numbers of insects are consumed by bats, which thus play an important part in the control of insect populations. A large colony of bats may eat a substantial weight of insects annually. A few species of bats catch and eat frogs, lizards, small rodents, and birds, or even other bats, although this is not an exclusive habit, since insects are also taken. A small number catch fish, skimming over the water and seizing their prey with their long claws and strong feet.

The three species of vampire are exceptional in that they live on the blood of other animals. Vampire teeth are highly specialized, with the upper incisors and canines enlarged and razor sharp to inflict a small wound from which the blood is lapped. Vampire bats are found only in Central and South America, where local populations of the common vampire are thought to have increased greatly since European colonization and the introduction of cattle and horses.

Fruits, flowers, nectar, and pollen form the staple diet of Old World fruit bats or Megachiroptera. In the New World, where there are no megachiropterans, these food habits have evolved independently in many species of the Microchiroptera. The post-canine or grinding teeth of fruit-eating bats are usually low and flat-crowned (in contrast to the ridged and strongly cusped teeth of insectivorous species) and are sometimes reduced in number and size, especially in nectar-feeding bats. Fruit-eating and nectar-eating bats are confined to the tropics and subtropics, where food is available throughout the year. They rely upon a wide variety of fruits and flowers, and play a vital role as pollinators and seed dispersers for many plant species of importance commercially, or as food. Many such plants are adapted for bat pollination — their flowers open at night, are often white, creamy or greenish with a musky or sour scent, are shaped to facilitate landing and entry by bats, and hang free so that they can be easily reached, an example of coevolution between plants and bats. A few species of Old World fruit bats are known to eat leaves occasionally but the digestive system of bats is not adapted for herbivory, and this seems, therefore, to be a rare habit.

Jane Burton/Bruce Coleman Ltd

▲ Evolution is largely a matter of seizing opportunities. In the absence of megachiropterans from the New World, some microchiropterans have evolved into omnivorous or fruit and nectar eating species, such as the Barbados fruit bat shown here.

Merlin D. Tuttle/Bat Conservation International

◄ The spear-nosed long-tongued bat, a member of the Phyllostomidae, is able to hover like a hummingbird while it sips nectar with its long tongue. In much the same way as a bee, it fertilizes the flowers of some trees in the American tropics.

▼ *Gould's long-eared bat, an Australian species, gives birth to twins. When young, they may be carried by the mother. These two are suckling from the mother's functional teats; when she is flying, they reverse their position and bite onto a pair of false teats in the groin.*

Kathie Atkinson

REPRODUCTION AND DEVELOPMENT

Reproduction is generally seasonal, with birth and development coinciding with periods of maximum food abundance. In the tropics the young of some species are born shortly before the onset of the rainy season, although there may be more than one reproductive cycle annually; in the temperate zones birth occurs at the beginning of the summer months. Temperate-zone bats mate in the fall and during or at the end of hibernation. Viable sperm can be stored in the female or male reproductive tracts of bats throughout hibernation or winter; ovulation and fertilization occur early in the following spring. Some tropical species also store sperm but, in addition, delayed implantation or retarded development of the fertilized ovum can be used to ensure that sperm production and birth occur at favorable times.

Gestation periods in bats vary from 40 to 60 days in small species to as long as eight months in larger species. Generally only one young is produced, but twins occur regularly in some species and up to four or five young have been recorded for a few species. At birth the young are helpless but have strong claws and hooked milk teeth with which they cling to the mothers. At birth baby megachiropterans are relatively large and hairy, and have their eyes open, but most microchiropterans at birth are relatively small and naked, and their eyes are closed.

The females of many temperate and some tropical species congregate in nursery colonies to bear and raise the young. The site of such a colony may be traditional and used year after year. Usually the young are left behind while the mother forages and feeds, although in small species the young can begin to make short flights within three weeks. Nursing lasts from one to three months while the young learn to fly, hunt, and feed. Infant bats emit loud calls with patterns that permit their mothers to reunite and nurse only their own infants; at close range, odor is also a likely component of the reunion. In some species the mother and infant call back and forth to each other, and the infant increasingly matches its mother's frequencies and in this way possibly learns some aspects of adult echolocation calls. Female bats have one pair of thoracic teats but a pair of false teats occur in the groin region in a number of species; these may provide an additional hold for the young, which grow rapidly and attain sexual maturity late in their first year or in their second year.

HIBERNATION FOR SURVIVAL

Many bat species can regulate their body temperature (heterothermy), allowing it to fall while roosting during the day to reduce their energy consumption, while others, such as large tropical fruit bats, maintain a relatively constant body temperature (homeothermy). In the temperate zones torpidity is extended into hibernation during the winter months when food is scarce or unavailable; body temperature and metabolism are reduced to very low levels and survival depends on fat stored during the fall. Hibernation is not necessarily continuous and hibernating bats may wake and move to a different part of the hibernaculum or even to another site.

Michael Fogden/Bruce Coleman Ltd

A number of temperate species migrate seasonally to avoid the extremes of winter, returning in the spring. The European noctule *Nyctalus noctula* travels as much as 1,000 to 1,600 kilometers (620 to 1,000 miles) or, exceptionally, 2,000 kilometers (1,200 miles) when migrating. Some fruit bats also appear to follow the seasonal flowering and fruiting of their food plants. (Accidental dispersal also occurs and occasionally bats are found far from their usual range, having been blown for long distances by the winds or accidentally transported in ships or aircraft.)

ECOLOGY AND BEHAVIOR

Bats occupy a wide variety of roosts. Many bats are cave dwellers, but they are found also in mines, tunnels, culverts, tombs, ruins, under shallow rock overhangs, in cracks and crevices, under loose rubble, or in buildings. Others live under bark, in hollow trees, or hang in the foliage of trees, shrubs, or bushes. A few New World bats of the genera *Uroderma, Artibeus,* and *Ectophylla* make a primitive, tent-like shelter by biting the supporting ribs of palm fronds so that these collapse. Many bats are strongly gregarious, sometimes forming large colonies of many thousands of individuals in

a single cave or cave complex. Others live in small groups and a few appear to be solitary. Some form breeding groups or harems that may persist throughout the year, the males using visual,

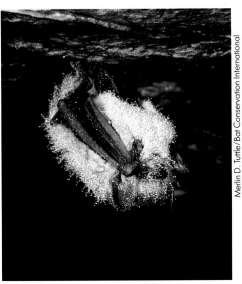

Merlin D. Tuttle/Bat Conservation International

▲ The tiny Honduran tent bat is one of the few white bats. Like other tent-building bats, they cut away the connection between the edges of a palm frond and its midrib, causing the leaf to curl. Several bats roost in the "tent" that is created.

◀ An eastern pipistrelle in hibernation. The bat has become so cool that moisture has condensed on its fluffed-up fur.

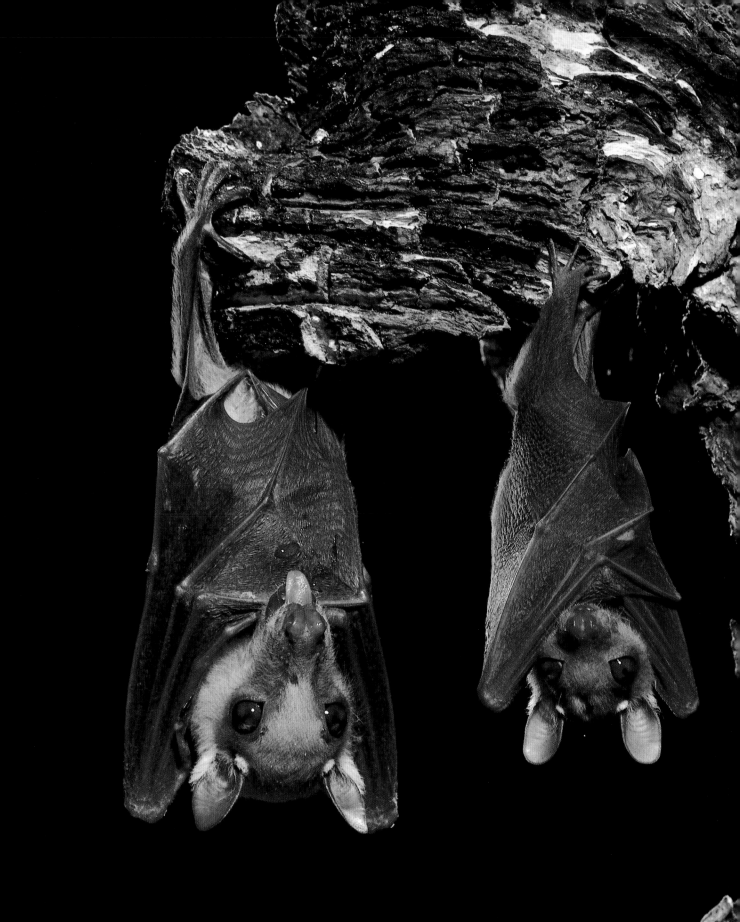

olfactory, or auditory signals to attract females or to discourage other males.

Bats appear to be relatively long-lived, with a maximum recorded age of 32 years in the little brown bat *Myotis lucifugus*. The European horseshoe bat *Rhinolophus ferrumequinum* has a known life-span of 26 years, and other species have been known to survive 10 to 20 years. The average life-span is, however, likely to be considerably less.

Bats are rarely a food item for any specific predator, but they can fall victim to a number of opportunistic hunters from large spiders to snakes, nocturnal birds, and mammals (even other bats).

CONSERVATION

Bats feature widely in mythology and folklore where, more often than not, they are quite unjustly associated with evil or, at best, the loathsome. Often there is an unjustified fear of bats, perhaps originating in their nocturnal and secretive habits or possibly from ignorance of these attractive and normally harmless animals. These unfounded fears and superstitions have adversely affected human attitudes towards bats and their conservation.

Occasionally fruit bats may cause local damage to crops but they are also important pollinating and seed dispersing agents and their destruction seems unjustified. The common vampire is a pest of domestic livestock in Central and South America, where it can also pose a health threat to humans. Insectivorous bats sometimes colonize roof spaces or other parts of buildings, but such colonies are usually harmless and, at worst, no more than a mild nuisance; often they are nursery colonies that will disperse at the end of the breeding season. If they have to be discouraged the only practical remedy is total exclusion by blocking off access points; extermination is useless since more bats move into the vacated area. In some countries, such as Great Britain, any unauthorized interference with bats is illegal. Bats can be

J.L.G. Grande/Bruce Coleman Ltd

encouraged by the provision of bat boxes on trees or in other suitable sites, or by other artificial roosts, and existing roosts can be protected: caves, for example, can be fitted with a carefully designed grille to prevent disturbance.

On the other hand, humans exploit bats. Large fruit bats are often cooked and eaten in parts of Africa, Asia, and the Pacific to the extent that the continued survival of some flying foxes of the genus *Pteropus* is seriously at risk. Some species are already extinct. In the past, bats have been a source of unusual remedies for various illnesses and, more recently, have been used in medical research. Bat manure has also been used for many years as a fertilizer. Perhaps the most bizarre involvement between humans and bats was a wartime project to use Mexican free-tailed bats *Tadarida brasiliensis* as self-propelled incendiary devices — the idea was hastily abandoned when, after some escaped, it became clear that they did not discriminate between friend and foe.

Bats are vulnerable to many factors that threaten their survival, many the result of human activity, including habitat destruction, the loss of roosting sites, the indiscriminate use of pesticides, or simple persecution. Adverse public attitudes, sometimes reinforced by ignorant and uninformed comment in newspapers and magazines, only serve to propagate the myth that bats are harmful and dangerous. Bats are protected by law in many countries but legislation alone is not enough. There is a need for public education and a realization that bats are an outstanding example of evolutionary adaptation and an irreplaceable part of the global ecosystem. In Great Britain such an approach, combined with legislation, has been remarkably successful in increasing public awareness of bats and of the need for their active conservation, and much progress is now being made in the United States.

▲ *Mouse-eared bats in a nursery cave. However difficult it is to believe, each mother recognizes its own infant in such nurseries, apparently by a distinctive call and, perhaps, odor.*

◄ *When a megachiropteran such as Wahlberg's epauletted fruit bat is roosting, it wraps the wings around its body and holds the head forward at a right angle to the chest. Microchiropterans, however, usually fold the wings at the side of the body and the head hangs down or is held at about a right angle to the bat's back.*

DO BATS CARRY RABIES?

Rabies virus or rabies-related viruses have been found in numerous species of New World bats; there are far fewer reports from Europe, Asia, or Africa. Their occurrence is confined chiefly to vampire bats (especially the common vampire) in Central and South America, where the sanguinivorous habits of these bats make them a potential risk. A very small number of human deaths over a long period in North America and still fewer elsewhere have been attributed to bat bites but, since the vast majority of bats do not normally bite humans, the likelihood of infection is remote. Sick or moribund bats should not be handled incautiously, especially in North and South America.

FLYING IN THE DARK

As long ago as 1793 the Italian Lazzaro Spallanzani discovered that blinded bats were able to avoid obstacles when flying. A Genevan scientist, Louis Jurine, found soon after this that bats with obstructed ears were unable to do so. Both scientists concluded that hearing was important in nocturnal bat flight but could not explain why. Their work was disregarded and fell into oblivion, but in the early part of the twentieth century it was suggested that high frequency sounds were involved, and the development of sufficiently sensitive apparatus enabled an American scientist, Donald Griffin, to establish in 1938 that this was so. Almost coincidentally the Dutch zoologist Dijkgraaf arrived at a similar conclusion after studying the faint sounds sometimes heard when bats are flying, but it was not until the Second World War ended that his work became known. Most of the sounds are ultra-sonic and therefore beyond the normal range of human hearing of 20 hertz to 20 kilohertz; most echo-locating bats use frequencies of 20 to 80 kilohertz, but some range as high as 120 to 210 kilohertz.

The use of high frequency echolocation for navigation, hunting, and catching prey is confined to the Microchiroptera. The sounds are produced in the larynx and emitted through the mouth or through the nostrils in the case of bats with noseleaves, the leaf apparently serving to modify, focus, and direct the beam of sound. Echoes received by the ears provide information that can be processed by the brain to provide the bat with an interpretation of its surroundings and the location of flying or resting prey. The Megachiroptera lack echolocation of this type, but a few species of rousettes of the genus *Rousettus* produce orientation sounds that are partially audible (5 to 100 kilohertz) by clicking the tongue. Unlike other

megachiropterans, these bats roost in the darker parts of caves and use this form of echolocation to navigate in and out of the caves, though they orientate visually in better light.

The ultrasonic pulses vary in duration from about 0.2 milliseconds to 100 milliseconds and may incorporate from one to five harmonics or overtones. By human standards of loudness the sounds may be very powerful and of high intensity. Frequency modulated (FM) signals sweep in a shallow curve or sharply through a range of frequencies, usually downwards; other signals are emitted at a constant frequency (CF). Most bats vary their acoustic repertoire, using both FM and CF patterns in varying degrees and combinations.

▼ *Pulses of ultrasound emitted by a bat spread outward from its head like ripples in a pond (below, right). The strength of the reflected vibrations gives information on the distance of the prey, while slight differences in the time taken for the reflections to reach each ear give information on its direction. Pulses of ultrasound are often complex. This sonogram of the relatively simple emission of a horseshoe bat (bottom) shows that the sequence begins with a series of constant tones, each of which glides rapidly to a lower frequency (through about a quarter of an octave). As the bat homes in on its prey, the pulses are emitted more rapidly, their frequency increases slightly, and the constant tone almost disappears.*

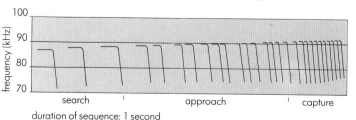

duration of sequence: 1 second

Stephen Dalton/NHPA

Some use CF signals of moderate to long duration to establish target movement and speed by means of frequency changes in the echo (Doppler effect). Typically, a hunting bat will search using an emission rate of 5 to 10 pulses a second. Once a flying insect has been detected the pulse rate increases to 15 to 50 a second, this approach phase being terminated by a stage in which the pulse rate may reach 200 pulses a second, this high rate providing the bat with continuous information about the target.

Some moths have a simple ear on the thorax that can detect bat echolocation sounds and, so warned, these moths are able to take evasive action. Other moths produce a train of ultrasonic clicks that mimic bat calls and so interfere with the bat's echolocation system. On the other hand, it is suspected that some long-eared bats can locate their prey by listening for its sounds; certainly the fringe-lipped bat *Trachops cirrhosus* is quite adept at discerning the social calls of the pond frogs that it hunts.

Megachiropteran bats also produce low frequency audible calls of many kinds. The epauletted fruit bat *Epomops franqueti* honks metallically during the night, and male hammer-headed fruit bats *Hypsignathus monstrosus* produce low, throaty, and intense metallic calls to attract females. Squeaks, squawks, and screams in the roost or around a fruiting tree may express annoyance or disputes over space.

▲ To be able to chase and capture a moth in the dark, like this horseshoe bat, seems an astonishing feat. Even more remarkable is the fact that the bat simultaneously recognizes all the other objects in its vicinity—the ground, trees, bushes, rocks, other bats, and owls. It "sees" by means of ultrasound.

PRIMATES

COLIN GROVES

The order Primates contains around 200 species of mainly arboreal, keen-sighted, intelligent animals: lemurs, monkeys, apes, and their relatives. They are a diverse lot—just how diverse we have only come to realize in the past thirty years, when field studies of primates really began to blossom. Most unexpected of all, perhaps, has been that completely new species of primates have continued to be discovered right up to the present.

DIVERSE MAMMALS

Surprisingly, for one of the most intensively studied of all orders of mammals, the primates are difficult to define. Nearly all have hands and feet modified for grasping; flat nails, rather than claws, on at least some of their digits; ridged friction pads on the undersurfaces of the ends of their digits; and highly sensitive nerve-endings in these digit pads. Males have testes that are permanently descended into a scrotal sac, and a penis that hangs free from the abdominal wall. Primates have long gestation periods, slow growth, long infant and juvenile dependency periods (during which there are opportunities for learning), and long life-spans. They have big brains. Most of these features are not unique to primates—arboreal marsupials, especially the opossums of the Americas, have the same adaptations of the hands and feet; fruit bats have a similarly complex visual system; carnivores, dolphins, and elephants have big brains—but taken together, these features do define primates.

Most primates are tropical animals—specifically, they live in tropical rainforests. But some species live in other forest types (sclerophyll forests, mangroves, even coniferous forests); some do not live in forests at all (baboons and other monkeys from the savannas and woodlands of Africa; langurs from the dry thorn-scrub of India); and some live in temperate regions where there is snow for part of the year (macaques in Japan, northern China, and the Atlas Mountains; snub-nosed monkeys from the mountains of southwestern China).

Specialists now generally classify the order primates into two suborders: Strepsirhini (lemurs and their relatives, with five or six families) and Haplorhini (tarsiers, monkeys, apes, and humans, with six to twelve families). The best classification is not by any means agreed on, as the inter-relationships of some groups, such as the New World monkeys and the lemurs, are still uncertain.

Members of the Strepsirhini, lemurs can easily be recognised among primates by their moist, dog-like snout (the rhinarium). Most of them have long, pointed snouts. Almost all strepsirhines possess two remarkable features: a "toilet claw", a long claw on the second toe of the foot, and a "dental comb", a row of incisor and canine teeth, six teeth in all, across the front of the lower jaw, compressed and pointing forward. Both specializations are used for grooming; whether they have any other functions in some species is unclear, but the dental comb is, in some bushbabies, also used for scraping resin, part of their diet, off the bark of trees.

The Haplorhini, which include monkeys, apes and humans, have a dry, sparsely haired nose instead of a rhinarium. Their eyes are also different. Most of them (not all) are non-seasonal breeders—the females have sexual cycles all year round. Their placenta is of a type that has a very close vascular connection between maternal and embryonic tissue, quite different from the placenta of strepsirhines. Each month, the wall of the uterus develops a special tissue called endometrium in preparation for implantation of an embryo, and if no fertilization takes place the endometrium is shed. Some haplorhines—apes, humans, and many Old World monkeys—have very extensive endometrium development, and its monthly shedding is accompanied by loss of blood: menstruation. No other mammals menstruate.

Humans, apes, and Old World monkeys are classified as Catarrhini, which is a major division of the Haplorhini. Members of this suborder have two, not three, premolars in each half of each jaw; nostrils that are close together and directed downwards; and opposable thumbs. Many other details of anatomy, physiology, and biochemistry unite the Catarrhini and distinguish them from the other suborder of the Haplorhini, the Platyrrhini or New World monkeys.

Living catarrhines divide easily into two superfamilies: one contains the Old World monkeys (Cercopithecoidea), and the other the apes and humans (Hominoidea). Hominoids have broad chests, short lumbar spines, and no tails; cercopithecoids have a narrow chest, a long lumbar region, and at least a short tail.

LEMURS AND ALLIES

As "lower primates", lemurs have for a long time been neglected by scholars in favor of monkeys and apes. For years the French primatologist Jean-Jacques Petter was working almost alone on the island of Madagascar (now the Malagasy Republic) studying the behavior and ecology of the lemurs found there, but, as their amazing diversity has become clear, lemurs have at last become fashionable. French, American, British, German, and, above all, Malagasy scientists have begun intensive work on lemurs. Many lemurs are increasingly threatened by habitat destruction for cultivation and cattle grazing; some are very localized in their distribution, and two species are so localized they were only discovered in 1987 and 1988, when they were already on the verge of extinction. Four families of lemurs are found on Madagascar; the fifth family of the Strepsirhini is Loridae, the lorises and bushbabies.

Dwarf and mouse lemurs

Dwarf and mouse lemurs, known as cheirogaleids, are, in the main, solitary; females occupy overlapping home ranges, and males defend territories that overlap with females' ranges. The smallest forms, such as the mouse lemurs *Microcebus*, which at about 50 to 60 grams (1¾ to 2 ounces) are the smallest of all primates, are mainly insectivorous; the larger ones, such as the dwarf lemurs *Cheirogaleus*, eat fruit as well as insects.

"True" lemurs

The ring-tailed lemur *Lemur catta* is about a meter in length—more than half of which is the tail—and weighs 2.5 to 3.5 kilograms (5½ to 7½ pounds). Ring-tailed lemurs live mainly on the ground, in open forest in the dry country of southwestern Madagascar, in troops of usually about 12 to 20, in which females are dominant to males—females can displace males in disputes, or from favorite foods or sitting-spots. They are territorial: the troop occupies an area that can be as little as 5 or 6 hectares (12 or 15 acres), which it defends against other troops. The males in a troop have glands on the wrist provided with horny spurs, which they rub onto small trees with an audible "click", leaving a slash in the bark impregnated with scent. As the scents are known to be characteristic for each individual, these spur-marks inform other lemurs, of the same and of neighboring troops, which male has been there and which troop "owns" that bit of land. Like all lemurs, ring-tailed lemurs are seasonal breeders; most mating occurs within a two week period during April, but females who do not conceive then may ovulate again a month later. The gestation period is 136 days, after which a single young is born. Young ride around clinging to their mother's fur at first, then gradually become independent, and are sexually mature at 19 to 20 months.

Walt Anderson/Tom Stack & Associates

Until 1988 the five recognized species of the genus *Petterus* were included in the genus *Lemur*, but they differ strongly in many ways, such as lacking the arm glands. They are all much the same size as the ring-tailed lemur, but tend to be brown or black, often with prominent cheek whiskers, ear tufts, and facial markings. They are distributed throughout Madagascar in both rainforest and dry forest; they tend to be more arboreal than *Lemur* and to live in smaller troops.

Sportive lemurs of the genus *Lepilemur* live throughout Madagascar in all habitats from rainforest to desert, where they cling to the stems

▲ *In the early morning, ring-tailed lemurs go to the tree-tops and sit facing the rising sun, arms spread, as if sun-worshipping. In fact, they are warming up after the cold night. In the heat of the day, as here, they shelter in the shade of the middle canopy. During the rest of the day they are mainly seen on the ground.*

Indrids

The black and white indri *Indri indri,* at over 6 kilograms (13 pounds) the largest surviving lemur, is also the only one that is virtually tailless. It has long hindlegs, and moves by jumping from tree to tree. A pair of indris live with their offspring in a large range of 18 hectares (44 acres); they share part of it with neighboring pairs, but keep most of it exclusive as a territory. Indris are entirely diurnal. Their loud wailing cries float through the forest and enable neighboring groups to avoid each other; they also rub their cheeks, which are presumably glandular, on branches. The other indrids, the diurnal *Propithecus* (the sifakas) and nocturnal *Avahi,* are smaller and have long tails.

The aye-aye

The aye-aye *Daubentonia madagascariensis* is the only strepsirhine that lacks the dental comb and toilet claw. The incisors, reduced to a single pair in each jaw, are huge and open-rooted, so that they grow throughout life; they are used for gnawing away the bark of trees to get at the grubs on which the ayes-ayes feed. Only the hallux (great toe) has a flat nail; all other digits have long curved claws. The middle finger is wire-thin and is inserted into crevices in trees to extract or pulp grubs. Aye-ayes weigh about 2.8 kilograms (6 pounds) and are about 800 millimeters (30 inches) long, half of which is the bushy tail; they are black, with coarse shaggy fur and huge ears and eyes. They are nocturnal, solitary, and secretive.

Lorises and bushbabies

These lemur-like creatures, the Loridae, are found in Africa south of the Sahara, Sri Lanka, South India, and Southeast Asia, but they differ little from the Malagasy lemurs (the aye-aye excepted).

The Loridae are divided into two subfamilies, Lorinae and Galaginae. They are very distinct and some authorities classify them as separate families. The Lorinae are short-limbed and short-tailed; they move quadrupedally in a rather slow, gliding fashion, using their strong hands and feet which grip round branches. They are the lorises of South and Southeast Asia, and the potto and angwantibo of Africa. The other subfamily is the Galaginae, the bushbabies or galagos of Africa; they have long bushy tails and long hindlegs, and move by jumping. Whereas Lorinae are entirely forest-living, Galaginae live in both forest and bush country, from Somalia and Senegal to the Cape of Good Hope. Most lorids seem to live solitary lives, though sometimes groups of females nest together and share a home range; the males occupy territories that overlap the range of females. The larger species, such as the potto and the bigger bushbabies, eat mainly fruit; the smaller ones, mainly insects. One species, the needle-clawed bushbaby *Euoticus elegantulus* of Cameroun and Gabon, eats the resin of forest trees.

▲ *The indri (top) is a very rare lemur. It is now protected in the mountain rainforest reserve of Perinet in northeastern Madagascar. Ruffed lemurs are the largest of the Lemuridae and live in the eastern rainforests of Madagascar; the brown ruffed subspecies (above, bottom) is confined to the Masoala Peninsula at the northern end of this range.*

of spiny succulents. The largest species has a head and body length of about 300 millimeters (12 inches), and weighs 900 grams (2 pounds); the smallest is 250 millimeters (10 inches) long and weighs only 550 grams (1 pound 3 ounces); their tails are longer than the head and body in some species, shorter in others. They live solitary, nocturnal lives, feeding on leaves and flowers in small territories that are vigorously defended against their neighbors of the same sex, but the territories of breeding males and females overlap.

TARSIERS

One genus of the Haplorhini that does not look much like a monkey at all is the tarsier (genus *Tarsius*). If anything, it looks rather like a rat-tailed bushbaby, with its long hindlegs and long tarsal region, long skinny fingers and toes, huge eyes, and big ears (which explains why in the past it was classified alongside bushbabies and other lemurs rather than with monkeys and apes, where it belongs). Tarsiers live in island Southeast Asia. There are four species: one on Sumatra, Bangka, Belitung, the Natuna Islands, and Borneo; one on Leyte, Samar, Bohol, Dinagat, and Mindanao in the Philippines; and two on the central Indonesian island of Sulawesi (one of them extending to the Sangihe Islands and Peleng). Tarsiers are nocturnal; their eyes are truly enormous, and in the most specialized species (*T. bancanus*, from the western Indonesian Islands) the orbits are so flared that the skull is actually broader than it is long.

Tarsiers live in pairs; they feed on insects, lizards, and other small vertebrates. They are inconspicuous, though not uncommon, especially in secondary forest. Their Indonesian name speaks volumes about their way of life and their general persona: *binatang hantu*—ghost animal.

Tarsiers form a subgroup of Haplorhini all by themselves. To the other subgroup, the Simiiformes, belong all the others, sometimes called "anthropoids": monkeys, apes, and humans. Their big brains and great intelligence distinguish them from other primates. The first division is between the New World monkeys (Platyrrhini) and the rest.

▲ *The nocturnal gray mouse lemur (top, left) is still common in the dry forests of western Madagascar. The aye-aye (top, center) is an enigmatic rainforest lemur; even its exact range is poorly known, but it is certainly very rare. The Philippine tarsier (top, right), a small, nocturnal, carnivorous primate from Southeast Asia, looks very much like a lemur but is in fact a primitive haplorhine. The angwantibo or golden potto (bottom) is a small lorid of West Central Africa, where it is nowhere common. It lives on insects and their larvae, including noxious caterpillars. It weighs up to 200 grams (7 ounces).*

A CONSERVATION CONCERN

Madagascar is ecologically diverse: there is a wide central grassy plateau, and down its eastern flank runs a belt of rainforest, whereas the western side of the island is covered with dry sclerophyll forest. In the south is a semi-desert region, where spiny succulents are the main vegetation; there is a small pocket of rainforest in the northwest; and in the far north is the forested Amber Mountain. Each of these zones has its own lemurs: most of the widespread genera have at least a rainforest species and a dry-forest species, and sometimes arid-zone, northwestern, and Amber Mountain species as well. But very few lemurs are at all common today. Mouse lemurs and other small forms survive well in small, remnant forest patches; ring-tailed lemurs and white sifakas are well-protected in a few strict nature reserves; black lemurs are revered as sacred in some villages in their range in the northwestern rainforest zone. The indri has a small range, but part of it is a nature reserve visited by tourists; *Lepilemur, Phaner, Cheirogaleus,* and some species of *Petterus* and *Hapalemur* are not endangered, though they are becoming more and more restricted as their forests are cut down.

But the conservation status of some lemurs is really desperate. Twenty years ago about half a dozen aye-ayes were caught and released on a small island to try to ensure their survival, as they are irrationally feared and so deliberately killed by some village communities; it is now known that the species still does survive in the eastern rainforest zone on the mainland, but very sparsely and in small numbers. The hairy-eared dwarf lemur *Allocebus trichotis* was long known only from three specimens, collected in the 1870s, and was feared extinct, until a fourth was found, kept as a pet, in 1965. Only in 1989 were specimens seen in the wild. The broad-nosed lemur *Hapalemur simus* was also feared to have been extinct since 1900, but in 1972 surviving populations, of not much more than 100 individuals, were found in the southeast; in 1985 it was discovered that about half of these were not of that species at all, but a new species, now known as the golden bamboo lemur *Hapalemur aureus*. Nearly as rare is the newly discovered Tattersall's sifaka *Propithecus tattersalli,* which has a few scattered populations of perhaps a couple of hundred individuals in the far northeast.

The Malagasy government is seeking urgent international aid to help it in ambitious education, village relocation, nature reserve, and conservation-for-development programs; the will is there, and the expertise is available, but the finance is critically lacking—and the exploding human population of Madagascar threatens to overwhelm even the best-protected areas.

▶ The white sifaka lives in the dry forests of western Madagascar in small family groups that generally consist of more males than females.

▼ The rare and mysterious aye-aye eats fruits by gnawing through the outer husk or rind and scooping out the pulp with rapid movements of its thin, wiry, long middle finger.

Frans Lanting/Minden Pictures

Frans Lanting/Minden Pictures

▲ The red uakari (top, left), the only
New World monkey without a long tail,
is a surprisingly agile leaper. The red
howler ((top, right), the largest of the
howler monkeys, lives in small territorial
groups high in the trees. The cotton-top

tamarin (bottom, left) is a now-rare
species of marmoset-like primate from
Panama. Dourocoulis or night monkeys
(bottom, center) belong to the only
genus of nocturnal monkeys. The
muriqui or woolly spider monkey
(bottom, right) is the largest and rarest
New World monkey. Only about 300
remain, in the dry forests of southeastern
Brazil.

NEW WORLD MONKEYS

Platyrrhines, or New World monkeys, are found in
South and Central America, almost entirely in
rainforest. Externally, they can be distinguished at
once by the nose: the nasal septum is broad, so that
the nostrils point sideways. Also, their thumbs are
not markedly opposable to their other fingers, and
some of them grip objects as readily between
forefinger and third finger as between thumb and
forefinger. The larger species have prehensile tails;
no other primate has a prehensile tail.

The problem of how these monkeys got to the

Americas is one that has puzzled paleontologists, because their remains first turn up in early Oligocene deposits of about 35 million years ago, when South America was already separated from Africa, where their closest contemporary relatives have been found.

Marmosets and tamarins

Marmosets and tamarins range from the tiny pygmy marmoset *Callithrix pygmaea,* at 125 grams (4½ ounces) the world's smallest living monkey, to the golden lion tamarin, which weighs 600 grams (21 ounces). They all have claws on all digits except the great toe (which bears a nail), making them unique among monkeys. Another unusual characteristic is that, except for *Callimico,* all marmosets tend to have twin births. The larger species eat mainly fruit, the small ones insects; many eat gums and nectar.

There is a curious puzzle about their typical social organization: in captivity, all species seem to be most easily kept in pairs, but the young will not breed until they are removed from their parents' cage. In the wild, pairs have been seen in some species, but other species seem more usually to live in quite large groups, of a dozen or more. What is more, the sex ratio in larger groups is often quite skewed, with more males than females.

It has recently been pointed out that the best analogies to marmosets' social systems are not from other primates, but from some birds, like Australian white-winged choughs and Florida scrub-jays, which exhibit what have been called "helper systems". The young stay with their parents well beyond maturity, and help in rearing the next clutch of offspring, and the next. What finally persuades them to leave is unclear, but in at least some birds that have this system the females tend to emigrate earlier than the males.

As nearly all marmosets bear twins, the potential load on the mother is enormous. When pairs are kept together in captivity, the male is the one who does most of the carrying of the infants, playing with them, and generally "babysitting"; in general, the female takes the young mainly for suckling. When the previous year's young are left in the group it is they, particularly the males, who perform more of the babysitting activities.

Night monkeys and titis

Whether night monkeys, of the genus *Aotus,* and titis, of the genus *Callicebus,* the other pair-living platyrrhines, also have "helper systems" is not known, but they do not have twins. A pair of titis will sit side by side, tails intertwined; the night monkeys seem less closely bonded. It used to be thought that there was only one species of night monkey, *Aotus trivirgatus,* but not only are there conspicuous differences in coloration from one area to another, there have also turned out to be very striking differences in chromosomes: those

Sullivan & Rogers/Bruce Coleman Ltd

from southern Colombia have 46 chromosomes, while those from the north of the same country have 54, 55, or 56. So nine or ten species of *Aotus* are now recognized. Similarly, evidence has been accumulating that *Callicebus* also has many more species than was once thought.

Other New World monkeys

Squirrel monkeys, of the genus *Saimiri,* are small greenish monkeys with white faces and black muzzles; females weigh 500 to 750 grams (18 to 26 ounces); males weigh a kilogram or more. These monkeys are unusual in that they are seasonal breeders. They live in large troops of 20 to 50 or more, with many females and only a few males. In the breeding season, lasting three to four months, the males put on fat and become very aggressive as they compete for mating opportunities.

The prehensile-tailed howler monkeys, of the genus *Alouatta,* are widespread from southern Mexico to northern Argentina. The "howls" are troop spacing calls. The howls have been known to carry, on the still evening air, as much as 5 kilometers (3 miles)! Different species loud calls sound different: zoologists Thorington, Rudran, and Mack, after studying the small black species *Alouatta palliata* in Panama, and the large reddish *A. seniculus* in Venezuela, said: "The howls of a distant troop of *A. palliata* sound rather like the cheers of a crowd in a football stadium, whereas the howls of *A. seniculus* are much like the sound of surf on a distant shore or a roaring distant wind." Most species of howlers live in troops of 10 to 30 consisting of several members of both sexes.

Spider monkeys, of the genus *Ateles,* are much more active animals than howlers; their troops are also much larger and split up into small foraging parties of unstable composition from day to day.

▲ *A group of red howler monkeys howl to defend their ranging area in forested areas of the Llanos of Venezuela.*

▼ The range of the diana monkey or diana guenon (below, top) is now confined to the diminishing high forests of West Africa. Male mandrills (bottom, left) are brightly colored; in females and the young, these colors are more muted. Mandrills live in Gabon and neighboring countries in Central Africa. The Celebes black "ape" or Celebes macaque (center, right) is one of four to seven species of macaques found on the island of Sulawesi, central Indonesia. In the gelada (bottom, right), only the adult male has the long mane, but both sexes have a bright red patch of bare skin on the chest.

OLD WORLD MONKEYS

The two major groups of Old World monkeys are usually referred to as subfamilies of a single family, Cercopithecidae, but many authorities now prefer to recognize two full families: the Cercopithecidae and Colobidae. Colobid monkeys tend to be specialized leaf-eaters. Like ruminants or kangaroos, they have stomachs with fermentation chambers containing bacteria that break down cellulose into short-chain fatty acids and so make extra nutrients available from the monkey's food. Cercopithecids have simple stomachs, and deep food-storage pouches in the lining of the cheeks.

Guenons

Cercopithecus monkeys, often called guenons, are a marvellously varied and brightly colored lot, divided into a number of species-groups. In rainforest regions members of three species-groups occur together, living high in the trees. In Gabon the local representatives of the three groups are the greater spot-nosed monkey *C. nictitans* (males weigh 6.6 kilograms, females 4 kilograms, 15 and 9 pounds); the much smaller moustached monkey *C. cephus* (males weigh 4 kilograms, females under 3 kilograms, 9 and 6½ pounds); and the intermediate-sized crested guenon *C. pogonias* (males weigh 4.5 kilograms, females a little more than 3 kilograms, 10 to 6½ pounds).

Each species lives in a small troop consisting of one adult male, three or four adult females, and their offspring. The troops are territorial, but normally a troop of monkeys consists of not just one of the species, but two or even all three. The troops of the three species have their usual social structure, but have more or less merged their separate identities together so that they move around, sleep, and feed together—perhaps for years. In some cases, the males of the two smaller species do not even bother to make their characteristic spacing calls, but rely on the deep booming call of the big spot-nose male to maintain the joint territorial space. Jean-Pierre and Annie Gautier have studied this remarkable symbiosis intensively in forests in northeastern Gabon. In one troop they studied, the male crested guenon was the leader: he made the spacing calls, and gave the warnings against birds of prey, whereas the male moustached monkey uttered the warning calls for terrestrial predators.

Baboons and their relatives

Other species of African monkeys have quite different social structures. The gelada *Theropithecus gelada,* a large baboon-like monkey from the grasslands of the Ethiopian plateau, is very sexually dimorphic: males weigh over 20 kilograms (44 pounds), females about 13 kilograms (29 pounds), and males have enormous manes. A troop consists of a male and one or a few females ("one-male groups" or "harems"), like the guenons, but in this species the troops associate together to form huge herds, sometimes of several hundred, which forage together on the cliff-tops.

The gelada is sometimes considered a kind of baboon. Other baboons include the hamadryas baboon *Papio hamadryas,* which like the gelada lives in one-male groups, and the savanna baboons *Papio cynocephalus* and relatives, which live in troops ranging from about ten to several hundred, with many adult males as well as females. Hamadryas and savanna baboons hybridize where their ranges meet, in northern Ethiopia, and there is some dispute whether they are better classified as separate species or not.

In savanna baboons, both males and females are organized into dominance hierarchies. Dominant animals get priority of access to scarce resources, such as favored foods, and they lead troop movements, sometimes taking the major role in the troop's protection; dominant males do most of the mating, and dominant females, too, appear to be more reproductively successful, as their offspring have a higher survival rate.

Among the baboons and their relatives, including the drill and mandrill *Mandrillus,* the mangabeys *Cercocebus* and *Lophocebus,* the talapoin *Miopithecus,* and the swamp monkey *Allenopithecus,* as well as some species of *Macaca,* the females develop periodic (generally monthly) sexual swellings—prominent fluid-filled swellings of the vulva and perineum, sometimes extending to the anus and under the base of the tail—which are at their maximum around the time of ovulation. At this time they are most attractive to the males, and exhibit the behavior patterns known as estrus: sexual solicitation, and cooperation in mating.

▼ The hamadryas baboon lives in rocky and arid areas in northern Ethiopia and southwestern Arabia. This group consists of a male — a rather young male — and three females, with offspring. This is the usual foraging unit.

Francisco Futil/Bruce Coleman Ltd

N. Tanaka, Orion Press/Bruce Coleman Ltd

▲ *The mountains of northern Honshu are snow-covered for more than half the year. In this harsh climate live Japanese macaques, also known as Japanese snow monkeys. During the winter they feed mainly on bark.*

Macaques

The macaques, of the genus *Macaca,* occupy in Asia the same ecological niches that the baboons and their relatives occupy in Africa. Like savanna baboons, most macaques live in multi-male troops; some are largely terrestrial, others mainly arboreal; some develop sexual swellings in the females, others do not. The crab-eating or long-tailed macaque *Macaca fascicularis* is one of the smaller species, weighing about 5 to 8 kilograms (11 to 18 pounds); it is widespread in tropical Southeast Asia, from Vietnam and Burma to Java, Timor, Borneo, and the Philippines. The closely related rhesus monkey *M. mulatta* replaces it in monsoon and deciduous forest regions, extending from northern India into China, as far north as Beijing. Even less tropical is the Japanese macaque *M. fuscata,* which in northern Honshu lives in regions that are snow-covered for more than half the year. The so-called Barbary "ape" *M. sylvanus* lives in North Africa, and a small colony of them on the

Rock of Gibraltar is kept going by fresh importations from Morocco every few years.

Colobid monkeys

The colobids have been separated from the cercopithecids since the Middle Miocene, about 12 million years ago, and the colobid digestive specializations are unique among primates. Whereas the cercopithecids are more diverse in Africa and have one genus in Asia, the colobids proliferate in Asia and are represented by only the colobus monkeys *Colobus* and *Procolobus* in Africa.

The mantled colobus *Colobus guereza* is a beautiful monkey with long silky black fur, varied with a white ring around the face, a white tail-tuft, and long white veil-like fringes along the sides of the body. It mostly lives in small one-male troops in forests in Kenya, Uganda, northern Tanzania, Ethiopia, and northern Zaire. The troops are territorial, the territories marked by the deep rattling roaring calls of the males. A study in the

Kibale Forest, Uganda, found that this species can survive for several months on little else but the dry, mature leaves of just one species of tree, the ironwood *Celtis durandii*; in Ethiopia it eats mainly *Podocarpus* leaves. A related species, the black colobus *C. satanas,* of the coastal forests of Cameroun, Gabon, and Equatorial Guinea, lives mainly on seeds (but of several tree species), avoiding the leaves which, in that region, are high in tannins and other toxic chemicals.

Colobus monkeys gallop, rather than walk, along branches and, at the end of the branch,

launch themselves into space without breaking stride and land in the neighboring tree, grabbing hold with their hook-like, thumbless hands.

In Asia, colobids are represented by the langurs; they differ from African colobus monkeys by having thumbs (though they are short), but their leaping skills are quite comparable. The entellus or sacred langur *Semnopithecus entellus* of India and Sri Lanka represents the monkey-god Hanuman of Hindu mythology, and consequently it is tolerated around Hindu temples and even in some towns, and is often fed by the faithful as a religious duty.

▲ The adult male proboscis monkey (above, left) has a rather grotesque appearance. This species is confined to riverine and mangrove forests in Borneo. The golden snub-nosed monkey (above, center) is a rare, gaudily colored monkey from the cool mountain forests of southwestern China. The eastern black and white colobus or mantled guereza (above, right), once hunted for its beautiful skin, is now common again in light forest in East Central Africa. It is even found quite close to Nairobi.

PRIMATE HANDS AND FEET

feet   hands

▲ The indri (top) clings to tree trunks; its great toe and thumb are stout and give a strong, wide grip. The aye-aye (second from top) has claws on all its digits except the great toe; it climbs by digging its claws into the bark, rather than by clinging like most other primates. The tarsier (third from top) has disc-like pads on its toes and fingers, to increase friction. The gorilla (bottom) has divergent thumb and great toe; the palm is broad to support its huge weight, and the great toe is stouter and less divergent than in other catarrhines, except humans.

► Muller's gibbon lives in Borneo. It is a close relative of the white-handed gibbon, from which it differs in color and in its longer, more trilling call. Gibbons have elongated arms and move mainly by brachiation — that is, they swing by their arms from branch to branch. They are smaller and have narrower chests and less mobile wrists than other apes.

Some langur troops are multi-male, others are one-male harems. In districts where one-male groups are usual, the surplus males form a bachelor band which, every two and a quarter years on average, invades a bisexual troop and ousts the troop male. The new troop male, one of the former bachelors, then tries to kill all the troop's unweaned infants. This unpleasant behavior trait appears on present evidence to be quite widespread among primates, especially in Old World monkeys, but is most conspicuous and best studied in these sacred langurs. Sarah Blaffer Hrdy, who has studied infanticide in langurs most intensively, points out that, when a female loses her infant, she stops lactating and so begins her sexual cycles again; the new troop can start breeding at once and so increase the reproductive output of the harem male before he, in turn, is ousted 27 months later.

Other langurs live in East and Southeast Asia. The forests of Malaysia and western Indonesia contain two genera, Presbytis and Trachypithecus. Mostly they live in the same areas, but whereas the silvery leaf-monkeys Trachypithecus auratus (of Java, Bali, and Lombok) and T. cristatus (from Sumatra, Borneo, and the Malay Peninsula) are relatively unvarying from place to place, the half dozen or so species of Presbytis are enormously diverse—every mountain bloc, offshore island, or forested region between two rivers has its own species or distinctive subspecies. But further north, where Presbytis does not occur, species of Trachypithecus may also show marked geographic variability. The capped langur T. pileatus of the Assam–Burma border region, and Francois's langur T. francoisi of the limestone hills of the Vietnam–Laos–Gwangxi border region, both vary enormously from one small area to another, and there is even a distinctive species, the beautiful golden langur T. geei, restricted to a small area in Bhutan.

In the mountains of Sichuan in China lives the bizarre and gaudily colored golden snub-nosed monkey Pygathrix roxellana. The males have fiery red-gold fur on the underside, flanks, and limbs, and around the face (the bare facial skin being pale blue), with long black fur on the back; they weigh 15 kilograms (33 pounds). The females are a drabber yellow-gold, again with black on the back, and weigh only 10 kilograms (22 pounds). Both sexes have curious, upturned, leaf-like noses. Two related species live in China, and another in Vietnam; a fifth species, the douc langur P. nemaeus, of southern Vietnam and Laos, has a flat nose but is even more gaudily colored.

The largest colobid is the brick-red-colored proboscis monkey Nasalis larvatus of Borneo. Males weigh up to 24 kilograms (53 pounds), females less than half that. The young have long, forward-pointing noses; in females these stop growing at maturity, but those of males carry on enlarging into enormous drooping Punch-like adornments. (A

nickname for them in Indonesia is belanda, which literally means "Dutchman"!) Proboscis monkeys live in troops in coastal areas, especially in and around mangroves, and along large rivers. They seem as much at home in the water as in the trees. It is one of nature's treats to see a huge male solemnly wade upright, arms held clear of the water, into a stream, lapsing into an energetic dog-paddle when he gets out of his depth.

Closely related to the proboscis monkey is the curious pig-tailed langur or simakobu Nasalis concolor (often placed in a separate genus, Simias) from the Mentawai Islands. Its nose is like a small version of the juvenile proboscis monkey's and it is the only colobid with a short tail.

APES: HAIRY AND NAKED

The apes, or hominoids (gibbon, orang utan, gorilla, chimpanzee, and human), differ from their sister-group, the Old World monkeys, in many ways. They have no tail: the few remaining caudal vertebrae, which make up the tail in other mammals, are variably fused together into a shelf-like small bone, the coccyx. The apes habitually sit (sometimes stand) upright, so it is no surprise that their lumbar vertebrae, being required to bear the weight of the upper body, instead of adding flexibility to the spine, are reduced in number (from seven or eight to four or five) and are much shortened and broadened. The thorax is broad—broader side-to-side than it is deep front-to-back—and the scapulae are on the back of the thorax, not on the sides, as in the Old World monkeys. The shoulders and wrists are very mobile. Apes, except for humans, have long arms, generally longer than their legs, but in fact in chimpanzees and gorillas it turns out that the arms are not especially long compared with the trunk—only in gibbons and orangs are the arms genuinely elongated.

There has been a lot of discussion about the evolutionary meaning of these anatomical features. Gibbons move mainly by brachiation, that is swinging by their arms under branches, and chimpanzees sometimes do this too, though orang utans clamber about with any old combination of arms and legs, and gorillas rarely climb at all and certainly never brachiate when they do. But, perhaps because of nineteenth century misunderstandings about the natural history of the apes, the idea that the hominoids' skeletal anatomy is "for brachiation" has become entrenched in both the specialist literature and the popular imagination, and it is almost impossible to get rid of. After all, did not our own ancestors "come down from the trees", where they had been swinging about by their arms? The facts suggest otherwise. First, gorillas are very largely terrestrial and chimpanzees are at least partly so—and these are our closest living relatives. Orang utans, the other great apes, do not brachiate. Gibbons (lesser

▲ *The siamang is the largest of the gibbons. It is seen here with an inflated throat sac, uttering the booming lead-in to its harsh, shrieking call.*

round to pull in the fruit. If brachiating ever was in our ancestry, it was surely only at a time when we shared a common ancestor with all other hominoids, and that common ancestor was small in size.

But even then the brachiating specializations do not go very far, for gibbons possess very many features, especially in the skeleton and muscles of their forelimbs, which are clearly valuable for their way of life but are not found in great apes or humans. When the human/great ape ancestor separated from the gibbon ancestor, it increased in size, and there was no more brachiation. When the human/gorilla/chimpanzee ancestor separated from the orang utan's ancestor, it came down from the trees. It is open-country existence, not terrestrial life as such, that distinguishes us from our closest relatives.

There are other hominoid characters too. An odd, and little-known, hominoid character is that we all possess a vermiform appendix, a thin projection on the end of the cecum (the blind gut at the junction of the small and large intestines), filled with lymphoid tissue. No other haplorhine primate possesses an appendix, so it must be an organ that the common ancestor of the hominoids developed, presumably to supplement the immune system. There is a popular belief that the human appendix is an evolutionary vestige (usually, it is suggested, of the cecum, and it is true that the cecum is very small in hominoids), but it is not.

Gibbons: arias and acrobatics in the tree tops

The smallest of the apes, the gibbons, are slenderly built, and have extraordinarily long arms and long narrow hands in which the thumb is deeply cleft from the palm, as far as the wrist; the legs are long, too, a fact often missed because the long arms overshadow everything else about them. They move easily and rapidly through the trees by brachiation: hand over hand over hand, in an up-and-down swinging gait, occasionally dropping easily to a lower branch, or jumping between trees.

All gibbons, as far as is known, live in pairs: a male and a female, with up to four offspring. The pair occupy a tree-top territory. From its center in the early morning, they launch into song, a duet: the female takes the major part, the male punctuating it at intervals with shorter, less coloratura phrases. The duet is primarily a spacing call, though by requiring the male and female to interweave their different contributions, it reinforces the pair-bond. Gibbons do not necessarily sing every morning, but when they do, the forest resounds with their soaring melodies as one pair after another takes up the theme, each from its own small territory. The wonders of gibbon acrobatics and operatics should not, however, obscure the fact that they are in fact only distantly related to the other hominoids.

The best-known species is the white-handed gibbon *Hylobates lar,* from northern Sumatra, the

apes), the ones that do brachiate, are not only small in size, but they also have the least hominoid anatomical specializations: the narrowest chests, the longest lumbars, and the least mobile wrists.

Observing apes in the wild suggests what the so-called "brachiating characteristics" are really all about. Great apes are lazy: they like to sit in amongst their favorite food plants and rake them in without moving much; to do this they sit upright and reach out, rotating their shoulders and wrists to pull in the vegetation or fruit. For a big-bodied animal it is a very worthwhile economy to be able to feed almost without moving. Gibbons practise a variation on this: they hang from the ends of branches, or sit on them, and again simply reach all

Malay Peninsula, Thailand, and parts of Burma and Yunnan. Weighing 5 to 7 kilograms (11 to 15 pounds), these small gibbons may be either buff or dark brown-black in either sex (the exact shade differs according to subspecies), both color types being found in the same population. They are remarkable, too, for their extremely dense fur, with more than 1,700 hairs per square centimeter (11,000 per square inch) of skin on the back—two or three times that in most monkeys. Related species live in Java, Borneo, southern Sumatra, and Cambodia: they differ in color and in vocalizations. Kloss's gibbon *H. klossii,* from the Mentawai Islands, which is completely black and has much less dense fur, is one of the few species in which the sexes sing separately and do not duet. In Burma, and extending into Assam, is found the larger hoolock gibbon *H. hoolock,* which weighs up to 8 kilograms (18 pounds). In this species, both sexes are black while juvenile (with a white brow-band), but on maturity—at about 7 years, as in all gibbons—the female turns buff-brown. The hoolock gibbon lives in monsoonal semi-deciduous forests, not rainforests like the *lar*-group, and it has a small inflatable throat-sac that gives it a much harsher call.

The concolor gibbon *H. concolor* lives in the Vietnam–Yunnan border region, and on Hainan Island. With age, the color changes in the female from black to buff, just as in the hoolock, although the two are not very closely related; the male, in addition, has an upstanding tuft of hair on the crown. Only the male has a throat sac; his voice is little more than a harsh shriek, while the female is melodious like members of the *lar*-group. Further south in Vietnam and Laos are species in which the black form has whitish cheek whiskers.

The largest gibbon is the siamang *H. syndactylus,* which weighs 9 to 12.5 kilograms (20 to 27½ pounds). Both sexes are black and have very large throat sacs, so that their calls (quite deafening at close quarters) are harsh barks and shrieks, punctuated by the boom of the throat sacs being inflated. They live in Sumatra and West Malaysia, in the same forests as *H. lar* and *H. agilis.* Compared with these smaller gibbons, siamangs eat more leaves and less fruit, vocalize less frequently, have smaller territories, and seem more closely pair-bonded. Remarkably, recalling those other pair-bonded primates, the marmosets, the adult male carries and babysits the infant, at least beyond its first year.

▼ *White-handed gibbons, the best known of all gibbons, are still quite common in the forests of the Malay Peninsula, and will survive as long as the forests survive. But how long will that be?*

Jean-Paul Ferrero/Auscape International

▲ The orang utan, the "man of the woods", is the least humanoid of the great apes but is still a remarkably intelligent and adaptable creature.

▶ Each night, the solitary orang utan makes a fresh sleeping nest high in the trees. Leaves and branches are pulled into a platform shape in the crown of a tree.

Orang utan: dignified tree-top mandarin

The orang utan *Pongo pygmaeus* is today restricted to Borneo and northern Sumatra. On Borneo it is still locally common in lowland forests, but in Sumatra, where it is not found south of about Lake Toba, it ascends to considerable altitudes. The name "orang utan" means simply "wild person", or "man of the woods", in Malay and Indonesian; indigenous people call it *maias* (in many Dayak languages) or *mawas* (in Sumatra).

Orangs are covered in sparse red hair through which the rough blue-grey skin can be seen. Bornean orangs *Pongo pygmaeus pygmaeus* tend to have thinner, more maroon hair, whereas in Sumatran orangs *P. p. abelii* the hair is lighter, more gingery, and fleecier, and often very long, especially on the arms. An orang's arms are very long, the legs appearing short by comparison; both the hands and feet are powerful and curved, with shortened thumb and great toe, and long second to fifth digits.

Orangs are sexually mature at about 7 years of age, after which time females more or less cease growing; adult females weigh 33 to 42 kilograms (73 to 92 pounds), and are 107 to 120 centimeters

(42 to 47 inches) high when standing bipedally. But the males carry on growing until they are 13 to 15 years old; a fully mature male weighs 80 to 91 kilograms (176 to 200 pounds)—though in zoos they become obese and can weigh much more—and stands normally 136 to 141 centimeters (53½ to 55½ inches) high, though a giant of 156 centimeters (61½ inches) has been recorded. In their teens, too, the males develop fleshy cheek-flanges; in Sumatran males, these tend to be smaller, and protrude sideways from the face, whereas those of a Bornean male are often very large and swing forward as he moves.

The red ape lives a solitary life. Females have overlapping home ranges of about 2 square kilometers (¾ square mile); a mature male occupies a much larger area, of 8 square kilometers (3 square miles) or more, which is not exactly a territory, as the area is too large to exclude others from it, but within it other males try to avoid the resident, who from time to time utters a succession of deep carrying roars: other males move away from the sound, whereas females may gravitate towards it. At night, every individual builds a nest, high in the trees; vegetation is packed securely into the crown of a tree, with a special rim placed around the main platform. When the orang lies in the nest, further vegetation is usually pulled down over it to make a cover. Occasionally a nest is reused, but generally a fresh one is made each night, even if the animal has not travelled far that day.

More than half the diet is fruit, with leaves, bark, and chance morsels like insects making up the rest. Rainforest trees come into fruit rather irregularly, depending not on season as such but on a combination of factors such as recent and current patterns of rainfall, cloudiness, temperature, wind, and so on. It has been pointed out that for a very large animal it is far more economic to be able to remember the locations of favorite fruiting trees over a wide area, and to be able to calculate the likelihood of a given tree being in fruit, than to go searching. And so, the theory goes, the high intelligence of the great apes is an almost inevitable consequence of being large and eating fruit.

There has been much discussion as to whether female orangs have restricted estrus, or are willing to copulate throughout their sexual cycle of 28 days. It now appears that they have only short estrus periods: for a few days around the time of ovulation, the female actively seeks out and solicits a mature, flanged male, and they associate together for a few days. But subadult males, not sought out by females, will chase and rape females at any time. It has been suggested that, as the penis is so short, a copulation will fail unless the female cooperates, so that most infants are born out of consortships with mature males, not from rapes by subadults. The gestation averages 245 days.

In the past, orangs were much in demand by

zoos, and as they were obtained by shooting mothers and collecting their infants (most of which would then die from inadequate substitute care), a dozen orangs would die for every one that reached a zoo. Now this is banned by international agreement: orangs breed well under good zoo management, and an increasing proportion of zoo orangs are now captive-born. The new threat to the orang is deforestation; even where, as in Indonesia, there is government concern for conservation, and logging is no longer regarded as the inevitable fate of every forest, the inexorable increase of the human population creates land hunger, and more and more inroads are made into reserves and national parks by cultivation.

▲ A mature male Sumatran orang utan. Sumatran males differ from Bornean in their flatter cheek-flanges, the smattering of downy hair on the face, and their long beards and moustaches.

▲ *A silverback male mountain gorilla reclines in the dense ground vegetation of the Virunga Volcanoes.*

▼ *A young mountain gorilla at play.*

Gorilla: Africa's gentle giant

The gorilla *Gorilla gorilla* is the largest living primate. Many people have a false impression of just how large a gorilla is. Reputed half ton, nine foot high monsters just do not exist: the record for an adult male in the wild is 219 kilograms (482 pounds) and 195 centimeters (77 inches), and even this is most unusual. The average adult male weighs on average 175 kilograms (385 pounds) and stands 156 centimeters (61½ inches) bipedally; a female weighs 85 kilograms (187 pounds) and stands 137 centimeters (54 inches) high. Like orangs, gorillas reach sexual maturity at about 7 years. At this time females cease growing, but males grow until they are about 12 or 13, and at full physical maturity develop a silvery-white "saddle" on the back, contrasting with the black hair (hence, a fully mature male is known as a silverback). The prominent brow ridges, smooth black skin, and flaring nostrils contrast with the high forehead, rough grey skin, and tiny nose of the orang. Gorillas have a pungent odor, emanating from the armpits: like humans, and also chimpanzees, they have a large cluster of specialized sweat glands there.

Gorillas are very largely terrestrial. They do climb, but not too commonly (especially not the big silverbacks), and they always travel long distances on the ground. They travel on all fours, with their hands not palm-down on the ground but flexed, so that the weight rests on the middle joint of the fingers: this is called knuckle-walking. Like orangs, they make nests every night—not such complex, well-made nests, and usually on the ground, though in a few areas there seem to be "traditions" of building nests in trees.

A gorilla troop consists of a silverback male (sometimes two, or even more), a few blackback (or subadult) males, and several females and young. The troop wanders over a large home range of 10 to 20 square kilometers (4 to 8 square miles) which overlaps that of other troops, with whom relations are normally peaceful. When a male matures, he has the option of staying in the troop, or leaving it. It has been suggested that he will stay in his natal troop if there are any females in it who are potential mates for him, otherwise he will leave and live alone for a while, sometimes following other troops and trying to "kidnap" females from them, usually by invading a troop when its silverback is off-guard and rounding up one or more females. In the fracas that ensues, there is often a fight, and infants may be killed—perhaps deliberately by the invading male, for an infant seems to be a bond between a female and her troop. Females may also take the opportunity to leave a weak male during an encounter of this nature, but if the new male in turn proves weak and cannot attract further females to him, then the females he already has will take the first opportunity to go elsewhere. The most stable kind of troop is probably one in which there is a second silverback ready to take over.

The gorilla has a fragmented distribution within Africa. There are three subspecies. The western lowland gorilla *G. g. gorilla* lives in West Central Africa. The much blacker eastern lowland gorilla (*G. g. graueri*) lives in Zaire, east of the Lualaba River, extending from the lowlands near the river up into the mountains. The black, long-haired, large-toothed mountain gorilla *G. g. beringei* lives in the Virunga Volcanoes, on the Zaire–Rwanda– Uganda border, and in the Impenetrable (Bwindi) Forest in southwestern Uganda.

Within lowland rainforest, gorillas seem to favor secondary growth especially; in the mountainous parts of their range they live in open forest and in bamboo thicket, up to 3,500 meters (11,500 feet). They feed in montane areas on bamboo shoots and stems of tall herbs, saplings and small trees, but in lowland areas more fruit is eaten, especially tough, somewhat woody fruits. This diet seems to have its drawbacks: 17 percent of gorilla skulls in museums have dental abscesses, and most mountain gorilla skulls show severe breakdown of the bone between the teeth—perhaps, it has been suggested, because of the packing of bamboo fibers, although this explanation is not the whole story, because other bamboo-eating gorilla populations seem not to show this condition. And finally, it may not be surprising in such a heavy-bodied, partially upright animal that 17 percent of skeletons in museums show traces of spinal arthritis.

▼ *As the population of mountain gorillas in the Virunga Volcanoes slowly recovers from near-extinction, the proportion of young is growing and is now nearly 40 percent.*

Yann Arthus-Bertrand/Auscape International

Helmut Albrecht/Bruce Coleman Ltd

▲ *Common chimpanzees spend much of their time on the ground, but are also active and efficient climbers.*

Chimpanzees: cunning and charisma

The chimpanzee, like the gorilla, is a specialized knuckle-walker; like the gorilla it is black, has large brow-ridges and a smooth skin which is black, at least in the adult. But most analyses, whether biochemical or morphological, tend to indicate that the chimpanzee is no less closely related to humans than it is to the gorilla—and it may even be closer. So perhaps this indicates that the characteristics it has in common with the gorilla were those of our own distant ancestor too.

The common chimpanzee *Pan troglodytes* is still widespread in West and Central Africa, reaching as far east as western Uganda and the extreme west of Tanzania. It lives not only in rainforest but also in montane forest, open forest, and even savanna woodland where it has not been exterminated. The very big Central African subspecies *P. t. troglodytes,* often called the bald chimpanzee, is very black-skinned from quite an early age, and adults go bald on the crown—males in a triangle, narrowing back from the forehead, females totally. Males weigh 60 kilograms (130 pounds) on average and average 120 centimeters (47 inches) in height when standing bipedally; females average 47.5 kilograms

(105 pounds) but are nearly as tall as the males, so they are more lightly built. The East African or long-haired chimpanzee *P. t. schweinfurthii* is much smaller: males in the famous Gombe Stream population of Tanzania average only 43 kilograms (95 pounds) and females 33 kilograms (73 pounds). They are less intensely black than Central African ones, have a more marked beard and cheek whiskers, and they do not go bald so completely or so early. The West African or masked chimpanzee *P. t. verus* is small, like the East African; it is very dark around the eyes and nasal bridge, but less so elsewhere on the face; it develops a long grey beard with maturity, does not go very bald, and when young has a central parting on the head.

But much more characteristically different, and generally considered a distinct species, is the pygmy chimpanzee or bonobo *Pan paniscus,* from south of the great bend of the Zaire River. Despite its name, it is about the same weight as the two smaller forms of the common chimpanzee, and stands 119 centimeters (46½ inches) high; it has longer legs, shorter arms, and a smaller head than the common chimpanzee, as well as long side-whiskers, rather small brow-ridges, and a face that is black from birth except for the contrastingly pinkish white lips.

Chimpanzees spend about half their time on the ground; in the trees, they knuckle-walk along branches, or brachiate beneath them. They feed on fruit, to a lesser extent on leaves, bark, insects, and even on vertebrate prey; some populations very regularly kill monkeys and other medium-sized mammals, which they hunt cooperatively (males do most of the hunting). At night they make nests in trees, like orangs, though their nests are not as complex; and, like gorillas, nests may also be made in the heat of the day for "siesta".

Chimpanzees live in large groups, which have been called communities by some authors and unit-groups by others: these number anything from 20 to 100 or more, but mostly its members associate in small parties, of varying composition, either bisexual parties roaming widely, searching for fruiting trees, or more sedentary nursery groups of mothers and infants. The community as a whole occupies a territory, whose boundaries are patrolled by the adult males against incursions by outsiders. The well-known field researcher Jane Goodall recorded an instance, in the Gombe National Park, of short-lived warfare between two neighboring communities, which ended in the virtual extermination of the smaller one.

Adult males seem almost always to remain in the communities in which they were born, whereas females seem always to leave them and join neighboring communities, sometimes changing communities more than once during their lives.

Whether this pattern applies to pygmy chimpanzees, or, indeed, to other common chimpanzees apart from the *schweinfurthii* populations of Tanzania and Uganda which have been best studied, is unclear, but it probably does at least as a rule. Pygmy chimpanzees do, however, appear to travel in groups of more consistent composition. Whereas in the East African chimpanzees males seem to form close bonds of association (perhaps because of their relatedness?), in pygmy chimpanzees it is females that form close ties, which appear to involve sexual interaction.

There is a well-marked difference between the two species in their heterosexual behavior. Females in both species have sexual swellings: enormous pink doughnut-shaped excrescences involving vulva, perineum, and anus. In the common chimpanzee, the swelling begins to appear after menstruation and gradually enlarges, reaching a maximum at the time of ovulation, after which it rapidly detumesces; the total cycle length is 35 days. The female solicits mating around the time of maximum swelling, though she will mate at other times: about 75 percent of copulations occur in the week or 10 days when she is fully swollen. In the pygmy chimpanzee, however, the cycle length averages 46 days in the wild (though in two captive females it was only 36 days); she has large sexual swellings for about half of this time, and never completely lacks them except when lactating. Most primatologists studying pygmy chimpanzees have found no restriction of copulation to any phase of the sexual cycle; only one has even reported a tendency for sexual activity to diminish during the times of lowest swelling. Moreover, common chimpanzees, like almost all other non-human primates (gibbons and orangs mating while hanging from trees may form an exception), mate dorso-ventrally, but pygmy chimpanzees very frequently mate ventro-ventrally, maintaining eye contact all the while.

When a female common chimpanzee has her maximum sexual swelling, males gather round her like flies and she willingly mates many times, apparently indiscriminately. But she may instead choose to form a consortship with one particular male; the pair slip away quietly for several days and sometimes the female will consort with the same male next time she comes into estrus. It is claimed that relatively more infants are born as a result of consortships than after promiscuous behavior. Pygmy chimpanzees also form consortships.

Gestation lasts about 230 days, shorter than the orang (245 days) or the gorilla (267 days). Like all great apes, lactation goes on for a long time—two or three years, or even more—and there is an interval of four years between live births. Sexual maturity is not reached until about 8 years of age and, unlike other apes, males as well as females seem to stop growing at this time. Because, like many other primates, including humans, there is a period when sexual cycles are non-ovulatory ("adolescent sterility"), a female chimpanzee may be 12 or more before she gives birth for the first time. Chimpanzees, like other apes, might live 35 to 40 years, sometimes more in captivity, and very aged females seem to show signs of menopause.

▼ Tool-use, even tool-making, is known in many populations of wild chimpanzees. Here a youngster, still not very proficient, pokes a stick into rotten wood to extract grubs.

Peter Davey/Bruce Coleman Ltd

modify twigs and grass stems the better to obtain termites, ants, or honey (in different populations); in one population of chimpanzees, in Ivory Coast, they select special hammer-stones to open *Cola* nuts, and leave them in known places, beside tree buttresses, which make suitable anvils. Alone of all mammals except humans, great apes readily (and spontaneously, given a few days exposure) come to recognize their own images in mirrors: they have, that is, a concept of self. Other animals—monkeys, gibbons, elephants—have the intelligence to use a mirror's reflection to find hidden food, and can recognize other individuals reflected in it, but never themselves.

Most staggering, perhaps, have been the language experiments. Using hand signs (based on

Dr Ronald H. Cohn/The Gorilla Foundation

▲ *All three great apes can be trained to use symbols to communicate, in a very rudimentary way, with their trainers, even sometimes with each other. Here Koko, a female gorilla, inspects a machine.*

Almost human?

It has long been known that great apes are more intelligent than any other animal (I almost forgot to add, "except humans"). They learn faster; they can learn a greater variety of special skills; and they can learn more complex tasks. They very clearly figure out how to perform the tasks set them in psychological tests, sometimes by apparent flashes of insight, sometimes, it would seem, by careful logic. They can learn to use implements, too. Gorillas are not so good at this, although they are very patient when faced with complex machinery, but chimpanzees and orang utans readily learn to modify objects to make tools, for example, to obtain food. All this has been known since the work of R.M. Yerkes in the 1920s, but the degree to which apes' minds resemble our own was never suspected then. It is now known that some groups of chimpanzees make tools in the wild: they

THE EVOLUTIONARY TANGLE

Gibbon (*Hylobates*), orang utan (*Pongo*), gorilla (*Gorilla*), chimpanzee (*Pan*), and human (*Homo*) are the five living genera of the Hominoidea. Perhaps more work has been undertaken to try to elucidate how these five genera are interrelated than on the whole of the rest of the primates put together. After all, we are involved; human pride is at stake.

Human chauvinists would prefer that we were the odd man and woman out (in other words, the sister-group of the rest); until 25 years ago, it was customary to place all the great apes together in one family, the Pongidae, separate from humans, Hominidae. Even many of those who can come to terms with scientific findings, and bring themselves to accept the reality of evolution, would like to see us separated from the other primates altogether. In the nineteenth century it was usual to place humans in a separate order, Bimana, from the other primates, which were called Quadrumana. In the 1920s and 1930s the great Australian anatomist and zoologist Frederic Wood Jones proposed his "tarsier hypothesis"—that our ancestor was a small long-legged beast related to the tarsier and so separate from the monkey/ape lineage since the Oligocene (over 25 million years ago), if not earlier.

In the 1950s a famous paper, "The Riddle of Man's Ancestry" by W. L. Straus, an American anatomist, proposed that the characters we share with the great apes are more or less illusory, and that we are at least as different from them as are the gibbons; his reasoning convinced many eminent authorities, and there was no surprise when, in the early sixties, the paleontologist Elwyn Simons claimed that a Miocene fossil ape, *Ramapithecus* (known only from fragmentary jaws!), was a human ancestor contemporary with *Dryopithecus,* the putative ancestor of living great apes—hence 20 million years at least must be envisaged for the time of separation of the two lines.

American Sign Language for the deaf, or Ameslan), chimpanzees learned to symbolize objects and actions, adjectives and sometimes fairly abstract concepts; in later experiments, both orangs and gorillas have learned the same. In the first flush of enthusiasm, many psychologists claimed that their apes could produce sentences with full syntax, but this now seems dubious. A continuing experiment at the Yerkes Primate Center in Atlanta, Georgia, uses computer consoles: chimpanzees learn to press different buttons for different words, and communicate in this fashion with their trainers, with "untrained" people (thus eliminating the danger of the "Clever Hans" effect, in which the trainers' unconscious body cues influence the animal's behavior)—and, to a degree, with each

The Gorilla Foundation/National Geographic Society

▲ *Koko with her pet kitten, All Ball. Koko was devoted to the kitten and, reportedly, grieved intensely when it was hit by a car and died.*

Biochemists such as Morris Goodman and Vincent Sarich began working on this problem in the early 1960s. While disagreeing on details, they all agreed that the blood proteins (albumin, transferrin, hemoglobin, and others) of the gibbons were most different from those of other hominoids, those of the orang utan next; while those of humans, chimpanzees and gorillas were very close to each other (some of their proteins being nearly or quite indistinguishable). This created an impasse: paleontologists insisting that they were working with the actual record of evolution, biochemists pointing out that the fossil material was incomplete.

The Gordian knot was cut in 1982 by Peter Andrews and John Cronin, who showed in a classic paper that, first, a fossil ape, *Sivapithecus,* which had previously been confused with *Dryopithecus,* is actually quite distinct; second, the by-now quite well-represented cranial remains of *Sivapithecus* are exactly what one would expect in the primitive ancestor of the orang utan; and third, the jaws assigned to *Ramapithecus* fall within the range of variation of *Sivapithecus.* The supposed special resemblances of *Ramapithecus* to humans (such as the thickness of the enamel on the molar teeth) turn out to be shared by *Sivapithecus,* too, and seem to be in any case just primitive features, retained from the common ancestor of humans and all great apes. This massive re-evaluation of the fossil data meant that there was no longer any evidence for an early separation of the human lineage; on the contrary, there was now good evidence for the early separation of the orang utan lineage, in agreement with the biochemical findings.

More recently, analysis of the morphological characters of the living hominoids has confirmed the picture as deduced long ago by Goodman and Sarich: the gibbon ancestors split off first, then the orang ancestor, while gorilla, chimpanzee, and human are very closely related to one another. Maybe the gorilla is the sister-group of chimpanzee plus human, or maybe gorilla plus chimpanzee form the sister-group to human—no one is sure.

other. Two young chimpanzees, for example, acquired the facility of requesting each other to share different types of tools for obtaining food. It has been claimed that pygmy chimpanzees do best of all at this kind of intellectual activity; so far there has been no independent test of this claim, but it is recorded that a pair of recently captured pygmy chimpanzees *spontaneously* made iconic hand-signs to each other in the context of their favorite activity, ventro-ventral copulation.

All this has, as might be expected, produced a great deal of soul-searching among biomedical researchers—people who would have no compunction about using monkeys for medical or "curiosity" experimentation—about whether it is really ethically justifiable to use apes for such purposes. Because orangs and gorillas are classified as being in danger of extinction, they are not used in any case, but until recently chimpanzees were not considered endangered and were used for terminal experiments, disabling surgery, or injection with infective agents without a second thought. This has all changed. Sadly, chimpanzees are now also put in the endangered category as their wild populations have declined catastrophically in the last 20 to 30 years, and only captive-bred chimpanzees may now legally be used in endangering research, at least in the United States; but it is very noticeable how thinking on caging, husbandry, and the sort of research that can be done on them has been altered by the cognitive studies of the past twenty years. As one physiologist has put it, we have discovered that they are close to us mentally as well as physically, and this carries ethical obligations with it.

Rod Williams/Bruce Coleman Ltd

▶ *The only nocturnal monkey is the douroucouli, or night monkey, widespread in the forests of South and Central America. It has enormous eyes, but no eye-shine.*

Rod Williams/Bruce Coleman Ltd

▲ *The smallest living monkey, the pygmy marmoset, lives in the Upper Amazon rainforests, where it feeds partly on tree sap and gum, which it extracts by gouging into the bark with its incisor teeth.*

SPECIAL EYES

Primates all have large eyes whose visual fields overlap, so that they have stereoscopic vision. Most primates also have color vision. These two features are not restricted to primates, but there is a complex sorting of the neuroanatomy of the visual system in the brain, which is shared between primates and just two other groups of mammals, the flying lemurs or colugos, and the fruit bats.

Lining the inside of the eyeball is the retina, where incoming light is intercepted. The retina has two types of receptor: rods and cones. Rods are capable of receiving very low-level light; cones interpret color. Day-living primates, as might be predicted, have a high proportion of cones; in the human retina there are 7 million cones and 125 million rods, which is indeed a high proportion (though about average for diurnal primates). Night-living primates have little use for color vision and therefore have nearly all rods—dwarf lemurs of the genus *Cheirogaleus,* for example, have 1,000 rods for every cone.

Many nocturnal or crepuscular mammals have eye-shine: when a bright light is shone into their eyes, a reflection comes back. The reflection is due to a crystalline layer behind the retina, called the tapetum lucidum. This acts to increase the amount of light that passes across the retina, and so it assists night vision.

Most strepsirhines are nocturnal, whereas nearly all haplorhines are diurnal; but all strepsirhines (even the diurnal ones) have a tapetum. With this eye-shine, strepsirhines seem basically adapted to lead nocturnal lives, and no monkey can compete with them at this, so in mainland Africa and Asia, where they share their range with monkeys, there is a clear split: all strepsirhines are nocturnal, all monkeys are diurnal. But in Madagascar, where the lemurs have it all to themselves, some have become diurnal and fill a "monkey-like" niche. And, interestingly, in Latin America, where there are no strepsirhines, one genus of monkeys, *Aotus,* has become nocturnal, filling a "lemur-like" niche (the other nocturnal haplorhine is the tarsier, of Southeast Asia). But no haplorhine has a tapetum, so *Aotus* and *Tarsius* compensate by having the largest eyes of all primates, and apparently they are the only ones to have no cones in their retinas.

Haplorhines have a special feature of the retina which strepsirhines lack: a macula, or cones-only region, with a depression in its center. This is an area of exceptionally high visual acuity. The two nocturnal haplorhines, *Aotus* and *Tarsius,* have some of the histological features of the macula, even though they have no cones.

Thus, primates retain in their eyes the stamp of their evolutionary past: all strepsirhines have a tapetum, whether they are nocturnal or not, because their common ancestor was nocturnal; all haplorhines have a macula, whether they are diurnal or not, because their ancestor was diurnal.

▼ *The diademed sifaka of eastern Madagascar (below, left) moves and feeds by day, but the structure of the retina discloses a nocturnal ancestry.*

▼ *Photographed at night, the lesser bushbaby's eyes (like those of all strepsirhines) reflect the light of the flash camera.*

Jane Burton/Bruce Coleman Ltd

Frans Lanting/Minden Pictures

SIZE
Smallest Least weasel *Mustela nivalis*, head and body length 15 to 20 centimeters (6 to 8 inches), tail length 3 to 4 centimeters (1 to 1½ inches), weight 100 grams (3½ ounces).
Largest Land: polar bear *Ursus maritimus*, head and body length 2.5 to 3 meters (8 to 10 feet), weight over 800 kilograms (1,750 pounds). Aquatic: southern elephant seal *Mirounga leonina*, head and tail length 490 centimeters (193 inches), weight 2.4 tonnes (5,300 pounds).

CONSERVATION WATCH
The following species are listed as endangered in the IUCN Red Data Book of threatened mammals: red wolf *Canis rufus*, simien fox *Canis simensis*, Baluchistan bear *Selenarctos thibetanus gedrosianus*, Mexican grizzly bear *Ursus arctos nelsoni*, Cameroon clawless otter *Aonyx congica*, black-footed ferret *Mustela nigripes*, Malabar large-spotted civet *Viverra megaspila civettina*, Barbary hyena *Hyaena hyaena barbara*, Asiatic cheetah *Acinonyx jubatus venaticus*, Florida cougar *Felis concolor coryi*, eastern puma *F. c. cougar*, iriomote cat *Felis iriomotensis*, Pakistan sand cat *Felis margarita scheffeli*, Spanish lynx *Felis pardina*, Asiatic lion *Panthera leo persica*, Sinai leopard *Panthera pardus jarvisi*, South Arabian leopard *P. p. nimr*, Amur leopard *P. p. orientalis*, Barbary leopard *P. p. panthera*, Anatolian leopard *P. p. tulliana*, tiger *Panthera tigris*, Siberian tiger *P. t. altaica*, snow leopard *Panthera uncia*, Japanese sealion *Zalophus californianus japonicus*, Mediterranean monk seal *Monachus monachus*, Hawaiian or laysan monk seal *M. schauinslandi*, Caribbean monk seal *M. tropicalis*. Many other species are listed as rare or vulnerable.

CARNIVORES

W. CHRIS WOZENCRAFT
AND JUDITH E. KING

From our fascination with man-eating tigers to our close association with domestic cats and dogs, beasts of prey have always been considered very special by humans, even though they represent only about 11 percent of all mammals. Carnivores have been seen historically as vicious predators, high on the food chain, and often as competing with humans. Within the order, however, are animals that live entirely on plants, invertebrates, large mammals, and even fruit. Nonetheless, within the Carnivora are some of the strongest and most formidable of all mammals.

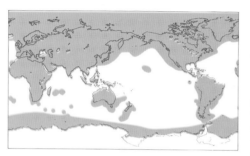

FASCINATING MAMMALS

In spite of our special fascination, we know relatively little about most of the wide diversity of species represented in the Carnivora. From the smallest, the least weasel *Mustela nivalis* at a mere 100 grams (3½ ounces), to the largest, the southern elephant seal at 2.4 tonnes (5,300 pounds), our knowledge of the biology of the group is scant. Within the group is the fastest land mammal, the cheetah *Acinonyx jubatus*, and perhaps some of the slowest: the earless seals (Phocidae), which are better adapted for the sea than the land. Some carnivores are extremely common where they occur, for example the North American coyote *Canis latrans* or the small Indian mongoose *Herpestes auropunctatus*; others, however, are quite rare, known only from a handful of museum specimens, such as the Columbian mountain weasel *Mustela felipei*.

Carnivores are found on every continent except Antarctica and in nearly every type of habitat, with species found in the oceans, the Arctic, tropical rainforests, prairies, temperate forests, deserts, high mountains, and even the urban environment. They are rarely a common or abundant part of any ecosystem, usually because they are found near the top of the food chain. They are the only order of mammals that have ocean-going, terrestrial, arboreal, and semi-fossorial species. Among the mammals, including known fossil groups, at least ten orders of mammals are primarily insectivorous and twenty are herbivorous, but only one modern and one extinct order are carnivorous: the Carnivora and the Creodonta. With such a wide diversity of kinds of mammals, one might ask why

these animals are grouped together and what they have in common.

WHAT IS A CARNIVORE?

Some carnivores are not classified in the order Carnivora; some members of the order Carnivora are not carnivores; and some carnivorous mammals are neither carnivores nor in the order Carnivora! So, just what is a carnivore? A carnivore is an animal that eats principally meat, but the word is most often used as a general term to refer to the members of the Carnivora. Being carnivorous means that an animal specializes in eating meat as the principal part of its diet; this is a food habit of some, but not all, of the members of the Carnivora. The order Carnivora consists of a unique group of mammals that all share a common evolutionary history, that is, the members of the Carnivora all share a particular set of ancestors. Members of each family within the order are also united by sharing a particular set of common ancestors, and these ancestors can be identified by certain unique morphological features. The first members of the Carnivora had adaptations that allowed them to eat meat more efficiently than other competing mammals around at the time (creodonts). These features are retained in nearly all of the present members of the order, despite the fact that many animals have modified them and now feed on many things besides meat. Some species specialize in eating fruit, insects, bamboo, worms, nuts, berries, fish, crustaceans, and seeds. Although these animals may specialize in a particular kind of food, for example, the giant panda *Ailuropoda melanoleuca* feeds on bamboo, nearly all are considered opportunists, that is, they will not turn down an easy meal when it is available. Consequently, many members of the Carnivora are probably more appropriately called omnivores—they eat a wide variety of plant and animal foods—and the portion of meat in their diet may range from nearly all, such as a cat, to hardly any, such as the giant panda.

HOW ARE CARNIVORES DIFFERENT?

Throughout their common evolutionary history, members of the Carnivora have acquired some unique features that distinguish them from other orders of mammals. These features relate to the basic eating habits of the early members of the order and can be grouped under these modifications. In order to survive by eating other animals, carnivores must not only be able to eat them efficiently, but must first be able to catch them. Modifications to find the animals they feed on, such as acute eyesight, hearing, or smelling, allow them to locate prey before it locates them. Second, speed and dexterity are essential, and modifications have allowed carnivores to catch prey quickly without undue expenditure of energy. Third, once the prey is caught, a carnivore must be able to kill and digest the animal efficiently and derive all its essential food requirements from it. And fourth, the carnivore must be smarter than the animal it is trying to catch!

The most widely accepted theory about the relationships among carnivore families is based primarily on modifications in the ear region and the sensitivity to hearing particular frequencies. Carnivores have a highly developed ear region, often with more than one inner ear chamber, which increases the sensitivity to certain frequencies and makes it easier to locate prey that makes those particular sounds. Skulls from each of the families of carnivores can be identified from their unique ear region alone.

Carnivores have adapted a variety of ways for feeding on everything from large mammals to fruit.

The single most important feature, and the one that is most often cited as uniting the entire order, is a unique modification of the teeth for eating meat. Carnivores have the fourth upper premolar and the first lower molar modified to form two vertical, sharp cutting surfaces, which slide against each other in a manner similar to the blades of a pair of scissors. At this location on the skull the jaws have their greatest force, making this pair of teeth, called the carnassial pair, extremely efficient as shears. However, some carnivores have evolved into non-meat-eaters and they have modified the carnassial pair to best utilize different food resources. Plant-eating and fruit-eating Carnivora have lost most of the vertical shear surface and increased the horizontal crushing surface (similar to humans). Insect-eating Carnivora have small teeth with small shearing surfaces that can pierce the exoskeleton of insects and get to the food inside.

THE DOG-LIKE CARNIVORES

The dog-like carnivores of the suborder Caniformia are believed to have evolved from an ancestor similar to the extinct family Miacidae, which lived during the Eocene. Caniforms are mostly terrestrial, although two families, the Phocidae and Otariidae (discussed at the end of this chapter), have become exclusively aquatic. The caniforms are the more numerous of the two suborders, both in terms of biomass and number of species, and they are found in all continents and oceans. The

CAT JAW SHOWING CARNASSIAL SHEAR

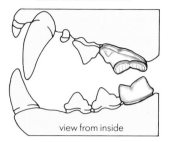

view from inside

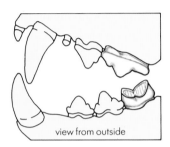

view from outside

▲ *Shading indicates the shearing facets on an upper premolar and a lower molar. (Cats lack the first upper and first two lower premolars.)*

▼ *Silver-backed jackals and a spotted hyena feeding on a wildebeest, killed by hyenas.*

terrestrial forms share a uniquely formed internal ear, different from that found in the feliforms. This group includes the largest (elephant seals) and the smallest (least weasel) of all carnivores.

Dogs, foxes, and jackals

The family Canidae is dominated by the genus *Canis*. The family is usually divided into roughly four groups: the foxes of the genera *Alopex, Urocyon,* and *Vulpes;* the dogs or wolves of the genera *Canis, Lycaon,* and *Cuon;* the interesting South American canids; and the more insectivorous canids, the raccoon dog *Nyctereutes procyonoides* and the bat-eared fox *Otocyon megalotis,* which are quite different from all other members of the family.

Canids live in open grasslands and show adaptations for the fast pursuit of prey. They originated in North America during the Eocene (57 to 37 million years ago) and are now distributed throughout the world. The familiar domestic dog

► ▼ Members of the Canid family. Above right, the bush dog of South America. Below: top, maned wolf of South America; middle, red wolf of North America (an endangered species); bottom, the African hunting dog.

and all of its breeds and varieties is believed to have evolved from a wolf-like ancestor.

Canids rely on hearing and smell to locate their prey. They are opportunists and highly adaptable, with a generalized dentition and digestive system that allow them to shift their diet from eating nothing but fruit at certain times of the year to being entirely carnivorous. Canids eat all types of vertebrates, insects, fruit, mollusks, and carrion. Some species, such as the wolf, African hunting dog, and red dhole *Cuon alpinus,* use cooperative hunting, which allows them to prey upon animals many times their own body size. Others, like the bat-eared fox, are insectivorous.

Canids have a long face and usually two molars in each jaw. The carnassial shearing teeth have the shearing surface at the front end of the tooth and then a crushing surface at the heel, usually with two cusps.

Canids are terrestrial carnivores with long, bushy tails, long legs, and slender bodies. Only one, the North American gray fox *Urocyon cinereoargenteus,* can climb trees easily. Many modifications reflect the pursuit way of life of canids, for example, the bones of the forelegs are interlocked to prevent accidental rotation during running. All canids have large pointed ears. The skin is almost completely without sweat glands and there is a dorsal scent gland at the base of the tail.

Some canids have had a long association with humans, and many cultures have a rich history of legends and folktales associated with these animals. Wolves are viewed as raiders of livestock and will attack humans when threatened. Coyotes may sometimes be seen around the edges of several urban communities; they are despised by landowners, who see them as predators of small livestock. The fox's attributes are legendary and the "fox raiding the henhouse" has even become a cliché. For these and other reasons massive predator control programs have been initiated in several areas and, combined with habitat destruction, have led to the demise of many species of canids from a large portion of their original range. Canids are economically important and are of high value in the fur industry. The raising of foxes on farms has in most cases eliminated the need for extensive hunting to supply the trade.

Foxes The foxes are small canids with acutely pointed skulls and bushy tails. The desert foxes are the lightest colored and the smallest, and have proportionately the largest ears. The fennec fox

Vulpes zerda (500 grams to 1.5 kilograms, 1 to 3 pounds weight) has an almost comical appearance, with enormous ears that help to dissipate heat and to locate prey. All of the foxes have basically the same eating habits: they eat whatever small vertebrates are in the area where they occur, but they also depend heavily on other food items such as insects and fruits. The Arctic fox *Alopex lagopus* (3 to 5 kilograms, 6½ to 11 pounds weight) occurs in two color phases, a brown form and a steel-gray or blue form. Both color types change to pure white in the winter. Most foxes are considered solitary hunters: cooperative foraging does not give much advantage to a predator that depends on small vertebrates such as mice. Some recent natural history studies indicate, however, that fox society may be much more complex. In some areas the basic social unit is usually the breeding pair and their offspring; however, recent studies suggest that in other areas the social unit consists of small groups, usually composed of one adult male and three to six females or vixens. The group appears to hunt in a specific area, although the individuals will disperse while foraging.

Dogs and wolves The true dogs and wolves are the most carnivorous of the canid family, with the African hunting dog being the only pure carnivore. Within this group are also the most social species. Wolf packs, for instance, have a well-developed hierarchy centering on the dominant breeding pair, known as the alpha male and female. Cooperative hunting by six to twelve individuals, seen in the wolf, Indian dhole, and the African hunting dog, allows the predator to specialize in very large prey such as zebra, antelope, elk, and deer. Despite the image of wolves running down moose in deep snow, or Indian dholes disembowelling a large deer, their success rate is actually quite low, almost always below 10 percent, and they usually have to go for long periods without eating. Coyotes and the four species of jackals (*Canis aureus, C. mesomelas, C. simensis,* and *C. adustus*) are smaller, more opportunistic foragers, and they concentrate more on medium-sized to smaller vertebrates. Coyotes can and will eat almost anything. Whereas the specialized, highly carnivorous *Canis* species are threatened by predator control and habitat destruction, coyotes do not seem to be suffering as

much from human intervention and they are increasing their range throughout North America.

South American canids The three kinds of South American canids are the long-legged maned wolf *Chysocyon brachyurus,* the "foxes" of the genus *Dusicyon,* which are morphologically and ecologically somewhere between the Northern Hemisphere *Canis* and *Vulpes,* and the unusual bush dog *Speothos venaticus.* The bush dog has the most carnivorous dentition of the group, but it is a most unusual-looking canid, with short legs, a rotund body, small ears, and a short tail. Its head and body length ranges from 55 to 75 centimeters (22 to 30 inches) and it weighs 5 to 7 kilograms (11 to 15 pounds). It will run in packs of five to ten individuals and feed on the larger capybaras. The maned wolf, so named because of its erectile dorsal crest of hair, has long, stilt-like legs adapted for tall grass prairie. It is a predator of medium-sized mammals such as rabbits, pacas, and armadillos.

Bat-eared fox This is one of the most peculiar of the canids, and the only one that is entirely insectivorous. It is unique among the Carnivora in having four to eight small extra molars, providing more chewing surfaces for feeding on insects. Like the fennec, it has relatively large ears that help both in locating prey and dissipating heat.

Raccoon dog Named because of its black face mask (similar to the North American raccoon), the racoon dog is the only canid that does not have any kind of bark. It is also distinguished from the rest of the family by living in dense undergrowth of forests. Raccoon dogs are commercially valuable in the fur industry and have been introduced into many areas.

▼ *More members of the Canid family. Top, gray fox of North America; bottom left, bat-eared fox of Africa; bottom right, raccoon dog of Asia.*

John Shaw/NHPA

▲ *Young polar bears at play. Largest of the terrestrial carnivores and well adapted to life in the Arctic, this species feeds almost entirely upon seals.*

▼ *The Asian black bear is an excellent climber.*

Jean-Paul Ferrero/Auscape International

Bears and pandas

The largest terrestrial carnivore, and the only bear that is exclusively a meat eater, is the polar bear. Males may weigh over 800 kilograms (1,750 pounds). Despite their Hollywood image, however, most bears do not feed regularly on large prey, and they are primarily herbivores or insectivores, although, like all carnivores, they are opportunistic and will take advantage of a good meal when they see it. The giant panda (75 to 160 kilograms, 165 to 250 pounds weight) is almost exclusively a bamboo feeder, whereas the sloth bear *Melursus ursinus* has long claws and powerful front teeth for breaking logs and getting insects and other invertebrates. Its long snout and long tongue are typical of insectivorous mammals. Another adaptation for insect eating is that it has lost its two upper middle incisors, providing a convenient channel for its long, prehensile tongue.

The ancestor to modern bears first appeared in the Oligocene, a red-panda-sized animal in the genus *Cephalogale*. At this time, all of the bear ancestors had long tails and were of this size. Throughout their evolutionary history, bears have increased in size, becoming the largest of the land-dwelling carnivores, with large massive heads, stocky bodies, and short tails (except for the red panda). Eurasia is believed to have been the center of the evolution of ursids.

There is a great similarity in general body form in all the modern bears save one, the panda. The anatomical features that define this family principally concern changes in dentition from the shearing–cutting form typical of carnivores, to a crushing–pulverizing form with broad, cuspidate molars. Bears have the largest molars of any carnivores. All the shearing function of the carnassials has been lost. Indeed, the first few premolars, typically shearing teeth, are rudimentary and often drop out at an early age.

Except for the giant and red pandas, the bears are all basically single colored, with some species showing striking color differences within the species. Nearly all show marks on the chest.

The members of this family are terrestrial and semi-arboreal omnivores; only the polar bear is semi-aquatic. Three species—red panda, sloth bear, and sun bear—are nocturnal, and the others are diurnal. They have delayed implantation, giving them a relatively long gestation. The northern temperate species are absent from their natural range for up to six or seven months of the year. During the winter their natural foods are not available, so they store up body fat just before retreating into a den or cave and go into a winter dormancy, distinguished from true hibernation because there is no drop in body temperature. During this period they do not eat and live entirely off their body fat. It is also during this time that one to five extremely small young—they weigh only about 1 percent of the mother's weight—are born in a protected, warm environment.

Brown and polar bears Also known as the grizzly, Kodiak, and Kamchatkan bear, the brown bear *Ursus arctos,* with a head and body length of 2 to 3 meters (6½ to 10 feet) and weight of 110 to 450 kilograms (240 to 990 pounds), is one of the largest ursids. Along with the Asian black bear *Ursus thibetanus* and American black bear *Ursus americanus,* they feed on tubers, berries, fish, and carrion. Polar bears live in an extremely harsh environment without such food resources, and they are adapted to feed almost exclusively on marine mammals, especially seals. Polar bears have large, long heads, and their skulls closely resemble those of their near relatives, the eared seals. They range in length from 2.5 to 3 meters (8 to 10 feet), and in weight, from 175 to 650 kilograms (375 to 1,430 pounds). Polar bears may travel on ice floes as much as 65 to 80 kilometers (40 to 50 miles) in a day in search of seals. The females do not breed until they reach five years of age. When pregnant, they isolate themselves in a den to have the young.

Pandas The giant panda and the red panda (3 to 5 kilograms, 6½ to 11 pounds, weight) are the only carnivores that feed almost exclusively on bamboo. Although they are not alike in appearance, both have some similar adaptations for feeding on bamboo that have led zoologists to call them "pandas". The giant panda has a special wrist bone that has enlarged and has an elongate form and serves like a thumb for grabbing bamboo. This adaptation is less developed in the red panda.

Philippa Scott/NHPA

▲ The giant panda feeds exclusively on bamboo.

▼ The red panda eats bamboo, other vegetation, and insects.

Joe van Wormer/Bruce Coleman Ltd

▲ *The common raccoon of Central and North America has adjusted to life in suburbs and farmland.*

▼ *The ringtail coati of South America uses its flexible snout to expose the insects and tubers that form much of its diet.*

Raccoons and coatis

Raccoons The inquisitive raccoon *Procyon lotor* (5 to 15 kilograms, 11 to 33 pounds, weight) is familiar to most people in North America. It is, however, just one of the more highly successful members of the procyonid family. This family is centered in the tropics of Central and South America, with only the raccoon, which can often be found associated with humans, increasing its range into northern North America. The family Procyonidae is entirely restricted to the New World. Although all of the members of this family are omnivores, some have become quite specialized and concentrate on eating fruit to an extent not seen in many other carnivores. All are nocturnal, except for the diurnal coatis.

The family is defined principally on morphological features. Among these are features of the skull, such as a deep posterior-oriented pocket in the external ear region, and the presence of inner ear sinuses, which are small pockets connected to the inner ear region where the ear bones lie. The carnassial shear, one of the basic carnivore traits, has been highly modified in this family. Only the ringtail has retained a sharp cutting edge, while the other species in the family have modified dentition with blunt cusps or even no cusps at all, such as the kinkajou *Potos flavus*. Except for the kinkajou and the olingo *Bassaricyon gabbii*, procyonids have a distinctive black facial mask and rings on their tail.

Raccoons are mainly omnivorous, but they prefer aquatic prey, such as frogs, fish, and crustaceans, and will also feed on various nuts, seeds, fruits, and acorns. They have the peculiar habit of washing their food with their highly dexterous forepaws. They are nocturnal, solitary, and found mainly in forested areas, where they live in dens in hollow trees or rock crevices. For this reason, the cutting down of old trees by many landowners has dramatically reduced raccoon populations.

Coatis The closely related coatis are diurnal inhabitants of woodlands and forests. Dramatic individual color variation is a common feature in coatis. One litter described in 1826 had a red, a gray, and a brown individual in the same litter;

black coatis and red coatis are not infrequent. Coatis are omnivorous, their food chiefly composed of insects and roots. Females and juveniles live in bands ranging from four or five to fifty individuals; they forage by spreading out over the forest floor and moving slowly along, digging, rooting, and investigating all possible sources of food. Their highly flexible snout allows coatis to forage in crevices and holes. Their tail acts as a balancing organ when the animal is in the trees.

Males are not allowed in the group except during the mating period. One dominant male will work its way into the foraging group and then mate with the females. Soon after mating, the males are expelled from the group. Adult females separate from the group to give birth.

Little is known about the behavior of the mountain coati *Nasuella olivacea,* apart from the fact that they are believed to feed predominantly on grubs and worms.

Olingos and kinkajous These are almost entirely arboreal and are mainly fruit eaters, but they may also feed on some birds and small mammals. With an elongate body form and uniformly colored fur, they have the general appearance of a primate. They are fast-moving, extremely agile and active, and travel constantly during the night. Kinkajous are 40 to 60 centimeters (16 to 24 inches) in length and weigh 1.5 to 2.5 kilograms (3 to 5½ pounds). The kinkajou was originally thought to be a lemur, and the manner in which it uses its feet and its prehensile tail, its arboreal habits, and its fruit diet may still lead observers to associate it with monkeys rather than carnivores. Kinkajous are the only New World carnivore with a prehensile tail.

Weasels, otters, skunks, and badgers

With nearly double the number of species of any other carnivore family, the family Mustelidae is clearly the most successful evolutionary group. They are found in every type of habitat, including both salt and fresh water, and on every continent except Australia and Antarctica. Despite their widespread and common occurrence, they are rarely seen by humans, because of their nocturnal, arboreal, or burrowing habits. The feature for which the family is most famous (or infamous!) is the scent gland located around the anus that can secrete a pungent, foul-smelling odor, sometimes for great distances.

The reproduction biology of the group is unique among the Carnivora: the males tend to be considerably larger than the females (5 to 25 percent) and the sexes live separately for most of their lives. Mating does not occur by mutual consent: the males violently accost the females, hold them down, often with a neck bite, and copulate vigorously for long periods of time. The stimulation of the copulation process causes the female to release an egg, which ensures successful

fertilization. In other mammals, after fertilization the egg implants on the uterine wall and begins to develop. In mustelids, however, the egg "floats" within the uterus and implantation is delayed for many months.

This family of fur-bearing animals has played an important historical role in the exploration of many areas. The fur of otters, minks *Mustela vison,* weasels, wolverines *Gulo gulo,* and martens of the genus *Martes* still brings high prices, and the trapping of these species is tightly controlled in many areas. The use of mink fur in producing fur coats is so great that this species has been successfully raised on farms.

Mustelids are distinguished from other carnivores by their musk gland: two modified skin glands located around the anus. They have five toes on each foot with non-retractile claws. The skull is low and flat, with a very short face. The teeth are distinctive: there is no second upper molar, or distinctive notch in the fourth upper premolar—found in all other carnivores. Most mustelids have an hourglass-shaped upper molar.

There are four subfamilies of mustelids: the almost purely carnivorous Mustelinae (weasels, wolverines, martens), the Melinae (badgers), which have insectivorous and carnivorous species, the Mephitinae (skunks), which concentrate heavily on insects, and the Lutrinae (otters), which eat fish and mollusks.

▲ *The kinkajou lives in tropical American forests, feeding mainly upon fruits and nectar. It is one of the very few carnivores to have a prehensile tail, which is a great aid to climbing.*

▼ *One of the smallest members of the raccoon family, the ringtail is also one of the most carnivorous. It preys upon lizards and small mammals but also eats fruits and nuts.*

Erwin & Peggy Bauer/Bruce Coleman Ltd

▲ An arboreal member of the weasel family, the South American tayra feeds mainly upon fruit. Its long tail is not prehensile.

▼ The wolverine (also known as the glutton) is a thickly furred carnivore that inhabits the colder parts of Eurasia and North America.

Weasels, wolverines, and martens The mustelines are one of the most successful groups of carnivores in the world. The terrestrial and semi-aquatic members of this family, referred to as weasels, have short round ears, extremely long cylindrical bodies, and short legs. They feed mostly underground and at night, chasing their prey, usually rodents, directly into their burrow systems. One species, the mink, is semi-aquatic and spends a large amount of its time foraging in the water for mollusks, crustaceans, and fish. The black-footed ferret *Mustela nigripes* was a specialized feeder on North American prairie dogs, but as landowners eliminated the prairie dogs when they converted their land to pasture for cattle, the ferret also became extinct in the wild. Martens are arboreal weasels, and their body size ranges are slightly larger. The largest marten, the fisher *Martes pennanti,* ranges in length from 45 to 75 centimeters (18 to 30 inches) and weighs 2 to 5 kilograms (4½ to 11 pounds). It is the only carnivore to feed heavily on porcupines. The marten accomplishes this by speed and dexterity. The porcupine's main defense is to keep its back, with the erect quills, toward the predator. The fisher will quickly move to the head, where there are no quills, and attack. Once the porcupine is dead, the marten flips the animal over and feeds from its underside. One South American arboreal marten, the tayra *Eira barbara,* depends on fruit for a large portion of its diet. The largest mustelid is the wolverine, which is 1,000 times heavier than the smallest species, the least weasel. Wolverines have robust bodies and short legs, and feed on small ungulates and medium-sized vertebrates. They are adept at killing relatively large animals in deep snow. Their winter fur, which has hollow hairs, is prized by Eskimos for its insulative properties.

Badgers Badgers are generally stocky, short-legged, large mustelids with powerful jaws that

Konrad Wothe/Bruce Coleman Ltd

Jeff Foott/Bruce Coleman Ltd

interlock and support a crushing dentition. Most are solitary, except for the Eurasian badger *Meles meles,* which lives in social groups and feeds mostly on earthworms and grubs. With a head and body length of 60 to 85 centimeters (24 to 34 inches), and weight of 10 to 15 kilograms (22 to 33 pounds), the Eurasian badger is slightly larger than the North American badger *Taxidea taxus.* The latter feeds on medium-sized vertebrates and uses its powerful front claws and limbs to burrow after rodents. Little is known about the natural history of the Southeast Asian badgers.

Otters The most highly successful aquatic carnivores, apart from the seals, are the otters, which spend nearly all of their time foraging in water for food. They occur in both fresh and salt water, and have webbed feet and stiff whiskers which they use as tactile sensors. Most otters are typical mustelids, leading solitary lives, but the sea otter is gregarious, and the males and females will form separate groups in different coastal areas. The giant otter *Pteronura brasiliensis* which ranges in head–body length from 85 to 140 centimeters (34 to 55 inches) and in weight from 20 to 35 kilograms (40 to 75 pounds), will sometimes forage in groups. The otter's muscular tail provides most of the force for swimming, and it attacks its prey with its teeth. The oriental short-clawed otter *Aonyx cinerea* uses its feet to grasp invertebrates to feed on. The sea otter often uses "tools", such as a flat rock, to break open mollusk shells.

Skunks The New World skunk subfamily, the Mephitinae, is perhaps the most infamous of the mustelids. The anal gland, present in all mustelids, is here modified to an organ that can project a foul-smelling liquid for a great distance. The skunk's coloration of vivid black and white patterns is a warning to most potential predators of what is to come. Skunks are omnivores, feeding on insects and occasionally small vertebrates, occupying a niche not unlike that of some mongooses.

▲ *Sea otter and young. This North Pacific species spends almost all of its life at sea, even sleeping on the surface.*

▼ *Striped skunk, swimming. Although not a habitual swimmer, the skunk—like most mammals—can swim when necessary.*

Jen & Des Bartlett/Bruce Coleman Ltd

▲ *The banded linsang, from the Indonesian region, is an arboreal civet that feeds on small invertebrates.*

▼ *The common genet, from the Mediterranean region, is a versatile predator on small vertebrates.*

THE CAT-LIKE CARNIVORES

The cat-like carnivores, of the suborder Feliformia, are believed to have evolved from an ancestor similar to the extinct family Viverravidae (which is also included in this suborder); these were more civet-like and lived during the Eocene. Feliforms are terrestrial; only the otter civet and the marsh mongoose *Atilax paludinosus* are semi-aquatic.

Civets and genets

An important commercial product for many centuries, civet oil is obtained from the civet gland of certain members of the Viverridae. This gland, located in the genital region, is unique to this family of carnivores. Civet oil, or just simply "civet", is refined and used as the basis for perfume.

Included within the Viverridae are some of the most diverse and interesting species of the Carnivora, but they are the least known and least studied. This possibly is because most occur in tropical regions, and are nocturnal and solitary, making observations difficult. Viverrids are found throughout Africa and southern Asia. Some species are quite common, like the common palm civet, sometimes called the toddy cat, because of its love for fermented palm sap (toddy). The common palm civet is also a notorious raider of coffee plantations. With the exception of a few species, we know little about the relative abundance or distribution of this family, our knowledge is limited to scattered records accumulated over the last century. With the destruction of the tropics occurring at such an alarming rate, there is the real danger that one of the most common components of the tropical rainforest, the family Viverridae, may suffer even more than others.

With the viverrids' occupation of such a wide diversity of ecological types, it is difficult to characterize the morphology of the whole family. The most common feature, the civet gland, is found in all genera except a few, although its exact

Clem Haagner/Ardea London

shape and position varies. Most viverrids have a very generalized dentition, with rather long faces, and a tail as long or sometimes longer than the head–body length. The ear region is diagnostic for this carnivore family: viverrids have an inner ear divided into two chambers, externally visible on the skull.

Because of the wide diversity in this family, it is useful to group its members by subfamilies. Here we will consider four: the palm civets (Paradoxurinae), Malagasy civets (Cryptoproctinae), true civets (Viverrinae), and banded palm civets (Hemigalinae).

True civets The most primitive subfamily, and the one with the most species, is the Viverrinae. Viverrines provide the basis for the perfume industry and are represented in Europe by the common genet *Genetta genetta*. The members of the diverse genus *Genetta* are arboreal, long-nosed cat-like animals that are found throughout Africa and into southern Europe. They prey on small vertebrates and invertebrates. The largest civet, the African civet, and the corresponding Indian civets, of the genera *Viverra* and *Viverricula*, are terrestrial —one of the few non-arboreal groups—and prey on medium-sized to small vertebrates.

Malagasy civets These civets share little in overall appearance. The fossa, the dominant carnivore on Madagascar, is extremely cat-like, and occupies a cat-like niche on the island. At 70 to 80 centimeters (28 to 32 inches) in length and 7 to 20 kilograms (15 to 45 pounds) in weight, it is about the same size as the African civet, making them the largest viverrids. The fossa occupies a niche similar to that of the clouded leopard of Southeast Asia. Fanalokas are rather fox-like in their ecology and appearance. Perhaps one of the most interesting of the Madagascar civets is the falanouc, a small (2 to 5 kilogram, 4½ to 11 pound), long-nosed animal with a bushy tail like a tree squirrel where it stores fat reserves. The falanouc and fanaloka are the only carnivores that produce young that are active immediately after birth.

Palm civets The palm civets demonstrate the evolution of a group of meat-eating carnivores into fruit-eating. The palm civets of Southeast Asia spend nearly all of their time in the tree canopy, and one species has been documented as feeding on over 30 species of fruit. Recently, the masked palm civet *Paguma larvata* has become a popular food item in restaurants in China, causing alarm over its conservation status and future.

Banded palm civets Like the Malagasy civets, the banded palm civets are a rather loose array of morphological types. This group is the least studied of the viverrids and occurs only in Southeast Asia. The banded palm civet appears to eat small vertebrates and insects. Both Owston's civet *Chrotogale owstoni* and the falanouc have long narrow snouts and feed on earthworms and other invertebrates. Both of these civets have strikingly

Frans Lanting/Minden Pictures

beautiful, broad transverse bands on their back. The otter civet resembles an otter and lives in the same habitat as an otter, but nothing is known of its natural history.

▲ The fossa is the dominant carnivore of Madagascar, an island that has no indigenous members of the dog or cat families.

▲ The African civet is the largest civet.

Frans Lanting/Minden Pictures

▲ The ring-tailed mongoose hunts in trees for small reptiles, mammals, and birds. It is one of five species, restricted to Madagascar, that are distinct from all other mongooses.

▼ Although related to the carnivorous hyenas, the aardwolf feeds almost exclusively on termites. Its teeth are degenerate but it is far less specialized for its diet than other mammalian "anteaters"

Mongooses

Mongooses, small carnivores of the family Herpestidae, have long slender bodies, well adapted for chasing animals down into burrows, though they mostly eat insects. Most mongooses are diurnal; none possesses the retractile claws necessary for adept tree climbing; and only a few species have external markings. Mongooses are basically a uniform color, though a few have stripes on the neck or different tail coloration. The banded mongoose *Mungos mungo* has black transverse bands on its body, and some Malagasy mongooses (Galidiinae) have longitudinal dark stripes running the length of their bodies. Mongooses are often the most abundant carnivore in an area, and they are found from deserts to tropical rainforests, from southern Africa through the Middle East, India,

and Southeast Asia, northward to Central China. Famous as "ratters", mongooses have been introduced to many islands in the Old and New World. The diurnal mongoose was introduced to many Caribbean islands to control the nocturnal rodents feeding on sugar cane, but they were not a great success, and they are now considered a pest species in most areas where they were introduced.

Mongooses are small, ranging from about 25 to 60 centimeters (10 to 24 inches) in head and body length, with short faces and two molars in each jaw. Their inner ear is divided into two chambers, but, unlike the other feliforms, these chambers are located one in front of the other, not in the typical side-by-side arrangement. All mongooses have an anal sac that contains at least two glandular openings. This is best developed in the African mongooses and least developed in the Malagasy mongooses. The glands deposit scent with the feces, which communicates the sexual condition and other characteristics to other mongooses.

Mongooses attain sexual maturity at 18 to 24 months. Litter size varies from species to species, but generally from two to eight young are born after a six to nine month gestation. The adult female solely raises the young in the solitary species; in the more social mongooses, the young may be cared for by several adult females.

Mongooses are essentially terrestrial and are poorly adapted for climbing, only the African slender mongoose *Herpestes sanguineus* and the ring-tailed mongoose *Galidia elegans* frequent the trees. One species, the marsh mongoose, is semi-aquatic and another, the crab-eating mongoose *Herpestes urva*, feeds heavily on aquatic crustaceans. Most are solitary or live in pairs. Members of one subfamily, the Mungotinae, are noted for their colonies and gregarious social systems. The dwarf mongoose *Helogale parvula* and banded mongooses will forage in groups of 5 to 20, and colonies of the yellow mongoose *Cynictis penicillata* have been found with 50 individuals. Both the yellow mongoose and the meerkat *Suricata suricata* sometimes live in burrow systems closely adjacent to the Cape African ground squirrel of the genus *Xerus*.

Hyenas and aardwolves

Few who have heard the cry of the spotted hyena *Crocuta crocuta* in the still air of the African night will forget the presence of one of the most fearsome predators in Africa. The hyenas at one time dominated the carnivore niche, and fossil hyenas have been found in North America, Europe, Asia, and Africa. Now the entire family consists of four species representing three genera. Hyenas are essentially carnivores of open habitats and are unknown in forested regions.

The common names are descriptive: the spotted, striped (*Hyaena hyaena*), and brown (*Hyaena brunnea*) are not easily confused. They range in size

Wardene Weisser/Ardea London

from 1 to 1.4 meters (40 to 55 inches) in head and body length and from 30 to 80 kilograms (65 to 175 pounds) in weight. The wolf-like or dog-like appearance of the aardwolf *Proteles cristatus,* which also has stripes, is clearly different. All members have a distinctive dorsal mane and they possess an anal pouch which is structurally unique in the feliforms and is used to deposit scent. The skull is distinguished from those of all other families of carnivores in the inner ear region. The inner ear consists of two separate chambers, which are also found in the cats, civets, and mongooses, but in the hyenas these chambers, rather than being arranged side by side, as in other carnivore families, are located one on top of the other.

The members of the two genera of true hyenas, *Crocuta* and *Hyaena,* are amongst the largest predators known: they have massive, long forelegs and short hindlimbs. The remaining member of this family, the aardwolf, although it retains many similarities to the hyenas, feeds almost entirely on termites and has a much more slender build and peg-like teeth.

True hyenas The true hyenas have large heads with impressive high bony crests on the top of the skull, which serve as the attachment for a massive jaw muscle that gives hyenas one of the strongest bites of any carnivore. With robust, bone-crushing teeth, these predators are best adapted for feeding on carrion, and they are superb scavengers. Hyenas, like other carnivores, are opportunists, but they derive a large portion of their diet from scavenging on large ungulate kills. Their digestive system crushes and dissolves nearly all of the scavenged kill, but they regurgitate hooves, antlers, and other matter that cannot be dissolved into pellets. The spotted hyena participates in group hunting of anywhere from 10 to 30 individuals, and different types of hunting strategies are employed for different types of prey. The females of the spotted hyena are larger than the males and have one to four young.

Aardwolf The aardwolf is a solitary forager, and for most of the year it depends on a few species of termites. Aardwolves are nocturnal; they use many dens and locate their prey primarily by sound. Teeth are relatively useless to an animal that feeds almost exclusively on termites, and their dentition is so reduced that the teeth are not much more than small stubs. They measure 63 to 75 centimeters (25 to 30 inches) in length and weigh 7 to 12 kilograms (15½ to 26½ pounds).

Lions, tigers, and cats

Of all carnivores, cats, of the family Felidae, are clearly the most carnivorous. Cats cannot crush food: their teeth consist almost entirely of sharp, scissor-like blades, with no flat surfaces for crushing. To masticate food, cats must turn their head to the side and use their sharp, rasp-like tongue to manipulate food around their carnassials

to cut it into swallowable pieces. The distinguishing features of the felids are the horny papillae on the tongue, and the presence of only one molar in each jaw. The felids have short faces, forward-pointing eyes, and high domed heads. They do not show the diversity of body form prevalent in other carnivore families. Cats' binocular vision, unique dentition, and other modifications have been successfully adapted to a wide variety of environments, from the deserts of the Sahara to the Arctic. Although they are principally meat eaters, at least two also rely on other types of food: the fishing cat *Felis viverrina,* which depends heavily on crustaceans, and the flat-headed cat *Felis planiceps,* which will eat fruit.

All cats have some capacity to climb trees. They are commonly stalk or ambush killers, and the final capture is achieved by a sudden pounce.

Traditionally, the cats have been divided into three groups: the large cats, called the pantherines, the small cats, referred to as the felines, and the cheetah. The principal difference, besides size, between the pantherines and the felines is that the small cats can purr but not roar, and the large cats can roar, but not purr.

Cheetah The cheetah, which ranges in weight from 35 to 65 kilograms (77 to 143 pounds), is the only true pursuit predator in the cat family, and its speed is famous: it has been clocked at 95 kilometers per hour (60 miles per hour). Unfortunately, during this burst of speed, the cat expends tremendous amounts of energy and builds

▼ *Famed as the fastest land animal, the cheetah is a sprinter rather than a long-distance runner: a chase seldom lasts more than 20 seconds.*

Anup & Manoj Shah/Planet Earth Pictures

A. Visage/Auscape International

▲ *Tigers have a predominantly tropical distribution but the habitat of the Siberian race is covered in deep snow during the winter.*

Jonathan Scott/Planet Earth Pictures

▲ *African lions, the most sociable of the big cats.*

up massive amounts of heat. Generally, a chase will only last for about 20 seconds. After a short time, the heat load becomes so great that, like a runner who has run too hard, the cat overheats and finds itself out of breath. After capturing its prey, often the cheetah must stop, cool down, and catch its breath, and it is during this time that other predators, taking advantage of the cheetah's vulnerability, will steal its prey.

Tiger The pantherine cats contain some of the most impressive carnivores, such as the tiger *Panthera tigris,* the lion, and the leopard *Panthera pardus.* The tiger, the largest cat (130 to 300 kilograms, 280 to 660 pounds), is the only carnivore that will occasionally supplement its diet by attacking humans. It feeds on very large prey—50 to 200 kilograms (110 to 440 pounds)—hunting alone and waiting in ambush to strike its prey. It is a solitary cat; the main type of social unit is the adult female and her immediate offspring. Females and males occupy home ranges that do not overlap with those of others of the same sex.

Lion By contrast, the lion is the most social member of the cat family and shows the most marked sexual dimorphism: the males have impressive manes and are 20 to 35 percent larger than the females. The main social unit of lions is the pride, usually consisting of five to fifteen adult females and their offspring, and one to six adult males. Lions, which range in weight from 120 to

▲ Small cats. Upper left, the caracal of Africa and Asia; upper right, the clouded leopard of Asia; lower left, the ocelot of the Americas; lower right, the jaguarundi of the Americas.

240 kilograms (260 to 520 pounds), hunt cooperatively and regularly kill prey above 250 kilograms (550 pounds). The large males, with their highly visibly manes, do little of the actual killing. The normal hunting strategy is for the adult females to fan out and surround the prey and slowly stalk the intended victim; the kill results from a final, short burst of speed from an ambush.

Leopard The extremely widespread leopard, and its ecological counterpart the jaguar *Panthera onca,* show a surprisingly high amount of melanism. The leopard ranges in weight from 30 to 70 kilograms (65 to 155 pounds), the jaguar, 55 to 110 kilograms (120 to 240 pounds). They are solitary nocturnal hunters that seek smaller prey than the lions or tigers, and they are adept tree climbers. The clouded leopard *Neofelis nebulosa* (15 to 35 kilograms, 33 to 77 pounds weight) hunts almost entirely in the trees, feeding on monkeys, small aboreal mammals, and birds. Snow leopards *Panthera unica* (25 to 75 kilograms, 55 to 165 pounds weight) are uniquely adapted for high altitude, and snow-covered areas; they have feet well covered with hair. They are solitary predators of large ungulates.

W. CHRIS WOZENCRAFT

▲ Leopards often rest in trees and may wedge dead prey in a fork to keep it out of reach of competitors and carrion-eaters.

149

Frans Lanting/Bruce Coleman Ltd

▲ A colony of Californian sealions.
Sealions, which swim with their fin-like
arms, hunt independently but form
dense associations when they come
ashore to breed.

PINNIPEDS: LIFE ON LAND AND AT SEA
Seals, sealions, and walruses show adaptations to
life in the water, with their streamlined bodies and
limbs modified to form flippers. They spend much
time in the water, but return to land to breed.
Different modifications of the hind flippers make
some more efficient movers on land than others.
Members of the family Phocidae are not able to
bend their hind flippers forwards at the ankle, so
their progression on land is a humping movement.
Phocids also lack an external ear pinna, and the
nails on the hind flippers are all the same size.
Phocids may be marine, estuarine, or fresh-water.

The members of the two families Otariidae
(sealions and fur seals), and Odobenidae
(walruses) are all marine and are able to bend their
flippers forwards at the ankle, so their movement

on land is much more "four-footed". The nails on
the hind flippers are very small on the two outer
digits, but longer on the middle three.

Sealions and fur seals have a small external ear
pinna, and slim cartilaginous extensions to each
digit of the hind flippers which can be folded back
so the middle three nails can be used for grooming.
Walruses lack the external ear pinna, and the
extensions to the hind digits are much smaller.
Many skull characters also distinguish the three
families of aquatic carnivores.

Interesting names are applied to pinnipeds. A
large group is a herd, but a breeding group is a
rookery. Adult males are bulls; females are cows;
and the newborn young are pups (calves for
walrus). Between about 4 months and 1 year the
young is described as a yearling; a group of pups is
a pod; and immature males are bachelors.

Most seals eat fish and squid, octopus,
crustaceans of all sizes, including krill, mollusks,
and lampreys. Leopard seals *Hydrurga leptonyx* and
some sealions are not averse to penguins and
carrion, and have been known to attack the pups of
other seals. The diet, however, varies with
geographic location.

Enemies vary according to locality and range
from sharks, killer whales, and polar bears to
humans and their pollution.

Sealions and fur seals
Both sealions and fur seals (otariids) and the walrus
came from ancestors that lived in the North Pacific
about 23 million years ago; these ancestors, in
turn, came from early bear-like ancestors.

Sealions The five species of sealions are still
primarily animals of the shores of the Pacific. They
are big animals: the adult males range from 2 to 3
meters (80 to 118 inches) long; the females are
smaller. Pups at birth are 70 to 100 centimeters (27
to 40 inches) long. Underfur hairs are few, so the
coat is coarse, and on adult males the heavy
muscular neck has a mane. The coat colour is a
darkish brown in males, often lighter to gray in
females, and in the male Australian sealion
Neophoca cinerea the top of the head is whitish.
Steller's sealion *Eumetopias jubatus* is the largest;
the Californian sealion *Zalophus californianus* is the
noisiest, barking almost continuously, and is the
seal most frequently seen in zoos and circuses. The
Australian sealion is restricted to Australian coastal
waters, and Hooker's sealion *Phocarctos hookeri* is
the most southerly, found on the Auckland Islands.

Fur seals There are two genera: *Arctocephalus*,
with eight species, and *Callorhinus*, with one.

Callorhinus, the northern or Pribilof fur seal, has
its main breeding area on the Pribilof Islands and
Commander Islands in the Bering Sea. Outside the
breeding season, the adult male Pribilof seals spend
the winter in the north, but the females and
juveniles range widely down the coasts of the
northern Pacific, as far south as Japan and

California. When the Pribilof Islands were discovered in 1786, there were over 2 million fur seals there, but the popularity of their soft underfur led to unrestricted sealing and drastic diminution of the herd. International agreement of the North Pacific nations led to research and supervised sealing, and the herds have now built up to their present numbers of nearly 2 million.

A month of commercial sealing is allowed in June, when a regulated number of young males of a specified length are taken. A close watch is also kept on the total numbers of the herd. The blubber is removed from the skins, which are then packed in salt. Many of the skins are processed in the United States, and 3 months and 125 operations separate the raw skin from the finished fur coat. The long rough guard hairs are removed, the soft thick underfur is dyed, and the skin is treated to make it supple. The seal carcasses are converted

into meal for chicken food, and the blubber oil is used in soap.

It is the soft underfur that is commercially desirable. The seal coat is composed of longer, stiffer, guard hairs covering a layer of fine, soft, chestnut-colored underfur hairs. Underfur hairs are relatively sparse in most seals, but abundant in fur seals. The more deeply rooted guard hairs are removed by abrading the inner surface of the skin.

Pribilof seals are dark brown in color with a characteristic short pointed snout. Adult males are about 2.1 meters (82 inches) in length; the females, slightly smaller. Pups are 66 centimeters in length and have a black coat.

Fur seals of the genus *Arctocephalus* are distributed mainly on the shores of Subantarctic islands. They are found on the coasts and offshore islands of Australia, New Zealand, South America, and South Africa, and on more isolated islands

▼ Cape fur seals. This species frequents African and southern Australian coasts. Like sealions, fur seals have visible ears and are able to use their hindlimbs when moving about on land.

such as Macquarie, South Georgia, Bouvet, and Marion. Those on Guadalupe, Galapagos, and Juan Fernandez islands are less known, but a reasonable amount is known about the others. All these fur seals suffered from indiscriminate sealing in the nineteenth century, and the populations are still, even now, building up their numbers. At the present time only the Uruguayan populations of the South American fur seal *A. australis* and the South African fur seal *A. pusillus pusillus* are taken on a carefully controlled commercial basis.

The nose to tail length of adult males of the *Arctocephalus* seals ranges from 1.5 to 2.2 meters (60 to 86 inches), with the Galapagos animals being the smallest and the South African and Australian ones the largest. The adult females are slightly smaller. The newborn pups range from 60 to 80 centimeters (24 to 32 inches) in length.

All these fur seals are very much the same in color: the adult males are a dark blackish brown, with longer hairs on the neck forming a rough mane. Females are brownish gray with a slightly lighter belly. Only the Subantarctic fur seal *A. tropicalis* has a distinctive creamy colored chest and face, and a brush of longer hairs on the head which rise to form a crest when it is agitated. Newborn pups have soft black coats which are molted for an adult coat at about 3 to 4 months.

Walrus

As the original Pacific walruses of the Oligocene died out, a group invaded the Atlantic 7 to 8 million years ago and developed into the modern animals. The 3 meter (10 foot), 1.2 tonne (2,600 pound) bulky body, the brown wrinkled sparsely haired skin and the long tusks make walruses unmistakable. They live in the shallower waters of the Arctic seas and, being social, haul out in large groups on moving pack ice.

The long tusks, present in both sexes, are the modified upper canines, which grow to 35 centimeters (26 inches) long, though record tusks of 1 meter (40 inches) are known. The ivory has a characteristic granular appearance, easily recognized in carvings. The tusks and strong whiskers are used to disturb the sediment to find mollusks, whose soft parts are sucked out of the shells and eaten.

With a gestation period of fifteen months, a walrus can produce only one calf every two years, and older females reproduce less frequently. The calves, which are 1.2 meters (43 inches) long, suckle and remain with their mother for two years, sheltering under her chest from the worst weather.

▼ *A bachelor herd of walruses. The tusks, present in both sexes, are used in seeking out mussels from sediment on the sea floor. Long tusks are also associated with dominance in males.*

Leonard Lee Rue/Bruce Coleman Ltd

The pharynx walls in the adult male walrus are very elastic and are expanded as a pair of pouches between the muscles of the neck. These pouches are inflated and used as buoys, and also as a resonance chamber to enhance the bell-like note produced during the breeding season.

Phocids

Phocids evolved round the margins of the North Atlantic from a stock of primitive mustelids (weasels and their relatives), but had become obvious phocids by the mid-Miocene (14 million years ago). The phocids, or true seals, are divided into 'northern phocids', and 'southern phocids', though this subdivision is not precise.

Northern phocids The northern phocids, four genera and ten species of seals, inhabit the temperate and Arctic waters of the Northern Hemisphere. Bearded seals *Erignathus barbatus* and ringed seals *Phoca hispida* live in the high Arctic: the ringed seal is the commonest seal of the Arctic, found up to the North Pole, anywhere there is open water in the fast ice. The bearded seal is so called because of its profusion of moustachial whiskers, which are curious in that they curl into spirals at their tips when dry.

The harp seals *P. groenlandica* and hooded seals *Cystophora cristata* live in the sea around Labrador and Greenland, the harp seal extending to the Arctic coast of Russia. The commercial exploitation of the appealing white-coated pups of the harp seal has incurred universal wrath. Hooded seal pups, known as blue-backs, are also very attractive in their first coats, which are steely blue dorsally and white ventrally. These are also hunted.

The hooded seal gets its name from the enlargement of the nasal cavity of the male. This forms a "cushion" on top of the head, which increases in size with age, and is inflated with air when the nostrils are closed. Hooded seals are also able to blow the very extensible membranous part of the internasal septum out through one nostril, forming a curious red "balloon". The inflation of the hood and balloon occurs both when the seal is excited and in periods of calm.

In the North Pacific, between approximately the Sea of Okhotsk and the Bering Sea, live the ribbon seals *P. fasciata* and larga seals *P. largha*. Harbor seals *P. vitulina* and gray seals *Halichoerus grypus* are found on the coasts on either side of both North Pacific and Atlantic Oceans; Caspian seals *P. caspica* are restricted to the Caspian Sea, and Baikal seals *P. sibirica* to Lake Baikal—the deep lake in eastern Russia.

Most phocids are grayish dorsally and lighter ventrally, plain or with spots, blotches or rings according to species. Only the harp and ribbon seals are strikingly marked. The harp seal has a distinctive black horseshoe mark on its back, while the ribbon seal has a dark chocolate brown coat with wide white bands round the neck, hind end,

Bruce Coleman Ltd

and each fore flipper.

Apart from most of the harbor seals, and the hooded and bearded seals, the northern phocids are born with a first coat of white woolly hair, which is shed after two to three weeks for a coat much like that of the adult. Bearded seals have a gray woolly first coat, and the others shed the first coat before they are born.

Most northern phocids range between 1.3 and 1.8 meters long (52 to 70 inches), and their pups are 65 to 90 centimeters (25 to 36 inches) long. Gray, hooded, and bearded seals are 2.2 to 2.7 meters (86 to 106 inches) long, with hooded and bearded seal pups measuring 1 to 1.3 meters (40 to 52 inches) long; gray seal pups are smaller, at 76 centimeters (29 inches) long.

▲ *Female harp seal and pup. Harp seals are typical phocids, lacking external ears and having the hindlimbs turned backward to form a tail fin. Locomotion on land is by an ungainly "caterpillar" crawl.*

▼ *The nasal cavity of a male hooded seal extends into an inflatable pouch of skin (the "hood") on the top of the head. By closing one nostril and exhaling, the male is able to evert the internasal membrane through the other nostril, forming a red "balloon". The function of this bizarre display is not understood.*

Frans Lanting/Bruce Coleman Ltd

▲ Phocids such as the monk seal shown here are so streamlined that they have no external ears. Nevertheless, they have excellent hearing. Many species use ultrasonic echolocation to navigate in dark waters and to find food.

▼ The crabeater seal does not eat crabs. When the jaws are closed, its teeth form a sieve that enables it to filter small crustaceans (krill) from water taken into its mouth—like a baleen whale.

Southern phocids Southern phocids are larger than northern seals, measuring 2.2 to 3 meters (86 to 128 inches) long; their pups measure 80 to 160 centimeters (32 to 62 inches). Many skull and skeletal characters also separate the northern and southern phocids, but the six southern genera are not entirely confined to the Southern Hemisphere.

The West Indian monk seal *Monachus tropicalis* is, unfortunately, probably extinct. It lived in the Caribbean, and, though it was seen by Columbus in the fifteenth century, there have been no undoubted sightings for the last thirty years.

The main center of the Mediterranean monk seal *Monachus monachus* is the Aegean Sea, with smaller numbers at suitable spots from Cyprus to Cap Blanc on the Atlantic coast of Mauritania. Early Greek writers such as Homer knew and wrote of this monk seal, but increasing human traffic and pollution in the Mediterranean has disturbed it to such an extent that its numbers are

declining and probably only 1,000 animals are now left. Officially it is protected, and education is alerting the inhabitants of the area to its plight.

On the western atolls of the Hawaiian Islands lives the laysan monk seal *M. schauinslandi*. It is also sensitive to disturbance, and air bases and tourism have reduced its numbers. The population of these monk seals is declining rapidly, and only about 700 animals may be left.

Monk seals are about 2.5 meters (98 inches) long and brown to gray in color, the Mediterranean seal frequently with a white ventral patch.

The Weddell, Ross, crabeater, and leopard seals are the Antarctic phocids. The Weddell seal *Leptonychotes weddelli* is the most southerly, found on fast ice close to land. Although reasonably common, these seals are not gregarious and tend to come singly to breathing holes in the ice. In spite of the hostile environment, these seals have been extensively studied, and it is known that they are particularly deep divers.

The crabeater *Lobodon carcinophagus* is probably the most abundant seal in the world. It is circumpolar and lives in the open seas, and is found on drifting pack ice. Its many-cusped cheek teeth interdigitate and sieve from the water the shrimp-like krill on which it feeds.

Ross seals *Ommatophoca rossi* are found in heavy pack ice, with the greatest numbers in King Haakon VII Sea. Little is known about its life, but it seems to be adapted for rapid swimming and fast maneuvers in order to catch squid and octopus.

The solitary leopard seal *Hydruga leptonyx* lives in the outer fringes of the pack ice, but immature animals wander far, and they are not infrequent visitors to the shores of New Zealand and Australia. The sinister reputation of this seal is not deserved. Like all carnivores, unless annoyed it is really only aggressive to its potential food, which can be anything from krill to fish, penguins, and even carrion, but its large three-pronged cheek teeth and big gape make an impressive display.

Elephant seals, at 4 to 5 meters (13 to 16 feet) long (pups measure 1.2 meters or 4 feet), the largest of the pinnipeds, are named from the pendulous proboscis that overhangs the mouth so that the nostrils open downwards. It is present only in males and attains its full size by the time the animal is 8 years old. The proboscis is most obvious during the breeding season and, when inflated, acts as a resonator for territorial roaring.

The southern elephant seal *Mirounga leonina* is circumpolar and found on most of the Subantarctic islands; the largest breeding populations are found on South Georgia, Kerguelen and Macquarie Islands. The northern elephant seal *M. angustirostris* occurs mainly on islands off the coast of California.

In the mid-1800s indiscriminate sealing for the oil of these seals nearly exterminated both species. The northern elephant seal has since made a

Francisco Erize/Bruce Coleman Ltd

remarkable recovery and numbers are increasing. The southern elephant seal did not suffer so much, and as its numbers increased, commercial licensed sealing started again. This stopped in 1964 and the seals are now fully protected.

Diving

All seals are able to stay under water for long periods, and some can dive to great depths. Weddell seals may stay under for 73 minutes and dive to 600 meters (2,000 feet), and recent research has found that southern elephant seals can dive to 1,200 meters (4,000 feet) and stay under for nearly 2 hours. Although these long dives are possible, 20 to 30 minutes is the average.

Seals have a lot of blood (12 percent of body weight, compared with 7 percent in humans), which carries a lot of oxygen. During diving, the peripheral arteries are constricted, so most of the blood goes to the brain, and the slowing of the heart rate keeps the blood pressure normal. Phocids can dive deep and long, but otariids, whose blood has a lower oxygen capacity, make shorter, shallower dives. Seals do not suffer from the bends, as they dive with very little air in their lungs and do not have a continuous air supply while diving.

Reproduction

Most seals produce their single pup in spring or summer, but there is considerable variation and one cannot generalize accurately.

Otariids spend much of the year at sea. At the beginning of the breeding season the dominant bulls arrive, select, and defend their chosen territory, and try to stop cows from leaving. The pups are born, head or hind flippers first, shortly after the females arrive, and about one to two weeks later, depending on species, the females mate again. There is a two to five month delay before the blastocyst becomes implanted and active gestation starts. The total gestation period is, therefore, about 11½ months, and the active gestation period, 7 to 8 months, depending on species. The pups start to suckle very shortly after birth, and most otariid pups are suckled for about a year and during this time become increasingly adventurous and start to find some solid food. The mother stays with her pup, on land, for about two weeks, and then goes to sea to feed, returning at intervals to suckle her pup. Pups frequently gather in groups, or pods, while their mothers are at sea, and mother–pup recognition is by call. Bulls ignore the pups, but will remain on land guarding their territory until the end of the breeding season, going without food for about two months, but details vary according to species. The Antarctic fur seal and the northern fur seal suckle for the shorter time of three to four months, and the walrus for two years. After a few months, the harem system breaks down and the animals disperse.

Dr Eckart Pott/Bruce Coleman Ltd

Amongst the phocids the territory system is found in gray and elephant seals, but in the others there may be concentrations of animals at breeding time, or just pairs coming together. Mating takes place from twelve days to seven weeks after birth, and the length of lactation is equally variable (twelve days to two months), depending on species. There is a delay of two to five months before the blastocyst is implanted. After lactation the seals disperse.

Seal milk is particularly rich in fat and protein: nearly 50 percent is fat (compared with 3.5 percent in cow's milk), and this is correlated with the fast rate of growth of the pup. Phocid pups have a high daily increase in weight, ranging from 1.3 kilograms (2¾ pounds) in gray seals, 2.5 kilograms (5½ pounds) in harp seals to 6 kilograms (13¾ pounds) in southern elephant seals. With the longer lactation period in otariids, the daily weight gain is lower (about 40 to 60 grams or 1½ to 2 ounces), though the two fur seals with shorter lactation periods (Antarctic and northern fur seals) have pups that put on weight faster (about 100 grams or 4 ounces a day). It is also necessary for all seal pups to lay down a layer of insulating blubber as soon as possible. There is virtually no lactose in the milk and as seals do not tolerate it, orphan pups must be given a specially designed formula, not cow's milk.

JUDITH E. KING

▲ A herd of northern elephant seals, consisting of a male with his harem and their offspring. Reaching a weight of 2.4 tonnes (5,300 pounds), male elephant seals are the largest of the pinnipeds and indeed the largest carnivores. The southern elephant seal has the amazing capacity to remain below the surface for nearly two hours, while reaching a depth of as much as 1,200 metres (3/4 mile).

Order Cetacea
9 families, 39 genera,
78 species

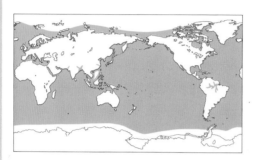

SIZE
Smallest Heaviside's dolphin
Cephalorhynchus heavisidii,
head and tail length 1.2
meters (4 feet), weight 40
kilograms (88 pounds).
Largest Blue whale
Balaenoptera musculus, head
and body length up to 30
meters (100 feet), weight up
to 130 tonnes (286,500
pounds).

CONSERVATION WATCH
The following species are
listed as endangered in the
IUCN Red Data Book of
threatened mammals:
Chinese river dolphin *Lipotes
vexillifer,* Indus dolphin
Platanista minor, blue whale
Balaenoptera musculus,
humpback whale *Megaptera
novaeangliae,* black right
whale *Balaena glacialis,*
bowhead whale *B. mysticetus.*
Many other species are listed
as vulnerable or of
indeterminate status.

WHALES AND DOLPHINS

M. M. BRYDEN

Whales and dolphins are found in all seas of the world, and in some rivers and lakes. Their streamlined bodies and their ability to dive for long periods and to great depths are just two of their many adaptations to life under water. The order Cetacea contains two extant suborders, the whalebone or baleen whales (Mysticeti), and the toothed whales (Odontoceti). A third suborder, the Archaeoceti, contains only extinct forms.

▼ *River dolphins have small, degenerate eyes. Because they usually live in turbid water, they navigate and find their prey by echolocation. The large "melon" on the forehead acts as a lens for ultrasonic vibrations emitted from the nasal cavity.*

Norbert Wu/Planet Earth Pictures

TOOTHED WHALES
There are six families of toothed whales. Many species are confined to relatively small areas of ocean, although the different species are found in a variety of habitats. Among the dolphins, some species are found only in rivers, some only in onshore waters, and yet others only in the open ocean. On the other hand some species, such as the killer whale and the bottlenose dolphin, are seen in both inshore and deep ocean waters.

Sperm whales
The family of sperm whales, Physeteridae, includes the largest and in some ways the most specialized of the toothed whales. Sperm whales have a very large head that contains a structure known as the spermaceti organ, and functional teeth only in the lower jaw, which is underslung.

White whales
The white whales, Monodontidae, have no beak; the dorsal fin is absent, rudimentary, or small; there are no grooves on the throat; and the neck vertebrae are all separate, providing some flexibility to the neck. Males of one member of the family, the narwhal, have a highly developed tooth in the upper jaw that forms a long, spiral tusk.

Beaked whales
The beaked whales, Ziphiidae, derive their common name from the characteristic long and protruding snout. A pair of throat grooves is present, and there is no notch in the middle of the trailing edge of the flukes. In all but one species there are two teeth in the lower jaw, but they protrude through the gums only in adult males.

Dolphins
Although often referred to as the "true dolphins", the family Delphinidae includes species that are not distinctly "dolphin-like", such as killer whales and pilot whales. Delphinids, the most abundant and varied of all cetaceans, are found from tropical to polar seas. Many of the smaller species have a beak, or elongation of the upper and lower jaws.

Porpoises
Porpoises, the Phocoenidae, are small toothed whales with the following characteristics: body length of less than 2.45 meters (8 feet); more than five neck vertebrae fused; and more than fifteen teeth in the back row of the upper jaw. None of the porpoises has a beak.

River dolphins
River dolphins, the Platanistidae, are characterized by an extremely long beak, and a long, low dorsal fin. The family includes four species that inhabit fresh-water rivers in Asia and South America, and one South American coastal species.

BALEEN WHALES

The three families of baleen, or whalebone, whales are found in all oceans, and most undertake extensive migrations which take them across and between oceans.

Right whales

Right whales, the Balaenidae, are characterized by long and narrow baleen plates, known as whalebone, and a highly arched upper jaw, which distinguish them from gray whales and rorquals. Other features include a disproportionately large head (more than a quarter of the total body length), a long thin snout, huge lower lips, and fused neck vertebrae. The skin over the throat is smooth, and there is no dorsal fin.

Gray whales

Gray whales, the Eschrichtiidae, have no throat grooves, although two to four furrows are present on the throat. There is no dorsal fin, but there are several humps on the upper surface of the tail stock. The neck vertebrae are not fused together.

Rorquals

Rorquals, the Balaenopteridae, are baleen whales that have relatively short, triangular baleen plates. The head length is less than a quarter of the body length; numerous grooves on the throat extend from the chin to the middle abdomen; and there is a dorsal fin, which is often small. The upper jaw is relatively long; the lower jaw bows outwards; and the neck vertebrae usually are not fused.

▼ *Killer whales cooperate in herding prey into shallow water; as well as eating large fishes, they prey upon warm-blooded animals and may patrol shores in search of unwary seals or penguins.*

Francois Gohier/Ardea London

▲ *Weighing up to 50 tonnes (110,000 pounds), the humpback whale is one of the largest filter-feeders. It eats small schooling fishes and krill—7 centimeter (3 inch) crustaceans that occur in dense populations in polar seas, particularly the Antarctic Ocean.*

TORPEDO-SHAPED FOR LIFE IN THE SEA

The torpedo-shaped bodies of cetaceans are very efficient for moving through water. The forelimbs are rigid paddles that help in balancing and steering. Most, but not all, species have a dorsal fin. There are no external ears.

The skin is smooth and virtually hairless, although fetuses bear a few hair follicles on the snout which in certain species persist throughout life. Sweat and sebaceous glands are lacking.

Immediately beneath the skin is an insulating layer of blubber, which contains much fibrous tissue interspersed with fat and oil. The color of whales varies from all white to all black, with varying shades of gray in between. Many species are dark gray or black along the back and top of the head and flukes, and light gray or white beneath. Some bear stripes, spots, or patches of black, gray, white, or brown. The most striking of these are the killer whale, with its distinctive black and white markings, the spotted and striped dolphins, and the common dolphin, which has an exquisite coloring of dark gray or black above, light beneath, and patches of gray and ochre in a figure-of-eight pattern along each flank. Toothed whales have numerous simple, conical teeth and telescoped, asymmetrical skulls. Baleen whales have no visible teeth but plates of baleen suspended from the hard palate, and symmetrical skulls. Baleen or whalebone is a horny substance that was formerly used for stiffening corsets. The real bones are spongy in texture and filled with oil. There is little or no visible neck, and the seven neck vertebrae are foreshortened. Some of them are fused in some species. There are no skeletal supports for the tail flukes or for the dorsal fin (if there is one).

The pelvic girdle is represented by two small rods of bone embedded in the muscles of the abdominal wall. They do not articulate with the backbone. In some species a projection from the pelvic bone represents the vestigal thigh bone.

Whales can attain great size because they live entirely in water and need not support their own weight. The blue whale, which grows to almost 30 meters (33 yards) in length, is the largest animal that has ever inhabited the earth.

Howard Hall/Planet Earth Pictures

◄ Dolphins swim very fast by relatively small movements of the horizontal tail fin. Their speed is facilitated by the streamlined body and by a layer of subcutaneous fat (blubber), which permits the skin to move in such a way that eddies (which cause friction) are minimized.

HOW WHALES BREATHE

Whales breathe less frequently than land mammals, and they can hold their breath for extraordinarily long periods during dives. Although their lung capacity is no greater than that of land mammals of equivalent size, whales take deeper breaths and extract more oxygen from the air they breathe, and exchange more air in the lungs with each breath. A whale's lungs are at least partially inflated when it dives, unlike the seal, which exhales before diving.

The nostrils are modified to form a blowhole at the top of the head. It is single in the toothed whales, double in the baleen whales. The skin immediately surrounding the blowhole has many specialized nerve endings, which are very sensitive to the change as the blowhole breaks the water surface. Breathing out and in again often occurs extremely rapidly, in the fraction of a second the blowhole is above the surface. The blowhole is closed when the animal is submerged by the action of muscles on a series of valves and plugs.

When a whale surfaces and exhales, a spout or "blow" is seen. This is composed of moist air exhaled under pressure, together with some condensation of water vapor as the pressure drops on release from the blowhole to the atmosphere. It probably also contains an emulsion of fine oil droplets from cells lining the nasal sinuses, and mucus from the respiratory passages.

▶ A blue whale "blowing". Baleen whales have two nostrils; toothed whales have a single blowhole.

Francois Gohier/Ardea London

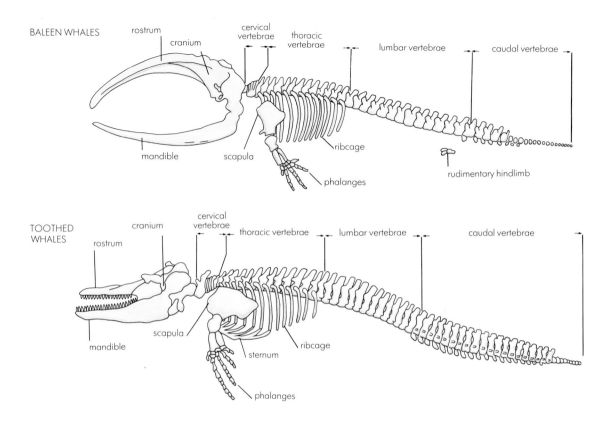

BALEEN WHALES
rostrum
cranium
cervical vertebrae
thoracic vertebrae
lumbar vertebrae
caudal vertebrae
mandible
scapula
ribcage
phalanges
rudimentary hindlimb

TOOTHED WHALES
cranium
rostrum
cervical vertebrae
thoracic vertebrae
lumbar vertebrae
caudal vertebrae
mandible
scapula
sternum
ribcage
phalanges

▲ *There are other differences between baleen and toothed whales besides the presence or absence of teeth, as these drawings of their skeletons show. The skull of a baleen whale is symmetrical; that of a toothed whale is asymmetrical. This is possibly related to the fact that toothed whales have a single blowhole. Sternal ribs are absent from the baleen whales but present in the toothed whales. But the most obvious external difference is size: few species of toothed whales approach the length of the smallest baleen whales.*

KEEPING WARM IN COLD SEAS

The large size of all whales, particularly sperm whales and most baleen whales, gives them significant metabolic advantages, especially in the cold seas where most of them spend part of their life. Blubber is a very effective insulator. It enables significant heat conservation when blood flow to the skin is reduced. The blubber varies in thickness seasonally within species, and there is great variation in its thickness among species. It has been recorded up to 50 centimeters (20 inches) thick in bowhead whales, whereas in bottlenose dolphins it is more in the region of 2 centimeters (1 inch).

When a whale is active, such as when diving and chasing prey, considerable metabolic heat is produced. Excess heat can be dissipated in the surrounding cold water by opening large arteries that pass through the blubber, thereby increasing the amount of warm blood flowing through the skin. Within the flippers, flukes, and dorsal fin is an elaborate arrangement of arteries and veins that provides fine control of heat loss or conservation in response to excess or deficiency.

ECOLOGY AND BEHAVIOR

The social and sexual behavior of whales is widely divergent among species. Most, but not all, baleen whales mate and breed in tropical seas in winter, and migrate to summer feeding areas in high latitudes. Sperm whales form nursery schools consisting of adult females, calves, and juveniles. As males mature they migrate annually to higher latitudes, and the largest males return and compete for access to the nursery schools during the breeding season.

Spinner dolphins rest during the day in groups of about 20 in inshore bays, moving offshore at night, where they unite to form groups of up to several hundred, and feed at great depth. Killer whales occur in all oceans and exhibit considerable variation in their behavior. They may hunt singly or cooperatively in groups. They tend to live in groups (or pods), which may join to form larger pods, either temporarily or permanently.

SWIMMING AND DIVING

The streamlined shape and smooth skin of cetaceans allow water to flow easily over the body surface. This, as well as the mode of locomotion and subtle changes in conformation of the skin, permit them to move through the water very efficiently. Streamlined flow over the body surface during movement means there is very little friction or drag, enabling some species to achieve great speed using relatively little energy.

Cetaceans propel themselves by up and down movements of the tail, in such a way that the flukes present an inclined surface to the water at all times. The force generated at right angles to the surface of the flukes is resolvable into two components, one

raising and lowering the body and the other driving the animal forward. The flippers are used for balancing and steering.

All species can dive to greater depths and remain submerged for longer periods than land mammals. Some species have exceptional diving abilities: for example sperm whales and certain beaked whales are known to dive to more than 1,000 meters (550 fathoms) and stay submerged for more than two hours. Baleen whales and many dolphin and porpoise species, however, generally make shallow dives of less than 100 meters (55 fathoms) and surface after ten minutes or less.

Whales' diving ability is due to anatomical and physiological modifications in the circulatory system, and chemical adaptations in the blood, muscles, and other tissues. The blood makes up a greater proportion of body mass and is capable of carrying significantly more oxygen than that of land mammals. The muscles contain large amounts of myoglobin, which permits increased oxygen storage and gives the muscles a characteristic dark color. When a whale dives, its heart rate slows down. Changes in the distribution of blood in the body mean that those tissues and organs that require oxygen most receive most of the blood. The whale is able to tolerate higher levels of carbon dioxide and other byproducts of metabolism than non-diving mammalian species, which means they can go for considerably longer between breaths

while still remaining active.

Whales have unusual blood vessels, known as retia mirabilia, in the upper part of the chest wall and extending along the vertebrae in the neck and towards the flippers. Although they are very extensive networks of small arteries, and very obvious on postmortem examination, their function is not known. They are present, particularly around the base of the brain, in some land mammals, but they are infinitely more extensive in whales. It has been assumed that they play some role in diving in whales, but precisely what is unknown.

KILLERS AND SIEVERS

The only toothed whale that regularly feeds on animals other than fish and cephalopods such as squid and octopus is the killer whale. It also eats the flesh of warm-blooded animals such as seals, sea lions, dolphins, and penguins. The false killer whale *Pseudorca crassidens* also eats warm-blooded animals occasionally.

The conical teeth of toothed whales are used to seize and hold prey but are not adapted for chewing. The teeth vary in size and number, and in some species, such as the sperm whales and Risso's dolphin *Grampus griseus*, they are absent from the upper jaw. Whereas Risso's dolphin has an average of only six to eight teeth in the lower jaw, the common dolphin has more than a hundred in each

▲ *The bottlenose dolphin is familiar to millions of visitors to marine mammal shows. Under natural conditions it leaps out of the water and, with training, it will do so on demand. Although dolphins have large brains, they are no more intelligent than many other mammals.*

Francois Gohier/Ardea London

▲ *Humpback whales feeding near the surface. They swim below a school of fishes and confuse them by emitting a multitude of small bubbles. They then swim upward with the mouth open, closing it as they break the surface. Water is squeezed out of the mouth, through the sieve of baleen plates, by raising the tongue. The concentrated mass of fish is then swallowed.*

jaw. The teeth have not come through the gums at birth; they emerge at several weeks or months of age, and those teeth remain throughout life. A thin lengthwise section of a whale's tooth reveals layers, which provide an accurate index of age.

The baleen of whalebone whales occurs as horny plates arranged along either side of the upper jaw, more or less like the leaves of a book, with the inner portion frayed into bristle-like fibers. They form a mesh that sieves the very small animals, collectively called plankton, that make up the diet of baleen whales. The composition of the plankton depends on the region in which whales feed, and to some extent different species feed on different kinds of plankton.

SIGHT, SOUND, AND TOUCH

Vision is considered to be quite acute in whales, although it is uncertain what its relative importance is. The eye is adapted to permit quite good vision both under water and in air, and at depths where light intensity is very low.

Whales produce a wide range of underwater sounds, some of which almost certainly are used for communication between individuals. The long and relatively complex song of humpback whales has attracted considerable interest, although its significance is not fully understood. It differs with location, and also within areas over time.

Toothed whales produce audible clicks, which are used in echolocation. They probably depend to a large extent on echolocation for orientation and to obtain food. It has been suggested that some toothed whales may stun prey with accurately

directed bursts of very high energy sound.

Whales have no external ear, but the middle ear receives sound through the tissues of the head and transmits it to the inner ear. The two ears are insulated by an enveloping layer of albuminous foam, allowing whales a three-dimensional perception of the direction and distance from which sound is coming. The ear canal of baleen whales contains a waxy plug, which in section shows layers that are used to determine the age of individual animals, just as tooth sections are used in toothed whales. Whales have no organ for smelling and therefore no sense of smell. They have some modified taste buds in the mouth, but their function is not known.

Whales have a highly developed sense of touch, which involves the entire body surface. Particular areas of the skin have specialized functions: for example, in toothed whales the skin in the region of the jaw may be used to "feel" and to detect sources of low-frequency vibrations.

REPRODUCTION AND DEVELOPMENT

Reproductive behavior is known to vary among species, although little is known about it in most. Breeding is seasonal in migrating species, but in other species it seems to occur through most, if not all, the year. The release of each egg leaves a permanent scar on the ovary of all species. This provides a permanent record of past ovulations, and has been used to estimate age in some large whales. Males of some species have a breeding season, while in others sperm are present all year.

The gestation period in most whale species is 11

to 16 months. A single young is born; twins are very rare. The newborn young is approximately one-third the length of the mother. (Body length is measured as the straight-line length from the tip of the upper jaw to the depth of the notch between the tail flukes.) The teats of the mammary glands lie within small paired slits one on either side of the genital opening. The mammary glands lie beneath the blubber over an extensive area of the abdominal wall. Milk is forced into the calf's mouth during suckling by contraction of muscle overlying the mammary glands. Sucking calves grow very rapidly, a fact that is associated with the high proportions of fat and protein in the milk.

CONSERVATION

For as long as humans have been on the Earth, whales have been considered an exploitable resource. We have killed many species of whales, particularly the large whales, for products such as oil, meat, baleen, and hormones.

As recently as the 1960s we became generally aware of the need to conserve the Earth's resources. The proposition that human use of nature should be placed on a rational basis became widely accepted, and we recognized the need to conserve species of whales. Gradually since then, measures have been introduced to try to achieve this end.

The measures taken have varied with the species and with the cause of the population decline. With a few exceptions, the size of whale populations is very difficult to assess. Some species, however, had obviously suffered drastic population declines through over-harvesting, and this problem was the first to be addressed. Drastic action was taken in some cases: for example the total ban on killing southern humpback whales introduced in 1963. Quotas for killing whales, often rather complex and, in some cases, of limited effectiveness for the most endangered species, were introduced. In very

recent times, whale harvesting has been reduced to a low level, and only those nations supporting so-called scientific whaling are involved.

New conservation problems have now arisen, however, affecting in particular some oceanic species of small toothed whales and dolphins. Many thousands of dolphins have been drowned in the huge gillnets, or driftnets, that are set to catch fish in the seas. Purse-seine nets for capturing tuna have been responsible for the deaths of millions of small toothed whales, particularly spotted and spinner dolphins, and, less frequently, common dolphins. The annual mortality of dolphins caused by these industries is monitored, and attempts have been made to modify the netting procedures so that the number of deaths is significantly reduced. But the world's most seriously endangered cetaceans include some of the river dolphins, notably the Chinese river dolphin that inhabits the Yangtze River. Netting, river pollution, and increased boat traffic have led to what could prove to be intolerable pressures on these animals.

▲ A minke whale being butchered in Iceland. Organized deep-sea whaling, using harpoon guns, began in the 1860s. A century later there were so few large whales left that some species were on the verge of extinction. Most nations now ban whaling.

WHY DO WHALES RUN AGROUND?

Like many other animals, from bacteria to mammals, whales appear to have a sensory faculty that can receive directional information from the Earth's magnetic field. Magnetite crystals have been observed in the tissues surrounding the brain of several species of toothed whales.

The total magnetic field of the Earth is not uniform, but is locally distorted by the magnetic characteristics of the underlying geology. In the floors of the oceans, the movements of the continents have produced series of almost parallel magnetic "contours", which could be used by whales in navigation. Evidence is mounting that whales do use this sense in navigation, and that the phenomenon of mass strandings of live whales results from navigational mistakes.

These sperm whales, disoriented in shallow water, were stranded at low tide off the coast of south-eastern Australia.

Order Sirenia
2 families, 2 genera,
4 species

SIZE
Dugong *Dugong dugon* has a
head to tail length up to 3
metres (10 feet) and weight of
400 to 500 kilograms (880 to
1,100 pounds).
Manatees range from head
and tail length 2.5 to 4.5
meters (8 to 14 feet) and
weight 350 to 1,500
kilograms (770 to 3,300
pounds). Steller's sea cow
(now extinct) was about 8
meters (26 feet) in length and
weighed up to 5.9 tonnes
(13,000 pounds).

CONSERVATION WATCH
All species are classified as
vulnerable and sanctions
apply to trade in their
products.

SEA COWS

HELENE MARSH

The forerunners of the modern sirenians, or sea cows, are believed to have been descended from an ancestor shared with elephants. Even though they have persisted since the Eocene epoch (57 million to 37 million years ago), there have never been many different types of sea cows. There are far fewer species of seagrasses, on which the sea cows feed, than terrestrial grasses and this is considered to be a major reason for the lack of diversity among sirenians. Only four species of sea cow survive today: one dugong (family Dugongidae) and three manatees (family Trichechidae). All are grouped in the obscure order Sirenia. A fifth species, Steller's sea cow, was exterminated by humans in the eighteenth century. Skeletal remains of this giant kelp-eating dugongid, the only cold-water sirenian, can be seen in many major museums.

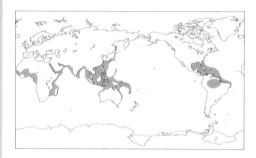

UNDERWATER GRASS EATERS
Manatees appeared during the Miocene (24 million to 5 million years ago), when conditions favored the growth of fresh-water plants in the rivers along the South American coast. Unlike seagrasses, these plants had high concentrations of silica, which rapidly wears away teeth. The manatees countered this problem by developing a system of continuous replacement of their molar teeth: throughout life worn teeth are shed at the front of the jaw and replaced at the back. Dugongs have a different method of coping with tooth wear: their two rear molars are open-rooted and continue to grow. However, the crowns of these teeth are worn continuously, which means that they are very simple and peg-like. In addition, the angle of the dugong's snout is more pronounced than that of the manatee, so the dugong is virtually obliged to feed on bottom-dwelling organisms, whereas the manatee can also feed on plants growing at or near the water's surface. These differences seem to have given manatees a competitive edge over dugongs, which they have displaced from the Atlantic area.

The four extant species of sea cows inhabit separate geographical ranges. All are restricted to the tropics and subtropics. The dugong's range spans the coastal and island waters of 43 countries in the Indo-Pacific region between East Africa and Vanuatu. In contrast to dugongs, which are restricted to the sea, manatees occur in rivers and estuaries. The West Indian and West African manatees occupy similar fresh-water and marine habitats on either side of the Atlantic. Their supposed common ancestor is believed to have migrated to Africa across the Atlantic. Amazonian manatees are restricted to fresh-water habitats in the Amazon basin.

As aquatic herbivores, sirenians have characteristics of both marine mammals and terrestrial herbivores. The dugong's body is shaped like a dolphin's but is a little less streamlined, and the tail flukes are like those of a whale or dolphin. The external ears also resemble those of whales and dolphins in being merely small holes, one on each side of the head. The dugong's hearing is acute. The nostrils close with valve-like flaps and are on top of the head so the dugong can surface without

its body breaking the water. But the dugong's head, with its vast upper lip covered with stout sensory bristles, is more like that of a pig, and inside the downturned mouth are horny pads like those a cow uses for grasping grass.

Manatees look very like dugongs. The most obvious difference is in the shape of the tail: manatees have a horizontal, paddle-shaped tail rather like that of a beaver or a platypus.

Sirenians are large animals. West Indian manatees can be well over 4 meters (13 feet) long and weigh over 1,500 kilograms (3,300 pounds). West African manatees are similar in size, while Amazonian manatees and dugongs range up to

▼ A female dugong is at least 10 years old before she bears her first calf, which is suckled for up to 18 months. Subsequent births are 3 to 5 years apart.

Jeff Foott/Bruce Coleman Limited

▲ *Manatees range from the sea to rivers. Their diet includes seagrasses, and vegetation floating on the water surface or overhanging the water's edge. The molar teeth continually move forward in the jaws, the oldest being shed as new teeth erupt at the back of the tooth-rows.*

about 3 meters (10 feet) and weigh up to 500 kilograms (1,000 pounds). Apart from their size, sea cows have few defenses. Mature male dugongs have tusks, which they apparently use in the fighting that precedes mating. Fortunately, dugongs and manatees have few natural predators.

ATTENTIVE MOTHERS

The life-span of sirenians is long and their reproductive rate low. The age of dugongs can be worked out by counting the layers laid down each year in their tusks, like growth rings in a tree. Individuals may live for seventy years or more, but a female does not have her first calf until she is at least ten years old, and then only bears a single calf every three to five years, after a gestation period of about a year. Information from manatees whose ages are known indicates that they may have a slightly higher reproductive rate, bearing their first calves at a minimum age of five years and having a minimum calving interval of two years.

Female dugongs and manatees are very attentive mothers, communicating with their calves by means of bird-like chirps and high-pitched squeaks and squeals. A young calf never ventures far from its mother and frequently rides on her back. Although it starts eating plants soon after birth, a calf continues to nurse from mammary glands near the base of each of its mother's flippers until it is up to 18 months old. The mammae look rather like the breasts of a human female, a probable reason for the belief that dugongs or manatees form the basis of the mermaid legend.

CONSERVATION

Dugongs and manatees have delicious meat, docile natures, few defenses, low reproductive rates, and geographic ranges that span the waters of many protein-starved developing countries. It is no

wonder that all are classified as vulnerable to extinction by the International Union for the Conservation of Nature.

If you ask Aborigines from the coastal regions of northern Australia to name their favorite food, they are likely to choose dugong. This choice is endorsed by coastal peoples throughout much of the tropical and subtropical regions of the world who prize the meat of sea cows. The meat does not taste of fish, but has been variously likened to veal, beef, or pork because, like the domestic land mammals we use for meat, sea cows feed primarily on plants.

Although successfully exploited by indigenous peoples for thousands of years, dugongs and manatees were not able to sustain the increased exploitation that often followed European colonization. Steller's sea cow was extinct within 30 years of its discovery by a Russian exploring party in the Commander Islands off the coast of Alaska in the eighteenth century. Today, dugongs and manatees are protected throughout most of their ranges, except in some areas where traditional hunting is permitted. But this protection is difficult to enforce, particularly in developing countries that cannot afford wildlife wardens. Even if it were possible to prevent all deliberate killings of sirenians, their preservation would not be assured. Dugongs and manatees often accidentally drown in gill nets set by commercial fishermen. In Florida, most manatees have distinctive scars inflicted by boat propellers; collisions with boats constitute the largest identifiable source of mortality.

The most serious threat to sirenians, however, comes from the loss and degradation of their habitat. This is most obvious in Florida, home to the only significant populations of the West Indian manatee, where there has been rapid growth in the human population. Despite attempts to reduce manatee mortality there, their future looks bleak.

The prognosis for the dugong is brighter, particularly in Australia, where human population pressure is low throughout its range, except in southeast Queensland. Dugong population estimates based on aerial surveys in Australia add up to more than 70,000 and not all the suitable habitat has been surveyed. Protected areas have been established in some important dugong habitats, particularly in the Great Barrier Reef region. However, because dugongs have such a low reproductive rate and because of the difficulties of censusing them, it will probably be a decade before it can be determined whether these conservation initiatives are working.

In the last decade we have learned a lot about the biology of dugongs and manatees. It remains to be seen if this knowledge can be applied to ensure their continued survival. The most effective action would be to identify and protect the habitats that still support significant numbers of these "gentle outliers in the spectrum of mammalian evolution".

ELEPHANTS

JEHESKEL SHOSHANI

Order Proboscidea
1 family, 2 genera, 2 species

SIZE Head and body length (excluding trunk) 300 to 500 centimeters (120 to 200 inches), trunk length 200 centimeters (80 inches), tail length 100 to 150 centimeters (40 to 60 inches), shoulder height 200 to 350 centimeters (80 to 140 inches) for the Asian elephant *Elephas maximus*, and 300 to 400 centimeters (120 to 160 inches) for the African elephant *Loxodonta africana*; weight 3 to 7 tonnes (6,600 to 15,000 pounds).

CONSERVATION WATCH
Both species are endangered.

Elephants are classified in the order Proboscidea, after the most distinguishing feature of these mammals — the proboscis or trunk. Ancestors of the modern elephants are believed to have lived 50 to 60 million years ago; they occupied extreme environments, from desert to tropical rainforest and from sea level to high altitudes. With the exception of Antarctica, Australia, and some oceanic islands, the proboscideans have at some time inhabited every continent on this planet.

AN IMPORTANT ECOLOGICAL ROLE

According to some authors, 352 proboscidean species have been identified; of these, 350 have become extinct. Parts of or complete carcasses of the extinct woolly mammoth *Mammuthus primigenius* have been discovered in the permafrost of the Arctic Circle. One of the best known is that of 'Dima', a 44,000-year-old mammoth calf whose intact frozen body, discovered in June 1977 in the Magadan region of Siberia, provided scientists with a unique opportunity to study both its anatomy and its soft tissues.

The living species belong to two genera, with one species each: the African elephant *Loxodonta africana* and the Asian elephant *Elephas maximus*. Elephants play a pivotal role in their ecosystem. Attributes that make the elephant an inseparable part of their environment include seed dispersal (for example, of acacia trees) through their fecal material, which promotes rapid germination; distribution of nutrients in their dung, which is carried below the ground by termites and dung

▼ *African elephants wallowing. The layer of mud that remains on the skin protects it from insects and sunburn. Elephants that wallow in clean water toss dust onto their backs to provide similar protection.*

Anthony Bannister/NHPA

D. Parer & E. Parer-Cook/Auscape International

▲ Young African elephant. In addition to their function in hearing, an elephant's ears are used as signals of emotional state and as radiators of excess body heat. African elephants have much larger ears than their Asian relatives.

beetles (soil aeration); providing water by digging waterholes which other species also use; trapping rainfall in the depressions from their footprints and bodies; enlarging existing waterholes as they plaster mud on their bodies when bathing and wallowing; making paths, usually leading to waterholes, which act as firebreaks; providing food for birds when walking in high grass by disturbing insects and small reptiles or amphibians; and providing protection for other species as they are tall and can see a long way and alert smaller animals to approaching predators.

THE SKELETON

As in all mammalian species, the skeleton of an elephant is divided into four major parts: the skull, vertebral column, appendages, and ribs and sternum. A side view of a skeleton shows the legs are in almost a vertical position under the body, like the legs of a table. This arrangement provides a strong support for the vertebral column, thoracic and abdominal contents, and the great weight of the animal. In other mammals — for instance, a dog or a cat — the legs are at an angle to the body.

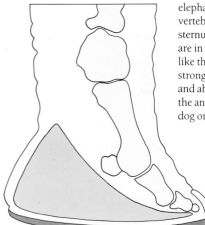

◀ Despite their great weight, elephants walk almost on tiptoe. The digits are supported from behind by a fibrous-fatty cushion and enclosed in a hoof-like structure of skin, with only the toes visible.

The number of teeth is identical for the two species within the lifetime of an individual. The total includes two upper incisors (tusks), no canines, twelve deciduous premolars, and twelve molars. Unlike most other mammals, elephants do not replace their cheek teeth, molars and premolars, vertically — that is, a new tooth developing and replacing the old one from above or below — but rather in a horizontal progression. A newborn elephant has two or three cheek teeth in one jaw quadrant and, as it ages, new teeth develop from behind and slowly move forward. At the same time the previous teeth are worn away, move forward, and fragment. And so, as in a conveyor belt system, new and bigger teeth replace the old. This happens six times in the life of an elephant.

SKIN, EARS, AND BRAIN

The old name of the order Proboscidea, "Pachydermata", refers to their thick skin. The thickness varies from paper thin on the inside of the ears, around the mouth, and the anus, to about 25 millimeters (1 inch) around the back and some areas of the head. Despite its thickness, the skin is a sensitive organ, owing to its sparse hair. The bodies of both species of elephant are usually gray; those of African elephants often seem brown or even reddish because the animals wallow in mudholes or plaster colored soil to their skin. Wallowing seems to be an important behavior in elephant societies; the mud, it appears, protects against ultraviolet radiation, insect bites, and moisture loss. Scratching against trees and bathing seem to be equally important behaviors for skin care.

Body heat can be lost through the ears: within a closely related group of mammals, the size of the ears is often related to climate (larger ears in hotter environments) and to size (proportionately larger ears in bigger animals). The African elephant, which is about 20 percent heavier than the Asian species, has ears of more than twice the surface area. That elephants' large ears function as cooling devices can be demonstrated by the large number of blood vessels at the medial side of the ears, where the skin is about 1 to 2 millimeters (1/16 inch) thick, and by the high frequency of ear flapping during warm to hot days when there is little or no wind. Large ears also trap more sound waves than smaller ones.

The brain of an adult elephant weighs 4.5 to 5.5 kilograms (10 to 12 pounds). It possesses highly convoluted cerebrum and cerebellum (in mammals, these control motor and muscle coordination, respectively). The temporal lobes, known to function as memory centers in humans, are relatively large in elephants. The digestive system is, for mammals, simple. The combined length of the small and large intestines may reach 35 meters (about 100 feet), and it takes about 24 hours to digest a meal. Elephants digest only about 44 percent of their food intake.

REPRODUCTION

The male reproductive system consists of typical parts found in other mammals. Unlike many mammals, however, the testes in elephants are permanently located inside the body, near the kidneys. The penis, in a fully grown male, is long, muscular, and controlled by voluntary muscles; it can reach 100 centimeters (40 inches) and has a diameter of 16 centimeters (6 inches) at its base. When fully erected, the penis has an S-shape with the tip pointing upwards.

The female reproductive system is also that typical for most mammals, but the cervix is not a distinct structure. The clitoris is a well-developed organ which can be 40 centimeters (15 inches) long, and the vulval opening is located between the female's hindlegs, not under the tail as, for example, in bovids and equines. The mammary glands are located between the forelegs, which enables the mother to be in touch with her calf while it suckles.

Elephants attain sexual maturity between the ages of 8 and 13. The estrous cycle may vary between 12 and 16 weeks, and estrus (ovulation and receptivity) lasts a few days; the egg is viable only about 12 hours. Mating is not confined to any season and the gestation period may last 18 to 22 months — the longest pregnancy of any known living mammal. Copulation may take place at any time in a 24-hour cycle; it is usually performed on land, but elephants have been observed to mate in water. The bull pursues the cow until she is ready to mate, at which time a short interaction period of trunk and body contact may take place, followed by the male elephant mounting the female in the usual quadruped position. The long penis of an adult bull plays an important role in the fertilization process; a young bull with a short penis may be able to mate with a cow in estrus, but it is unlikely that he would impregnate her. The entire copulating process lasts about 60 seconds, at the end of which the two elephants separate and may remain near each other for a short period.

Elephants usually bear a single young. Newborn calves weigh 77 to 136 kilograms (170 to 300 pounds) and are 91 centimeters (3 feet) tall at the shoulder. Young are hairy compared with adult animals; the amount of hair reduces with age. Calves may consume 11.4 liters (3 gallons) of milk a day. Weaning is a very gradual process, which begins during the first year of life and may continue until the seventh or sometimes the tenth

THE ELEPHANT'S TRUNK

The trunk of an elephant is its single most important feature; it is an indispensable tool in everyday life. Anatomically, it is a union of the nose and the upper lip. Early naturalists described the trunk as "the elephant's hand" or as "the snake hand". Its flexibility and maneuverability are certainly extraordinary. It is said that an elephant can pick up a needle from the ground and bring it to its trainer. This belief is probably apocryphal, but elephants are capable of picking up objects as small as a coin. A 25-year-old female Asian elephant has been observed using her trunk with an amazing dexterity: she cracked peanuts with the back of her trunk, blew the shells away and ate the kernels.

The early nineteenth century French anatomist G. Cuvier and his colleagues examined the trunk of an elephant and estimated the number of muscles in it at about 40,000. A recent study has shown that the number of muscle units that manipulate this highly sensitive organ may total over 100,000, and that the trunk appears to have a more complex internal structure than previously thought. It has no bones or cartilage; it is composed of muscles, blood and lymph vessels, nerves, some fat, connective tissue, skin, hair, and bristles. The nostrils continue as separate openings from the base of the trunk to its tip. Each is lined with a membrane.

Measurements show that the trunk of an adult Asian elephant can hold 8.5 liters (2.2 gallons) of water, and a thirsty adult bull elephant can drink 212 liters (56 gallons) of water in 4.6 minutes. Trunks are used for feeding, watering, dusting, smelling, touching, lifting, in sound production and communication, and as a weapon of defense and offense.

Anthony Bannister/Oxford Scientific Films

▲ *African elephants in mutual caress. This behavior strengthens social bonds between members of a herd.*

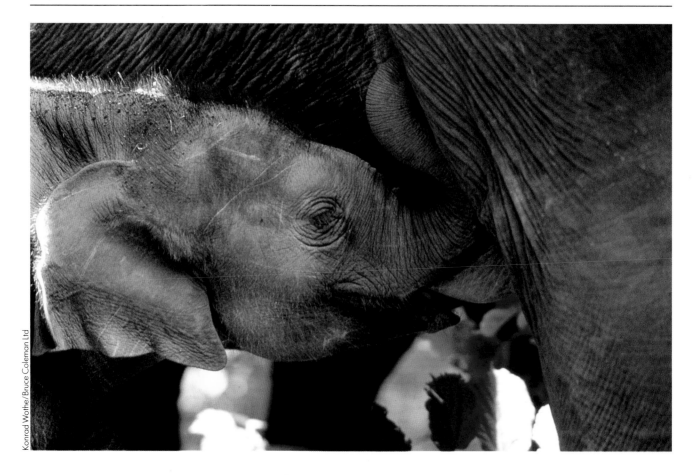

▲ *Asian elephant calf suckling: the trunk has to be lifted back so that its mouth can reach one of the mother's nipples. Gestation lasts about 20 months; the infant suckles for 8 to 10 months, after which the weaning process begins. Females reach sexual maturity when aged about 10 years but may not bear young until very much older.*

year. In her lifetime, which may be 50 to 80 years, a female elephant has the potential to give birth to 7 or 8 offspring under optimum environmental conditions. This potential is rarely realized.

MUSTH

Musth is a periodical phenomenon known to occur in both Asian and African elephants and is associated with physiological and behavioral changes. Elephants in musth — mostly males — can become uncontrollable, and those in captivity have been known to attack and kill their keepers. In Hindi, the word musth means intoxicated. During musth, secretion oozes from the musth gland, which is also known as the temporal gland because of its location — beneath the skin midway between the eye and the ear on each side of the elephant's head. It is not found in any other living mammal. The exact function of the secretion is not known, but it might be associated with sexual activity or communication. Musth, however, is not rut. In rut, there is a period when the males of a species have a heightened mating drive and aggressiveness, and the females are receptive. Musth does not necessarily involve heightened mating drive, nor are the females receptive when the males are in musth.

ECOLOGY AND BEHAVIOR

Eighteen to twenty hours of an elephant's daily cycle is devoted to feeding or moving towards a food or water source: they consume 75 to 150 kilograms (165 to 330 pounds) of food and 83 to 140 liters (27 to 30 gallons) of water a day. Their diet is strictly herbivorous. Feeding constitutes about 80 percent of total behavior, and the balance is filled with activities such as bathing, playing, sleeping, and reproducing.

Elephants are highly social animals. The basic unit is a matriarchal family of five to ten animals; when families join they become a herd. The matriarch is the oldest and usually the most experienced female in the herd. During a drought, for instance, the matriarch will lead her family and relatives to the best possible foraging habitats. The rest of the herd learns and accumulates knowledge through close relationships; it is this tight bonding and herd experience, passed from one generation to the next, that has contributed to the survival of the elephant species.

Calves are tended not only by their mothers but also by other females in the herd. Males leave the herd at maturity, at about 13 years, and are sometimes joined by other males to form bachelor herds. Adult bulls join cow herds when a female is

AFRICAN AND ASIAN ELEPHANTS: WHAT'S THE DIFFERENCE?

Stephen J. Krasemann/NHPA

The most obvious difference between African and Asian elephants is the size of their ears — the African's are much larger. There are, however, many other differences; only some can be given here. African elephants are generally heavier and taller: a bull African can weigh 7 tonnes (14,000 pounds) and reach as much as 4 meters (over 12 feet). The maximum recorded weight and height for Asian elephants are 5.4 tonnes (12,000 pounds) and 3.35 meters (11¼ feet).

The African elephant has a concave back, whereas the Asian has a convex (humped) or level back. Tusks are usually present in both sexes of the African elephant; among Asians, mostly males possess them.

The trunk of the African elephant has more folds of skin in the form of "rings" or annulations, and the tip possesses two instead of one finger-like projection. The trunk of the African appears to be "floppy", while that of the Asian seems slightly more rigid. This can best be seen when they raise their trunks towards their foreheads.

Comments such as "Asian elephants are more easily trained than African" or "the trunk of the Asian is more versatile than that of the African" are partially or wholly correct, and can be explained in evolutionary context. The differences have anatomical bases: the more advanced and specialized of the two species, the Asian elephant, has a higher degree of muscle coordination and therefore is able to perform more complicated antics than its cousin the African species.

▲ African elephants. Note the large ears. African elephants are bigger than Asian elephants and both sexes develop large tusks.

Adrian Warren/Ardea London

Joanna Van Gruisen/Ardea London

▲ There are three subspecies of the Asian elephant, respectively from Sri Lanka, Indo-China, and Sumatra. This female from Sri Lanka lacks tusks and, like all Asian elephants, has an arched (sometimes straight) back; African elephants are sway-backed.

◀ A male Asian elephant from Malaysia. The tusks of this individual are of about average size for a mature male: the largest known tusks of this species were 3 meters (nearly 10 feet) long.

Gerald Thompson/Oxford Scientific Films

▲ *African elephants (above) feed largely on the leaves and branches of trees, particularly acacias. Asian elephants eat more grasses and shrubs. When food is scarce, African elephants push over trees to reach the topmost twigs.*

in estrus. There is no evidence of territoriality; home range is 10 to 70 square kilometers (4 to 27 square miles) or more, depending on the size of the herd and the season of the year.

Acuteness of the senses of elephants appears to change with age. Generally speaking, sight is poor, but in dim light it is good; hearing is excellent; smell is acute; sense of touch is very good; and taste seems to be selective. On balance, the African elephant feeds more on branches, twigs, and leaves than the Asian elephant, which feeds predominantly on grassy materials. Elephants are crepuscular animals — that is, they are active mostly during early morning and twilight hours. When the sun is at its zenith and until early afternoon, elephants have their siesta.

An intriguing aspect of elephant behavior emerged from the study of K.B. Payne and her colleagues, who demonstrated the "secret" language of elephants. This infrasonic communication, a vocalization not audible to the human ear, appears to carry for long distances. Elephants up to 2.4 kilometers (1½ miles) from the source of the sound seem to recognize signals emitted by other elephants and respond accordingly. Males move toward a sound produced by a female in estrus; other elephants at different locations seem to synchronize their behavior in response to stress signals.

IVORY AND CONSERVATION

In elephants, milk incisors or tusks are replaced by permanent second incisors within 6 to 12 months of birth. Permanent tusks grow continuously at the rate of about 17 centimeters (7 inches) a year and are composed mostly of dentine. Like all mammalian teeth, elephant incisors have pulp cavities containing blood vessels and nerves; tusks are thus sensitive to external pressure. On average, only about two-thirds of a tusk is visible externally, the rest being embedded in the socket within the cranium.

The longest recorded tusks of an African elephant measured 3.264 meters (10 feet 8½ inches), and the heaviest weighed 102.7 kilograms (226½ pounds). Comparable published figures for an Asian elephant are 3.02 meters (10 feet) and 39 kilograms (86 pounds). A live 45-year-old Asian elephant, Tommy, carries a pair of tusks measuring about 1.5 meters (5 feet) each, which are estimated to weigh at least 45 kilograms (100 pounds) each.

Elephants use their tusks to dig for water, salt, and roots; to debark trees; as levers for maneuvering felled trees and branches; for work (in domesticated animals); for display; for marking trees; as weapons of defense and offense; as trunk-rests; as protection for the trunk (like a car's bumper bar); and tusks may also play a role as a "status symbol". Just as humans are left or right handed, so too are elephants left or right tusked; the tusk that is used more than the other is called the master tusk. Master tusks can easily be distinguished since they are shorter and more rounded at the tip through wear.

In a cross section, a tusk exhibits a pattern of criss-cross lines that form small diamond-shaped areas visible to the naked eye. This pattern is unique to elephants. The term "ivory" should be applied only to elephants' tusks.

The hardness, and therefore carvability, of ivory differs according to country of origin, habitat, and the animal's sex. Once it is removed from an elephant's body, ivory soon dries and begins to split along the concentric lines unless it is kept cool and moist. But it also deteriorates if kept too moist; the water-absorbing properties of ivory are well known to certain African tribes, who use ivory as a rain predictor by planting it in the ground in selected locations.

As long as there is a demand for ivory, the slaughter of elephants in Africa will continue. It is estimated that every year about 70,000 elephants are killed to supply the insatiable demand for the white gold. True, some of this ivory comes from legal sources, such as culling operations, but about 80 percent of the tusks come from poached, brutally killed elephants. Poachers indiscriminately shoot males and females, young and old, and many

Silvestris/Australasian Nature Transparencies

◀ African elephants in ritual combat. Such struggles seldom involve serious injury: an aggressive display by a dominant male is usually sufficient to discourage a rival with smaller tusks. The tusks of very old male African elephants occasionally grow so long and heavy that they become an encumbrance.

of the animals killed are in their reproductive prime. As poaching continues, the average size of tusks on the offspring of the surviving elephants is getting gradually smaller. In 1982 the average tusk weight of an African elephant was 9.7 kilograms (21 pounds); in 1988 the average was 5.9 kilograms (13 pounds). In Asia, where the size of tusks is much smaller in males and most females have no tusks, elephant poaching is almost non-existent.

Poaching is not, however, the only problem elephants face. The rapid growth of human populations leads to encroachment on wildlife habitat, so that elephants are 'compressed' into a given locale. In this limited area the vegetation is soon over-exploited, and damage to the habitat follows. Over all Africa in 1989, the wild elephant population was estimated at fewer than 750,000, which is half the number of ten years ago.

Various conservation measures are taken to protect wildlife in Africa, public education in the form of lectures, seminars, and travelling museums being the most important. Incentives have also been offered. For example, compensation is given to farmers whose crops are damaged or eaten by elephants. Protecting elephants may prove to be a very costly operation, yet it is imperative for the protection of other wildlife in the environment because, by preserving a large area for elephants to roam freely, one provides suitable habitat and protection for many other animal and plant species of the same ecosystem.

Because of the long generation time and continuous decline in numbers and habitat, elephants — Asian and African — cannot maintain healthy and reproducing populations. The present rate of elephant mortality is alarming. The Asian elephant *Elephas maximus* was declared an endangered species throughout its range in 1973, and is listed in Appendix I of the Convention on International Trade in Endangered Species of Wild Fauna and Flora (CITES). The African elephant *Loxodonta africana* was declared a threatened species and listed in Appendix II in 1978. Because

of recent decimation, however, the African elephant was upgraded in January 1990 to the status of an endangered species (Appendix I of CITES).

What must be done to ensure the survival of wild elephants in Africa? Four strategies are put forward here as essential to the conservation battle: continue with public education for wildlife conservation, especially among young students; enforce existing laws; create natural corridors to connect isolated populations so that they can exchange genes; and establish international cooperation with neighboring countries to ensure uninterrupted elephant habitats, especially where they are known to cross the borders.

▼ Ivory poaching became so intense in the latter half of the 1980s that there is a real danger of extinction of the species over much of its range in East Africa. Since it has not been possible to control poaching, there is increasing international pressure to put an end to the ivory-carving industry, thereby reducing the market for tusks.

Gerald Cubitt/Bruce Coleman Ltd

Order Perissodactyla
3 families, 5 genera,
18 species

SIZE
Smallest Mountain tapir
Tapirus pinchaque, head and
body length 180 centimeters
(71 inches).
Largest White rhinoceros
Ceratotherium simum, 370 to
400 centimeters (145 to 160
inches), tail length 70
centimeters (27½ inches),
weight up to 2.3 tonnes
(5,000 pounds).

CONSERVATION WATCH
The following species are
listed as endangered in the
IUCN Red Data Book of
threatened mammals: African
ass *Equus africanus,* Grevy's
zebra *Equus grevyi,* Syrian
wild ass *Equus hemionus
hemippus,* Indian wild ass *E. h.
khur,* Przewalski's horse *Equus
przewalskii,* Malayan tapir
Tapirus indicus, northern
white rhinoceros
Ceratotherium simum cottoni,
Sumatran rhinoceros
Dicerorhinus sumatrensis,
Javan rhinoceros *Rhinoceros
sondaicus,* great Indian
rhinoceros *Rhinoceros
unicornis.* Many other species
are listed as vulnerable.

ODD-TOED UNGULATES

NORMAN OWEN-SMITH

The Perissodactyla represent the older of two orders of hoofed mammals or ungulates. They have retained functionally either three toes or a single toe on each limb, and hence are referred to as the odd-toed ungulates. The three families in the order are the Equidae (one genus), the Tapiridae (one genus) and the Rhinocerotidae (three genera). While the horses, asses, and zebras of the equids are clearly related, their association with the rare and strange tapir and the lumbering rhinoceros is more surprising.

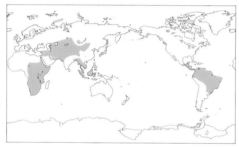

SPECIALIZED GRAZERS

Horses represent the group of ungulates that first adapted to running on the tips of their toes across the grassy plains that developed around 20 million years ago. Tapirs have remained forest animals, feeding on fruits as well as leaves. Rhinoceroses represented the dominant form of large herbivore between 25 and 40 million years ago, but they subsequently declined in diversity, although not necessarily in abundance. All these animals are distinguished from other ungulates not only by their number of toes, but also by the techniques they have evolved for digesting the cellulose content of plant cell walls. They are all specialized grazers and are found in grasslands in parts of Africa, Asia, and South America.

THE EQUID FAMILY

Horses and asses originated in North America, but the last of the American equids became extinct towards the end of the Pleistocene Ice Ages, about 12,000 years ago. The horse *Equus caballus* no longer exists in the ancestral wild form in Eurasia, having been domesticated by 2000 BC. Przewalski's horse *E. przewalski,* which is believed to be closely related to ancestral horses, persisted on the plains of Mongolia until recent times, but may now survive only in zoos. Domesticated horses have, however, established feral populations in all six continents. Two species of wild ass exist: the Asiatic ass *E. hemionus* in parts of the Middle East and southern Central Asia, and the African ass *E. africanus* in semi-desert areas of northeast Africa.

All equids are grazers, although they may include small amounts of leaves, twigs, and succulents in their diet during the lean season. All equids have a gestation period of about 12 months, so that they cannot produce a foal every year and still maintain synchrony with the seasonal cycle. As a result, reproductive rates are lower than in antelope and deer.

Horses

The original wild horses were more stockily built than the slender-limbed breeds prevalent today. Horses are the largest of the equids, weighing up to 700 kilograms (1,500 pounds). Unlike most horned ruminants, males are not much larger than females. This may be related to their fighting technique, which involves attempts to bite the legs of opponents and so depends more on agility than on strength.

The domestication of the horse was an important step in human cultural development, providing not only a beast of burden but also a vehicle for warfare. Foot-slogging armies had little answer to the mobility of attacking hordes mounted on their steeds, which led to the early conquests of European tribes by the Huns. Subsequently the near extermination of bison in North America and certain antelope in South Africa

▼ *Feral horses in New South Wales, Australia. Being descended from various breeds of domesticated stock that strayed or were let loose, these animals display great variability in color, size and form.*

Jean-Paul Ferrero/Auscape International

Liz & Tony Bomford/Ardea London

▲ Przewalski's horse, from Mongolia, may no longer exist in a wild state. However, this stocky species, which is close to the ancestor of all the domesticated horses, is breeding well in an international program that involves many of the world's best zoos.

was aided by the ability of hunters to ride in amongst herds on horseback, from where they could shoot large numbers of animals within a short time. Perhaps most African antelope were saved from the fate of American bison only because of the prevalence of horse sickness in much of tropical and subtropical Africa. Horses survive well without human care in the prairie region of North America as well as in the Camargue delta in France, while feral populations also exist in the Southern Tablelands of Australia, and northern and Central Australia, and in southwest Africa.

Asses

The donkey, or burro, is derived from the African wild ass. Used formerly as beasts of burden by gold prospectors, they have formed feral populations in Death Valley, California, and in the region of the Grand Canyon in Arizona. The African wild ass is now rare in the wild, persisting only in desert regions of Ethiopia and neighboring countries. They are shy and difficult to approach. Asiatic wild asses are somewhat more horse-like, and also quite rare. They are likewise desert inhabitants, dependent on their ability to outrun predators.

Jean-Paul Ferrero/Auscape International

▲ The African or Somali ass is the ancestor of the domestic donkey; it is now rare. Asses and donkeys are better adapted to dry conditions than are horses.

◄ A distinct species, the Asiatic ass exists in four subspecies, all of which are rare in the wild. The subspecies shown here is the onager.

Joanna Van Gruisen/Ardea London

177

▲ *Grant's zebra is a subspecies of Burchell's zebra. This plains-dwelling form is the only zebra that remains common.*

Zebras

Zebras are still abundant and widespread in Africa. There are three species: the plains or Burchell's zebra *E. burchelli,* which occurs throughout Africa in a number of races; Grevy's zebra *E. grevyi,* occupying northeast Africa; and the mountain zebra *E. zebra,* which has one race in Namibia and another in the southern Cape mountains. The latter two species have rather ass-like bodies.

Zebras are distinguished by their coloring of black stripes on a white background. Grevy's zebra and mountain zebra have finely divided patterns of black on white, while different races of plains zebra show varying degrees of brown shading between the stripes. A zebra with black stripes against a dull brown background, the quagga, formerly occurred in the Cape, but was exterminated by white settlers. Genetic evidence obtained from the hides of the two surviving museum specimens suggests that it may have been only an extreme color variant of the plains zebra. Weights vary from 235 kilograms (520 pounds) in some races of plains zebra to 400 kilograms (880 pounds) in Grevy's zebra. Zebras do not have much stamina, and early hunters found them easy to ride down on horseback. For the same reason, attempts to domesticate them were abandoned.

Grevy's zebra and mountain zebra occur in semi-desert areas, while plains zebra migrate through grassy plains and savanna regions.

Mountain zebra hooves have a rubbery texture to resist the bare rocks of arid ranges.

Plains zebras commonly form mixed herds with various antelope. While they tend to be outnumbered in any local area by one or another species of antelope, these zebras are more widely distributed through habitat types and regions than any single antelope species. They occur in tight-knit groups of several mares accompanied by a stallion, as do feral horses and mountain zebras. Young females are abducted out of their family at the time they become sexually receptive. This is probably a mechanism to ensure outbreeding, in circumstances where stallions may remain attached to the same breeding group for many years. In contrast, in Grevy's zebra and wild asses, females form loose affiliations, while males occupy large territories.

Mixed herds may give zebras some additional degree of protection against predators. Adults commonly fall prey to lions, while hyenas and African wild dogs take a toll of foals. Stallions may actively try to protect their young against attacks by hyenas and wild dogs, while mares may even try to distract lions from killing foals. As non-ruminants, zebras are able to feed on the coarser, more stemmy fractions of grasses, including seed-bearing stems, so they do not compete strongly for food with grazing antelopes, even though they favor many of the same grass species.

Jen & Des Bartlett/Bruce Coleman Limited

▶ *The mountain zebra is now uncommon and regarded as vulnerable. The individual shown here, readily identified by the broad black and white stripes on the rump, was photographed in the extremely arid Namib Desert.*

TERRITORIAL SOCIAL SYSTEMS

Solitary animals may nevertheless have a highly developed social organization. For instance, white rhinos have a system of territories held by a proportion of the adult males. Each territory-holding male restricts his movements almost entirely to an area of about 2 square kilometers (¾ square mile). Within this area, he is supremely dominant over all other males. Neighboring territory holders express the balance of power in ritualized horn-to-horn confrontations at territory borders. Males scent-mark their territories by spraying their urine on borders and rhino trails crossing the territory, and also scatter their dung over the middens where other rhinos concentrate their feces. This means that any rhino entering a territory can soon detect that it is occupied, and perhaps even recognize the holder by his scent should there be a meeting. This does not necessarily deter other rhinos from entering. Females and young animals wander in and out of the territories. Adult males that have been unable to claim a territory of their own settle within the territory of another male and accept subordinate status by avoiding spraying their urine or scattering their dung. They noisily demonstrate their submission whenever they are confronted by the holder.

The significance of the territories becomes evident when a rhino female is sexually receptive. Territory-holding males attach themselves to such females and prevent them from crossing out of the territory by blocking their movements at boundaries. The result is that, when females permit courtship advances and eventually the lengthy mating, this takes place without interference from other males. Territories seem especially beneficial where a species attains high local densities and so many males are competing for limited mating opportunities. Territoriality is less clearly expressed or is absent in other rhino species, which occur at lower densities than those typical of white rhinos.

Male Grevy's zebra and both species of wild ass also occupy territories, but these can be as large as 10 to 15 square kilometers (4 to 6 square miles). Perhaps in semi-desert environments animals can maintain visual surveillance over large areas, but stallions also mark boundaries with dung piles. Male territories are commonly found among African antelope species. Where female groups are highly mobile, as is the case in plains zebra and in horses, it is more effective for males to attach themselves to particular herds of females and accompany them in their movements. Hence the grouping patterns and mobility of the female segment of the population determine what type of social system males adopt.

▼ *All equids are social, and males engage in combat that is far from ritual. Grevy's zebra, shown here, is no more closely related to the other zebras than it is to horses or asses. We can regard zebras as horses that happen to be striped, and horses as unstriped zebras: stripes do not signify kinship.*

K.W. Fink/Ardea London

Kevin Shafer/Tom Stack & Associates

▲ *Baird's tapir, from Central America, is a swamp-dweller. Like other tapirs, it forages for leaves and shoots with its flexible proboscis-like snout.*

TAPIRS

Tapirs are squat animals, weighing about 300 kilograms (660 pounds), that are characterized by a short mobile trunk at the tip of the snout. There are three South American species: the Brazilian tapir *Tapirus terrestris*, the mountain tapir *T. pinchaque*, and Baird's tapir *T. bairdii*; and a single Asian species, the Malayan tapir *T. indicus*. While the American species are a dull brown, the Malayan tapir displays a contrasting black and white pattern. Newborn tapirs exhibit mottled cream flecks or stripes, evidently a camouflage for when they lie hidden while their mothers are absent.

The tapir's way of life has not changed much from that of ancestral forms of 20 million years ago. They browse forest shrubs and herbs, but also include much fruit and seeds in their diet when they are available. By feeding on large fruits, they may play an important role in disseminating the seeds of many forest trees. Tapirs seem to spend much time in water, and they are largely nocturnal and difficult to observe. They appear to be largely solitary in their habits. Male tapirs scent-mark their home areas by spraying urine. Because of the orientation of the flaccid penis, the urine is ejected backwards between the hindlegs (this feature is shared with rhinoceroses as well as with cats).

Tapirs are threatened by widespread logging and clearing of tropical forests in both Asia and South America.

Jany Sauvanet/NHPA

◄ All young tapirs are patterned with pale stripes and spots on a dark background. This juvenile Brazilian tapir blends marvellously with the background of dappled light in its woodland habitat.

▼ The Malayan tapir is very conspicuously marked with a disruptive pattern of black and white that provides effective camouflage in the dense rainforest where the few remaining animals survive.

Alain Compost/Bruce Coleman Limited

RHINOCEROSES

Rhinoceroses are characterized by the horns they bear on their snouts. Unlike the horns of antelope, those of rhinos (as the animals are commonly called) lack a bony core. They consist of a mass of hollow filaments that adhere together and are attached fairly loosely to a roughened area of the skull. Early in their evolutionary history rhinos developed a tendency towards large size. An extinct hornless form, *Indricotherium,* attaining 5.5 meters (18 feet) at the shoulder and weighing perhaps 20 tonnes (over 40,000 pounds), was the largest land mammal ever. Like elephants, male rhinos have no scrotum, the testes remaining internal. Five species of rhinoceros have survived. They fall into three distinct subfamilies, one restricted to Africa and the other two to Asia, although they formerly extended into Europe.

African rhinoceroses

The African two-horned rhinos are represented by two species, the black or hook-lipped rhinoceros *Diceros bicornis* and the white or square-lipped rhinoceros *Ceratotherium simum.* Although assigned to separate genera, the two species have a common ancestor that existed about 10 million years ago. They are distinguished most importantly by their feeding habits: the white rhino is a grazer, with a long head and wide lips designed for cropping short grasses, and the black rhino is a browser, with a prehensile upper lip for drawing branch tips into its mouth. There is in fact no clear distinction in skin color between the two species, despite their popular names. Both are basically grey, but with the precise tinge modified by local soil color. The name white rhino has been ascribed to a corruption of a Cape Dutch name, *wijd mond* (wide mouth) rhinoceros, but this is a myth. Probably the first white rhinos encountered by European explorers in the northern Cape had been wallowing in pale, calcium-rich soil.

The white rhino is by far the larger of the two species, with adult males reaching weights of about 2.3 tonnes (5,000 pounds). It is regarded as the third-largest land mammal alive today, after the two species of elephant. Other distinguishing features include the shape of the neck, back, ears, and folds on the skin. The white rhino occurs in two distinct races. The southern form *C. s. simum* was distributed through southern Africa south of the Zambezi River. The northern race *C. s. cottoni* was found west of the upper Nile River in parts of Sudan, Zaire, and neighboring countries. Differences in appearance between the two races

◄ *The white rhinoceros is only a little paler than its "black" (actually gray) African relative. It is better referred to as the square-lipped rhinoceros, in reference to its wide, straight, non-hooked upper lip. It is the only grazing rhinoceros.*

▼ *The black rhinoceros is better referred to as the hook-lipped African rhinoceros, in reference to the short prehensile projection of the upper lip, which is used to pull leaves into the mouth. Like the rhinoceroses of Asia, it is a browser.*

P. Jeans/Australasian Nature Transparencies

▲ *The Sumatran rhinoceros has a sparse covering of long hair. In this respect, and in the retention of incisor and canine teeth, it is more primitive than the African rhinoceroses.*

The white rhino is mild and inoffensive in temperament and was soon hunted toward extinction in southern Africa when guns capable of piercing its thick hide became available. The species is more social than other rhinos, forming groups of three to ten made up of young animals and females without calves. Although these groups, and females accompanied by calves, move independently, they approach one another to engage in playful horn wrestling when they meet. Adult males are solitary and territorial.

The black rhino attains weights of up to 1.3 tonnes (nearly 3,000 pounds). It is renowned for its aggressive charges at human intruders and their vehicles, making puffing noises that sound like a steam engine. This behavior, coupled with the thick bush the animals commonly inhabit, proved fairly effective deterrents to human hunters until recently. Hence the species was widely distributed through much of savanna Africa, from the Cape to Somalia, and even into desert regions in Namibia and northern Kenya. Various races have been distinguished. The large Cape form *D. b. bicornis* has become extinct in the Cape, but may still be represented by Namibian animals. A smaller southern form *D. b. minor* has a distribution

are small, but they occur in different kinds of savanna and appear to be fairly distinct genetically. The existence of the northern race was only confirmed this century, such is the remoteness of the region. Fossil remains indicate that the white rhino was once more widely distributed through East and North Africa.

Alain Compost/Bruce Coleman Limited

extending from the eastern parts of South Africa through Zimbabwe and Zambia into southern Tanzania. In Kenya it is replaced by another form, *D. b. michaeli*. Other races may occur in Somalia and in West Africa, but too few specimens remain to confirm their distinctiveness. In some regions black rhinos are characterized by peculiar festering sores behind their shoulders. These are caused by a filarial roundworm, which lives in the wounds. White rhinos never develop these sores. The two species of African rhino usually ignore one another when they meet, but sometimes appear mildly curious about each other, and may even approach to rub horns.

Asian rhinoceroses

The Asian two-horned rhinos are represented by a single species, the Sumatran rhinoceros *Dicerorhinus sumatrensis*. This is a small (for a rhino) and somewhat hairy animal, which weighs up to 800 kilograms (1,750 pounds). It is a browser occupying mountainous forests not only in Sumatra, but also in other parts of Southeast Asia. Other rhinos of this subfamily occurred in Europe and Asia during the period of the Ice Ages. They included the famous woolly rhinoceros *Coelodonta antiquitatis*. Specimens preserved in ice

in Siberia confirm that it was a grazer, like the white rhino.

The Asian one-horned rhinoceroses are represented by the great Indian rhinoceros *Rhinoceros unicornis*, which rivals the white rhino in size, and by the smaller Javan rhinoceros *R. sondaicus*. Both bear a single horn on the tip of the snout. A prominent characteristic of the Indian rhino is the armor-like folds on the skin, which feature in a famous tale by Rudyard Kipling. The Indian rhino does not use its horns in fighting, as the African rhinos do, but lunges at opponents with its tusk-like incisor teeth. The Indian rhino is largely a grazer, although not as exclusively as the white rhino. It favours tall swampy grasslands in northern India and adjacent Nepal, and spends much time wallowing in ponds. Although animals may congregate around ponds, they are largely solitary in their movements. Indian rhinos exhibit noisy mating chases, with females making honking sounds and males squeaky panting noises. They are famous for their prolonged matings, copulation commonly persisting for as long as 60 minutes. The Javan rhino is a browser occupying lowland forests, formerly occurring through much of Southeast Asia. Little is known of its behavior, and there are no specimens in zoos.

▲ *Much smaller than its Indian relative but also with a very thick skin and a single horn, the Javan rhinoceros has a projecting, prehensile upper lip, which is used to grasp leaves and twigs. It is often encountered in water.*

◄ *The Indian rhinoceros has such extremely thick skin that it is arranged in plates, like a jointed suit of armor.*

THE TRADE IN RHINO HORNS

Rhinos of all species are being pushed towards extinction on account of the very high prices fetched by their horns. These horns are used primarily as a medicine in China and neighboring countries of the Far East. Horns are ground into a powder and swallowed as a potion. They were also taken as an aphrodisiac in a region of west India, but this practice is no longer of much importance. Another use is for making dagger handles, which are especially prized as a symbol of manliness in North Yemen. In fact the substance of rhino horn, a protein called keratin, is no different from that making up the outer covering of the horns of cattle and antelope, hooves, and human fingernails.

Although the efficacy of rhino horn as a drug may be quite mythical, the fact remains that horns fetch extremely high prices (exceeding that of gold per unit weight), and the value tends to rise as rhinos become rarer. This spurs increasing endeavors by unscrupulous entrepreneurs to acquire horns, paying peasant farmers prices to hunt the animals that are vastly in excess of the honest wages these people could otherwise earn. The result is that conservation agencies in Africa and Asia have been fighting a losing battle to protect the vanishing remnants of the herds of these great beasts.

There is no easy solution to this problem. As horns will regrow if removed, some conservationists advocate farming rhinos by repeated cropping of their horns, thereby producing a steady supply of revenue to impoverished communities through honest channels. However, the relationship between remaining numbers of rhinos and potential markets in the east is such that the temptation to hunt will continue, forcing rhino populations ever downwards. Another proposal, already instituted in Kenya, is for rhinos to be handed over to private owners who have the wealth for the fencing and patrolling to protect the animals. A third suggestion is for as many of the remaining rhinos as possible to be moved into zoos, there to be bred awaiting more favorable circumstances for them to be reintroduced into the wild. However, by the time

Anthony Bannister/NHPA

memory of the illegal horn trades has faded, the rhinos' former habitats may have become settled by peasant farmers desperately trying to eke out an existence. Some conservationists believe that every attempt should be made to keep rhinos in the wild, however hopeless the situation may seem.

Rhinos can serve as potent symbols for marshalling international aid for parks and other conservation action in impoverished Third World countries. While the value of rhinos to humans today may be largely symbolic, they may serve to gain protection for many lesser species that are also threatened but attract little notice. The ultimate value to humankind of these inconspicuous species, such as plants and insects, may be vastly greater than that of spectacular mammals like rhinos.

▲ Horns of the black rhinoceros, confiscated by African wildlife authorities. It has been seriously suggested that the last of the world's black rhinoceroses might be saved by government programs of de-horning the animals, making them valueless to poachers.

▶ Symbols of death. Daggers with rhinoceros-horn shafts on sale in North Yemen.

CONSERVATION

Rhinos have been the center of much attention from conservationists. The Javan rhino is among the rarest species alive today, with a surviving population of about 60 animals confined to a single reserve at the western tip of Java. The Sumatran rhino is only slightly better off, as the 800 or so animals remaining are thinly scattered through Malaysia and adjacent countries, and are vulnerable to both poaching and logging. Remnants of the Indian rhino are protected in reserves in northeast India and Nepal, but number under 2,000. The southern white rhino was

reduced to perhaps 50 animals in the Umfolozi Reserve in South Africa at the turn of the century, but under careful protection its number has increased steadily to over 4,000 now. This has enabled the species to be reintroduced into many areas from which it had been exterminated by early white hunters, for example in Zimbabwe, Botswana, and Mozambique. The northern white rhino maintained its abundance until the late 1960s, but has now been almost exterminated. Only about 20 animals remain, all confined to the Garamba National Park in northern Zaire. Regular monitoring of these few individuals has halted

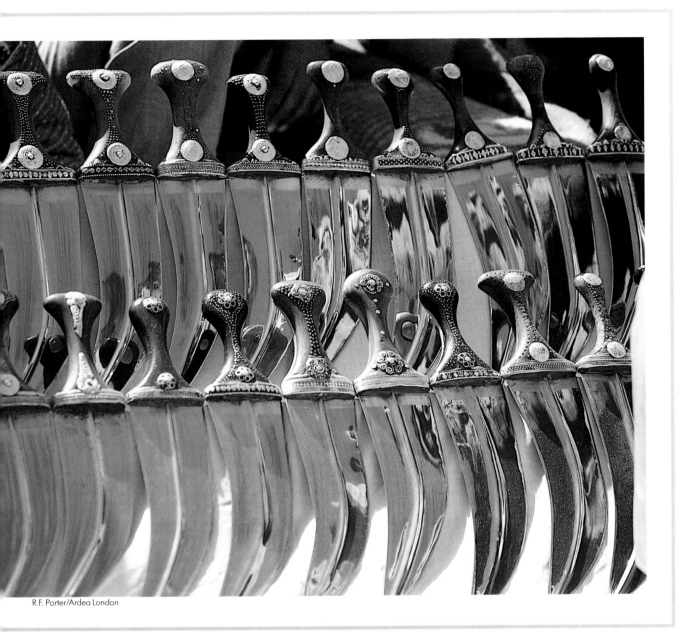

R.F. Porter/Ardea London

further losses. The black rhino was formerly the most abundant and widespread of the five species, with a total population of over 60,000 in 1970. This situation has changed drastically in recent years: widespread poaching has caused its numbers to plummet to perhaps 3,500 at present.

The decline in the abundance of rhinos is entirely due to humans, through hunting coupled with habitat destruction. Adult animals are large and formidable, and hence rarely fall prey to carnivores, although young animals are vulnerable. Hence adult mortality is low and is balanced by slow reproductive rates. The gestation period is 16 months in white rhinos and Indian rhinos, and 15 months in black rhinos. The birth interval is about two and a half to three years, so that the maximum rate of population growth is under 10 percent a year. Once levels of the harvest by human hunters exceed this threshold, populations are driven inexorably downwards. This happened to the white rhino between 1860 and 1890, when an abundant and widespread species was reduced to a remnant of a few score animals. Today unscrupulous hunters are being lured by the high prices fetched by rhino horns in markets in Asia and the Middle East.

Order Hyracoidea
1 family, 3 genera, 4 species

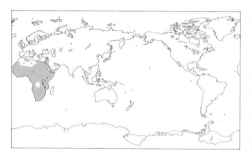

SIZE Head and body length
30 to 60 centimeters (12 to
24 inches), tail length up to 3
millimeters (⅛ inch),
shoulder height 30
centimeters (12 inches),
weight 2 to 5 kilograms (4 to
11 pounds).

CONSERVATION WATCH
Tree hyraxes, genus
Dendrohyrax, are the most
endangered owing to habitat
destruction and hunting.

HYRAXES

JEHESKEL SHOSHANI

Sometimes mistaken for rodents or rabbits, hyraxes or dassies are small but robust mammals that resemble a huge guinea-pig. However, hyraxes are really so different from all other mammals that zoologists place them in an order by themselves — the Hyracoidea. This order has one extant family, Procaviidae, which includes three living genera: *Dendrohyrax,* the tree hyrax, *Heterohyrax,* the yellow-spotted rock hyrax, and *Procavia,* the rock hyrax. The classification of *Procavia* is still provisional: there are at least four species — more, according to some zoologists.

▼ The southern tree hyrax. Hyraxes represent an evolutionary experiment that all but failed. Forty to fifty million years ago a wide range of hyraxes of various sizes were the dominant grazing mammals of North Africa, but they were later displaced by antelopes, cattle, sheep, and goats. The surviving hyraxes are small herbivores that eke out an existence in arid, rocky country.

ROBUST ROCK-DWELLERS

The earliest fossil hyracoids date to the Eocene epoch, about 50 million years ago. Living hyraxes are all approximately rabbit-sized, but extinct species reached the size of a tapir. It appears that hyraxes, extinct and extant, were adapted to a variety of ecological herbivorous niches, though today they are restricted to tropical and subtropical habitats, in altitudes from 400 meters (1,300 feet) below sea level, at the shores of the Dead Sea, to mountains 3,500 meters (11,500 feet) above sea level in East Africa.

For their size, hyraxes are robust, stocky animals. They have a muscular, massive short neck; a long and arched body, stubby legs, and a tail is not visible. They have large eyes, medium-sized ears, and a truncated snout with a cleft in the upper lip. Sexual dimorphism is evident: the males are heavier, more muscular, possess longer upper incisors, and behave more aggressively than the females. The coat is dense and consists of short underfur and long tactile guard hairs. General body color is highly variable among the genera and species, from light gray or yellowish brown to dark brown; the flanks are lighter and the underparts are buff. In most species, there is a light, thin band of hair along the dorsal edge, and a tuft of long hair a different color from the surrounding area is present at about the center of the back. This tuft covers a scent gland that secretes a sticky, smelly substance, believed to have a communicatory function. The skin is relatively thick.

Hyraxes walk on the soles of their feet, which are unique, with large, rubbery-soft elastic pads which are kept moist by secretory glands; these adaptations assure them a firm grip on rocks and trees. The forefoot has five digits but only four are functional, while the hindfoot has four digits with three functional. Except for the inner toe of the hindfoot, all digits terminate in short, flat, hoof-like nails. That of the inner hind toe is a long, curved, claw-like nail, which is used for grooming. Grooming is also accomplished by the spatula-shaped lower incisors in combination with the upper teeth. A hyrax grooms itself, and can usually reach most parts of its body, but occasionally hyraxes engage in social grooming.

REPRODUCTION AND TERRITORY

An animal the size of a hyrax (for example, a rabbit) usually has a gestation period of about one month, but pregnancy for hyraxes has been reported to last from six to eight months. One possible explanation for such an extremely long gestation period is that the hyrax is an evolutionary vestige of much larger ancestral animals.

During the mating season, *Procavia,* which mark rocks with their dorsal scent glands, may exhibit territoriality-related behavior. The odor may also serve as a form of social communication among the colony members and help the young identify or locate their mothers. The mating season in the wild varies among the genera, but most have their

John Shaw/NHPA

young during the spring season. *Dendrohyrax*
usually produce one or two offspring and *Procavia*
and *Heterohyrax*, two to four. The newborn are
highly precocial, like all living ungulates. Young
Procavia often climb on their mothers' backs,
engage in "play" behavior, and are more active than
adults. Sexual maturity is attained at one year.

GREGARIOUS VEGETARIANS

Hyraxes can live to be 10 years old. Colonies of the
gregarious rock hyraxes, numbering up to 80
animals, consist of a few to several families, each of
which is headed by a male. All species are
vegetarians, but they sometimes eat invertebrates.
Rock hyraxes feed mostly on the ground, and tree
hyraxes feed on the ground as well. Usually, one or
more individuals keep watch and warn other
members of their group of an approaching
predator or potential predator with a loud warning
cry. All the animals suddenly disappear among the
rocks and boulders, but once the danger is over
they gradually reappear and resume their activities.
Natural predators of hyraxes include leopards, wild
dogs, eagles, and pythons, while natives of some
African countries hunt them for their flesh and
skins. In places where their predators have been
reduced or eliminated, hyrax numbers increase
and they become a nuisance.

Most hyraxes are active during the mornings and
evenings, when they bask in the sun — often close
together in a group — and feed. The basking,
passive behavior presumably increases their
metabolic rate and prepares them to forage. Hot
hours of the day are spent in the shade. Hyraxes
have a habit of staying in one spot and staring in
one direction for a long time.

Hyraxes do not burrow, although they have
been observed digging; their rock outcrops provide
them with most of the shelter they need. On
occasion, they have been known to inhabit the
burrows of aardvarks and meerkats.

No other animal sounds like a hyrax. Their
vocabulary changes and increases throughout life;
the young ones utter long chatters which increase
in their intensity from beginning to end of each
bout. It has been reported that *Procavia* infants
make only five of the twenty-one sounds made by
adults. The best-known calls are those of
Dendrohyrax. They begin to vocalize soon after
dark in a series of croaks that end in a loud scream.
Travelers on safari in Africa are often warned of
these sounds and told not to confuse them with
those of bandits.

Peter Davey/Bruce Coleman Ltd

▲ *Yellow-spotted rock hyraxes on an old
termite mound. Although they look like
guinea pigs, hyraxes are most closely
related to elephants, sea cows, and the
aardvark. They live in groups with a
complex social structure, ever on the
alert for predators. Unusually for small
mammals, they are active during the
day.*

S. Krasemann/NHPA

◄ *Yellow-spotted rock hyraxes basking.
Living in a habitat with sparse food,
hyraxes are adapted to a low-energy
diet. They are unable to regulate their
body temperature effectively but they
conserve heat by huddling together and
warm themselves by basking in the sun.*

Order Tubulidentata
1 family, 1 genus, 1 species

SIZE
Aardvark *Orycteropus afer*, head and body length 100 to 160 centimeters (40 to 65 inches), tail length 50 to 80 centimeters (20 to 30 inches) shoulder height 60 to 65 centimeters (24 to 26 inches), weight 40 to 60 kilograms (90 to 160 pounds).

CONSERVATION WATCH
Not endangered, but sanctions apply to trade in their products.

AARDVARK

JEHESKEL SHOSHANI

The aardvark — a name derived from the Dutch, through Afrikaans, for "earth pig" — is one of the strangest of all living mammals. It is classed in an order of its own, the Tubulidentata. The order consists of two families, only one of which, Orycteropodidae, is extant. The only surviving species is the aardvark *Orycteropus afer*. Tubulidentata probably originated in Africa at some time in the Paleocene or earlier, about 65 to 57 million years ago, and are believed to have evolved from an early hoofed-type mammal. Because they exhibit primitive mammalian anatomical and molecular characters, aardvarks can be thought of as "living fossils".

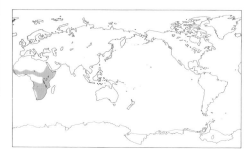

FIERCE AND PIG-LIKE

Although aardvarks eat ants they are not "ant-eaters" in the strictest zoological sense. Early naturalists classified aardvarks together with pangolins and armadillos, but as it became evident that the similarities between them are due to convergence rather than affinities, this view was abandoned and today each one of these groups is assigned to a different order.

The aardvark is a medium-sized pig-like animal with a stocky body, a short neck, and an arched back; it has a long snout, large ears resembling those of a donkey, and a long muscular tail that is thick at its base. The skin is thick and sparsely covered with bristly hair, light in color on the body (though usually stained by soil) and darker on the limbs. The head and tail are whitish. Females are slightly smaller than males; their heads are lighter and the tail has a bright white tip.

The snout contains a labyrinth of nine to ten thin bones — the highest number of all mammals. The nostrils can be closed by means of muscular contraction of hair, an adaptation that prevents ants and termites from entering the snout. The eyes are reduced and the animal is color blind, but hearing and smell are acute. The mouth is small with a long, narrow tongue; food particles are taken in with the tongue and scraped off by ridges on the palate as the tongue protrudes again. The jaw muscles are weak and the mandible is slender.

An aardvark's forelimbs are short, powerful digging tools that have four shovel-shaped claws on each foot. The hindlegs, which have five clawed toes, are long and function as support and as a springboard for the body. With its massive skeleton, thick skin, and sharp claws, an aardvark is well armed — humans and hyenas are among the few predators that will attempt to hunt a healthy adult. The young, the old, and the sick, however, are preyed on by lions, leopards, cheetahs, hunting dogs, and pythons. When attacked, aardvarks are fierce and will kick and slash with their legs and claws. With great speed they can somersault and stand on their hindlegs and tails to defend themselves with their forefeet.

FEEDING AND NESTING

Unlike those of other mammals, the teeth of aardvarks are oval or figure-of-eight-shaped, flat on the top, and columnar. They consist of hexagonal prisms with many fused minute dentine columns and pulp cavities that appear as tubes on the chewing surface of the tooth. The teeth grow continuously, are without enamel, are covered with cement, and are rootless. Adult aardvarks have no incisors or canines, and only eight premolars and twelve molars. The cheek teeth can easily crush the hard exoskeletons of ants and termites, which constitute their main diet. Aardvarks also eat locusts, and grasshoppers and insects of the family Scarabidae. The stomach is one-chambered, and

▼ *The aardvark feeds on ants and termites by digging into their nests with its powerful, clawed forelegs. Insects are picked up on a long, sticky tongue and swallowed without being chewed. The nostrils, on the tip of the snout, can be closed to prevent the entry of insects.*

Gary Milburn/Tom Stack & Associates

the cecum is large for an insectivorous mammal. Aardvarks have been observed eating vegetable matter, especially the fruits of a wild plant known in South Africa as "aardvark cucumber"

Generally speaking, wherever termites are found one can expect to find aardvarks: open canopy forests, bush veldts, and savannas are among their favorite habitats. As a well-equipped digger, an aardvark can excavate a termite mound in a few minutes where a human would need to use an ax or other heavy-duty tool. Termites and ants are ingested by means of a sticky, 25 centimeter (10 inch) extensible tongue. Most termites and ants are unable to hurt aardvarks, whose thick hide protects them. On one occasion, however, I observed a female aardvark that had been stung by large ants; she rolled about in a frenzy for three minutes, rubbing her body against the hard ground, rocks, and logs.

While foraging, the animal moves in a zigzag path and continuously sniffs the ground in an area about 30 meters (33 yards) wide. It may travel 10 kilometers (6 miles) or even up to 30 kilometers (18 miles) a night in search of food.

Aardvarks dig three types of burrows: for food, for temporary shelter, and for permanent residence. Food and temporary burrows vary from shallow to deep holes excavated by the animals, but they sometimes use abandoned termite mounds. A permanent residence is about 2 to 3 meters (2 to 3 yards) long, usually with one entrance descending at an angle of about 45°. It may be a single main straight tunnel or an extensive tunnel system with many access holes. Sleeping chambers are wide enough for the animal to turn around: aardvarks enter and leave their burrows head first. Unused aardvark burrows are important for the survival of many small species such as hyraxes, which occupy them as dens or refuges in case of fire.

REPRODUCTION AND TERRITORY
The uterus of a female aardvark is similar to that found in rodents and rabbits. Nipples are in two pairs, one pair in the lower abdomen and one in the groin; the number of milk ducts varies from one to three. The breeding season appears to relate to latitude: in habitats away from the Equator breeding begins earlier than in habitats close to the Equator. Births occur from May to July in South Africa and Ethiopia, and later in the year in Zaire and other equatorial countries. Estrus is signalled by swelling of the vagina, and sometimes a discharge is visible.

The one offspring, which usually weighs about 2 kilograms (4 pounds 6 ounces) and measures 55 centimeters (22 inches) is born after a seven month gestation period. The newborn are partly precocial: their eyes are open and their claws are well developed. They join their mothers on nocturnal foraging trips at the age of 2 weeks and remain

Clem Haagner/Ardea London

with them until the next mating season. Sexual maturity is attained at about 2 years. In captivity, a female has given birth to 11 young in 16 years, and a male has sired 18 offspring by the age of 24.

Males are more vagabond than females and spend most of the year in separate burrows, except during the breeding season. Evidence for territoriality is inconclusive: in areas where density is high, several animals may occupy and feed in the same or overlapping home ranges. Aardvarks visit watering holes frequently and are good swimmers.

AARDVARKS AND HUMANS
The economic importance of aardvarks may be measured by their control of termites. In areas where aardvarks and other insectivorous animals have been exterminated, crops have suffered extensive damage. Grazing by wild and domestic ungulates creates suitable conditions for termites, which in turn are consumed by aardvarks; but aardvark burrows damage farmland and are hazards to vehicles and galloping horses. Bushmen, Hottentots, and other residents still hunt aardvarks for their meat and hide, as well as for medicines, amulets, and for sport. The meat is very tasty and is similar to that of a pig.

Aardvarks are rarely observed in the wild. They are nocturnal, solitary, and elusive and their habitat is shrinking as more land is cultivated.

▲ During the day, an aardvark usually sleeps in its burrow: it is rare for one to be active by day, as in the photograph above. Although it has not been proved, it seems likely that the large ears act as radiators that assist in temperature regulation, as in such other desert dwellers as the fennec, jack rabbit, and bilby.

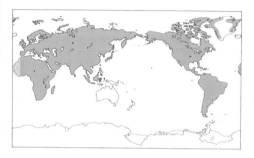

SIZE
Smallest Lesser mouse deer
Tragulus javanicus, head and
body length 44 to 48
centimeters (17 to 19 inches),
weight 1.7 to 2.6 kilograms
(3¾ to 5¾ pounds).
Largest Tallest is the giraffe
Giraffa camelopardalis,
average head and body length
4.2 meters (13¾ feet), average
height to horn tips 5 meters
(16½ feet), average weight
1.35 tonnes (3,000 pounds).
Largest in mass is the
hippopotamus *Hippopotamus
amphibius,* average head and
body length 3.4 meters (11¼
feet), average weight 2.4
tonnes (5,300 pounds).

CONSERVATION WATCH
The following species are
listed as endangered in the
IUCN Red Data Book of
threatened mammals: pygmy
hog *Sus salvanius,* swamp deer
Cervus duvauceli, brow-
antlered deer *Cervus eldi,*
South Andean huemul
Hippocamelus bisulcus, Fea's
muntjac *Muntiacus feai,* wild
yak *Bos grunniens,* kouprey
Bos sauveli, lowland anoa
Bubalus depressicornis,
tamaraw *B. mindorensis,*
mountain anoa *B. quarlesi,*
Jentink's duiker *Cephalophus
jentinki,* Edmi gazelle *Gazella
cuvieri,* rhim *G. leptoceros,*
Arabian tahr *Hemitragus
jayakari,* Arabian oryx *Oryx
leucoryx,* Mediterranean
mouflon *Ovis musimon.* Many
other species and subspecies
are classified as rare,
vulnerable, or of
indeterminate status.

EVEN-TOED UNGULATES

VALERIUS GEIST

The Artiodactyla, or cloven-hoofed mammals, are the younger of two orders of hoofed mammals, or ungulates. They have two or four weight-bearing toes on each foot, hence the name even-toed ungulates. Artiodactyls are a successful group of herbivores: they have a high species diversity; they inhabit a wide range of habitats, from tropical to polar; and they have a large biomass and wide geographic distribution, occurring naturally in all continents except Australia and Antarctica.

RUNNING TO SAFETY

The Artiodactyla include non-ruminating forms such as the pigs of the Old World (family Suidae); the pig-like peccaries of the New World (Tayasuidae); and hippos, now restricted to Africa (Hippopotamidae). The ruminants include camels (Camelidae); the chevrotains or mouse deer (Tragulidae); the musk deer (Moschidae); antlered deer (Cervidae); giraffes (Giraffidae), two species, now restricted to Africa; the lone American pronghorn (Antilocapridae); and numerous hollow-horned bovids (Bovidae).

Artiodactyls are herbivores, though some pigs are omnivores, and all are prey of carnivores. Consequently, their biology is dominated by means of evading predators. The foot structure of the artiodactyls, evolved for sprinting, reflects this. The bones of the soles of their feet are large, to absorb the stress of running, and fuse into a common cannon bone in all but the pigs, hippos, and the front legs of the peccaries. The hooves, or large claws in the case of camels, ensure immediate traction on almost any earth surface when power is suddenly and massively applied. Only the hippos use water to defeat predation, and so they differ in build from other artiodactyls.

While some pigs have simple stomachs, the majority of artiodactyls have complex stomachs and enlarged ceca that serve in the anaerobic fermentation of plant food. Even some pigs, and especially the peccaries and hippos, have complex stomachs. Fermentation allows a percentage of the otherwise undigestible cellulose in the cell walls of plants to be digested into metabolizable products, namely short-chain fatty acids. These, once liberated, are directly absorbed into the bloodstream through the walls of the stomach. The fermentation of plant food and the absorption of fatty acids is followed by the digestion of the bacteria and associated organisms in the true stomach. That is, ruminants culture their own protein source. This not only supplies the protein, but also most of the vitamins as well.

The fermentation vat also accepts inorganic matter, such as sulfur, which the bacteria incorporate into their bodies, creating sulfur-bearing amino acids, so vital for growing connective tissue and hair, horns, and hooves. For this reason mineral licks are sought out not only for their vital minerals, such as magnesium, sodium, and calcium, but also for sulfur. In fact, in the beautiful white Dall's sheep, of northern Canada and Alaska, females and young make a mineral lick the centre of their summer activity, for as long as the females are heavily lactating.

EVOLUTION

The artiodactyls appeared early in the age of mammals, in the Eocene (57 to 37 million years ago), as small, forest-adapted omnivores. They were barely the size of rabbits and, like them, apparently depended for survival on hiding plus a swift getaway by hopping rapidly along the ground and over obstacles. Proceeding by leaps and bounds is costly in energy, as the runner must generate considerable lift with each jump, and therefore tires rapidly. For all that, however, it must be an effective way to escape predators, because for small ungulates it has remained the primary mode of escape from predators.

In tropical forests only a small amount of easily digestible food, such as fallen fruit or shoots growing on the ground, is available to ground dwellers. The density of such herbivores is, consequently, low. Artiodactyls blossomed into great diversity during the middle of the Tertiary (65 to 2 million years ago), when savanna and steppe replaced much of the earlier forests. The

Hans Reinhard/Bruce Coleman Ltd

artiodactyls now had access to grasslands which, unlike tall forests, could be totally exploited for food. With that arose a multitude of highly gregarious, large-bodied runners that lived in high density on open plains.

Gregariousness and speedy, enduring running leave their mark on body shape and weapons. The chest cavity enlarges to house large lungs and heart; the shoulders become more muscular; and the legs equalize in length and become slim and light. Some species become "short-legged" runners that make good their escape across level, even, hard, and unobstructed ground. These species, such as saiga antelopes, reindeer, and addax, run with their heads low. They minimize costly body lift and convert almost all their energy into forward propulsion.

Then there are the "long-legged" runners, which run with their heads held high. They are specialized to run over uneven, broken terrain,

▲ *Giraffes run with a "pacing" gait, both legs on a side moving more or less together. The forelegs are very long and must be spread apart to allow the mouth to be brought low enough to drink.*

▲ *The babirusa (top, center), unlike the other pigs, has tusks that seem to be purely ornamental. In the males, they usually jut through the top of the snout. The red river hog (above, right) is probably the most colorful of the pig family. It has several color forms, not all of them red. The pig-like chacoan peccary (above, left) was discovered in 1975. It is the largest and rarest of the three species of peccary.*

where they need an early warning of obstructions to come and the ability to shift their hooves quickly within each bound to avoid obstacles. Examples are the dama gazelle, the argali sheep, and the pronghorn. And there are runners that specialize in throwing a stream of obstacles into a pursuing predator's path. Examples are the mule deer, with its unique bounding gait, or the moose that trots smoothly and fast over low obstacles that pursuing bears or wolves must cross in costly jumps.

ATTACK AND DEFENSE

Not only the body form changes with gregariousness; so do weapons. In bounding runners from tropical thickets, the weapons are long combat teeth, and short and sharp horns, or outgrowths on the head that allow the head to be used as a club. These types of weapons are associated with the defense of material resources within a territory. Weapons that damage the body surface, and which will not stick in an opponent's body, combined with tactics of surprise attacks, are highly adaptive. Such attacks aim at quickly provoking great pain, making the intruder run away with a most unpleasant memory of the event and the locality. Sharp pain discourages retaliation that would very likely damage the attacking territory-owner. So-called "knives and daggers" are universally associated with the defense of resource, but not of mating, territories.

In the so-called "selfish herd" on open plains, damaging weapons are a liability. By grouping in the open, each individual gains from the presence of the others: the more there are, the more secure each individual. A predator striking may kill one, and the chances of being that one decline with companions. When fleeing, an animal has a much better chance in a large than in a small group of not being last: it's the last one that gets caught! By running first and attracting others to follow one is pretty sure of not being the last. Consequently, it pays to have a "follow me" marker on the tail. That appears to be the origin of the large, species-specific rump patches in gregarious species.

Weapons must allow the animal to defeat an opponent, but their use must not normally lead to retaliation or wounding of either participant in a fight. Blood attracts predators and it matters little whose blood it is. Consequently, with gregariousness a new type of weapon arises: the horns or antlers change into twisted, rugose, or branched "baskets" that readily catch and bind the opponent's head, permitting head-wrestling. With weapons that can be used harmlessly, sporting engagements can flourish, and they do. Called "sparring", these are much engaged in by large and small, particularly outside the rutting season.

Gregarious species that defend mating territories retain fairly simple grappling horns, but in species with serial or harem defence of females, the horns may enlarge. Huge antlers or horns may evolve, as in reindeer, the extinct Irish elk, and the water buffalo.

What artiodactyls do, no matter where they may live, is a logical consequence of generating security, minimizing competition for resources, maximizing the extraction of nutrients and energy from fibrous, toxic plants, and maximizing the chances of successful, frequent reproduction. Their lives illustrate how the same problems have been diversely and ingeniously solved.

Jeff Foott/Bruce Coleman Ltd

▲ The distinctive patches on the rumps of many gregarious animals act as signals to other members of the herd: their message is "follow me". The patch is particularly prominent in pronghorns, shown here.

Jonathan Scott/Planet Earth Pictures

◀ Vast herds of wildebeests (brindled gnu) migrate annually across the plains between Kenya and South Africa, grazing as they go.

▲ *Domestic pigs are direct descendants of the wild boar of Eurasia and North Africa. The young of the wild form have a camouflage pattern of broken white stripes. Feral descendants of European breeds of pigs revert to an appearance very like that of the wild boar shown here but their young are plain colored.*

PIGS, PECCARIES, AND HIPPOS

The pigs of the Old World (eight species) and the unrelated pig-like peccaries (three species) of the New World have evolved in similar ways and are externally similar to one another. The peccaries have more complex stomachs and fused cannon bones in the hindlegs, indicating a longer history of herbivory and of sprinting from predators. Pigs and peccaries exploit, in part, concentrated food sources below the surface of the ground, such as roots and tubers. Pigs may also take carrion, birds' nests, newborn mammals, and small rodents.

Peccaries differ from pigs by living in closely knit groups that jointly defend territories; these are usually much smaller than the home ranges roamed over by pigs. Group size varies, but may number over one hundred in the white-lipped peccaries. As a consequence of putting priority on defense, male and female have both evolved as "fighters": they have sharp canines and are alike in external appearance and size. They also save on reproduction by bearing few and very small young. Unlike pigs, peccaries expose their newborn young to the environment and so do not inhabit cold climates. Peccaries range in head and body length from 80 to 120 centimeters (30 to 48 inches) and weigh 17 to 43 kilograms (38 to 95 pounds).

Females of wild boar shelter their tiny but numerous young against cold, snow, and rain by building domed nests and by "brooding" the young, so wild boar can give birth at any time of the year, including winter. Moreover, wild boar, freed from the need to synchronize birth with warm seasons, may bear several litters a year.

Pigs are a diverse group that evolved in Africa. Here giant species appeared early in the Ice Ages, and even today peculiar species with huge tusks, such as the wart hog or giant forest hog, exist. These large tusks are used in fighting, much like short, curved horns in bovids, as a means of defense to hold and control the opponent's head. An aberrant tropical island species with large, ornate tusks is the babirusa; another aberrant form is the pygmy hog from the eastern Himalayan foothills. Compared with the giant forest hog, which may reach 275 kilograms (610 pounds), the pygmy reaches only 10 kilograms (22 pounds).

Pigs rely for security, in part, on confronting and attacking predators. Otherwise they hide in dense vegetation. Large males tend to be solitary and a hierarchy controls access to estrous females.

Jane Burton/Bruce Coleman Ltd

▲ With a weight of about 230 kilograms (500 pounds), the pygmy hippo is only about a tenth of the mass of the "true" hippopotamus. Inhabiting dense tropical forests, it is solitary and less aquatic than its larger relative.

Short, sharp tusks for attack are matched in dominant wild boar males by thick dermal shields on the shoulders and sides for defense. Small males tend to roam in bachelor groups. Females live in mother–daughter kinship groups within large home ranges; pigs are not known to defend territories, but they do share resources. Home ranges are marked with glandular secretions. The gregarious nature of pigs and the lack of territorial defense make them ideal for domestication. They have been domesticated in many cultures since the early Neolithic. Domestic pigs have reinvaded natural habitats as feral populations in North America, Australia, and New Zealand.

◄ Hippopotamuses spend the day together in water. At sunset, each individual moves out along an established path (marked at intervals by piles of its dung) to a grazing area. They return around sunrise. A hippopotamus is rarely aggressive to humans unless it is provoked or encountered on its marked path.

Jonathan Scott/Planet Earth Pictures

Gunter Ziesler/Bruce Coleman Ltd

▲ The vicuña, a wild camel from the high Andes of South America, is the smallest member of the camel family. Like the related llamas, it has a long neck and legs, but no hump.

A close relative of pigs, the hippopotamus is entirely herbivorous, fermenting grasses in a large, complex stomach. The food habits of the pygmy hippo, which lives in moist, tropical forests, are not known. Like the pigs, hippos do not have cannon bones. Rather, the four weight-bearing toes each have a metatarsal or metacarpal bone. This is a primitive feature. It suggests that pigs, peccaries, and hippos evolved, at an early stage, a way of life that did not require speedy running on hard surfaces. Hippos illustrate this well: they seek security within groups in water, but wander out at night to feed on closely cropped patches of grass. An aquatic existence, or life in high humidity, is

apparently dictated by a skin that offers little protection against dehydration. Grouping may protect the young against crocodiles, but the mortality of young is high.

Hippos use tusks in combat and have a thick dermal armor to protect themselves. Large males place a territory over areas with aquatic groups, but tolerate submissive males. Hippos have a long life and low birth rate.

During warm periods in the Ice Ages hippos were found as far north as England. They also colonized Mediterranean islands, where they shrank greatly in size. These island dwarfs, along with dwarf elephants and dwarf deer, fell victim to

CAMELS

Camels, which evolved in North America, diversified in the Tertiary from tiny gazelle camels to huge giraffe-like browsers. They died out in North America during the Ice Ages, but some had emigrated and survived in South America, Eurasia, and North Africa. The two species of Old World camels, whose ability to withstand dehydration is legendary, are desert-adapted social mammals that bear large, well-developed young after long gestation periods. They also have red blood cells that contain a nucleus, an anomaly among mammals. They are able to feed on dry, thorny, desert vegetation and move long distances when they detect distant rainfall and green pastures. Both the dromedary and Bactrian camels average in height at the hump 210 centimeters (85 inches) and in weight 550 kilograms (1,200 pounds). Like their small South American cousins, the Old World camels are domesticated. The last of the wild camels in the cold deserts of northwest China and Mongolia are greatly endangered.

There are still natural populations of wild South American camels such as the guanaco and vicuña—the latter very vulnerable. The South American camels are much smaller than the Old World species and have no humps. The vicuña, the smallest of the South American species, averages 91 centimeters (36 inches) in shoulder height, and 50 kilograms (110 pounds) in weight. South American camels are actually North American in origin and have been in South America only since the early Pleistocene (about 2 to 3 million years ago). Males tend to defend a territory which is used by a group of bonded females and their young.

All camels have remarkable combat teeth, in which not only the canines but also the first premolars have been formed into caniforme fighting teeth.

▲ *There are about 15 million domesticated camels in the world. A few wild Bactrian (two-humped) camels survive in the Mongolian region and more than 25,000 dromedaries (one-humped camels) thrive in a feral state in the deserts of Australia. These latter are descendants of domesticated animals that were abandoned early this century.*

Lee Lyon/Bruce Coleman Ltd

the first wave of Neolithic settlers about 8,000 years ago. The pygmy hippo of Liberia and Ivory Coast is distinctly less specialized for aquatic life than the hippo. It has relatively longer legs, less webbing between the toes, and less protrusion of the eyes to allow better vision above water. It lives apparently in family groups in moist forests and swamps, and depends on wallows much as the larger and more specialized hippo. The pygmy hippo has an average head and body length of 1.6 meters (64 inches) and weight of 230 kilograms (500 pounds), in contrast to the hippo's average length of 3.4 meters (125 inches) and weight of 2,400 kilograms (5,300 pounds).

▲ The water chevrotain, which inhabits the rainforests of Western Africa, is about the size of a rabbit. Lacking antlers, it defends itself with sharp canine teeth. It often seeks shelter in hollow trees, within which it climbs with agility.

MOUSE DEER AND MUSK DEER

The most primitive artiodactyls are the four species of tropical tragulids, the mouse deer of Asia, and the water chevrotain of Africa. These are the remnants of a family abundant in the early Tertiary. Though superficially similar to small deer, tragulids are in many respects closer to pigs. They have a simpler rumen than the others, are diverse in food habits, and are territorial resource

defenders armed with ever-growing, sharp, fighting canines. The premolars still have conical crowns, much like those of the primitive ungulates from the early Tertiary. Water chevrotains lack cannon bones in the front legs. As in other small tropical resource-defenders, the female may be larger than the male. Water chevrotains average 75 centimeters (30 inches) in head and body length, 35 centimeters (14 inches) in shoulder height, and 10 kilograms (20 pounds) in weight.

Close to tragulids are the three species of musk deer, which can be viewed as "improved and enlarged" cold-adapted tragulids. In fact, the Siberian musk deer penetrates further north than the large wapiti. The musk deer differs enough from tragulids and deer to warrant separate family status. It lives in temperate or cold climates within continental Asia, where it feeds on small bits of highly digestible plant matter, including shoots, lichens, fruit, and soft grasses. It runs like a rabbit in long jumps, propelled by powerful hindlegs. It is also a surprisingly good climber of rocks and large, well-branched trees. Larger than most tragulids, it rivals the roe deer in size: it averages 90 centimeters (36 inches) in head and body length, 60 centimeters (24 inches) in height, and 12 kilograms (26 pounds) in weight.

Only the males carry long combat canines. They are strongly territorial and the apple-sized musk

▼▶ The Indian muntjac (below) is a small deer, weighing about 22 kilograms (48 pounds). Males have very simple antlers and well developed canine tusks. Muntjacs are either solitary or move about in pairs or small family groups. Most of the world's fallow deer (right) now live under some degree of protection in deer parks or reserves. Males may weigh more than 100 kilograms (220 pounds). The prominent Adam's apple is characteristic.

gland, which is situated between the penis opening and the umbilicus, may serve in olfactory communication. The spotted young are very small compared to those of other deer; breeding is seasonal. Much is to be learned yet about the musk deer, whose valuable musk has made it a target for commercial exploitation. It is heavily snared and hunted, which has led to local scarcity or extinction. The musk of the male is used in the folk medicine of the Far East as well as by perfume manufacturers. Experimental projects to keep musk deer in captivity and periodically remove the musk of males are still beset by difficulties. The musk deer is just one of the many species of wildlife whose existence is threatened by the economic rewards placed on their dead bodies.

DEER

The Cervidae or deer family (39 species) consists of the Old World deer, the New World deer, and the aberrant, antlerless water deer of Korea, *Hydropotes inermis*. Deer are characterized by bone antlers that are grown and shed annually. Antlers have reached the monstrous dimensions of a 3.5 meter (12 foot) span in the extinct Irish elk and a weight of just under 50 kilograms (110 pounds). Deer are advanced ruminants that are tied to woody vegetation, though one species, the highly gregarious reindeer, has adapted to tundra. Deer never invaded sub-Saharan Africa, but they reached tropical South America just before the Ice Ages and there diversified into many small-bodied species.

▼ One of the largest of the living deer, weighing up to 450 kilograms (990 pounds), the wapiti occurs naturally in north-western North America. Unusually for an animal living seasonally in snow, its coat is darker in winter than in summer. Unlike its close relative the red deer, which roars in the rutting season, the wapiti "bugles"—a high-pitched whistling sound.

◀▼ *Antlers are very different from horns. Horns are permanent structures made of the same substance as hair or claws. Antlers consist of bone and are shed and regrown each year. While they are growing, antlers have a covering of skin, called "velvet"; when growth is complete, the skin peels off or is rubbed off. There is a broad correlation between the size of a deer and the size and complexity of its antlers. The water deer has no antlers. Small species such as the pudu (below, left)—which is in fact the smallest species of deer, with a shoulder height of 35 to 38 centimeters (14 to 15 inches)—have simple, unbranched antlers. Large species such as the moose (top, left)—the largest species of deer, with a shoulder height of 168 to 230 centimeters (5 feet 7 inches to 7 feet 8 inches)—and reindeer or caribou (below, right) have enormous, many-branched antlers, often with webs of bone between the branches. The reindeer is the only species of deer in which both sexes have antlers.*

WHY ARE ANTLERS SO LARGE?

This question was addressed by Charles Darwin. Quite correctly he saw here a parallel to the showy feathers of pheasants and peacocks. Darwin suspected that these organs somehow evolved by female choice, but he did not resolve how. The mystery surrounding large horns is lifting, but the explanation at first sight borders on the incredible: large horns and antlers are an indirect consequence of the security adaptations of newborns.

In a species adapted for running, the young must soon run as fast and with as much staying power as the mother. The dangerous post-birth period can be shortened by making the young as large and as highly developed at birth as possible, and by supplying it with much or rich milk so that it grows rapidly, and this adaptation has occurred.

A mother following these rules must be very good at saving nutrients and energy from her maintenance costs and body growth towards the growth and feeding of the young. But what about father? If he spares nutrients from growth and maintenance they at once increase horn or antler growth. That, however, should be an advantage, because now the male carries a symbol of his biological success on his head. Females, then, should choose males with big antlers to ensure large neonates for and copious milk production by their daughters. After all, the bigger the antlers, the better

Hans Reinhard/Bruce Coleman Ltd

▲ Shed antler of a roe deer.

▼ Male red deer with antlers in velvet.

the male was at obtaining surplus resources, and the more efficient he was at maintenance. In addition, symmetry of antlers is an excellent indicator of health. We therefore expect that in species with larger horns or antlers the males will be more likely to show off with them in courtship, but not in dominance displays to other males.

These predictions work out: the antler size of the father correlates with the size of the young at birth, and with the percentage of solids in the milk of female descendants. The larger the antlers, the more they are shown to females in courtship, but not to males in display. In small-antlered animals, where female choice is expected to be minor, we expect males to gain access to females by violence. In that case males are better at shunting their surplus energy and nutrients into body growth as body size is vital to victory. That is, if males are small-horned and not territorial, we expect them to grow larger in body size year by year; in large-antlered species we expect males to plateau soon in body size. That also appears to be the case. Not surprisingly, the extinct Irish elk, with its huge antlers, turns out to be the most highly evolved runner among deer. And the reindeer, which has the largest antlers and is the most highly evolved runner among living deer, has highly developed, relatively large young and the richest milk among deer.

Hans Reinhard/Bruce Coleman Ltd

▲ Because they feed upon the leaves of trees, giraffes have a year-round food supply and therefore can breed at any time, unlike other hoofed animals of the African plains which must adjust their breeding to seasonal variation in the abundance of ground vegetation.

GIRAFFES

The family Giraffidae is now reduced to two species: the tropical, forest-adapted okapi, a species similar to primitive giraffes of the mid-Tertiary; and the largest ruminant of all, the giraffe of the open African plains. Giraffe species were more numerous in Tertiary times. They also evolved several lineages of grazing giraffes with ox-size bodies and large "horns", such as the sivatheres. In giraffes, as in bovids, the horns are formed from ossicones, but giraffes grow a covering of hair, not horn, from the skin over the ossicones. The ossicones continue to get a cover of bone throughout life, so that a bull's skull grows increasingly massive with age. In both species of giraffe the skull is used like a club in fighting and the skin is enlarged into a thick dermal armor to counter blows from the mace-like head.

Giraffes also share a unique courtship behavior with bovids, suggesting a common descent. The giraffe is a highly specialized, very successful foliage feeder which, by obtaining green foliage year round, has escaped the limitation experienced by ground-level grazers in the savanna: a seasonal supply of easily digested green vegetation. It can thus reproduce year-round. Giraffes, like many plains ungulates from productive landscapes, give birth to very large, highly developed young. Being freed from the constraints of a seasonal food supply, giraffes can also grow very large. Male giraffes have an average head and body length of 4.2 meters (14 feet), an average height to the horn tips of 5 meters (16 feet), and an average weight of 1,350 kilograms (3,000 pounds). Females average 4.2 meters (14 feet) in height and 870 kilograms (1,900 pounds) in weight.

Giraffes mature relatively slowly, bear few young, but have a long potential life-span. Bulls gain access to cows by means of a dominance hierarchy. Despite their odd body shape, giraffes are good runners.

The okapi, a forest giraffe, is also a foliage feeder, and in its biology appears to share many similarities with its advanced relative from the open plains. It has an average head and body length of about 2 meters (7 feet), an average height of 36 centimeters (15 inches), and an average weight of about 230 kilograms (500 pounds). Today it is highly localized geographically and, though protected, shares the scourge of most species of large mammals: illegal killing for commerce in meat and parts.

PRONGHORNS

The pronghorn is a gazelle-like ruminant, the only species of the many peculiar indigenous North American ruminants to survive to the present day. Like other large North American mammals to survive the great extinction at the end of the Ice Ages, pronghorns are ecological opportunists and have high reproductive rates. These strikingly colored, keen-eyed "antelopes" live gregariously in open plains as speedy, enduring runners; they are very light in build and deposit little fat on their bodies. Their average weight is 60 kilograms (130 pounds); their head and tail length averages 140 centimeters (55 inches); and shoulder height 87 centimeters (34 inches). By forming breeding territories in summer, pronghorns have not only a long gestation period, but also a breeding system surprisingly similar to that of the forest-adapted roe deer.

Pronghorn bucks also share social signals with gazelles. The branched horn sheath regrows annually, but from the tips of the horn cores only. Females may also carry short horns. The horn sheaths are shed annually right after the mating season. Female pronghorns superovulate so that a handful of embryos implant in the uterus. Here the embryos kill one another by growing long outgrowths through the bodies of rival fetuses until only two survive to grow to term. Like other gregarious plains runners, pronghorns have large young. They mature rapidly, but have a short life-span as adults. Twin births are the norm.

▼ *The okapi (below, left) was not known to Europeans until 1900. It is a short-necked giraffe which browses in equatorial rainforest. Its color pattern provides excellent camouflage in its natural environment. The pronghorn (below, right), the only member of the Antilocapridae family, lives in North America. Often referred to as an antelope, it is not really a member of that group. Pronghorns are the fastest-running mammals in North America.*

▲ *Thick, shaggy hair and a densely matted undercoat provide the yak with insulation against the cold of the Himalaya mountains.*

THE BOVID FAMILY

The largest group of artiodactyls, with about 107 species, is the family Bovidae. It includes cattle; goat-antelopes, shrub and musk oxen, true goats and sheep; gazelles and their relatives; the primitive duikers; twisted-horned antelopes, such as the eland and kudu; reed and water bucks; roan and sable antelopes; gnus and hartebeests; the dwarf antelope; plus several odd species such as the four-horned antelope and the blue bull of India.

Bovids have an Old World origin and are currently distributed from hot deserts and tropical forests to the polar deserts of Greenland and the alpine regions of Tibet. No indigenous bovid species are found in South America or Australia. They are characterized by non-deciduous horns that grow from an ossicone on the forehead. This ossicone forms in the skin of the forehead and attaches to the frontal bone. It is largely hollow inside, giving rise to the designation of bovids as hollow-horned ruminants. The diversity of horn shape and size is striking, as is the diversity of ecological adaptations and body sizes.

Bovids form the bulk of the tropical grazers and desert dwellers and they supply most species of important livestock such as cattle, sheep, and goats. Extinctions at the end of the last Ice Age have affected mainly large-bodied species in Africa, Europe, and North America; most of the smaller species survived to the present. Domestication of sheep and goats began early in the Neolithic era. The small wild sheep of Corsica, Sardinia, and Cyprus, as well as the wild goats on Crete, appear to be non-native species derived from early Neolithic transplants by people. On some islands feral goats have gravely damaged indigenous vegetation and associated animals. Species of interest to hunters have also been widely distributed by human hand. Such "exotics" are currently a threat to native wildlife in North America.

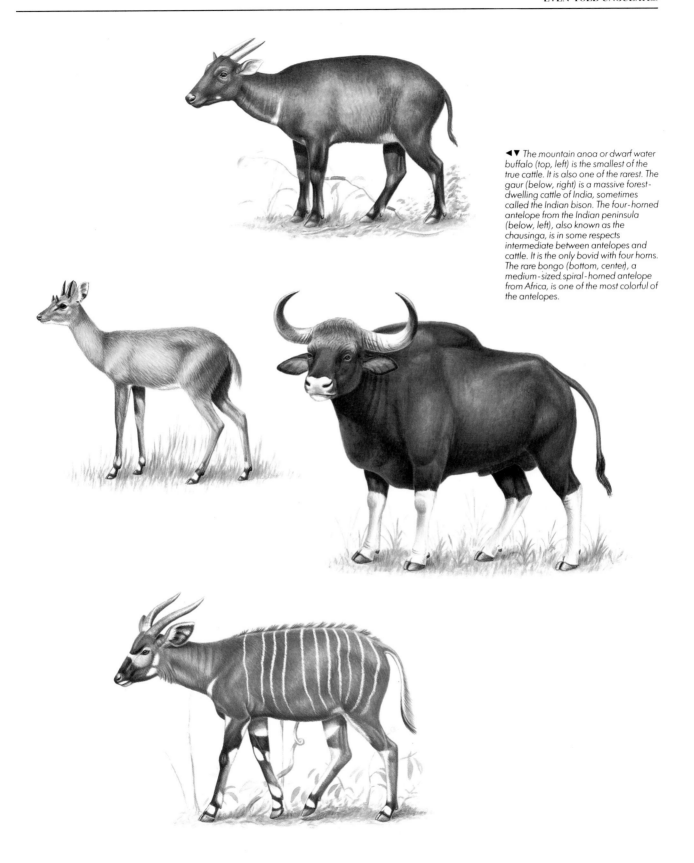

◄▼ The mountain anoa or dwarf water buffalo (top, left) is the smallest of the true cattle. It is also one of the rarest. The gaur (below, right) is a massive forest-dwelling cattle of India, sometimes called the Indian bison. The four-horned antelope from the Indian peninsula (below, left), also known as the chausinga, is in some respects intermediate between antelopes and cattle. It is the only bovid with four horns. The rare bongo (bottom, center), a medium-sized spiral-horned antelope from Africa, is one of the most colorful of the antelopes.

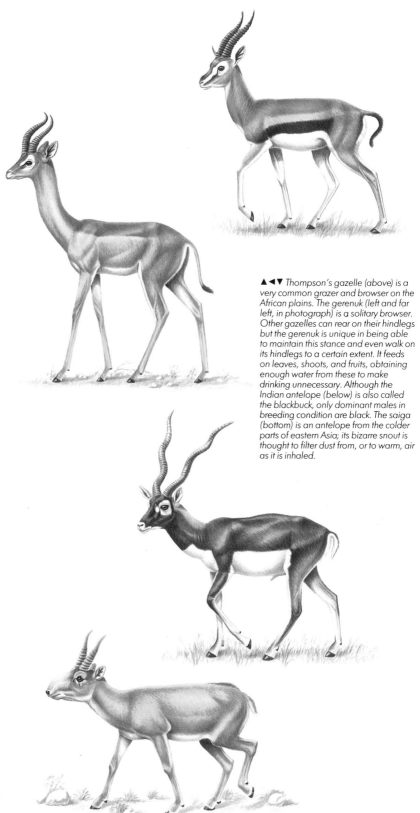

▲ ◄ ▼ *Thompson's gazelle (above) is a very common grazer and browser on the African plains. The gerenuk (left and far left, in photograph) is a solitary browser. Other gazelles can rear on their hindlegs but the gerenuk is unique in being able to maintain this stance and even walk on its hindlegs to a certain extent. It feeds on leaves, shoots, and fruits, obtaining enough water from these to make drinking unnecessary. Although the Indian antelope (below) is also called the blackbuck, only dominant males in breeding condition are black. The saiga (bottom) is an antelope from the colder parts of eastern Asia; its bizarre snout is thought to filter dust from, or to warm, air as it is inhaled.*

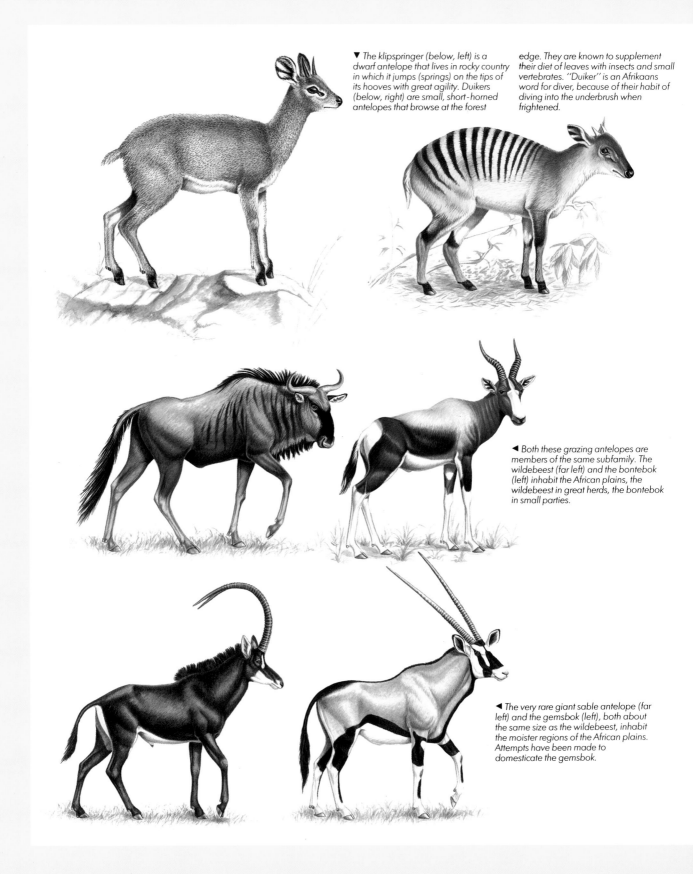

▼ The klipspringer (below, left) is a dwarf antelope that lives in rocky country in which it jumps (springs) on the tips of its hooves with great agility. Duikers (below, right) are small, short-horned antelopes that browse at the forest edge. They are known to supplement their diet of leaves with insects and small vertebrates. "Duiker" is an Afrikaans word for diver, because of their habit of diving into the underbrush when frightened.

◄ Both these grazing antelopes are members of the same subfamily. The wildebeest (far left) and the bontebok (left) inhabit the African plains, the wildebeest in great herds, the bontebok in small parties.

◄ The very rare giant sable antelope (far left) and the gemsbok (left), both about the same size as the wildebeest, inhabit the moister regions of the African plains. Attempts have been made to domesticate the gemsbok.

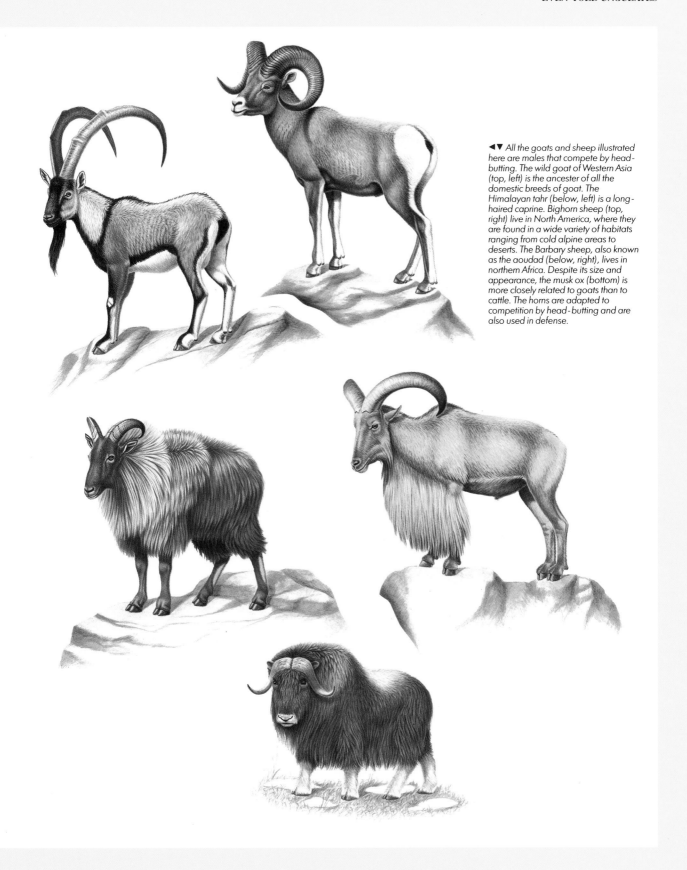

◄▼ *All the goats and sheep illustrated here are males that compete by head-butting. The wild goat of Western Asia (top, left) is the ancester of all the domestic breeds of goat. The Himalayan tahr (below, left) is a long-haired caprine. Bighorn sheep (top, right) live in North America, where they are found in a wide variety of habitats ranging from cold alpine areas to deserts. The Barbary sheep, also known as the aoudad (below, right), lives in northern Africa. Despite its size and appearance, the musk ox (bottom) is more closely related to goats than to cattle. The horns are adapted to competition by head-butting and are also used in defense.*

PANGOLINS

R. D. STONE

Order Pholidota
1 family, 2 genera, 7 species

SIZE
Smallest Long-tailed
pangolin *Phataginus
tetradactyla* head and body
length 30 to 35 centimeters
(12 to 14 inches); tail 50 to
60 centimeters (20 to 24
inches); weight 1.2 to 2
kilograms (2½ to 4½ pounds).
Largest Giant pangolin
Phataginus gigantea head and
body length 80 to 90
centimeters (30 to 36 inches);
tail 65 to 80 centimeters (25
to 30 inches); weight 25 to 35
kilograms (55 to 80 pounds).

CONSERVATION WATCH
Not endangered, but
sanctions apply to trade in
these animals or their
products.

At one time classified together with the anteaters, sloths, and armadillos in the order Edentata, the pangolins, or scaly anteaters, have now been placed in their own distinctive order, the Pholidota. The pangolins are a single family, Manidae, and are represented today by two genera, the Asian pangolins, *Manis,* and the African pangolins, *Phataginus.* Distinguishable from all other Old World mammals by their unique covering of horny body scales, pangolins are truly strange-looking animals that, at a glance, appear more reptilian than mammalian. These body scales, which grow from the thick underlying skin, protect every part of the body except the underside and inner surfaces of the limbs. With a tail that, in some species, is twice the length of the entire body, small limbs, a conical head, and short but powerful limbs, this is one of the most extraordinary-looking mammals of the Old World tropics.

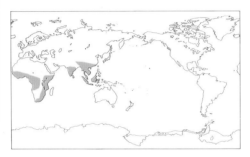

MAMMALS WITH SCALES
Pangolins occur in much of Southeast Asia and in tropical and subtropical parts of Africa. The three Asian species possess external ears, have a scale-clad tail, and also have hairs at the base of the body scales. The four African species do not have external ears; the rear part of the breastbone is very long; and the tail has no scales beneath. Two species, the giant (*Phataginus gigantea*) and the Cape (*P. temmincki*) pangolin, are terrestrial, while the long-tailed (*P. tetradactyla*) and small-scaled (*P. tricuspis*) tree pangolins are, as their names suggest, arboreal. The giant pangolin has a body length of 80 to 90 centimeters (31 to 35 inches) and a tail measuring 65 to 80 centimeters (26 to 31 inches); it weighs 25 to 35 kilograms (56 to 78 pounds). The body of the long-tailed pangolin ranges from 30 to 35 centimeters (12 to 14 inches) in length, the tail, from 50 to 60 centimeters (20 to 24 inches); it weighs 1.2 to 2 kilograms (2½ to 4½

▼ *Pangolins are the only mammals to have a covering of scales. These are made of horn.*

C.B. Frith/Bruce Coleman Ltd

pounds). In all species, the male is much larger than the female—the male Indian pangolin may be as much as 90 percent heavier than the female.

All species have short but powerful limbs, which are used for digging into termite mounds and anthills. The terrestrial species also use their claws for scooping out underground burrows where they conceal themselves during the day. The arboreal species seek refuge in tree hollows, curling themselves up for protection when asleep. The tail, though covered with scales, is fairly mobile, and, in some forms, even prehensile. In addition, the tail has two other important functions: it is very sensitive at the tip and, even when not prehensile, can be hooked like a finger over a solid support. If threatened, a pangolin can also lash its tail at an adversary, using the razor-sharp scales with devastating effect. This action may also be supplemented by spraying an attacking animal with a foul-smelling fluid from the anal glands.

Pangolins feed exclusively on insects—basically termites and ants—catching them with their long, proboscis-like tongue. Housed in a special sheath attached to the pelvis, the 70 centimeter (27 inch) tongue is coiled up in the animal's mouth when at rest. Viscous saliva secreted onto the tongue by special glands in the abdomen traps its prey when the tongue is flicked into the chambers of the mound. Pangolins have no teeth and all food is crushed in the lower section of the stomach leading to the intestines. This region usually contains small pebbles and seems to function by grinding food in the same manner as the gizzard of a bird. When it is feeding, thickened membranes protect the pangolin's eyes, and special muscles seal its nostrils to shield it from the bites of ants.

The degree of sensory development among these different species is directly related to the animal's diet and way of life. Largely nocturnal, pangolins have a poor sense of vision and only average hearing. The sense of smell, however, is exceptionally acute, and this probably plays a major role in communication. Pangolins are largely solitary animals that do not appear to actively defend a fixed territory from neighboring animals of the same species. However, by repeatedly marking selected trees and rocks with secretions from the anal gland, a pangolin notifies neighboring and potentially intruding animals that the area is already occupied.

CONSERVATION

In Africa, large numbers of pangolins are killed each year for meat, while in Asia the Chinese have traditionally attributed medicinal values to the scales and for that reason the animals have always been relentlessly hunted. Pangolins are unlikely to be replaced by more adaptable competitors, but the greatest single threat to the survival of these strange-looking creatures is the destruction of their habitats, which will certainly

have a severe effect on animals with such highly specialized, restricted feeding behavior. Formerly extensive tracts of tropical rainforests are being severely eroded and irreparably damaged each day. Without human action to protect what remains of their habitats, these specialized species are unlikely to survive.

▲ Although the Chinese pangolin climbs with agility it feeds mainly on the ground, digging for termites with its strongly clawed feet.

▼ The tree pangolins of Africa climb with the aid of a very prehensile tail; the single young uses its tail to cling to its mother.

M.P.L. Fogden/Bruce Coleman Ltd

Keith & Liy Laidler/Ardea London

RODENTS

M.J. DELANY

Order Rodentia
29 families, 395 genera, 1,738
species

SIZE
Smallest Pygmy jerboa
Salpingotulus michaelis, head
and body length 36 to 47
millimeters (1⅖ to 1⁹⁄₁₀
inches), tail length 72 to 94
millimeters (2⅘ to 3⁷⁄₁₀
inches).
Largest Capybara
Hydrochoerus hydrochaeris,
head and body length 106 to
134 centimeters (42 to 53
inches), weight 35 to 64
kilograms (77 to 141
pounds).

CONSERVATION WATCH
The following species are
listed as endangered in the
IUCN Red Data Book of
threatened mammals:
Vancouver Island marmot
Marmota vancouverensis,
Delmarva fox squirrel *Sciurus
niger cenereus*, Morro Bay
kangaroo rat *Dipodomys
heermanni morroensis*, salt-
marsh harvest mouse
Reithrodontomys raviventris,
Cabrera's hutia *Capromys
angelcabrerai*, large-eared
hutia *C. auritus*, dwarf hutia
C.nanus, little earth hutia
C. samfelipensis. Many other
species are listed as rare,
vulnerable, or of
indeterminate status.

The success of rodents can hardly be in doubt. They make up just under 40 percent of all mammal species and have a worldwide distribution. They have adapted to habitats from the high Arctic tundra to tropical deserts, forests, and high mountains. Rodents have also reached—often with human help—some of the most isolated oceanic islands. Their relationships with humans are often close and frequently deleterious: historically, they have spread fatal diseases on an enormous scale, and they consume crops and stored products.

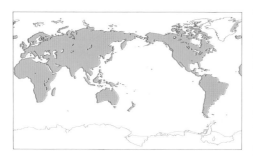

INGREDIENTS OF SUCCESS

To what do these animals owe their great success? Three main factors come to mind. First, although there was major evolutionary diversification during the Eocene (about 57 to 37 million years ago), one family, the Muridae, did not appear until the Pliocene (about 5 million years ago) and is therefore relatively young. With more than 1,000 species, this family is still maximizing its genetic diversity. Throughout its evolution this group has remained relatively unspecialized.

Second, rodents are generally small: most weigh less than 150 grams (5 ounces), though there are many exceptions, with the capybara weighing up to 66 kilograms (145 pounds). Small size affords a good opportunity for the exploitation of a wide range of microhabitats.

Third, many rodents are reproductively prolific: short gestation periods, large litters, and frequent breeding are characteristic, permitting survival under adverse conditions and rapid exploitation under favorable ones.

The combination of evolutionary flexibility, small size, and high production have permitted relatively modest structural and functional modifications to be sufficient to produce the diverse array of contemporary species.

ADAPTATION FROM A SIMPLE PLAN

Rodents have a remarkably uniform mouse-like body plan that has been subject to modification principally of the teeth and digestive system, the limbs, and the tail. All rodents have a single pair of open-rooted, continuously growing sharp incisors to gnaw into food. Behind these is a space or diastema whose presence permits the lips to be

brought together to exclude unwanted particles of gnawed material. There are no canines. At the back of the mouth is a row of molar, and sometimes premolar, teeth, usually used for grinding food. Molars and premolars vary in number from 4 in the one-toothed shrew mouse *Mayermys ellermani* to 24 in the silvery mole rat *Heliophobius argentocinereus;* there are commonly 12 or 16. The chewing surface may consist of ridges or cusps.

Although most rodents are herbivorous and have a relatively large cecum to house the bacterial flora used in cellulose digestion, many have alternative feeding habits and appropriately adapted digestive systems. Examples include omnivory in the house rat *Rattus norvegicus* and the Arabian spiny mouse *Acomys dimidiatus;* insectivory in the speckled harsh-furred rat *Lophuromys flavopunctatus;* and carnivory in South American fish-eating rats of the genera *Anotomys* and *Ichthyomys* and the Australian water rat *Hydromys chrysogaster.*

Most rodents walk on the soles of their feet. In jumpers, such as the jumping mice and jerboas, the hindfoot is greatly elongated and only the toes reach the ground surface. Climbers have opposable big toes (for example, the palm mouse *Vandeleuria oleracea*), hands or feet, or both, broadened to produce a firmer grip (for example, Peter's arboreal forest rat *Thamnomys rutilans*), or sharp claws (tree squirrels). For rapid running, the agouti *Agouti paca* has elongated limbs with only the fingers and toes reaching the ground. Aquatic forms can have long and slightly splayed hindfeet, as in the African swamp rats of the genus *Malacomys,* or webbing, as in the Australian water rat.

The bushy tails of squirrels serve for balance and, in the African ground squirrel *Xerus erythropus,* when recurved over the body, for shade; aquatic forms may have a horizontally flattened tail for swimming (as in the beaver, *Castor* species), a laterally flattened tail to act as a rudder (muskrat *Ondatra zibethicus*), or a longitudinal fringe of hairs running under the tail and adding to its surface area (earless water rat *Crossomys moncktoni*). The rapid runners and jumpers commonly have a long balancing tail, often with a distinct tuft of hairs at its tip, as in jerboas. In the harvest mouse *Micromys minutus* the tail is a grasping organ used in climbing. The tail may also be used for communication. The elaborate fan of Speke's gundi *Pectinator spekei* is for both balance and social display. The smooth-tailed giant rat *Mallomys rothschildi* has a dark tail with a white end which probably has a behavioral function.

The three major groups of rodents are separated on the basis of their jaw musculature. The muscles involved are the deep and lateral branches of the masseter which pull the lower jaw forward (in gnawing) and close the lower on the upper jaw. Their functions vary in different groups, depending on the position of the muscle branches. In the Sciuromorpha or squirrel-like rodents the lateral masseter brings the jaw forward and the deep masseter closes the jaw; in the Caviomorpha or cavy-like rodents the lateral masseter closes the jaw and the deep masseter provides the gnawing movement; and in the Myomorpha or mouse-like rodents both branches of the masseter are involved in gnawing. The last provides the most effective gnawing action. The parts of the skull associated with these different muscle functions are quite distinct for each of the three groups.

SQUIRREL-LIKE RODENTS

Apart from the sciuromorph jaw musculature, the seven families of this group have little in common and probably diverged early in rodent evolution. Squirrels account for almost 74 percent of the species. With a worldwide distribution, except Australia, Polynesia, southern South America, and the Sahara and Arabian deserts, squirrel-like rodents are to be found in most habitats. They range in size from the pocket mouse *Perognathus flavus*, which weighs 10 grams (⅓ ounce) to the beaver *Castor canadensis*, which weighs 66 kilograms (145 pounds).

Mountain beaver and beavers

The mountain beaver *Aplodontia rufa* is the most primitive of living rodents. It weighs 1 to 1.5 kilograms (2 to 3 pounds) and lives in coniferous forests in North America, where it constructs an elaborate burrow inhabited for much of the year by one animal. The mountain beaver has difficulty regulating its temperature and conserving body moisture, which makes hibernation and summer torpor impracticable.

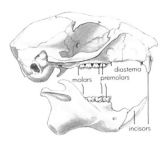

▲ The skull of a rodent can be recognized by a pair of continuously growing, chisel-edged incisors in the upper and lower jaws, and a long gap between these and the grinding teeth.

RODENT SKULL MUSCULATURE

SCIUROMORPHA

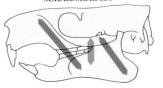

MYOMORPHA

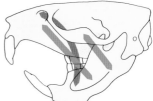

CAVIOMORPHA

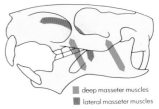

■ deep masseter muscles
■ lateral masseter muscles

▲ The major jaw muscles (masseters) are arranged differently in the three suborders of rodents. In the diagrams above, the orientation of the lateral masseter is shown in red and that of the deep masseter in blue. In the squirrel-like rodents (top), the lateral masseter pulls the lower jaw forward and the deep masseter closes the jaws. In the mouse-like rodents (center), the lateral and deep masseters work together to pull the jaw forward. In the cavy-like rodents (bottom), the lateral masseter closes the jaws and the deep masseter pulls the jaw forward.

◄ Rodents are specialized as gnawing animals, none less so than beavers, which fell substantial trees with their teeth. Pictured is the European beaver.

The North American beaver *Castor canadensis* and European beaver *Castor fiber* are large herbivores, highly adapted to a semi-aquatic life with streamlined body, flattened tail, and webbed feet. They live in closed, hierarchical family units consisting of an adult pair and the offspring of up to several previous years; they have one litter each year. Using their incisor teeth, beavers cut down trees for food and for building dams across streams to impound water and create ponds. They build conical lodges in the ponds with access to the living chamber through an underwater tunnel.

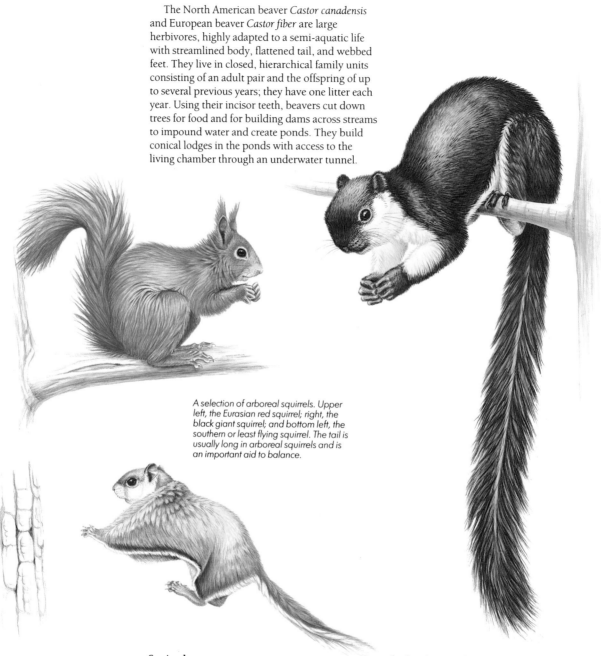

A selection of arboreal squirrels. Upper left, the Eurasian red squirrel; right, the black giant squirrel; and bottom left, the southern or least flying squirrel. The tail is usually long in arboreal squirrels and is an important aid to balance.

Squirrels

Squirrels are a fairly uniform, little-specialized group that have successfully adapted to many habitats from rocky cliffs and semi-arid deserts to temperate and tropical grasslands and forests, in all continents except Australia. They can be divided into diurnal arboreal species, nocturnal arboreal gliders, and diurnal ground-dwellers. They range in size from the African pygmy squirrel of about 10 grams (⅓ ounce) to the Alpine marmot of up to 8 kilograms (17½ pounds).

It is probably the bushy, spectacular tail, active habit, bright eyes, and general alertness that people find endearing about squirrels. With their excellent eyesight and wide vision, tree squirrels can precisely appreciate distance in three dimensions. Movement through the trees is facilitated by sharp claws on the digits and a counterbalancing tail. When the animal jumps the limbs are outstretched, the body flattened, and the tail slightly curved, presenting as broad a surface as possible to the atmosphere. On the ground, movement is by a series of arched leaps.

Flying squirrels glide rather than fly, using a membrane down each side of the body, called the patagium, as a parachute and the tail as a rudder.

Direction is controlled by the legs, tail, and stiffness of the patagium. On landing, flying squirrels brake by flexing the body and tail upward. They are less agile climbers than tree squirrels.

Ground squirrels are widespread and include prairie dogs of the genus *Cynomys* in America, marmots of the genus *Marmota,* and sousliks of the genus *Spermophilus* across the Northern Hemisphere, and the African ground squirrels of the genus *Xerus.* Many inhabit burrows where young are reared, food stored, and protection provided from predators. Many diggers have strong forelimbs and long claws. In some species, for example *Cynomys* and *Marmota,* the tail is much reduced. Many temperate species hibernate.

Most arboreal squirrels are herbivores and feed on fruit, nuts, seeds, shoots, and leaves, though this diet can be supplemented by insects. Terrestrial species often consume grasses and herbs in the immediate vicinity of their burrows. Squirrel feeding habits can affect the environment: red squirrels damage forestry plantations by eating young conifer shoots, and bark stripping by grey squirrels *Sciurus carolinensis* destroys some deciduous trees. On the other hand, failure to recover cached acorns serves as a dispersal mechanism, and the preferential feeding of prairie dogs on herbs favors growth of grama grass.

Social organization is most highly developed in terrestrial species. Prairie dogs live in social units (coteries) consisting of an adult male, several adult females, and associated young. No dominance exists in the coterie, resources being shared within the territory. Considerable aggression is displayed between adjacent coteries, although there is some seasonal relaxation in spring and summer. Numbers of coteries with interconnecting burrow systems form extensive "towns". New territories are established when adult animals move out of the coterie. Cooperative alarms are raised at danger.

Marmots have a similar social structure, but the mountain-dwelling species, experiencing a shorter growing season, have the highest degree of sociality. Although grey squirrel home ranges may overlap, dominance hierarchies occur in both sexes, and low-status animals are forced to emigrate at times of food shortage. This system may be widespread among arboreal species as it ensures optimal use of resources, minimal competition, and social contact with other animals of the same species.

Francois Gohier/Auscape International

▲ Most ground squirrels live in small social groups, sharing a burrow system. The social unit of prairie dogs includes a male, several females, and their young. No adult is dominant over the others.

A selection of ground squirrels. Left, the Columbian ground squirrel; top center, the hoary marmot; right, the least chipmunk. The tail is relatively short in these burrowing rodents.

Pocket gophers and pocket mice

Both of these families have "pockets" or cheek pouches on either side of the mouth for holding and carrying food.

The pocket gophers, which weigh 45 to 400 grams (1½ to 14 ounces), occur in various habitats in North and Central America. They are highly adapted burrowers with tubular bodies and short powerful limbs; the hand may be broad with strong nails, as in the large pocket gopher *Orthogeomys grandis*. The incisors assist the forelimbs in burrowing. Pocket gophers eat surface vegetation and underground roots and tubers; in the process they can be serious agricultural pests.

The pocket mice are granivores of arid to wet habitats of North, Central, and northern South America. Up to six species have been recorded from the same habitat in arid regions, whereas far fewer are found in tropical rainforest. This is attributed to seeds having better chemical protection in rainforest. Most species are quadrupedal and range close to their burrows. The bipedal kangaroo rats of the genus *Dipodomys* and kangaroo mice of the genus *Microdipodops* extend over wider areas, being partially protected from predators by very efficient hearing.

Scaly-tailed squirrels

These African rainforest dwellers are, with the exception of one species (*Zenkerella insignis*), excellent gliders. They have a well-developed patagium and, beneath the basal region of the tail, triangular scales with outwardly projecting points, which possibly serve to help the animals to grip branches. The two pygmy species, *Idiurus macrotis* and *I. zenkeri,* weigh about 20 grams (¾ ounce) and live in colonies of up to a hundred in tree cavities. These and the larger species are herbivorous, eating fruits, nuts, leaves, and flowers. They are all probably mainly nocturnal.

The jumping hare of the African veldt

The spring hare or springhaas *Pedetes capensis* inhabits dry savanna and semi-desert in eastern and southern Africa. A nocturnal grazer, it spends the day underground in a complex burrow system with several entrances. This large rodent, which weighs 3 to 4 kilograms (6½ to 9 pounds) and has an upright posture, is an excellent jumper, having long hindlegs. The female produces one young about three times a year.

MOUSE-LIKE RODENTS

Occurring throughout the world, the mouse-like rodents or Myomorpha, which account for over one-quarter of all mammal species, have successfully adapted to all habitats except snowy wastelands, notably in Antarctica. They are generally small, the largest species being the African maned rat *Lophiomys imhausii*, which weighs 2.5 kilograms (5½ pounds).

Rats and mice

The vast assemblage that constitutes the family Muridae accommodates all but 59 of the myomorph species.

The New and Old World rats and mice belong to two distinct groups, the Hesperomyinae and Murinae, respectively. They evolved separately but have come to occupy comparable niches in their two land masses. Very closely related to the murines are the Australian water rats, Hydromyinae; they range from Australia to New Guinea and the Philippines. All three groups arose from early Oligocene (about 37 million years ago) hamster-like stocks inhabiting North America, Europe, and Asia. The establishment of a land bridge between North and South America during the Pliocene (5 to 2 million years ago) gave the ancestral hesperomyines their greatest opportunity as they entered and exploited new environments in South America. For the murines Southeast Asia was an important center for evolution. They probably evolved from an ancestral stock arriving there during the Miocene (24 to 5 million years ago). They then diversified and spread westward and eastward to establish secondary evolutionary centers in Africa and Australia. Thus in both the Old and New Worlds most speciation has taken place relatively recently, in the Pliocene and

▶ *European harvest mice are tiny seed-eating rodents that climb in grasses and shrubs, aided by a very prehensile tail. American harvest mice, which have a similar way of life, belong to a different subfamily.*

▼ *The spring hare is a very large burrowing rodent which bounds on its hind legs like the much smaller jerboas and hopping mice.*

G. Cubitt/Bruce Coleman Ltd

Pleistocene (5 million to 10,000 years ago).

The absence of insectivores and lagomorphs from much of South America provided a major opportunity for the ancestral invaders to radiate. That this was achieved is illustrated by the fact that only 46 of 359 living Hesperomyinae occur in North America. In Central and South America their diversification includes omnivorous field-mice (*Akodon*); insect-eating, burrowing mice (*Oxymycterus*); arboreal fruit-eating and seed-eating climbing rats (*Tylomys*); and grain-eating pygmy mice (*Scotinomys*). There are also fish-eating rats (*Daptomys, Ichthyomys*), mollusk and fish-eating water mice (*Rheomys*), and riverbank-dwelling water rats (*Kunsia*). Numerous high altitude species occur, including the herbivorous paramo rats (*Thomasomys*), forest mice (*Aepomys*), the diurnal leaf-eared mice (*Phyllotis*), and several rice rats (*Oryzomys*).

Like the New World rats and mice, those of the Old World have much higher species densities in the tropics. For example, in Europe there are only 9 species of murine, whereas Zaire, astride the Equator, has 41 species. Even so, certain temperate genera, such as *Apodemus*, are extremely successful, as is its North American counterpart, *Peromyscus*.

The murines can be divided into two major geographical zones, Africa and Indo-Australia. Both areas are rich in species with only one genus (*Mus*) common to both. They are separated by the arid Saharo-Sindian zone. This points to relatively recent, independent evolution in the two areas.

African murines are generally small, with the African swamp rats, *Malacomys*, among the largest. These animals have a head and body length reaching about 18 centimeters (7 inches) and a weight of 130 grams (4½ ounces). In contrast, in Southeast Asia there are 10 species in the Philippines with head and body lengths over 20 centimeters (8 inches) and 6 species in New Guinea of more than 30 centimeters (12 inches). The largest representative is Cuming's slender-tailed cloud rat *Phloeomys cumingi*, over 40 centimeters long (16 inches), in the Philippines.

Three species of murine, the roof rat *Rattus rattus*, from India–Burma, the brown rat *Rattus norvegicus*, from temperate Asia, and the house mouse *Mus musculus*, from southeastern Soviet Union, have, through their close and successful association with human beings, become widespread throughout the world. In the process, the cost to humans in health, damage, and destruction has been prodigious.

Rats and mice are renowned as prolific breeders: many have gestation periods of 20 to 30 days, repeated breeding, and litters of 3 to 7. Typically, the young are born without fur, with closed eyes, and are relatively inactive. Breeding often takes place within a few months of birth. But not all conform to these characteristics. The spiny mouse *Acomys*, for example, has a gestation period of 36

NHPA/Australasian Nature Transparencies

to 40 days and usually gives birth to one or two large well-haired young. These have open eyes and are mobile within a few hours. The multimammate rat *Praomys natalensis* has an average litter size of 12, a gestation period of 23 days, and reproduction starting at 55 days; it has been estimated that one pair and their progeny could produce over 6,700 animals in 8 months.

Many murines cause havoc to agricultural crops and stored products. Among the most striking examples are the depredations of the introduced house mouse in Australia. Here populations build up to unbelievable levels: one farmer recorded 28,000 dead mice on his veranda after one night's poisoning and 70,000 killed in a wheat yard in an afternoon. In the Pacific basin and Southeast Asia three rats (*Rattus exulans, R. rattus* and *R. norvegicus*) do considerable damage to sugar cane (particularly in Hawaii), coconuts, oil palm, and rice. In Kenya, a 34 percent loss of wheat and 23 percent loss of barley crops has been reported. The species responsible were the multimammate rat, the Nile rat *Arvicanthis niloticus,* and the four-striped grass-mouse *Rhabdomys pumilio.* The first two have also seriously damaged crops of cotton in the Sudan and maize in Tanzania. Extensive

damage by these pests is not necessarily annual, but is often irregular, and, in some cases, several years may elapse between outbreaks.

Voles and lemmings
Voles and lemmings are north temperate rodents belonging to the subfamily Microtinae. They are attractive animals characterized by a blunt snout, small ears, small eyes, and a short tail. The molar teeth have flattened crowns and prisms of dentine surrounded by enamel and are highly adapted for grinding the toughest vegetation, including grasses, sedges, mosses, and herbs. Fruits, roots, and bulbs are also eaten, and seasonal changes in diet reflect availability. Food is often in shortest supply in spring and early summer.

Most of the 121 species are small, weighing less than 100 grams (3½ ounces); an exception is the semi-aquatic muskrat *Ondatra zibethicus* which can weigh 1.4 kilograms (3 pounds). With very few exceptions, they are surface dwellers and burrowers occurring in tundra, grasslands, scrub, and open forest. Some species have periodic cyclical fluctuations in numbers, sometimes attaining extremely high densities. They are active throughout the winter, surviving beneath the snow

RODENTS AND HUMAN HEALTH

▼ *The black, or ship, rat appears to have originated in western Asia and to have reached Europe in the thirteenth century. It has spread to most parts of the world on ships. A closely related species from eastern Asia is almost identical in appearance and habits.*

John Markham/Bruce Coleman Ltd

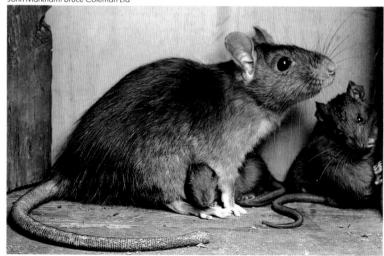

Many species of rodents carry microorganisms harmful to human health. These include bacteria, rickettsias, viruses, fungi, and protozoa. Between them, these microorganisms cause many diseases. Transmission can be through an intermediate organism (vector) or direct.

Some rodent-borne diseases have been known for many hundreds of years to cause devastating mortality in human populations. One such disease

is plague, which is known to be carried by over 200 species of rodent. Transmission to humans is mainly through fleas. A particularly serious threat is posed by those rodents that live near humans: the black rat *Rattus rattus* is notorious. In Central Asia two gerbil species are carriers and in southern Africa a complex transmission route exists through the main reservoir species, a gerbil, and the more commensal multimammate rat. Outbreaks of the plague are often irregular or cyclic, and are currently known to occur in Africa, North and South America, and much of temperate Asia.

Tularemia is another naturally occurring widespread bacterial disease of rodents. In Central Asia and eastern Europe voles are the main reservoirs. Transmission to humans is by various routes, including blood-sucking arthropods (mites, ticks, fleas, etc.), contaminated water, and inhaled dust containing fecal material.

The Lassa fever virus of West Africa occurs in the multimammate rat and is transmitted by human ingestion of its excreta. Similarly, the Bolivian hemorrhagic fever virus is deposited in the urine of the mouse *Calomys callosus,* a common house species in parts of South America.

For a number of diseases, protection can be assured by maintaining high levels of hygiene, including keeping rodents out of dwellings. Fortunately, transmission is automatically limited because the ecological habits of humans and many infective rodent species have little or no overlap.

Jane Burton/Bruce Coleman Ltd

in elaborate tunnel systems.

Microtines have complex social structures. Scent is used to define male and female territories in the breeding season. According to the species, male and female territories may be separate or overlap; females may have separate or shared territories. Males, which can be monogamous or promiscuous according to the species, commonly establish dominance hierarchies. Dispersal is an important facet of microtine behavior. In some species, for example the meadow vole *Microtus pennsylvanicus,* the dispersers have a different genetic makeup, more suitable for establishment in new locations, from the stay-at-homes. An expression of dispersal is the mass migrations of lemmings. In the Norway lemming *Lemmus lemmus,* these usually occur in summer or fall after a population buildup. Migrations start as small modest movements from upland areas. Initially, movement is random. As the descent progresses, more and more animals become incorporated into the migration until finally, with many thousands of animals involved, mass panic sets in. Movement becomes more reckless and undirectional, culminating in mass deaths from drowning in lakes, rivers, and even the sea.

▲ *The European water vole (usually known as the "water rat" in England) is primarily herbivorous, like other voles, but also eats insects, mice, and small birds. Related to rats and mice, voles have smaller eyes, a blunter snout, and a shorter tail.*

▲ *The crested rat, which is the only member of the subfamily Lophiomyinae, inhabits arid, rocky areas of northeastern Africa. In response to threat, it has the very unusual ability to "part" tracts of hair along its body to reveal bold lateral stripes and scent glands.*

More African rats and mice

While most African rats and mice are in the subfamily Murinae, a further 51 species, excluding gerbils, are not included in this group. They are a diverse assemblage, ranging in size from the crested rat *Lophiomys imhausii* at 2.5 kilograms (5½ pounds) to Delany's swamp-mouse *Delanymys brooksi* at 5 to 7 grams (³⁄₁₆ to ¼ ounce). They are widespread throughout the continent south of the Sahara. One group of eleven species, the Nesomyinae, is restricted to Madagascar. Probably derived from a single ancestor, there are now species resembling mice, rats, voles, and rabbits.

The crested rat has long hair, a bushy tail, and the ability to part its hairs along the length of its body to expose its lateral stripes and scent glands. The other large species are the giant rats, *Cricetomys* species, which can weigh 1.2 kilograms (2½ pounds); they have cheek pouches, dull brown or gray dorsal fur, and a bicolored tail. Smaller species include the climbing mice, *Dendromus* species. Their small size—up to 18 grams (⅔ ounce)—and prehensile tails enable them to climb grasses and herbaceous vegetation. The swamp-rats (*Otomys*) inhabit grasslands, particularly in wet areas, where they can be extremely numerous. They are vole-like in appearance, are grass eaters, and, like voles, leave small piles of grass clippings in their runs.

Subterranean rats

There are 22 species of vegetarian, subterranean murids in Eastern Europe, Africa, and Asia. They inhabit burrows constructed with the aid of their protruding front teeth and strong forelimbs. Their food generally consists of roots, rhizomes, and bulbs, and underground storage is common. "Mole hills" are often constructed. The most highly adapted are the eight species of blind mole-rats, the Spalacinae, which have no external ears, no tail, and eyes permanently beneath the skin. The zokors, of the subfamily Myospalacinae, are less extremely adapted and have external ears, tails, and very small eyes. The root rats, of the subfamily Rhizomyinae, include the east African mole rats and the Asian bamboo rats, the latter showing the least underground specialization. Several species are serious agricultural pests.

Desert dwellers

Within this grouping are the hamsters, of the subfamily Cricetinae, the gerbils, of the family Gerbillinae, and the jerboas. The last are in a separate family, the Dipodidae, which has 11 genera and 30 species. With the exception of the hamsters, which also have a limited range in Europe, they are confined to Asia and Africa. They are found from temperate rocky mountains, steppes, and deserts to dry tropical grasslands. The jerboas differ from the hamsters and gerbils in having greatly extended hindlegs, and the main bones of the foot fused to form a single cannon bone. This adaptation enables jerboas to rapidly traverse remarkably long distances: a single animal has been recorded as covering over 12 kilometers (7½ miles) in one night.

Living in places with scant vegetation poses problems of predator avoidance. These problems are partially solved by exceptionally acute hearing through a highly developed middle ear; activity confined to the night; a burrow into which to bolt; excellent vision, often covering a broad field; and a body color that blends with the background. The desert dwellers are primarily granivores and vegetarians, though they often eat insects and other animals too.

In hot arid climates water intake and retention are particular problems. Gerbils and jerboas minimize water loss by sleeping in deep cool burrows during the day; in jerboas the depth of burrow varies seasonally according to the external temperature, the lower levels being favored in hot weather. Plugging the entrance prevents evaporation and maintains a moist internal environment. Water loss is also minimized by highly adapted kidneys that produce concentrated urine and digestive systems that produce dry feces.

▼ Most gerbils live in deserts or very arid country. They sleep in a deep, blocked-off burrow during the day and feed at night, mostly on seeds. They do not need to drink, obtaining sufficient water from their food and conserving this by excreting only small amounts of highly concentrated urine.

Rod & Moira Borland/Bruce Coleman Ltd

Dormice

Dormice occur in Africa south of the Sahara, from western Europe and North Africa to central temperate Asia and Arabia, and in Japan. With their agility, bushy tails, and typically arboreal habits, they share many characteristics with squirrels. The temperate species hibernate from about October to April. Mating takes place soon after emergence, with the first litters appearing in May. As they have no cecum, cellulose presumably plays little part in their diet. Mainly vegetarian, particularly the edible dormouse *Glis glis* and the common dormouse *Muscardinus avellanarius,* they eat fruits, nuts, seeds, and buds; the garden dormouse, *Eliomys* species, tree dormouse *Dryomys nitedula* and African dormouse, *Graphiurus* species, eat more animal material. Dormice can be very vocal: the sounds they make have been variously described as twitters, shrieks, clicks, growls, and whistles. Such behavior probably has territorial, mating, and hierarchical functions. *Selevinia betpakdalensis* is a desert dormouse of eastern Kazakhistan in the Soviet Union. Two species of spiny, oriental dormice are included in the murids.

Jumping mice and birch mice

Jumping mice are widespread through North America and Europe, and birch mice are found in Central Asia. They have poorly developed cheek pouches, very long tails and, in the jumping mice, particularly elongated hindfeet. Hibernation is prolonged, lasting from six to nine months. Body temperature may then fall to a little above freezing and bodily functions operate at minimal levels. The birch mice of the *Sicista* species, with their partly prehensile tails, are good climbers and spend much of their lives in bush and thicket. Jumping mice are ground dwelling.

◀ *The European dormouse.*

UPS AND DOWNS OF RODENT POPULATIONS

Rodent populations are subject to considerable fluctuations in numbers. Among the best-known examples are the cyclical peaks of north temperate voles and lemmings, which occur every three to four years. In the increase phase of the cycle, the breeding season is protracted and breeding activity at a maximum. At the peak, the breeding season is abbreviated and juvenile mortality high. Food becomes short; stress sets in; and the population soon declines. A gradual recuperation of the population follows.

The underlying causes of these fluctuations are not firmly established. Two possibilities have been proposed. One is that growth and decline are attributable to exploitation and exhaustion of the food supply. The other is that behavioral changes within the population through the cycle maximize the species' ability to exploit its environment as well as colonize new areas.

In the semi-arid Sahel region of west Africa, the granivorous gerbil *Taterillus gracilis* underwent population fluctuations from less than 1 per hectare in 1971–73 to 143 per hectare in 1976. In this precarious environment, which has a low annual rainfall, climatic variations greatly influence plant productivity and rodent resources. Below-average rainfall in 1971 and a drought in 1972 reduced numbers. But above-average rainfall in 1972–76 meant more food and a longer breeding season. Another Sahelian species, the Nile rat, is normally present at low densities. It reached the incredible figure of 613 per hectare in 1975.

▲ *The introduced house mouse erupts in plagues periodically in countries like Australia.*

CAVY-LIKE RODENTS

The Caviomorpha derives its name from the South American guinea pig or cavy, which has a number of features in common with other members of the group, particularly in South and Central America. The characteristic jaw muscles, large size, robust body, large head, slender limbs, and short tail are widespread attributes, though only the first is ubiquitous.

The cavy-like rodents differ from squirrel-like and mouse-like rodents in having the largest number of families and the smallest number of species. In fact, four families are represented by only one species each. The reason for the more striking and fundamental diversification of the cavy-like rodents probably lies in their early evolution. It has been suggested that the first caviomorphs evolved from a Northern Hemisphere ancestor during the Eocene (57 to 37 million years ago), and that entry into South America took place across a sea barrier during the late Eocene or early Oligocene (37 to 24 million years ago). After North and South America reconnected in the Pliocene (5 to 2 million years ago) only one caviomorph, a porcupine, moved north and successfully established itself.

New World cavy-like rodents

Porcupines American porcupines (family Erethizontidae) are found from northern Argentina to Alaska and Canada. They have much shorter protective spines (about 4 centimeters, 1½ inches) than their Old World counterparts. Herbivorous, unaggressive, and slow moving, they spend much of their time in trees. Some, such as the prehensile-tailed and tree porcupines (*Coendou* species) are more arboreal than others. Normally, the spines lie smoothly along the back but at times of danger are erected by muscles in the skin. Quills are easily detached and remain embedded in an attacker by means of small barbs along much of their length. In the thin-spined porcupine *Chaetomys subspinosus* of the Brazilian forests the sharp quills are restricted to the head.

Domestic guinea pigs These guinea pigs *Cavia porcellus* take their common name from Guyana (where cavies occur) and from their chunky bodies and short limbs which give them a vaguely pig-like appearance. They are also edible and have been reared for food in South America for several centuries. The domestic guinea pig no longer occurs in the wild, but other cavies (Caviidae) are numerous and widespread in South America, where they typically occupy grassland, open pampas, bushy and rocky areas, and forest edges. All are herbivorous, and some climb in search of leaves. Sexual maturity is attained between 1 and 3 months; gestation lasts 50 to 75 days; litters are small; and the young born highly precocial. These attributes apply generally within the caviomorphs.

▲ Prehensile-tailed porcupines (top) are arboreal leaf-eaters in tropical American forests. New World and Old World porcupines belong to different families and it seems that spines evolved independently in each group. The South American family Echimyidae, generally known as spiny rats (bottom), includes some soft-furred species but most have hair that is stiff and spiny. Although not at all closely related to porcupines, the spiny rats suggest how quills may have evolved from ordinary hairs.

▲ *Long-legged South American rodents. Top, the Patagonian mara or hare, a monogamous species that rears its young in a communal crèche. Center, the plains viscacha. Males of this species are twice the weight of females. Bottom, the golden agouti, a social species that shares a common burrow system.*

food is abundant they may congregate in groups of up to 100, the pair-bonds being maintained.

Capybara The capybara *Hydrochoerus hydrochaeris* is the only living member of an apparently fairly recent family (Hydrochoeridae) whose fossil record only goes back to the Pliocene (5 million years ago). It is the largest living rodent and can weigh up to 66 kilograms (145 pounds), although the fossil members of the family were all considerably bigger than this extant species. Semi-aquatic, its adaptations include nostrils, eyes, and ears all high on the head. They graze at water's edge and are social animals living in groups of up to about 15, which may temporarily aggregate into larger assemblages. Capybaras are widespread east of the Andes from Panama to Argentina. In places they are cropped for food and leather. Another large herbivorous rodent, the coypu *Myocaster coypus* (family Myocastaridae), which weighs up to 10 kilograms (22 pounds), is more aquatic in its habits than the capybara. It burrows into river banks. Introduced into farms in various parts of the world for their fur, escaped coypus have established damaging feral populations.

Chinchillas and viscachas These members of the family Chinchillidae are medium-sized grazers that have dense coats and well-furred tails. The forelimbs are short and the hindlimbs long. With their large ears the mountain viscachas, of the genus *Lagidium,* look like rabbits. The plains viscacha *Lagostomus maximus* has short ears and is sexually dimorphic, the males being twice the size of the females. The smaller chinchillas, which weigh up to 800 grams (1¾ pounds), are prized for their fur.

Agoutis The agoutis (Dasyproctidae), which weigh up to about 2 kilograms (4½ pounds), have a short tail and long slim limbs adapted for rapid running. The forefeet have four toes and the hind three toes; the claws are blunt and almost hoof-like in some species. Diurnal forest-dwellers, they live in excavated burrows and feed on fallen fruits.

Cavies These medium-sized rodents weighing 300 to 1,000 grams (10 to 35 ounces) are nocturnal and often live in colonies. Exceptions are the maras or Patagonian hares *Dolichotis patagonum* and *D. salinicolum*, which are much larger (8 to 9 kilograms, 17 to 20 pounds). With their long limbs they are capable of rapid running. In *D. patagonum*, male and female are monogamous for life. The male defends the female but does not assist in rearing the pups. Up to 15 pairs can deposit their young in a communal den where they remain for up to 4 months. Adults do not enter the den, but the pups leave it periodically for suckling and grazing. Communal suckling does not take place. Generally adults are dispersed in pairs but when

▲ *Capybaras, the largest living rodents.*

Hutias The 13 species of hutias (Capromyidae) are restricted to the West Indies, where they have evolved to fill a number of niches. They are heavy, thickset rodents, the largest being the forest-dwelling hutia *Capromys pilorides,* which weighs up to 7 kilograms (15 pounds). Although arboreal it has, like other hutias, an underground den.

Tuco-tucos, degus, and coruros The tuco-tucos (Ctenomyidae) of the lowland pampas and upland plateaus are the South American equivalent of the pocket gophers. Both are adapted for digging. The fur is dark to pale brown; the limbs are short with large digging claws and fringes of hair on the toes and limbs to assist in moving soil; the tail is short; and the eyes and ears small. Leaves, stems, and roots are eaten. Burrowing is also common in the degus (Oetodontidae), which occur from sea level to 5,000 meters (16,000 feet). One of the more specialized species is the coruros *Spalacopus cyanus* of Chile. Groups of animals occupy a common burrow system where they feed entirely underground, largely on the tubers and stems of a particular lily. When the food supply is exhausted the colony moves to a new site.

Spiny rats The spiny rats (Echimyidae) are an assemblage of 55 species of medium-sized herbivores covered with spiny or semi-spiny bristles. Superficially similar to the murids, they were established in South America several million years before this group entered. Within their range from Nicaragua to Paraguay they can be very common in both forest and grassland. The species with long tails and slender bodies tend to be tree dwellers; the more solid short-tailed species are ground-dwelling burrowers; and the intermediate forms exploit both situations. These animals readily lose their tail owing to structural weakness in the fifth caudal vertebra—doubtless a unique adaptation to escape a predator.

Old World cavy-like rodents
As in their South American counterparts, there is a considerable range of form and size within this group. The smallest is the naked mole-rat of East Africa *Heterocephalus glaber,* which weighs 30 to 40 grams (1 to 1⅓ ounces), and the largest, the crested porcupine *Hystrix,* which weighs up to 24 kilograms (53 pounds). The large African cane rats *Thryonomys* species, which weigh up to 9 kilograms (20 pounds), have a covering of short, harsh, bristly hairs. They prefer thick grassland close to water and the larger species, *T. swinderianus,* is semi-aquatic. They feed mainly on grasses but are renowned as pests of maize, sugar cane, and other agricultural crops. On the other hand, they are very popular as a source of food.

Porcupines Included in the Old World porcupines (Hystricidae) are three species with long tails and a distinct terminal tuft of bristles. In the African (*Atherurus africanus*) and Asian (*A. macrourus*) brush-tailed porcupines, each bristle

consists of a chain of flattened discs, enabling the tail to be rattled, probably as a warning to predators. These relatively small (up to 3.5 kilograms, 7¾ pounds), forest-dwelling species lie up by day in burrows, crevices, and caves. Roots, tubers, and fallen fruit are their main foods. The long-tailed porcupine *Trichys lipura* of Malaya, Sumatra, and Borneo has a terminal brush of hollow bristles. It is small (1.5 to 2 kilograms, 3 to 4½ pounds) and, with a body covered in short bristles, looks like a spiny rat.

The remaining porcupines, *Hystrix* species, have shorter tails and longer quills. They occur from Java and Borneo, through southern Asia to southern Europe, and through most of Africa. They are vegetarian, favoring roots, bulbs, and fruit. They excavate burrows, which they share with their progeny. If they are attacked, the quills of the back are raised and those on the tail rattled as a threat. The porcupine then attacks by running sideways or backwards, trying to drive its spines into its adversary.

Gundis Of the family Ctenodactylidae, gundis are small rodents, weighing up to 180 grams (about 6 ounces), that live in broken, rocky country in northern and northeast Africa. They are agile and dexterous rock climbers, aided by the sharp claws and pads on the soles of their feet. Their soft fur insulates the body from excessive exposure to the sun—an important adaptation for species foraging during the warm times of the day when the temperature ranges from 20° to 30°C (68° to 90°F). At the hottest part of the day they shelter from the sun on shaded rock ledges. Their sharp nails are unsuitable for grooming the soft coat; rows of stiff bristles arising from the two inner toes of the hindfeet are used instead. Gundis live in groups within which are family territories. They are opportunistic herbivores with a limited capacity to concentrate their urine when dry food is eaten.

▲ *The crested porcupine. Like other Old World porcupines, it is terrestrial and has a relatively small and non-prehensile tail. The long spines (quills) are erected and rattled in response to threat, and the porcupine then rushes backward or to the side, attempting to drive them into its attacker.*

John Visser/Bruce Coleman Ltd

▲ Like most mammals that live underground, mole-rats are blind. The legs are short and powerful and the body is cylindrical.

Mole-rats As their common name suggests, mole-rats (Bathyergidae) are rats that have assumed habits similar to moles. The nine species, all occurring in Africa, display considerable adaptation to subterranean life, often paralleling their murid counterparts. They differ from the insectivorous moles in being vegetarian. Mole-rats spend almost their entire lives underground, inhabiting burrow networks of varying complexity. The lips close behind the incisors to prevent soil entering the mouth. The soil is then pushed back by the hindfeet, and when a pile accumulates it is removed by the rat to the outside to appear as a "mole hill" or to be displaced elsewhere in the burrow network. Further adaptations for a subterranean life include cylindrical bodies, short tails, small eyes which no longer serve the function of sight but have instead become sensitive to air currents in the burrow, loss of outer ears, and keen senses of hearing and smell. Mole-rats range from 30 to 60 grams (1 to 2 ounces) in the hairless, naked mole-rat *Heterocephalus glaber,* to about 800 grams (28 ounces) in the Cape dune mole-rat *Bathyergus suillus.* Mole-rats feed on roots, tubers, and the like obtained during the course of their excavation. Their extensive burrowing has been known to cause sagging of railway lines.

THE NAKED MOLE-RAT: A HIGHLY SOCIAL RODENT

The fascinating social organization of the naked mole-rat is unique among mammals. These almost hairless rodents live in colonies of up to 40 (and possibly more) individuals in the drier, warmer savannas of East Africa. They inhabit an elaborate network of foraging tunnels plus a breeding chamber. In the naked mole-rat colony there are clear divisions of labor, presumably to make the community function more efficiently, which have parallels only in the social insects such as bees, wasps, and ants.

Within the colony a single breeding female produces recognizable castes of the more numerous "frequent workers" and the less numerous non-workers. The former are the smallest members of the colony; they dig, transport soil, forage, and carry food and bedding to the communal nest. Digging is performed cooperatively: each animal excavates its soil, then pushes it backwards under the forward-moving animals who have already disposed of their load. Underground food supplies are exposed during the course of this digging.

The larger non-workers spend most of their time in the nest with the breeding female. The females are not sexually active, but they have the potential to be so if the breeding female is removed. The breeding female produces 1 to 4 litters a year with an average of 12 young, though up to 27 have been recorded. The young are suckled only by the breeding female, and at weaning eat food brought to the nest and feces of colony members. After weaning growth is slow. The juveniles join the worker caste, but some eventually grow larger to join the non-worker caste. It is probable that a hierarchy exists within the non-workers, dominated

Jane Burton/Bruce Coleman Ltd

▲ Except for sensitive bristles on the snout, the naked mole-rat has no hair. Its naked condition may have evolved as a means of losing heat from the body.

and controlled by the breeding female. Chemicals (pheromones) secreted by her probably inhibit breeding in non-workers. Chemical influence by the breeding female is also witnessed immediately before she gives birth, when all colony members develop teats and male hormone levels drop. This is probably to prepare the colony to care for the young.

LAGOMORPHS

M. J. DELANY

Order Lagomorpha
2 families, 12 genera,
59 species

SIZE
Smallest Steppe pika *Ochotona pusilla*, head and body length 18 centimeters (7 inches), weight 75 to 210 grams (2½ to 7½ ounces).
Largest European hare *Lepus europaeus*, head and body length 50 to 76 centimeters (20 to 30 inches), tail length 7 to 12 centimeters (2¾ to 4¾ inches), weight 2.5 to 5 kilograms (5½ to 11 pounds).

CONSERVATION WATCH
The following species are listed as endangered in the IUCN Red Data Book of threatened mammals: Assam rabbit *Caprolagus hispidus*, Ryukyu rabbit *Pentalagus furnessi*, volcano rabbit *Romerolagus diazi*. The Sumatran short-eared rabbit *Nesolagus netscheri* is rare.

The order Lagomorpha includes hares, rabbits, and pikas — a familiar group of animals with large ears and eyes set wide on their heads. Lagomorphs have a widespread distribution, either naturally or as a result of introduction by humans, but they are absent from parts of Southeast Asia, including several of the larger islands. Although typically animals of tundras, open grasslands, rocky terrains, and arid steppes, some occur in temperate woodlands, cold northern forests, and tropical forests.

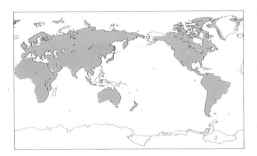

PIKAS

Pikas, which belong to the family Ochotonidae, are the smallest lagomorphs, weighing from 80 to 300 grams (3 to 11 ounces). They have prominent rounded ears and no tail. They occur in steppes and rocky areas, particularly slopes, up to 6,000 meters (20,000 feet) in western North America (two species) and northeastern Asia (twelve species). Pikas retire to rock crevices or burrows according to the terrain. Some species, for example the Afghan pika *Ochotona rufescens* and the daurian pika *O. daurica*, are found in both types of terrain; others, such as the steppe pika *O. pusilla*, are burrowing steppe-dwellers. Vertical segregation of species is common, particularly on high mountains; even on low ground there is little evidence of two species occupying the same habitat.

Pikas are opportunistic vegetarians. They eat any plants available near their homes. During summer and autumn they harvest plants and hoard them under rocks or in burrows for the winter. These hay piles are stored within a territory, which in the case of the northern pika *O. hyperborea* is held by a male and a female throughout the year.

HARES AND RABBITS

Hares and rabbits belong to the family Leporidae. They range in size from the pygmy rabbit *Sylvilagus idahoensis*, which weighs up to 300 grams (11 ounces), to the Arctic hare *Lepus timidus*, which weighs up to 5 kilograms (11 pounds).

In several parts of the world, members of this family are best known for the damage and depredations wrought by the European rabbit *Oryctolagus cuniculus*, which was introduced by the Normans into Britain, and from there by eighteenth and nineteenth century settlers into Australia and New Zealand. Once the European rabbit was established in the wild in these new environments, its numbers increased and livestock pastures and crops were devastated. In Australia, the near extinction of certain marsupials can be attributed to competition with rabbits.

Rabbits, including the European rabbit, can be prolific breeders. One doe can produce up to thirty young in a single breeding season, from January to August. Even though rabbits are born blind, inactive, and with a thin covering of fur, they are capable of breeding at three and a half months.

Some species, however, exist in relatively small numbers. The volcano rabbit *Romerolagus diazi* is restricted to two volcanic sierras close to Mexico City; the Amami rabbit *Pentalagus furnessi* to two of the Japanese Amami islands; and the little-known greater red rockhare *Pronolagus crassicaudatus* is found only in parts of Natal and Cape Province, South Africa.

▲ Although they resemble rodents such as guinea pigs or hamsters, pikas are most closely related to rabbits and hares. Like these, they have two pairs of upper incisors and a slit between the upper lip and the nostrils ("hare lip"). Pikas store leaves and grass during the summer to be eaten, as hay, in winter.

Martin W. Grosnick/Ardea London

John Cancalosi/Auscape International

▶ In addition to their function in hearing, the enormous ears of the black-tailed jack rabbit are radiators that help to control the temperature of this desert animal. When the blood vessels of the ears are engorged with blood, the jack rabbit loses heat; when they are constricted, body heat is conserved.

▼ The "boxing" that occurs between European hares seems to be mainly a matter of a female rejecting an over-amorous male.

The European rabbit can be active at any time of the day or night. It lives in groups with a highly organized social structure maintained, particularly at high densities, by aggressive conflict. Dominant bucks and does occupy the whole territory, while subordinates are restricted to smaller areas within

it. During breeding, dominant does have their nesting chambers in the main warren while the lowest-ranking females have to nest in less-favored locations. Territories are marked by secretions from a gland under the chin and the anal glands, and by urine. These are important components of communication by smell.

In Europe, foxes and stoats are among the main predators of adult rabbits; the young are vulnerable to a wider range of predators including birds of prey, weasels, badgers, and domestic cats. Alarm signals are provided by the rabbit's flashing white tail when it runs, and by thumping with the feet both above and below ground. Both rabbits and hares indulge in tooth-grinding, which may serve as a warning signal. With their long, strong hindlegs, hares and rabbits are adapted for rapid running. Agility, plus their protective coloring, keen sense of smell, and long mobile ears which give them acute hearing, are their main defenses against predators. In contrast to the highly social rabbit, many hares, including the European brown hare, are solitary. They are seen in pairs or small groups only during the breeding season.

Hares and jack rabbits of the *Lepus* genus differ from the remaining nine genera (some of which are confusingly called hares) in several respects. They do not burrow much and have developed improved running ability. Hares typically give birth to small litters of well-furred, mobile young,

Stefan Meyers/Ardea London

Hans Reinhard/Bruce Coleman Ltd

Hans Reinhard/Bruce Coleman Ltd

which have open eyes. Litters are deposited on a form, that is, the hare's bed, shaped by the animal's body. The European brown hare *Lepus capensis* can breed throughout the year, although the main reproductive activity is in April and May. A female can produce three or four litters a year, totalling about seven to ten young — a much lower rate of production than that of the rabbit.

The spectacular "boxing matches" and chasing seen in brown hares in spring has given rise to the phrase "mad as a March hare". This activity can hardly be related to breeding, which starts appreciably earlier than March and continues well into winter. Much of the "boxing" is probably the doe repelling an over-ardent buck. And, as chasing, fighting, and mating go on for much of the year, it appears that the concept of "March madness" has arisen from limited observation!

The Arctic hare is well adapted to living in the far north, as its white winter coat provides excellent concealment. In summer, the coat changes to brown or grey. It is one of a few species of hare that will dig a burrow—into snow or into

earth. It can be used as a bolt-hole for the young, as the adults rarely enter. Another species of the north with a white winter coat is the snowshoe hare *Lepus americanus*. While the Arctic hare extends across much of the Northern Hemisphere, the snowshoe hare is restricted to North America. The value of this animal's pelt resulted in its systematic trapping for over two centuries.

Two species of hare are widespread in Africa, occurring everywhere but in tropical rainforest areas and the most arid parts of the Sahara. They are the Cape hare *Lepus capensis* and the scrub hare *Lepus saxatilis*. Both are nocturnal, lying up in their forms during daytime, and both are grass eaters. There is, however, a fairly sharp distinction in the habitat preferences of the two species. The Cape hare is found in more arid, open terrain and the scrub hare in scrub and woodland adjacent to grassland. That the former is well adapted to arid country is confirmed by its widespread occurrence in Arabia. Here it is found in sand desert, steppe, scrubland, and mountains, wherever there is an adequate food supply.

▲ *The Arctic or alpine hare has a brown or gray coat in summer (above, left), which changes to white in winter (above, right). The snowshoe hare undergoes the same seasonal change, which aids concealment. However, a similar change in the coats of the Arctic fox and snowy owl also reduces the visibility of these predators.*

ELEPHANT SHREWS

R. DAVID STONE

Elephant shrews are a highly distinctive group of animals. They are so named because of the resemblance—vague though it is—between the trunk of an elephant and the extended, highly flexible snouts of these animals. All the elephant shrews are characterized by long, thin legs, large eyes, external ears, long, rat-like tails, and, of course, the proboscis-like snout.

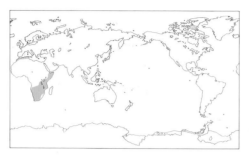

SPECIALIZED BOUNDERS

In comparison with the insectivores, elephant shrews have a well-developed brain. In body form they resemble the oriental tree shrews more closely than any other living group, but they have specialized in a totally different way of life. Almost all of their time is spent on the ground, where their long legs enable them to jump and run at impressive speeds. Alone among insect-eating mammals they are specialized for bounding, which is probably a useful anti-predatory tactic that serves to confuse pursuing predators.

Elephant shrews occupy a considerable variety of habitats, including open plains, savannas, thornbush, and tropical forests. Although many of the species are active during at least part of the day, and are therefore fairly easy to see, their behavior and ecology remain poorly known.

TYPES OF ELEPHANT SHREW

As with the true insectivores, the classification of the elephant shrews has had a checkered history. At one time or another, elephant shrews have been associated with primates, insectivores, tree shrews, and even ungulates. Now, however, elephant shrews have been firmly installed in their own distinct order, Macroscelidea, which has one family, the Macroscelididae. Within this family there are four genera: the *Macroscelides* (one species), *Petrodromus* (one species), *Elephantulus* (nine species), and *Rhynchocyon* (three species).

Body form varies considerably according to species. The aptly named golden-rumped elephant shrew *Rhynchocyon chrysopygus,* for example, is a large species, measuring 76 centimeters (30 inches) from the nose to the tip of the tail and weighing approximately 540 grams (19 ounces). These animals are a striking dark amber color with black legs and feet and a distinct gold-colored rump patch. The short-eared elephant shrew *Macroscelides proboscideus,* in contrast, is a dull gray-brown color and is the smallest species of the group, measuring only 35 centimeters (14 inches) and weighing about 45 grams (1½ ounces). There is no sexual dimorphism within a given species.

TERRITORY AND NESTS

Elephant shrews have a clearly defined monogamous mating system. Animals display a strong degree of fidelity toward their partners—they usually mate for life—and, together, each pair defends a fixed territory against other elephant shrews. When faced with intruders, the resident male usually engages in contest if the intruder is a male, the female of the pair if the intruder is a female. Direct interactions between the male and female of a pair are rare, but contact is

▼ *Elephant shrews are unrelated either to elephants or to shrews: rabbits may be their closest kin. They have large eyes and a long, sensitive and mobile snout with which they probe in leaf litter for invertebrate prey. This spectacled elephant shrew is eating a cricket.*

Jane Burton/Bruce Coleman Limited

achieved through depositing scent at strategic parts of the shared territory.

At least one species, the spectacled elephant shrew *Elephantulus refescens,* builds and maintains an elaborate system of trails traversing the territory. The male of the pair may spend as much as 25 percent of his active time clearing debris such as leaves off the trail system by pushing it away with his forepaws. Grasses and small branches are chewed until they break into pieces small enough to move. The reason for such fastidious housework appears to be that it provides clear escape routes within the home territory if any family member is surprised by predators. The trail system may often extend for more than several hundred meters and remains relatively stable throughout the year. All activities, such as resting, sleeping, grooming, or foraging, occur at favorite parts of the trail system.

The spectacled tree shrew does not construct a complete nest, but some other species create elaborate nests and burrows. Nest sites are dispersed throughout the range, usually at the base of a tree or under a heap of branches. When they sleep, elephant shrews rarely lie on their sides; instead, they keep their limbs beneath the body in a position that allows for a quick escape.

FINDING FOOD
Apart from serving as a communications highway, the trail system is used extensively for foraging. The principal food items are ants and termites, although beetles, worms, leaf litter, invertebrates, and, occasionally, small lizards are captured.

The various species have different techniques for locating prey: species of the genus *Rhynchocyon,* for example, appear to detect potential prey by smell as they forage among leaf litter, continuously poking their long snouts into the debris and thereby seeking out beetles, grasshoppers, and spiders. In contrast, the spectacled elephant shrew prefers to feed on termites and ants, breaking into their tunnel systems by a combination of clawing and biting. Because of the dispersed, unpredictable nature of leaf-litter invertebrates, there is little advantage for an animal feeding on them to defend such a resource. Thus, the golden-rumped elephant shrew spends about three-quarters of its active time looking for food within a territory of almost 2 hectares (5 acres), but the spectacled elephant shrew, which feeds primarily on termites, a localized food resource, spends less time searching for prey and so needs to defend a smaller territory, usually less than half a hectare (1 acre).

BREEDING BEHAVIOR
Elephant shrews breed throughout the year. At a carefully chosen site, females give birth to one or two rather precocial young which are nursed only at 24-hour intervals. Because the infants are so well developed they require minimal parental attention in terms of feeding, warmth, protection, and

▲ *Strictly terrestrial, elephant shrews have long, slender legs that enable them to avoid predators by running fast or bounding like miniature antelopes. Species range in size from that of a mouse to that of a half-grown rabbit.*

Jen & Des Bartlett/Bruce Coleman Limited

grooming, and thus do not serve as a focus for maternal and paternal behavior. This behavior is probably intended to reduce any attention from potential predators. Juveniles remain at the nest site for about two weeks, after which they begin to forage independently, with only infrequent nursing. Unlike insectivores such as moles, shrews, or desmans, weaned elephant shrews are permitted to remain within the adults' territory until they reach sexual maturity, when they must seek a vacant territory.

The most vulnerable stage of the life cycle is probably that experienced by the newly independent subadult animals, as they do not have an established territory. The main predators of elephant shrews are snakes, owls, and small carnivores. In some areas of East Africa, species of the genus *Rhynchocyon* are eaten by local people and are hunted with dogs for this purpose. Few animals live to more than four years of age, when tooth wear becomes a serious handicap to feeding.

CONSERVATION
Most of the extant elephant shrews are under no immediate threat of extinction. However, nearly every species is associated with specific—often narrow-ranged—habitats. For example, the rock elephant shrew *Elephantus mymyurus* is found almost exclusively on rocky outcrops, whereas the largest species, the golden-rumped elephant shrew, frequents the periphery of dry deciduous forest. In such environments, which are subject to rapid change as a result of human interference, these and other species may not be adaptable enough to survive the destruction of their native habitats.

PART TWO
BIRDS

▲ *Birds are characterized by feathers. The intricacy of these structures offers the potential for a wide range of colors and patterns among birds, from muted browns and grays to the gorgeous iridescent hue of this lesser double-collared sunbird.*

INTRODUCING BIRDS

JOSEPH FORSHAW

Scientists divide the animal kingdom into several major groups for classification purposes. By far the largest group is the invertebrates: it contains about 95 percent of the millions of known species of animals, including sponges, mollusks, arthropods, and insects. Groups of vertebrates, or animals with backbones, contain the other 5 percent of known species. They can be divided roughly into fishes, amphibians, reptiles, birds, and mammals. The class Aves—birds—consists of approximately 9,000 species, grouped (in the classification system adopted by this book) into 24 orders. One order, the Passeriformes (known as passerines or songbirds), contains more than half of the known bird species. The remaining orders are known collectively as non-passerines. Birds come in all shapes and sizes, from the ostrich *Struthio camelus,* standing about 2.5 meters (8 ¼ feet) tall, to the bee hummingbird *Mellisuga helenae,* which measures less than 6 centimeters (2 ⅓ inches) from tip of bill to tip of tail and possibly is the smallest bird. There are large birds that cannot fly, small ones that can hover or fly backwards, and just about every conceivable intermediate. But it is the possession of feathers that immediately differentiates birds from other animals. All birds have feathers.

PLUMAGE

Feathers constitute the plumage of a bird. As well as providing mechanical and thermal protection, the plumage assists in streamlining the body, thereby reducing friction during flight, or when moving on the ground or through water.

There are several types of feathers, but the most important are contour feathers, which constitute the ordinary visible plumage, and down feathers, which form a hidden underlayer in most adult birds and usually constitute the plumage of newly-hatched chicks. In most species, feathers other than down grow from definite tracts of skin (the pterylae), and the intervening areas (the apteria) are bare. A bird's plumage is replaced regularly by molting.

Plumage coloration and pattern helps birds to recognize other individuals of their own species and features prominently in their displays. The dazzling array of plumage colors exhibited by birds are the result of the structural and pigmentary colors found in feathers. Structural colors are due to either interference with light (the result being iridescence) or the scattering of light, and the responsible structures are in the barbs and barbules of feathers. Pigmentary colors are widespread in birds and are due to pigments, the most common of which is melanin, the same dark pigment found in human skin, hair, and eyes. Many colors result from a combination of two or more pigmentary colors or from a combination of pigmentary and structural colors.

BILLS AND FEET

The bill and feet are other major elements of a bird's external appearance, and the enormous variation in the structure and shape of these appendages contrasts with a relative uniformity in plumage. The bill is the projecting jaws of a bird encased in horny sheaths, and it must serve the bird in the same way as our hands function for us. It is used in nest building, for preening feathers, or as a weapon in combat, but its primary function is for food-gathering. Different foods require different bill shapes.

Structural modifications are present also in the legs (tarsi) and feet. Birds that spend most of their time on the ground have long legs, well suited to walking and running, whereas tree-dwelling birds have short, stout legs to facilitate climbing and perching. Some birds, especially those occurring in cold climates, have feathered legs and even feathered toes.

No bird has more than four toes; some have three, and the ostrich is unique in having only two. In most species, three toes point forward and the first toe is turned backward, but there are exceptions, ranging from all four pointing forward, as in swifts (family Apodidae), to having two pointing forward and two turned backward, as in parrots (order Psittaciformes). There are structural

► Birds range in size from the tiny bee hummingbird, so small it could fit comfortably into a matchbox, to the ostrich, towering much taller than a human being; the variety of their colors and patterns is kaleidoscopic. But this variety masks a fundamental similarity in body plan, and birds vary far less in size and basic body structure than most other groups of animals.

PARTS OF A BIRD

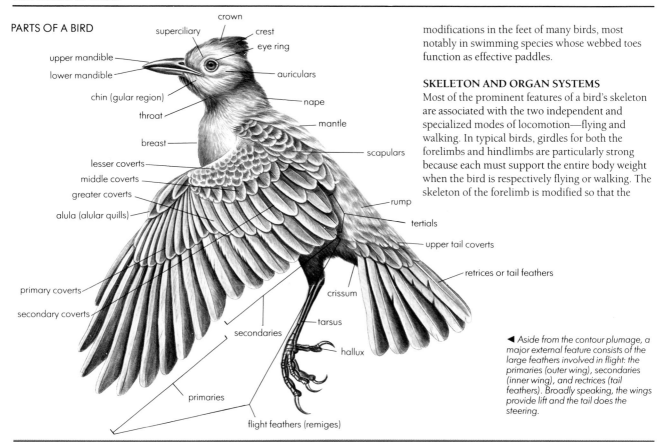

crown
superciliary
crest
eye ring
upper mandible
lower mandible
auriculars
chin (gular region)
throat
nape
mantle
breast
scapulars
lesser coverts
middle coverts
greater coverts
alula (alular quills)
rump
tertials
upper tail coverts
retrices or tail feathers
primary coverts
secondary coverts
crissum
secondaries
tarsus
hallux
primaries
flight feathers (remiges)

modifications in the feet of many birds, most notably in swimming species whose webbed toes function as effective paddles.

SKELETON AND ORGAN SYSTEMS

Most of the prominent features of a bird's skeleton are associated with the two independent and specialized modes of locomotion—flying and walking. In typical birds, girdles for both the forelimbs and hindlimbs are particularly strong because each must support the entire body weight when the bird is respectively flying or walking. The skeleton of the forelimb is modified so that the

◄ *Aside from the contour plumage, a major external feature consists of the large feathers involved in flight: the primaries (outer wing), secondaries (inner wing), and rectrices (tail feathers). Broadly speaking, the wings provide lift and the tail does the steering.*

THE FEATHERS OF A BIRD

The structure of a feather is an extremely successful and very efficient combination of lightness and strength. The feather itself is merely an elaborate, highly specialized product of the outer layer of skin, and it consists almost entirely of a very strong substance called keratin—the same substance that constitutes the hair and fingernails of humans, the fur and claws of mammals, and the scales of reptiles.

In a contour feather there are two clearly discernible components: the central spine or rachis (often referred to as the quill) and the vane or web on each side. The rachis has strong keratin walls and is filled with keratinized cells vaguely resembling the pith of some plants. Similar material fills the laterally branching barbs, which together make up the vane. Along the sides of each barb are hooked barbules, which interlock with barbules on the adjoining barb to give the feather its strength. If the interlocking barbules become unhooked, they can be re-engaged simply by stroking the feather from the base towards the tip, as is done by a bird in the act of preening.

In down feathers only the calamus or rachis base is retained. Parallel barbs project from around the top of the calamus, like spikes atop a coronet, and these give down its characteristic fluffy texture.

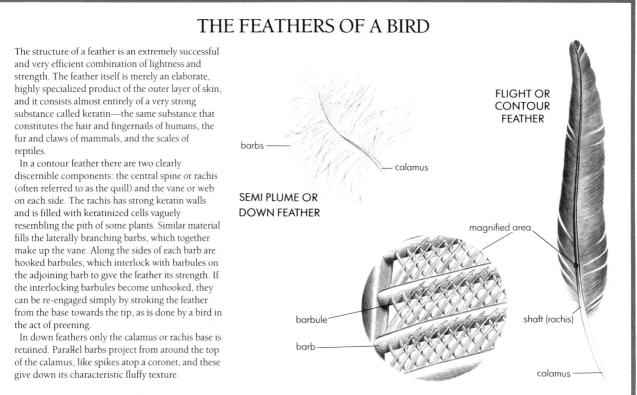

barbs
calamus

SEMI PLUME OR DOWN FEATHER

FLIGHT OR CONTOUR FEATHER

magnified area

barbule
barb

shaft (rachis)

calamus

arm, forearm, and hand support the wing. The hindlimb also is specialized, in that some metatarsals (bones of the foot) are fused and lengthened so that the leg appears to contain an extra segment.

Except for the neck, the vertebral column is comparatively immobile, and many of the vertebrae are fused. The skull is light, with a compact rounded brain-case, which is filled quite tightly by the brain. The bill is attached to the skull quite flexibly, so allowing movement of the upper mandible to increase the gape; this mobility of the upper mandible is especially marked in parrots but is very slight or absent in the ratites and a few other birds. The skeleton of many birds is lightened by extensive pneumatization, a condition in which bones are hollow and contain air-sacs. Even though some bones are hollow, they are structurally very strong, having internal trusslike reinforcements which add to their strength. Lightness and strength are essential for flight.

Dominating the circulatory system is a four-chambered heart, which is different in arrangement from the four-chambered heart of mammals. Thick-walled arteries convey blood at high pressure from the heart to all parts of the body. Thin-walled veins carry the blood through a capillary network within the organs and tissues before returning it to the heart, and fluid exudates pass into the lymphatic vessels. Birds capable of sustained flight have a proportionately large heart in comparison to poor fliers or those that fly only short distances. Like mammals, birds are "warm-blooded". Although this capability of maintaining body temperature above that of the surroundings was acquired independently by the two groups, the physiology of thermoregulation for mammals and birds is remarkably similar, and is a striking example of parallel evolution.

Birds have a highly specialized respiratory system. The small lungs comprise only about two percent of the body volume, but connecting air-sacs are well developed, and in total may be up to 20 percent of the body volume. These air-sacs are located in various parts of the body, and they play an important role in the through passage of air.

The digestive tract of a bird is basically the same as that of other vertebrates and consists of a coiled tube or gut leading from the mouth to the anus. Food passes from the mouth into the gullet and then to the crop, which is a thin-walled distensible pocket of the gullet where food is stored for subsequent digestion or feeding of the young by regurgitation. The crop is well developed in grain-eating and many flesh-eating birds, less developed in other species, and absent altogether in some insect-eating birds. The proventriculus and the ventriculus or gizzard together correspond to the stomach in mammals, and again are well developed in grain-eating species. From the gizzard food passes to the duodenum and intestines, where

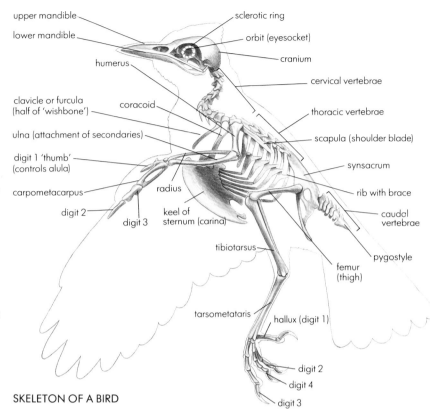

SKELETON OF A BIRD

upper mandible
lower mandible
humerus
clavicle or furcula (half of 'wishbone')
coracoid
ulna (attachment of secondaries)
digit 1 'thumb' (controls alula)
carpometacarpus
digit 2
digit 3
radius
keel of sternum (carina)
sclerotic ring
orbit (eyesocket)
cranium
cervical vertebrae
thoracic vertebrae
scapula (shoulder blade)
synsacrum
rib with brace
caudal vertebrae
pygostyle
tibiotarsus
femur (thigh)
tarsometataris
hallux (digit 1)
digit 2
digit 4
digit 3

digestion is completed before waste is excreted through the anus. Birds have no urinary bladder, so nitrogenous wastes are excreted in the form of urea, a semi-solid paste-like substance, after water has been absorbed in the cloaca. The cloaca is a common opening through which the products of the reproductive, digestive, and excretory systems are passed. Some birds, such as owls, eliminate the indigestible components of their food in the form of pellets regurgitated through the mouth.

THE SENSES

The general structure of the bird eye is similar to that found in all vertebrates. However, the extremely well-developed, efficient eyes possessed by almost all birds, especially the large eyes of some birds of prey and nocturnal species, lead ornithologists to conclude that vision is of the utmost importance.

Attempts to ascertain the level of hearing possessed by birds have met with only partial success, but the few auditory functions that have been measured are almost as sensitive as they are in humans. A higher proficiency was detected in the ability to recognize different sounds repeated so rapidly that to the human ear they become inextricably fused. Birds apparently possess adequate olfactory organs, but in some species the sense of smell seems to be poorly developed and plays little part in their lives.

▲ The skeleton of a bird, although broadly similar to that of other vertebrates, is highly modified to support powered flight. The bones of the hand, for example, are fused to form what is essentially a single digit, which supports the main flight feathers. Other typically avian features include the backward-pointing tabs ("uncinate processes") on each rib, and a prominent keel-like structure on the sternum or breastbone, which serves to anchor the enormous pectoral muscles supplying power to the wings.

Cross section of bone

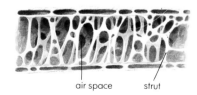

air space   strut

▲ The long bones of the limbs (and in some birds many other bones in the body) are thin-walled and hollow, but intricately braced and strutted inside to provide the maximum strength and rigidity with the least premium in weight.

A brood of baby lesser black-backed gulls hatches at a colony in Wales:

▶ *A baby bird breaks free of the eggshell with the help of the "egg-tooth" on the tip of the bill. Obvious only at hatching, this feature will be shed or re-absorbed within a day or so.*

▶ *Damp and bedraggled, the newly-hatched chick rests briefly after its exertions. Chicks of some species utter piping calls while still within the egg, and it has been shown that these calls hasten the hatching of other chicks in the brood, improving the odds that all will emerge together.*

▶ *Birds like this, whose chicks hatch down-covered, open-eyed, and mobile, are known as precocial. Some can even feed themselves immediately. Even so, it may be several days before the baby bird is fully capable of regulating its own body temperature, and it will require frequent brooding by its parents to shield it from rain and extremes of hot and cold weather.*

HOW BIRDS REPRODUCE

All birds lay eggs, within which development of the embryo subsequently takes place, but of course this form of reproduction is prevalent in other groups of animals. As in the majority of vertebrates, the adult male has testes and the female has ovaries, although in nearly all bird species, only the ovary on the left side is functional. During copulation the cloaca of both sexes is everted so that sperm can transfer from male to female, but in some birds (for example, many ducks and the ratites) part of the cloaca of the male is modified to form a penis. Fertilization of released ova takes place in the upper oviduct, then as each egg passes along the oviduct, layers of albumen are deposited on it. In the wider and greatly distensible uterus, the shell and pigment are added to complete the egg, which finally passes through the vagina and cloaca to be expelled into the nest.

The egg must be maintained at the correct temperature for embryonic development. This is usually brought about through contact with the body of a parent, and the adults of many species develop brood-patches—areas denuded of feathers and richly supplied with blood vessels. The parent bird settles on the nest so that its brood patch or patches cover the egg or eggs; and with regular changeovers of the parents or short breaks away for feeding, the eggs are incubated until the chicks break out of the shell. Incubation periods vary and may be as long as 80 days for a large albatross or as brief as 10 days for some small passerines. Some species do not incubate their own eggs but parasitize other birds by laying their eggs in the host's nest, and the megapodes make use of natural sources of heat such as solar radiation or the fermentation of decaying vegetation to maintain the temperature of eggs buried in a mound or sand.

▶ *This brood of Eurasian robins is almost ready to fledge (that is, become fully feathered and capable of flight). The Eurasian robin is an altricial bird: the chicks at hatching are naked, blind, and helpless, and dependent on their parents for food and care for several weeks before they can leave the nest.*

CLASSIFYING BIRDS

WALTER J. BOCK

The diversity of a group of animals such as birds could not be covered easily in this book without the use of a well-established classification. It enables biologists to summarize a vast amount of biological information in an efficient fashion. For example, categorizing birds as members of the animal kingdom, the phylum Chordata, the subphylum Vertebrata, and the class Aves informs us that each individual bird possesses, among other things, gill slits, a dorsal hollow nerve cord, a vertebral column, a neural crest, feathers, and forelimbs modified into wings.

AVIAN CLASSIFICATION

The class Aves is divided into orders—24 in the classification adopted for this book, although there are additional orders containing only extinct birds. Knowing that a bird is an owl, order Strigiformes, informs us that it possesses a bony arch on its radius bone, among other features. And if a bird is a hornbill, family Bucerotidae, we know that the first two cervical vertebrae are fused into a single unit. Each order is subdivided into families (about 165), each family into genera (slightly over 2,000), and each genus into species (slightly over 9,000).

It must be emphasized that considerable disagreement still exists about the limits of some groups and the relationships between orders and between families included in an order. However, most of the family-level groups covered in this book are well substantiated and will retain their identity even when we learn more about the relationships of birds to each other.

WHAT IS A SPECIES?

A species is defined as a group of actually or potentially interbreeding populations of organisms, which are genetically isolated from other such groups. Species maintain their separation from each other by the possession of intrinsic isolating barriers which prevent the exchange of genetic material among them.

The scientific name given to a species is made up of two words derived from Greek or Latin—for example, *Falco peregrinus*. This name is used by ornithologists around the world, no matter whether they are Dutch taxonomists working in Egypt, Russian researchers in Siberia, or Spanish-speakers in South America. However, the vernacular or common name that people give to this species can vary from place to place; thus the peregrine falcon may also be called the black-cheeked falcon in English, and different names in many other languages, but they all refer to one species, *Falco peregrinus*.

Where populations of a species are separated geographically they may develop slightly different details of size or plumage color, and can be identified as separate races or subspecies. A third word is then added to the scientific name. For example, *Falco peregrinus tundrius* is one of the North American subspecies, and *Falco peregrinus calidus* one of the European subspecies. An adult female should be able to breed successfully with an adult male of a related subspecies where their geographic ranges overlap and the habitat conditions are suitable.

From time to time biologists have to decide whether geographic representatives should be considered as subspecies of a species or as separate species. Such decisions are largely arbitrary, as there are no objective tests for judging the specific status of geographic representatives whose ranges do not meet.

THE BASIS OF CLASSIFICATION

Biological classification, the work of systematists, has two main goals. The first is recognition of the basic units of biological diversity—species and their subunits (mainly subspecies); this then establishes the extent of diversity throughout all organisms, living and extinct. The second is the arrangement of these basic units into a system of increasingly higher-level groups, providing the foundation for summaries of biological knowledge. When two or more species are quite similar in their morphology, physiology, behavior, and ecology, they can be classified in the same genus; all species in a genus are presumed to be descendants of a common ancestor.

Biologists are interested in a single natural classification suitable for all comparative analyses, one which reflects the past evolutionary history of organisms by summarizing the amount of evolutionary change along each lineage and the splitting of lineages. These aspects would be revealed by the number of taxa (groups) at successive levels—such as species, genus, family, order, class—indicating the degree of relationships among species and higher-level taxa.

Several different types of biological classification have been used in recent years. One type is known as phenetic. The only aspect of evolutionary

Morten Strange

correlated with each other, so evolutionary taxonomists must decide which aspect should be given greater importance in particular cases. The order Passeriformes, for example, includes more than half of the known species of living birds, for its members show much less diversity than the rest of the orders of birds combined. In contrast, the order Coliiformes (mousebirds) contains a single genus with only six species but it is quite distinct from other orders. Evolutionary classifications contain the greatest amount of information for biologists, and provide the best all-purpose general reference system. The evolutionary classification used in this book follows closely the system advocated in the 16 volumes of the recently completed Peters' *Check-list of Birds of the World*.

ESTABLISHMENT OF A CLASSIFICATION

The establishment of a biological classification is a two-step process. First is the formulation of hypotheses about the classification of groups—for example, that the kingfishers (family Alcedinidae) and the hornbills (family Bucerotidae) are members of the same group, the order Coraciiformes. This hypothesis is tested scientifically against taxonomic properties of characters, of which the most important is homology. The second step is the formulation of hypotheses about the taxonomic properties of characters—homology, for example—which are tested against empirical observations. This second step, character analysis, is the most important part of classifying organisms into higher-level groups.

The words homology, homologous and homologue come from the Greek *homologos* meaning "agreeing, corresponding". In biological usage a homologue is a feature in two or more organisms that stems phylogenetically from the same feature in the immediate common ancestor of these organisms. Thus the hypothesis that the fused first and second cervical vertebrae are homologous in species of hornbills means that this feature was inherited from such fused vertebrae in the immediate ancestor of all known hornbills. Hypotheses about homologous features are tested by comparing them and ascertaining their similarities. These similarities are assumed to be paternal ones—descriptive of the feature in the immediate common ancestor and remaining unchanged during the evolution to each descendent organism. Thus, homology of the fused cervical vertebrae in hornbills would be tested by establishing similarities in the structure of this feature in diverse hornbill species. This is the only available valid way to test hypotheses about homologies.

Unfortunately this test is frequently not very robust and often does not provide correct answers. Hence further analyses are needed to establish a degree of confidence in each homologue. This involves functional and adaptational analyses of the

▲ *Birds which superficially resemble each other are not necessarily related, while differences in outward appearance often mask a common lineage. Thus this rhinoceros hornbill shares many features with kingfishers despite its different appearance, and both are grouped in the order Coraciiformes.*

change reflected in such a classification is the amount of evolutionary change expressed as the similarity among taxa. A second type is cladistic (from the Greek *klados* meaning "branch"). The only evolutionary aspect reflected in this type of classification is the splitting of phyletic lineages (branching patterns). A third type, evolutionary classification, attempts to summarize both the amount of evolutionary change in phyletic lineages and the splitting of these lineages; however, these two evolutionary aspects are not absolutely postulated homologues, and estimates of the probability that two similar features evolved independently ("convergent evolution").

The possibility of being fooled by independent evolution of unrelated organisms subjected to similar demands from their environment can be reduced by studying various features so that at least some of the features will be independent of similar selective demands. For this reason, the scientist will attempt to use a diversity of features, choosing them carefully to include those associated with different aspects of the life of the organisms. So although both grebes and loons have webbed feet, the presence of different types of webbing suggests that these two groups are not closely related despite being foot-propelled diving birds. Systematists are more confident in the correctness of a classificatory hypothesis if it is supported by a variety of homologous features. But each feature must be carefully and independently analysed.

Ornithologists have used this approach, but with varying success. The major problem appears to be a great emphasis placed on finding new taxonomic characters—biochemical ones during the past two decades—but in the absence of functional/adaptational analyses to establish how much confidence should be given to taxonomic characters in different groups of birds.

Moreover, the tendency has been for each systematist to emphasize the classification supported by the characters he or she used. After all, most of the classic morphological characters used to establish the currently accepted classification have never been properly analysed. And neither have the newly established biochemical and genetic (DNA) characters.

WHAT IS A SEQUENCE?
What is a classification and what is a sequence? Why does the sequence of birds vary in different books? Before addressing these questions, we should consider the difference between a classification and a sequence.

Classifications are systems expressing the evolutionary relationships of taxonomic groups

arranged in an inclusive, non-overlapping hierarchy. In any one taxon, all members descend from a single common ancestor. The taxa in this type of hierarchy are arranged in a series of categories at different levels; for birds, the class Aves is the highest categorical level, followed by orders, families, genera, and species. Intermediate levels such as superfamilies, subfamilies, and tribes are also used.

Sequences are arrangements of the taxa to suit books and other data banks with similar linear restrictions. Rules do exist for the establishment of sequences—such as more primitive groups being listed before more advanced groups—but other equally-valid sequences could be established from the same classification. Broadly-accepted standard sequences are important because they permit greater ease of communication. For this book we have adopted not only the basic classification but also the general sequence used in Peters' *Check-list* because it is the most standard recent sequence for birds of the world.

RULES FOR SCIENTIFIC NAMES
The International Code of Zoological Nomenclature is concerned with names for groups at different levels, from subspecies to families, with the goal of establishing a stable universal set of taxonomic names for all animals.

Priority means that the valid name of any taxon is the oldest name applied to it. If new studies reveal that two species are members of the same genus but were formerly classified in separate genera, they should both be given the generic name that was published first. However, priority is only one of the rules used to achieve stability and universality in zoological nomenclature. Long-term established usage regulated through plenary powers of the International Commission on Zoological Nomenclature is another.

Special care has been used in this volume to use the valid name for each avian taxon, especially those advocated in the recently developed list of names for bird families.

Leo Meier/Weldon Trannies

▲ *Most Coraciiformes, like this Smyrna kingfisher, have large bills, colorful plumage, and nest in cavities. But truly revealing links are often far less obvious. For instance, one structural feature suggesting their common lineage is the fact that the three forward-facing toes are fused together along part of their length, an unusual condition known as syndactyly.*

ORDERS AND FAMILIES OF BIRDS
The following list is based on *Check-list of Birds of the World* by J.L. Peters, E. Mayr, J.C. Greenway Jr., *et. al.,* 1931–1987, 16 volumes, Cambridge, Massachusetts, Museum of Comparative Zoology.

CLASS AVES

ORDER		ORDER		ORDER	
STRUTHIONIFORMES	**RATITES AND TINAMOUS**	**PROCELLARIIFORMES**	**ALBATROSSES AND PETRELS**	**SPHENISCIFORMES**	**PENGUINS**
Struthionidae	Ostrich			Spheniscidae	Penguins
Tinamidae	Tinamous				
Rheidae	Rheas	Diomedeidae	Albatrosses	ORDER	
Casuariidae	Cassowaries	Procellariidae	Shearwaters	**GAVIIFORMES**	**DIVERS**
Dromaiidae	Emu	Hydrobatidae	Storm petrels	Gaviidae	Divers (loons)
Apterygidae	Kiwis	Pelecanoididae	Diving petrels		

Order
PODICIPEDIFORMES — GREBES
Podicipedidae — Grebes

Order
PELECANIFORMES — PELICANS AND THEIR ALLIES
Phaethontidae — Tropicbirds
Pelecanidae — Pelicans
Phalacrocoracidae — Cormorants, anhingas
Sulidae — Gannets, boobies
Fregatidae — Frigatebirds

Order
CICONIIFORMES — HERONS AND THEIR ALLIES
Ardeidae — Herons, bitterns
Scopidae — Hammerhead
Ciconiidae — Storks
Balaenicipitidae — Whale-headed stork
Threskiornithidae — Ibises, spoonbills

Order
PHOENICOPTERIFORMES — FLAMINGOS
Phoenicopteridae — Flamingos

Order
FALCONIFORMES — RAPTORS
Cathartidae — New World vultures*
Accipitridae — Osprey, kites, hawks, eagles, Old World vultures, harriers, buzzards, harpies, and buteonines
Sagittariidae — Secretarybird
Falconidae — Falcons, falconets, caracaras

Order
ANSERIFORMES — WATERFOWL AND SCREAMERS
Anatidae — Geese, swans, ducks
Anhimidae — Screamers

Order
GALLIFORMES — GAMEBIRDS
Megapodiidae — Megapodes (mound-builders)
Cracidae — Chachalacas, guans, curassows
Phasianidae — Turkeys, grouse, etc
Opisthocomidae — Hoatzin

Order
GRUIFORMES — CRANES AND THEIR ALLIES
Mesitornithidae — Mesites
Turnicidae — Hemipode-quails (button quails)
Pedionomidae — Collared hemipode (plains wanderer)
Gruidae — Cranes
Aramidae — Limpkins
Psophiidae — Trumpeters
Rallidae — Rails
Heliornithidae — Finfoots
Rhynochetidae — Kagus
Eurypygidae — Sunbittern
Cariamidae — Seriemas
Otididae — Bustards

Order
CHARADRIIFORMES — WADERS AND SHOREBIRDS
Jacanidae — Jacanas
Rostratulidae — Painted snipe
Dromadidae — Crab plover
Haematopodidae — Oystercatchers
Ibidorhynchidae — Ibisbill
Recurvirostridae — Stilts, avocets
Burhinidae — Stone curlews (thick knees)
Glareolidae — Coursers, pratincoles
Charadriidae — Plovers, dotterels
Scolopacidae — Curlews, sandpipers, snipes

Thinocoridae — Seedsnipes
Chionididae — Sheathbills
Laridae — Gulls, terns, skimmers
Stercorariidae — Skuas, jaegers
Alcidae — Auks

Order
COLUMBIFORMES — PIGEONS AND SANDGROUSE
Pteroclididae — Sandgrouse
Columbidae — Pigeons, doves

Order
PSITTACIFORMES — PARROTS
Psittacidae — Parrots

Order
CUCULIFORMES — TURACOS AND CUCKOOS
Musophagidae — Turacos, louries (plaintain-eaters)
Cuculidae — Cuckoos, etc

Order
STRIGIFORMES — OWLS
Tytonidae — Barn owls, bay owls
Strigidae — Hawk owls (true owls)

Order
CAPRIMULGIFORMES — FROGMOUTHS AND NIGHTJARS
Steatornithidae — Oilbird
Podargidae — Frogmouths
Nyctibiidae — Potoos
Aegothelidae — Owlet nightjars
Caprimulgidae — Nightjars

Order
APODIFORMES — SWIFTS AND HUMMINGBIRDS
Apodidae — Swifts
Hemiprocnidae — Crested swifts
Trochilidae — Hummingbirds

Order
COLIIFORMES — MOUSEBIRDS
Coliidae — Mousebirds

Order
TROGONIFORMES — TROGONS
Trogonidae — Trogons

Order
CORACIIFORMES — KINGFISHERS AND THEIR ALLIES
Alcedinidae — Kingfishers
Todidae — Todies
Momotidae — Motmots
Meropidae — Bee-eaters
Coraciidae — Rollers
Upupidae — Hoopoe
Phoeniculidae — Wood-hoopoes
Bucerotidae — Hornbills

Order
PICIFORMES — WOODPECKERS AND BARBETS
Galbulidae — Jacamars
Bucconidae — Puffbirds
Capitonidae — Barbets
Ramphastidae — Toucans
Indicatoridae — Honeyguides
Picidae — Woodpeckers

Order
PASSERIFORMES
Suborder Eurylaimi — BROADBILLS AND PITTAS
Eurylaimidae — Broadbills
Philepittidae — Sunbirds, asitys
Pittidae — Pittas
Acanthisittidae — New Zealand wrens

Suborder Furnarii — OVENBIRDS AND THEIR ALLIES
Dendrocolaptidae — Woodcreepers
Furnariidae — Ovenbirds
Formicariidae — Antbirds
Rhinocryptidae — Tapaculos

Suborder Tyranni — TYRANT FLYCATCHERS AND THEIR ALLIES
Tyrannidae — Tyrant flycatchers
Pipridae — Manakins
Cotingidae — Cotingas
Oxyruncidae — Sharpbills
Phytotomidae — Plantcutters

Suborder Oscines — SONGBIRDS
Menuridae — Lyrebirds
Atrichornithidae — Scrub-birds
Alaudidae — Larks
Motacillidae — Wagtails, pipits
Hirundinidae — Swallows, martins
Campephagidae — Cuckoo-shrikes, etc
Pycnonotidae — Bulbuls
Irenidae — Leafbirds, ioras, bluebirds
Laniidae — Shrikes
Vangidae — Vangas
Bombycillidae — Waxwings
Hypocoliidae — Hypocolius
Ptilogonatidae — Silky flycatchers
Dulidae — Palmchat
Prunellidae — Accentors, hedge-sparrows
Mimidae — Mockingbirds, etc
Cinclidae — Dippers
Turdidae — Thrushes
Timaliidae — Babblers, etc
Troglodytidae — Wrens
Sylviidae — Old World warblers
Muscicapidae — Old World flycatchers
Maluridae — Fairy-wrens, etc
Acanthizidae — Australian warblers, etc
Ephthianuradae — Australian chats
Orthonychidae — Logrunners, etc
Rhipiduridae — Fantails
Monarchidae — Monarch flycatchers
Petroicidae — Australasian robins
Pachycephalidae — Whistlers, etc
Aegithalidae — Long-tailed tits
Remizidae — Penduline tits
Paridae — True tits, chickadees, titmice
Sittidae — Nuthatches, sitellas, wallcreeper
Certhiidae — Holarctic treecreepers
Rhabdornithidae — Philippine treecreepers
Climacteridae — Australasian treecreepers
Dicaeidae — Flowerpeckers, pardalotes
Nectariniidae — Sunbirds
Zosteropidae — White-eyes
Meliphagidae — Honeyeaters
Vireonidae — Vireos
Emberizidae — Buntings, tanagers
Parulidae — New World wood warblers
Icteridae — Icterids (American blackbirds)
Fringillidae — Finches
Drepanididae — Hawaiian honeycreepers
Estrildidae — Estrildid finches
Ploceidae — Weavers
Passeridae — Old World sparrows
Sturnidae — Starlings, mynahs
Oriolidae — Orioles, figbirds
Dicruridae — Drongos
Callaeidae — New Zealand wattlebirds
Grallinidae — Magpie-larks
Artamidae — Wood swallows
Cracticidae — Bell magpies
Ptilonorhynchidae — Bowerbirds
Paradisaeidae — Birds of paradise
Corvidae — Crows, jays

*Recent genetic and morphological evidence suggests that the New World vultures are more closely related to the storks (family Ciconiidae, order Ciconiiformes).

BIRDS THROUGH THE AGES

JOEL CRACRAFT

The biology of living birds is better known than for any other group of vertebrates. Our understanding of the evolutionary history of birds is not as far advanced, however, in part because fewer fossils have been found. But many spectacular fossil finds in recent years, along with comparative studies of the anatomy and genetic structure of species alive today, are providing valuable evidence with which to reconstruct the pattern of avian evolution, especially the early history of birds.

"FEATHERED DINOSAURS"

Although birds arose more than 150 million years ago, the first 50 million years of avian history has yielded relatively few fossils. Those that do exist have provided important information about the modernization of the avian body plan, especially as it pertains to the evolution of flight.

During the past decade, paleontologists have reassessed the relationships between birds and two-legged theropod dinosaurs, a group that includes perhaps the most famous dinosaur of them all, *Tyrannosaurus rex*. The realization that birds are "feathered dinosaurs" arose as a result of the discovery of new specimens of the oldest known bird, *Archaeopteryx lithographica*. Indeed,

the two beautiful, nearly complete specimens of *Archaeopteryx* now housed in museums in London and Berlin were among the very first fossil birds to be described.

All six specimens of *Archaeopteryx* were found in the Solnhofen limestones of Bavaria, in southern Germany, and had lived during the late Jurassic period, 200 to 145 million years ago. The close relationship of *Archaeopteryx* and other birds to theropod dinosaurs is significant, because it means that the most distinctive characteristic of birds—flight, and all the structural modifications necessary for flight—arose in ancestors that were swift two-legged predatory creatures designed to exploit terrestrial environments rather than trees.

Although *Archaeopteryx* is very clearly related to birds, much of its skeleton recalls that of a theropod. Most of the bones of the forelimb, for example, had none of the special modifications developed in modern birds, such as the fusion of the bones in the wrist and hand facilitating the attachment and fine manipulation of the flight feathers. The bones of the shoulder girdle—the scapula and coracoid—were like those of theropods, and the two clavicles were fused into a U-shaped furcula, or "wishbone". It was long thought that the furcula was a distinctive avian feature, bracing the two shoulder girdles during flight, but we now know that it was present in

◀ About the size of a crow or a pigeon, the earliest known bird, Archaeopteryx lithographica, *lived in Europe over 145 million years ago.*

some advanced theropods, where it also probably functioned as a brace, but for arms used in prey capture not flight.

From the days of its discovery in 1861, *Archaeopteryx* was considered a bird because its body was covered with feathers arranged in feather tracts similar to those of modern birds. Yet despite the presence of feathers, paleontologists have debated whether *Archaeopteryx* was capable of powered flight or was simply a glider. The detailed shapes of the flight feathers suggest the former, for they were asymmetric—the leading edge was very much narrower than the trailing edge. Such a configuration appears to be correlated with a well-developed aerodynamic function and thus suggests that *Archaeopteryx* was capable of strong flight. In addition, the brain of *Archaeopteryx*, especially the portions associated with motor activity and coordination, was enlarged and also indicates a major refinement in locomotor behavior.

THE CRETACEOUS PERIOD

Some of the oldest forms of the Cretaceous period (145 to 65 million years ago) are only a little bit younger than *Archaeopteryx* itself. Despite this, all Cretaceous birds are much more like modern

▶ *A number of birds inhabited the seas of the late Cretaceous period some 70 million years ago, among them several species of Hesperornis, a fish-eating bird that may have been the ancestor of modern grebes. This bird had lost the ability to fly and, like* Archaeopteryx *but unlike all modern birds, it had teeth.*

birds, and less like theropods, and most of them were undoubtedly capable of strong flight. At the same time, their anatomy provides evidence of their dinosaur ancestry. Among the most primitive were two species whose fossils have recently been discovered near Las Hoyas, Spain. Both were the size of small songbirds and had forelimbs and shoulder girdles that were relatively similar to modern birds, whereas the skeletal elements of the pelvis and hindlimbs were more similar to *Archaeopteryx* and theropods. These birds also had a pygostyle—a fusion of the most posterior caudal vertebrae, providing a central point of attachment for the tail feathers—and an ossified sternum, which indicate improvements in the ability to fly.

Several other early Cretaceous birds are interesting because they suggest that anatomically more advanced forms had already evolved by this time. One of these, *Ambiortus*, found in Mongolia, was the size of a small crow. *Ambiortus* possessed a well-developed keel on the sternum and a very modern shoulder girdle, and (unlike the Las Hoyas fossils) the bones of the wrist were fused to form a single element, the carpometacarpus, just as in modern birds. Although nothing is known about the pelvis and hindlimbs of *Ambiortus*, another early Cretaceous bird fills in some of the gaps in our knowledge. This form, named *Gansus* after the Gansu Province of China where it was discovered, is known only from the distal part of the leg, but those bones are of an entirely modern aspect and thus are much more advanced than the Las Hoyas birds. *Gansus* was apparently a small shorebird-like species.

Perhaps the most famous of all Cretaceous fossil birds were *Hesperornis* and *Ichthyornis* from the late Cretaceous of North America. Hesperornithiform birds are also known from the early Cretaceous of England and the late Cretaceous of South America, which suggests they were successful and widespread. *Hesperornis* and *Ichthyornis* are notable because they retained the primitive theropod condition of having teeth on their upper and lower jaws, as did *Archaeopteryx*. Although many other Cretaceous birds probably had teeth, no evidence for this has yet been found. *Hesperornis* and its allies were flightless, foot-propelled divers. As such they had lost the keel on the sternum, greatly reduced the size of wing, and developed non-pneumatic bones. *Ichthyornis*, in contrast, was a strong flier, reminiscent in some respects of modern gulls, but much more primitive and unrelated to them.

THE RISE OF MODERN BIRDS

Modern birds, called the Neornithes ("new birds"), are divisible into two well-defined groups. The first of these is the paleognaths ("ancient jawed", in reference to their somewhat primitive skull) including the tinamous of South and Central America, and large flightless ratite birds such as the ostrich of Africa, the rheas of South America, the emu and cassowaries of Australia–New Guinea, and the kiwis of New Zealand. Paleognaths have had a long history that predates the breakup of Gondwana and the drift of the southern continents. Several fossils from the late Cretaceous period have been found in Mongolia and Europe. In the Paleocene (65 to 57 million years ago) of Europe and North America there existed numerous species of relatively small paleognaths, most of which were capable of powered flight.

The second group of modern birds is the neognaths ("new jawed", in reference to their more advanced skull anatomy). More than 99 percent of all species alive today are neognaths. We know that most of the major groups were represented in the Eocene (57 to 37 million years ago) and Oligocene (37 to 24 million years ago), but few of them have a fossil record from as early as the Cretaceous, primarily because of the scarcity of sediments containing fossils. Paleontologists are probably justified in inferring that many of these groups, or their ancestors, extended well into the Cretaceous.

One of the more primitive lineages of neognaths includes the galliform birds (chickens, pheasants, quails) and the anseriform birds (ducks, geese, swans). Both are first known from fossils in the Eocene, and each apparently had a worldwide distribution by that time. This radiation also included a small group of very large birds, some more than 2 meters (6½ feet) tall: the flightless diatrymas of the North American, European, and Asian Eocene. Although it was assumed for a long time that they were fierce predators, recent studies suggest they were herbivores, which would be consistent with their apparent relationship to anseriforms. It is also possible that another group of flightless giants, the dromornithids or "Mihirung birds" of Australia, are members of this lineage.

Recent evidence suggests that waterbirds such as penguins, loons, grebes, pelicans, cormorants and their allies, and the albatrosses, shearwaters and their allies comprise a distinct evolutionary lineage. Many of these groups have fossil representatives in the Eocene, and so it is reasonable to assume this radiation had its beginnings in the Cretaceous. Penguins are well represented in the fossil record of Australia, South America, and New Zealand, where they live today, and even in the Eocene they were already specialized for "flying" through water.

Few lineages of birds have as interesting a fossil record as the pelecaniforms. The most bizarre group was the pseudodontorns ("false-toothed birds"), a diverse assemblage of albatross-like gliders. All had jaws with teeth-like bony projections, which were presumably used for capturing prey while skimming the ocean's surface. Some pseudodontorns were truly gigantic, with a wingspan of as much as 6 meters (20 feet), far larger than any living albatross.

The most fascinating order of birds from a

paleontological perspective is the Gruiformes, which includes rails and cranes, as well as a number of morphologically distinct families. The gruiforms have perhaps the best fossil record of any order of birds. One lineage now represented only by the trumpeters and seriemas of South America was much more diverse; it included several closely related families that radiated extensively in Europe and North America during the Eocene and Oligocene. Another lineage radiated in South America as the spectacular phorusrhacoid birds. These included a large number of gigantic species, most, if not all, of which were flightless and roamed the savannas and pampas as fierce predators. They survived to the end of the Pliocene (5 to 2 million years ago). Inexplicably, one of the largest members of the group, *Titanis walleri*, has also been discovered from Pliocene fossil records in Florida, the only

▲ *New Zealand was the home of the moas, a group of large to very large wingless birds that probably existed from the Pleistocene to within 200 to 300 years ago. About a dozen species are known; Dinornis maximus, portrayed here, stood an estimated 3 meters (10 feet) tall.*

▲ Fossils of the Pleistocene period (2 million to 10,000 years ago) include many species that are alive today, but many others are now extinct. The large, vulture-like Teratornis merriami lived in western North America, and many specimens have been found at the Rancho La Brea tar pits in California.

certain record of these birds north of Brazil.

Because they inhabit aquatic environments, charadriiform birds—shorebirds, gulls, terns, and their allies—are well represented in the fossil record. At least four extinct families of the late Cretaceous of North America are tentatively placed in this order, thus attesting to the ancient origins of this group. Many contemporary families, including sandpipers, plovers, avocets, puffins, and auks, were present by the Eocene.

Another lineage of aquatic forms includes flamingos, storks, and ibises, and they too have a relatively good fossil record. All three were present and widely distributed by 45-50 million years ago.

The two great groups of raptorial birds include the falconiforms (hawks, eagles, falcons, and vultures) and the owls. Whether they are all closely related has been the subject of intense debate. Although the fossil record does not help solve this problem, it documents that both groups have been in existence for at least 50 million years. Hawks and eagles had diversified on most continents by the Eocene, and other falconiform groups such as ospreys and secretarybirds are nearly as old. The "New World" cathartid vultures, whose evolutionary relationships are still uncertain, also have an extensive fossil record, including a number of forms in the Eocene and Oligocene of Europe

and Asia as well as in North and South America. But unquestionably the most spectacular vulture-like birds—their relationships are also obscure—were the teratorns ("wonder birds") of the Miocene (24 to 5 million years ago) and Pliocene (5 to 2 million years ago) of South America and the Pleistocene (2 million to 10,000 years ago) of North America. *Teratornis merriami* found in the Rancho La Brea tar pits in California was quite large, having a wingspan of perhaps 3.8 meters (12½ feet). But soaring across the Miocene skies of the South American pampas was *Argentavis magnificens,* the largest-known flying bird. *Argentavis* had a wingspan that may have reached 7.5 meters (24½ feet) and a body weight of 80 kilograms (176 pounds). Present evidence based on skull morphology suggests this giant was largely a predator rather than a scavenger like most vultures.

The second group of raptors, the owls, was remarkably diverse: no fewer than three families, now extinct, are known from the Paleocene and Eocene of North America and Europe, and numerous extinct genera of barn owls are known from the same deposits. Owls in the modern family Strigidae are first known from the early Oligocene, and since then have radiated nearly worldwide.

Other orders and families of birds, while not having as extensive paleontological records as the preceding groups, are represented by fossils which indicate they too had their origins more than 50 million years ago. Included among these are nightjars and their allies (Caprimulgiformes), cuckoos (Cuculiformes), parrots (Psittaciformes), swifts (Apodiformes), and kingfishers, rollers, and their allies (Coraciiformes).

The largest order of birds is the Passeriformes, or songbirds. More than 70 families and thousands of species are alive today, but because virtually all of them are small tree-dwelling birds, their fossil record is relatively poor; the fossils that have been found are only tens of thousands of years old at most. Although a small number of species have been described from deposits that are at least 40 million years old, their relationships are not clear. But genetic distances among their most divergent families are much greater than among many non-passerine groups that were present 50-60 million years ago. This suggests that songbirds arose long before that time.

All of the evidence leads ornithologists to infer that most orders and many families of living birds probably had their origins in the Cretaceous, more than 65 million years ago, and then subsequently radiated. It may be difficult to document this conclusion directly until we find more fossils in non-marine sediments of the late Cretaceous. Nevertheless, recent years have seen a remarkable growth in the field of avian paleontology, which is certain to expand our knowledge of avian evolution in the very near future.

HABITATS AND ADAPTATIONS

HUGH A. FORD

The habitat of a bird can be loosely defined as the environment it occupies, particularly the climate and vegetation. Its habitat must provide food, foraging sites, cover from predators and the weather, and nesting sites. Birds have adapted to habitats as diverse as the Arctic tundra, the Sahara Desert, the Amazonian rainforest, and the middle of the oceans. They have carved out niches from the available resources. Most have survived ice ages and periods of great aridity, causing the expansion and contraction of their favored habitats. It is true that some became extinct, but others evolved to take their place. Each habitat has a characteristic array of species, many of which will display morphological adaptations to that particular habitat.

HABITAT REQUIREMENTS

A bird's habitat may be restricted by geographical barriers; for example, numerous families of songbirds such as cotingas, mannakins, antbirds, and woodcreepers are restricted to South and Central America; bowerbirds, fairy-wrens, and lyrebirds are found only in the Australia-New Guinea region. More often, habitat restriction comes about because a species requires a particular resource to be present. This may be food, such as the seeds of spruce trees for the common crossbill *Loxia curvirostra* in Scandinavia, or protea flowers for the Cape sugarbird *Promerops cafer* in South Africa. Often it is a safe nesting site, such as a hole in a living pine tree for the red-cockaded woodpecker *Picoides borealis* in southeastern United States, and termite mounds for the golden-shouldered parrot *Psephotus chrysopterygius* in northern Australia.

Birds may be generalized or specialized in their habitat. The peregrine falcon *Falco peregrinus* and the barn owl *Tyto alba* occupy a wide range of habitats around the world. Kirtland's warbler *Dendroica kirtlandii* is an example of a highly specialized species, living only in jack pine woodlands recovering from fire that burnt through them six to thirteen years previously. Some species occupy different habitats in different parts of their range. The horned or shore lark *Eremophila alpestris* breeds in the high Arctic, the cold deserts of Central Asia, and the mountains of the Balkans, Morocco, and the Middle East. In North America the species is widespread in tundra, mountains, and deserts, as well as fields and grasslands. There is even a small isolated population in the Andes.

In many regions the presence of other animals, particularly predators, parasites, and competitors, may deter birds from an otherwise suitable habitat. Introduced predators such as stoats and rats have eliminated many native birds from the main islands of New Zealand, so the stitchbird *Notiomystis cincta* and the saddleback *Philesturnus carunculatus,* for example, are now restricted to tiny offshore islands. An introduced mosquito which carries avian malaria has forced the endemic honeycreepers of the Hawaiian Islands to retreat to the mountain forests of each island. Evidence for competitive exclusion is more difficult to find, but many species expand their habitat in the absence of a similar species. For example, the horned lark is probably so widespread in North America because there are no other true larks there.

▼ The golden-shouldered parrot is a rare bird that inhabits dry savanna woodlands on Cape York peninsula, Australia. It feeds almost entirely on seeds on the ground, and builds its nest only in termite mounds. Much sought after by aviculturists, its future is threatened by illegal trapping.

G. Longford

Tom & Pam Gardner

▲ *Pardalotes are largely restricted to the eucalypt forests and woodlands of Australia, foraging for insects in the canopy foliage. This is the most widespread species, the striated pardalote. It nests in tree-cavities or in tunnels on the ground.*

▶ *(Opposite page) The seas are vast but suitable nesting islands for seabirds are small and scattered; many species congregate to nest in huge, crowded colonies, like these king penguins on Macquarie Island.*

Many migratory birds occupy quite different habitats in the breeding and non-breeding seasons. Seabirds roam across the oceans but breed on islands and cliffs. Perhaps the most remarkable is the marbled murrelet *Brachyramphus marmoratus* which nests in the crowns of forest trees 50 metres (150 feet) high, inland from the Pacific coast of North America.

ECOLOGICAL NICHES

The ecological niche is the role an animal occupies in its habitat—its relation to food, shelter, and enemies. Generally it is the feeding behavior that tends to determine a species' niche. For example, the great spotted woodpecker *Picoides major* of Eurasia is a forest-dwelling, bark-gleaning and probing insect- and seed-eater; and the eastern meadowlark *Sturnella magna* of North America is a grassland-inhabiting, ground-gleaning insect-eater. Of course these are little more than sketches of the birds' lives. We could add that the woodpecker requires rotten trunks to excavate its nest sites, it sometimes takes nestling birds, and occasionally feeds on the ground. But the ecological niche is usually used as a quick way of pigeonholing a

species' position in its community.

Niches are perhaps best seen when comparing the species living in any one habitat. In a forest, for example, there are seed-eaters, fruit-eaters, numerous insect-eaters, sometimes nectar-eaters, and consumers of vertebrate prey. Some species combine different foods, such as fruit and invertebrates eaten by thrushes in North America and Europe, or insects and nectar by honeyeaters in Australia. Different species taking the same type of food often forage on different levels within a habitat, known as microhabitats; for example, among insect-eaters there are different species feeding on the ground, among foliage, on bark, and capturing prey in the air. Species may take similar foods in the same place but of different sizes.

Scientists studying the ways in which birds share or partition the resources in similar communities around the world have noticed two things. First, there are often similar-looking (but possibly unrelated) birds filling similar niches on different continents; this is known as convergence. Second, there are differing numbers of species present in any given habitat.

SPECIES DIVERSITY

A birdwatcher in spring in a North American, European, or southern Australian forest might expect to see 40 or 50 bird species on a good morning. More could be added in subsequent visits, but an observer would be very pleased if their bit of forest attracted as many as 100 species during the year. However, a small area of rainforest in New Guinea may be the home of more than 200 bird species, and the best parts of Amazonia may support 300 to 500 species. Why are tropical rainforests so rich? It seems that there are several answers. Rainforests hold a greater variety of resources, such as a vast array of fruits and flowers and large insects, and foraging opportunities in vine tangles and palm fronds. Birds such as parrots, fruit-doves, jacamars, motmots, and oropendolas exploit these niches. In South America there is a whole range of species that follows army ants, capturing the insects they displace.

Islands typically have fewer kinds of birds than continents. Generally the smaller the island and the more distant it is from the mainland the fewer species it possesses. The reason that larger islands have more species is because they have more habitats and hence more ecological niches. Also each species that finds a suitable niche can become common enough to be reasonably safe from extinction. Remote oceanic islands may have few species partly because not many birds have reached them. However, recent research on fossil remains on the Pacific islands has revealed that they had many more species when they were settled by Polynesians than when Europeans arrived. So their impoverished wildlife today is partly because the island peoples exterminated many local forms.

sharp-billed or vampire finch — blood

tree finch — insects

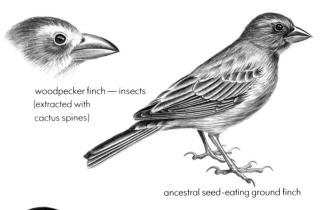

woodpecker finch — insects
(extracted with
cactus spines)

warbler finch — insects

ancestral seed-eating ground finch

large cactus ground
finch — cacti

large ground finch — seeds

tree finch — plant material

dislodge prey from inaccessible positions. A further species, the warbler finch *Certhidea olivacea*, adopted the way of life of a warbler and gleans small insects from foliage and twigs. Perhaps most remarkable of all, one of the finches, the ground or sharp-billed finch *Geospiza difficilis*, now feeds on blood from seabirds. It perches on nesting boobies, pecks at the base of their wing and tail feathers, and laps up the blood that oozes out.

The differences in appearance of the finches on this arid, remote archipelago led Charles Darwin to the belief that species were not immutable, but can change over time—in other words, evolve. The process by which a single species diversifies into a whole array of forms exploiting different niches is known as adaptive radiation. It can be seen on many groups of oceanic islands. The Hawaiian islands provide perhaps the supreme example. Again, a finch arrived some millions of years ago, perhaps from Asia, and found an environment with endless opportunities. The result was about 40 species of honeycreepers or sicklebills (family Drepaninidae), which exploited all sorts of seeds, insects, and nectar over all the main islands. Sadly, the familiar story of extinction under the impact of Polynesian and European settlement followed; direct predation, habitat clearing and burning, and introduced predators and parasites have severely depleted the number of honeycreepers.

Just as adaptive radiation can occur on islands, so too can it occur on continents. It is just more complex on continents, with many colonizations from other continents having taken place.

▲ *Arising from a single ancestral species that probably looked very much like the bird in the center, Darwin's finches have evolved into a number of species that now occupy most islands and most habitats in the Galapagos Islands.*

ADAPTIVE RADIATION

Island birds include some remarkable forms. These have evolved as a result of long periods of isolation from mainland relatives in a strange environment, but also because many other mainland forms are absent. Frequently one kind of bird has diversified in ecology and appearance, so that there are now many, and these occupy different ecological niches.

This can be seen best in Darwin's finches of the Galapagos Islands. There were probably no other land-birds when the first finches arrived from the neighboring South American mainland. But there were insects, fruits, flowers and buds, as well as the seeds that make up the main diet of finches. Some of the finches continued to occupy seed-eating niches, so the larger islands have large-, medium-, and small-billed species (*Geospiza magnirostris, G. fortis* and *G. fuliginosa*), which eat large, medium, and small seeds respectively. One species evolved a longer bill and feeds on cactus flowers and fruits as well as seeds (*G. scandens*). Another evolutionary line led to the tree-dwelling finches (genus *Camarhynchus*), which feed on fruit, buds, seeds, and insects. The woodpecker finch exploits insects from beneath bark and in rotten wood; it does this by using a small stick or cactus spine to locate and

▶ *Just over half of the total body length of the sword-billed hummingbird is made up of the extraordinary bill. An inhabitant of the high Andes in South America, this hummingbird gathers nectar from a number of plant species characterized by very deep trumpet-shaped flowers.*

John S. Dunning/Ardea London Ltd

BILLS, FEET, WINGS, AND TAILS

Birds have characteristic sizes and shapes of their body parts, which have been adapted to suit their feeding behavior and the niches they occupy. Birds' bills display adaptation best. The long dagger-shaped bill of the herons, the huge pouches on the pelican's bill, the hooked bill of predatory birds, and the deep, heavy bills of seed-eating parrots and finches are all adaptations to their food. Differences between bills of related species reflect differences in ecology. The blue tit *Parus caeruleus* in Europe has a deeper bill than the coal tit *P. ater* as it takes insects from broad-leaved trees, while the latter feeds in conifers; a fine bill offers an advantage when probing into pine needles. A similar pattern is shown in North America, with the plain titmouse *P. inornatus* having a deeper bill for deciduous trees and the chestnut-backed chickadee *P. rufescens* having a finer bill for coniferous trees.

Feet, too, show adaptations to a bird's feeding behavior and environment. The talons of raptors for gripping large prey, the webbed feet of ducks for swimming, and the extremely long toes of the jacana for walking on waterlilies are good examples. Ground-feeding birds usually have long legs and toes, whereas tree-creeping birds have very long toes and claws but short legs. Aerial birds such as swallows and swifts have small feet; indeed the scientific name for one group of swifts is *Apus,* meaning "no feet". Legs may also be short in birds in very cold climates, to reduce heat loss; the tundra-dwelling ptarmigans (genus *Lagopus*) have very short, feathered legs and feet.

Mark Newton/Auscape International

▲ *Birds that seek food by wading in shallow water generally have long legs, long necks, and long bills, like this black-winged stilt. Stilts are found in shallow wetlands of all kinds in temperate and tropical regions throughout the world.*

Wings and tails can be important too. Long wings provide economy during flight. The house martin *Delichon urbica* uses far less energy when flying than similar-sized but more terrestrial birds. Short wings give maneuverability, however, and are found in birds living in dense habitats or those that indulge in aggressive aerobatics. A comparison of two species, of which one is a migrant and the other is not, shows that the former invariably has longer wings. Long tails can also provide maneuverability and are shown by most flycatching birds. They can also be important in display, as in pheasants.

Of course, many birds change their diet and even their habitat during the year, so their bills, legs, and wings have to be compromises. Many birds eat seeds and insects, but the former require a deep bill and the latter a fine bill. It seems that birds are adapted morphologically to the time when food is in shortest supply. The chaffinch *Fringilla coelebs* and the great tit *Parus major* feed on insects in the summer but nuts and seeds during winter. They have fairly deep bills, adapted to the time when food is scarce. The long bills of many waders are poorly adapted for taking insects, which they eat in the breeding season, but are ideal for probing estuarine mud in winter.

Natural selection operates on birds' bills, legs, and wings by favoring those with the most efficient size and shape. These birds will survive best and leave the most offspring, who will have inherited their parents' advantageous characteristics. Natural selection takes place in this way over thousands and millions of years, and this evolution allows species to adapt to changing conditions.

BIRD BEHAVIOR

PETER J.B. SLATER

The behavior of birds is governed primarily by their senses of vision and hearing. In this respect they are very like humans, which probably goes part of the way—along with their beautiful plumage and striking songs—toward explaining why birds are so attractive to us. Although the behavior of birds is wonderfully varied, all of them must find their way from one place to another, find food, avoid being eaten by predators, breed with a mate of their own kind, and rear young which are well equipped to achieve all these feats in their turn. The senses play an important role in all of these activities.

EYES AND EARS LIKE HUMANS

The eyesight of most birds is rather like our own, although recent evidence suggests that some of them see very much better in ultraviolet light than we do. Likewise, their hearing range is similar to ours, but some, such as owls, have special abilities that are remarkable. The barn owl can home in on and kill a mouse in a pitch-black room within seconds because its ears are adapted for extremely accurate sound location. Most birds also resemble us in having a poor sense of smell, but again there are some exceptions: the New Zealand kiwis are noted for their ability to smell out prey.

The other main factor to be considered as a background to discussing bird behavior is their movement patterns. Birds are extremely mobile. The power of flight enables them to travel long distances in pursuit of food or mates. Those that breed in higher latitudes need not hibernate or eke out a precarious existence during the short and cold winter days, when many foods are absent or in short supply. They can travel to more equable climates where living is easier.

FINDING FOOD

The ability to find a reasonably constant supply of food is obviously very important to a flying animal that must be light and therefore cannot store large reserves. Whether they are diving for fish, probing at the water's edge for crabs, gleaning insects from the forest canopy, or searching for seeds in the undergrowth, most birds spend a high proportion of their waking hours in pursuit of food.

Not many birds cooperate in the search for food. They may be solitary hunters, like hawks or owls, or they may gather in groups where food is in abundant supply, as do finches or penguins, but they do not often assist each other to catch it.

▼ A river kingfisher emerges triumphantly from the water with its catch. This colorful species lives along wooded rivers and streams, intently scanning the water for prey from a series of favorite low perches along the banks.

Australian Picture Library/ZEFA

Pelicans swimming in formation, and cormorants diving in synchrony, probably help to round up fish shoals, but close cooperation to track down and kill a single prey, like that of wolves or lions, is rare among birds. The social hunting displayed by Harris's hawks, where several birds combine to catch large prey such as a rabbit, may be an example. Members of a mated pair may forage together, and there is evidence that some birds that roost together at night may benefit by gaining information from each other about where best to feed. But, when feeding, most birds look after themselves alone.

Small birds, which do not have extensive food reserves, must feed very actively through most of the daylight hours. Indeed, some very small ones, such as hummingbirds, have so little in reserve that they lose heat overnight and rely on the warmth of the sun to get them going again in the morning.

It is of great benefit to birds to find food as quickly and economically as possible, especially if they feed in the open where their searching may expose them to the danger of being eaten themselves. A good deal of evidence suggests that birds do indeed feed in this way. A thrush that has just found a worm will search in the same area more carefully—a good strategy, given that worms usually occur in groups. A flycatcher, which eats small insects in the treetops early in the day, moves closer to the ground later on when large flies become active, as these yield more energy for the work expended in catching them.

Individual birds may also develop different feeding skills and concentrate on the foods to which they are best adapted. Some gulls may feed on the shore, eating crabs and other invertebrates, while others search for food on agricultural land, and yet others specialize in the spoils to be found

▲ *Talons extended, a barn owl is captured in its final approach for a landing. A cosmopolitan species, this owl has remarkably keen hearing: laboratory experiments have shown that it is quite capable of catching a mouse in utter darkness, guided solely by the tiny sounds made by its prey in breathing and moving about.*

on rubbish dumps. Even with a single type of food, techniques may differ. It is not easy to prize apart the two shells of a mussel and so gain the meat inside, but oystercatchers have various different ways of doing it. Some specialize in "stabbing", inserting their bill between the valves and cutting the muscle that holds them together. The favored technique for others is "hammering", by which they break their way in through the shell. Some gulls and crows have yet another method: they fly up into the air and drop the mussel repeatedly until it breaks. They may even choose hard surfaces on which to do this, so that it is more likely to be successful.

Dropping shells onto a hard surface is just one stage removed from the use of tools. Song thrushes use special stones, their "anvils", on which they repeatedly smash snail shells until they break. Egyptian vultures take the opposite approach to break the very thick-shelled eggs of ostriches, gaining access by hitting them with a heavy stone thrown from their bill. Even more subtle is the behavior of the woodpecker finch from the Galapagos Islands. It uses a cactus spine held in the bill to extract grubs from holes in trees.

Some birds have overcome the lack of a constant food supply by storing it. A marsh tit that finds a rich supply of seeds will hide them one at a time in the surrounding area, remembering their locations and returning to eat them one at a time during the next couple of days. Other birds, such as the acorn woodpecker, use food storage as a longer-term strategy to tide them over the months when the nuts that they eat are scarce.

AVOIDING BEING EATEN

As well as feeding, birds must avoid being fed upon. A lone finch foraging in the open is easy prey to a cat or hawk. It is probably largely for this reason that many small birds feed in flocks where they can benefit from the warning provided by more pairs of eyes. In a large group one of them is bound to have its head up, looking around for danger. The first to spot a predator often produces an alarm call and so warns the others. Each bird may be able to spend more time feeding and less looking out for predators simply because of the safety in numbers. An ostrich, for example, feeding on its own, will raise its head and look around more often than when it is in a group.

Solitary birds have several ways of avoiding being preyed upon. The snipe, a secretive wading bird, sits tight until one is almost upon it, and then darts into the sky with a zig-zag flight that would be very hard for a predator to follow. Its plumage, streaked in various shades of brown, matches the long grass of the marshes where it lives, and like many cryptically colored birds its main defense comes from the difficulty predators have in detecting it. The burrowing owl in North America has an even better trick. It lives down the burrows of ground squirrels, and if one of these should chance upon it, it has a call that closely resembles that of a rattlesnake. The squirrel does not stay around to find out who produced the call!

COURTING AND MATING

The breeding behavior of birds is wonderfully diverse. Most birds are monogamous, and the male often defends a territory in which sufficient food may be found for the pair and their young. In songbirds the male may sing to attract a mate and to keep rivals out of his territory; he will threaten male intruders and court female ones with displays that often show off brightly colored parts of his plumage. But this general picture hides a wealth of variety. For example, in some species, such as phalaropes, which are small waders nesting high in the Northern Hemisphere, it is the female that courts the male. In some species that are colonial, such as gannets, the territory is only large enough to contain the nest, and feeding is done elsewhere. In some species males form "leks"—groups of very small territories on which they display to attract females to mate, then the females themselves nest elsewhere. Some males may mate with several females on the same territory, some may have several territories with a different female on each. In some songbirds, such as the house sparrow, the

▼ Many birds use courtship displays of various kinds to cement their pair bonds. These complex rituals are often beautiful or spectacular. Here a pair of western grebes perform their serene and graceful "weed dance".

Gary Nuechterlein

Brian Chudleigh

male has no real song, whereas in others, such as the superb lyrebird, the nightingale, or the brown thrasher, he may have hundreds or even thousands of phrases. These lavish songs, like the tail of a peacock, are thought to have evolved because females find the variety attractive, and the bigger and better the display the more mating success the male achieves.

After pairing comes nest-building, mating, and egg-laying. The nest may be an elaborate affair, such as those made by weavers, whose construction involves a complex sequence of movements and carefully chosen materials, or it may be a simple scrape. Some birds build no nest at all, whereas others create a vast pile. The mound of vegetation, several meters across, amassed by the male brush turkey is the most notable, especially as the eggs are incubated by the warmth of its fermentation rather than by either of the parents. In more conventional cases incubation may be by the male, by the female, or by both. Either or both of the sexes may carry out care of the young as well, with the added variant that many species have now been found to have helpers at the nest: additional birds, often the offspring from earlier broods, which help the parents to feed their chicks

and so to raise more young. Helping like this is perhaps at its most bizarre in the white-rumped swiftlet, where the female lays two eggs several weeks apart. By the time the second is laid the first has hatched and is old enough to incubate it.

Birds with young or eggs have a particular problem in keeping predators away from their brood. Special alarm calls produced when a hawk is spotted may serve to warn the mate and young so that they keep quiet and still until the danger has passed. Some small birds on the nest hiss and rattle their wing feathers in frightening similarity to a nasty predator such as a snake. Wading birds often produce distraction displays. When a dangerous intruder such as a human comes near the nest, they run away with wings dragged along the ground as if badly wounded. When the predator has been tempted to chase them far from the nest, they rise in the air and fly strongly away. The predator is distracted from one meal by the prospect of another, and as a result loses both.

Predators may also be driven off. The fulmar, a seabird from the North Atlantic which eats foods such as squid and jellyfish, as well as offal thrown from trawlers, defends its nest by spitting its last oily and half-digested meal at an intruder. It can

▲ *An Australasian gannet returns to the colony with nesting material. Most birds build nests in which to cradle and protect their young; some nests are extraordinarily intricate structures taking days or even weeks to complete, while others may be little more than a rough heap of litter on the ground.*

Belinda Wright

▲ *Solicitous and protracted care of their young is a very conspicuous feature of bird behavior. Here a painted stork brings food to its nestlings in India.*

▼ *The downy young of western grebes spend much of their early lives snugly riding on the back of one or other of their parents. In grebes both sexes co-operate in parental care, but male involvement is by no means universal among birds. In many species the female rears her brood unaided.*

Gary Neuchterlein

and must be fed in the nest until they have developed sufficiently to fledge. This can often be several weeks. In other birds (for example, ducks and gulls) the young are well developed at birth, can stand within a short period, and leave the nest within a day of hatching. They rapidly learn to feed themselves, and the parents provide them with only shelter and protection.

Experience plays an important role in the development of all birds, particularly after they leave the nest. A young gull learns the call of its own parent and will perk up to beg only when it hears the call of that particular adult returning to the colony. It also learns to peck at its parent's bill to obtain food. In chickens and ducks, where the young follow the mother in search of food, they learn her features also. At first a young chick will follow any large object it sees, but after a few days it has imprinted on its mother and will follow only her. Other large objects are now frightening to it and if one should appear, instead of following it the young bird will run to its mother for safety.

By experience with their parents and their siblings young birds also learn the more general features of their own species, and, when they are more mature, it is for these that they look when seeking a mate. This is true of songbirds too, even though they remain in the nest for longer. If a young male zebra finch is reared by a pair of Bengalese finches, it will prefer to court and mate with a Bengalese when it becomes adult.

As they grow, young birds develop many of the skills required to survive and reproduce by trial and error and by interactions with their parents and others around them. Young mammals, particularly predatory ones such as cats, often romp around in rough-and-tumble play. Play is less obvious in birds, but young predators such as falcons will often fly at each other and grapple in dazzling aerobatic displays. One theory is that such predators, in their play, are learning the complex skills required to capture and subdue their prey.

score a hit at up to 2 meters (6½ feet), and its aim is deadly, so intruders are best to keep away. They are also well advised to keep clear of the territories of skuas. These gull-like seabirds dive-bomb large animals that stray near their nests, swooping down from above to approach at high speed. Though they seldom hit, only those with a steely nerve remain close by.

GROWING UP

In some birds (for example, songbirds and birds of prey) the young hatch small, naked, and helpless,

BIRD BRAINS

Much of the learning shown by birds is more a case of special abilities matched to a way of life than a sign of wide-ranging intelligence. A brown thrasher will master more than a thousand song phrases; a marsh tit can memorize the locations of hundreds of hidden seeds. But neither could manage the other task, nor many other tasks that to us seem simple. Natural selection has endowed them with special abilities where they need them.

This is not to argue that birds are stupid. They have large brains like those of mammals, and their behavior is, as we have seen, elaborate and varied. The evidence for their learning by imitation is as good as that for any mammal besides ourselves and our closest relatives. This is just one of the many striking features of bird behavior that make it a fascinating subject for study.

ENDANGERED SPECIES

ALISON STATTERSFIELD

More than a thousand bird species are threatened with extinction today. This is the startling conclusion from recent research by the International Council for Bird Preservation (ICBP). In fact, the situation is worse than even this figure implies, for many more are declining or are potentially vulnerable and could soon be threatened with extinction too.

THE GLOBAL THREAT

All types of birds are at risk—passerines, non-passerines, big birds, small birds, landbirds, seabirds—but the family with by far the most threatened species, more than 70 species in total, is the parrot family, Psittacidae. Beautiful birds like the world's largest parrot, the hyacinth macaw *Anodorhynchus hyacinthinus* from central Brazil,

◄ *Worldwide habitat destruction and poaching have reduced many parrots to critically low populations. This is especially true of the large macaws, like this hyacinth macaw – largest of all parrots – of South America. Large birds generally need large territories, and the combination of spacious habitat requirements, ease of capture, high cash value, and inadequate protection make it unlikely that these beautiful birds can survive in the wild.*

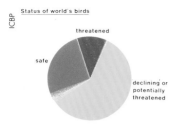

Status of world's birds

threatened

safe

declining or
potentially
threatened

ICBP

▲ The precarious status of the world's
birds: about 11 percent of the world's
total avifauna must be regarded as
threatened.

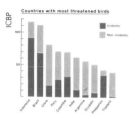

Countries with most threatened birds

■ Endemic
■ Non-endemic

ICBP

▲ The political connection: many of the
world's poorest nations also have the
highest total of threatened bird species.

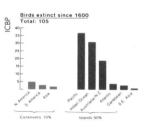

Birds extinct since 1600
Total: 105

ICBP

Continents 10%   Islands 90%

▲ The geography of extinction: the grim
catalog to date suggests that birds of
small oceanic islands are most at risk,
closely followed by tropical rainforest
species.

▶ One of the largest and most
impressive of all birds of prey, the harpy
eagle is threatened by widespread
destruction of the forests of Central and
South America. It is now very rare.

eastern Bolivia, and northeastern Paraguay, or the world's only flightless parrot, the kakapo *Strigops habroptilus,* now confined to a couple of islands off New Zealand, could well disappear soon.

Threatened birds come from all walks of avian life—some live in wetlands, others in deserts, many inhabit tiny islands—but the habitat that is home to the highest number of threatened species is tropical rainforest. Many of the threatened parrots, pigeons, pheasants, birds of prey, and hornbills are forest-dwelling, and some families such as the antbirds (Formicariidae) of Central and South America, the broadbills (Eurylaimidae) of Africa and Asia, and the birds of paradise (Paradisaeidae) of Indonesia, New Guinea, and Australia, are restricted almost entirely to forests.

Birds are threatened in all corners of the globe, on all continents, and on most islands. But two countries stand out from the others: Indonesia and Brazil together account for more than 25 percent of all threatened bird species.

Since 1600 at least 100 species, mostly island birds, have died out. More than 30 have disappeared this century. Of course extinction is a natural process: species have always died out, and others have diversified in response to a changing environment. As a consequence of evolution, the birds of today bear little resemblance to their *Archaeopteryx*-like ancestors which flew on earth some 140 million years ago. The big difference between the species that have become extinct in historical times, and those that disappeared in more ancient eras, is that recent extinctions can be attributed almost wholly to human beings.

This extinction crisis is not confined to birds. Their demise signifies that other wildlife in the same ecosystems—mammals, reptiles, invertebrates—are likely to be in trouble too. We are in fact facing the possibility of a massive loss of global biodiversity within the next few generations, unless we act now to conserve the natural world.

WHY BIRDS ARE ENDANGERED

Causes of extinction and current threats vary, depending on the type of bird, the habitat it lives in, and the region it inhabits. The greatest cause of bird extinction has been the effect of introduced species on island birds. The greatest overall threat to birds today, however, is undoubtedly habitat destruction, which affects both island species and continental species.

Habitat destruction

Almost all major habitat types have been affected by encroachment: grasslands have been widely plowed or subjected to intensive overgrazing by livestock; wetlands have been drained and converted to farmland; tropical forests have been degraded, chopped down, and burnt. Already, tropical forests have been reduced by 44 percent of their original area, and the destruction continues

apace. Until recently the most authoritative estimate of deforestation in the tropics was 11.4 million hectares (44,000 square miles) per year, but it is now thought that deforestation is much greater than previously estimated in Brazil, Costa Rica, India, Myanmar (formerly Burma), the Philippines, and Vietnam. Forest clearing has also increased sharply in Cameroon, Indonesia, and Thailand. If these new studies are accurate, the world is losing up to 20.4 million hectares (78,700 square miles) of tropical forest each year.

In some areas there has been almost complete deforestation—for example, on the islands of Cebu in the Philippines and on Sangihe in Indonesia, which have consequently lost their forest-dependent endemic birds, the four-colored flowerpecker *Dicaeum quadricolor* (Cebu, extinct 1906) and the cerulean paradise-flycatcher *Eutrichomyias rowleyi* (Sangihe, presumed extinct 1978). In other areas the forest has become fragmented. Some species can of course survive partial clearance of their habitats, or may even make use of cutover secondary forest. However, there are many more species that are unable to adapt and that rely on pristine habitat for their continuing survival.

The world's most powerful bird of prey, the harpy eagle *Harpia harpyja,* is threatened by the clearance and fragmentation of forest. This magnificent bird lives in the tropical forests of Central and South America, from Mexico to eastern Bolivia, southern Brazil, and extreme northern Argentina. It has been estimated that a pair of harpy eagles need a territory of undisturbed forest of 100 to 200 square kilometers (57 to 77 square miles) to satisfy their feeding requirements. Thus the minimum area of intact rainforest to ensure the harpy eagle's survival could be as large as 37,500

Norman Tomalin/Bruce Coleman Ltd

square kilometers (14,500 square miles) if it is assumed that a population of 250 pairs is necessary for perpetual viability. But as harpies tend not to overlap territory with their closest competitor, the crowned eagle *Harpyhaliaetus coronatus* (also threatened), an even larger area may be necessary for the required 500 birds. At a very rough guess, 60,000 square kilometers (23,000 square miles) might be sufficient, an area equivalent to the tract of Amazonian forest destroyed in 1988. This species may survive in the long term only if the exponential rate of forest destruction is brought under control throughout its range, and a network of large undisturbed reserves is established.

Introduced species

Introduced species, especially predators such as cats, rats, and mongooses, have been the major cause of extinction of island birds. Many of these birds had evolved without any predator pressure and were unable to cope with introduced aliens which stole eggs and nestlings, or even hunted adults, especially of species that had lost the ability to fly. One such bird was the Stephen Island wren *Xenicus lyalli* (extinct 1894). Stephen Island is a tiny speck of land totalling 2.6 square kilometers (1 square mile) lying in the Cook Strait between the North and South Islands of New Zealand. The wren may have had the smallest natural range of any bird, as well as being the only passerine truly incapable of flight. But its main claim to fame was the remarkable manner of its extinction—by the lighthouse keeper's cat. This single feline discovered the species then destroyed the entire population in just a few months.

Introduced herbivores such as rabbits and goats can be as deadly as introduced predators because of changes they wreak by browsing on vegetation. For example, the Laysan duck *Anas laysanensis* declined to near extinction on Laysan Island (a coral atoll belonging to the Hawaiian Islands) in the 1920s after the introduction of rabbits and the consequent denudation of the island's foliage. Two further species, the Laysan rail *Porzanula palmeri* (extinct 1944) and the Laysan millerbird *Acrocephalus familiaris* (extinct between 1912 and 1923) were unable to survive, whereas the Laysan finch *Telespyza cantans* was able to persist by changing its diet and feeding on seabird eggs.

Species introduced in the past still severely threaten a number of birds today. For example, the brown tree snake may have been responsible for the population crash of all forest birds on the island of Guam in the Pacific, and in particular the endemic Guam rail *Rallus owstoni* and Guam flycatcher *Myiagra freycineti;* and exotic fish in Madagascar have caused a decline in waterlilies and other aquatic plants, and so threaten the Madagascar little grebe *Tachybaptus pelzelni.*

Nowadays efforts are made to keep islands free of introduced predators, and in some places

Brian Chudleigh

measures have been developed to eradicate the invaders. Cats had been present on Little Barrier Island, off Auckland, New Zealand, for over 100 years, but an eradication program between 1977 and 1980 resulted in their complete elimination and a marked improvement in a number of threatened bird species. For example, Cook's petrel *Pterodroma cooki* and the black petrel *Procellaria parkinsoni,* both heavily preyed on (100 percent of black petrel fledglings were killed from 1974 to 1975), have shown a significant recovery; the stitchbird *Notiomystis cincta,* once widespread in the forests of New Zealand but surviving only on Little Barrier Island, was declining but is now increasing; and the saddleback *Creadion carunculatus,* extinct on Little Barrier Island, has been successfully reintroduced.

Human predation

Hunting has been another major cause of extinction, and includes hunting for food (both birds and eggs), for feathers, and in some cases for

▲ *Introduced predators such as feral rats and cats have precipitated the extinction of many small birds. Once widespread in the original forests of New Zealand, the stitchbird shown here now survives only on a few tiny offshore islands, where constant vigilance is required to prevent the accidental introduction of predators.*

individuals wintering at a Moroccan wetland. It seems likely that uncontrolled hunting over a long period of time has reduced the population year by year, and the breeding success (or lack of it) of the species has been unable to compensate for the hunting losses. If this species is to survive, a ban on all shooting of curlews and godwits (because of misidentification by hunters) in all these countries is urgently needed.

International trade

The wild-bird trade is a growing threat to birds. Some species are particularly sought after, and the rarer they become, the higher their market value. Thus it is very difficult to control the trade in prized birds because trappers and dealers are tempted by rich rewards. The red siskin *Carduelis cucullata* is one popular cage-bird. Trade in the species began in the nineteenth century, and demand escalated this century when it was discovered that it was possible to hybridize the red siskin with the domestic canary *Serinus canarius* (in the same subfamily of finches) and so introduce the genes for red plumage. The siskin once occupied a continuous range across northern Venezuela into northeastern Colombia, with isolated populations on the islands of Gasparee, Monos, and Trinidad. Thousands of birds have been trapped and numbers have declined drastically, so that now there are confirmed localities in only a few states in Venezuela and perhaps as few as 600 to 800 birds.

Parrots are the most popular of all cage-birds, and the trade in some species has had devastating effects. Spix's macaw *Cyanopsitta spixii* has been illegally trapped down to the very last bird in the wild. Few ornithologists have ever seen the species, and little is known about its way of life. It comes from a remote region of northeast Brazil, and in June and July 1990 a search for the bird was carried out over a large portion of its expected range. The survey team located only one bird. There were reports that Spix's macaw had persisted in another nearby location until the previous year, but residents thought that the last individuals had been taken by trappers. Further work indicated that the preferred habitat of the species was mature woodland along watercourses and that only three small patches currently remain in the state of Bahia. It is becoming clear that the range of this parrot is much smaller than previously thought, and the species is now virtually extinct because of trappers exploiting an already tiny population in a diminishing habitat. The only hope for the species lies with the 15 or so individuals held in captivity, and a carefully planned captive breeding program.

Pesticides and pollution

The use of pesticides and the pollution of the environment threaten many bird species throughout the world. During the 1960s scientific

▲ *Red in the plumage is a quality that long eluded canary fanciers, until it was found that controlled cross-breeding with the red siskin of South America would introduce the necessary genetic material; as a consequence of this discovery red siskins were trapped to the point that they are now a critically endangered species.*

museum specimens. For example, the Hawaii mamo *Drepanis pacifica* (extinct 1899) was hunted for its brilliant yellow rump feathers, and 80,000 birds were sacrificed to make the famous royal cloak worn by Kamehameha I. Perhaps the most famous case of hunting to extinction is that of the passenger pigeon *Ectopistes migratorius* (extinct 1914). Once vast numbers darkened the North American skies, and hunting competitions were organized in which more than 30,000 dead birds were needed to claim a prize. The supply appeared inexhaustible, yet within a century the incredible multitudes were reduced to small dwindling bands, and in 1914 the sole survivor died in Cincinnati Zoo.

Many species are still hunted today, and while some are able to withstand the losses, others are seriously reduced. One species that has suffered particularly is the slender-billed curlew *Numenius tenuirostris,* one of the rarest and most poorly known species of the Western Palearctic region (Europe, USSR, Middle East, and North Africa). This species is believed to nest in the USSR, and on migration it visits Turkey, Romania, Hungary, Yugoslavia, Greece, Italy, Tunisia, Morocco, and Iran. It appears to have been common in much of its range up to the end of the nineteenth century—observations from Algeria describe "incredible flocks ... as big as starling flocks"—yet in 1989 the only records of the species were of three proof emerged that more than 20 bird species across Europe and North America were suffering disastrous breeding failures because of malformed and broken eggs. The peregrine falcon *Falco peregrinus,* Eurasian sparrowhawk *Accipiter nisus,* golden eagle *Aquila chrysaetos,* osprey *Pandion haliaetus,* and brown pelican *Pelecanus occidentalis* were a few of the birds affected. These birds, at the top of their food chains, all showed high levels of DDT contamination, and this was linked beyond doubt to their unusually thin eggshells. Chlorinated hydrocarbons (of which DDT is one) persist in the environment for many years and, because they are soluble in fats and oils, accumulate in the fatty tissues of top predators with detrimental reproductive effect.

The use of chlorinated hydrocarbons is now restricted and the populations of affected species are recovering, but new classes of chemicals have taken their place and these too can have deadly side-effects. For instance, there is a general feeling that the use of pesticides in Africa may represent a serious threat to many European migrants (as well as to intra-African migrants and resident species). It is also claimed that locust and bird control campaigns—for example, of the quelea, which is a pest to agriculture—have caused casualties among an array of species of migratory birds including storks, herons, and birds of prey. Furthermore, there is circumstantial evidence that pesticides, especially rodenticides, are responsible for the decline of resident bird populations in Egypt.

As well as pesticides and insecticides, there is a host of other synthetic chemicals which now pollute the atmosphere, the land, and the water, and which are likely to be harmful to birds (and to other wildlife and humans). High levels of polychlorinated biphenyls (PCBs) have been recorded in the carcasses of the threatened white stork *Ciconia ciconia* in Israel and in addled eggs in Germany and Holland. These chemicals, produced by the combustion of plastics and other waste materials from the electronics industry, are known to severely reduce avian breeding success, and so could be a factor contributing to the decline of the white stork.

Climatic catastrophes

Changes in climate may also play a part in the decline of some bird species. For example, it has been suggested that the great auk *Alca impennis* of the North Atlantic (extinct 1844) was never very numerous and was restricted to a relatively narrow climatic zone. It seems likely that a period of severe weather coincided with increased human predation, and these combined factors eventually overwhelmed the species. But climatic change is not just a phenomenon of the past. We are currently experiencing a climatic change which may have far-reaching results—the so-called greenhouse effect. If the use of fossil fuels such as

coal, gas, and oil cannot be reduced, scientists speculate that the average global temperature could rise by 3°C (5°F) within the next 50 years. The world would be warmer than it has been for 10,000 years. Even this small change is likely to affect the ranges of many species, disrupt natural communities, and contribute to the extinction of some vulnerable species.

Some climatic changes can be sudden and dramatic. Hurricane Gilbert was the most powerful storm recorded in the Caribbean this century. When Gilbert hit Jamaica on September 12, 1988, gusts in excess of 220 kilometers per hour (137 miles per hour) were registered as winds tore across the island causing enormous destruction, especially to the montane and mid-level forests. It was feared that six endemic species largely confined to these habitats could have been badly affected. In fact the birds survived the hurricane surprisingly well. As hurricanes are not infrequent visitors to Jamaica, and the forests are probably always in some state of recovery, the avifauna is doubtless adapted to this dynamic system. But there is concern that continued forest clearance and charcoal extraction could reduce the forests to tiny areas, and that hunting could reduce the bird populations to such low numbers, that next time they may not recover from the impact of a storm as fierce as Gilbert.

BIRDS WITH RESTRICTED RANGES

Some birds are considered threatened because they have tiny ranges—for example, birds confined to single small islands, like the Ascension frigatebird *Fregata aquila.* This seabird breeds only on Boatswainbird Islet (3 hectares, or 7 acres) just off the coast of the remote South Atlantic island of Ascension. It once bred on Ascension itself, but egg collecting, disturbance, and the introduction of

▼ *Pollution has exacted its toll on wild bird populations. One of the effects of DDT on peregrine falcons is to interfere with the body chemistry needed by the female to properly form the shell around her egg. As a result, eggs are fragile and break easily in the normal stresses of incubation, killing the chick inside. This effect almost wiped out the peregrine, but with the banning of DDT, populations began a gradual recovery.*

R.T. Smith/Ardea London Ltd

► *One of the most encouraging success stories in recent bird conservation is the saving of the Lord Howe woodhen, confined to a tiny island in the Tasman Sea. Reduced to a critical 20 birds in the late 1970s, a captive breeding and reintroduction program was instituted, with such success that the bird now occurs commonly throughout the island.*

Tom & Pam Gardner

▼ *Confined to tussock swamps in the most remote high mountains of the South Island of New Zealand, the takahe was considered extinct until dramatically rediscovered in 1947. It remains critically endangered.*

Brian Chudleigh

alien mammals, especially cats, have forced the frigatebird onto cat- and people-free Boatswainbird Islet. Despite the contraction of its range it still appears to be a fairly numerous species with a population of about 10,000 birds. Nevertheless, the fact that the entire population is confined to such a minute area renders it a species that could be easily and quickly threatened should cats, for example, get a foothold on the island.

Many continental species also have very restricted distributions, confined to single mountaintops or tiny patches of forest. Bannerman's turaco *Tauraco bannermani* and the banded wattle-eye *Platysteira laticincta* are two such species, found only in the shrinking forest of Mount Kilum (also known as Mount Oku) in the Bamenda Highlands of Cameroon. The black-breasted puffleg *Eriocnemis nigrivestis* is another species at risk because of its small range, being known only from two volcanoes in north-central Ecuador, and threatened by habitat destruction because of proximity to the capital city, Quito. It is vital that birds like these are given adequate protection before it is too late.

One restricted-range species that has been given the necessary protection is the noisy scrub-bird *Atrichornis clamosus*, a bird of swampy eucalypt forest in the extreme southwest of Western Australia. At one stage it was feared extinct, not being recorded since 1889, but in 1961 a small population of 40 to 45 singing males was rediscovered at Two Peoples Bay, east of Albany. The disappearance of the scrub-bird from most of its former range has been attributed to the intense fires started by European newcomers to encourage

the growth of grasses more suitable for cattle-grazing. The Two Peoples Bay population probably survived because the area was rugged and unsuitable for agriculture.

However, the rediscovery of the noisy scrub-bird presented a problem, for the area where the bird was living was precisely where the Western Australia government had proposed establishing a new town. To the government's credit the site was cancelled, and in 1966 Two Peoples Bay Nature Reserve was declared. In 1987 the population of the noisy scrub-bird had risen to about 500, and other populations had been established in other nearby reserves.

MYSTERY BIRDS
Some birds are included on lists of threatened species because they are virtual mysteries—they have been seen so few times that we assume they must be extremely rare and therefore endangered. The Fiji petrel *Pterodroma macgillivrayi*, for example, was formerly known from one specimen collected on Gau (an island in the Fiji group) in 1855. However, in 1984 an adult was captured on Gau and released, and in 1985 a fledgling was found there. So it is likely that the species breeds on the island, although there have been no further sightings since.

Mystery birds keep turning up. One recent discovery in 1988 was that of a new species of shrike from Somalia. This bird (yet to be described and named, at the time of writing) was found in dense *Acacia* bushland and had presumably escaped detection because of the impenetrable nature of its habitat. Other mystery birds are the orange-necked partridge *Arborophila davidi,* known from one specimen collected in 1927, 60 kilometers (37 miles) east of Ho Chi Minh City, Vietnam, and the Red Sea cliff swallow *Hirundo perdita,* known from one specimen found dead at a lighthouse off Port Sudan in 1984.

BACK FROM THE BRINK

Of the 1,000 or so species threatened with extinction there are some with so few individuals that they are on the very brink of extinction. And yet it may be possible to conserve even these species. The whooping crane *Grus americana* exemplifies such a species, being the center of a remarkable USA-Canadian conservation effort which has seen the total population rise from an all-time low of 21 wild birds in 1944 to 200 birds in wild and captive flocks in the late 1980s. The whooping crane has probably never been very numerous in historical times—possibly as few as 500 birds existed in the late eighteenth century. Species with naturally low numbers are especially vulnerable to change, but low numbers coupled with small (and shrinking) breeding and wintering grounds, and a hazardous (because of hunting and power lines) migration route over 3,000 kilometers linking the two, had stacked the odds against the chances of the whooping crane's survival. The story of the whooping crane recovery is a complicated one, involving much ferrying of eggs and birds from one location to another in a massive attempt to maximize the reproductive rate, but the enormous effort and expense has paid off.

The majority of threatened birds, however, live in developing countries where funding for conservation takes second place to the needs of people, and it is therefore unlikely that many species can be saved once they reach such a critical state. Instead it is vital that appropriate conservation action is taken at a much earlier stage.

Nothing more clearly illustrates the urgent need for swift action to save species on the brink of extinction than the story of Gurney's pitta *Pitta gurneyi,* a stunningly beautiful bird with brilliant blue crown and chrome-yellow flanks. The pitta is a resident of southern Burma (now Myanmar) and peninsular Thailand, and though common at the start of this century it had not been encountered in the wild since 1952. Much of its original lowland forest habitat had been cleared and settled by a burgeoning human population. A search in 1984 and 1985 revealed no birds, but in 1986, following a lead from a bird dealer, the quest for the species was rewarded with the discovery of a nesting pair in a tiny patch of forest at Khao Noi Chuchi in the far south of Thailand. Subsequent intensive survey work has shown that virtually the entire population of fewer than 30 pairs is confined to this one forest. A petition to the Minister of Agriculture drawing attention to this priceless forest fragment (apart from the pitta, it still harbors

a spectacularly diverse range of other animals and plants) has brought protection just in time, for Khao Noi Chuchi was destined to be logged and encroachment by local people had already started. Now a major conservation project has been launched to help protect the forest and at the same time to aid the nearby farmers so they have no need to destroy the forest. Without this action Gurney's pitta would almost certainly be extinct today.

CONSERVATION IN THE FUTURE

The species of birds at risk of extinction, and those that are declining or potentially vulnerable, must be saved. Different species have different conservation needs, but some combination of preserving habitat, eliminating introduced species, banning hunting, controlling trade, and preventing pollution will help to ensure their survival. But assessing the required action and executing it takes time and money, and both are in short supply. While conservation action to save flagship species like Gurney's pitta can be very successful—not just for the species itself but also for associated wildlife—conservation in the future is perhaps best focused on key sites where several threatened species occur together. For it is in these "hotspots" that there will be the greatest return for the conservation effort invested. If we preserve such sites we will make a major contribution to maintaining the earth's biological diversity.

Brian Chudleigh

▲ The black robin was snatched from the brink of extinction by determined last-minute measures. Confined to Chatham Island in the far southwestern Pacific, it was reduced to about seven individuals before a complex cross-fostering program began to offer some hope that its numbers might be rebuilt.

▼ Generally, birds of lowland tropical rainforest seem most at risk: this is the habitat of the very rare and beautiful Gurney's pitta.

M.D. England/Ardea Photographics

SIZE
Smallest Little tinamou
Crypturellus soui, total length
15 centimeters (6 inches);
weight 450 grams
(16 ounces).
Largest Ostrich *Struthio
camelus*, height 2.75 meters
(9 feet); weight 63 to 105
kilograms (140 to
230 pounds).

CONSERVATION WATCH
The following species are
listed in the ICBP checklist of
threatened birds: little spotted
kiwi *Apteryx owenii*, solitary
tinamou *Tinamus solitarius*,
black tinamou *T. osgoodi*,
yellow-legged tinamou
Crypturellus noctivagus,
Magdalena tinamou *C.
saltuarius*, Taczanowski's
tinamou *Nothoprocta
taczanowskii*, Kalinowski's
tinamou *N. kalinowskii*, lesser
nothura *Nothura minor*, and
dwarf tinamou *Taoniscus
nanus*.

RATITES AND TINAMOUS

S.J.J.F. DAVIES

Ratites, the giants of birds, and their small relatives, the tinamous, are the living representatives of the order Struthioniformes. Their distribution on southern continents has long encouraged the view that they represent a primitive group of southern origin that did not penetrate the Northern Hemisphere, but recent discoveries of ratite fossils in Europe indicate that their southern distribution may be all that remains of a once wide-ranging group. Members of this order—the ostrich, rheas, cassowaries, emu, and kiwis—provide spectacular evidence that birds can evolve into large flightless vertebrates comparable with the large herbivorous mammals.

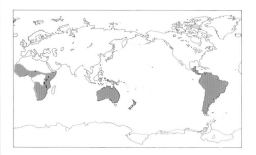

AN UNKEELED BREASTBONE
The order contains six living families:
Struthionidae represented by the ostrich of Africa,
the world's largest bird; Tinamidae, about 45
species of tinamous in South and Central America;
Rheidae, two species of rhea in South America;
Casuariidae, the cassowaries, three species in New
Guinea and Australia; Dromaiidae, the emu of
Australia; and Apterygidae, three species of kiwi in
New Zealand. In the historical past the
Diornithidae, at least twelve species of moas,
inhabited New Zealand, and individuals may have
survived into the nineteenth century. The final date
of extinction of another family, the Aepyornithidae
(elephantbirds), confined to Madagascar, was
about 1650.

All "true" ratites are flightless, and they have a
flat sternum (breastbone) without the keellike
prominence of most flying birds. (The Latin word
ratis means "raft"). The tinamous have a keeled
sternum, and also have the ability to fly, but they
resemble the other Struthioniformes in many
unusual anatomical characters such as the
structure of the palate. Their plumage is loose
compared with the feathers of most other birds.
Cassowaries and the emu have plumage that hangs
like hair from their bodies; each feather has two
shafts of equal length. The ostrich, rheas,
tinamous, and kiwis have feathers with one
main shaft, but the barbules, if present, do not
interlock closely so the birds appear shaggy.

THE OSTRICH: A KING AND HIS HAREM
Ratites are running birds. They gain a mechanical
advantage by having long thin legs to support the
body's weight well above the ground, in a similar
way to the ungulates (horses, cows, and their
relatives). The number of toes has been steadily
reduced in the course of ratite evolution. Most
birds have four; in ratites three is the common
number, but the ostrich has gone further and has
only two toes, a feature unique among living birds.
In the ostrich, unlike most ratites, males are larger
than females: males grow to 2.75 meters (9 feet)
tall, females to 1.90 meters (6¼ feet) tall. The
ostrich's wide range once included the Middle East,
North Africa, and Africa south of the tropical
rainforests, but it is now extinct in the Middle East
and most of North Africa. Almost all southern
African populations are confined to national parks.
In South Australia there is a small feral population.

Ostriches use their large wings in courtship, and
each cock collects a harem of two to five hens. The
male usually builds the nest in which all his hens
lay, then normally the dominant hen incubates by
daylight and the cock by night during the
incubation period of 39 to 42 days. Both sexes
guard the chicks, which may remain as a family for
twelve months. Sexual maturity sometimes appears
in the second year, although breeding is usually
delayed until the third or fourth year. Their
breeding behavior is very variable, depending on
the density of the population, the habitat, and the
climatic conditions. Some nests are the property of
a monogamous pair, especially where ostriches are
scarce. In dense bush, nests may be closer to each
other than the usual 2 kilometers (1¼ miles) and
on the plains may be more widely spaced than that.
Large concentrations of ostriches occur daily
around water or where food is abundant, and
immatures are found in flocks of up to 100 birds.
The diet is a selection of fruits, seeds, succulent
leaves, and the growing parts of shrubs, herbs, and
grasses. They also take small vertebrates, and ingest
gravel to assist with the maceration of their food.

Mitch Reardon/Weldon Trannies

◄ The ostrich was once widespread across Africa and the Middle East, but it is now essentially confined to various national parks in Africa. Until about the turn of the century the ostrich was widely farmed for its feathers, which were used in the fashion industry; the industry is not entirely defunct, and captive flocks in South Africa still total about 60,000 birds. In the Arab world the shells of ostrich eggs are credited with magical powers, and are often used on the roofs of Muslim homes or Egyptian Orthodox churches with the intention of deflecting lightning.

THE SOUTH AMERICAN FAMILIES

The two representative families of ratites in South America, the rheas and the tinamous, total 47 species and inhabit a variety of habitats from forest, to the high antiplano (puna) of the Andes.

Rheas

Rheas are sometimes called South American ostriches. The greater rhea *Rhea americana* stands about 1.5 meters (5 feet) tall and weighs 20 to 25 kilograms (44 to 55 pounds). The lesser rhea *Pterocnemia pennata* is smaller. Both species have gray or gray-brown plumage, with large wings that cover the body like a cloak. When they run rheas sometimes spread their wings, which then act as sails, but the birds are unable to fly.

The original distribution of the two species was unusual. The greater rhea lived on the plains from northeastern Brazil to central Argentina, but its range has been dissected by agricultural

▲ A one-wattled cassowary. Protein being scarce for many native peoples in the rainforests of New Guinea, a common practice is the capture and raising of young cassowaries, which are slaughtered for food when they reach maturity.

▼ The tinamous of South America are in many respects the ecological equivalents of quail and partridges elsewhere in the world. All are terrestrial, though many roost in trees. Some species, like this highland tinamou shown at its nest, inhabit rainforest.

development. The lesser rhea has two separate populations: one on the pampas (grasslands) of Patagonia, known as Darwin's rhea; and the other in the high Andes of southern Peru and northern Chile, the puna rhea.

Rhea males fight for territories, and once a male has established his domain he builds a nest, a scrape on the ground lined with leaves and grass. To this he attracts females, often as a small flock. Each female lays an egg in the nest, returning to do so every two or three days until the male, responding to the size of the clutch, drives them away and begins to incubate. Before and after the females lay in the nest, however, they lay eggs on the ground in the vicinity, some of which the male rolls beneath him; the others rot. Once he sits, the female flock goes off to attend another male and may serve half a dozen nests in a season. The cock leads the chicks, which grow quickly and are of adult size in about six months but do not breed until they are two years old.

Tinamous

The 45 species of tinamou vary from the size of a quail to that of a large domestic fowl. While they show close relationships to other Struthioniformes in their anatomy, their eggwhite proteins, and the structure of their genes, they differ in some conspicuous ways. Many, perhaps all, species can fly, although they seldom do so; they usually escape predators by stealing away through cover or freezing. Most species have three toes, a common ratite number, but some have four. Members of the genus *Tinamus* roost in trees; all other tinamous roost on the ground. In *Tinamus* species the back of the tarsus is roughened to give the birds a good grip on the branch when at rest.

Many species feed on vegetable matter, but some (for example, *Nothoprocta* species) take much

animal food, and the red-winged tinamou *Rhynchotus rufescens* digs for roots and termites. In some species (such as the ornate tinamou *Nothoprocta ornata*) a single male and female establish the nest, but in most a male associates with several females at nesting. Tinamous nest on the ground, lining a depression with grass and leaves, and like most ratites the male undertakes the incubation (19 days for *Eudromia*) and looks after the family.

Tinamous are found in many habitats, including rainforest and the high and barren Andes. On the open tablelands the martineta tinamou *Eudromia elegans* lives in flocks of up to a hundred. Tinamous are diverse and abundant, an impressive achievement considering that they are now thought likely to be close to the ancestral stock of all ratites.

EMU AND CASSOWARIES

The Australian emu *Dromaius novaehollandiae* lives a nomadic existence, continually moving to keep in touch with its food—not that the food moves, but rather that abundances of flowers, fruits, seeds, insects, and the young shoots on which it feeds appear in random sequence in the Australian deserts. Emus, standing 2 meters (6½ feet) tall and weighing up to 45 kilograms (100 pounds), move over vast distances, usually as monogamous pairs, stopping when they find abundant food and moving again when it is exhausted. Only when the male undertakes the eight weeks of incubation is it impossible for him to move to find food. During incubation he does not eat, drink, or defecate, living instead on the fat reserves he has built up in the previous six months. If conditions have not allowed the pair to store fat before the winter breeding season, they do not breed, or if eggs are laid the male may desert them before they hatch. The male guards the chicks and leads them for their first seven months and sometimes longer. The female may remain nearby, or move far away in search of food, or mate with another male. Seldom do the pair re-form for a second season.

Emus live throughout southern Australia, not just in deserts. They become less common in the north, although a few birds can be found as far north as Darwin and Cape York. Their numbers, currently estimated at about 500,000, can rise or fall rapidly in correlation with wet and dry seasons. They are common in coastal scrub, in eucalypt woodland, and on saltbush plains. A few venture into alpine heath, and many are still present in farming areas provided some bushland remains.

Cassowaries favor jungle. Three species live in New Guinea, and one of these, the double-wattled cassowary *Casuarius casuarius,* also lives in the tropical rainforests of far northeastern Queensland, Australia. In New Guinea the original distribution of the three species is uncertain because humans have transported them and released them beyond their natural range. It is likely that the double-

C.A. Henley

wattled cassowary favored mid-level rainforest, the one-wattled cassowary *C. unappendiculatus* low-level rainforest, and the dwarf cassowary (or moruk) *C. bennetti* the highlands, perhaps even the montane grasslands. Several islands around New Guinea have cassowary populations, usually of one species—for example, the double-wattled cassowary on Ceram and the moruk on New Britain. The double-wattled cassowary stands 1.5 meters (5 feet) tall and may weigh more than 55 kilograms (120 pounds). Its glossy black plumage grows after the first year; before that the young birds are clad in a sober gray. All species have throat wattles, brilliant red and blue in adults but less colorful in immatures. The moruk does not have a distinct casque, but a casque adorns the heads of the other two species. Cassowaries depend on forest fruits for food—fruits from more than 75 species of tree in northern Queensland. Individuals seem to maintain a territory of 1 to 5 square kilometers (⅓ to 2 square miles), moving around it to gather fruit as different trees ripen. The territory is occupied by pairs during the winter breeding season, and the clutch of six to eight eggs is incubated for about two months by the male, who also looks after the young chicks. Cassowaries are not abundant anywhere, and their survival will be imperiled if the diversity of the forests in which they live is reduced by logging.

KIWIS, THE BURROWING RATITES

Three species of kiwi remain in New Zealand: the brown kiwi *Apteryx australis* is the largest, 55 centimeters (21 inches) long, with females

weighing 3.5 kilograms (7¾ pounds); the great spotted kiwi *A. haastii,* intermediate in size; and the little spotted kiwi *A. owenii,* 35 centimeters (14 inches) long, weighing 1.2 kilograms (2½ pounds). The brown kiwi is still found on North, South and Stewart Islands, the great spotted only on South Island, and the little spotted on Kapiti Island, where it was introduced in 1913.

Kiwis are nocturnal, feeding on invertebrates which they find mainly by scent, probing with their long and sensitive bills. Pairs form during the late winter/spring breeding season, and the male excavates a burrow in which the female deposits one to three white eggs. Each egg is equivalent to 25 percent of the female's body weight, proportionately the largest egg laid by any bird. The eggs are incubated by the male for 78 to 82 days, and the chicks appear to be independent almost from the time they emerge from the burrow.

The call of the male brown kiwi is a shrill whistle with a long ascending phrase and a short descending one at the end, giving rise to the name "kiwi". Females have a hoarse, low cry. Two anatomical features set kiwis apart from other ratites. Firstly, the wings—small in the emu and cassowaries—are vestigial in kiwis. Secondly, female kiwis have paired, functional ovaries—in most birds usually only the left is functional, and the right is absent altogether. All species live in native podocarp (southern conifer) forest, but the brown kiwi has survived in farmland and in pine forest, although it is still unclear if such populations are self-sustaining outside these natural forests.

▲ *Emus sometimes form wandering parties when not breeding, but after mating and the eggs are laid, the male incubates and raises his brood of young alone. The chicks are covered with down when hatched, and can follow their father within a few hours. He indicates food to his chicks, but they normally feed themselves.*

▼ *The kiwi is New Zealand's national symbol, though in fact there are three very similar species. Restricted to New Zealand, the group has no close relatives anywhere in the world. All three kiwis are flightless, nocturnal, and live in burrows. The great spotted kiwi, shown here, occurs only on the South Island.*

Brian Chudleigh/National Wildlife Centre

SIZE
Smallest Least storm petrel
Halocyptena microsoma, body
length 12½ to 15 centimeters
(5 to 6 inches); wingspan 32
centimeters (12½ inches);
weight 28 to 34 grams (1 to
1⅓ ounces).
Largest Wandering albatross
Diomedea exulans, body length
1.1 to 1.4 meters (3⅗ to 4⅗
feet); wingspan 3.4 meters
(11 feet); weight 6 to 11
kilograms (13 to 24¼
pounds).

CONSERVATION WATCH
The following species are
listed in the ICBP checklist of
threatened birds: Amsterdam
albatross *Diomedea*
amsterdamensis, short-tailed
albatross *D. albatrus,*
Mascarene black petrel
Pterodroma aterrima, black-
capped petrel *P. hasitata,*
cahow *P. cahow,* Beck's petrel
P. becki, magenta petrel *P.*
magentae, gon-gon *P. feae,*
freira *P. madeira,* dark-
rumped petrel *P. phaeopygia,*
Cook's petrel *P. cooki,*
Chatham Island petrel *P.*
axillaris, Defilippe's petrel *P.*
defilippiana, Pycroft's petrel *P.*
pycrofti, Fiji petrel *P.*
macgillivrayi, black petrel
Procellaria parkinsoni,
Westland black petrel *P.*
westlandica, pink-footed
petrel *Puffinus creatopus,*
Heinroth's shearwater *P.*
heinrothi, Newell's shearwater
P. newelli, Townsend's
shearwater *P. auricularis,*
Guadalupe storm petrel
Oceanodroma macrodactyla,
Markham's storm petrel *O.*
markhami, ringed storm petrel
O. hornbyi, and Peruvian
diving petrel *Pelecanoides*
garnoti.

► *The wedge-tailed shearwater*
inhabits tropical and subtropical waters
of the southern Pacific and Indian
oceans.

ALBATROSSES AND PETRELS

J.P. CROXALL

Birds of the order Procellariiformes, known collectively as petrels or tubenoses, are highly adapted to a marine way of life, spending much of their time at sea and feeding on the larger zooplankton, cephalopods (squid), and fish. They can be found throughout the world's oceans, but it is thought that their ancestors evolved in the southern seas. Their fossil record goes back some 60 million years.

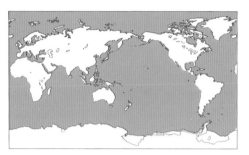

BIRDS OF THE OPEN OCEAN

A number of characteristics are shared by the Procellariiformes, notably their long, external, tubular nostrils — hence the vernacular name, tubenoses. This special feature is associated with a comparatively (for birds) well-developed olfactory part of the brain, and it is likely that the Procellariiformes locate their food, breeding sites, and each other by their sense of smell. All species have a distinctive musty body odor, which persists for decades in museum specimens.

Other characteristics include a deeply-grooved, markedly-hooked bill, and the ability of most species (not the Pelecanoididae, diving petrels) to store large quantities of oil in their stomachs. This oil, derived from the energy-rich food stores laid down by the marine organisms on which they feed, is regurgitated to feed the young or ejected to deter would-be predators. It is one of a number of adaptations for the highly pelagic life, with vast oceanic ranges, typical of the group. Recent satellite tracking studies have shown that the wandering albatross *Diomedea exulans* covers between 3,600 and 15,000 kilometers

R. Drummond

Graham Robertson/Auscape International

(2,200 to 9,300 miles), flying at speeds of up to 80 kilometers (50 miles) per hour, in a single trip while its partner takes a 10-day incubation shift.

The Procellariiformes can be divided into four families: albatrosses (family Diomedeidae), shearwaters (family Procellariidae), storm petrels (family Hydrobatiidae), and diving petrels (family Pelecanoididae).

ALBATROSSES

When fifteenth-century Portuguese navigators first ventured down the coast of Africa into the windy south Atlantic they encountered large black-and-white seabirds with stout bodies and long pointed wings. They called these strange birds *alcatraz,* the Portuguese word for large seabirds. English sailors later corrupted alcatraz to albatross.

Albatrosses are the supreme exponents of gliding flight, sometimes gliding for hours without a single wing-beat. They are typically associated with the belt of windswept ocean lying between the Antarctic and the southern extremities of America, Africa, and Australia, but they also breed in the more temperate waters of the Southern Hemisphere. Three species breed in the north Pacific, and one species breeds at the Galapagos Islands on the Equator.

Albatrosses are distinguished from the other Procellariiformes by the position of their tubular nostrils, which lie at either side of the base of the bill, rather than being fused on the top of the bill. They eat a variety of food, often scavenging behind ships, but fish, squid, and crustaceans are favorite items and are frequently caught at night. They catch prey mainly from the surface of the sea, but occasionally catch it from just beneath the water by plunge-diving

into the waves with bent wings.

Albatrosses are long-lived birds with an average life span of 30 years. But what they gain in longevity they lose in productivity, with most species not breeding until they are 10 to 15 years old, and several species breeding only biennially. Most albatrosses breed in closely-packed colonies, sometimes numbering thousands of pairs, but sooty albatrosses of the genus *Phoebetria* are solitary or nest in small groups on cliff edges. For several species the nest is a heap of soil and vegetation, although the tropical species make do with a scanty nest and the waved albatross *Diomedea irrorata* doesn't prepare one at all.

Albatrosses usually pair for life, and "divorce" occurs only after several breeding failures. They are famous for their spectacular courtship displays, standing with wings open and tails fanned, while stretching out their necks and throwing back their heads until the tip of the bill is buried in the plumage of the back. All this is accompanied by gurgling and braying sounds, and in the case of the great albatrosses, which have the most complex rituals, by the rattling of their bills.

In common with all Procellariiformes, albatrosses lay a single white egg which is incubated in alternate shifts of several days by both parents, from about 65 days in the smaller species to 79 days in the royal albatross *Diomedea epomophora.* The chick is initially brooded for three to four weeks, but thereafter the parents remain ashore only long enough to transfer a meal. Fledging takes from 120 days for the black-browed albatross *D. melanophrys* and yellow-nosed albatross *D. chlororhynchos,* to 278 days for the wandering albatross. The extremely long nesting period of the latter species, and of five others (royal

▲ *The wandering albatross differs from most other tubenoses in the bewildering series of different immature plumages worn by each individual through its lengthy adolescence. Old males have largely white upper wings, unlike these rather young birds at their breeding ground on Macquarie Island. Albatrosses mate for life, and greet each other on their return at nesting time with spectacular displays involving outstretched wings and trumpeting calls.*

271

Graham Robertson

▲ *The yellow-nosed albatross is typical of the mollymawks, a group of rather small albatrosses with plain dark upperwings and colorful bills. Like all albatrosses it is a bird of the open ocean, seldom approaching land except to nest. This species breeds annually (not in alternate years like the great albatrosses) on several islands in the Indian and southern Atlantic oceans.*

albatross, Amsterdam albatross *D. amsterdamensis,* gray-headed albatross *D. chrysostoma,* light-mantled sooty albatross *Phoebetria palpebrata,* and sooty albatross *P. fusca*) means that these species, if they are successful in rearing a chick, breed only in alternate years.

The 14 albatross species can be conveniently divided into three groups.

Great albatrosses

The great albatrosses are at once recognizable by their huge size, the average wingspan being about 3 meters (10 feet). There are three species: the wandering albatross (breeding at most subantarctic islands) and the royal albatross, which are both circumpolar in the southern oceans, although the royal albatross is far less common away from its breeding islands around New Zealand; and the Amsterdam albatross, from the French subantarctic island of Amsterdam in the Indian Ocean, a recently described and nearly extinct species, with about five pairs breeding each year.

Mollymawks

This group includes nine smaller species of albatross with an average wingspan of 2.2 meters (7 feet). The term mollymawk comes from the Dutch word *mollemok* meaning "foolish gull". Three species breed in the north Pacific: the short-tailed or Steller's albatross *Diomedea albatrus,* from the tiny volcanic island of Torishima off Japan; the black-footed albatross *D. nigripes,* from the northwest Pacific; and the Laysan albatross *D. immutabilis,* from the Hawaiian archipelago. The waved albatross is the only equatorial species of albatross, breeding on the Galapagos and Isla de la Plata off Ecuador, and wintering in the rich waters of the cool Humboldt Current off Ecuador and Peru.

Sooty albatrosses

These are similar in size to the mollymawks and include two species only: the light-mantled sooty albatross and the sooty albatross. Both species breed on temperate and subantarctic islands in the Atlantic and Indian oceans, south and north respectively of the Antarctic Polar Front, and co-occur at a few sites. They are dark-plumaged birds with more slender wings and longer pointed tails than other albatrosses.

SHEARWATERS

The shearwaters are the most diverse family of the Procellariiformes, ranging in size from the giant petrels (the size of albatrosses) to diminutive prions, but typical members are about the size of small gulls. They get their name from some species' habit of skimming just over the surface of the sea. They have the widest distribution of any bird family, with species nesting 250 kilometers (155 miles) inland in Antarctica (the snow petrel *Pagodroma nivea*) and as far north as there is land in the Arctic (the northern fulmar *Fulmarus glacialis*). Despite this great variation, the family can be divided into four groups which have some common attributes.

Fulmars

The fulmars are a cold-water group, only venturing into the subtropics along cold-water currents. All of them have large, stout bills and feed on larger zooplankton and fish; most are proficient scavengers and characteristically associate with fishing vessels. Of the seven species, six are confined to the Southern Hemisphere. The single northern representative is the northern fulmar, which has shown a remarkable recent increase in range and numbers. Its southern counterpart is the Antarctic fulmar *Fulmarus glacialoides.* Three species of fulmar belong to genera with only one species in each genus: the Antarctic petrel *Thalassoica antarctica,* the Cape petrel *Daption capense,* and the only small all-white petrel, the snow petrel.

In contrast to these medium-sized species, which have an average wingspan of 1 meter, or 3¼ feet, the northern giant petrel *Macronectes halli* and southern giant petrel *M. giganteus* (both wide-ranging in southern oceans) are heavier than some albatrosses though have a wingspan of only 2 meters (6½ feet). Giant petrels take a wide variety of prey, but are notable as specialist scavengers at sea and ashore, where they exhibit vulture-like behavior at seal carcasses.

Prions

Prions are another southern group of petrels breeding chiefly on subantarctic islands. They are small (wingspan about 60 centimeters, or 2 feet) and are all very similar in size and appearance, blue-gray above and white below with distinctive W markings across their wings. They were once known as whalebirds because they were frequently seen in the presence of whales. There are about six species of prions (the status of some subspecies is still uncertain), which form two groups. Four species excavate underground burrows and have comb-like outgrowths from the palate to filter surface zooplankton from seawater. The other two species, the fulmar *Pachyptila crassirostris* and the fairy prion *P. turtur,* often nest among boulders, are less widespread (breeding mainly in Australian waters), and have less specialized bills.

Gadfly petrels

The gadfly petrels are a difficult group to classify and identify. Many species are very similar in coloration and markings, and there is controversy as to the number of distinct species (about 30). There is also debate as to whether the blue petrel *Halobaena caerulea* belongs to this group or to the prions. Gadfly petrels mostly nest underground in large colonies, but many species breed on remote islands where predators are absent.

They are medium-sized birds (wingspan about 80 centimeters, or 2½ feet), whose bills are short and stout with a sharp cutting edge and strong hook for gripping and cutting up small squid and fish. They are widely distributed in tropical and subtropical seas, particularly in the Pacific. Some have a very restricted distribution and breed on single islands, such as the cahow *Pterodroma cahow* from Bermuda, and the freira *P. madeira* from Madeira. Many species are very poorly known: the Mascarene petrel *P. aterrima,* for example, known only from Réunion by four specimens collected in the nineteenth century and three birds found dead in the 1970s; the magenta petrel *P. magentae,* recently rediscovered on a small stack off the coast of Chatham Island, New Zealand, 111 years after the unique type specimen was collected in the south Pacific Ocean; and Jouanin's petrel *Bulweria fallax,* described only in 1955, its breeding grounds still remaining unknown.

True shearwaters

The fourth group of shearwaters are the true shearwaters of the genera *Procellaria, Calonectris,*

and *Puffinus,* about 23 species in total. They vary in size, with a wingspan from 60 centimeters (2 feet) to 1.5 meters (5 feet). They are widespread and very mobile. Shearwater bills (except *Procellaria*) are proportionally longer and thinner than those of fulmars, prions, or gadfly petrels, and are mostly used for seizing fish or large zooplankton underwater. The four *Procellaria* species are the largest members of the group and are restricted to the southern oceans, although the black petrel *P. parkinsoni* is known to winter off western Mexico. The two *Calonectris* species are geographically isolated, Cory's shearwater *C. diomedea* breeding in the north Atlantic and the streaked shearwater *C. leucomelas* breeding in the north Pacific; both migrate south during northern winters. The genus *Puffinus* comprises about 17 species of small to medium-sized shearwaters; the short-tailed shearwater *P. tenuirostris,* which breeds off southern Australia, is sometimes referred to as the muttonbird because chicks are harvested commercially for meat and oil. Most shearwaters nest in vast colonies of burrows which they visit at night; they are highly social, often forming dense feeding "rafts" at sea. Several of the southern species are transequatorial migrants, wintering in the north Pacific.

STORM PETRELS

Storm petrels are small birds (average wingspan 45 centimeters, or 1½ feet) and are found in all oceans. Many species have striking black plumage

▲ A sooty albatross at its nest on Gough Island in the southern Atlantic Ocean.

▼ The tubenoses are most strongly represented in the southern hemisphere, but the northern fulmar is common in colder parts of the northern Atlantic and Pacific oceans, where it has increased dramatically in range and number this century. Fulmars are scavengers, and often congregate around fishing boats for offal.

Jean-Paul Ferrero/Auscape International

▶ Diving petrels use their wings more for propulsion underwater than for flight in air; their bodies are somewhat tubby and their wings small and rigid. In their general appearance and behavior they show some remarkable parallels with the auks, which do not occur in the Southern Hemisphere. This is the common diving petrel, the most widespread of the four species.

Graham Robertson

with a white rump, and all are colonial and nest underground, mostly on isolated islands. Long migrations are often undertaken. For example, Wilson's storm petrel *Oceanites oceanicus* migrates from its Antarctic breeding grounds to the subarctic oceans of the Northern Hemisphere, and Matsudaira's storm petrel *Oceanodroma matsudairae* appears to migrate along a west–east axis from the Pacific to the Indian Ocean.

The northern group of storm petrels, about 13 species in total, have pointed wings and short legs. They swoop down and pluck food from the surface of the sea. The 11 species of the genus *Oceanodroma* have moderately forked tails.

The seven species of southern storm petrels are characterized by short, rounded wings and long legs, which are often held down as the birds bounce or "walk" on the sea surface while feeding.

DIVING PETRELS

As their name implies, diving petrels dive for their food, simply flying into the water and disappearing, and then suddenly reappearing from the depths, traveling through air and water with equal ease. Their wings are small and broad, an adaptation for swimming underwater, and their flight is whirring rather than gliding.

Their tubular nostrils open upwards rather than forwards, which is presumably an adaptation to their diving habit. All four species, which belong to one genus, *Pelecanoides,* are found in the southern oceans. They are small, with an average wingspan of 30 centimeters (1 foot). Unlike many other petrels, diving petrels do not appear to make extensive movements even outside the breeding season and are usually seen in the waters near their breeding areas.

THREATS TO SURVIVAL

Traditionally the albatross was thought to be a bird of ill omen, embodying the souls of drowned sailors. To kill one was bad luck. Despite this superstition, made famous by Coleridge's poem *Rime of the Ancient Mariner,* the albatross hasn't escaped persecution, being caught by many seafarers to relieve the monotony of their diets on long voyages.

Discovery of the remote island breeding sites favored by albatrosses and other Procellariiformes has led to losses through egg-collecting, and almost wholesale slaughter at some colonies for their luxuriant feathers, much in demand as soft bedding and popular with ladies' fashions at the turn of the century. The short-tailed albatross suffered particularly at the hands of the plume-hunters. Once abundant in the north Pacific, by 1953 it was reduced to about 10 pairs on the volcanic island of Torishima (eruptions in 1902 and 1939 also contributed to its near-extinction). The population is slowly recovering, with 146 adults and 77 fledglings recorded in 1986, and breeding success has improved since grass transplantation to stabilize nesting areas on the precarious volcanic slopes.

The biggest threats today are from predators such as cats and dogs, introduced by humans to the remote islands where petrels breed, and from commercial fisheries. In addition to actual and potential competition for food, there is increasing evidence of substantial mortality from gill and drift-net fisheries, which mainly affect shearwaters, and from long-line tuna fisheries, which mainly involve albatrosses. This is particularly serious for birds such as Procellariiformes which have low reproductive rates.

PENGUINS

FRANK S. TODD

enguins form a distinct group of highly specialized, social, flightless pelagic seabirds, widely distributed throughout the cooler waters of the southern oceans. The greatest concentrations and largest number of species occur in the subantarctic between latitudes 45° and 60°S, with the greatest diversity of species in the New Zealand area and around the Falkland Islands. Only two species live south of latitude 60°S in the Antarctic. Penguins are absent from the Northern Hemisphere, although the Galapagos penguin *Spheniscus mendiculus* sometimes ranges slightly north of the Equator. Other species inhabit the mainland coasts and offshore islands of Australia, New Zealand, Patagonia, Tierra del Fuego extending northward to Peru, and offshore islands of southwestern Africa. Although penguins inhabit regions free of terrestrial predators, at sea they must contend with such efficient aquatic predators as carnivorous leopard seals and killer whales. In some areas, skilled aerial predators such as skuas take enormous numbers of eggs and chicks.

Order Sphenisciformes
1 family, 6 genera, 17 species

SIZE
Smallest Little blue or fairy penguin *Eudyptula minor,* total length 39 to 41 centimeters (15 to 16 inches); weight 0.7 to 2.1 kilograms (1½ to 4½ pounds).
Largest Emperor penguin *Aptenodytes forsteri,* total length 115 centimeters (45 inches); weight 19 to 46 kilograms (42 to 100 pounds).

CONSERVATION WATCH
The yellow-eyed penguin *Megadyptes antipodes,* Humboldt penguin *Spheniscus humboldti,* and black-footed or jackass penguin *S. demersus* are listed in the ICBP checklist of threatened birds.

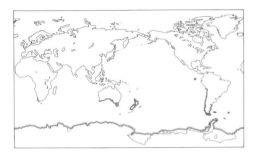

UNDERWATER SWIMMERS

Although flightless, penguins evolved from flying birds but have remained essentially unchanged for at least 45 million years. The most aquatic of all birds, some species may spend up to three-quarters of their life in the sea, coming ashore only to breed and molt. Many are migratory. With wings that have evolved into stiffened, flattened, paddle-like flippers, penguins are supreme swimmers, and the only other avian groups that swim underwater using their wings rather than feet are auks and their allies (in the Northern Hemisphere the counterparts of penguins) and diving petrels. While brief sprint speeds of up to 24 kilometers per hour (15 miles per hour) are attainable, penguins generally swim at 5 to 10 kilometers per hour (3 to 6 miles per hour). They are by far the most accomplished of avian divers. An emperor penguin *Aptenodytes forsteri* was recorded diving to 265 meters (870 feet) and remained submerged for

▼ *Penguins are often popularly associated with the icy wastes of Antarctica, but the Galapagos penguin is restricted to an archipelago lying almost exactly on the Equator.*

18 minutes, and even much smaller species such as the gentoo penguin *Pygoscelis papua* dive to depths exceeding 150 meters (500 feet). Penguins feed on fish, krill, and other small invertebrates, and cephalopods (squid), which are captured and consumed underwater.

All penguins are faced with thermoregulatory challenges: the polar penguins must conserve heat, whereas the temperate and tropical species have to shed excess heat. Thus the well-insulated south polar penguins generally have relatively smaller appendages, and feathering may extend well down on the bill. Conversely the tropical penguins have larger appendages and bare skin about the face which, when flushed, provides a mechanism for dissipating excess heat. No other group of birds is forced to endure air temperatures ranging from -60°C (-75°F) during the dark Antarctic winter to more than 40°C (105°F) at the Equator. Penguins depend on the insulative quality of their overlapping feathers to maintain their body temperature, the dense waterproof layer effectively trapping warm air. During the annual molt, when all feathers are lost simultaneously, a penguin is not waterproof and must come ashore or onto the ice. Molting birds cannot enter the sea to feed and therefore during the molt period of three to six weeks the fasting birds may lose a third or more of their body weight.

Like most seabirds, penguins tend to be rather long-lived, although juvenile mortality may be high. In the breeding season most species are highly territorial, but the emperor penguin forms large "huddles" during the winter. Some species do not become accomplished breeders until their tenth year. Upon hatching, the chicks are down-covered but are dependent on the adults for warmth and protection. Chicks are fed via regurgitation, and in surface colonies the parents recognize their young by voice.

THE SIX GENERA
The two largest and most colorful species, the king and emperor penguins, are both included in the genus *Aptenodytes*. Unlike other penguins which typically produce two-egg clutches, both lay a single egg which is incubated on top of the feet and covered by a muscular fold of abdominal skin. The emperor penguin is unique in that it breeds during the height of the dark Antarctic winter; only the males incubate for the entire incubation period of 62 to 67 days; and colonies are typically located on the annual fast ice, thus the emperor penguin is the only bird (under normal conditions) never to set foot on solid ground. The fasting period of 110 to 115 days endured by incubating males is the longest for any bird.

The smaller but more colorful king penguin *A. patagonicus,* of subantarctic regions, weighs nearly 20 kilograms (44 pounds) and is capable of producing only two chicks in a three-year period.

Graham Robertson

A group of emperor penguins crosses an ice field in characteristic single file.

◀ Most penguins walk on land, but the rockhopper penguin (far left) often bounds along with feet together, like someone in a sack-race. Second-largest of penguins, the king penguin (near left) breeds in large dense colonies, many of which exceed 100,000 pairs. Severely persecuted by humans last century, the king penguin is now increasing rapidly in number. Both species inhabit subantarctic islands.

The chicks spend the winter in large groups known as crèches where they are fed sporadically, and many perish. The chicks require nine to thirteen months to fledge, the longest fledging period of any bird. Formerly exploited for their oil, most king penguin colonies have recovered since being given legal protection.

The Adelie, gentoo and chinstrap penguins are collectively referred to as the long-tailed penguins. The Adelie penguin *Pygoscelis adeliae* is essentially restricted to the Antarctic and may be the most numerous of the penguins with a population of many millions. The chinstrap penguin *P. antarctica* occurs in an area known as the Scotia Arc, extending from the tip of the Antarctic Peninsula and including the South Shetland, South Orkney and South Sandwich islands, and South Georgia. The gentoo penguin *P. papua* inhabits mainly the subantarctic although some breed along the north coast of the Antarctic Peninsula. All three species nest during the southern spring and summer, October to February. The migratory Adelie penguin winters in the pack ice, whereas wintering chinstraps favor open water. Some gentoo penguin populations remain near their colonies year round, but others may disperse widely.

The six species of thick-billed crested penguins (genus *Eudyptes*) are essentially circumpolar throughout the subantarctic, and adults are characterized by prominent orange or yellow crests. All species lay dissimilar-sized two-egg clutches, and although some (at least three species) may hatch out two chicks, none is capable of fledging both young. Several species have very restricted breeding ranges—for example, the royal penguin *E. schlegeli* at Macquarie Island, and the Snares Island crested penguin *E. robustus* only at Snares Island.

The four species of basically non-migratory temperate and tropical penguins within the genus *Spheniscus* span the greatest latitudinal range: the New World species extend from the Galapagos Islands south to the tip of South America and the Falkland Islands. The black-footed penguin *S. demersus* lives in South African and Namibian waters. Most spheniscids are burrow- or crevice-nesters, but at some crowded colonies they may nest on the surface.

The burrow-nesting little blue or fairy penguin *Eudyptula minor* of Australia and New Zealand is the smallest of the penguins—up to 30 would be required to equal the weight of one large emperor penguin. The yellow-eyed penguin *Megadyptes antipodes* of New Zealand is the most endangered of all the penguins, with possibly fewer than 4,500 individuals still extant, a decline of nearly 80 percent since the 1950s and 1960s. Unlike most other penguins, nesting yellow-eyed penguins are not social, and if pairs are not visually isolated from one another they will fail to rear offspring. The loss of suitable nesting habitat and the introduction of terrestrial predators have had a damaging impact.

SIZE
Smallest Red-throated diver
Gavia stellata, total length
(extended) 53 to 70
centimeters (20 to 27 inches);
weight 1.1 to 1.7 kilograms
(2 ½ to 3¾ pounds).
Largest White-billed diver
Gavia adamsii, total length
(extended) 109 centimeters
(43 inches); weight up to
6.5 kilograms (14⅓ pounds).

CONSERVATION WATCH
Divers have declined in
numbers but are not at risk of
extinction.

**Order Podicipediformes
1 family, 6 genera,
c. 19 species**

SIZE
Smallest Least grebe
Tachybaptus dominicus, total
length 24 to 28 centimeters
(9½ to 11 inches); weight 110
to 150 grams (4 to 5 ounces).
Largest Great grebe *Podiceps
major*, total length (extended)
up to 70 centimeters (27
inches); weight 1.5 kilograms
(3½ pounds).

CONSERVATION WATCH
The following species are
listed in the ICBP checklist of
threatened birds: hooded
grebe *Podiceps gallardoi*, Junin
flightless grebe *P. taczanowskii*,
Madagascar little grebe
Tachybaptus pelzelnii, and
Aloatra grebe *T. rufolavatus*.
The giant pied-billed grebe
Podilymbus gigas and the
Colombian grebe *Podiceps
(nigricollis) andinus* are
probably extinct.

▶ *A great northern diver. Members of
this family of waterbirds are known as
divers in Europe, loons in North America.*

DIVERS AND GREBES

JON FJELDSÅ

Divers (known as loons in North America) and grebes (including the smaller
dabchicks) are superbly streamlined aquatic birds about the size of a duck or
small goose. Although classified in two separate orders—the Gaviiformes and
Podicipediformes, respectively—they were once thought to be closely related, but the
similarities are now considered to be the result of convergent evolution, involving
independent specializations for diving. Genetic comparisons suggest that divers are
related to penguins and petrels, which use their wings under water; and although today
the divers are foot-propelled, some anatomical details suggest that they had wing-
propelled ancestors. The affinities of grebes are still unclear, but certain anatomical
features suggest relationships with sungrebes or rails.

DIVERS, OR LOONS

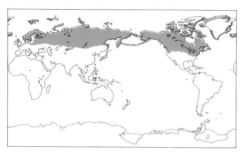

Divers are shy birds of lakes in northern forests and
tundra. Their wild hollering and yodelling calls,
audible for several kilometers, conjure up for many

people the essence of these lonely northern
wildernesses. After the breeding season the divers
fly to their winter quarters at sea, where they molt
their wing-feathers in October–December. They
travel rapidly under water in pursuit of fish and in
extreme cases can dive to a depth of 70 meters
(230 feet). Although preferring rather shallow
water, they usually stay well off the coasts, so the
best chance to see them is when they migrate over
certain projecting points along northern coasts.

In winter they are a rather uniform sooty gray
above and white below, and the species are hard to
tell apart, but for the breeding season their
plumage is adorned with the most elaborate
patterns. The red-throated diver *Gavia stellata*
acquires its rufous throat-patch and a pattern of

Wayne Lankinen/Bruce Coleman Ltd

Gary Neuchterlein

black and white lines on the hind-neck. The black-throated diver *G. arctica,* and its sibling species *G. pacifica* in eastern Siberia, Alaska, and Canada, have black and white lines on the sides of the neck and four white-checkered areas on their black back. The great northern diver *G. immer* of Iceland and North America, and the white-billed diver *G. adamsii* of Arctic North America and Siberia, are mostly black with their backs decorated by white square patches almost like a chessboard.

As they are unable to walk properly, divers place their nest at the lake's edge. This permits the bird to slip easily into the water. Most nest-sites are on peninsulas or small islands, places with a good view yet exposed to flooding from waves. Nests are usually just shallow scrapes in the boggy ground, but some pairs assemble reeds and waterweeds. The female lays two eggs on average, which are incubated for a month. The young are clad with sooty brown down and have two successive downy plumages (like the young of penguins and petrels). They can swim and dive at once, but the young of larger species prefer to ride on their parents' backs at first. The red-throated diver breeds mainly in small remote ponds and flies to larger lakes or out to sea for food. The other species prefer lakes large enough to satisfy the family's needs until the young fledge at 10 to 12 weeks of age.

GREBES AND DABCHICKS

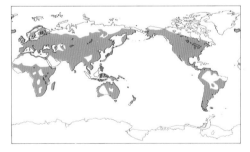

Slim, elegant, and adorned with crests and ruff, some grebes are considered the most fascinating birds on the lake. The best-known species is the great crested grebe *Podiceps cristatus,* widely distributed in temperate parts of Eurasia, Africa, and Australia. In North America its place is taken by the red-necked grebe *P. grisegena* and the western grebe *Aechmophorus occidentalis.* The smaller grebes, often called dabchicks, are less spectacular, resembling ducklings with small pointed bills. The feet, flexible and compressed, with semi-webbed toes, can work almost like a propeller, giving great maneuverability and speed when the birds pursue fish and invertebrates in the water.

▲ *Grebes are notable for their spectacular courtship displays, which involve both sexes in a variety of highly ritualized movements. These western grebes skittering over the surface on rapidly pattering feet are performing a distinctive display known as "rushing".*

Morten Strange/Flying Colours

▲ *A great crested grebe presents a feather to its chick. Feather-eating is common in several grebe species, but its purpose remains enigmatic: perhaps it is related in some way to a diet of fish.*

Grebes spend almost their entire lives in water. They even nest there, making floating platforms of plant material, which they usually conceal among reeds. The chicks can swim at once, but being sensitive to cold water they spend most of the time under the parents' wings.

Grebes have diverged considerably in their feeding adaptations. The primitive type of grebe is small and dumpy, living secluded among the reeds, taking water insects and making quick darts for small fish; such grebes exist worldwide, even in small creeks and ephemeral ponds. The large and streamlined grebes prefer larger lakes, for they are specialized for taking fish. The elegant western grebe can spear fish with its slender bill. Other dumpy and small-billed grebes are specialized to pick tiny arthropods from among the waterweeds. While fish-stalking grebes feed singly, the latter type are social feeders. Diet is largely determined by the anatomy of the bill, and where several closely related species occur in the same geographic area they have the most differently sized bills. Such "character displacement" reduces food competition between species.

Grebes usually avoid danger by diving, so flying is unnecessary for their daily life and they can therefore molt all wing-feathers at the same time. Some species congregate in very productive lakes when molting—750,000 black-necked grebes *Podiceps nigricollis* spend the fall at one lake in California. To become airborne the birds need a running start. They fly fast but, unable to make fast turns, may be vulnerable to birds of prey and therefore prefer to migrate at night. Northern

Hemisphere species spend the winter on icefree lakes or at sea. Grebes living under stable climatic conditions may not need to fly at all. A couple of species have entirely lost the power of flight and remain at one lake for their whole life: the Junin flightless grebe (puna grebe) *P. taczanowskii* in a highland lake in Peru, and the Titicaca flightless grebe (short-winged grebe) *Rollandia micropterum* in lakes of the Peruvian–Bolivan high plateau.

The grebes' courtship rituals are complex, and especially in the larger species the two partners often face each other for long bouts of mutual head-shaking. They also "dance", maintaining their bodies vertically almost out of the water, or they rush side by side in upright positions. The sexes look alike, but each sex is distinguished by fine voice differences.

THREATS TO SURVIVAL
The major threat to the survival of divers and grebes is the destruction of their freshwater habitat. Being specialized feeders, they are extremely sensitive to changes in their lake ecosystems — divers being adversely affected by acid rain, and grebes by pollution, and draining of their shallow breeding habitat. Human activities also have a detrimental affect on the breeding success of many species, especially that of the divers, who will readily surrender their nests if disturbed. Despite these threats, divers are not at risk of extinction, although numbers have declined in recent years. Two species of grebes, however, may already be extinct, and three species are listed in the ICBP checklist of threatened birds.

PELICANS AND THEIR ALLIES

E.A. SCHREIBER

**Order Pelecaniformes
5 families, 7 genera,
56 species**

SIZE
Smallest White-tailed
tropicbird *Phaethon lepturus,*
head-body length 35 to 40
centimeters (14 to 16 inches);
central tail feathers 35 to 40
centimeters (14 to 16 inches);
weight about 460 grams (1
pound).
Largest Dalmatian pelican
Pelecanus crispus and
Australian pelican *P.
conspicillatus,* total length 160
to 180 centimeters (63 to 71
inches); weight about 15
kilograms (33 pounds).

CONSERVATION WATCH
The following species are
listed in the ICBP checklist of
threatened birds: spot-billed
pelican *Pelecanus philippensis,*
Dalmatian pelican *P. crispus,*
Abbott's booby *Sula abbotti,*
New Zealand king cormorant
Phalacrocorax carunculatus,
pygmy cormorant *Halietor
pygmeus,* Galapagos flightless
cormorant *Nannopterum
harrisi,* Ascension frigatebird
Fregata aquila, and Christmas
frigatebird *F. andrewsi.*

O ne family in the order Pelecaniformes is particularly well known around the
world—the pelicans of the family Pelecanidae. These ungainly comic characters
with their large feet, waddling gait, long bill and huge pouch have attracted
attention throughout recorded history. The other four families are tropicbirds (family
Phaethonitidae), gannets and boobies (family Sulidae), cormorants and anhingas (family
Phalacrocoracidae), and frigatebirds (family Frigatidae). The one characteristic that
unites these diverse families is their totipalmate feet: all four toes are connected by a
web. The most familiar feature of the group is probably the large naked throat (gular)
sac, which can be spectacular in pelicans and male frigatebirds, but is entirely absent in
the tropicbirds.

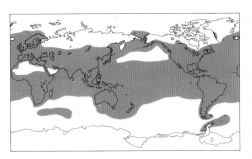

NESTING IN COLONIES
While primarily marine birds, pelicaniforms are
found in all types of water environments from the
open ocean and sea coasts to lakes, swamps, and
rivers. Pelicans and cormorants may switch
between fresh water and salt water during the year.
For example, the American white pelican *Pelecanus
erythrorhynchos* nests inland and migrates to coastal
waters during the non-breeding season. Most
species inhabit tropical and temperate regions, but
several species of gannets and cormorants live in
Antarctic and subantarctic waters.

Pelecaniforms commonly live more than 20
years, and many return faithfully to the same nest
site each year, mating with the same individual. All
species are colonial to some extent, nesting with
members of their own type, in isolated places such
as on islands. Cormorants and gannets often build

Graham Robertson

◄ *A colony of Cape gannets in South
Africa. Gannets nest in large dense
colonies in which all stages of the
breeding cycle are closely synchronized.
The birds vigorously defend their
territories, but these are so small as to be
effectively bounded by the reach of the
owner's bill while sitting on its nest.*

their nests within a few inches of each other. Aggressive interactions between neighbors are frequent in these dense colonies as each pair defends a space around their own nest. In contrast, masked booby colonies are often loose associations of birds nesting 6 to 60 meters (20 to 200 feet) apart. Frequently, a single colony will have two or three, or more, different pelecaniform species nesting in it.

In all species, both parents share incubation and chick-rearing duties. An individual's shift of sitting on the egg(s) varies from a few hours to more than a week, according to the nesting locale and the species. When the mate returns from feeding, the pair may go through a brief greeting display, allowing mate recognition and reinforcing the pair bond. If the incubating bird does not relinquish its duties, the returning bird, seemingly impatient to begin, may sit beside the bird on the nest and begin gently pushing it off. Incubation takes four to seven weeks depending on the species, and the adults either wrap the eggs in their large webbed feet or tuck them into the breast feathers to provide warmth. They do not have a brood patch. Chicks hatch naked in all pelecaniforms except the tropicbirds, whose single chick is covered by thick, fluffy grayish down when it hatches. Chicks must be carefully shaded from the hot sun during the day and kept warm at night. After two to three weeks, all chicks are covered with down and are frequently left alone while both parents forage for food. Adults regurgitate meals directly into the chicks' mouths—birdwatchers may wonder how any of the chicks get fed, when three nestling pelicans all have their long bills jammed into one parent's throat. Pelecaniform birds feed primarily

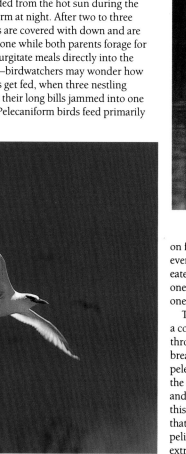

▼ A white-tailed tropicbird soars over its breeding colony in the Seychelles. Except when breeding, tropicbirds are solitary inhabitants of the open ocean, distributed as their name implies. They are characterized by a pair of slender tail streamers equal in length to the rest of the body.

C.A. Henley

on fish, but squid and other invertebrates and even other birds' eggs and small chicks may be eaten. In years when food is plentiful, more than one chick is raised in species that lay more than one egg.

The air-sac system, which enables birds to have a continuous supply of oxygenated air passing through the lungs both when breathing out and breathing in, is very extensive in many pelecaniforms. It also branches extensively across the chest and lower neck, making this area feel soft and cushioned to the touch. The suggestion is that this padding provides a shock absorber to species that dive into the water from the air (tropicbirds, pelicans, gannets, and boobies) and may provide extra buoyancy.

C.A. Henley

TROPICBIRDS

Tropicbirds differ in many characteristics and behaviors from other pelicaniforms, while sharing the one defining character of the order: totipalmate feet. In many ways they seem closely allied to gulls and terns (order Charadriiformes). They are widely distributed in tropical and subtropical seas, occurring in the Caribbean, Atlantic, Pacific, and Indian oceans, and are not seen on land during the non-breeding season. The three species (genus *Phaethon*) are similar in appearance, being white, with some black on the head, back, and primaries. They all have two long central tail feathers.

Amazingly, tropicbirds cannot walk—their legs are located too far back on the body—and to move on land they push forward with both feet and plop on the belly. They perform their courtship rituals in the air, flying around in groups of two to twelve or more, squawking loudly. A pair will land and push their way into their potential nest site, a hole in a cliff or under a bush, where they sit and squawk at each other, somehow deciding whether or not to form a pair. Tropicbirds are often referred to as boatswain birds, presumably because their harsh call is reminiscent of a boatswain's yell. They are plunge divers, diving into the water from varying heights to catch their fish or squid prey.

PELICANS

All seven species of these fascinating birds have the characteristic pouch, long bill, and long neck. The adults of five species are primarily white, with

▲ *A group of Australian pelicans glides in for a landing. In terms of body weight, pelicans are among the largest of all flying birds.*

► The blue-footed booby, like other gannets and boobies, feeds on fish. The birds plunge in spectacular dives from a height of 15 meters (50 feet) or so above the sea. The impetus of the dive takes it beyond the fish, which it snatches from below as it returns to the surface.

D. Parer and E. Parer Cook/Auscape International

some black in the primary and secondary feathers. The gray or spot-billed pelican *Pelecanus philippensis* is light gray with some black primaries. The brown pelican *P. occidentalis* has a very complicated sequence of annual adult plumage changes: it is silver gray on the back, with a black belly, white head, and white neck, but at different times of the year, depending on the breeding stage, the neck is chocolate brown and/or the head is yellow, and the pouch varies in color.

Most pelicans nest near and feed in fresh water, although all species are able to feed in either fresh or salt water. The brown pelican is the only marine species and is the only one to dive from varying heights into the water to catch a meal. The other species all feed as they sit on the surface, dipping down into the water with their bill and extensible pouch. Groups of pelicans will herd fish into shallow water where they are more easily caught. Brown pelicans especially are known to scavenge for meals around fishing piers and boats, unfortunately contributing to their decline in many places as they easily become entangled in fishing line which later snags on rocks or trees.

Adult pelicans are essentially voiceless, and courtship involves the use of "body language" (visual displays) rather than vocalization. Males pick a nest site in the colony—pink-backed *P. rufescens* and spot-billed pelicans mainly in trees; Australian *P. conspicillatus*, great white *P. onocrotalus*, Dalmatian *P. crispus*, and American white pelicans on the ground, and brown pelicans either in trees or on the ground—where they display by posturing, primarily with the head and neck, as females fly over. When a female is attracted, the two go through a series of coordinated displays. Once mated, the male brings nest material to the female who builds the nest. The number of birds in a colony ranges from as few as five pairs to several thousand. A large pelican colony is very noisy because chicks do have a voice and use it loudly to beg for food from their parents.

▲ The brown pelican (left) is unusual among pelicans in that it plunge-dives for its prey in a manner similar to that of gannets and boobies. Second rarest of the world's boobies, the blue-footed booby (right) is restricted to the eastern Pacific Ocean.

Leo Meier/Weldon Trannies

GANNETS AND BOOBIES

The three species of gannets live primarily in temperate regions, while the six species of boobies range throughout the tropical and subtropical regions of the world. Most boobies and gannets have a white head, neck, and underside, and are white with varying degrees of brown or black on the back. Colors of the soft parts of the bill and feet vary from bright sky-blue in the blue-footed booby *Sula nebouxii* to vivid red in the red-footed booby *S. sula*. Plumages are similar in both sexes, except the eastern Pacific subspecies of brown booby *S. leucogaster* in which females have brown heads which match the body and males have lighter brown heads. Male gannets are larger than females, but the reverse is true for boobies.

It is thought that the name booby is derived from the Spanish word *bobo* meaning clown or stupid fellow. Courtship can involve much parading around, lifting the feet up high, mutual preening, fencing with bills, and tossing of heads. When seen, it does make the name booby seem appropriate. Only Abbott's *S. abbotti* and red-footed boobies nest in trees, building a nest of twigs which may be lined with some leafy vegetation. The others lay their egg(s) on bare ground or build a nest of twigs or dirt.

Gannets and boobies dive like missiles, often from amazing heights, into the water to capture fish and squid, feeding alone or in flocks. Sometimes gannets may even pursue fish underwater, moving with powerful feet and half-opened wings. A few boobies are reported to be kleptoparasitic, chasing other boobies until they regurgitate, then stealing the meal.

CORMORANTS AND ANHINGAS

Most of the 28 species of cormorants (also known as shags) and four species of anhinga (or darters) live in tropical and temperate areas, but some inhabit colder Antarctic and Arctic waters. Some species are solely freshwater, others solely marine, and some are found in both habitats. One member of the family, the Galapagos cormorant *Nannopterum harrisi,* cannot fly. It hops in and out of the water, scrambling up rock ledges to roost or get to its nest site on predator-free islands. Tree-nesting species construct nests of twigs, whereas species that nest on rock islands or cliff ledges use seaweed, or even guano and old bones, to build their nests. Cormorants form some of the largest and densest seabird colonies in the world which, as might be expected, produce great quantities of excreta. This guano is mined in some areas of the world for fertilizer.

Most cormorants and anhingas are black and may have an iridescent green or blue sheen, while others have striking white markings. Their diet is mostly fish but includes smaller amounts of squid, crustaceans, frogs, tadpoles, and insect larvae. They generally pursue their food by swimming underwater. The legs and feet, placed far back on the body, may not make walking easy but they make great propellers. The fish-catching ability of

▲ *Cormorants lack any waterproofing agent in their plumage. This reduces the energy needed to remain underwater in pursuit of fish, but it also means that eventually the plumage becomes too waterlogged to continue. Cormorants therefore spend much of their time loafing at the water's edge while their plumage dries out, like these pied cormorants.*

Zefa/Wisniewski

and sit in groups with other males, all with the large red gular sac expanded like a huge red balloon. When a female flies over, the males begin bill-clattering and fluttering the wings. Females land and display with various males until a pair determines they are "compatible". The male then begins bringing nest material to the female, frequently stealing twigs from unwatchful neighbors. A single egg is laid.

Frigatebirds probably have the longest chick-rearing period of any bird. The young begin to fly at five to six months but return to the nest to be fed until they are up to a year old. Flying juveniles are often seen "playing" with sticks or other items around the colony. One will dip down, picking up a stick from the ground, and others will pursue it, agilely coming at the young bird from all directions trying to grab the stick. The bird drops the stick, which may be picked up by another juvenile, often before it hits the ground, and the chase is on again. They are no doubt learning an important skill for catching their own meals.

Walking is impractical for frigatebirds because they have very short legs and small feet, and the very long tail drags if they land on the ground. Some ornithologists have suggested that they do not sit on the water because their feathers are not well waterproofed. With their large wings (wingspan up to 2.5 meters, or 8 feet), small feet, and small body, they do have a very difficult time taking off from water, getting the wings at the correct angle for flapping and eventual flying. Frigatebirds therefore feed on the wing, grabbing fish or squid from near the surface of the water or catching flying fish. With their huge wing area and light weight, they have the lowest wingloading of any bird measured and can remain in the air for days. They also have incredible maneuverability and often chase and harrass other birds, particularly boobies, causing them to regurgitate and then stealing the meal.

THREATS TO SURVIVAL

The severest threats to pelecaniforms are caused by humans: disturbance of nesting colonies, and destruction of nesting habitat. The taking of birds for food causes the loss of many individuals. Predators such as rats, cats, and pigs introduced to islands continue to destroy breeding colonies; ground-nesters are particularly susceptible. Organochlorine pesticides such as DDT, which cause pelicans and other fish-eating birds to lay thin-shelled eggs that crush during incubation, are legally restricted, but the organophosphate pesticides now used are also highly toxic to birds. Oil spills and other forms of water pollution cause local mortality. Pelicans are the most endangered of the pelecaniforms, probably because they live in close proximity to humans. Safe nesting habitats and unpolluted food sources are critical to their survival.

▲ A male greater frigatebird with its bright red gular pouch inflated. The frigatebird's courtship display ranks among the most bizarre of all avian activities. Males perch in bushes on their breeding islands: as females fly overhead, each male inflates his pouch, clappers his bill, and waves his wings wildly about in an attempt to coax the female to land beside him, whereupon mating takes place.

cormorants has been exploited by humans since the sixth century AD, and a few fishermen in Asia still keep trained flocks of cormorants. A collar is tied around the neck of each bird to prevent it from swallowing fish; then with a long line attached, the fishermen let the birds dive and catch fish, pulling them back to the boat and taking the fish when they resurface.

FRIGATEBIRDS

The five species of frigatebird range widely over tropical oceans during the non-nesting season and nest on isolated islands. All five are very similar in size and appearance: adult males are all black, except for the Christmas frigatebird *Fregeta andrewsi* and the lesser frigatebird *F. ariel* which have some white on the ventral side. Females are the larger sex, and all have some white markings on the underside, except the Ascension frigatebird *F. aquila* which is dark. Males pick out a nest site

HERONS AND THEIR ALLIES

K.W. LOWE

Order Ciconiiformes
6 families, 43 genera,
110 species

SIZE
Smallest Little bittern
Ixobrychus minutus, total
length 27 to 36 centimeters
(10½ to 14 inches); weight 46
to 85 grams (1½ to 3 ounces).
Largest Goliath heron *Ardea
goliath,* total length 1.4 to 1.5
meters (4½ to 5 feet).

The order Ciconiiformes includes the herons, storks, ibises and spoonbills, whereas the order Phoenicopteriformes is made up of the flamingos. These two orders, plus the family Gruidae (the cranes), are usually known as the large waders because of their long legs and their generalized habit of feeding when wading in water. In contrast the small waders belong to the order Charadriiformes and are also known as shorebirds.

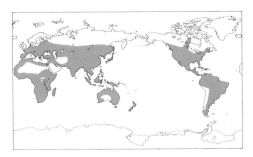

IDENTIFICATION BY BILL SHAPE
The feeding behavior and thus the bill shapes of the large waders are the best keys to their identification. The heron family shows the greatest evolutionary radiation with 61 species; they are recognized by their short straight bills, except for one or two species with unusually shaped bills for specialized feeding. Storks (17 species) are very large birds with correspondingly large bills of various shapes. Ibises (26 species) and spoonbills (6 species) belong to one family; ibises have sickle-shaped, downcurved bills, and the appropriately named spoonbills are easy to identify. The five species of flamingos have short, thick, curiously downcurved bills which immediately suggest an unusual feeding style.

Recently, the New World vultures (family Cathartidae) have been included within the order Ciconiiformes on the basis of genetic and some morphological studies. Before this, they were considered to be members of the order Falconiformes, and detailed information about them is included in that chapter of this book. There is considerable debate about the taxonomic status of these vultures.

Large waders are found throughout the world, except near the poles, and in all habitats. Many species are gregarious, and frequently huge groups

CONSERVATION WATCH
The following species are listed in the ICBP checklist of threatened birds: zigzag heron *Zebrilus undulatus,* Japanese night heron *Gorsachius goisagi,* white-eared night heron *G. magnificus,* slaty egret *Egretta vinaceigula,* Chinese egret *E. eulophotes,* Madagascar heron *Ardea humbloti,* white-bellied heron *A. imperialis,* shoebill *Balaeniceps rex,* milky stork *Mycteria cinerea,* Storm's stork *Ciconia stormi,* oriental white stork *C. boyciana,* lesser adjutant *Leptoptilos javanicus,* greater adjutant *L. dubius,* white-shouldered ibis *Pseudibis davisoni,* giant ibis *P. gigantea,* waldrapp *Geronticus eremita,* southern bald ibis *G. calvus,* crested ibis *Nipponia nippon,* dwarf olive ibis *Bostrychia bocagei,* and black-faced spoonbill *Platalea minor.*

Belinda Wright

◀ *A little egret at its nest. Around the world, egrets were once slaughtered in vast numbers for their plumes. These specialized feathers, worn only during the breeding season, were known in the trade as "aigrettes" and were used to decorate women's hats.*

287

▲ The painted stork (far left) is a large colorful Asian stork commonly seen in marshland reserves. The waldrapp or northern bald ibis (center left) is a cliff-nesting species of ibis from northern Africa. Highly endangered, it is the subject of an intensive conservation program. The green-backed heron (center right) is a common small species of the Americas, while the purple heron (far right) is a common large species of Europe and Asia.

of several species can be seen feeding, roosting, or nesting together. When species are solitary they occupy ecological niches that seem to favor a lone existence. Most species are migratory and cover long distances to escape cold weather, probably because the weather readily affects the availability of their food and also because their large bodies consume a lot of energy to keep warm.

HERONS

Perhaps the most confusing thing about herons is their names; what are herons and what are egrets? The word egret is derived from the term "aigrette" which describes the filamentous breeding plumes found in six species of white heron. The definition has since been broadened by popular usage to cover several other heron species of other colors and lacking the fancy plumes.

Herons are among birds that have specialized feathers, called powder down, which are never molted but fray from the tip and continually grow from the base. The tips fray into a fine powder, which the birds use to remove slime and oil from their feathers.

Within the family Ardeidae there are three main types: typical herons, which generally are familiar

to most people; night herons, which are mostly active at night; and bitterns, which are rarely seen but often heard producing their characteristic "booming" call.

Typical herons

Typical herons are a diverse group varying in size, color, plumage pattern, and feeding behavior. Possibly the best-known species is the cattle egret *Bubulcus ibis* which has successfully invaded most areas of the globe. This heron has specialized its feeding behavior to benefit from catching insects disturbed by the grazing of large animals. This habit must have developed in association with wild animals such as buffalo, rhino, hippopotamus, giraffe, and elephant, but nowadays their link with domestic stock such as cattle and horses and their habit of following working plows have assisted their near-cosmopolitan spread.

Most typical herons are simple variations of one theme. They vary in size from the rufous heron *Ardeola rufiventris* to the giant Goliath heron *Ardea goliath;* they are monochrome (white, black, gray) or a mix of two or more colors including rufous, green, maroon, gray, and a huge array of shades; and some feed by standing still and waiting for prey to come to them, while others actively pursue prey by running or by flying and pouncing onto prey on the surface of water.

There are, however, some bizarre forms, such as the boat-billed heron *Cochlearius cochlearius,* which ranges from Mexico to Argentina. The bill of this bird has been described as slipper-like; it is much broader and deeper than the usual heron-bill and is flat above and curved below. The shape is accentuated when the bird displays its backward-drooping fan-shaped crest. The bill seems to be an adaptation to increase the area and number of prey sensors and thus the chances of snapping shut onto prey when the bird feeds at night. Some have suggested that the boat-billed heron is a weird

variant of the black-crowned night heron.

The black heron *Egretta ardesiaca* of southern Africa has a remarkable feeding method. It inhabits shallow saline and fresh waters, where it bends its body forward, raises its wings above its back, and extends the wing tips downward to create an umbrella, casting a shadow on the water's surface. The bird's head is below this canopy of wings, which seems to act as a false refuge under which aquatic prey are tricked into sheltering.

Night herons
The night herons are so named because they feed mostly at night. They are short dumpy herons with rather thick bills and comparatively short legs. One

◄ *The yellow-billed stork is common and widespread over much of Africa. A desultory feeder and only mildly gregarious, it wades in shallow water catching frogs and other small aquatic animals.*

species, the black-crowned night heron *Nycticorax nycticorax*, is perhaps the most common heron in the world. They take a wide range of food items, including fish, frogs, snakes, small mammals, spiders, crustaceans, and insects. During the day most night herons roost quietly, usually high in trees, where the astute observer may see them.

Bitterns

Bitterns are inconspicuous, quiet creatures which inhabit densely vegetated wetlands and hence are rarely seen. They adopt cryptic postures, including standing fully erect with the bill pointed skyward, thus resembling the reeds and other thin vertical vegetation of their homes. They will even wobble slowly to resemble the movement of vegetation

blown by the wind. They have a curious adaptation whereby the eyes are placed widely on the head which allows them to see across a wide field of vision even when they are sky-pointing. All of this reflects their dependence on camouflage rather than flight to avoid predators.

STORKS

Storks are best known through tales about just one of the 17 species. The white stork *Ciconia ciconia* of Eurasia is prominent in folklore because of its image of natality. Its habit of nesting on buildings, often on chimneys, and its return to the breeding grounds in spring after wintering in Africa have long been associated with human birth. But inspiring as this species is, other members of the stork family are also intriguing. For example, one unusual variation is shown by the shoebill *Balaeniceps rex*. Imagine a huge dark gray bird that seems to be wearing a Dutch clog-shoe on its face! Well, it's true; this bird is found in wetlands in Africa. Unfortunately, despite its immense appeal, it has not been studied in any detail. The two species of openbill storks (genus *Anastomus*) are also easy to recognize. Their huge bills seem to have been bent about half-way along when they were biting on something hard. In fact they use this gap to grasp and crush freshwater snails, which seem a strange food supply for such a large animal.

Storks as a group have been more intensively studied than most of the other large waders. This can be attributed to a single researcher, Phillip Kahl, who spent more than a decade traveling the world to study all 17 species. Kahl's main interest was to try to tease out the evolutionary relationships between the species. This could be achieved with storks because they have elaborate and complicated courtship displays which they perform conspicuously. The results of Kahl's work have been presented in a fascinating series of eminently readable scientific papers. Perhaps more importantly, Kahl's other passion is photography, and through an extensive series of photo articles published by his main sponsor, the National Geographic Society, general awareness of the need to protect wetlands and wetland birds has been heightened.

IBISES

Ibises are another successful group that has radiated into most habitats and can be recognized by most people. They are strongly linked with wetlands, where they probe at and into soft sediments or water for their food. They also nest in wetlands, often in huge tightly packed colonies. However, several species have adopted a different lifestyle. One species, the hadadah *Bostrychia hagedash* of central and southern Africa, utilizes wet and dry forests for feeding and breeding. The bald ibis *Geronticus calvus* is confined to southern Africa in an area where large trees are rare because

▼ *The extraordinary shoebill is restricted to the interior of Africa, especially the Sudd region of southern Sudan. It resembles a heron rather than a stork in much of its behavior and, like herons, retracts its head back between its shoulders in flight.*

David C. Houston/Bruce Coleman Ltd

◄ Spoonbills are found on all continents except Antarctica. All six species are very similar in general appearance, and all but one (the roseate spoonbill) have white plumage like this African spoonbill.

frequent lightning strikes have burnt much of the native forest; this adaptable ibis now nests on sandstone cliffs, often beside waterfalls that are inaccessible to predators.

Unfortunately, ibises have more endangered species than the other families of large waders. The waldrapp *G. eremita,* a close relative of the bald ibis, is now restricted to a small population at only a few sites. The Japanese crested ibis *Nipponia nippon* was once widespread in Japan, but now only a handful of birds survives there in the wild, with another small population in China.

SPOONBILLS

These wonderfully unusual birds seem to be well liked throughout the world. The elongated and flattened bill resembles a spoon, which is whipped (partly opened) from side-to-side in sweeping strokes through the water until a prey item is touched. Acutely sensitive nerve endings lining the bill register the vibrations caused by the prey as it is touched, and the bill snaps shut in an instant. The bird then throws its head back, releasing its grip on the prey which falls down into the throat. Watching spoonbills feed has caused me to wonder

Order Phoenicopteriformes
1 family, 1 genus, 5 species

SIZE
Smallest Lesser flamingo *Phoenicopterus minor*, total length 100 centimeters (40 inches); weight 1.9 kilograms (4⅕ pounds).
Largest Greater flamingo *Phoenicopterus ruber*, total length 140 centimeters (55 inches); weight 2.1 to 4.1 kilograms (4⅖ to 9 pounds).

CONSERVATION WATCH
The Andean flamingo *Phoenicoparrus andinus* and puna flamingo *P. jamesi* are listed in the ICBP checklist of threatened birds.

why they don't get headaches or feel ill from all these jerking movements of their heads. Intensive studies of spoonbills are few, but work in Australia on the royal spoonbill *Platalea regia* has shown that these birds can be voracious feeders, sometimes consuming hundreds of prey items per day. In some marine ecosystems where they concentrate on a shrimp diet, their selective feeding on large, mainly female, shrimps seems to change the sex ratio in the shrimp population, so the number of shrimps that the spoonbills consume has an important effect on the whole ecosystem. This species can be found alongside the yellow-billed spoonbill *P. flavipes* at inland sites in Australia, where contrast between the birds and their feeding behavior is strong: the royal has a shorter, fatter body and bill than the yellow-billed; the former's bill is moved more rapidly through the water as the bird wades forward. These differences are reflected in the type of prey that the species catch at the same place: the royal catches the faster-moving prey such as fish.

Apart from the European spoonbill *P. leucorodia*, other spoonbills are not so well known. The beautiful African spoonbill *P. alba* is poorly named because all spoonbills have predominantly white plumage; rather it should have been named after its pink legs, bill, and face. The black-faced spoonbill *P. minor* inhabits the Orient but is not commonly reported. The roseate spoonbill *P. ajaja*, which inhabits the coastal swamps of the Americas, is well named and presents an exciting splash of color in its drably-colored habitats.

FLAMINGOS

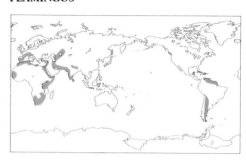

The bright pink of the flamingo's feathers and the strongly hooked bill instantly identify these birds. At first sight the birds look strange and awkward because their necks and legs are longer in proportion to their bodies than in other birds.

THE SACRED IBIS AND THE WISDOM OF THOTH

The sacred ibis *Threskiornis aethiopicus* had great importance in the everyday life of the ancient Egyptians. To these people certain animals symbolized individual gods, and the animals were worshiped as divinities. The sacred ibis was a symbol of Thoth, the god of writing and wisdom. This marvelous papyrus painting depicts the ibis-headed Thoth at the after-life ceremony involving weighing the heart of a deceased person. If the heart outweighed the feather of truth and judgment, it would be devoured by another animal-god; the result was recorded by Thoth, who reported it to the assessor-gods. Live birds were kept in temples and were mummified after death—huge collections, sometimes numbering half a million mummified birds, have been found in animal necropolises near the wetland breeding sites. Possibly, many of these were nestling birds that died of natural causes and were collected by the animal-protectors as a sign of respect. Unfortunately, the ibises' importance in human culture did not stop this species becoming extinct in Egypt by the mid-nineteenth century, although it thrives elsewhere.

▶ *This Egyptian papyrus painting dating from about 1250 BC depicts Thoth, the ibis-headed god of writing and wisdom, recording the result of the heart-weighing ceremony.*

C.M. Dixon

But their beauty and grace are soon revealed.

Flamingos are or were found on all continents, although only fossils are present now in Australasia. They seem to prefer salty or brackish waters through which they drag their bills upside-down. The upper mandible is lined with rows of slits, and the tongue is covered with fine tooth-like projections. The bill is opened, and then as the lower mandible is closed, water and mud are pumped out through the bill-slits. The residue probably contains a mix of microscopic food,

which is swallowed.

The best-known species is the greater flamingo *Phoenicopterus ruber* of Eurasia, Africa, and Central and southern America; it occasionally wanders from the Caribbean islands to Florida in the USA. All flamingos nest on lakes in shallow mud-cups scraped together so that they project above the water level. Only one egg is laid, and it takes about 30 days to hatch. The young birds leave the nests and herd together in large crèches, and are able to run and swim well at an early age.

▲ *A group of lesser flamingos feeding in the shallow waters of Lake Nakuru, East Africa. Lesser flamingos are more commonly found feeding by night in waters so deep that the birds swim rather than wade.*

RAPTORS

PENNY OLSEN

The Latin word raptor means "one who seizes and carries away", a term that immediately calls to mind the archetypal eagles, the most powerful of avian predators. Yet the order Falconiformes—one of the largest and most fascinating of all avian groups—includes an unexpected range of form and habit, from swift, bird-catching falcons, among the fastest of birds, to huge, ugly, carrion-eating vultures. Their great strength and remarkable powers of flight and sight have inspired, enthralled, and terrified humans for thousands of years. So potent is their image that even today they appear on the crests of many nations, as military insignia and the logos of many businesses, and terms such as "eagle-eyed" have universal meaning.

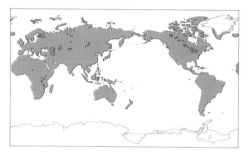

HOOKED BILLS AND SHARP TALONS

Raptors are characterized by hooked bills, strong feet, sharply curved talons, and large eyes. Their nostrils are surrounded by a fleshy cere straddling the base of the bill. Typically, they hunt in the daytime—which is why they are often called diurnal birds of prey—and eat live prey caught with their feet. Also characteristic is the crop (a pouchlike extension of the gullet) to store freshly eaten food.

Differences in the basic design are adaptations to the extraordinary variety of raptor behavior. The vultures soar on long, broad wings and have strong feet and relatively straight talons; they must search vast distances to find carrion, then hold it while they tear with their bills rather than subdue it. In general, the falcons have a muscular body, long, pointed wings and long toes—all features necessary for swift, agile flight and capture of airborne prey. Many of the kites have a relatively slim body and weak fleshy feet, reflecting their generally less predatory habits and their scavenging or collecting of easily-caught prey. Long-legged harriers and sparrowhawks reach into bushes or grass; sturdy-legged falcons hit prey in the air, often with great force. Some of the large forest eagles have massive legs and talons, and capture monkeys, sloths, and other large mammals from the trees. Falcons kill their prey with a blow or by biting the prey's neck and severing the spinal chord; hawks and other members of the family Accipitridae kill by the force of their grip, often compulsively squeezing the victim. All raptors use

their feet to hold prey and their bills to dissect it.

The plumage of most birds of prey is brown to chestnut, dark gray, or black, often with mottled, barred, or streaked undersides. The Australian race of the variable goshawk *Accipiter novaehollandiae* can be pure white. A few are adorned with crests, which they can raise in emotion. Some species have a distinctive juvenile plumage, and in a few the adult male differs in color from the female. Females are almost always larger than males, particularly in the agile bird-catching species; the diminutive male Eurasian sparrowhawk *A. nisus* is about half the weight of the female. In contrast, raptors that depend on immobile prey are the least dimorphic; the male of these species is about the same size as the female.

Raptors are renowned for their remarkable powers of flight and sight. From a distance of 1.5 kilometers (1 mile) a wedge-tailed eagle *Aquila audax,* a similar species to the golden eagle *A. chrysaetos,* can distinguish a rabbit from its surroundings; a human must approach to within 500 meters (550 yards). All raptors locate their prey primarily by sight.

Not least because they are prominent predators, raptors have suffered from pesticide contamination, habitat loss, and persecution, and some have long been coveted for falconry. Yet their image and appeal has also been to their advantage: considerable effort has gone into various programs to conserve species.

HABITATS

Raptors can be found in almost any habitat: from Arctic tundra to equatorial rainforest; arid desert to damp marshland; farmland to city. Because structural features of the habitat, rather than plant type, are most important to raptors, woodlands around the world tend to support a range of several different species. And raptors are not evenly distributed around the world: more than 100 species breed in the tropics, but only four in the high Arctic. Some habitats support raptors only at certain times of the year. For example, the rough-legged hawk *Buteo lagopus* of North America and

Tom & Pat Leeson

◄ National symbol of the United States of America, the bald eagle is now much reduced in number, and remains common only in Alaska.

Eurasia breeds above the treeline on open Arctic tundra, and then the entire population moves south to winter in farmlands and marshes. Eleonora's falcon *Falco eleonorae* deserts its barren, rocky nesting islands in the Mediterranean and coastal Morocco each winter and migrates to the humid woodlands and forests of Madagascar.

Some raptors can occupy a wide range of habitats, as long as there is suitable prey to catch. At the other extreme are the highly specialized species, dependent on a particular type of habitat. The snail kite *Rhostrhamus sociabilis* eats only snails, collecting them in the freshwater lowland

marshes of Florida, Cuba, and Mexico, south to Argentina.

Isolated islands often support lone species. For example, the Galapagos hawk *Buteo galapagoensis* is the sole raptor on the semi-arid Galapagos Islands. In New Zealand there are two diurnal raptors: the New Zealand falcon *Falco novaezeelandiae* in the forests and the swamp harrier *Circus approximans* in the open grasslands. The raptor with the most limited distribution is the endangered Hawaiian hawk *Buteo solitarius,* confined to one island.

Habitat and form are linked. Forest-dwellers such as goshawks tend to have short rounded

Francois Gohier/Auscape International

▲ A group of Andean condors congregates at a carcass. These are even-tempered birds, and squabbling while feeding is rare. Weighing up to 14 kilograms (31 pounds) and with a wingspread of 3 meters (10 feet), the Andean condor is the largest of the world's raptors.

wings for buoyant flight among the trees; open-country species have either long pointed wings for rapid flight (falcons), or long broad wings for effortless soaring (buzzards and eagles).

THE RANGE OF RAPTOR TYPES
The raptors alive today can conveniently be divided into 14 general types.

Family Cathartidae
New World vultures These vultures are typically vulturine in appearance and habit, but their resemblance to the Old World vultures is almost certainly superficial—the result of convergent evolution to suit them to a similar lifestyle—and they may more properly belong with the storks, in the order Ciconiiformes. Like the storks, but unlike the raptors, they have perforate nostrils (open from one side of the head through to the other) and squirt excreta onto their legs to cool themselves. All cathartids have a bare head and neck; the king vulture *Sarcorhamphus papa* has an extraordinary corrugated and wattled head in brilliant colors of red, orange, and yellow. The immense California and Andean condors are among the largest of all flying birds. Turkey

vultures *Cathartes aura* are medium-sized and slightly proportioned; they can locate carrion by smell alone, even rejecting carcasses too badly decayed.

Family Accipitridae
Osprey The osprey *Pandion haliaetus* is cosmopolitan in distribution, although in the Southern Hemisphere it breeds only in Australia. Its nostrils close as it makes spectacular dives into water for fish, sometimes submerging completely, either on the seacoast or on inland lakes. A reversible outer toe, long curved talons, and rough spiny toes help to grasp the slippery prey. Pesticides and persecution have decimated some populations, but local recovery programs have had good success.

Honey buzzards and white-tailed kites These birds are quite varied in appearance and habits. Some are extemely specialized: honey buzzards (especially the genus *Pernis*) eat the grubs and nests of wasps; the bat kite *Machaerhamphus alcinus,* which occurs in Africa and from Burma to New Guinea, catches bats at dusk; the hook-billed kite *Chondrohierax uncinatus,* of Mexico south to Argentina, neatly extracts land snails from their shells with its deeply curved bill; bazas (genus *Aviceda*) eat insects, much preferring green grasshoppers to brown, and occasionally figs. The small gray and white kites (genus *Elanus*), with an almost cosmopolitan distribution, are rodent-eaters and accomplished hoverers.

True kites and fish eagles   The black kite *Milvus migrans* eats many different foods but, with its taste for carrion, excreta and other refuse, can mass in hundreds around abattoirs and villages. Its large cousins, the sea eagles *Ichthyophaga* and *Haliaeetus* (including the bald eagle *H. leucocephalus*) snatch live fish from near the surface of water, catch rabbits, and scavenge. They are also aggressive pirates, stealing prey from other birds, even other raptors.

Old World vultures   Vultures are scavengers that rarely kill prey. They are incapable of sustained flapping flight and depend on rising air currents to keep them soaring aloft. Because of the vast distances they can travel in search of carcasses, they are exceptionally efficient and important scavengers. Most species have bare areas of skin on their head and neck, thought to reduce fouling of feathers when feeding and perhaps to help with heat regulation. The seven species of griffon vultures (genus *Gyps*) are usually numerous, and several hundred may gather at a food supply. They spread themselves in the sky and keep an eye on their neighbors; when one bird spots a carcass and descends, many others follow. They gorge on muscle and viscera deep within a beast, and their ample crops can hold about one-quarter of the bird's body weight. The cinereous vulture *Aegypius monachus* is usually dominant over other vultures bickering at a carcass; its powerful bill enables it to eat coarse tissue, tendon, and skin. The Egyptian vulture *Neophron percnopterus* cracks eggs and eats their contents; it picks up stones and hurls them at eggs too large to pick up and drop. Bones and tortoises are dropped from a height repeatedly by the bearded vulture *Gypaetus barbatus* which swoops down to eat the fragments; its habitat is

Jean-Paul Ferrero/Auscape International

◄ With its large size and heavy, powerful bill, the lappet-faced vulture usually dominates the various vulture species that quickly congregate at any carcass on the African plains.

▼ The white-bellied sea-eagle is an Australasian species. It is not confined to the coast, being common along major rivers and around larger lakes in the interior. It catches fish by spectacular plunges from a height above the water surface, but it also patrols shorelines for carrion.

▲ The extraordinarily flexible legs of the African gymnogene can be bent up to 70° behind and 30° from side to side, enabling it to grope in awkward tree hollows for the nestling birds and similar small animals on which it feeds.

▼ As a group, the harriers are birds of open country. Most species nest on the ground, and the spotted harrier of Australia is the only harrier that nests in trees.

open mountainous country from Spain to Central Asia. The unusual palmnut vulture *Gypohierax angolensis* of Africa is dependent on the fruit of the oil palm.

Snake eagles Powerful feet and short rough toes help the members of this group to grasp snakes and other reptiles. They resemble the Old World vultures in some characters and, like them, spend long hours soaring. Most live in Africa, but the short-toed eagle *Circaetus gallicus* occurs also in southern Europe and eastwards to India. Although placed with this group, the bateleur *Terathopius ecaudatus,* an all-but-tailless bird of Africa, eats mostly carrion.

Harriers A uniform group (genus *Circus*) of long-winged and long-tailed hawks. They characteristically fly low and slow over open country and drop on small reptiles, birds, and mammals. Cosmopolitan in distribution, the 13 species do not overlap geographically except in Australia, where two species occur, the spotted harrier *C. assimilis* and the swamp harrier. The spotted harrier breeds in trees like most of the raptors. The other species nest on the ground in long vegetation. Some are polygynous; one male will pair with up to six females. When not breeding, some species are quite gregarious and roost communally on the ground.

Sparrowhawks and goshawks The genus *Accipiter* is the largest of the raptor genera, with

about 48 species in woodlands and forests worldwide. Typically round-winged, long-tailed and long-legged, the largest is the northern goshawk *A. gentilis* (up to 1.3 kilograms, or almost 3 pounds), a powerful predator of large birds and mammals up to the size of hares. The smallest is the little sparrowhawk *A. minullus* (less than 100 grams, or 3½ ounces), which feeds on insects and small birds.

Typical buzzards and their allies Buzzards are medium to large hawks found in many habitats on all continents except Antarctica and Australia. Most are brown or gray in color. The 25 species of the genus *Buteo* soar readily on long broad wings and expansive rounded tails and catch mammals and reptiles on the ground. Harris's hawks *Parabuteo unicinctus* live and breed in small groups: perhaps four birds will hunt cooperatively, and after a jackrabbit is located it will be flushed by one bird towards the others waiting in ambush.

Harpy eagles These are very large, powerful eagles with unfeathered legs. The most powerful is the harpy eagle *Harpia harpyja* of tropical South America, followed by the monkey-eating eagle *Pithecophaga jefferyi* of the Philippines. Unusual among raptors, the latter has blue eyes.

True or booted eagles Legs that are feathered to the top of their feet distinguish these true eagles (including the genera *Aquila, Hieraaetus, Spiziastur* and *Spizaetus*), from the other eagles. They are

Tom & Pam Gardner

classic eagles in form and, in general, are active predators of mammals and ground-dwelling birds.

Family Sagittariidae

Secretarybird The secretarybird *Sagittarius serpentarius* is the only living representative in the family and has no obvious close relatives. It may not belong with the raptors, but is placed with them because it is eagle-like in appearance and in some habits. Large (standing about 1 meter, or 3 feet tall) and semi-terrestrial, it strides over the African grasslands. Prey are subdued with a kick from the long legs, short stout toes, and nail-like claws.

Family Falconidae

True falcons About 38 species of the genus *Falco* and seven falconets comprise the true falcons. The largest are powerful birds capable of swift flight, which often seize or kill their prey with a mid-air blow. Measured in a stoop a peregrine

falcon *F. peregrinus* reached 180 kilometers per hour (112 miles per hour), the fastest of all birds; it has the widest natural distribution of any bird. The gyrfalcon *F. rusticollis*, of the open Arctic tundra, captures ptarmigans flushed from the ground. Smaller kestrels habitually hover in search of ground prey; some are gregarious and live and breed in colonies. Most falcons breed as solitary pairs. The pygmy falcons (genus *Polihierax*), one in Africa and one in Southeast Asia, capture small birds and large insects on the wing.

Caracaras and others The caracaras are odd falconids. One species, the red-throated caracara *Daptrius americanus,* eats wasps and their nests, plus forest fruits and seeds. The other seven eat carrion. Unlike the true falcons, caracaras build a nest. Also of Central and South America are the forest falcons (genus *Micrastur*), which are five little-known primitive falcons superficially resembling harriers and goshawks.

▲ Kestrels of one species or another occur on all continents except Antarctica; illustrated here is the North American representative, the American kestrel (top left). The crested caracara (lower left) is one of a group of 9 species confined to Central and South America. They are chicken-sized birds that spend much of their time on the ground, hunting small animals or carrion. The wings of the bateleur (right), like those of a modern sailplane or glider, are unusually long and narrow, enabling it to swiftly cover large amounts of territory in search of food. It spends most of the day on the wing. In most other raptors the tail functions as a rudder, but the bateleur's relatively very small tail means that it must bank into turns in a highly distinctive manner.

▲ *The secretarybird of Africa differs from most other raptors in that it hunts on the ground, not from the air. It stalks about in open country, attacking prey ranging from grasshoppers to snakes and small mammals with powerful blows from its bill, or by stamping on them with its feet.*

naumanii typically nests in colonies of about 20 pairs, even as many as a hundred pairs, on a cathedral in Europe, an old fortification in southern Asia or, less often, a well-recessed cliff. The kestrels feed on insect swarms and, like other colonial species, hunt, roost, and breed together.

Most raptors are ostensibly monogamous, although some new evidence indicates that cuckoldry is quite common (the female mates with several males but only her permanent partner raises the nestlings). A few raptors, such as the harriers, are polygamous. Almost all raptors have courtship displays, often with mock aerial battles. The male bald eagle *Haliaeetus leucocephalus* commonly dives at the female flying below. In response, she rolls and raises her legs to him. Occasionally, she grasps his feet and the pair tumble spectacularly earthwards. Typically male raptors offer food to the females (courtship feeding). The accipitrids (family *Accipitridae*), the secretarybird, and caracaras build a nest; the falcons appropriate the nest of another species or use a hollow in a tree, or a cliff ledge.

Typically there is a clear division of labor during breeding: the male hunts, and the female carries out most of the incubation and brooding. She tears up food and offers it to the nestlings for the first few weeks after hatching. Unlike most other raptors, the secretarybird and honey buzzards regurgitate food from their crop to feed their young nestlings; and like the vultures, they also share nest duties more than is typical—behavior that is presumably related to the ease with which they can gather food.

Small species tend to have larger clutches, and shorter incubation and nestling periods than large species: a kestrel or falcon may lay four eggs, have an incubation period of four weeks, and the young are in the nest for another five weeks. In contrast, the cinereous vulture *Aegypius monachus* lays one egg, incubates it for eight weeks, and the nestling is in the nest for 17 weeks. A few species lay two eggs but raise only one young—the first-hatched chick kills the second. Smaller species mature at an earlier age. For example, a kestrel can breed at the end of its first year, whereas a vulture must wait until it is five or six years old.

At the end of the breeding season some species migrate. Most prefer to migrate over land, and therefore they funnel along land bridges such as the isthmus of Panama. Between September and November each year a vast stream of raptors passes over Panama on the way from North to South America for the winter: two million birds, mostly of two species (Swainson's hawk *Buteo swainsonii* and broad-winged hawk *B. platypterus*), have been counted, although many more go undetected. They travel as far as 11,000 kilometers (7,000 miles), taking as long as two months. Other raptors remain in their breeding area year-round, and some are nomadic and wander to wherever prey is available.

TERRITORIES AND MIGRATION

Generally territorial, in suitable habitat raptor pairs space themselves fairly evenly—about 1 square kilometer (⅓ square mile) for each kestrel pair, and 100 square kilometers (116 square miles) for each golden eagle pair. For some species the territory is permanent, but for others it is occupied and defended only during the breeding season; the latter may build a nest wherever prey is temporarily abundant. If a territory has a good nest site it may be occupied by successive pairs of raptors for centuries. The lesser kestrel *Falco*

WATERFOWL AND SCREAMERS

FRANK S . TODD

Collectively known as waterfowl or wildfowl, ducks, geese, and swans are among the most beautiful of all birds. They are included in the family Anatidae and thus are often referred to as the anatids. All species are remarkably similar in form. While the center of distribution is in the Northern Hemisphere, their distribution is worldwide except Antarctica. Among the first birds to be domesticated, ducks and geese have been raised as a food source for more than 4,500 years, and all domestic varieties have been derived from the mallard, muscovy, graylag, and swan goose. The three species of South American screamers are the closest surviving relatives of the anatids, although superficially screamers more closely resemble gallinaceous birds such as the domestic fowl.

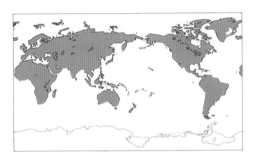

FLYING, SWIMMING, FORAGING

Most species are powerful fliers—although several are flightless—and many northern forms are highly migratory. Flying with continuous wing beats their speeds can exceed 110 kilometers per hour (68 miles per hour), but usually are considerably slower. Topographic barriers such as mountain ranges and oceans are not major factors in determining their distribution because most species can fly over them. Bar-headed geese *Anser*

◀ *The hooded merganser inhabits forested lakes and swamps in North America.*

Derrick Hamrick

**Order Anseriformes
2 families, 45 genera,
151 species**

SIZE
Smallest Indian pygmy goose *Nettapus coromandelianus*, total length 30 to 37 centimeters (12 to 14½ inches); female's weight 185 to 255 grams (6½ to 9 pounds); male's weight 255 to 310 grams (9 to 11 pounds).
Largest Trumpeter swan *Olor (Cygnus) buccinator*, total length up to 150 centimeters (59 inches); wing span up to 3 meters (10 feet); weight 9.4 to 11.9 kilograms (20 to 26 pounds).

CONSERVATION WATCH
The following species are listed in the ICBP checklist of threatened birds: northern screamer *Chauna chavaria*, West Indian whistling duck *Dendrocygna arborea*, lesser white-fronted goose *Anser erythropus*, Hawaiian nene goose *Branta sandvicensis*, red-breasted goose *B. ruficollis*, freckled duck *Stictonetta naevosa*, Baikal teal *Anas formosa*, New Zealand brown teal *A. aucklandica*, Hawaiian duck *A. wyvilliana*, Laysan duck *A. laysanensis*, marbled teal *Marmaronetta angustirostris*, scaly-sided merganser *Mergus squamatus*, white-headed duck *Oxyura leucocephala*, white-winged wood duck *Cairina scutulata*, Madagascar teal *Anas bernieri*, Baer's pochard *Aythya baeri*, Madagascar pochard *A. innotata*, and Brazilian merganser *Mergus octosetaceus*. The pink-headed duck *Rhodonessa caryophyllacea* and the crested shelduck *Tadorna cristata* are considered to be extinct.

301

indicus have been observed flying near the summit of Mount Everest at an altitude of 8,500 meters (28,000 feet).

With comparatively short legs and strongly webbed front toes, most waterfowl are excellent swimmers. Exceptions are the Hawaiian nene goose *Branta sandvicensis,* the Australian magpie goose *Anseranas semipalmata,* and the Cape Barren goose *Cereopsis novaehollandiae,* which have adapted to more terrestrial lifestyles and have less-webbed feet. The usual swimming speed is about 3 to 5 kilometers per hour (2 to 3 miles per hour), but faster speeds can be attained if the birds are pursued. Some species typically come ashore to roost, but others, such as the sea ducks and stifftails, generally doze on the water.

Most waterfowl have relatively long necks and flattened, broad, rather spatulate bills. However, the six species of mergansers differ, in that the bill is long, hard, and slender, with tooth-like serrations along the edges to aid in grasping slippery fish. In general, variations of bill shape reflect different feeding techniques; waterfowl feed on a wide variety of food items, and while grass, seeds, cultivated grain, and aquatic vegetation are favored by some, others will seek fish, mollusks, crustaceans, and insects. Aquatic invertebrates are particularly important for many ducklings and goslings, even though as adults they may feed primarily on plants. There are three main methods of foraging: grazing, surface feeding, and diving. True geese are noted for grazing and spend a great deal of time foraging ashore. Surface feeding is the most typical, not only for the least specialized of wildfowl, but also for some of the most specialized such as the Australian pink-eared duck *Malacorhynchus membranaceus* which filter-feeds mainly on blue-green algae. Dabbling ducks are primarily surface feeders, typically ingesting material from the surface of the water, but they also upend in order to obtain food from the bottom. Diving ducks submerge for their food and rarely leave the security of the water because the rearward position of their legs makes it difficult to move on land. Most rarely dive deeper than 3 to 6 meters (10 to 20 feet) although there are exceptions. The oldsquaw or long-tailed duck *Clangula hyemalis* and the king eider *Somateria spectabilis* can dive to at least 55 meters (180 feet). Sea ducks feed chiefly in salt water and have evolved efficient salt glands, located above the eyes, to facilitate removal of excess salt.

All wildfowl, except for the fish-eating mergansers, have a highly functional, sensitive, fleshy tongue which is lined with many small spiny projections. The action of the tongue, working

▼ A gallery of waterfowl. The northern screamer (below, left) is a member of a group of three species restricted to South America and isolated in the family Anhimidae. Other waterfowl belong to the cosmopolitan family Anatidae: top right, the red-breasted goose of western Asia; far right, the Carolina or wood duck of North America, and bottom right, the king eider, an inhabitant of the high Arctic.

against the rows of horny lamellae that line the mandibles, functions as an efficient food-sifting mechanism. The peculiar sound emitted during feeding is known as "chattering", and except for the flamingos (which some ornithologists believe are closely related) no other group of birds feeds in this specialized manner.

All anatids are densely feathered with compact waterproof plumage, as well as a thick coat of insulating down beneath. Of all birds thus far examined, the tundra (whistling) swan *Cygnus colombianus* has the greatest number of feathers—more than 25,000—most of which are on the head and neck. While a number of species are dull and nondescript, many others are brightly colored and patterned. Males are generally more ornate and vividly colored, but with some birds such as the Siberian red-breasted goose *Branta ruficollis* both sexes are colorful. During the annual wing-molt the flight feathers are shed simultaneously, resulting in flightlessness for a period of three to four weeks. Only the Australian magpie goose (and possibly also the continental population of South American ruddy-headed geese *Chloephaga rubidiceps*) has a graduated wing-molt, so it avoids a vulnerable flightless period. The drakes of many northern ducks change their colorful nuptial plumage to a cryptic female-like "eclipse" plumage during the summer.

Waterfowl vocal abilities vary considerably in tone, intensity, and quality. Most people are familiar with the typical duck quack, but wildfowl also honk, hiss, trumpet, grunt, bark, squeak, cluck, and coo. In many cases the sexes have different voices, the female's generally being lower. No species is totally mute, not even the so-called mute swan. A number of species also produce mechanical sounds: ruddy duck *Oxyura jamaicensis* drakes, for example, have specialized throat sacs that can be inflated and beaten upon with the bill to create a distinct drumming sound.

COURTSHIP AND NESTING
Well known for their highly ritualized courtship behaviors, waterfowl are unlike most birds that require only cloacal contact for fertilization, and males have a distinct erectile penis. Copulation typically occurs on the surface of the water, but there are several exceptions. Most anatids are monogamous, and polygamous behavior is confined to only a few forms. The duration of the pair bond is variable: lifelong for swans and geese; almost nonexistent for the muscovy duck *Cairina moschata,* a polygamous species of the neotropics. Many nest in the open, while others seek the security of cavities. Nest construction tends to be rather rudimentary because waterfowl do not carry nesting material, but use only material that can be reached from the nest site. Some species such as the snow goose *Anser caerulescens* and common eider *Somateria mollissima* nest in colonies. The

S. Nielsen/Bruce Coleman Ltd

most unusual nester is the black-headed duck *Heteronetta atricapilla* of South America which lays its eggs in the nest of a host species—other ducks, or gulls, rails, ibises, herons, coots, and even snail kites. In all but a dozen species the female alone incubates; the period varies according to the species, for ducks about four weeks on average. Most waterfowl line their nests and cover the eggs with a thick layer of down plucked from the breast. Except for the Australian musk duck *Biziura lobata* and the magpie goose, which are fed by their parents, waterfowl young must feed themselves from the very beginning.

WATERFOWL
Magpie goose
The magpie goose, or pied goose, of northern Australia is the most atypical of all the wildfowl: a gradual wing molt, long toes only partially webbed, polygamous breeding, onshore copulating, and adult feeding of young are not typical waterfowl traits. Almost exclusively herbivorous, this striking bird has a specialized strongly-hooked bill that is used to dig out food from the hard-baked clay during the dry season.
Whistling ducks
The eight species of essentially tropical, long-legged whistling ducks are noted for shrill whistling calls, often uttered in flight. Gregarious

▲ *The white-faced whistling duck is widespread in Africa and in South America. Like most whistling ducks it is strongly gregarious, and spends most of its time in groups or flocks at the margins of swamps, lakes and rivers. The sexes are alike in appearance.*

C.A. Henley

▲ *The black swan is common in most wetland habitats across southern Australia, but it tends to avoid the northern tropics. Recent studies have revealed an unusual flexibility in its breeding habits. It may nest whenever conditions are appropriate, and sometimes raises several broods in succession. Birds are ready to breed at 18 months of age. Desertion, divorce, and various kinds of polygamy are common in young breeders, but pair bonds become increasingly stable as the partners mature.*

for the greater part of the year, most are sedentary, but those in temperate regions are at least partly migratory. Unlike most anatid species, pairs of whistling ducks preen each other and the pair bond is very strong, possibly lifelong. In some regions, the ducks are considered detrimental to crops. One of the most cosmopolitan of all birds, the fulvous whistling duck *Dendrocygna bicolor* lives in North and South America, Africa, and India, with no significant variation in size or color.

Swans

Largest and most majestic of the waterfowl, the seven species of swan are indigenous to every continent except Africa and Antarctica. The northern swans are noted for unbelievably loud voices which is reflected in their descriptive vernacular names such as trumpeter, whooper, and whistler. Most swans construct huge nests, and that of the trumpeter swan *Cygnus buccinator* of North America may be a floating structure. Both sexes care for the young. Several species carry their young (cygnets) on their backs.

True geese

The 14 species of true geese are highly gregarious and are confined to the Northern Hemisphere, with most breeding in Arctic or subarctic latitudes.

Geese can be long-lived, and some captive birds have attained ages of nearly 50 years. Many, but not all, are somberly colored. The true geese are not sexually dimorphic, but the lesser snow goose *Anser caerulescens* has a distinct dark color morph known as the blue goose, which at one time was considered a separate species. The familiar Canada goose *Branta canadensis* has no less than 12 races ranging in size from scarcely larger than a large duck up to the size of a small swan.

Shelducks and sheldgeese

These birds live mainly in the Southern Hemisphere. They are all highly pugnacious, particularly during the breeding season, and many of the larger species have hard bony knobs at the bend of the wing which are used effectively in combat. Other than the northern swans, the adult kelp goose *Chloephaga hybrida* gander is the only all-white-plumaged waterfowl.

The Cape Barren goose of Australia has traditionally been regarded as an aberrant sheldgoose, but may represent a transitional link to the true geese of the Northern Hemisphere.

Steamer ducks

Often linked with the shelducks, all but one of these South American and Falkland Islands ducks

C.A. Henley

◀ Native to central Asia, the mute swan has been raised in a semi-domesticated state since the time of the Ancient Greeks. It is now common on ornamental waters across Europe, with a substantial feral population. In England, all mute swans by law belong to the Crown, a relationship extending back to the twelfth century. Numbers are now declining in England as many die of lead-poisoning through accidental ingestion of lead weights lost or discarded by anglers. One recent survey found 228 such weights per square meter of sediment.

▼ The Canada goose originated in North America, but it has been widely introduced elsewhere and is now common in a feral state in, for example, New Zealand and Great Britain.

are flightless, but are capable of "steaming" over the surface of the water with flailing wings at speeds up to 18 to 28 kilometers per hour (11 to 18 miles per hour). Excellent divers, the aggressive steamer ducks feed primarily on shellfish and other marine organisms.

Perching ducks

This is a widely varying group, ranging in weight from a mere 230 grams to 10 kilograms (8 ounces to 22 pounds). As implied by their name, perching ducks are more arboreal than the other anatids and most are cavity nesters. Included in the group are the pygmy geese (*Nettapus* species) and the spur-winged goose *Plectropterus gambensis*. Drakes of the American wood duck *Aix sponsa* and Asian mandarin duck *A. galericulata* have such complex patterns and bright colors that they are considered to be among the most beautiful of all birds.

Dabbling ducks

In this, the largest group, which has some of the most familiar of all ducks, the genus *Anas* alone consists of 38 species including the mallard, widgeons, pintail, shovelers, and many teal. Also known as river or puddle ducks, the dabblers prefer freshwater habitats, but may frequent salt water, particularly during migration. An iridescent,

Stephen J. Krasemann/Bruce Coleman Ltd

► *Most widespread, versatile and numerous of the dabbling ducks, the mallard occurs naturally across much of the Northern Hemisphere, but it has been widely introduced elsewhere. This is of some concern to conservationists because the mallard hybridizes freely with local species in many parts of the world, compromising their genetic integrity. It is the ancestor of the domestic farmyard duck.*

Andy Purcell/Bruce Coleman Ltd

▼ *The common pochard belongs to a cosmopolitan group closely related to the dabbling ducks, but which dive for their food. Although flocks of some species often congregate in sheltered bays and inlets of the sea, on the whole they favor fresh water over salt water. This species is abundant across much of Europe and Asia.*

Richard T. Mills

metallic, mirror-like speculum on the wing secondaries is typical of both sexes of most *Anas* species.

Most dabblers breed during their first year and lay relatively large clutches of eggs. A new mate is courted each year. Best known is the mallard *A. platyrhynchos,* which ranks among the most successful of all avian species. No wild duck is more tolerant of human pressure and disturbance, and they have been able to adapt well to a rapidly-changing human-dominated world. Mallards and their near relatives are worldwide in distribution, and even occupy some remote islands.

Pochards
The pochards are rather closely related to the dabblers, but most are excellent divers. Unlike the dabblers, which can become airborne instantly from the water, the pochards (along with eiders and sea ducks) have to run over the surface of the water for a considerable distance before taking flight. Some species such as the greater scaup *Aythya marila* form up in huge rafts, principally at sea. The largest pochard, the canvasback *A. valisineria,* is the prime target for most North American duck-hunters because of its succulent taste. Sadly, the spread of agriculture over much of the prairie pothole country, as well as the drought of the late 1980s, has had a damaging impact on this magnificent duck and many other anatid species that breed in this habitat.

Eiders
The four eider species are specialized diving ducks, all of which inhabit the Arctic and subarctic. The famous eider down is thick and heavy, with the best thermal quality of any natural substance; in some regions such as Iceland, eiders are still "farmed" for their down. Drakes in nuptial plumage are among the fanciest of ducks, although the female plumage is a well-camouflaged brown. Three of the eider species are extremely heavy-bodied birds, but are strong fliers nonetheless. Much of their time is spent at sea, frequently in rough waters, eating aquatic vegetation and large quantities of mollusks, sea urchins, crabs, and other crustaceans.

Sea ducks
Scoters, goldeneyes, bufflehead, and mergansers are among the most accomplished of anatid divers.

With the exception of the rare Brazilian merganser *Mergus octosetaceus,* all sea ducks inhabit the Arctic and temperate regions in the Northern Hemisphere. While many do spend considerable time at sea, most nest inland near fresh water. The exquisite harlequin duck *Histrionicus histrionicus* nests along fast-moving streams (often well inland), but gravitates to the sea during the winter. Its shrill whistling calls are clearly audible above the rushing roaring water. The oldsquaw is unique in that the drakes undergo three molts annually: the basically white plumage of winter is followed by an elegant brown with white for spring and early summer (it nests in the far north), and during the summer a dull-colored "eclipse" plumage is assumed. The largest merganser species, *Mergus merganser,* is known as the goosander in Britain because of its goose-like size.

Stifftails

Named for their distinctive stiff tails which the drakes often cock jauntily up into the air, the dumpy stifftails are so adapted to an aquatic, diving lifestyle that they can scarcely walk. During the breeding season the bills of the drakes of most species become a brilliant blue. Stifftails are essentially vegetarian, although the musk duck, a large Australian stifftail, is carnivorous. It is also the only stifftail that does not exhibit distinct sexual dimorphism in plumage; the male, however, is considerably larger than the female.

SCREAMERS

The three distinct species of neotropical screamers bear no superficial resemblance to typical waterfowl but are apparently closely related through a number of anatomical affinities. Like the magpie goose, they undergo a graduated wing-molt and thus do not have a flightless period. They are unique among living birds in lacking the uncinate process, the overlapping rib projection which strengthens the rib cage by serving as a cross-strut, a feature shared only with *Archaeopteryx,* one of the earliest-known avian fossils. In addition, screamers lack feather tracts; the only other birds that have feathers growing randomly are ratites, penguins, and African mousebirds.

Large, heavy-bodied birds with small heads and long legs, screamers are strong fliers and can soar for hours. Their very long toes exhibit only a trace of webbing, but despite this they are excellent swimmers although they do not dive. The long toes increase the surface area of the foot, distributing the weight and allowing them to walk over floating vegetation. The bill is distinctly fowl-like, and each wing is armed with two long, sharp, bony spurs. The largest species, the horned screamer *Anhina cornuta,* has a peculiar frontal "horn" that curves forward from the forehead toward the bill.

CONSERVATION

Despite the enormous pressures brought about by the increase in human population, it is heartening to note that most waterfowl species are holding their own, at least for the present. However, the extended drought of the late 1980s and early 1990s has severely reduced the numbers of North American species and the loss of both breeding and wintering habitat continues to be of major concern. Oil spills and other forms of pollution also have a detrimental impact on waterfowl populations.

▼ *In summer harlequin ducks inhabit rushing mountain streams in North America and Asia, but they spend the winter months at sea, congregating in bays, inlets, and fiords. The male is among the most striking of waterfowl, but the plumage of the female is mainly dark dingy brown.*

Tim Fitzharris

GAMEBIRDS

MICHAEL R.W. RANDS

Order Galliformes
7 families, 76 genera, 264 species

SIZE
Smallest Asian blue quail *Coturnix chinensis*, total length 14 centimeters (5½ inches); weight 43 to 57 grams (1¾ to 2 ounces).
Largest Blue peafowl *Pavo cristatus*, total length 2 to 2.3 meters (6½ to 7½ feet); weight 3.8 to 5 kilograms (8½ to 11 pounds).

CONSERVATION WATCH
There are 68 species listed in the ICBP checklist of threatened birds. They include the following: malleefowl *Leipoa ocellata*, rufous-headed chachalaca *Ortalis erythroptera*, Cauca guan *Penelope perspicax*, Djibouti francolin *Francolinus ochropectus*, Elliot's pheasant *Syrmaticus ellioti*, crested argus *Rheinartia ocellata*, white-breasted guineafowl *Agelastes meleagrides*, and ocellated turkey *Agriocharis ocellata*. The Himalayan quail *Ophrysia superciliosa* is probably extinct.

The galliforms or gamebirds are familiar to almost everyone, not necessarily because of their ornithological characteristics but more likely because of their nutritional value. "Chickens", which are domesticated forms of the jungle fowl of central Asia, and to a lesser extent turkeys from North America, are to be found on dinner plates around the world. Other gamebirds are known to hunters, notably the pheasant and several species of partridge, grouse, and quail, and yet others such as the peafowl are noted for their stunning beauty and ornate displays. Despite wide variation in size, all galliforms are characteristically stocky and have relatively small heads, plump bodies, and short broad wings. Flight is fast and usually low. Only the quail *Coturnix coturnix* is truly migratory, although several species are quite nomadic and some mountain dwellers make altitudinal migrations with the seasons.

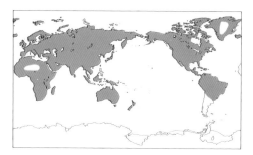

MEGAPODES

Members of the megapode ("large-footed") family have unique nesting habits. They lay eggs not in conventional nests, but in burrows or mounds. Thereafter parental care is restricted to maintaining the mound in such a way as to ensure stable temperatures for the eggs. The eggs are incubated by the sun, by organic decomposition of plant material or even by volcanic activity, and when the chicks hatch they are immediately able to fend for themselves. Of the 12 species distributed in Australasia and some Pacific islands, most inhabit rainforest and monsoonal scrub, but the best-known member of the family, the malleefowl *Leipoa ocellata* of southern Australia, inhabits semi-arid eucalypt woodland, where its numbers have declined markedly. All megapodes are vulnerable to human exploitation because the adults are conspicuous, largely ground-dwelling, tasty, and easy to shoot, and their eggs are large, highly nutritious, and easy to find. With widespread forest destruction, no less than eight species are threatened with extinction.

Several species build mounds as nests, the largest exceeding 11 meters (36 feet) in diameter and 5 meters (16 feet) in height. They are

▶ *Ready to lay, a female malleefowl approaches the mound while the guardian male looks on; he will drive her away if conditions in the mound are unsuitable for laying. Maintained entirely by the male, mounds remain in use year after year. Several females each lay several eggs in the mound at intervals throughout the nesting season, and the accumulated total may reach 30 or more.*

composed of leaf litter and soil and are used year after year. The malleefowl is strongly territorial, and the male maintains a mound of rotting vegetation throughout the year. Eggs are laid from September to January, in holes excavated in the organic matter by the male, who then re-covers the mound with sand and regulates its temperature at a constant 33°C (91°F) by changing the depth of sand as necessary. Other species on islands where soils are heated by volcanic activity lay their eggs in burrows where hot streams and gases provide the heat for incubation. Still more remarkable is the Niuafo'ou megapode *Megapodius pritchardii,* confined to Niuafo'ou island in the Tonga archipelago, which uses hot volcanic ash to incubate its eggs. The maleo *Macrocephalon maleo,* of Sulawesi in Indonesia, lays its eggs in tropical rainforest close to hot springs or on beaches exposed to the sun; egg collecting used to be supervised by rajas (kings) or other local authorities, and nesting grounds were leased and limits set for the number of eggs that could be taken. Unfortunately the loss of this traditional control has resulted in drastic overexploitation. Efforts are now being made to establish national parks and other protected areas where the maleo can survive alongside the local human population.

CHACHALACAS, GUANS, AND CURASSOWS
The Cracidae, or cracids, are found in the tropics and subtropics of the Americas, where they are largely arboreal (tree-dwelling). Within minutes of hatching the chicks show a remarkable instinct to climb and seek refuge in trees, and in a matter of days they can flutter from branch to branch in the canopy almost with the agility of adults. The smallest are the nine chachalaca species, which are the only ones not confined to humid forest, preferring the low woodland thickets more typical of the drier tropics. The guans (22 species) are medium-sized cracids and the most arboreal, while the curassows (13 species), which are the largest members of the family, spend up to half the day on the ground. In general the family is highly vocal, its members noted for the variety of their songs and cries echoing through the forest—the chachalaca name comes from its loud "cha-cha-la-ca" cry. In some species, notably guans, the windpipe is adapted for amplifying sound in the dense forest, resulting in the most far-reaching calls of all birds. Some species, in particular the curassows, perform elaborate courtship displays and courtship feeding; and many of the guans have a wing-whirring or drumming display (unique to the family), performed in flight.

TURKEYS
Two species of turkey exist: the common turkey *Meleagris galloparo,* native to the southern United States and Mexico; and the ocellated turkey *Agriocharis ocellata,* an endangered relative

L.C. Marigo/Bruce Coleman Ltd

confined to fragments of lowland rainforest in Mexico, Guatemala and Belize. It is believed that Mexican Indians were the first to domesticate the common turkey, which was subsequently introduced to Europe by the Spanish in the sixteenth century. Turkeys are large (males on average weigh about 8 kilograms, or 17 pounds; females 4 kilograms, or 8½ pounds), forest-dwelling birds with strong legs, large spurs and almost completely bare heads and necks. To attract females as mates, the males gather on strutting grounds known as leks. Here each male performs an elaborate display, spreading his tail fans, lowering and rattling his flight feathers and swelling up his head wattles. In this posture the males strut around, uttering their "gobble gobble" call. Because turkeys are polygynous (one male mating with several females), the dominant males achieve most of the successful matings at a lek.

▲ *The wattled curassow is widely kept in zoos and similar collections, but little is known of its behavior in the wild. Curassows inhabit the rainforests of Central and South America. They feed mainly on the ground and roost in trees.*

A willow ptarmigan in its pure white winter plumage is superbly adapted to deep snow and bitter cold, right down to its feathered toes.

Mark Newman

GROUSE

The more temperate zones of the Northern Hemisphere are the home of 17 species of grouse. Some live in coniferous forests, such as the spruce grouse *Dendragapus canadensis* and blue grouse *D. obscurus* of North America; others frequent more open habitats, such as the willow ptarmigan *Lagopus lagopus* of Europe's moorlands and mountainsides. In many northern areas, grouse are the largest year-round food source for predators, and their own numbers play an important part in determining the population size of some birds of prey. Some grouse undergo regular cyclic fluctuations in numbers, which have been studied extensively. Studies of the red grouse (a subspecies, in Britain, of *Lagopus lagopus*), for example, have shown that these cycles may be explained in some circumstances by fluctuations in parasite numbers. Other studies reveal variations in behavior of the grouse: aggressive birds hold large territories and so limit or decrease population growth; passive individuals tolerate crowding and so allow the population to rise again. Despite intensive investigations of blue grouse in North America and red grouse in Great Britain, the mechanism for such behavior is not yet fully understood. Humans also have an influence on gamebirds and their habitats: the desire to hunt game and therefore to maintain unnaturally high numbers leads people to control predators, provide additional food, and even release large numbers of artificially reared birds, which are all factors likely to distort natural population dynamics.

There is a wide range of reproductive social behavior. For example, the willow ptarmigan/red grouse is strictly territorial and forms a strong pair bond between male and female for the breeding season. In contrast, the sage grouse *Centrocercus urophasianus* males gather at large leks, each performing dazzling displays to attract females for a few moments, and the dominant male may well mate with more than 75 percent of the females in a "territory" not much bigger than two grouse.

NEW WORLD QUAILS

These are typically plump, rounded little quails, often boldly marked and with distinctive forward-pointing crests. All species studied so far are strictly monogamous and territorial during the breeding season; outside the breeding season they are gregarious, forming coveys or flocks. The subfamily includes three species of tree quail that inhabit montane forests in Mexico and Central America; the barred quail (*Dendrortyx* species) of Mexican arid scrub and woodland; the mountain quail *Oreortyx picta* of the woodlands of western USA; four grassland-dwelling North American species of crested quail, including the California quail *Lophortyx californica* (of which hunters shoot more than 2 million every year); four colins or bobwhites; and 14 species of wood quail

◄ *Despite its name, the California quail is common and widespread in western North America from Vancouver Island to the tip of Baja California. It has been introduced to many other parts of the world. Both sexes wear the unusually-shaped topknot, but the female is otherwise much duller and browner than the male.*

(*Odontophorus* species), medium-sized forest-adapted birds of Central and South America, two of which are threatened by deforestation in Colombia

OLD WORLD QUAILS AND PARTRIDGES

There are more than a hundred species in this group, occurring naturally from Africa east to New Zealand and north throughout most of Eurasia. The majority of species, however, are found in tropical Asia and in Africa south of the Sahara. A few have been introduced into North America.

Old World quail are small, rounded birds with short legs and relatively pointed wings. The common quail *Coturnix coturnix* is the only gamebird that regularly migrates from breeding grounds in Europe and central Asia to winter in Africa and India. Two other species, the harlequin quail *C. delegorguei* and the blue quail *Excalfactoria adansonii,* are nomadic and will move into areas in large numbers following rain; they inhabit grasslands and appear to show marked fluctuations in their population sizes.

The partridges are a much more diverse group: the very large, mountain-dwelling snowcocks which live in alpine zones from the Caucasus to Mongolia; the almost-unknown monal-partridges of China; seven species of red-legged partridges of Europe and the Middle East, adapted for life in arid regions; the larger francolins, predominantly of Africa; the spurfowl of Asia; the poorly known group of Southeast Asian species that inhabit tropical rainforests; and the so-called "typical" partridges, which were originally natural-grassland dwellers of northern Asia and Europe. The best-known is undoubtedly the gray partridge *Perdix perdix,* a native of Europe and parts of the Soviet Union but successfully introduced into North America. It has the largest clutch of any bird: 20 eggs. Throughout its range this bird is a prize

▶ The green peafowl is an even more handsome bird than the familiar blue peafowl, but it is somewhat more difficult to maintain in captivity, and hence not as widely known. It inhabits Southeast Asia.

C.A. Henley

▼ A male red junglefowl. This bird, the ancestor of the domestic chicken, was living in a semi-domesticated state in the Indus valley at least 5,000 years ago. It has since been taken by humans to virtually every corner of the globe, and it may well now be the most numerous bird on earth.

quarry for hunters, and although it initially adapted well to the intensification of agriculture in Europe it has since suffered an estimated 90 percent decline in population there, especially because of pesticides.

Almost all of the quails and partridges are monogamous, and pairs with their immediate offspring form the basic social unit, a covey. In the relatively few species that inhabit forests, adults live in pairs or alone year-round, whereas in species that occupy open habitats (snowcocks, red-legged partridges, common quail), coveys tend to amalgamate during winter months.

TRUE PHEASANTS, PEAFOWL, AND JUNGLE FOWL

All of the species in this group, apart from the tragopans, are ground-dwelling forest birds; and with the notable exception of the Congo peafowl *Afropavo congensis,* they all inhabit Asia. All males are spectacularly colored and adorned with a rich variety of vivid plumes for use during elaborate courtship displays. The crested argus *Rheinartia ocellata* has the longest tail feathers in the world. While each species has its own remarkable courtship routine, perhaps the most impressive is that of the great argus pheasant *Argusianus argus.* The male clears a hilltop in the forest as a dance floor. From here he gives loud cries soon after first light to attract females. When a potential mate

Morten Strange

appears he begins to dance around her, suddenly throwing up his wings into two enormous fans of golden decorated "eyes" which appear three-dimensional. The display culminates in an attempt to mate, after which the female will go off to nest and bring up a family, the male playing no further part. Most pheasants are either polygynous (one male mating with several females) or promiscuous. In some species, such as the ring-necked pheasant *Phasianus colchicus,* males defend a group of hens as a harem, forming bonds with them until the eggs are laid. This social system, which also occurs in the jungle fowl, is very rare in birds although it often occurs in mammals.

Pheasants have a close association with humans. A number of the most beautiful species have been released by (or have escaped from) collectors around the world. Feral populations of the golden pheasant *Chrysolophus pictus* and Lady Amhurst's pheasant *C. amherstiae* have established themselves in Britain.

The three peafowl species are famous for their train of decorated feathers, raised and fanned during courtship. Blue peafowl *Pavo cristatus* have adapted easily to living with humans, not only in

► *Reeves's pheasant is restricted to hill forests in central China. For many centuries its plumage (especially the tail feathers) was used by the Chinese as a decorative, ceremonial, or religious motif, a fact noted by Marco Polo in his writings.*

their native India—male peacocks and female peahens are also found in parks throughout Europe. The green peafowl *P. muticus,* however, is threatened by hunting and the destruction of its forest habitat in Southeast Asia. The Congo peafowl was discovered only in 1936 by J.P. Chapin, in the equatorial rainforest of what is now eastern Zaire, Central Africa. How this one species became isolated from all other members of the pheasant group which live in Asia is a real mystery.

The closest association between any bird and humans occurs with the jungle fowl. Chickens are

▼ *Tragopans are a group of five species restricted to high altitudes in the Himalayas from Kashmir to central China. They live in bleak, damp forests and spend much of their time in trees rather than on the ground. Males are characterized by rich red plumage, but females are dull brown. Pictured is the satyr tragopan of Nepal.*

Belinda Wright

▶ The vulturine guineafowl inhabits arid scrublands in East Africa, but six other species of guineafowl occur across Africa in most habitats from dense rainforest to open savanna. In guineafowl the sexes do not differ obviously in appearance.

kept wherever people live, including on the high seas. The red jungle fowl *Gallus gallus*—which still lives in the wild in India and Southeast Asia—was domesticated at least 5,000 years ago and since then has provided food (eggs and meat) to almost every human race on earth.

GUINEA FOWL

The seven species of guinea fowl all inhabit Africa, mainly in semi-arid habitats, sometimes in forests. They have largely naked and pigmented heads, with wattles and usually a crest or casque on top. Variation in the size and shape of these adornments, and the extent of naked skin, is believed to help the birds to regulate their brain temperature in different climates. All species are monogamous but congregate in flocks outside the breeding season. They feed opportunistically on the ground but roost, whenever possible, in trees. The best-known and most widespread (both naturally and as a result of domestication) species is the helmeted guinea fowl *Numida meleagris*. It is the only species that naturally occurs outside the Afrotropics—in both Morocco and the Arabian peninsula. Outside the breeding season, vast flocks of up to 2,000 birds gather towards dusk at waterholes; during the day, smaller groups regularly walk 20 to 30 miles while foraging.

HOATZIN

The hoatzin of tropical South America is an unmistakably prehistoric-looking bird. Superficially, an adult looks like an elongated pheasant: long-necked, long-tailed, and with a long, frizzled rufous crest. The body is predominantly brown above and rusty-red to reddish-yellow below. The face is bright blue with a prominent red eye. It has an extraordinarily large crop which is a highly specialized adaptation for grinding its food —the leaves, buds, fruits, and flowers of only a few species of tree and shrub. Breeding occurs year-round. A few days after hatching, the chicks leave the nest and are able to clamber in the trees using their feet, bill, and specially adapted wings that each have two "claws" on them. These claws disappear as the chick grows, and they are absent from adults.

This remarkable species is in its own genus. Its bizarre appearance has led to speculation that it provides the "missing link" between the reptiles and the birds, especially since the claws of the chick are quite similar in form to those of *Archaeopteryx*, one of the earliest-known bird fossils. However, the highly adapted crop and other specializations suggest that it is not a primitive species at all, and even the claws may in fact be the product of recent evolution and not a relic.

D. & R. Sullivan/Bruce Coleman Ltd

The hoatzin is restricted to flooded forests along the banks of quiet rivers in Amazonia. For much of the year it lives in groups of 10–20 or more, but during the rainy season smaller groups split off to breed.

CRANES AND THEIR ALLIES

GEORGE W. ARCHIBALD

Order Gruiformes
12 families, 61 genera,
220 species

SIZE
Smallest Hemipode-quails
11.5 to 19 centimeters
(4½ to 7½ inches).
Largest Sarus crane *Grus
antigone*, total length
153 centimeters (5 feet).

CONSERVATION WATCH
There are 48 species listed in
the ICBP checklist of
threatened birds. They
include the following: plains
wanderer *Pedionomus
torquatus*, black-breasted
buttonquail *Turnix
melanogaster*, red-crowned
crane *Grus japonensis*,
whooping crane *G. americana*,
New Caledonian rail
Tricholimnas lafresnayanus,
corncrake *Crex crex*, masked
finfoot *Heliopais personata*,
kagu *Rhynochetos jubatus*,
great bustard *Otis tarda*, and
Bengal florican *Houbaropsis
bengalensis*.

The gruiforms are predominantly ground-living birds that have a much greater propensity for walking and swimming than for flight. Given the isolation of an island habitat, some have become flightless. They usually nest on the ground or on platform-nests in shallow water, and the young are active immediately after hatching. Most have loud vocal displays, and in many species the male and female duet. Many have elaborate nuptial dances. From a ground-loving shorebird ancestor, the 12 families of gruiforms have radiated into a diversity of niches around the world.

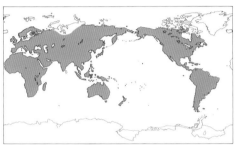

aberrant in that it has a flute-like voice and, at least in captivity, does not breed until it is five to seven years old. In contrast, other cranes have raucous trumpet-like calls and can breed in their third year.

Cranes are symbols of long life and good luck in

FLIGHTLESS ANCESTORS
Among the early gruiforms the gigantic *Diatryma* of the northern continents and *Phororhacos* of South America stood more than 185 centimeters (6 feet) tall, were flightless but possessed powerful legs and hooked bills. *Diatryma's* skull was almost as large as that of a horse, and *Phororhacos* was believed capable of chasing down fast-running mammals. Their closest surviving relatives are the two species of long-legged and flight-capable grassland birds of South America, the seriemas. Other ancient flightless gruiforms perhaps gave rise to ostriches; shorebirds of the order Charadriiformes are believed to be the closest order to the Gruiformes. The oldest species of living bird is the North American sandhill crane *Grus canadensis* whose bones are known from deposits dating back some 10 million years. Other species of now-extinct cranes, rails, and bustards date back 40 to 70 million years.

CRANES
Found in North America, Europe, Asia, Africa, and Australia, the crane family (Gruidae) includes 15 species of large wading birds. Cranes are best-known for their loud calls, spectacular courtship dances, monogamy, and the care they lavish on their young. The crowned crane *Balearica pavonina* of Africa is considered to be closely related to the ancestral stock that subsequently gave rise to the 14 other species. Within this latter group, the Siberian crane *Grus leucogeranus* is somewhat

▶ The distinctive crest and the fact that the birds frequently roost in trees are but two of the features that set the crowned crane apart from other cranes.

Stanley Breeden

▲ Close to extinction over much of Southeast Asia, the sarus crane is the subject of a vigorous program of reintroduction in Thailand, the Philippines, and elsewhere.

▶ Veiled by falling snow, a group of red-crowned cranes bugle in concert. Proclaimed as a "special national monument", a population wintering in Hokkaido, Japan, is gradually rebuilding its numbers from a low of 33 in 1952.

period. Cranes of the northern continents migrate thousands of kilometers between breeding and wintering areas. Juveniles learn the route by accompanying their parents throughout the autumn migration.

Hunting on the northern continents and, more recently, poisoning in Africa, in addition to the destruction of their wetland and grassland habitats, have eliminated these magnificent and space-demanding birds from much of their former range. Seven species are listed as endangered, and unless immediate conservation action is taken, several other species may soon be added. The rarest is the whooping crane *Grus americana* of North America. Reduced to as few as 15 individuals in the wild in 1941, the flock now numbers about 150. Cooperation between Canada (where the cranes breed) and the USA (where they winter) has been important in implementing a conservation program. Similarly, Asian nations are now working together to help five species of endangered migratory cranes. Captive propagation centers for cranes have been established in China, Germany, Japan, Thailand, the USA, and the USSR. Programs are underway to use captive stock to reintroduce sarus cranes *Grus antigone* into Thailand, whooping cranes into southeastern USA, and to bolster the western population of Siberian cranes. Cranes that have been reared in captivity in visual and vocal isolation from human keepers, manipulated through hand-puppets and crane-costumed persons, readily join wild cranes and thrive.

If protected, most species of cranes are remarkably adaptable to surviving in close proximity to humans. The tallest crane, the sarus of the Indian subcontinent, is protected by Hindus and thrives in densely populated areas, breeding at the edge of village ponds and foraging in the agricultural fields. Mongolians protect the demoiselle crane *Anthropoides virgo*, and on the vast grasslands the cranes often nest and forage near settlements. Likewise in regions of Africa where the crowned crane is protected, cranes roost at night on trees within the villages.

The most abundant species is the North American sandhill crane. More than half a million birds gather each March along 65 kilometers (40 miles) of the Platte River in central USA before continuing north to their northern breeding grounds. Thousands of tourists travel to Nebraska in spring to see one of Earth's greatest wildlife shows: the staging of the hordes of sandhill cranes.

LIMPKINS AND TRUMPETERS
Closely related to cranes are the limpkins of the family Aramidae, and the trumpeters of the family Psophiidae. Both families are native to the New World. There is but a single species of limpkin, *Aramus guarauna,* and it is found in the tropics. It uses its long curved bill to extract snails from their shells.

the Orient, and the attractive red-crowned crane *Grus japonensis* is prominent in their art. Cranes are, in fact, long-lived birds—one captive Siberian crane died at the age of 83 years, having successfully fathered chicks in his 78th year!

A crane pair will defend a large acreage of wetland and grassland as their breeding territory. Intruders are buffeted by trumpeting duets of the defending pair. In a secluded area, a platform nest is constructed in shallow water, and typically two eggs are laid. Both sexes assist in incubation and care of young; then the crane chicks remain with their parents until the onset of the next breeding

Jeff Simon/Bruce Coleman Ltd

▲ *The limpkin feeds almost entirely on large water snails, especially apple-snails of the genus* Pomacea. *It favors forested swamps but also sometimes occurs in marshes and scrub. In some parts of its range it is popularly known as the "crying bird" because of its eerie nocturnal wails and screams.*

In contrast to the water-loving limpkin, the three species of trumpeters are forest birds with soft feathers and short, somewhat curved bills. Like the crowned crane, trumpeters roost in trees and have velvet-like feathers on their heads and crane-like courtship dances. However, trumpeters are found only in South America, whereas the crowned crane is endemic to Africa.

RAILS

Perhaps the best known of the gruiforms is the family Rallidae, including 133 species of rails, gallinules or moorhens, and coots, although 14 species are probably extinct. Of all living groups of birds, the rails are the most likely to lose the ability to fly. One-quarter of the rail species endemic to islands cannot fly, and the introduction of rats, snakes, and domestic cats have brought about their demise. Captive breeding has been helpful for several species, and as predation and habitat problems are resolved, reintroductions have been possible.

One of the largest and most unusual members of the family is the takahe *Notornis mantelli*, a flightless purple bird from New Zealand's Fiordland. It was known from fossils, so the discovery of living birds was a great surprise. But grazing competition from introduced deer, and predation by introduced stoats, have reduced the takahe to perhaps fewer than 180 birds. Captive-

bred birds have been released and are now breeding on Maud Island.

Although many members of the Rallidae are threatened, others are remarkably successful. Almost every major wetland on all continents and many islands has a species of rail, gallinule, or coot. The New Zealand weka *Gallirallus australis,* when introduced to other, smaller islands, inflicted serious damage to vegetation and through predation brought about the demise of a petrel.

BUSTARDS

The family Otididae includes 22 species of long-legged, short-toed, broad-winged birds of the deserts, grasslands, and brushy plains in the Old World. The majority of the bustards are native to Africa; five species occur in Eurasia, and one is found in Australia.

Predominantly brown plumage provides cryptic coloration, and when alarmed, bustards often crouch and are difficult to see. Males are usually much larger than females and have elaborate ornamental plumes used in display. Their nuptial displays are bizarre—the males in some species inflate gular sacs and elevate their elongated and predominantly white neck feathers. To attract females, the male florican bustard *Houbaropsis bengalensis* of India and Southeast Asia makes helicopter-like vertical flights within its breeding territory. One male will mate with many females,

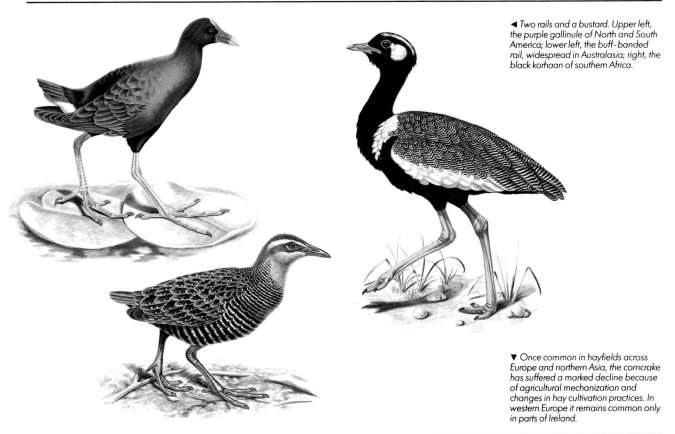

◄ Two rails and a bustard. Upper left, the purple gallinule of North and South America; lower left, the buff-banded rail, widespread in Australasia; right, the black korhaan of southern Africa.

▼ Once common in hayfields across Europe and northern Asia, the corncrake has suffered a marked decline because of agricultural mechanization and changes in hay cultivation practices. In western Europe it remains common only in parts of Ireland.

and incubation and rearing duties are usually the sole responsibility of the hen. Three to five speckled dark eggs are usually laid in a scrape on the ground.

Although loss of grassland nesting habitat to agriculture is the primary problem for the conservation of bustards, falconry is also a threat to several species. Falconry is a cherished tradition in several Arab cultures, and as a probable consequence bustards have been wiped out in much of the Middle East. Parties of falconers have recently traveled to India, Pakistan, and eastern Asia to continue their age-old sport. Concerned to keep bustards available for hunting, an elaborate captive-breeding center for bustards has been set up in Saudi Arabia. Unfortunately, bustards are difficult to breed in captivity.

In Hungary, before a bustard nest is plowed under, farmers collect the eggs for a government-supported hatching and rearing center. Later in the summer, juvenile bustards are released from rearing pens, and they join the local flocks of wild birds before migration.

UNUSUAL GRUIDS
Finfoots
The three species of finfoots (family Heliornithidae) are aquatic birds of the subtropics and tropics: one species in Africa, one in Southeast Asia, and one in Central and South America.

Richard T. Mills

▲ *The crested seriema and one other species constitute the family Cariamidae, restricted to South America. Seriemas are terrestrial, inhabit savanna, arid scrub, and thorny woodland, and feed on a variety of small animals, including snakes.*

▼ *Sunbitterns forage inconspicuously along the margins of forest rivers and streams in Central and South America. The sexes are alike in plumage.*

Generally grebe-like in external appearance, the finfoots have an unusual mixture of features that suggest affinities to cormorants, darters, ducks, coots, and rails. Drably colored and equally agile on land and in the water, these little-known birds frequent brushy edges of lakes and streams. They perch on branches over the water, seldom fly, but flap across the water in coot-like manner when disturbed. Two to seven cream-colored eggs are laid in nests built in the reeds or on low branches over the water.

Sunbittern

In the family Eurypgidae, the single species of sunbittern *Eurypyga helias,* like the finfoots, prefers heavily vegetated zones around streams and ponds. Found only in Central and South America, it is a teal-sized bird with intricately barred, soft brown, gray and black plumage, a graceful neck, and light and graceful flight. Nuptial behavior includes an elaborate display: while perched at a conspicuous spot on a branch, it spreads its wings and tail rigidly to form a continuous fan-like bank of attractively-banded feathers. Sunbitterns construct a bulky nest of sticks and mud in a tree, and both sexes assist in the incubation of the two or three nearly oval, blotched brown or purplish eggs. The young, although active immediately after hatching, remain in the nest for several weeks.

Kagu

The kagu *Rhynochetos jubatus* (sole member of the family Rhynochetidae) is found on the island of New Caledonia in the Pacific Ocean and is considered to be closely related to the sunbittern. It is slightly larger than the domestic chicken but with longer legs. It is incapable of flight but is able to glide. Pale gray with an elaborate crest, the kagu is a forest-floor bird. Like cranes, the male and female kagu have a loud and elaborate duet, and they engage in bizarre courtship displays in which they whirl around holding the tip of the tail or of a wing in the bill. They build a stick nest on the ground and produce a single egg. Both male and female assist in rearing the young. Eggs and young have fallen prey to introduced cats, dogs, pigs, and rats, and much of the kagu's forest habitat has been destroyed by humans in the course of farming and open-pit nickel-mining. It now survives only in a few inaccessible mountain valleys.

THE TINY GRUIDS
Mesites

The three species of mesites (family Mesitornithidae) of Madagascar are an ancient group of isolated lineage. Superficially resembling large thrushes but with the gait of a pigeon and the behavior of a rail, mesites are forest-floor species that seldom fly. The stick nest is constructed on a

Australian Picture Library

low branch, two or three eggs are laid, and there is some evidence of polyandry (the female mating with two or more males in the breeding season). As the forests in Madagascar continue to be destroyed, and if their primary egg-predator, the brown rat, continues to flourish, the mesites can be expected to decline in numbers.

Hemipode-quails
Polyandry and a three-toed foot are salient features of the hemipode-quails or buttonquails (family Turnicidae). With 15 species widely distributed in southern Asia, Africa, and Australia, these terrestrial birds, although true gruiforms, have some similarities with sandgrouse and pigeons. Females are larger and more brightly colored than males, and they have a loud booming call. They circulate among several males, and they are particularly aggressive in expelling other females from their breeding territory. Both sexes build the

ground-nest, but only the male incubates the clutch of four eggs. The incubation period is about 12 or 13 days, and within two weeks of hatching the young are capable of flight. First breeding can take place when hemipode-quails are only three to five months old.

Collared hemipode
The collared hemipode or plains wanderer *Pedionomus torquatus* (sole member of the family Pedionomidae) inhabits Australia's open grasslands and plains. It closely resembles its relatives the hemipode-quails; however, its persistent hind toe, egg shape, and paired carotid arteries suggest it deserves family status. The bird is compact, about 16 centimeters (6¼ inches) long, and the female's chestnut-colored breast distinguishes her from the male. The male incubates the four-egg clutch and rears the young. They seldom fly and prefer to run and hide when afraid.

▲ *The red-chested buttonquail inhabits rank grasslands in northern and eastern Australia, but it becomes progressively more scarce towards the south boundary of its range. As in other buttonquail, the male assumes sole responsibility for incubation and the care of the young.*

Order Charadriiformes
14 families, 82 genera,
292 species

SIZE
Smallest Least sandpiper
Calidris minutilla, total length
11 centimeters (4½ inches),
wingspan 33 centimeters (13
inches); weight 23 to 37
grams (¾ to 1⅓ ounces).
Largest Great black-backed
gull *Larus marinus,* total
length 64 to 78 centimeters
(25 to 30 inches), wingspan
150 to 163 centimeters (5 to
5⅓ feet); weight to 18.7
kilograms (41 pounds).

CONSERVATION WATCH
There are 33 species listed in
the ICBP checklist of
threatened birds. They
include the following:
Chatham Island oystercatcher
Haematopus chathamensis,
sociable plover *Chettusia
gregaria,* piping plover
Charadrius melodus, St Helena
plover *C. sanctaehelenae,* New
Zealand shore plover
Thinornis novaeseelandiae,
black stilt *Himantopus
novaezealandiae,* Jerdon's
courser *Cursorius bitorquatus,*
Eskimo curlew *Numenius
borealis,* slender-billed curlew
N. tenuirostris, spotted
greenshank *Tringa guttifer,*
Tuamotu sandpiper
Prosobonia cancellatus, Asian
dowitcher *Limnodromus semi
palmatus,* relict gull *Larus
relictus,* Kerguelen tern *Sterna
virgata,* black-fronted tern *S.
albostriata,* Chinese crested
tern *S. bernsteini,* and
Japanese murrelet
*Synthliboramphus
wumizusume.* The Canarian
black oystercatcher
Haematopus meadewaldoi and
Javanese wattled lapwing
Vanellus macropterus may be
extinct.

► *Not quite as strongly migratory as
many waders, the Eurasian curlew
breeds on moorland and winters on
coastal mudflats and river estuaries.*

WADERS AND SHOREBIRDS

COLIN J.O. HARRISON

The shallow waters of the world—pools, puddles, and the shores of lakes, rivers, and sea—seem always to have been rich in tiny creatures. They are therefore valuable habitats, and a whole order of birds, the waders or shorebirds, Charadriiformes, has evolved to exploit them. These places do, however, have some fundamental limitations: birds must wade or swim after food; and the shallow waters are vulnerable to rapid change, drying-up in hot areas, freezing in cold ones, so the birds must be able to fly far and fast should the need arise.

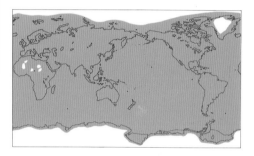

BIRDS OF COASTAL AND INLAND SHORES
This order of birds comprises several groups and subgroups that have diverged to take advantage of different resources. The typical waders are perhaps those of shallow waters and shores. Their legs are long and slender for wading, necessitating long slender necks and bills for feeding, and their wings are long and narrow for fast flight. Species that prefer estuaries and seashores tend to be a little heavier in build, with a wider range of bill sizes and shapes for snatching, probing, or prizing the more varied prey.

Gulls, a smaller and more uniform group, are mainly seabirds, although a few species use fresh waters. Legs tend to be stouter and feet webbed for swimming, enabling them to exploit the surface of deeper water as well. Their bill is generally thicker, and hooked a little at the tip, for they are scavengers as well as predators. They spend much more time on the wing in search of food.

The terns, a small group probably derived from gulls, have shorter legs and smaller feet, but long narrow wings—flight being more important than walking or swimming. Their tail is long and usually

Richard T. Mills

forked for quick maneuverability in flight, and the bill is tapered, for they are mostly plunge-divers, usually at sea, exploiting deeper water layers than gulls.

Finally the auks, the smallest main group, have moved out to exploit the deeper seas, swimming underwater after their food, and in consequence becoming penguin-like. The body is compact, with webbed feet set well back, and the wings are used as underwater flippers as well as in flight. Heads tend to be largish, and bills stout and strong. While much smaller and less specialized in structure than penguins, which live in the Southern Hemisphere, auks appear to be evolving to fill a similar place in the avifauna of the Northern Hemisphere.

WADERS, OR SHOREBIRDS

Waders are generally birds of open places, many sociable in their nesting. Nests are often sketchy or non-existent, and the eggs are relatively large, producing downy young that can be active from the moment they hatch. The typical waders, of which there are some 216 species, form the major part of this order.

Curlews, sandpipers, and snipe

The family Scolopacidae includes 88 species. They tend to have cryptically colored, brown or grayish, streaked plumage, but may assume brighter orange or black patterns when breeding. They share a similar basic structure—a long body, narrow wings, and long legs—but in size they vary from the eastern curlew *Numenius madagascariensis* at about 66 centimeters (26 inches) to the least sandpiper *Calidris minutilla* at 11 or 12 centimeters (4½ inches).

There are oddities, such as the short spatulate bill of the spoonbilled sandpiper *Eurynorhynchus pygmaeus,* but most bills are narrow, varying in length and curvature. They range from the long decurved bills of curlews, which probe deep into worm burrows in mud, and long straight ones for godwits searching mud under water, to shorter probes, and the small stubby bills of stints. The bills have sensory nerve endings near the tip, enabling the birds to sense food as they probe blindly. Most bills seem designed to exploit part of the range of organisms in the mud and sand of the birds' winter quarters. On rich mudflats large

▲ *The Charadriiformes constitute a diverse group. Top, the black skimmer, like other skimmers, has a highly asymmetrical bill for its habit of flying just above the water and dragging the lower bill in the water to catch food. The banded lapwing (center, below) is a common Australian plover, while the lesser golden plover (far left) migrates regularly from western Alaska and Siberia to as far south as Australia. During the breeding season the bill of the male Atlantic puffin (far right) develops bright colors.*

► It is characteristic of the Scolopacidae family that various species associate freely together on their wintering grounds, especially at high-tide roosts. At least three species are portrayed at this roost in northwestern Australia: the bird stretching its wings is a black-tailed godwit; most others of equal size are bar-tailed godwits, while the smaller birds are great knots.

Brian Chudleigh

flocks can feed together, and may move and maneuver in close rapid flight.

Almost all of these species nest in the Northern Hemisphere, many using the short summer of the Arctic tundra when waterlogged ground overlays the permafrost. The breeding grounds may not seem ideal for the adults' bill-shapes, but they have abundant insects that soon become food for the short-billed young. The open areas used by many for nesting encourage aerial displays, which are accompanied by the musical piping calls typical of these birds, and the dark-bellied breeding plumage of some species may be a visual advantage. The eggs, usually three or four, are often laid in bare scrapes—no nest material is carried, and only what is nearby is used. The downy young usually feed themselves from the earliest stage, with a few exceptions. Snipe and woodcock have proportionally the longest bills, and probe deep into the soft mud of inland marshes. The large laterally-placed eyes with almost panoramic vision watch for predators, not prey. These species bring food to their short-billed chicks.

The breeding season may be short, and when it ends the birds migrate south to other shallow waters or coasts, some even going as far as Australasia. Although often territorial when breeding they may be sociable at other times, with loud calls for maintaining contact.

Within this family there are many subgroups—and vernacular names such as whimbrel, redshank, greenshank, yellowlegs, willet, tattler, woodcock, dowitcher, surfbird, knot, sanderling, ruff—and many have specialized adaptations. The two turnstones (*Arenaria* species), with small plump bodies, stubby bills and strong necks, find food by flipping over stones and seaweed. The seemingly delicate phalaropes (*Phalaropus* species) swim and rotate to pick tiny creatures from the water's surface, and two species spend the winter far out at sea; the third, Wilson's phalarope *P. tricolor*, winters around South American inland waters.

In the phalaropes, and in some other families of waders such as jacanas, painted snipe, the Eurasian dotterel, and some sandpipers, the female is larger than the male, more brightly colored and dominant. She is polyandrous, leaving several males to care for clutches of eggs and the subsequent young. The ruff *Philomachus pugnax* goes to the other extreme: the breeding male grows a great individually-colored and patterned ruff, and ear-tufts, and these are displayed in ritual posturing on a communal "lek" or display ground where females mate with the most conspicuous and impressive males.

Oystercatchers

The oystercatchers, family Haematopodidae, are birds of shores, estuaries, and stony rivers. They are heavier in build, and the long bill has a blunt laterally-flattened tip that opens shellfish and chips limpets off rocks, as well as taking other prey. They are pied or black, loud-voiced, and aggressive. They bring food to their small young.

Plovers and dotterels

The 66 species of the family Charadriidae also show bold and bright color patterns. They have short, stout bills and feed in a "run and snatch" fashion on visible prey. They need rather bare open habitats, so coastal beaches and bare watersides support many smaller species. The golden and gray plovers winter on coasts, but nest on tundra or bare northern uplands. The boldly-colored and broader-winged lapwings and wattled plovers are birds of moister grasslands. The habitats of the others vary from the semi-deserts of Australia's interior to the Arctic/Alpine mountaintops of Eurasia.

They are territorial when breeding, usually with conspicuous aerial displays—the Eurasian lapwing *Vanellus vanellus* being famous for its erratic broad-winged tumbling.

Richard T. Mills

One small plover, the New Zealand wrybill *Anarhynchus frontalis*, is unique among birds in having a bill that bends sideways, to the right. It appears to be an adaptation to extracting insects from undersides of stones in riverbeds.

Thick-knees, or stone curlews

The stone curlews, family Burhinidae, are like very large, round-headed, and large-eyed plovers of

▲ *Oystercatchers of one species or another are found along seashores almost world-wide. The plumage of some species is entirely black; others are boldly pied, like this Eurasian oystercatcher.*

Klaus Wernicke/Silvestris GmbH/Frank Lane Picture Agency

◀ *Outside the breeding season the male ruff looks little different from the female (known as a reeve) except for his substantially larger size. In spring, however, males don a cape or "ruff" of long loose feathers around the neck and on the crown. This finery is used in courtship displays and differs widely in color and pattern among individuals. The color varies from pure white through a range of reds and browns to black; the feathers may be barred, spotted or plain, and no two birds are precisely alike.*

▲ *A Eurasian stone-curlew. Members of the Burhinidae are widely known as stone-curlews in Europe, thick-knees in North America, and dikkops in Africa. They have large yellow eyes, an oddly furtive manner, and loud wailing cries uttered usually at night. Several species inhabit arid scrub-lands.*

bare inland places and stony riverbeds. Cryptically striped, they tend to rest by day and are active at dusk and by night when their big staring eyes, pied wing-patterns, and loud wailing and whistling calls become apparent. The seven small *Burhinus* species have short stout bills, but in the two larger *Esacus* species these become longer and more massive. The beach thick-knee *E. magnirostris,* of Oriental and Australasian coasts, has the largest bill and feeds on crabs and marine invertebrates, living and nesting on the shore.

Crab-plover

The crab-plover *Dromas ardeola* is like a very large plover or stone curlew, with a massive bill as long as the head. It has a conspicuous black and white plumage and is a shorebird of Indian Ocean coasts and islands, feeding on crabs. It breeds in colonies on sandy areas, digging a tunnel more than 1.5 meters (5 feet) long, and laying one white egg. The young remains in the tunnel while the adults bring food, and it is still fed after it leaves the nest. Adults fly slowly and low, often in chevrons, with the head supported on the shoulders.

The adaptive radiation of the waders that produced these various groups of birds has also resulted in the evolution of more-specialized small groups variously adapted to wetter conditions, to deserts, to mountains, to aerial hunting, or to life on the Antarctic borders.

Ibisbill

The ibisbill *Ibidorhynchus struthersi,* a stoutish gray bird with black on the face and breastband, is of

▶ *The crab-plover is restricted to tropical shores of the Indian Ocean, where it frequents sandy beaches and mudflats, feeding largely on crabs. Compared to other waders, its breeding habits are extraordinary: it breeds in dense colonies in sand dunes, digging burrows in which the single, pure white egg is laid. The young are dependent on their parents for food and, as in the oystercatchers, this dependence persists to some extent even after fledging.*

Geoff Longford

doubtful affinity, tentatively linked with both stilts and oystercatchers. Its heavy build, shortish legs, and curved bill may be adaptations for feeding in the stony mountain streams where it probes for food. It is confined to southern Asian uplands.

Painted snipes

On lower wetlands there are two painted snipe (genus *Rostratula*), already mentioned for their sex-role reversal. They are birds of tropical lowland swamps, one species in the Old World, the other in South America. They have slightly curved long bills and broad wings, and feed on insects and worms, nesting within the swamps. They are furtive and rather silent birds, active at dusk and dawn, and as a result little is seen of the beautiful intricately-patterned plumage that is shown in flight and in spread-wing displays.

Stilts and avocets

Stilts and avocets, family Recurvirostridae, are more cosmopolitan in distribution and occur around more open waters. They are slender and long-legged, and unlike other waders they have webbed feet. The banded stilt *Cladorhynchus*

leucocephalus has the smallest range, on saline waters of southern Australia. Black, chestnut, and white, with long thin legs and bill, it feeds in shallow waters, mainly on brine shrimps; it is nomadic to take advantage of rain at inland lakes, and at suitable sites will form large breeding colonies of shallow scrapes near water. The black-winged or pied stilt *Himantopus himantopus* is seen on every warm continent, in forms varying in the black coloration on the head and neck. The neck is long; the bill long, thin, and straight; and the legs so long that it must tilt its body to reach the ground. It looks absurd, but when wading belly-deep in water, picking food from the surface, appears natural. It is a freshwater species, nesting in loose colonies, and making a quite substantial nest on a tussock in the water. It has a yelping call and flies with long legs trailing behind. The feet are only slightly webbed.

The avocets (genus *Recurvirostra*), one species on each warm continent, have pied plumage with heads varying from black and white to chestnut, buff or all-white. Their very thin, long bill is upturned, and they feed in shallow brackish waters, sweeping the bill from side to side in search of tiny prey.

▲ *As a breeding bird, the banded stilt is characteristic of the most inhospitable deserts of Australia, and few nesting colonies have ever been found. In its behavior it shows some remarkable similarities to the flamingos: like them, it is strongly gregarious, favors arid salt lakes, feeds largely on shrimps, often swims as it feeds, and breeds erratically in very large dense colonies.*

► *Like other jacanas, the lotusbird or comb-crested jacana prefers deep, permanent lagoons with an abundance of floating vegetation. This species occurs in Indonesia, New Guinea, and Australia; others inhabit tropical wetlands around the world. During the breeding season the birds are strongly territorial, but they often congregate in flocks at other times of year.*

Belinda Wright

Jacanas

The family Jacanidae has gone a stage further as waterbirds. There are five species in warmer climates, rail-like with rounded wings and sharp bills; four have forehead plates or wattles. The large legs are long and strong, with long splayed long-nailed toes that can support the bird's weight as it runs over floating vegetation or waterlily leaves. They live on still inland waters. The skimpy nest

rests on water plants. The adults can move the boldly-scribbled eggs or downy young a little way by gripping them under the closed wings.

Coursers and pratincoles

The desert waders are classified as "coursers" in the family Glareolidae. The little Egyptian plover *Pluvianus aegyptius* is an odd species of river sandbars and banks. It is boldly patterned, and ploverlike in appearance and feeding. With rapid kicking, it buries its two to three eggs in warm sand for incubation, and cools them with water from wet belly feathers; it cools chicks in the same way and buries them too if danger threatens. The eight coursers are larger birds with long legs and short toes, and small curved bills. They occur in dry country and desert borders of Africa, Arabia, and India. They are well camouflaged and can run fast, tending to crouch in alarm, but able to fly well and for long distances.

This family also contains the aerial waders, the pratincole species, which occur through warmer parts of the Old World. These show some swallow-like adaptations for catching insects on the wing, although they can also run fast on long legs. They have more pointed wings, and forked tails, and the bill is short but with a very wide gape. They live on large rivers or marshes and are highly gregarious, feeding in flocks with constant high-pitched calls. Their flight is swift and agile. The Australian pratincole *Stiltia isabella* has longer wings and legs and is a migrant, nesting in inland gibber (stony) deserts within reach of water.

Seedsnipes

High-latitude grassland and lower southern plains in South America support the four species of seedsnipe (family Thinocoridae). These are curious gamebird-like waders resembling partridge or quails. They are plump, mottled brown, with short stout legs and short deep bills. Their flight is snipelike and they feed on seeds and plants.

Sheathbills

Least wader-like are the two sheathbills (*Chionis* species) of the Antarctic shores. Like large white pigeons or small domestic fowl, they have short strong legs and run well. Their flight is pigeon-like but lazy. The posture is upright, and they have a rounded head and a very stout deep bill, with bare wattles around the bill-base and the eyes. They are scavengers, haunting the breeding colonies of penguins, cormorants and seals, taking eggs, carrion and feces, and at other times feeding on small shoreline creatures or around rubbish dumps. They nest in scattered pairs in crevices and rock cavities and usually rear two young.

GULLS AND GULL-LIKE BIRDS
Skuas

The skuas or jaegers (family Stercorariidae), consist of seven gull-like species, but are brown in color with white wing-patches. The largest species (genus *Catharacta*) are like big, heavily built, broad-winged gulls. Far-ranging, they breed on shores of cold seas, with three in the south to Antarctic regions, and one in the North Atlantic. They scavenge around seabird colonies, taking eggs and young, and killing weak birds. Strong and fast on the wing, they are piratical as well as raptorial,

forcing other species to disgorge food.

The three smaller species (genus *Stercorarius*) are more slender and long-winged, with the two central tail-feathers of adults elongated, tapering or rounded at the tip. They breed on the Arctic tundra but at other times range widely over the seas into the Southern Hemisphere. They are fast-flying and very agile, and find much of their food by harrying smaller birds such as terns, forcing them to disgorge fish in aerial attacks. When breeding they also hunt lemmings and voles, kill small birds, and take carrion.

Gulls

The gulls of the family Laridae are a readily recognizable group, solidly built, with long strong webbed-footed legs for running and swimming, and long wings for the steady sustained flight that goes with aerial scavenging. Buoyant gliding and soaring is often involved, as food is hunted over water as well as land.

With a few dusky exceptions, the plumage is mainly white with gray or black across the back and wings. There is little conspicuous patterning, but some species have a brown or black hood that combines with a red or yellow bill to act as a display pattern. Color and patterns are similar in both sexes, and young birds have mottled brown plumage with dark bills and a dark tail band.

The bill is slender in the smaller inland species that feed mainly on insects, but in larger species it is stout with a tip for tearing —adapted for feeding on carrion, particularly fish—but gulls will feed on small marine life exposed by tides or take any small creatures, and the larger gulls will attack and kill young or smaller seabirds. (They are probably responsible for the nocturnal nesting habits of many smaller seabirds.) Gulls tend to be sociable in

Jean-Paul Ferrero/Auscape International

▲ *The yellow-billed sheathbill inhabits Antarctica and several subantarctic islands.*

▼ *A southern skua raids a gentoo penguin colony. Agile, strong, and piratical, skuas take the place of raptors on many subantarctic islands.*

Francisco Erize/Bruce Coleman Ltd

Geoff Longford

▲ Many calls and postures of gulls around the world are highly ritualized, and differ only in detail from species to species; this feature is useful in unraveling their relationships. These kelp gulls are uttering the so-called "long call", common to many of the larger gulls. The kelp gull occurs in Australia and New Zealand, South Africa, and southern South America.

nesting, resting, and feeding; and they respond quickly when the behavior of individuals indicates that food has been found.

The 48 species are spread around the shorelines and islands of the world, and to the larger inland waters. Food is taken from land or from the surface of the water, and smaller species sometimes hunt insects on the wing. As scavengers they have readily learnt to use discarded waste from human activity, following boats, searching recently plowed land, and feeding on rubbish tips.

Nesting is usually colonial, using sites varying from cliff ledges to level ground, and from the Chilean desert to polar shores. The small North American Bonaparte's gull *Larus philadelphia* is an exception, nesting inland in conifer trees. A gull's nest is usually a substantial cup of carried vegetation, normally with two or three spotted eggs. The dark-mottled downy young are fed by the adults, but are mobile and even when feathered may run and hide if threatened. On the nests of kittiwakes (*Rissa* species), which are trampled drums of mud and plants fixed to small projections of precipitous cliffs, the young are adapted to making little movement on the site.

The voice of gulls is usually harsh—wailing, cackling or sqawking calls—and colonies are noisy.

Terns

Terns are gull-like but adapted for more specialized feeding such as plunge-diving. The plumage tends to be pale gray and white, with black on the crown or nape, but a few species are mainly brown or black. Young are brown-mottled on the back and wings. Wings are long and slender, the tail distinctly forked, the bill strong, thin and tapering, and the legs rather small and short, with webbed feet.

Perhaps the least modified are *Chlidonias* species: the two black terns and the whiskered tern. Mainly black or gray when breeding, they have thin bills and less forked tails. These so-called "marsh terns" are birds of inland waters of warmer climates and are very light on the wing, feeding mainly by briefly dipping to snatch tiny creatures from the surface. They will also hunt insects over land. They have sharp, short, high-pitched call-notes. Their nests are in colonies on inland marshes and lakes, flimsy structures usually on floating vegetation, sometimes in shallow water. Adults bring food, but after a few days the active downy young may leave the nest to hide in nearby vegetation. After breeding they disperse rather than migrate, keeping mainly to inland waters.

The typical sea terns, 34 species, have compact and streamlined bodies, a heavier head with a tapering dagger bill, long and narrow wings, forked tails for quick braking and maneuvering, and relatively small legs and feet, since the birds rely mainly on flight. Most are white with gray backs and wings, and a black cap or nape when breeding, but a few are brown or black on the upper parts and head, with a white forehead. The Inca tern *Larosterna inca* differs in being deep gray with an elongated and exquisitely curled mustache, and is also peculiar in nesting in cliff cavities. In size, the sea terns range from the big Caspian tern *Sterna caspia,* about the size of some larger gulls at 53 centimeters (21 inches), to the little tern *S. albifrons* at 24 centimeters (9½ inches). They fly low over oceanic or inland waters, spotting prey, hovering briefly, and then plunging for food, often sending up a plume of spray on impact. The sooty tern *S. fuscata* may snatch small fish or squid at or above the surface when larger fish pursue them, and may fish at night.

Sea terns are highly sociable nesters, making a collection of scrapes, unlined unless material is to hand, on beaches, sandbanks, and low, sparsely-vegetated shores. Even where space is available the nests are closely packed, often a bill-stab apart, and there is frequent bickering. Displaying males parade around their mates with plumage sleeked and a small fish held high. Calls are usually harsh single notes, with higher-pitched alarms. Colonies are noisy and are fiercely defended by aerial attacks on intruders. Birds nesting on hot tropical beaches

may shade the nest at times rather than incubate, and return with wet belly feathers to moisten eggs or chicks. The older young of some species assemble near the water's edge to await returning adults. Migrating young leave the colonies while still relying on parents for food.

As strong fliers, sea terns spend much of their time on the wing, some species migrating long distances. The Arctic tern *S. paradisaea* is famed for traversing the world from one polar sea to the other twice yearly.

An exception among these species is the gull-billed tern *Gelochelidon nilotica.* It has a shorter, thick bill, less deeply forked tail, and longer legs than other terns. It is a bird of open inland waters, coastal lagoons, estuaries, and salt-marsh. It tends to hunt on the wing, swooping to snatch insects on land, and small fish, crabs or frogs from the surface of water, but may also hunt on foot. It nests colonially on small raised islands in shallow water.

The noddy terns consist of the three blackish-brown *Anous* species and the pale blue-gray noddy *Procelsterna cerulea,* all with whitish foreheads. The bill is slender, the wings long, the tail broader and wedge-shaped with a blunt fork at the tip, and the legs are short. Terns of tropical and subtropical seas, they are coastal and oceanic birds, mainly around islands, and generally do not range widely. They are not plunge-divers but surface-dippers, hovering to snatch from the surface or catch in the air smaller fish or squid trying to escape

▼ *A pair of Arctic terns greet each other. Arctic terns show strong fidelity both to their mates and their nest sites of previous years, but there is little evidence that they remain together during their winter migrations. Rather, Arctic terns tend to congregate at roosts immediately on their return to the breeding grounds and for a few days before the nesting colonies themselves are reoccupied; pair bonds are re-established in courtship flights at the roost and above the nesting colony.*

lower mandible compressed laterally to a flat blade projecting beyond the upper mandible.

They usually feed by flying low over the water with the tip of the longer lower mandible plowing a little below the surface. When prey is touched, it is snapped up. The young have even-lengthed mandibles and can feed more normally. The flight is light and graceful with steady, shallow beats of raised wings.

AUKS

The 22 auk species show a certain uniformity. Medium-sized to small, they are plain-colored, varying from black or dark above and pale below to all-dark species, some with pale wing or shoulder patches. Breeding may bring vivid colors on the bill and gape, and odd feather adornments on the heads of some species; but the voices are unexceptional and calls are mostly short and deep hoarse notes.

With largish heads, thick waterproof plumage on a compact body, and a tendency to ride high in the water, they look stout and stubby when

▲ *The white tern breeds on oceanic islands in the tropical Pacific, Indian, and Atlantic oceans. It builds no nest: instead it balances its single egg in any available crevice on the upper surface of the limbs of trees.*

▶ *Weighing around 5 kilograms (11 pounds), the great auk was by far the largest of the alcids. It could not fly, and was restricted to colder regions in the northern Atlantic Ocean. It was driven to extinction by sailors and fishermen, mainly to provision boats, and the last birds were killed in Iceland in 1844.*

underwater predators. The birds will flap onto the surface in a shallow plunge, or swim to seize surface prey. They swim well, and rest on the water. They feed sociably, sometimes in large numbers, and the brown noddy *A. stolidus,* which is more migratory, may occur well out at sea in huge flocks. Noddies nest colonially in bushes or trees such as mangroves, building untidy platforms of seaweed, sticks, and other debris, but in some regions they may nest on cliff ledges or even on the ground. The white forehead is used in head-nodding displays, and they have other terrestrial and aerial posturings.

The other odd type of tern is the fairy or white tern *Gygis alba.* A tropical coastal tern, it is small, delicate-looking and pure white, slender-billed and with large black eyes. It often feeds at dusk, snatching small fish and squid at or above the surface, in similar fashion to the noddy terns. It makes no nest, balancing its single egg on a small precarious surface such as a tree branch, rock pinnacle or similarly hazardous place. The chick has strong feet with sharp claws, and from the outset must cling to stay on the site.

Skimmers

The skimmers (genus *Rhynchops*) are superficially tern-like birds, white with black wings and crown. There are three species, one each in Africa, India, and the Americas. They are highly sociable at all times, tending to feed, rest, and breed in flocks. Relying wholly on flight for feeding, and resting on the sandbars or gravel of shores and rivers at other times, they have long tapering wings, short tails with a fork, and shortish legs with webbed feet. The unusual bill is long and tapering, with the

floating or paddling at the surface. Under water they are more streamlined and move with agility, swimming with strong strokes of the wings held out from the body, and steering with the feet. The larger species take mostly fish, but the smaller ones

C.A. Henley

may rely on plankton. The wings are short, small, strong, and firm-feathered for swimming. Flight is usually direct and low over the water, with continuously and rapidly beating wings. Having small tails they tend to use their splayed webbed feet in maneuvering.

The guillemots or murres—larger auks up to 46 centimeters (18 inches) long—are birds of colder and arctic seas, with narrow, tapering bills. They breed on rocky coasts. The common guillemot *Uria aalge* and the thick-billed or Brunnich's guillemot *U. lomvia* huddle close-packed on broader ledges, making no nest; the very varied color and pattern of their single egg may aid recognition by the owners. As with some other auk species, the young flutter down from the ledge before they are fully fledged and go out to sea escorted by the parents. The smaller *Cepphus* guillemots breed in deep crevices, laying two eggs, and the young leave more fully fledged.

These auks and the puffins breed in the north of both the Atlantic and Pacific oceans. Two puffins (*Fratercula* species) have stout deep bills which in breeding birds grow temporary bright bill-sheaths producing a great parrot-like bill, with wattles on the bill-base and eyelids completing a clown-like face. The tufted puffin *Lunda cirrhata* has a simpler pattern ornamented with big swept-back tufts of blond plumes. Puffins burrow into the softer soil-covering of cliffs and islands, and they gather nest material. The serrated bill-edge allows them to pack in slender fish, held crossways, to carry to the single young.

The North Atlantic razorbill *Alca torda* is a guillemot with a deep, blunt, and laterally flattened bill, resembling the common guillemot in habits. Its close relative was the great auk *A. impennis,* which used to nest in colonies on low islands in the North Atlantic; being flightless it was easily killed, and it was finally exterminated in 1844.

The smallest of the Atlantic auks is the little auk *Alle alle,* a tiny bird with big head and stubby bill. It winters out at sea except when driven inshore by storms, and nests in huge swarms in the cliff screes of arctic islands. It is more active and agile on the wing than larger species.

In the North Pacific the smallest members of the auk family are called auklets and murrelets, six species each, breeding mainly in colonies on coastal islands and archipelagos. For safety they tend to nest in burrows or rock crevices; they usually have a single egg. The auklets are the more heavily built and go farther out to sea when not nesting. Like puffins, some develop brightly colored bill-sheaths which are shed after each breeding season, and some have fine projecting feather tufts on the forehead and cheeks.

Murrelets are smaller and tend to stay closer inshore. They have small fine bills, stubby or slender. The two *Brachyramphus* species are peculiar in having mottled brown breeding

Dean Lee/Weldon Trannies

plumage, and they may fly inland to nest. Both are little known, but Kittlitz's murrelet *B. brevirostre* has been found using a nest-hollow in bare stony ground above the tree level on mountains. The marbled murrelet *B. marmoratus* sometimes occurs in huge numbers but very few nests have been found: two were on bare tundra; and two on large level branches or lodged debris high in conifer trees in forest, in one instance more than 8 kilometers (5 miles) inland. The young bird must try to reach the sea in its first flight.

▲ The horned puffin is among the most numerous birds along the coast of Alaska. It is named after the small fleshy appendage over each eye, which the bird can raise or lower at will.

Order Columbiformes
2 families, 45 genera,
320 species

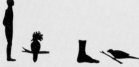

SIZE
Smallest Diamond dove
Geopelia cuneata, head-body
length 190 to 215 millimeters
(7½ to 8½ inches), tail 87 to
107 millimeters (3½ to 4½
inches); weight 23 to 37
grams (¾ to 1¼ ounces).
Largest Victoria crowned
pigeon *Goura victoria*, head-
body length 71 to 80
centimeters (28 to 31½
inches), tail 25 to 30
centimeters (10 to 12 inches);
weight 1.7 to 3 kilograms
(3¾ to 6½ pounds).

CONSERVATION WATCH
There are 50 species listed in
the ICBP checklist of
threatened birds. They
include the following:
Mauritius pink pigeon
Nesoenas mayeri, giant
imperial pigeon *Ducula
goliath*, ochre-bellied dove
Leptotila ochraceiventris,
Society Islands imperial-
pigeon *Ducula aurorae*, gray
woodpigeon *Columba
argentina*, Nicobar pigeon
Caloenas nicobarica, Victoria
crowned-pigeon *Goura
victoria*, tooth-billed pigeon
Didunculus strigirotris, and
Marianas fruit dove *Ptilinopus
roseicapilla*.

PIGEONS AND SANDGROUSE

FRANCIS H.J. CROME

There are two quite different families of birds within this order, and they may not even be closely related. Pigeons and doves of the family Columbidae are basically seed- and fruit-eating, tree-dwelling, terrestrial birds, occurring throughout the world except in the high Arctic and classified in more than 300 species. Less well known are the 16 species of sandgrouse of the family Pteroclididae, which are desert-dwellers of Africa and Eurasia. Scientists argue frequently about whether they are related to pigeons at all, some suggesting they are waders (Charadriiformes), others that they should form their own order. A third family, Raphidae, now extinct, consisted of the dodo and two related species of the Mascarene Islands in the Indian Ocean.

SANDGROUSE

Sandgrouse are medium-sized birds, specialized for a life in the deserts and semi-arid regions of Africa and Eurasia. They have dull, well-camouflaged brown, gray or khaki-colored plumage and are compact and streamlined, with small heads and short necks. The tops of their feet are densely clothed in feathers. Males and females have different plumages. Their adult feathers are similar to those of pigeons; and like pigeons (and

► The Nicobar pigeon (left) feeds on the ground in rainforest and mangrove swamps; it inhabits Indonesia and New Guinea. The superb fruit dove (top right) occurs in New Guinea and northeastern Australia; the pin-tailed sandgrouse (bottom right) inhabits southern Europe and North Africa.

M.P. Kahl/Auscape International

unlike most other birds) they can drink by sucking, although less expertly than pigeons—they have to raise their heads to swallow. They differ from pigeons, however, in several important ways: they cannot produce crop milk; they have a pair of large functional ceca (pouches, or blind tubes, forming the beginning of the large intestine); they have a different syrinx or voice-box; and they have oil-glands that produce an oil which is used to preen the feathers.

They are very strong and fast fliers, the Namaqua sandgrouse *Pterocles namaqua* of the deserts of southern Africa being able to outpace a falcon in level flight. They need to drink regularly and may have to commute over 60 kilometers (37 miles) to water. A typical species is the pin-tailed sandgrouse *P. alchata* of North Africa, Spain, the Middle East, and Central Asia west of the Caspian Sea. Flocks of at least 50,000 have been recorded at waterholes in Turkey. In the breeding season, March to May, they break up into pairs, then gradually congregate again in flocks after the young fledge in September.

Sandgrouse nest on the ground, and the young leave the nest a few hours after hatching. The chicks take every opportunity to shield themselves from the hot sun and even shelter under their parents while moving. As a protection against predators they half-bury themselves in the sand under the shade of bushes. Sandgrouse have a unique way of watering their young: the male flies

to a waterhole and wades in with his feathers lifted; the central feathers soak up water, then the male returns to his family where the chicks drink from his wet feathers.

PIGEONS AND DOVES

About three-quarters of the 304 species of pigeons and doves live in tropical and subtropical regions. The term "pigeon" is used for larger species, and "dove" for the smaller, more delicately built ones. They are medium-sized to small birds which feed on seeds and fruits. Most are tree-dwelling, yet some of the common species feed in huge flocks on the ground. The plumage is soft and dense, and the feathers have characteristically thick shafts and fluffy bases. They have no, or very small, oil-glands—instead, special plumes disintegrate to produce powder that cleanses and lubricates the plumage. The most specialized feature of the family is the ability to produce crop milk: when the birds are breeding, special glands in the crops of both male and female enlarge and secrete a thick milky substance which is fed to the young.

The sexes are usually similar in appearance, but some species have different male and female plumages. Pigeons and doves have characteristic sexual and advertising displays, such as bowing, and special display flights. Their calls are usually pleasant cooing notes. All species build flimsy nests of a few sticks in trees, or on the ground or on

▲ The African green pigeon is common and widespread over much of the African continent; closely related species also occur in tropical Asia. These pigeons are strongly arboreal and feed mainly on fruit.

Frithfoto/Australasian Nature Transparencies

▲ *Nearly turkey-sized, the three species of crowned pigeons, or gouras, are the largest of all pigeons. They are terrestrial and live in lowland rainforest on the island of New Guinea. Hunted widely for food, they now remain common only in remote areas. Males use their glorious crests in bowing displays during courtship. The Victoria crowned pigeon, shown here, is the smallest of the three and inhabits the northern part of the island.*

of fruit dove (genus *Ptilinopus*) of the Indo-Pacific region. These smallish plump pigeons are spectacularly colored with bright greens, brilliant reds and oranges, purples and pinks, blues and golds. They live high in the canopy of the rainforests, and some species are found only on single islands in the Pacific Ocean. Fiji, for example, has three very specialized species collectively called golden doves: the orange dove *P. victor* is fiery orange; the golden dove *P. luteovirens* is metallic greenish-gold; and the yellow-headed dove *P. layardi* is green with shining gold fringes to the feathers. Fruit doves eat only the fruits of rainforest trees and have specialized digestive systems—features they share with the 36 species of imperial pigeons (larger birds, in the genus *Ducula*) which have the same distribution. Most birds take grit that lodges in the crop and helps grind up food, but fruit doves and imperial pigeons do not take grit and have a thin gizzard with horny knobs which gently strips the flesh from the seeds. The seeds are defecated whole and so these birds act as important dispersers for rainforest trees.

The New Guinea rainforests are the home of some spectacular birds such as the three species of crowned pigeon in the genus *Goura*. The size of small turkeys, they are the largest pigeons and are characterized by big filmy crests and subtle purple and gray plumage. They forage in small groups on the forest floor and, despite their size, nest up to 15 meters (50 feet) in trees. Elsewhere in the Pacific and Indian oceans pigeons have adapted well to life on islands. They have evolved into many distinctive species, and several frequently fly long distances between the islands in their range. The white and black Torresian imperial pigeon *Ducula spillorrhoa* migrates from New Guinea to northern Australia in August and breeds in huge colonies, mostly on the offshore islands of the Great Barrier Reef. While there, flocks of several thousand birds fly to the rainforests of the mainland every day to feed; in March they return to New Guinea.

In Asia and Africa, the aptly-named green pigeons (genus *Treron*) replace the fruit doves and imperial pigeons as the arboreal fruit-eating species in the tropical forests. They lack the specialized gut, however, and grind up the seeds of the fruits they eat. Elsewhere, in Eurasia and the Americas, the various pigeons are less specialized and less spectacularly plumaged. Many pigeons are gregarious and form small to large flocks. Flocks of up to 100,000 woodpigeons *Columba palumbus* have been recorded in Germany, and the eared dove *Zenaida auriculata* of South America breeds in huge colonies of tens of thousands of birds. The flock pigeon *Phaps histrionica* of semi-arid north and central Australia occasionally irrupts in huge flocks of thousands of birds; explorers in the nineteenth century described the noise of the flocks as deafening, like "the roar of distant thunder".

ledges. One or two plain white eggs are laid, and the chicks are cared for by both sexes; they leave the nest in 7 to 28 days, depending on the species.

The common street pigeon or rock dove *Columba livia* has adapted well to agriculture and towns. It originally occurred in Eurasia and North Africa, where it nested colonially on cliffs, but it has easily made the transition from cliffs to buildings and now occurs in almost all the world's cities. But this rather drab species gives no indication of the diversity and brilliance of plumage in the family. For instance, the snow pigeon *C. leuconota* from the high plateaus of the Himalaya mountains is a striking white, black, and gray, and the Seychelles blue pigeon *Alectroenas pulcherrima* is deep metallic blue with silver-gray foreparts and a red head with naked wattles.

But perhaps the most beautiful are the 47 species

DRIVEN TO EXTINCTION

Nothing symbolizes human treatment of wildlife and the need for conservation better than the tragic extermination of the dodo *Raphus cucullatus*. It was discovered in 1507 and exterminated by 1680. The dodo of Mauritius was one of three species of massive, flightless, highly aberrant birds on the remote Mascarene Islands, east of Madagascar in the Indian Ocean. Presumably, they derived from pigeon-like ancestors that flew to the islands.

The strange dodo was ash gray with a reddish tinge to its black bill and weighed about 23 kilograms (50 pounds). Its wings were reduced to useless stubs and it had no defense against, or means of escape from, the seafarers who killed it for food, sport, and because they thought it abominably ugly. The pigs, cats, rats, and monkeys that were introduced to the islands may have contributed to its extinction, but basically it fell victim to human persecution.

The very similar white solitaire *R. solitarius* of neighboring Réunion Island was wiped out in the same way by about 1750, but the Rodriguez solitaire *Pezophaps solitaria* managed to survive until perhaps 1800. There are few accounts of the behavior and biology of these species and, indeed, few specimens in museums. They supposedly laid one egg each year, were vegetarian, used their stubby wings for fighting, and were agile runners despite their size and gross proportions.

Like the dodo, the North American passenger pigeon *Ectopistes migratorius* was hunted to extinction. But unlike the dodo, the passenger pigeon was found over much of a continent and was incredibly abundant, possibly the most numerous bird in the world. When white people first came to North America there may have been three to five thousand million or more of the species.

Passenger pigeons underwent irregular migrations within their huge range and were not abundant every year. In good years, however, flocks reached staggering proportions. In 1871 one flock seen over Wisconsin occupied 2,000 square kilometers (850 square miles) and contained 136 million birds; in 1810 a single flock of two and a quarter *billion* birds was seen in Kentucky; and in Ontario a flight of birds moving north from the United States in 1866 was 480 kilometers (300 miles) long and 1.6 kilometers (1 mile) wide and continued for 14 hours. Possibly there were three billion birds in it.

Forest clearing obviously hastened the decline of this species, but the passenger pigeon was exterminated by relentless slaughter just as the bison almost was. Birds were shot, trapped, and poisoned in millions; at one nesting colony in Michigan alone 25,000 birds were killed daily for market during the breeding season of 1874. Over 700,000 a month! Even the commonest bird in the world could not sustain such obscene carnage indefinitely. By the 1880s the species was close to extinction, and the last passenger pigeon died in Cincinnati Zoo on September 1, 1914. Thus passed one of the greatest wildlife spectacles witnessed by modern man.

The dodo, the solitaire, and the passenger pigeon have not been the only species in this order to suffer extinction. The Mauritius blue pigeon *Alectroenas nitidissima*, Norfolk Island dove *Gallicolumba norfolciensis*, tanna ground dove *G. ferruginea*, bonin wood pigeon *Columba versicolor*, and the silver-banded black pigeon *C. jouyi* have all been exterminated. The small, beautiful, Solomon Islands crowned pigeon *Microgoura meeki* is probably extinct, and many other species, perhaps all those on the small islands of the Pacific where forests are being cleared, are endangered. For such an inoffensive group of birds, pigeons have suffered badly at the hands of humans.

◄▼ *Two notorious extinctions: the passenger pigeon (left) once migrated across North America in hordes that darkened the skies, but the last one died in the Cincinnati Zoo in 1914; the dodo (below) was confined to the island of Mauritius in the Indian Ocean, but was exterminated by early explorers before 1680.*

PARROTS

JOSEPH FORSHAW

SIZE
Smallest Buff-faced pygmy parrot *Micropsitta pusio* and allied species, total length 85 millimeters (3⅜ inches); weight 10 to 15 grams (½ ounce).
Largest Hyacinth macaw *Anodorhynchus hyacinthinus*, total length 1 meter (3¼ feet).

CONSERVATION WATCH
There are 71 species listed in the ICBP checklist of threatened birds. They include the following: scarlet-breasted lorikeet *Vini kuhlii*, ultramarine lorikeet *V. ultramarina*, salmon-crested cockatoo *Cacatua moluccensis*, golden-shouldered parrot *Psephotus chrysopterygius*, orange-bellied parrot *Neophema chrysogaster*, night parrot *Geopsittacus occidentalis*, Mauritius parakeet *Psittacula eques*, hyacinth macaw *Anodorhynchus hyacinthinus*, Spix's macaw *Cyanopsitta spixii*, thick-billed parrot *Rhynchopsitta pachyrhyncha*, blue-chested parakeet *Pyrrhura cruentata*, Puerto Rican amazon *Amazona vittata*, and the kakapo *Strigops habroptilus*. The paradise parrot *Psephotus pulcherrimus* and the glaucous macaw *Anodorhynchus glaucus* are probably extinct.

Probably no group of birds is more widely known to the general public than the parrots. Indeed, one species—the budgerigar from inland Australia—rivals goldfish as the most popular pet animal in the world. The popularity of keeping parrots as pets dates from early recorded history: rose-ringed parakeets were known to the ancient Egyptians, and it was probably Alexander the Great who introduced tame parrots from the Far East to Europe. Today, the international trade in live parrots has reached alarming proportions, and there is virtually no city or town without a pet shop selling budgerigars, cockatiels, or lovebirds.

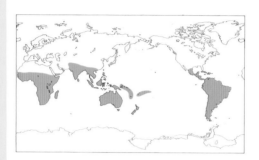

▼ *Two Australasian parrots: the black-capped lori(left) is one of the most conspicuous parrots in New Guinea, and the eastern rosella (right) is common in open eucalypt woodland over much of eastern Australia.*

BRIGHT BIRDS IN BOLD PLUMAGE
Parrots belong to a very distinct order of ancient lineage and are strongly differentiated from other groups of birds. Some distinguishing features are obvious to even a casual observer; most prominent is the short blunt bill with a downcurved upper mandible fitting neatly over a broad, upcurved lower mandible. This unique design enables parrots to crush the seeds and nuts that constitute the diet of most species.

Another conspicuous characteristic is the typical parrot foot, with two toes pointing forward and two turned backward. Parrots show remarkable dexterity, using their feet for climbing or for holding food up to the bill. The skull is broad and relatively large, with a spacious brain cavity. The

extremely muscular tongue is thick and prehensile, and in lorikeets of the subfamily Loriinae it is tipped with elongated papillae for harvesting pollen and nectar from blossoms.

Parrots are renowned for the generally brilliant coloration of their plumage. There are plain or dull-colored parrots, such as the two *Coracopsis* species from Madagascar, but these are few. Green predominates in most species, and is effective as camouflage amidst foliage in the rainforest canopy where many species live. Bold markings, mainly of red, yellow, and blue, are prevalent on the head or wings, and many species have brightly-colored rumps. Unusual plumage patterns are present in the spectacular *Anodorhynchus* and *Cyanopsitta* macaws from South America and some *Vini* lories from the Pacific Islands, all of which are entirely or almost entirely blue, while bright yellow predominates in the plumage of the golden conure *Aratinga guarouba* from Brazil, and the regent parrot *Polytelis anthopeplus* and yellow rosella *Platycercus flaveolus* from Australia. Some *Ara* macaws from Central and South America are almost entirely red, as are some lories from the Indonesian archipelago.

The sexes generally are alike, though females may be appreciably duller, but the eclectus parrot *Eclectus roratus* of Australia is notable in that the bright green males are strikingly different from the predominantly red females.

VARIATIONS ON A THEME
Despite the homogeneity of their basic features, parrots come in all shapes and sizes. Tails may be long and pointed, as in the long-tailed parakeet *Psittacula longicauda* from Malaysia and the princess parrot *Polytelis alexandrae* from Australia, or short and squarish as in the short-tailed parrot *Graydidascalus brachyurus* and some *Touit* parrotlets from South America, or there may be ornate feathers as in the Papuan lory *Charmosyna papou* from New Guinea or the *Prioniturus* racket-tailed parrots from Indonesia and the Philippines. Wings can be narrow and pointed, as in the swift parrot *Lathamus discolor* and the cockatiel *Nymphicus hollandicus* from Australia, or broad and rounded, as in the *Amazona* parrots from South America. Cockatoos of the subfamily Cacatuinae have prominent, erectile head-crests, while other parrots may have elongated feathers on their crowns or hindnecks.

Even in the characteristic bill there are variations in shape, which represent modifications for different feeding habits. A curved, less elongated upper mandible enables the slender-billed corella *Cacatua tenuirostris* of Australia to dig up roots and corms, while similarly-shaped bills of the slender-billed conure *Enicognathus leptorhynchus* of South America and the red-capped parrot *Purpureicephalus spurius* of Australia seem to be ideal for extracting seeds from large nuts. Parrots that feed extensively on pollen, nectar, or soft fruits tend to have narrow, protruding bills.

DISTRIBUTION AND HABITATS
Parrots live mainly in the Southern Hemisphere, and are most prevalent in tropical regions. Once the Carolina parakeet *Conuropsis carolinensis* of North America became extinct in the early part of this century, the northernmost species became the

▲ *A blue-and-yellow macaw. These splendid birds were once common in forests across much of South America, but they have been much reduced because of illegal trafficking for the cage bird trade. They prefer tall palms growing along watercourses. It seems probable that they mate for life, and even when congregating in large flocks the pairs remain in close contact.*

CYG/The Photo Library

▲ *Brilliantly plumaged rainbow lorikeets are commonly seen feeding in suburban gardens throughout their range in northern and eastern Australia. The species also occurs widely in the New Guinea region and on many islands in the southwestern Pacific. Strongly gregarious, rainbow lorikeets forage in parties of up to 50 or so, but at night they often congregate in thousands to roost.*

▶ *The blossom-headed parakeet inhabits hill forest from Bengal to Indochina. In places it remains quite common, and is usually encountered in small flocks or family parties that hurtle through the forest with remarkable speed and agility. It roosts communally and feeds on seeds, nuts and flowers.*

slaty-headed parakeet *Psittacula himalayana,* in eastern Afghanistan. The most southerly parrot is the Austral conure *Enicognathus ferrugineus,* which reaches Tierra del Fuego. The strongest representation of parrot species is in Australasia and South America, with 52 species recorded from Australia and 71 from Brazil, but in South America there is a marked uniformity of types. There are parrots in Asia, mainly on the Indian subcontinent, and in Africa, but representation in these regions is surprisingly low. The most widely distributed species is the rose-ringed parakeet *Psittacula krameri,* which occurs in Asia and northern Africa and has been introduced to parts of the Middle East and Southeast Asia. Most of the species with restricted ranges are confined to quite small islands. With a total area of some 21 square kilometers (8 square miles), the Antipodes Islands south of New Zealand are inhabited by two *Cyanoramphus* parrots; one of these, the Antipodes green parakeet *C. unicolor,* is endemic.

Parrots are particularly plentiful in lowland tropical rainforest, although in Australia and parts of South America open country is preferred by many species. While some species, especially those

C.A. Henley

restricted to rainforest, show little capacity to withstand interference with their habitat, others have adapted remarkably well to the human impact and are commonly seen in parks, gardens, or even trees lining city streets. In Australia the galah *Eolophus roseicapillus* is plentiful in many towns and cities, while in downtown São Paulo, Brazil's largest city, small flocks of plain parakeets *Brotogeris tirica* can be seen in parks surrounded by towering buildings.

Parrots tend to be less common at higher altitudes, and species that occur there normally are absent from or are rare in neighboring lowlands.

Distinctive highland forms include the Johnstone's lorikeet *Trichoglossus johnstoniae* in the Philippines, the Derbyan parakeet *Psittacula derbiana* in Tibet, and the yellow-faced parrot *Poicephalus flavifrons* in Ethiopia. Possibly the most interesting of highland parrots is the kea *Nestor notabilis* from the Southern Alps of New Zealand; it is a species that has been much maligned because of its alleged sheep-killing habits.

FAMILIAR AND UNFAMILIAR SPECIES

Parrots are difficult to observe in the wild. Most are predominantly green and live in the rainforest canopy, so sightings usually are little more than momentary glimpses of screeching flocks flashing overhead. Species that inhabit open country or are plentiful near human habitation are conspicuous, and there is much more information on their habits. We know a great deal, for example, about the habits of species such as the eastern rosella *Platycercus eximius* and red-rumped parrot *Psephotus haematonotus* in Australia, Meyers parrot *Poicephalus meyeri* in Africa, and the monk parakeet *Myiopsitta monachus* in South America, but virtually nothing about many forest-dwelling species in New Guinea, Central Africa, and South America.

The Australian ground parrot *Pezoporus wallicus* is largely nocturnal, and there are reports of some normally diurnal species being active on moonlit nights. Migration of swift parrots and *Neophema* species across Bass Strait, in southern Australia, usually takes place at night. Patterns of daily activity in most parrots is typical of birds in tropical regions: peak periods in the morning and towards evening, and low activity during the heat of the day.

Many parrots are gregarious. Pairs and family parties come together to form flocks, which in arid areas may build up to enormous sizes after breeding brought on by favorable conditions. At such times, massive flocks of bare-eyed corellas *Cacatua pastinator* and budgerigars darken the skies of inland Australia.

◄ The hanging-parrots, so called because of their extraordinary habit of roosting, bat-like, upside down, constitute a group of ten species best represented in Indonesia. They seem closely related to the lovebirds of Africa and, like them, transport material for their nests among the feathers of the rump. This is the blue-crowned hanging-parrot, which is widespread in Malaysia, Sumatra, and Borneo.

Morten Strange

▶ *Pesquet's parrot (left) inhabits hill forest in New Guinea, where it is widespread but rare. It feeds on soft fruit and nectar, and the largely naked head is probably an adaptation to avoid having plumage matted with sticky juice and nectar. The hawk-headed parrot (right) of Amazonia has a striking ruff of long, colorful, erectile feathers on the nape. It is a noisy, conspicuous parrot that is usually encountered in small parties.*

The flight of most parrots, especially the smaller ones, is swift and direct. Some have a characteristically undulating flight produced by wing beats being interspersed with brief periods of gliding. In the larger species it is variable; macaws are fairly fast fliers and their wing beats are shallow, but the buoyant flight of *Probosciger* and *Calyptorhynchus* cockatoos is conspicuously slow and labored. The kakapo *Strigops habroptilus* is the only flightless parrot.

The distinctly metallic call-notes of most parrots are harsh and unmelodic, generally being based on a simple syllable or combination of syllables. Variation comes primarily from the timing of repetition. Larger species normally have raucous, low-pitched calls, while small parrots give high-pitched notes. The mimicry of captive parrots is well known, so it is surprising that there are no convincing reports of wild birds imitating other species.

FOOD AND FEEDING
Most parrots eat seeds and fruits foraged from treetops or on the ground. Lories and lorikeets of the subfamily Loriinae are strictly arboreal, and

feed on pollen, nectar, and soft fruits. Insects are often found in crop and stomach contents; wood-boring larvae are an important food item for some of the black cockatoos from Australia. Mystery still surrounds the diet of *Micropsitta* pygmy parrots, which seem to take lichen from the trunks and branches of trees, but at times they have been observed foraging for termites.

When feeding, a parrot makes full use of its hooked bill; while climbing among foliage it often uses the bill to grasp a branch onto which it then steps. Many species use a foot as a "hand" to hold food up to the bill, and with the bill they expertly extract kernels from seeds and discard the husks.

BREEDING BEHAVIOR

The age at which parrots reach sexual maturity varies, but in general it is three or four years in larger species and one or two years in small birds. As far as can be ascertained from observations, most species are monogamous and the majority remain paired for long periods, perhaps for life. Notable exceptions are the kea, which is polygamous, and the kakapo, a lek-display species with males almost certainly taking no part in incubation or care of the young. Pairs and family groups are usually discernible within flocks. Courtship displays are simple, with bowing, wing-drooping, wing-flicking, tail-wagging, foot-raising, and dilation of the eye pupils being the more common actions.

Parrots usually nest in hollows in trees or holes excavated in termite mounds, occasionally in holes in banks, or in crevices among rocks and cliff-faces. The ground and night parrots from Australia and some populations of *Cyanoramphus* parakeets on New Zealand islands nest on the ground, usually under or in grass tussocks. Monk parrots from South America gather twigs to build a huge communal nest in a tree, and each pair has its own breeding chamber.

Eggs are normally laid every other day, and clutches vary from two to four or five, sometimes up to eight for small parrots. Incubation starts with or immediately after the laying of the second egg, but there is mounting evidence that this can vary

◄ Few sights are more evocative of the arid interior of Australia than a tree festooned with little corellas, or a massed flock of these birds rising from the ground in a roar of wings and a cacophony of harsh screeches. They are intensely gregarious, and flocks may sometimes number thousands of birds. They roost communally, invariably near waterholes. This is a habit that was exploited by many of the early desert explorers, who followed the gathering flocks at dusk to be led to water.

Leo Meier/Weldon Trannies

individually. Generally the female alone incubates. The duration of incubation for small parrots is from 17 to 23 days, but for the large macaws it can be up to five weeks. Newly-hatched chicks are blind and naked or have sparse dorsal down, which in most species is white.

Young parrots develop slowly, and remain in the nest for three to four weeks in the case of the smallest species, and up to three or four months for the large macaws. After leaving the nest, young birds are fed by their parents for a brief time while learning to fend for themselves; young black cockatoos are fed by their parents for up to four months after leaving the nest.

Juveniles generally resemble females or are duller than either adult sex. There are species, such as the crimson rosella *Platycercus elegans* from Australia and some *Psittacula species,* that have a distinct juvenile plumage. A striking difference between adults and juveniles occurs in the vulturine parrot *Gypopsitta vulturina* from Brazil: in adults the bare head is sparsely covered with inconspicuous "bristles", but in juveniles the head is well covered with pale green feathers. The time taken for juveniles to attain adult plumage varies greatly between species; it may be within months of leaving the nest, or possibly up to three or four years.

CONSERVATION OF PARROTS

Ten extinct species of parrot are represented by specimens in museums, while others are known from subfossil material or reports in the writings of early explorers. Probably the best known of these extinct species is the Carolina parakeet; the last living bird died in the Cincinnati Zoo on February 21, 1918. Even for this species the causes of extinction will never be fully understood, but the loss is a warning that should be heeded if parrots are to be protected from the serious threats they now face in virtually all parts of their range. Habitat destruction is by far the most serious threat, especially the widespread clearing of tropical forest. Of special concern are parrots confined to small islands, where the habitat is finite and cannot be extended.

Perhaps more than any other group of birds, parrots suffer from exploitation for the live-bird trade. Methods of capture are wasteful and often inhumane, and there are signs that the level of trapping is having an adverse effect on wild populations.

THE PARROT THAT CANNOT FLY

In just about every aspect of its biology, the kakapo or owl parrot *Strigops habroptilus* of New Zealand is unique. It is much heavier than other parrots. It is nocturnal. And although its wings are well developed, there is no sternal keel for attachment of the wing muscles, and hence the bird cannot fly! Its method of feeding is peculiar, and produces telltale signs of its presence in any given area. The parrots chew the leaves or stems, extract the juices, and leave behind, hanging on the plant, tight balls of macerated fibrous material, which are then bleached white by the sun. But its strange breeding behavior sets the kakapo apart from all other parrots most decisively. Male courtship is a lek-display involving social displaying and "booming" from inflated gular air-sacs at excavated bowl-like depressions or "courts" on arenas or traditional display grounds. Females respond to the booming and come to the courts, where mating takes place. Males take no part in nesting, leaving the female to excavate a burrow under rocks or tree roots, in which she lays one or two, rarely three, eggs, and where she alone rears the chicks. Tragically, the kakapo is now seriously endangered. A major conservation effort by New Zealand wildlife authorities is underway, with funding coming from both government and corporate sources.

► *A kakapo in display.*

Don Merton

TURACOS AND CUCKOOS

S. MARCHANT

Order Cuculiformes
2 families, *c.* 40 genera,
c. 150 species

SIZE
Smallest Little bronze-cuckoo *Chrysococcyx malayanus*, total length 14 to 15 centimeters (5½ to 6 inches); weight about 30 grams (1 ounce).
Largest Great blue turaco *Corythaeola cristata*, total length about 90 centimeters (3 feet); weight 1 to 1.2 kilograms (2¼ to 2⅔ pounds).

CONSERVATION WATCH
There are 12 species listed in the ICBP checklist of threatened birds. They include the following: Bannerman's turaco *Tauraco bannermani*, Cocos cuckoo *Coccyzus ferrugineus*, red-faced malkoha *Phaenicophaeus pyrrhocephalus*, banded ground-cuckoo *Neomorphus radiolosus*, short-toed coucal *Centropus rectunguis*, and green-billed coucal *C. chlororhynchus*. The snail-eating coua *Coua delalandei* may be extinct.

T he order Cuculiformes consists of two quite different families—the turacos (louries or plantain-eaters) of the family Musophagidae, and the cuckoos of the family Cuculidae—which are united chiefly because their egg-white proteins are similar. Some ornithologists classify them in two separate orders because of their many anatomical differences, totally different juvenile development, different patterns of molt, and the different lice living among their feathers. Both groups have a long evolutionary history, being known as fossils from about 40 million years ago. At present, the turacos are confined to Africa south of the Sahara. The cuckoos are virtually cosmopolitan but are best represented in the tropics and subtropics.

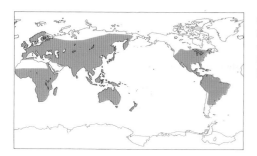

TURACOS

Turacos are rather long-necked birds with long tails, short rounded wings, and erectile, laterally compressed crests, except in one species. The great blue turaco is the largest, but all the other species are roughly the same size, about 45 centimeters (17¾ inches) in length and weighing 250 to 350 grams (9 to 12½ ounces). Their feet are semi-zygodactylous (the fourth or outer toe is at right angles to the main axis of the foot and is capable of being directed backwards or forwards). They are gregarious, noisy birds, going about in parties of five to ten in forest and savanna, flying rather weakly from tree to tree with gliding and flapping flight, but running and bounding nimbly among the branches within a tree, communicating with harsh loud barking calls that can be heard from afar and may be recognized by other creatures such as antelopes as signals of alarm. In general they are sedentary, or at least non-migratory. Five species are dull-plumaged, grayish, brownish or whitish birds that inhabit the savannas—for example, the gray go-away bird *Corythaixoides concolor*. The rest are brightly colored in greens, reds, purples, and blues, which are formed by special pigments such as turacin, apparently unique in the animal kingdom. These brightly colored species typically live in forests. Turacos eat mostly fruit, leaves, buds, and flowers but also take insects, especially when breeding. Their nests, always in trees, are

simple frail platforms of sticks. The young hatch at an advanced stage of development, and have thick down and open eyes (or nearly so). They are fed by regurgitation, and usually scramble out of the nest long before they can fly.

▲ *The white-cheeked turaco has a restricted distribution in Ethiopia and Somalia in northeastern Africa. It frequents dense brush and forest edges, feeding largely on fruit.*

Richard T. Mills

▲ *Cuckoos are notorious for their habit of laying their eggs in the nests of other birds, to be raised by the unwitting fosterers. However, it is not true that all cuckoos do this, nor are cuckoos the only birds that practice this strategy. The European cuckoo lays its eggs in the nests of a wide range of hosts, including warblers, wrens and pipits. This is a juvenile bird.*

CUCKOOS

For many people, the name "cuckoo" will conjure up a symbol of summer—a rather sleek, long-tailed grayish bird with long pointed wings, a well-known call, and the habit of laying eggs in the nests of other species, from which its chicks evict the eggs or young of the host. That image is based on the performance of one species, the European cuckoo *Cuculus canorus*, around which a huge body of myth and speculation has been built up over the centuries. In fact, of the 130 or so species in the family Cuculidae, only about 50 are truly parasitic.

The family is subdivided into six groups of great diversity, classified more on breeding habits and geographical distribution than on structural characters, though all have zygodactylous feet (two toes pointing forwards and two backwards).

True cuckoos Members of the subfamily Cuculinae have parasitic habits like those of the European bird and are confined to the Old World. Mostly they are drab, black and white, or black birds, differing more or less between the sexes, with long pointed wings. The smaller bronze-cuckoos of the tropics and Southern Hemisphere may, however, be brightly colored—for example, the emerald cuckoo *Chrysococcyx cupreus* of Africa, which is green and gold. Cuckoos generally seem to be solitary because the males are often very conspicuous and noisy with persistent bouts of loud, striking, and rather monotonous calls, even at night, whereas the females are unobtrusive and have different, even muted, calls. Probably few species, if any, form simple pairs for breeding; the male's territory may overlap with those of more than one female, and they may therefore be promiscuous. Species are to be found in a variety of habitats, from open moorland in northwestern Europe to tropical rainforest. Most species (at least, outside the tropics) are strongly migratory. Hairy caterpillars form a large part of their diet.

In spite of all sorts of claims, these cuckoos do lay their eggs directly into the nests of other birds, however improbable or impossible this may seem when the nest that they parasitize is small and enclosed. The female cuckoo usually removes an egg of the host when laying her own; her eggs generally hatch earlier, and the young cuckoo evicts unhatched eggs or young of the host within three to four days of hatching. However, some large species, such as the channel-billed cuckoo *Scythrops novaehollandiae*, often lay several eggs in the host's nest; rather than evicting the host's eggs or young, the young cuckoos out-compete them for food and they die. Cuckoos' eggs often mimic those of the host, especially if the host makes an open cup-shaped nest, and different females within one species of cuckoo may each be adapted to parasitize one species of host.

Nest-building cuckoos The subfamily Phaenicophaeinae comprises the malkohas of tropical Asia, quite large, long-tailed birds with bare, brightly colored faces; large, lizard-eating cuckoos (genus *Saurothera*) and other diverse species of the Caribbean; and the Coccyzus cuckoos of the New World.

Anis and guiras The anis and guiras of the subfamily Crotophaginae are a small New World group of gregarious, short-winged, long-tailed cuckoos that breed communally and maintain group-territories. Characteristically their bills are deep and laterally compressed.

Road-runners and ground cuckoos The members of the subfamily Neomorphinae of the

Americas, with one genus (*Carpococcyx*) in Asia, are terrestrial birds, some of which rarely fly and some of which inhabit arid regions. The striped cuckoo *Tapera naevia*, pheasant cuckoo *Dromococcyx phasianellus*, and pavonine cuckoo *D. pavoninus* are parasitic like the true cuckoos but are classified here because they appear to be so like road-runners.

Couas The subfamily Couinae comprises 10 non-parasitic species confined to Madagascar; one species may now be extinct. They are long-legged birds, the size of pigeons. Some are brightly colored and have naked skin on the head.

Coucals The coucals of the subfamily Centropodinae are terrestrial cuckoos with long straight hind-claws (thus sometimes called "lark-heeled cuckoos"), which build domed nests, lay white eggs, and are confined to the Old World. Their persistent hooting monotonous calls—praying for rain or giving thanks for fine weather, according to local folklore in Africa—proclaim their presence, although most species tend to skulk in thick undergrowth in woodlands.

R. Drummond

▲ The pheasant coucal is common in rank grasslands in New Guinea and in northern and eastern Australia.

▲▶ Two aberrant cuckoos. The channel-billed cuckoo (above) is unusual among cuckoos in that it feeds largely on fruit; it breeds in northern and eastern Australia and spends the winter in Indonesia and New Guinea. The original inspiration for the famous cartoon character, the roadrunner (right) inhabits deserts of the American southwest. As its name suggests it is terrestrial, and feeds on lizards, snakes, and other small animals.

Order Strigiformes
2 families, 24 genera,
c. 162 species

SIZE
Smallest Least pygmy owl
Glaucidium minutissimum,
total length 12 to 14
centimeters (4¾ to 5½
inches); weight less than 50
grams (1¾ ounces).
Largest Eurasian eagle owl
(European race) *Bubo bubo*,
total length 66 to 71
centimeters (26 to 28 inches);
weight 1.6 to 4 kilograms (3½
to 8¾ pounds).

CONSERVATION WATCH
There are 21 species listed in
the ICBP checklist of
threatened birds. They
include the following: Congo
bay owl *Phodilus prigoginei*,
Blakiston's fish owl
Ketupa blakistoni, and rufous
fishing owl *Scotopelia ussheri*.
Flores scops owl *Otus alfredi*
and Madagascar owl *Tyto
soumagnei* may be extinct.

OWLS, FROGMOUTHS, AND NIGHTJARS

PENNY OLSEN

The Strigiformes (owls) and Caprimulgiformes (frogmouths, nightjars, and their allies), are both well-defined groups, and even for people with little ornithological training the members of each are instantly recognizable. The two orders share many characteristics and are thought to be distantly related. Both are crepuscular (twilight-active) and nocturnal (night-active). Their soft plumage is typically in "dead-leaf" and "mottled-bark" colors and patterns, which are most refined in nightjars and frogmouths. Immobility and posture add to the effectiveness of their camouflage. With flattened feathers, bill tilted skyward, and eyes closed to a slit, a disturbed frogmouth is indistinguishable from the broken branch of a tree. So cryptic are the nightjars as they crouch, roosting on the ground, that photographs of them become "find-the-hidden-bird" puzzles. The owls also flatten their plumage when slightly disturbed but with their longer legs and wider eyes they are more obvious; they roost in more hidden places and flush more readily than many of the caprimulgiforms.

OWL DISTRIBUTION

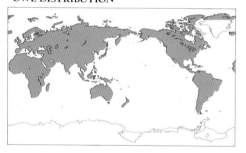

▼ *Many night-birds, like this common potoo in South America, can be located by their eye-shine reflecting back from a powerful torch.*

ADAPTED TO DIM LIGHT

These birds are more often heard than seen; their distinctive calls, described as startling, strange, or weirdly beautiful, often carry across the countryside. Their calls and mysterious nocturnal habits have been the basis for much superstition, from shrieking ghosts to the ancient belief that the nightjars steal milk from goats, hence one of their common names, goatsucker. In fact, they flit around goats and other livestock in pursuit of the insects attracted to them.

Life in dim light has led to some remarkable sensory adaptations: large eyes with good vision in poor light; and, in total darkness, navigation by echolocation (oilbirds) and hunting by exceptional hearing (barn owls).

They all have rather large heads. Most species have large eyes: forward-facing in the owls for increased binocular vision; more laterally placed in the caprimulgiforms. Their eyes are specialized for vision in poor light, with more rods (light-sensitive elements) than diurnal (day-active) birds. Nevertheless, most species appear to need some light before they are able to hunt. While most species habitually hunt in poor light, they can see well by day and some occasionally hunt in daylight (for example, the barn owl *Tyto alba*, the burrowing owl *Athene cunnicularia*, and the barking owl *Ninox connivens*); the northern hawk-owl *Surnia ulula* is largely diurnal.

The remarkable barn owls (genus *Tyto*), with rather small eyes, have the most exceptional hearing. They are able to catch prey in total

Richard Cannings

◀ In many owls, the young leave the nest well before they can fly; the parents continue to feed them for weeks, and in a few cases for months, thereafter. These young saw-whet owls have already donned the distinctive juvenile plumage but have still not reached independence. So-called because one of its calls resembles the sounds produced in sharpening a saw, the saw-whet owl inhabits dense coniferous forests in North America.

OWL SKULL

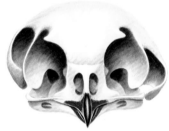

▲ In owls the ears are asymmetrical in both size and shape. This enhances the "stereo" effect — the subtle difference between the sound signal reaching one ear relative to that reaching the other ear — and enables very precise location of prey.

darkness, guided by sound alone. They and several other owls have facial masks to catch sound, and some have asymmetrical ear openings. Either the external feathering or the skull itself is modified so that sounds reach one ear at a slightly different time to the other; by turning its head the owl can locate the source of a low sound, such as a mouse chewing grain, very precisely. Oilbirds also have a remarkable adaptation for night navigation. They nest and roost gregariously, deep in caves. At night a mass of birds navigates through the cave by making audible (to humans) clicks and using the echoes that return to their ears to guide them from the cave; once outside they cease clicking.

Most species have soft, loose plumage, with frayed trailing edges to the flight feathers of their wings and tail, for noiseless hunting flight. Exceptions are the fishing owls (genus *Scotopelia*) of Africa and the oilbirds, which are hunters of fish and gatherers of fruit, respectively, presumably with little need for silent flight; both have firmer feathers.

The two groups differ most obviously in their bill and feet. The owls have a sharp, hooked bill and strong legs and feet, with sharp curved talons for their predatory lifestyle. The nightjars and their allies have a broad flattened bill, an enormous gape, and small, weak feet and legs. Both orders have reversible outer toes and can perch with two toes forward, two back. The barn owls and the caprimulgiforms have a serrated edge on the talon of their middle toe, perhaps as an aid to grooming.

OWLS

Currently, the owls are split into two families. All have rather long, broad wings.

Barn owls

Members of the family Tytonidae are medium-sized owls with heart-shaped faces, inner toes as long as their middle toes, and long bare legs. The bay owls (genus *Phodilius*) are currently placed in this family but may resemble barn owls (genus *Tyto*) only superficially.

Hawk-owls or true owls

The family Strigidae are small to large owls with rounded heads, large eyes, stout, sometimes feathered legs, and the inner toe shorter than the middle toe. Some show little sign of a mask; others are partially masked; and some have a full, rounded mask. Several species have two tufts of erectile feathers or "ears" which they can raise in emotion and which may help with concealment by disguising the owl's outline. One species, the maned owl *Jubula lettii* of west Africa and the Congo, has voluminous crown and nape feathers.

The female of most owl species is larger than the male, sometimes considerably so: female Tasmanian masked owls *Tyto novaehollandiae* weigh an average of 965 grams (34 ounces), males a mere 525 grams (18½ ounces). But, in some of the *Ninox* species, it is the male that is larger: for example, the female barking owl weighs 510 grams (18 ounces), the male 680 grams (24 ounces).

SIZE
One of smallest Donaldson-Smith's nightjar *Caprimulgus donaldsoni,* total length 19 centimeters (7½ inches); weight 21 to 36 grams (¾ to 1¼ ounces).
Largest Papuan frogmouth *Podargus papuensis,* total length 50 to 60 centimeters (20 to 23½ inches); weight 300 to 570 grams (10½ to 20 ounces).

CONSERVATION WATCH
The following species are listed in the ICBP checklist of threatened birds: Dulit frogmouth *Batrachostomus harterti,* satanic nightjar *Eurostopodus diabolicus,* Puerto Rican whip-poor-will *Caprimulgus noctitherus,* white-winged nightjar *C. candicans,* Vaurie's nightjar *C. centralasicus,* Salvadori's nightjar *C. pulchellus,* long-trained nightjar *Macropsalis creagra,* and sickle-winged nightjar *Eleothreptus anomalus.* The long-tailed potoo *Nyctibius aethereus,* white-winged potoo *N. leucopterus,* and rufous potoo *N. bracteatus* are very rare. The Jamaican pauraque *Siphonorhis americanus* is probably extinct.

NIGHTJARS AND THEIR ALLIES

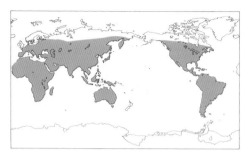

The Caprimulgiformes are divided into five families.

Oilbird
The oilbird *Steatornis caripensis* is the sole member of the family Steatornithidae. It has a fan-like tail, long broad wings, and is dark brown with white spots and black bars. Adult size is about 30 centimeters (12 inches). In common with the other caprimulgiforms, it has a strong hook-tipped bill, a wide gape surrounded by bristles, and large eyes.

Frogmouths
Members of the family Podargidae are the largest of the caprimulgiforms. They have been described as the most grotesque of birds, with a great flat shaggy head dominating the body, which tapers from it. The massive bill, surrounded by large tufts of facial bristles, as wide as it is long and heavily ossified (hardened like bone), acts as a heavy snap-trap. Their legs are short and weak.

Potoos
The potoos (family Nyctibiidae) resemble frogmouths in their arboreal roosting habit and color pattern. Yet their broad, weakly ossified bill, which is surrounded by relatively few bristles, and their aerial hawking behavior, ally them with the nightjars.

Owlet-nightjars
These birds (family Aegothelidae) are somewhere between a nightjar and an owl in appearance, but their closest relatives are the frogmouths. They have a broad flat bill almost hidden by bristles. Their feet are slightly stronger than those of the other caprimulgiforms, and their legs longer, perhaps because they run about more.

Nightjars
The nightjars (family Caprimulgidae) are a large group and comprise about half of the species in the Caprimulgiformes. They have long pointed wings and swift flight, a wide gape, stubby bill, and brightly colored mouth (usually pink) shown in threat. Their legs and feet are weak, much reduced and rarely used. The nightjars are fairly uniform in appearance, but variations on the basic form include the standard-winged nightjar *Macrodipteryx vexillaria* and pennant-winged nightjar *M. longipennis* which have extraordinary trailing feathers used in courtship, and the long-tailed nightjar *Caprimulgus climacurus* with a long gradated tail; these three live in Africa. In North America, members of the subfamily Chordeilinae are called nighthawks.

▶ Roosting in trees, a frogmouth usually spends the day in a distinctive posture, more or less upright and oriented along the branch; in this position its resemblance to a broken-off stub is extraordinary. But the tawny frogmouth is a sociable bird, and sometimes pairs and even families cuddle up together on a branch.

Robert Cook/Weldon Trannies

◄ Mostly eyes and fluff, an Australian owlet-nightjar pauses at the entrance to its nest hollow. This bird is common in woodlands across Australia; several other species occur in New Guinea. Owlet-nightjars are unusual among caprimulgids in that they roost and nest in cavities in trees. The birds pair for life but roost apart.

Tom & Pam Gardner

DISTRIBUTION AND HABITATS

Owls are cosmopolitan in distribution. They occur on all continents except Antarctica and are absent from some oceanic islands. *Strix* and *Otus* are widespread genera, the latter mostly in tropical areas; they do not occur in the Australia–Papua New Guinea region, where they are replaced by the genus *Ninox*. Some species such as the barn owl and the short-eared owl *Asio flammeus* are among the most widely distributed of all birds. In contrast, the Palau owl *Pyrroglaux podargina* is found only on the Palau islands in the western Pacific Ocean. Habitat destruction and introduced animals have taken their toll and pushed some of these island owls toward extinction.

The majority of owl species inhabit woodlands and forest edges. A few species prefer treeless habitats: for example, the snowy owl *Nyctea scandiaca* of Arctic tundra regions, and the elf owl *Micrathene whitneyi* of the southwestern deserts of the USA. Some, long-legged, species are terrestrial and live in flat grasslands (the grass owl *Tyto capensis* of Africa, and India to Australia) or

marshes (the marsh owl *Asio capensis* of Africa), or among rocks. Various species can be found in most habitats, from tundra and desert to rainforest and swampland, from wilderness to the suburbs; the great horned owl *Bubo virginiatus* occurs in most habitats of North America. Some are more specific in their habitat requirements than others: the northern hawk owl lives in the northern conifer forests (taiga) of North America and Eurasia; the white-throated owl *Otus albogularis* is found in the cloud forests of the Andes; and the Peruvian screech owl *O. roboratus* likes habitat with mesquite and large cacti in arid parts of Peru.

The nightjars and their allies have a similar distribution to owls but are not found at such high latitudes or on as many small islands. Nor are they found on the islands of New Zealand. The oilbird is found only in the neotropics (Guyana to Peru and Ecuador, and Trinidad); it depends on suitable caves, and ranges out from them to forage locally —most are in the mountains, but in Trinidad sea-caves along the rocky coast are used. The potoos prefer open woodland of cultivated areas such as

► *Largest of Australian owls, the powerful owl roosts by day in trees, usually selecting places that are fairly open all around but with a screen of foliage above. This one is roosting near its offspring. Feeding largely on possums, the powerful owl frequently holds the remains of its night's meal, tail dangling, clamped in its talons throughout the following day, often polishing it off as a sort of wake-up snack the next evening before going off to hunt again.*

Geoff Longford

coffee plantations in tropical Central and South America. Frogmouths are arboreal inhabitants of forest, woodland and forest edge: *Batrachostomus* species occur from India to Malaysia; and the *Podargus* species in New Guinea, the Solomon Islands, and Australia. The tawny frogmouth *P. strigoides* is found throughout Australia in most habitats, but avoids dense rainforest and treeless desert. Owlet-nightjars live in rainforest or open forest and woodland in New Guinea, Australia, and New Caledonia. The nightjars are widely distributed in warmer parts of the world and have a diversity of habitat preferences. In Africa, for example, there is a different *Caprimulgus* species in almost every habitat.

FEEDING AND BREEDING

Owls drop down from a perch to catch mammals (from mice to hares, depending upon the size of the owl) or insects on the ground. They also snatch insects from foliage, and large species grab arboreal mammals while smaller species hawk insects in the air. They catch prey with their feet and may reach down and dispatch it with a few bites. Small prey is often lifted to the bill with one foot, in the manner

of a parrot; it is swallowed whole. Large prey is held in the feet and dissected with the bill. Once or twice a day owls regurgitate a pellet containing the fur, most bones, chitinous insect remains, and other indigestible parts of their prey.

The caprimulgiforms do not produce a pellet. Oilbirds hover to pluck a variety of fruits and seeds, which they locate by sight and scent. Frogmouths pounce from a perch to catch non-flying animal prey in their massive gape and heavy bill; they batter prey to soften it before swallowing it whole. Owlet-nightjars take some insects and frogs on the ground and dart out to snatch termites, moths and other insects from the air. Most aerial of all are the fast flying nightjars and nighthawks which commonly trawl, open-mouthed, for airborne insects.

Since they are more often heard than seen, it is not surprising that many of these birds are named for the calls they make. All owls call, especially as the breeding season approaches. They have a variety of shrieks, hoots, and barks, which are typical of individual species. Several owls sing quite musically. Some screech-owls (genus *Otus*) duet, the male and female each taking turns to complete their section of a song. Caprimulgiforms also have a variety of far-carrying calls: low drumming in frogmouths; churring in owlet-nightjars; and various other screams and shrieks, hence the name "nightjar". However, some, like the whip-poor-will *Caprimulgus vociferus* of North America, make quite melodious whistles, and one African species, the fiery-necked nightjar *C. pectoralis*, sings "Good Lord, deliver us".

Owls nest in a hole in a tree or cliff, an old building, or the old stick nest of a crow or raptor. The burrowing owl takes over a gopher hole, eagle owls sometimes dig a nest cavity in the side of an anthill, and the snowy owl nests on the open ground in a scrape to which it adds a little lining. A woodpecker hole, drilled in a cactus, is used by the the elf owl. Some caprimulgiforms build a nest. Regurgitated fruit, which sets firm, forms the oilbird's nest, which is built in recesses in the cave and added to each season. Frogmouths build a flimsy nest of sticks on a horizontal branch (*Podargus* species) or use a pad of down from the birds themselves, plus spider webs and lichen (*Batrachostomus* species). Owlet-nightjars nest in a hollow in a tree or occasionally a bank; potoos use a depression on a branch; nightjars and nighthawks nest unceremoniously on bare ground or occasionally on an epiphyte.

Perhaps because they are nocturnal, many species do not seem to have elaborate courtship displays. The male owls feed the females during courtship. During the breeding season, the second primary in each wing of the male standard-winged nightjar projects about 35 centimeters (14 inches) beyond its neighbors and is shown to effect in slow aerial breeding displays. The bird nips them off after the displays cease.

Some owl species lay a similar-sized clutch of eggs each year during a regular breeding season (for example, two to three eggs for *Ninox* species); others vary the start of breeding and the clutch size quite dramatically according to seasonal conditions. The snowy owl will lay up to 14 eggs in a year when its lemming prey are abundant, but two to four when prey is scarce. Incubation takes between four and five weeks depending upon the species, and the young must be brooded for a few weeks. The male forages and the female incubates, then both parents feed the young by offering them food. The nestlings often leave the nest to perch nearby when still downy, and in some species the family may stay together for several months.

The caprimulgiforms have a small clutch of between one and four eggs. Because the risk of predation is high, some ground-nesting nightjars have short incubation times (17 or 18 days in the Northern Hemisphere, but longer in Australia) and the young are semi-precocial—within hours they can totter around. They stay with their parents until migration. Oilbirds feed their nestlings on regurgitated oily fruit for 120 days, until they reach almost adult size. They have long been collected by South American Indians for their fat, which is used for cooking and in lamps.

Typically, owls and caprimulgiforms are solitary; the gregarious oilbirds are an exception. In both orders some species are resident, others partially or totally migratory. For example, some European populations of *Otus* owls migrate to Africa for the winter. Most temperate-zone species of nightjar spend winter in the tropics and some tropical species are partial or total migrants. Rather than departing the North American winter, however, each year the common poorwill *Phalaenoptilus nuttallii* hibernates, clinging to the sides of a rock crevice. Its heart rate and respiration drop to almost unmeasurable levels, and its temperature falls from about 41°C (105°F) to about 19°C (66°F). It is one of the very few birds that hibernate regularly.

▼ *The great gray owl inhabits the vast subarctic coniferous forests of the Northern Hemisphere. It is one of the largest of owls, but its bulk is mainly feathers and in fact it weighs less than half as much as many other owls more or less its equal in size. It nests in the abandoned nests of other birds of prey, never adding to them, although it will often rearrange the material.*

Jeff Foott/Auscape International

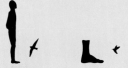
SWIFTS AND HUMMINGBIRDS

CHARLES T. COLLINS

Swifts, crested swifts, and hummingbirds are generally classified as three families (Apodidae, Hemiprocnidae, and Trochilidae, respectively) in the order Apodiformes. They share some anatomical features, particularly the relative length of the bones of the wing, which is related to their rapid wing beats and flight behavior. The connection between swifts and crested swifts seems clear, but the inclusion of the very dissimilar hummingbirds in this order has often been challenged. Any communality of ancestry is indeed old.

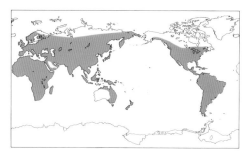

SWIFTS

Swifts, with their narrow swept-back wings, have a well-deserved reputation for being among the fastest flying birds. They range in size from the pygmy swiftlet *Collocalia troglodytes* of the Philippines and pygmy palm swift *Micropanyptila furcata* of Venezuela, which weigh less than 6 grams (¼ ounce), up to the white-naped swift *Streptoprocne semicollaris* of Mexico and purple

needletail *Hirundapus celebensis* of the Philippines, both of which approach 200 grams (7 ounces). All are predominantly dark brown or sooty, with some areas of white or gray, and they have short legs with strong claws. All swifts pursue and capture their food, mostly insects, on the wing and stay aloft throughout the day, perching only at their overnight roosts. Sometimes the food ball or bolus taken to a nestling will contain mainly swarming insects such as termites, mayflies or aphids, as well as winged ants, wasps, and bees. At other times up to 60 different kinds of insects and spiders and several hundred individual prey items can be found in a single bolus.

Although most numerous in the tropical areas of the world the 80 or so species of swifts are widely distributed and even occur in Scandinavia, Siberia, and Alaska. The common swift *Apus apus* and alpine swift *A. melba* of Europe, the white-throated needletail *Hirundapus caudacutus* of Siberia and the chimney swift *Chaetura pelagica* of eastern North

▶ *The black swift of western North America nests in rock crevices in cliffs, often behind waterfalls.*

Tom Ulrich/Oxford Scientific Films

George D. Lepp/Comstock

◄ *A female Allen's hummingbird feeds her brood of young. As in most hummingbirds, nest-building, incubation, and care of the young is left entirely to the female. The usual clutch is two, and two or three broods are raised each season. This hummingbird is common in coastal California, often nesting in city parks and gardens.*

America all make long migration flights, often over stretches of ocean, to Southern Hemisphere wintering grounds. Even on the breeding grounds some swifts regularly spend the night on the wing.

Many swifts use secretions of their salivary glands in nest building. (Members of the New World subfamily Cypseloidinae do not do this, however, and their nests of mosses, ferns, and other plant material are placed near or behind waterfalls). The salivary glands enlarge during the breeding season to produce a sticky material which, in the genus *Chaetura*, is used to glue together small sticks to form the nest and also to

attach it to the vertical wall of a hollow-tree nesting site; while in flight, the birds break dead twigs from the tops of trees. The use of saliva in nest building is most highly developed in some of the smaller cave-inhabiting swifts, known as cave swiftlets, of Southeast Asia, where saliva makes up the bulk of the nest. It is sometimes mixed with plant material and feathers (black nests) or forms the entire nest (white nests). These nests are collected by men who climb rickety bamboo scaffolding or vine ladders to reach the high ceilings of caves where tens of thousands of these swiftlets nest. Although white nests are considered the most valuable, as the main ingredient of bird's nest soup, both white and black nests are harvested and have become a major economic resource in that part of the world. Harvesting is controlled to protect the birds and keep this a renewable resource. Occasionally swiftlets, as well as other species of swifts, nest in close association with humans and use buildings and bridges as nest sites; chimney swifts now nest more commonly in chimneys than in hollow trees.

Researchers have found that several species of swifts and their nestlings are able to survive short periods of inclement weather by entering a semi-torpid state with a lowered body temperature. There are anecdotal accounts of what appears to be true hibernation in the chimney swift, but this needs further study.

Some cave swiftlets (genus *Aerodramus*) can nest and roost in total darkness deep in caves, sometimes more than a kilometer from any light. These birds make a series of audible clicks or rattle calls, and the returning echoes enable them to navigate within a cave and locate their own nest or roost site. (The only other bird that uses echolocation is the oilbird *Steatornis caripensis* of northern South America and Trinidad.) Non-echolocating swiftlets nest in the twilight zone and near the entrance of caves where there is still sufficient light for visual flight.

Both sexes participate in nest building,

incubation, and provisioning of the chicks; incubation requires 19 to 23 days, and the nestling period may last as long as six to eight weeks. For species that nest in colonies there often is much social activity in the form of grouped flights and vocalizing. Their calls vary, from short sharp chips to long-drawn-out buzzy screes or screams.

CRESTED SWIFTS
The four species of crested swifts (genus *Hemiprocne*) are distributed from peninsular India eastward through Malaysia and the Philippines. All have frontal feathered crests, various degrees of forked tails, and patches of brighter colors. They are far less aerial in their behavior than the true swifts, in that often they alternate between perching on prominent treetops and making graceful flights in pursuit of flying insects. Nowhere as abundant as swifts, crested swifts tend to be more solitary and sparsely distributed. Their nest is tiny, consisting of a small cup of plant material and lichens barely large enough to hold the single egg; it is fastened to a small lateral twig, and the brooding bird straddles the nest and supports itself on the underlying branch. Because their nests are typically high in the outer branches of large trees, many aspects of the breeding biology of crested swifts remain to be studied.

HUMMINGBIRDS
Hummingbirds are known for their small size, bright iridescent colors, and hovering flight. This diverse New World family, with 320 species in 112 genera, is most abundant in the warm tropical areas of Central and South America, but some are also found from Alaska to Tierra del Fuego and from lowland rainforest to high plateaus in the Andes. The average weight of these tiny birds is between 3.5 and 9 grams (less than ⅓ ounce)—the bee hummingbird is perhaps the smallest living species of bird, at about 2.5 grams (¹⁄₁₀ ounce) —but a few are larger, the giant hummingbird *Patagona gigas* being almost 20 grams (⅔ ounce).

The shape of the bill clearly reflects the type of flowers each species visits for nectar and insects. Hummingbird foraging takes two major forms: territoriality, in which floral nectar sources are vigorously defended; and trap-line foraging, where rich but more widely dispersed sources of nectar are regularly visited. The tongues of hummingbirds are brush-tipped to aid in nectar acquisition, but insects provide a needed source of protein and are a major component of their diet.

The extremely rapid wing beat (22 to 78 beats per second), coupled with a rotation of the outer hand portion of the wing and a powered upstroke, permits hummingbirds to hover adroitly in front of flowers during foraging. They also make vigorous acrobatic flights during territorial chases, and some species make elaborate aerial courtship displays. Longer flights to follow seasonal flowering patterns

▼ *A male and a female black-chinned hummingbird consort briefly at the same flower. In most hummingbirds there is no pair bond, and the sexes come together only casually or when actively courting.*

Robert A. Tyrrell

are also undertaken, and some species migrate over several thousand kilometers from nesting areas in temperate zones to wintering grounds in the tropics. During its migration, the ruby-throated hummingbird *Archilochus colubris* flies 1,000 kilometers (620 miles) across the Gulf of Mexico.

The breeding season of most species is keyed to the local flowering cycle, although it avoids seasons of intense rain. In most species the female alone builds the nest and incubates the eggs. Nests are typically small inconspicuous cups of plant material held together with spider web and sometimes adorned with bits of moss or lichen. More bulky nests, sometimes attached to the underside of a leaf blade, are typical of the hermit hummingbirds and some cave-nesting species.

▲ A hummingbird drains the nectar from a flower. Hummingbirds need constant access to flowers to fuel their prodigious metabolism, by far the highest measured among birds. An intricate relationship exists between many hummingbirds and the flowers at which they feed, and some flowers rely on hummingbirds for pollination.

**Order Coliiformes
1 family, 2 genera, 6 species**

Size
Head-body length
10 centimeters (4 inches); tail
20 to 24 centimeters (8 to
9½ inches); weight 45 to
55 grams (1½ to 2 ounces).

Conservation Watch
These species do not appear
to be threatened.

**Order Trogoniformes
1 family, *c.* 8 genera,
c. 37 species**

Size
Smallest Black-throated
trogon *Trogon rufus*, total
length 23 centimeters
(9 inches).
Largest Resplendent quetzal
Pharomachrus mocinno,
body length 33 centimeters
(13 inches); tail 70
centimeters (24 inches).

Conservation Watch
The resplendent quetzal
Pharomachrus mocinno, eared
trogon *Euptilotis neoxenus*,
and Baird's trogon *Trogon
bairdii* are listed in the ICBP
checklist of threatened birds.

MOUSEBIRDS AND TROGONS

G.R. CUNNINGHAM-VAN SOMEREN

Mousebirds (order Coliiformes), also known as colies, are drably colored, small-bodied birds of Africa. The name "mousebird" comes from the curious way in which the birds creep and crawl among the bushes, clinging upside down, with the long tail high in the air. Among the trogons (order Trogoniformes) the resplendent quetzal *Pharomachrus mocinno* of Central America is perhaps the best known, but all of these birds of the tropical woodlands and forests have brilliantly colored plumage.

MOUSEBIRDS, OR COLIES

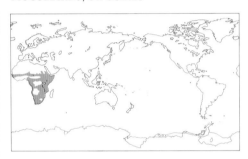

Mousebirds are distributed throughout Africa south of the Sahara. The species inhabit a wide range of country, from almost-dry bushland to the edge of forests (but not in the forest), from low altitudes to 2,000 meters (6,500 feet). Some have taken up residence in people's gardens and smallholdings, where they are justifiably regarded as pests. Mousebirds are vegetarians and their diet consists of foliage, buds, flowers, wild or cultivated fruits, and even seedlings.

The nest is well hidden in a bush, and is of rough construction but has a softly lined cup. The

► Confined to Africa, the mousebirds are superficially unremarkable in appearance, but they have a number of odd characteristics. Like the other species, the speckled mousebird habitually perches in a distinctive fashion with the feet more or less level with the shoulders. Its feathers have very long aftershafts, which contributes to an unusually fluffy-looking plumage.

John Shaw/NHPA

eggs (usually four to six) are laid at daily intervals, and incubation begins with the first, with the result that nestlings of different sizes will be found. The nestlings are naked when newly hatched but soon grow down and then feathers. Older nestlings, if disturbed or in full sun, may clamber out and hide—but then may be cannibalized by their parents. Nest building, incubation, and feeding by regurgitation are undertaken by both parents. A family may remain together for a considerable time or may join others to form groups, the leader going from bush to bush while others follow. Their flight is often swift and direct, with whistle contact-calls. When resting or roosting the family or group will clump together haphazardly, some clinging to vegetation, others on the backs of those below. Their legs and feet are curiously articulated.

Mousebirds dislike rain and cold, and even huddling together they may become torpid. They seldom drink, but groups are often seen dust-bathing and sunning themselves with wings and tail well spread. The feathers grow randomly, and the birds are often parasitized by flies, lice, fleas, and ticks.

TROGONS

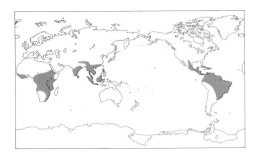

Trogons are pantropical birds of Central America and the West Indies, Africa, and Asia. They are forest dwellers, not very well known, and there is still disagreement among taxonomists about details of their classification. Generally the head, breast, and back are metallic iridescent green with some reflections of blue and yellow. The belly may be carmine, red or pink, orange or yellow. Females are usually duller than the males, and Asian species are not as colorful. The tail is long and slightly graduated, usually with white or black bars. Quetzals (genus *Pharomachrus*) of Central America are well known for their long, drooping upper tail-coverts. The lower part of the leg is feathered, and the toes are heterodactylous (toes 1 and 2 directed backwards, toes 3 and 4 forwards).

Secretive and territorial, a trogon will perch quietly scanning for a food item, especially insects. With a slow undulating flight it may snatch a caterpillar from a leaf or twig, then return to its perch. Occasionally small lizards are taken; and some South American species also eat fruit. The male calls to attract a mate to a suitable nest site.

▶ ▼ *The national bird of Guatemala, the resplendent quetzal (right) was considered divine by the Aztecs, and killing one was a capital offence. The bird lost this protection after Cortes laid waste the Aztec Empire, and its numbers have declined steadily ever since. It inhabits cloudforest above 1,300 meters (about 4,000 feet); females differ from males most conspicuously in lacking the tail streamers. The violaceous trogon (below) inhabits rainforest edges and clearings from Mexico to Amazonia.*

Females may answer; some remain silent. At each note the tail is depressed downward. They nest in a cavity in a tree; some excavate nest holes in dead trees. In the species that have been studied, two or three rounded somewhat glossy eggs are laid. Nestlings are naked and helpless when hatched, but soon acquire down. Both parents take part in the incubation and tending of nestlings and fledglings. Males may indulge in a display in which several gather to chase each other through the trees. Their skin is delicate and tears readily; soft feathers may fall out.

Order Coraciiformes
9 families, 47 genera,
206 species

SIZE
Smallest Puerto Rican tody
Todus mexicanus and allied
species, total length 10
centimeters (4 inches); weight
5 to 6 grams (⅕ ounce).
Largest Southern ground
hornbill *Bucorvus cafer,* and
allied Abyssinian ground
hornbill *B. abyssinicus,* total
length 80 centimeters (31
inches); weight 3 to 4
kilograms (6½ to 8¾
pounds).

CONSERVATION WATCH
There are 20 species listed in
the ICBP checklist of
threatened birds. They
include the following: Blyth's
kingfisher *Alcedo hercules,*
cinnamon-banded kingfisher
Halcyon australasia, Mangaia
kingfisher *H. ruficollaris,*
Tuamotu kingfisher
H. gambieri, Marquesas
kingfisher *H. godeffroyi,*
mustached kingfisher
Acterioides bougainvillei,
short-legged ground roller
Brachypteracias leptosomus,
long-tailed ground roller
Uratelornis chimaera, rufous-
necked hornbill *Aceros
nipalensis,* wrinkled hornbill
A. corrugatus, plain-pouched
hornbill *A. subruficollis,*
Sumba hornbill *Rhyticeros
everetti,* blue-capped wood
kingfisher *Halcyon hombroni,*
Biak paradise kingfisher
Tanysiptera riedelii, keel-billed
motmot *Electron carinatum,*
scaly ground roller
Brachypteracias squamiger,
and rufous-headed ground
roller *Atelornis crossleyi.*

KINGFISHERS AND THEIR ALLIES

JOSEPH FORSHAW
AND ALAN KEMP

Kingfishers, todies, motmots, bee-eaters, rollers, hornbills, the hoopoe, and wood-hoopoes are linked by details of their anatomy and behavior. They have generally small feet with a fusing of the three forward toes, and an affinity of the middle ear bone and of egg proteins. Many species are brilliantly colored birds which make their cavity nests by digging holes in earth-banks or rotten trees.

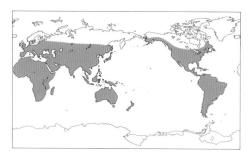

KINGFISHERS
The most visible distinguishing features of the kingfishers (family Alcedinidae) are their bill and feet. The legs are very short, and the toes are syndactyl, the third and fourth being united along most of their length, and the second and third joined basally; in some species the second toe is much reduced or absent altogether. A "typical" kingfisher bill is proportionally large, robust, generally long and straight, with a sharply-pointed or slightly hooked tip.

Two predominant bill shapes reflect the two main ecological groupings of kingfishers: in the subfamilies Cerylinae and Alcedininae the laterally compressed, pointed bill is for striking at and grasping prey, especially fishes, whereas in the subfamily Daceloninae the rather broad, slightly hook-tipped bill is more suited to holding and crushing prey. Some species have intermediate bills, and the bills of a few others are modified for specialized foraging techniques. As its name implies, the hook-billed kingfisher *Melidora macrorrhina* from New Guinea has a broad bill with a strongly hooked tip, apparently an adaptation for digging in the ground to procure prey, whereas the shovel-billed kingfisher *Clytoceyx rex,* also from New Guinea, uses its short, wide, and extremely robust bill to shovel down into moist soil in search of earthworms.

Kingfishers vary in size, from the laughing kookaburra *Dacelo novaeguineae* of Australia or the giant kingfisher *Ceryle maxima* of Africa, with a total length of more than 40 centimeters (16 inches) and weight of about 400 grams

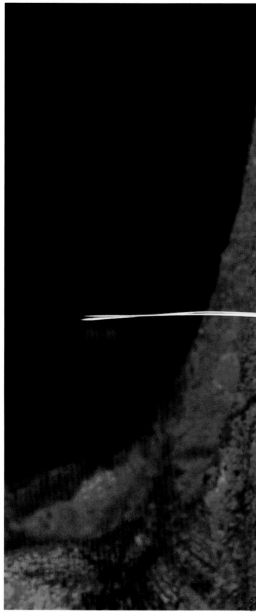

(14 ounces), to the black-fronted pygmy kingfisher *Corythornis lecontei* of Central Africa at 10 centimeters (4 inches) and little more than 10 grams (⅓ ounce). Their plumage is commonly of bright colors, often with a metallic brilliance, in blues, greens, purple and reddish or brown tones which are frequently offset with white patches or dark markings. The wings tend to be short and rounded, while the tail varies from extremely short, as in the *Ceyx* species, to very long in the beautiful *Tanysiptera* kingfishers.

Kingfishers are distributed worldwide, except the highest latitudes and remote islands. The centre of abundance is Southeast Asia and New Guinea, but tropical Africa also has strong representation. The northernmost species, summer-breeding migrants, are the Eurasian kingfisher *Alcedo atthis* in southern Scandinavia and western Russia and the belted kingfisher *Ceryle alcyon* in Alaska and Central Canada. The southernmost limit is Tierra del Fuego, which is the southern extremity of the summer breeding range of the ringed kingfisher *C. torquata*. The Eurasian kingfisher is the most widely distributed species with a range extending from western Ireland to the Solomon Islands. Most of the species with very restricted ranges are confined to small islands or groups of islands.

◄ Kingfishers of the genus *Tanysiptera* are very closely related to the widespread *Halcyon* kingfishers but are characterized by greatly elongated central tail feathers. All eight species inhabit the New Guinea region, but one, the white-tailed or buff-breasted paradise-kingfisher, shown here, migrates across Torres Strait to breed in the extreme northeastern part of Australia. It nests in arboreal termite mounds, often close to the ground.

Tom & Pam Gardner

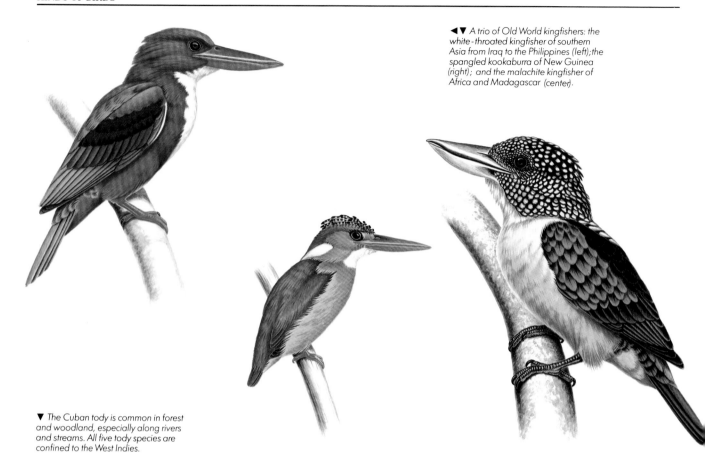

◄▼ A trio of Old World kingfishers: the white-throated kingfisher of southern Asia from Iraq to the Philippines (left); the spangled kookaburra of New Guinea (right); and the malachite kingfisher of Africa and Madagascar (center).

▼ The Cuban tody is common in forest and woodland, especially along rivers and streams. All five tody species are confined to the West Indies.

D. & M. Zimmerman/Vireo

Kingfishers inhabit a wide variety of aquatic and wooded habitats, though tending to avoid open country where there is a dearth of trees. There are noticeable habitat preferences among fish-eating kingfishers, with some species showing marked preferences for littoral habitats, especially mangroves, whereas others favor large open waterways. Lowland tropical rainforest is the habitat favored by many "non-fishing" kingfishers.

Regular seasonal migration is undertaken by some species occurring at higher latitudes, particularly in the Northern Hemisphere. The belted kingfisher of North America, the Eurasian kingfisher of Europe and western Russia, and the sacred kingfisher *Halcyon sancta* of Australia are migrants whose seasonal departures and arrivals are well known. Some continental populations of the Eurasian kingfisher are known to migrate over 1,500 kilometers (930 miles) between breeding and wintering areas.

Because the best-known species, the Eurasian kingfisher, feeds mainly on fish captured by diving into the pond or stream from an overhanging branch, the general concept is that all other kingfishers do likewise. In fact, most species are generalized predators that take a wide variety of invertebrates and small vertebrates from the ground or in water. The usual "sit and wait" foraging technique involves quiet surveillance from a vantage perch, then a swoop or dive down to seize the prey, which is brought back and, while firmly grasped in the bill, is immobilized by being struck repeatedly against a branch. Undigested food is regurgitated as pellets.

Territorial and courtship displays include loud calling and conspicuous flights high above the treetops. The normal breeding unit is a monogamous pair, which may nest solitarily or in loose colonies, but cooperative breeding, involving helpers at the nest, also occurs. With the blue-winged kookaburra *Dacelo leachii* and the laughing kookaburra in Australia, members of the family group participate in all facets of breeding, and more than one female may lay in the same nest.

Nests are in natural hollows in trees or in burrows excavated by the birds in earth-banks, termite mounds, or rotten tree-stumps. Both sexes excavate the burrow by prizing loose the soil or rotten wood with the bill, and then by using their feet to kick out the dislodged material. At the end of the burrow there is an enlarged chamber, where the eggs are laid on the unlined floor. During the course of nesting, the burrow becomes fouled with excrement, regurgitated pellets, and food remains, resulting in a characteristically pungent odor. Chicks of small fish-eating species such as the Eurasian kingfisher become independent almost immediately after fledging, whereas chicks of the larger woodland kingfishers are dependent on their parents for up to eight weeks after leaving the nest.

TODIES

Todies are very small, stocky birds with a proportionately broad head, fairly long, much-flattened bill and a short, slightly rounded tail. Their striking plumage has brilliant green upperparts and an almost luminescent red throat-patch. The sexes are alike, but young birds are noticeably duller than adults and lack the red throat-patch or have it only slightly indicated. Differences in the color of the flanks and the presence or absence of bluish subauricular patches are the main distinguishing features of the species.

All five species belong to the genus *Todus* and are confined to the West Indies, where the only island with two species is Hispaniola: the narrow-billed tody *T. angustirostris* in wet montane forests, and the broad-billed tody *T. subulatus* in dry forests or scrublands at a lower altitude. On other islands the single species shows a wider habitat tolerance. Todies remain common throughout much of their restricted range and seem to have coped remarkably well with the two major threats that affect so many island birds, loss of habitat and introduction of predators or competitors.

The typical method of capturing their insect prey is to take it from the underside of a leaf in the course of an arc-like flight from one perch to

D. Wechsler/Vireo

another. The flattened bill seems well suited to this technique and also to catching insects in flight. The nest is in a burrow excavated by the birds in an earth bank, and the two or three white eggs are laid in a slight depression in the floor of an enlarged chamber at the end of the burrow. Incubation is shared by the sexes, and chicks are fed by both parents. At some nests, extra adults have been recorded assisting in the feeding of nestlings.

MOTMOTS

The 10 species in the Momotidae family are confined to continental tropical America, from northern Mexico to northern Argentina. They are birds of forests and woodlands in the lowlands and foothills, a notable exception being the blue-

▲ Despite its colorful plumage, the rufous motmot is an unobtrusive bird of dense rainforest, more often heard than seen. It perches quietly, capturing prey on the ground or near it in brief sallies. Most prey is brought back to the perch to be subdued and swallowed. Motmots have a distinctive habit of persistently swinging the tail from side to side like a pendulum.

Hilary Fry

▲ *In typical bee-eater stance, a cinnamon-chested bee-eater sits alert for prey. This species is common in the highlands of East Africa.*

throated motmot *Aspatha gularis* which is restricted to cloud forest and mixed woodlands in the highlands of Central America.

Motmots are robust birds with a distinctive overall appearance that comes from a broad head with a proportionately long, strong bill, as well as proportionately long legs and, in most species, an acutely gradated tail with elongated central feathers. Variation in size is quite marked, from the tody motmot *Hylomanes momotula*, total length 19 centimeters (7½ inches) to the upland motmot *Momotus aequatorialis,* total length 53 centimeters (21 inches). The plumage is a combination of green and rufous, and most species have striking head and facial patterns in black and shades of brilliant blue. The stout, slightly decurved bill varies from rounded and laterally compressed to extremely broad and flattened, and in all species except the tody motmot the cutting edges are serrated.

It is the tail that most typifies the motmots, for in some species the elongated central tail-feathers are denuded part-way along their length but end in fully-intact tips that form flaglike spatules. Newly-acquired feathers are without these bare shafts, and it may be that the barbs are loosely attached and therefore fall away as the bird preens and also in consequence of rubbing against vegetation. The tails of females are slightly shorter than those of males, but otherwise the sexes are similar. Young birds resemble adults, though the head and facial markings are usually duller.

Motmots fly forth from vantage perches to capture prey on the ground, amid foliage, from the surface of a branch or tree trunk, or in the air.

Usually, captured prey is brought back to be struck repeatedly against the perch before being swallowed. Motmots have been seen with other birds following army ants to capture insects fleeing from the marauding columns.

The nests are burrows excavated by the birds in earth-banks, and some species nest in colonies of just a few pairs or 40 or more pairs. The eggs are laid in an enlarged chamber at the end of the burrow, and the incubation (about 20 days) is shared by the sexes. Both parents feed the chicks, which leave the nest about 30 days after hatching. Motmots remain locally common in much of their range, but their future is uncertain as the destruction of forests and woodlands continues unabated.

BEE-EATERS
Bee-eaters are particularly graceful birds, both in appearance and in their actions. A slim body shape is accentuated by a narrow, pointed bill and a proportionately long tail, the central feathers of which are elongated in many species. Bright green upperparts are a prominent feature of most species, and there are distinctive head or facial patterns. The predominantly red plumage, with little or no green, of the northern carmine bee-eater *Merops nubicus,* southern carmine bee-eater *M. nubicoides,* and rosy bee-eater *M. malimbicus* clearly sets apart these three spectacular species. In all species the sexes are alike or differ very slightly, and young birds are generally duller than adults. In size the adults vary from 18 to 32 centimeters (7 to 12½ inches).

Being specialist predators of stinging insects,

▶ *The swallow-tailed bee-eater is common in scrub and savanna woodland across much of Africa. Markedly less gregarious than some bee-eaters, it is usually encountered in pairs or small parties. Bee-eaters feed on flying insects such as bees and butterflies, capturing them in marvelously dextrous sallies from high exposed perches.*

Nigel Dennis/NHPA

◄ The blue-crowned motmot (far left) is the most widespread member of this small American family of birds; in Brazil it is known locally as the hudu, expressive of its most persistent call. The rainbow bee-eater (near left) is the only Australian representative of its family, but it is common and widespread across much of the continent. It is strongly gregarious, and most populations are migratory.

bee-eaters have developed an effective technique for devenoming their prey. A captured insect is struck repeatedly against the perch, and then is rubbed rapidly against the perch while the bird closes its bill tightly to expel the venom and sting. Bees, wasps, ants, and their allies predominate in the diet of most species, and insects nearly always are captured in the air. As specialized insectivores, bee-eaters may be vulnerable to the effects of chemical pesticides, although no species are at present in danger of extinction.

The 27 species occur only in the Old World, where the center of abundance is in northern and tropical Africa. Most live in open, lightly wooded country, often in the vicinity of waterways. Along routes through the Middle East and the Arabian Peninsula, a conspicuous seasonal phenomenon is the passage of large numbers of the European bee-eater *Merops apiaster* and the blue-cheeked bee-eater *M. persicus* migrating between their breeding areas in Europe and Asia and the wintering grounds in Africa. There are also intracontinental migrations in Africa. Bee-eaters are common in Southeast Asia, where there is not only seasonal movement over shorter distances but also long-distance regular migration—for example, the rainbow bee-eater *M. ornatus* which reaches Australia for breeding in spring.

The burrow nests excavated by bee-eaters in an earth-bank or flat sandy ground may be solitary (as for the *Nyctyornis* species), in groups of two or three (as for the black-headed bee-eater *Merops breweri*), in loose aggregations spaced regularly or irregularly along a bank, or in large colonies containing up to a thousand or more holes. Incubation is shared by the sexes and both parents feed the chicks, which vacate the nest 24 to 30 days after hatching. The role of additional adults as helpers at the nest has been confirmed for some species and is suspected to be present in others—it has been determined that participation by helpers increases the fledging success rate quite dramatically.

ROLLERS

Rollers take their name from the twisting or "rolling" actions that characterize the courtship and territorial flights undertaken by some of the 12 species in the family Coraciidae. Typical rollers are fairly large stocky birds, 25 to 37 centimeters

(10 to 14½ inches) long, with robust bills and rather short legs. The squarish tail is proportionately short, though some African species have tail streamers. Different foraging techniques are reflected in the bill shape: the eight *Coracias* species capture prey largely on the ground, and their projecting bills are narrower and laterally compressed; the four *Eurystomus* species take insects in the air, and their short, flattened bills are very broad at the base. Bills are blackish in *Coracias,* but bright yellow or red in *Eurystomus.* Plumage is brightly colored, with shades of blue predominating, and young birds resemble the adults, though are noticeably duller.

The European roller *Coracias garrulus* is a regular intercontinental migrant, leaving its breeding range in Eurasia to overwinter in Africa. Rollers are especially prevalent in Africa, where both resident and overwintering species are conspicuous in open and lightly wooded habitats. They range east to Southeast Asia, where there are markedly fewer species, and only one species extends to Australasia, the red-billed roller or dollarbird *Eurystomus orientalis;* the Australian populations of this species migrate north to overwinter in New Guinea.

Raucous calling, bright plumage, and frequent aerobatics make rollers very conspicuous. They are particularly active and vocal at the onset of breeding, and often dive at intruders, including humans, in the vicinity of the nest. From a low vantage perch a *Coracias* roller will drop to the ground to capture a large insect or small vertebrate, whereas *Eurystomus* rollers sally forth from high exposed branches to take flying insects.

Nests are in hollows in trees or in burrows excavated by the birds in earth banks, or

▲ One of the most conspicuous of African birds, the lilac-breasted roller is usually encountered in pairs in open country, often perched high in a dead tree.

sometimes in crevices under the eaves of houses or in adobe walls. Both sexes incubate the eggs for 17 to 20 days and feed the chicks in the nest for up to 30 days.

There have been dramatic declines in numbers of the European roller in many parts of Europe,

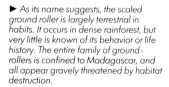

▶ As its name suggests, the scaled ground roller is largely terrestrial in habits. It occurs in dense rainforest, but very little is known of its behavior or life history. The entire family of ground-rollers is confined to Madagascar, and all appear gravely threatened by habitat destruction.

O. Langrand/Bruce Coleman Ltd

Belinda Wright

possibly because of loss of woodland habitat and the widespread use of chemical pesticides. Two species, the azure roller *Eurystomus azureus* and Temminck's roller *Coracias temminckii,* have very restricted ranges in Southeast Asia and so are vulnerable to habitat disturbance.

GROUND ROLLERS
A group of highly-specialized, terrestrial ground rollers is found only in Madagascar. The five species are stout-bodied birds, varying in length from 25 to 40 centimeters (10 to 15¾ inches), and are immediately distinguished from other rollers by their long sturdy legs, obviously an adaptation for life on the ground. Plumage coloration is variable, but in most species is characterized by finely striped or flecked patterns. Four species frequent the eastern rainforests of Madagascar, where a preference for secluded shaded areas, coupled with their secretive habits, make detection difficult, and so they remain poorly known. The long-tailed ground roller *Uratelornis chimaera* inhabits arid woodland in the southwest and is common in areas of undisturbed habitat. All are insect-eaters, foraging quietly in the leaf litter, and some species are known to feed at dusk and into the night. The birds excavate burrow nests in the ground, but little is known of their breeding behavior.

THE COUROL, OR CUCKOO-ROLLER
The courol *Leptosomus discolor* is confined to Madagascar and the Comoro Islands. It is a fairly large bird, about 50 centimeters (20 inches) long and weighing up to 240 grams (8½ ounces). There is a marked difference in plumage coloration of the sexes: in the adult male the upperparts are grayish-black, strongly glossed with metallic green and mauve-red, while the underparts and prominent frontal crest are ash-gray; the adult female is almost entirely rich rufous, broadly barred and spotted with black. Young birds resemble the adult female.

The courol is generally common in forests and wooded areas, and is usually encountered singly or in pairs during the breeding season but in groups of a dozen at other times of the year. The birds are confiding, generally allowing a close approach, and their presence is particularly noticeable when they call loudly while flying back and forth above the treetops. Amid the branches of trees and shrubs, they capture large insects and small reptiles, especially chameleons. The nest is in a hollow limb or hole in a tree, where up to four white eggs are laid.

THE HOOPOE
A large erectile crest and a long decurved bill are unmistakeable features of the hoopoe *Upupa epops,* the sole member of the family Upupidae. The

▲ *The Indian roller perches on telephone wires or fence posts, periodically descending, parachute-like, to snatch prey on the ground. It has been known to snatch fish from the surface of water, or even dive in after them like a kingfisher. Rollers are named for their flamboyant aerial courtship displays, which often involve both sexes.*

Guy Robbrecht/Bruce Coleman Ltd

plumage is pink-buff to rich cinnamon or rufous, relieved by black and white banded wings and tail. The adult bird weighs about 55 grams (2 ounces) and is about 28 centimeters (11 inches) long.

Widely distributed in Africa, Eurasia, and Southeast Asia, the hoopoe inhabits a variety of lightly timbered habitats where open ground and exposed leaf litter are available for foraging. It feeds mainly by probing with the long bill into soft earth, under debris, or into animal droppings for insects and their larvae—even in grazing pastures and well-watered gardens. During the breeding season it is usually encountered singly or in pairs, but at other times in family parties or loose flocks of up to 10 birds. The hoopoe is migratory in northern parts of its range, and partially migratory or resident elsewhere. Most migrants from Europe probably overwinter in Africa south of the Sahara Desert, where they are not distinguishable in the field from local residents. Populations from central and eastern Siberia migrate to wintering grounds in southern Asia.

Hoopoes nest in hollow limbs or holes in trees, in burrows in the ground, crevices in walls of buildings, or under the eaves of houses, and will use nest-boxes. They reuse some sites year after year. Only the female undertakes incubation, which lasts 15 to 20 days. Newly-hatched chicks are brooded and fed by the female, but later are cared for by both parents. The nesting hollow becomes fouled with excrement and remnants of food as the chicks grow, and eventually is vacated by the brood some 28 days after hatching. Though the species remains generally common throughout much of its range, there has been a steady decline in numbers in Europe during most of this century.

WOOD-HOOPOES

Endemic to Africa south of the Sahara Desert, the eight species of wood-hoopoe (family Phoeniculidae) are small to medium-sized arboreal birds with long tails and proportionately long, pointed bills, which in some species are strongly decurved. The plumage is black, usually with pronounced iridescent sheens of metallic green, blue, or violet, generally with white bars or spots on the wings and tail, and in some species the head is white or brown. Females usually have shorter bills and tails than the males, and young birds are noticeably duller than adults.

The forest wood-hoopoe *Phoeniculus castaneiceps* and the white-headed wood-hoopoe *P. bollei,* which is larger, favor primary and secondary forest, where they keep to the higher branches and canopy. Other species inhabit mainly savanna, open woodland, and dry thornbush country. All are active birds and clamber along tree trunks or amid foliage with much agility, continually probing in crevices or under bark with their long bills in search of insects and their larvae. As with woodpeckers and woodcreepers, the tail is used as a brace, and often a bird will hang upside down to reach its prey. Larger species tend to be gregarious, noisy, and conspicuous, especially when displaying the white markings on their wings and tails as they fly from one tree to the next.

Wood-hoopoes nest in hollow limbs or holes in trees, often holes originally excavated by barbets or woodpeckers. The purple wood-hoopoe *P. purpureus* is a cooperative breeder, with up to 10 helpers being recorded at a nest; usually these are previous offspring from the nesting pair. Incubation lasts 18 days and is only by the female, then the young birds fledge at 28 to 30 days. The nesting habits of other species are largely unknown.

JOSEPH FORSHAW

HORNBILLS

Hornbills are one of the most easily recognized families of birds, the Bucerotidae. They are conspicuous for their long down-curved bills, often with a prominent casque on top, their loud calls and, at close range, their long eyelashes. All but two species are also notable for sealing the entrance to their nest cavity into a narrow vertical slit, the incarcerated female and chicks being fed by the male. Hornbills are confined to the Old World regions of sub-Saharan Africa and of Asia east to the Philippines and New Guinea. Toucans, some with relatively larger and more gaudy bills than hornbills, are their New World equivalents.

The large bill of hornbills serves a variety of functions including feeding, fighting, preening, and nest-sealing. The head is supported by strong neck muscles and strengthened by fusion of the

W.S. Paton/Bruce Coleman Ltd

◀ Widely distributed in Africa, the purple wood-hoopoe is a noisy, gregarious species that lives in small parties. The birds move through forest and woodland, following each other from tree to tree. The wood-hoopoe feeds mainly on insects, using its curved bill to probe for them in crevices in the bark. It climbs actively about limbs and branches, often clinging to the trunk using its tail, woodpecker-like, as a prop.

first two neck vertebrae, unique among birds. The casque, barely developed in species such as the southern ground hornbill *Bucorvus cafer* and red-billed hornbill *Tockus erythrorhynchus*, appears to function primarily as a reinforcing ridge along the crest of the bill. However, in many species it is modified secondarily: as a hollow resonator in the black-casqued hornbill *Ceratogymna atrata*; as a special shape in the wrinkled and the wreathed hornbills *Aceros corrugatus* and *A. undulatus*; or, most extremely, as a block of solid ivory in the

◀ (Opposite page) A hoopoe arrives at its nest. Widespread in the warmer parts of Europe, Africa and Asia, the hoopoe nests in cavities in trees, among rocks, or burrowed in earthen banks. The female incubates alone while food is brought to her by the male. After the young hatch, the female joins the male in bringing food to the nest, but normally the male merely fetches, handing the food to the female who in turn presents it to the young.

◀ A yellow-billed hornbill. All but one of the 14 species in the genus *Tockus* are confined to Africa. They differ from most other hornbills in their ground-foraging habits, and in feeding largely on insects rather than fruit.

helmeted hornbill *Rhinoplax vigil,* so that the skull is 11 percent of the body weight.

The plumage of hornbills consists of areas of black, white, gray, or brown, with very few special developments apart from a loose crest, as in the bushy-crested hornbill *Anorrhinus galeritus,* or long tail feathers as in the helmeted hornbill. However, the bill, bare facial skin, and eyes are often brightly colored, with reds, yellows, blues, and greens, and the throat skin may form wattles, as in the yellow-casqued hornbill *Ceratogymna elata,* an inflated sac, as in the wreathed or ground hornbills of the genus *Bucorvus,* or the whole head and neck may be exposed as in the helmeted hornbill.

In all species, development of the bill and casque indicates the age and sex of individuals, supplemented in many species by differences in colors of the plumage, bare facial skin, or eyes. Males are slightly larger than females, and the shape of their more prominent casques is often indicative of the species, such as in the rhinoceros hornbill *Buceros rhinoceros.* Each species also has loud and distinctive calls—among the most obvious in their environment—from the clucking of von der Decken's hornbills *Tockus deckeni* or the booming of Abyssinian ground hornbills *Bucorvus abyssinicus* in Africa, to the roars of great Indian hornbills *Buceros bicornis* or the hooting and maniacal laughter of the helmeted hornbill in Malaysia. Several species conduct conspicuous displays while calling, and most larger species are audible in flight, the air rushing between the base of their flight feathers where lack of stiff underwing coverts is another feature of the family.

▶ Largest of all the hornbills, the African ground hornbill lives in open country in small family parties that daily patrol well-defined territories. Maintaining contact with frequent loud, deep booming notes, the birds spread out and methodically search through grassland, feeding on a wide variety of ground-living animals, including large insects, snakes, and frogs.

Mitch Reardon/Weldon Trannies

Stanley Breeden

The 45 species occupy habitats varying from arid steppe to dense tropical rainforest. All but one of the 12 species that inhabit savanna and steppe occur in Africa, the exception being the Indian gray hornbill *Meniceros birostris*. These include the extremes of size found within the family: from 150 grams (5¼ ounces) to 4 kilograms (8¾ pounds). Most hornbills are sedentary and live as mated pairs within defended territories, which range in size from 10 hectares (25 acres) up to 100 square kilometers (39 square miles). However, species of arid habitats are often forced to move locally during dry seasons, sometimes on a regular migration. Rainforest species may also range widely in search of fruit-bearing trees, especially members of the genus *Aceros* such as the wreathed hornbill, which will daily cross open sea between islands. A number of species reside in groups where all members defend the group territory, and most help a dominant pair to breed.

Hornbills nest in natural cavities in trees or rock faces, and in most species the female closes the entrance hole to a narrow vertical slit, using the side of the bill to apply mud, droppings, or food remains as a sealant. In some species, such as the great Indian hornbill, the male assists with delivering mud and sealing the exterior; and in others the male swallows mud to form pellets in his gullet which are then regurgitated to the female. The entrance is also resealed by half-grown chicks in those species where the female emerges before the end of the nesting cycle. The exceptions are the two species of ground hornbill, which neither seal the nest nor squirt their droppings out of the entrance for sanitation, and which may excavate their own holes in earth-banks or even use old stick nests of other birds.

In all species the male delivers food to his mate, and later to the chicks, assisted by group members in cooperative species. The food is carried as single items in the bill tip of small insectivorous species, as a bolus of several small animals in the carnivorous ground hornbills, or as a load of fruits in the gullet of many forest species, which are then regurgitated one at a time at the nest and passed in to the inmates.

Small hornbills lay up to six eggs, have an incubation period of about 25 days, see the female emerge when the eldest chick is about 25 days old, and have a nestling period totaling about 45 days. Large species lay two eggs, incubate for about 45 days, and leave the chick alone when it is a month old even though the nestling period extends to about 80 days. In most large forest species the female remains in the nest until the chick is fledged, a total period of incarceration of four or five months. Females of most species molt while in the nest, casting all flight feathers within a few days of beginning to lay eggs, and regrowing them before emerging from the nest.

ALAN KEMP

▲ *The great Indian hornbill feeds largely on fruit, especially figs; these are usually plucked from among the foliage, but the bird will sometimes descend to the ground to gather fallen fruit. Most often found in small parties that roost communally, this hornbill habitually visits several fruiting trees in succession on a regular daily basis.*

Order Piciformes
6 families, 65 genera,
378 species

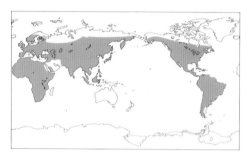

Size

Smallest Asian rufous piculet *Sasia abnormis* and South American scaled piculet *Picumnus squamulatus*, total length 77 millimeters (3 inches); weight 7 grams (¼ ounce).

Largest Black-mandibled toucan *Ramphastos ambiguus* and white-throated toucan *R. tucanus*, total length 61 centimeters (2 feet); weight 500 grams (17½ ounces).

Conservation Watch

The following species are listed in the ICBP checklist of threatened birds: three-toed jacamar *Jacamaralcyon tridactyla*, white-mantled barbet *Capito hypoleucus*, toucan-barbet *Semnornis ramphastinus*, black-banded barbet *Megalaima javensis*, white-chested tinkerbird *Pogoniulus makawai*, yellow-footed honeyguide *Melignomon eisentrauti*, yellow-browed toucanet *Aulacorhynchus huallagae*, Cuban flicker *Colaptes fernandinae*, red-cockaded woodpecker *Picoides borealis*, helmeted woodpecker *Dryocopus galeatus*, ivory-billed woodpecker *Campephilus principalis*, red-collared woodpecker *Picus rabieri*, and Okinawa woodpecker *Sapheopipo noguchii*. The imperial woodpecker *Campephilus imperialis* is possibly extinct.

WOODPECKERS AND BARBETS

LESTER L. SHORT

The order Piciformes contains six families of cavity-nesting birds: jacamars (family Galbulidae), puffbirds (Bucconidae), barbets (Capitonidae), toucans (Ramphastidae), honeyguides (Indicoridae), and woodpeckers (Picidae). Although they look different, they share certain anatomical features such as zygodactyl feet (two toes in front, two behind) with associated tendons that flex them, they generally lack down feathers (except Galbulidae), and lay white eggs. Most species live in tropical regions. Most are colorful, and some are very gaudy.

BILLS FOR SPECIAL PURPOSES

The largest family (with more than half the 378 species) are the well-known woodpeckers, which are distributed around the world except on oceanic islands, Australasia, Madagascar, Antarctica, and Greenland. Woodpeckers and barbets excavate roosting and nesting cavities, which are later taken over and used by hole-nesting species of other families and orders. Hence, many other birds depend on them.

There is much variation in feeding habits and therefore in bill structure. Jacamars have longish, pointed bills and are insect-eaters. Puffbirds too are insectivores, but have large, often hooked bills. Barbets generally are fruit-eaters, with a largish, sometimes notched bill. Toucans have remarkably large colorful bills, somewhat resembling those of hornbills, and feed mainly on fruits. The woodpeckers are unique in their strong, tapering, often chisel-tipped bills and strong tail feathers used as a prop against the bark of trees, where they glean, pry, probe, and excavate for insects. Honeyguides eat insects and particularly wax obtained from beehives

► *A purple-necked jacamar. Jacamars are exclusively American, while the bee-eaters are restricted to the Old World; although the two groups are unrelated they show some remarkable similarities in diet, appearance, and general behavior.*

J. Dunning/Vireo

◄ The gaudy red-and-yellow barbet (far left) lives in groups in East African scrublands, nesting in termite mounds. The rufous-tailed jacamar (near left) is one of the most widespread members of its family, inhabiting clearings, farmland, and forest edges from Mexico to northern Argentina.

or the exudate of certain insects, using their relatively unspecialized short bills; one African species "guides" humans to the location of beehives, giving rise to the name of the family.

JACAMARS

The male rufous-tailed jacamar *Galbula ruficauda* (length 25 centimeters, or 10 inches) is long-billed and long-tailed, metallic green in color with elongated central tail feathers, and generally rufous or chestnut below with a white throat. The female is similar but buff-colored below and on the throat. They live a solitary life except when breeding. Then they excavate a nesting tunnel in a termite mound or earthen bank, and for about three weeks both parents incubate the eggs (two to four) during the day, the female incubating alone at night. The young hatch bearing whitish down feathers, and are fed insects, especially butterflies, which the parents catch by flying out from a perch. Nestlings make trilling calls, weak versions of the adults' notes. When leaving the nesting cavity about 24 days after hatching, the fledglings closely resemble their parents, sex for sex. This species is less forest-dependent than most other members of this strictly American family, and it may be found even in grassland with scattered trees. Perhaps that explains its very extensive distribution from Mexico to Argentina.

Some ornithologists consider that the jacamars and the puffbirds are not related directly to the other four families in the order Piciformes.

PUFFBIRDS

Like jacamars, the puffbirds sally forth from a perch to catch insects, but in contrast to the giant hummingbird-like appearance of the colorful jacamars, puffbirds are stout-billed, short-necked, big-headed birds of subdued colors. Most are solitary species, but the nunbirds (genus *Monasa*) are social and even breed cooperatively. The black nunbird *M. atra* digs a tunnel nest into the ground, then covers the entrance with a pile of sticks, under which a horizontal "tunnel" leads to the actual nest chamber; adults in a cooperatively breeding group utter a loud chorus of "churry-churrah" notes, answered by neighboring groups. The swallow-winged puffbird *Chelidoptera tenebrosa,* a small-billed, mainly black bird (white rump, rufous belly) of northern South America, makes long flycatching sweeps, often repeated, hence is quite aerial. Its nest is excavated straight into the forest floor and lacks a stick-covering or "collar" of twigs or leaves around the entrance.

BARBETS

Tropical Asian, African, and American forests and woodlands and (in Africa) arid scrublands, are home to the brightly colored and patterned barbets. These

Belinda Wright

▲ Not much bigger than a house sparrow, the coppersmith barbet is a colorful, arboreal bird that obtains most of its insect food from the bark of trees. It takes its name from its monotonous, metallic call-notes.

have generally heavy bills, some with a notch on either side, with which they seize and bite into fruits such as figs. They excavate holes in dead trees with what at first glance appears to be an ineffective bill that is broad but pointed. A red cap, white facial stripes, olive back, and orange throat mark the male scarlet-crowned barbet *Capito aurovirens* (length 19 centimeters, or 7½ inches), which lives along the base

of the Andes, from Colombia to Bolivia. Females are duller, with a whitish crown. Repetitive croaking notes are uttered when breeding, but generally they are inconspicuous.

African barbets are the most diverse, ranging from highly social (up to 50 or more pairs nesting in one dead tree), dull-colored brown or olive species, to the group-social, duetting, multicolored species of the genus *Lybius,* and the solitary, tiny (length 7 to 8 centimeters, or 3 inches) tinkerbirds of the genus *Pogoniulus.* Unusual among them are the ground-barbets which nest in holes excavated straight into the ground, or in termite mounds. The gaudy red-and-yellow barbet *Trachyphonus erythrocephalus* (length 20 centimeters, or 7¾ inches) is marked with black, yellow, white and red spots, stripes and blotches. Nesting in groups, they excavate holes in banks or termite mounds. A group of five to eight chorus together regularly, all year, uttering a melodious series of notes sounding like a repetitive "red-'n-yell-oh", but the primary pair maintain the highest perches among the group, carry on displays with the head and tail, and sustain the chorus as a duet of their own. The bright orange-red bill is long and tapering, and with it this barbet seizes insects, fruits, and even young birds it encounters in the scrublands of Ethiopia south to Tanzania.

Tropical Asian barbets tend to be larger than those in Africa or tropical America and are large-billed,

▶ A rufous-necked puffbird. Named for their unusually loose, fluffy plumage, the puffbirds are rather inconspicuous, inactive, solitary birds of the middle levels of South American rainforests.

L.C. Marigo/Bruce Coleman Ltd

large-headed, short-tailed birds. The numerous species generally are green with sexually-differing intricate patterns of reds, blues, yellows, and blacks concentrated on the head and throat. Most belong to the genus *Megalaima* and are solitary (though they may join other barbets, often of several species, to feed in a fruiting tree), but highly vocal, uttering repetitive ringing "towp", "chook", or "pop" notes during much of the year. The coppersmith barbet *M. haemacephala* is named for its ringing, oft-repeated notes—it is one of several diverse species called "brain-fever birds" because of their monotonous unceasing song. Pairs excavate a nesting cavity in a dead tree trunk or branch, often on the sloping underside, in which they lay their eggs and raise their young.

TOUCANS

The huge-billed, comical-looking toucans of tropical America might be termed "glorified barbets", but of course their large size, frilled tongue, colorfully patterned and often serrated bill, and their inability to excavate a nest-cavity on their own, render them distinctive. Most are black, blue, green, brown, yellow or red on the body, with brilliantly colored stripes or patches on the bill, and often bright, bare skin around the eyes. They adroitly feed on fruits, moving long distances from feeding tree to feeding tree, but can also pluck untended eggs or baby birds out of nests. In flight they appear ungainly and conspicuous, but like parrots they are surprisingly inconspicuous when perched, unless they are calling or displaying.

The collared aracari *Pteroglossus torquatus* (length 41 centimeters, or 16 inches) lives in pairs and social groups, from Mexico to northern South America. Its serrated yellow and black bill is a quarter of its length; the body is greenish above, with a black head, a rufous band around the nape, and bare red skin around the eye, and yellowish below with two red bands, blackish in their center. Fast-flying compared with larger toucans, these aracaris range through forests feeding largely on fruits, calling "ku-sik" repetitively when moving about. They roost in old woodpecker holes, shifting from one to another at intervals, and also nest in such cavities. Sleeping in a group in one "dormitory", they are able to fold the tail over and onto the back. Their blind and naked young find themselves on a pebbly floor of egested seeds after hatching, and both parents as well as up to three helpers join them in the nest for the night.

HONEYGUIDES

Predominantly African in distribution, honeyguides are dull olive-green or grayish birds. Most species have a stubby or pointed bill with raised nostrils, a very thick skin, and an ability to locate beehives. However, only one species, the greater honeyguide *Indicator indicator* (length 20 centimeters, or 7¾ inches), exhibits guiding behavior with a series of calls and flight displays that entice the local people to

◀▼ *Widespread in South American forests, the sulphur-breasted toucan (left) lives in small parties in the treetops. The northern flicker (below) is one of the commonest of North American woodpeckers. It spends much of its time on the ground, feeding mainly on ants.*

follow them to a honey source. The birds are able to locate and enter some hives on their own to feast on beeswax. The male, black-throated and golden shouldered, sings a repetitive song, fending off other males and attracting various pale-throated females with which it mates. The female can breed and lay several clutches a year—using the nests of about 60 diverse host species that have made cavity nests or covered nests—one egg to a nest. She may destroy or remove an egg of the host when she lays; and when the honeyguide egg hatches, the hatchling has a pronounced bill hook with which it regularly strikes out, injuring and killing the hosts' young. When it leaves the hosts' nest the aggressive young bird, in its distinctive yellow and brown plumage, seeks out wax in cavities of trees (abandoned beehives) and soon develops the ability to "guide". Other species of honeyguides monitor noisy humans and their fires in woodlands, watching for the opening of beehives, but they do not "guide" like the greater honeyguide.

WOODPECKERS

Specialists at clinging to the bark of trees and foraging beneath it, woodpeckers excavate several roosting cavities, and a nesting cavity, sometimes into live wood. Their stiffened tail, chisel-tipped bill, bony- and muscle-cushioned skull, and long-clawed strong toes enable them to live anywhere there are trees in

the Americas, Eurasia, and Africa. The smaller species, such as the downy woodpecker *Picoides pubescens* of North America, which is mainly black and white (length 14.5 centimeters, or 5¾ inches), and the great spotted woodpecker *P. major* (length 21 centimeters, or 8¼ inches) of Eurasia, are common visitors to birdfeeding-stations, to the delight of many people who put out suet for them in the cold winter months.

The soft-tailed wrynecks (genus *Jynx*) of Africa and Eurasia are the only members of the woodpecker family that do not excavate their nests; they use natural cavities or old woodpecker or barbet holes. The tiny piculets (genera *Picumnus, Sasia, Nesoctites*) of tropical America, Africa and Asia have short soft tails not used as a brace in "woodpecking", but these woodpeckers drum with the bill to communicate and do excavate their own nesting cavities. Other unusual woodpeckers include some of the flickers (genus *Colaptes*) which live in open country, walk instead of hop, and nest in holes excavated in the ground or in termite nests. Some species of *Melanerpes,* mainly the American acorn woodpecker *M. formicivorus,* are social and nest cooperatively; this species also stores acorns in tiny holes in special storage trees, which are fiercely defended so that the food will be there for the lean times of winter. The sapsuckers (genus *Sphyrapicus*) of North America make rows of small holes in the trunks of certain trees where they regularly feed on the sap that accumulates and on insects attracted to it. Some sapsuckers and flickers are the only members of the order Piciformes that undertake extensive migrations.

Large woodpeckers include the pileated woodpecker *Dryocopus pileatus* (length 42 centimeters, or 16½ inches), widespread in wooded North America, and formerly decreasing in numbers. It now seems to be adapting and will even nest in trees within sight of New York's skyscrapers. The body is black with white marks; the male has a red crown, crest and "mustache", whereas the shorter-billed female shows red only on the crest. Members of a pair usually feed alone in a large territory, excavating in rotting trees and fallen logs for tunnels of their chief food, ants, which are extracted by the elongated, frilled-edged, and very sticky tongue. The male and female maintain contact with occasional loud series of "wuk" calls, and they call and drum in a rapid cadence, loudly, to establish and maintain their territory, which is defended sex by sex (the female driving away trespassing females, and the male evicting other males) throughout the year. Each uses several roosting cavities alternately. Both excavate a new nest annually, usually in a partly rotten live tree at 10 meters (33 feet) or more above the ground. The male incubates at night, and both alternate by day, during the 18 days to hatching. Adults take turns feeding the young almost hourly, by regurgitating large volumes of ants, their larvae, and eggs. The young leave the nest at about four weeks of age, and are fed and led about by their parents for two months or more,

after which they leave or are forced from the territory.

Woodpeckers, especially larger species, require large trees and some dead ones. Forest clearing has reduced many populations, and some face extinction. Numerous other bird species, which are partly dependent on old woodpecker holes for their own nesting, such as large toucans and hornbills, are thus brought under threat.

Donald D. Burgess/Ardea London Ltd

▲ *Despite its size (about that of a crow or raven) and its vivid red crest, the pileated woodpecker is generally a rather quiet and inconspicuous bird. It inhabits coniferous forests across North America.*

WOODPECKER'S TONGUE

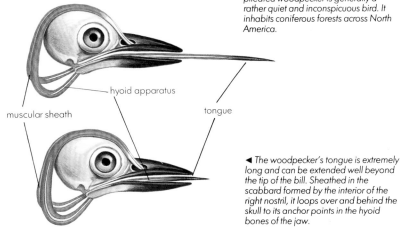

hyoid apparatus

muscular sheath

tongue

◄ *The woodpecker's tongue is extremely long and can be extended well beyond the tip of the bill. Sheathed in the scabbard formed by the interior of the right nostril, it loops over and behind the skull to its anchor points in the hyoid bones of the jaw.*

◄ *(Opposite page) A female great spotted woodpecker leaves her nest in a tree hollow, bearing a fecal pellet. As in many nestling birds, the excrement is ejected wrapped in a gelatinous coating for easy removal by the parents. In woodpeckers these pellets quickly gather a coating of sawdust from contact with the floor of the nest cavity.*

BROADBILLS AND PITTAS

H. ELLIOTT McCLURE

Order Passeriformes
Suborder Eurylaimi
Family Eurylaimidae
8 genera, 14 species
Family Philepittidae
2 genera, 4 species
Family Pittidae
1 genus, 23 species
Family Acanthisittidae
2 genera, 3 species

Left to right: broadbill, asity, pitta, New Zealand wren.

SIZE
Broadbills (Eurylaimidae)
total length 13 to 28
centimeters (5 to 11 inches).
Asitys (Philepittidae) total
length 15 centimeters
(6 inches).
Pittas (Pittidae) total length
15 to 25 centimeters
(6 to 10 inches).
**New Zealand wrens
(Acanthisittidae)** total length
8 to 10 centimeters
(3 to 4 inches).

CONSERVATION WATCH
The following species are
listed in the ICBP checklist of
threatened birds: African
green broadbill
Pseudocalyptomena graueri,
Schneider's pitta *Pitta
schneideri*, whiskered pitta
P. kochi, bar-bellied pitta
P. ellioti, Gurney's pitta
P. gurneyi, fairy pitta
P. nympha, superb pitta
P. superba, Steere's pitta
P. steerii, Solomons pitta
P. anerythra, small-billed
false-sunbird *Neodrepanis
hypoxantha*, and bush wren
Xenicus longipes.

A mong the 5,000 or more species of passerines (or songbirds) grouped in 50 to 74
families—depending upon the opinions of the ornithologists making the
grouping—the broadbills are considered the most primitive. They are just over
the line dividing passerines from non-passerines. The pittas and two families of
primitive species (Philepittidae and Acanthisittidae) are classified vaguely near each
other because of the musculature of the syrinx (the organ of voice or song), but
geographically and ecologically these groups are widely separated.

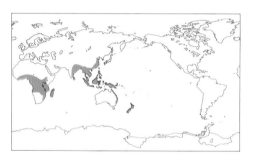

BROADBILLS

The 14 species in the family Eurylaimidae are very
colorful, with striking plumage and some having
brilliant red, green, or yellow eyes. They are
chunky birds with broad heads, broad flattened
bills, and short legs. Their feet have three toes
forward and one behind. They differ from other
passerines in having 15 neck vertebrae instead of
14, and two front toes that may be partially joined
at the base; the 11 "typical broadbills" have

▼ *Two colorful birds of Southeast Asian
rainforests: the banded pitta (left)
searches leaf litter on the forest floor for
food, while the green broadbill (right)
catches insects in the middle layers of
foliage.*

11 primary feathers in each wing, instead of the nine or ten found in other passerines.

The three *Smithornis* species are small dusky birds of forests in Central Africa. During courtship flights they display areas of color; they have loud calls, and hawk insects like flycatchers. The fourth African species is known by only a few specimens from forests near Lake Tanganyika.

All the other species live in India or Southeast Asia. Although most of them are mainly insect-eaters, hawking aerial insects in the lower canopy like clumsy flycatchers or catching them on the ground, they will also take small lizards and frogs. The diet of the three superbly-colored "green broadbills" (genus *Calyptomena*) is predominantly soft fruits and buds. Most species appear to be gregarious and move about the forest in small flocks, which could be family groups. In Malaysia, the long-tailed broadbill *Psarisomus dalhousiae* has been seen in flocks of 20 in association with "bird-waves" totaling more than a hundred birds of seven or eight species. The dusky broadbill *Corydon sumatranus* also joins bird waves; they have been observed sitting quietly on twigs, looking for insects while the flock passes, then flying on to join it and repeating the procedure.

Broadbills' nests are masterpieces of camouflage. The nest is usually attached to a vine suspended in the open, often above a stream, where it appears as debris caught there at high water. This appearance is further amplified by a trailing tail of fibers or debris hanging below it, as well as decorations of lichen and spider webs. Access to the nest hollow is through a hole in the side, which may have a short roof or vestibule. The common Malaysian species, the black-and-red broadbill *Cymbirhynchus macrorhynchos,* which is spectacularly plumaged, nests above streams in dense forest, but also along creeks through farmland. It also suspends its nest above roadsides from power or telephone lines.

ASITYS

The family Philepittidae comprises four poorly known species restricted to Madagascar. They are solitary birds of dense forest undergrowth, small but stout in structure, short tailed, with long legs, and a bill broad at the base and shorter than the head. The velvet asity *Philepitta castanea* of the lowland forests of eastern Madagascar is dimorphic (sexes unalike): the female is greenish-olive, whereas the male has yellow-tipped black body feathers which, when the yellow has worn away, leaves it a velvety black. He also has a long greenish wattle above each eye. The diet appears to be small fruits. They build a bulky nest suspended from twigs in the lower or middle canopy, which has a side entrance with a small projecting roof. Three whitish elongate eggs make up the usual clutch, but incubation and nesting behavior seem not to have been reported.

Schlegel's asity *P. schlegeli,* also a forest species

Morten Strange/Flying Colours

but in western Madagascar, is yellower, and the male is black on the top of the head and has black eye wattles. This species probably has habits similar to those of the velvet asity, but appears to be more active and less restricted to ground cover.

The small-billed false-sunbird *Neodrepanis hypoxantha* is known only by a few specimens from eastern Madagascar. In contrast, the wattled false-sunbird *N. coruscans* is found over much of the island; it is a tiny creeper-like bird only 10 centimeters (4 inches) long, which feeds on insects in bark crevices and also probes flowers with its long downcurved bill. The male is iridescent blue above and yellow below, with a large wattle about each eye. The female is dark green above, yellowish below, and lacks the wattle.

PITTAS

These beautiful birds resemble thrushes—indeed, are sometimes called jewel thrushes—but are separated from them by their syrinx musculature

▲ *The black-and-yellow broadbill inhabits lowland forest edges from Burma to Malaysia. In broadbills the sexes are distinguishable but not very different.*

▲ *Despite their brilliant plumage, pittas are surprisingly inconspicuous in the gloom of the rainforest floor. They feed quietly, and seldom call except when breeding. They roost in trees. Here a noisy pitta brings food for its young.*

and the structure of the tarsus (leg). The 23 species, all in the genus *Pitta,* are distributed from Africa to the Solomon Islands and from Japan through Southeast Asia to New Guinea and Australia. The probable geographic origin of the species is in the Indo-Malaysian region, which has about 20 species. They are medium-sized insect-eating birds, terrestrial forest inhabitants which, when disturbed, prefer to walk or run rather than fly, yet some species are migratory and travel long distances at night. Most have similar nesting and feeding habits, but the distinctive and vivid coloring of the various species make them some of the world's most brilliant birds. The female is usually duller in color than the male.

Large bulky nests resembling piles of plant debris built on the ground or in low vegetation have a spacious brood chamber and a side entrance at which the brooding bird rests facing out; the color pattern on the head is such that it blends with the surrounding plant debris, so there appears to be no opening. Both parents care for the eggs and attend the young. All species have loud melodious double whistles, usually heard in early morning and evening.

The African pitta *P. angolensis* breeds in Tanzania and southward to the Transvaal, and when non-breeding migrates north to the Congo basin, Uganda, and Kenya. The green-breasted pitta *P. reichenowi* is a rare forest species of equatorial Africa; it breeds from June into November, and is non-migratory. In Australia, the noisy or buff-breasted pitta *P. versicolor,* a snail-eating forest species, is found from Cape York to New South Wales and breeds from October to January. The northern black-breasted or rainbow pitta *P. iris* favors coastal scrub and mangrove swamps, breeds from January to March, and is thought to be sedentary. The red-breasted pitta *P. erythrogaster* from New Guinea, Java, Malaysia and the Philippines, migrates to northeastern Australia in October; it prefers dense scrub environments and breeds from October to December.

The blue-winged or Moluccan pitta *P. brachyura/moluccensis* (taxonomists are not in agreement as to whether this is one or two species) is the most widely distributed. It is found from India and Sri Lanka east through Southeast Asia to New Guinea and Australia, as well as north through southern China and east into Taiwan, Korea and southwestern Japan. Segments of the population are migratory; pittas that breed in northern India migrate to Sri Lanka, arriving in September and October, overwintering in forests, and leaving in April or May. In Southern Japan they arrive in April or May, breed there, and leave in October. In Malaysia they are permanent residents of mangrove and coastal forests and breed from May into August. The blue-winged pitta incubates for about 18 days, the nestlings fledge in 18 to 20 days, and there is only one brood. This is probably standard for the family.

NEW ZEALAND WRENS

The three species alive today in the family Acanthisittidae (or Xenicidae) may be survivors from an ancient colonization of New Zealand. They have a primitive syrinx structure, weak songs, and are weak fliers. Four species were already widespread before humans arrived with their animals. The rifleman *Acanthisitta chloris* and bush wren *Xenicus longipes* were originally common on both North and South Islands of New Zealand. In recent years the bush wren has been recorded at only one North Island locality and is rare in the South. The rock wren *X. gilviventris,* restricted to the South Island, was first described in 1867 and is an inhabitant of Alpine and subalpine scrubland, where it spends much time on the ground searching for insects. Both bush wrens and rock wrens bob when alighting from a short flight. The Stephen Island wren *X. lyalli,* discovered on that island in 1894, was quickly exterminated by the lighthouse keeper's cat.

The rifleman forages for insects on tree trunks like a creeper and is common in beech forests on both islands from sea level to an altitude of 350 meters (1,150 feet). It breeds from August to January and builds loosely-woven nests of moss and plant debris, with a side entrance, in tree hollows or behind loose bark and sometimes on the ground in protected places. After the chicks fledge, the family remains together for several weeks. The bush wren has somewhat similar nesting habits, but builds its nests in low shrubbery from August to December. As its name denotes, the rock wren nests among rocks in exposed areas, from September to December.

◄ *A female rifleman visits her nest in a tree cavity. The New Zealand wrens are an isolated group with, apparently, no near relatives; they are unusual among passerines in that females are substantially larger than males, although males are more brightly colored.*

M.F. Soper

Order Passeriformes
Suborder Furnarii
Family Furnariidae
34 genera, c. 222 species
Family Dendrocolaptidae
13 genera, 49 species
Family Formicariidae
52 genera, 42 species
Family Rhinocryptidae
12 genera, 30 species

Left to right: ovenbird, woodcreeper, tapaculo.

Size
Ovenbirds (Furnariidae)
total length 9 to 27
centimeters (3½ to 10⅗
inches).
**Woodcreepers
(Dendrocolaptidae)** total
length 14 to 35½ centimeters
(5 ½ to 14 inches).
Antbirds (Formicariidae)
total length 8 to 35½
centimeters (3⅕ to 14
inches).
Tapaculos (Rhinocryptidae)
total length 11.5 to 25.4
centimeters (4½ to 10
inches).

Conservation Watch
There are 54 species listed in
the ICBP checklist of
threatened birds. They
include the following:
mustached woodcreeper
Xiphocolaptes falcirostris,
white-bellied cinclodes
Cinclodes palliatus, stout-
billed cinclodes *C. aricomae,*
Apurimac spinetail *Synallaxis
courseni,* Alagoas foliage-
gleaner *Philydor novaesi,*
henna-hooded foliage-gleaner
Automolus erythrocephalus,
orange-bellied antwren
Terenura sicki, fringe-backed
fire-eye *Pyriglena atra,* gray-
headed antbird *Myrmeciza
griseiceps,* giant antpitta
Grallaria gigantea, crescent-
faced antpitta *Grallaricula
lineifrons,* hooded gnateater
Conopophaga roberti, and
Stresemann's bristlefront
Merulaxis stresemanni.

▶ *Looking much like old-fashioned clay
ovens, the nests of the rufous hornero
are a common sight along the roadside
fences of the grasslands of South
America.*

OVENBIRDS AND THEIR ALLIES

EDWIN O. WILLIS

Some of the commonest birds in South and Central America are the Furnarii—more than 222 rust-colored ovenbirds (family Furnariidae), 49 trunk- climbing woodcreepers (Dendrocolaptidae), 30 skulking tapaculos (Rhinocryptidae), and 242 brightly patterned antbirds (Formicariidae). Several new species are recorded every decade.

BIRDS OF WOODLAND AND FOREST
The rufous hornero *Furnarius rufus* and some of its relatives are familiar birds near houses, stalking about lawns and singing noisy duets; the barred antshrike *Thamnophilus doliatus* sometimes nests in bushy backyards. However, most Furnarii live hidden in dense shade, wherever humans have not yet changed the tropics with cars, cows, cane, or cocaine.

These birds prefer dense foliage, gleaning their prey from leaves, tree trunks, and leaf litter in the undergrowth; because they tend to walk rather than fly, the closer the leaves and accompanying

Gunter Ziesler/Bruce Coleman Ltd

insects and spiders, the better. Antbirds, and some woodcreepers, are less restricted to dense foliage because they can fly for prey. A few species, such as the jaylike giant antshrike *Batara cinerea* of southern regions, vary the usual diet with small frogs and snakes, eggs, and the nestlings of other birds.

OVENBIRDS
Ovenbirds received their name because the rufous hornero and a few of its relatives build a mud nest like a traditional clay oven, with an entrance at the left or right and the nest chamber inside the remaining half. Most others make oven-shaped nests of twigs or moss, although some hide cup nests inside burrows, crevices, or even in dense foliage. Each family group of the firewood-gatherer *Anumbius annumbi* works together for weeks each year to make two or three wheelbarrow-load nests where all can roost to avoid the cold southern nights. In spring one nest will be occupied by the adult pair for breeding.

The rufous-fronted thornbird *Phacellodomus rufifrons* builds family apartments of sticks at the tips of drooping limbs of savanna trees. The wrenlike rushbird *Phleocryptes melanops* glues wet marsh leaves together to make a cardboard-like oven nest. Closed nests are hard to build, but they do protect against sun, cold, and the birds' predators. Using an oven nest with a narrow entrance poses no problem for these birds, because they are adapted to walk and creep in narrow places when they forage. In contrast, stick nests are not safe from the striped cuckoo *Tapera naevia,* which lays eggs that the unsuspecting hosts will incubate and raise. Some nests attract persistent nest-robbers, notably the oriole *Icterus icterus,* which takes over the nests of thornbirds.

The unusual names of several ovenbirds refer to their behavior or morphology. The campo miner *Geositta poeciloptera* pumps its short tail as it walks over newly burnt Brazilian savannas, and excavates nest holes in armadillo burrows. *Cinclodes* species are sometimes called shaketails because they jerk their tails upward on landing. The wiretail *Sylviorthorhynchus desmursii* has tail filaments more than twice as long as its body, and the softtail

Thripophaga macroura has a yellow train. Foliage-gleaners and treehunters (many *Philydor, Automolus,* and *Thripadectes* species) clamber in, or dig into, dense epiphytes or dead leaves, following more alert species that call the alarm if a predator appears. Leaftossers (various *Sclerurus* species) toss or rake leaves on the forest floor. The streamcreeper *Lochmias nematura* tosses leaves to uncover insects along limpid creeks, but a few individuals at sewer outlets above Rio de Janeiro give the whole species the local name "President of Filth". Nuthatch-like bark-probing birds include *Xenops* species in the forests, the recurvebill *Megaxenops parnaguae* in dry woods of northeastern Brazil, and the treerunner *Pygarrhichas albogularis* of woodlands in Patagonia.

We know little about the reproductive behavior of these birds. Male and female build a nest, incubate a few whitish or bluish eggs for 15 to 22 days, and raise nestlings for 13 to 29 days; fledglings receive food for unknown periods and stay with parents for up to a year. One adult buffy tuftedcheek *Pseudocolaptes lawrencei,* studied by Alexander Skutch in Costa Rican cloud forest, built a nest, incubated, and fed its young alone.

WOODCREEPERS

Until the 1930s the family Dendrocolaptidae ("tree hewers") included both woodcreepers and ovenbirds. The former are essentially ovenbirds with strong toes and curve-tipped tails used when climbing tree trunks. Voices of the two groups are similar—long chatters or sharp cries—but only some ovenbirds make a duet. Widely confused with woodpeckers in Latin America, woodcreepers are easily distinguished by their normal feet (woodpeckers have two toes forward and two back) and by their brown or pale-streaked plumage. Unlike ovenbirds, woodcreepers do best in tall-trunked equatorial forests.

Differences between species are greatest in their bills and behavior. Scythebills (*Campylorhamphus* species) use their thin semicircular bills to probe in small holes. The long-billed woodcreeper *Nasica longirostris* of Amazonian riversides, uses its long straight bill to probe in epiphytes growing on horizontal tree limbs. *Drymornis* and *Xiphocolaptes* species have strong and slightly decurved bills, used to rummage among ground litter, rotten logs, and epiphytes growing on vertical trunks. Less-strong bills, some decurved, are a feature of the trunk-climbing *Xiphorhynchus* or *Lepidocolaptes* species. The wedgebill *Glyphorynchus spirurus* wedges insects from scales of bark. Tiny straight bills help *Sittasomus* and *Deconychura* species to peck small insects on tree trunks or capture them in short flights. Stronger straight bills of *Dendrocincla, Dendrocolaptes* and *Hylexetastes* species denote birds that wait and then sally to the ground or to distant trunks or foliage, especially when army ants are flushing prey. The cinnamon-throated woodcreeper *Dendrexetastes rufigula* uses its straight bill in foliage at the tips of branches, pecking apart wasp nests or lunging for prey like a foliage-gleaner. Surface-pecking species can outmaneuver, even chase, the larger but slower sallying species.

Woodcreepers nest in natural tree cavities, or occasionally in wider holes made by woodpeckers, with little nest lining. Male and female incubate one to three white eggs for 15 to 21 days and care for young in the nest for 19 to 24 days, and afterwards for one month or longer—although in *Dendrocincla* species and a few others, the female works alone.

Among the ant-following woodcreepers there are several different behaviors. Big and dominant *Hylexetastes* individuals form pairs and even let their single offspring follow them for a year. Females of the medium-sized *Dendrocolaptes* attack their terrified mates if food is preempted by large

Luiz Claudio Marigo

▲ Though unrelated to treecreepers elsewhere in the world, the woodcreepers of South America are similar in their general appearance and in their tree-creeping behavior. Like woodpeckers they use their stiffened tail feathers as a prop to brace themselves against the trunks of trees. This is the wedge-billed woodcreeper.

A. Greensmith/Ardea London Ltd

▲ In its general appearance and habits, the Andean tapaculo resembles a wren. It usually frequents dense undergrowth and shrubbery, where it creeps mouse-like with its tail often cocked over its back. It is very difficult to observe, and has a loud penetrating song.

▼ Antbirds are so-called because of their habit of following columns of army ants through the forest. Here a white-tufted antbird perches low in typical stance, alert for insects flushed from the ground by the predatory ants.

birds or faster ones such as antbirds. Small *Dendrocincla* males stay away from their females: either the females seem wary of predators and the increased danger to pairs; or they don't have enough foraging room to let males stay near if dominant competitors are present. The extreme of social isolation is found in the solitary and retiring tyrannine woodcreepers *Dendrocincla tyrannina* of the equatorial Andes, where ant swarms are rare and predators are well hidden in cold and mossy forests; one sex (probably the male) sings for the other like a bowerbird every morning from a notch in a ridge, so that its long, stirring trill is amplified by the sides of the notch over a wide radius.

TAPACULOS

The name "tapaculo" either refers to "cocked tails" (in very risqué Spanish), or to the repeated "tap-a-cu" of the Chilean white-throated tapaculo *Scelorchilus albicollis*. The resemblance of tapaculos to wrens—in size, colors, and methods of rummaging for insects in dense vegetation—is belied by the big feet and spooky repetitive whistles of these non-songbirds. The covered nostrils of the *Rhinocryptidae* species may help foraging in dense debris. They resemble ovenbirds in making covered or burrow nests and in their nearsighted foraging, and like the ovenbirds they rarely follow army ants; however, they also

D. Wechsler/Vireo

resemble antbirds in color patterns, and may be the surviving members of an intermediate lineage.

Now mostly restricted to cold forests atop the Andes and in Patagonia and Chile, tapaculos are skulking and wary, hiding in vegetation, but appearing if the observer waits quietly. One of the 14 *Scytalopus* species—birds that tick on and on, or snore like frogs, and sneak like little gray mice in tangled undergrowth—reaches Costa Rica, and another the central Brazilian tablelands. Four bright-colored antbirdlike crescentchests (genus *Melanopareia*) hop on the ground in dense savanna grass on both sides of the Andes and across central Brazil. Three thrush-sized turcas and huet-huets (genus *Pteroptochos*) walk low in Chilean–Argentine beech forests and chaparral. The rest of the family is a diverse scatter of perhaps ancient forms: the brightly colored rusty-belted tapaculos *Liosceles thoracicus*, like cactus-wrens on the floor of Amazonian forests; the tinking gray males and rufous females of two tuft-nosed bristlefronts (genus *Merulaxis*) in the litter of eastern Brazilian forests; white-specked rufous ocellated tapaculos *Acropternis orthonyx* in bamboo atop the equatorial Andes; two gallitos ("little roosters", genera *Rhinocrypta* and *Teledromas*) that run on the ground in the semi-deserts of interior South America; and the spotted bamboo-wren *Psiloramphus guttatus,* giving mysterious screech-owl sounds in southern Brazil.

Nesting and social behavior are little known, except for a few burrow or oven nests with white or blue eggs. Small families and pairs are registered in a few species. They hide in foliage rather than join mixed-species flocks.

ANTBIRDS

Antbirds are so-named because some species cling to vertical stems just above the jaws and stings of moving army ant swarms, darting after fleeing insects and other prey. They may be accompanied by woodcreepers, tanagers, ground-cuckoos, and a motley crowd of occasional followers, including lizards. Several antbird species may compete above a single swarm, the larger species taking the center and supplanting the smaller ones peripherally. In compensation, most small species can find enough prey away from ants, or at small swarms. Large species, which get enough food only at the biggest swarms, have become very rare and subject to extinction. One speedy Amazonian species, the small white-tufted antbird *Pithys albifrons*, specializes at darting among larger birds to grab a contested bite. However, it has to fly long distances to find an ant colony with few large birds, and each individual needs an area about 3 kilometers (1¾ miles) in diameter. Specialized birds like this, and the large ant followers, tend to vanish when only small areas of forest are preserved.

The bright blue or red bare faces of several species appear to imitate the eyes of forest cats and therefore frighten predators or competitors. Patterns are often bright black and white or rust, but many are gray or black, and females may be a less-colorful brown. White or buff back patches, wing corners, and tail tips, which are usually concealed, may be spread in fights for territories. White-bearded antshrikes *Biatas nigropectus* are colored like the white-collared foliage-gleaners *Philydor fuscus* with which they associate, probably to avoid being singled out if a hawk attacks. Most antbirds fly only short distances in the understory or foliage of the forest; none appear in bright sunlight, and different species often occur on different sides of large rivers in the Amazon region.

Antthrushes (*Chamaeza* and *Formicarius*) walk on the forest floor like bantams, pounding their short upraised tails as they go. They rarely follow army ants and their spooky whistles and white eggs laid in cavities resemble those of tapaculos. Antpittas (*Pittasoma, Grallaria, Hylopezus, Myrmothera,* and small *Grallaricula*), which are big-headed, stub-tailed, and mottled, hop on low vines and the forest floor giving owl-like calls, especially in the Andes. Nests tend to be low platforms with two greenish eggs. Only the two *Pittasoma* species, found west of the northern Andes, regularly "bound" like kangaroos around ant swarms.

The eight species of gnateaters (*Conopophaga*) are small short-tailed brown and black birds, and males often have a white tuft behind the eye. They peck insects nearby or flutter to snap them from the forest floor or low vegetation. Nests are bulky cups with two speckled brownish eggs.

Other antbirds, including most ant followers, are among the commonest birds in Amazonian and trans-Andean lowland forests, and in certain Peruvian localities there are more than 40 species. In cutover or second-growth zones, few survive. Some hop or walk on the forest floor, but most peck or sally short distances for prey in the mid-level foliage, while a few species act like warblers in the subcanopy. They are specialized for moderately dense foliage, unlike creeping ovenbirds of dense foliage or sallying manakins of open foliage.

Antwrens (*Myrmotherula*, plus other genera) are small, short-tailed, and warblerlike. Different species have separate niches in the Amazon: commonly one lives near the ground, three or more in the understory (at least one investigating dead leaves, and one in dense tangles), one or more in the canopy, plus others in streamside or forest-edge. Often they join understory flocks of up to 50 different species of birds, including woodcreepers and ovenbirds. In each flock may be one or more slightly larger antvireos (*Dysithamnus*) and various thick-billed antshrikes (*Thamnophilus, Taraba,* and *Sakesphorus*), from sparrow to jay size, though these often live apart in thickets. Occasionally with the flocks may be various "antbirds", including ant followers, plus long-tailed tangle living *Cercomacra* or short-tailed undergrowth-haunting *Myrmeciza*.

▲ *A little-known bird of forest undergrowth, the black-spotted bare-eye is distributed east of the Andes from Colombia to Bolivia.*

Luiz Claudio Marigo

The most unusual species are little gray or brown antcatchers (*Thamnomanes*), which are actually good flycatchers, perching upright and snapping up insects flushed by the mixed flock. Like drongos in Old World flocks, they faithfully sound the alarm if a hawk or other danger appears, and for this service they attract others. (Sometimes, however, they give a false alarm and catch insects in the confusion!) Amazonian Indians, and later Brazilians, thought the "uirapurus" very potent medicine because lots of birds followed them.

The advantages of joining a mixed flock—allowing many rummaging specialists like ovenbirds to seek food while alert flycatching antcatchers keep a lookout for predators—may limit the density of some species in the forest because the one pair of *Thamnomanes* antcatchers in a flock has a large territory and each pair of other species has to take the same large area. If humans continue to cut down the trees leaving only small areas of forest, the resulting loss of *Thamnomanes* species can then cause the predation and loss of other species as a "domino effect".

The songs of antbirds tend to be series of whistles, with rough sounds for enemies and soft ones for family; in some species the female sings a different song in reply to her mate. When courting and before copulation, the male feeds the female. Both build the nest, a small cup or, less often, oven shape. At night the female incubates or broods, but both sexes incubate the two (rarely more) spotted whitish eggs for 15 to 20 days, then care for nestlings for 9 to 18 days and for one fledgling each for a month or so. The association of juvenile birds with parents lasts for between one month and a year.

Order Passeriformes
Suborder Tyranni
5 families, 160 genera,
514 species

SIZE
Smallest Short-tailed pygmy-tyrant *Myiornis ecaudatus,* total length 6 centimeters (2⅖ inches); weight 5 grams (⅕ ounce).
Largest Amazonian umbrellabird *Cephalopterus ornatus,* total length 45 centimeters, (17¾ inches); weight 400 grams (14 ounces).

CONSERVATION WATCH
There are 33 species listed in the ICBP checklist of threatened birds. They include the following: swallow-tailed cotinga *Phibalura flavirostris,* gray-winged cotinga *Tijuca condita,* white-cheeked cotinga *Ampelion stresemanni,* turquoise cotinga *Cotinga ridgwayi,* long-wattled umbrellabird *Cephalopterus penduliger,* black-capped manakin *Piprites pileatus,* white-tailed shrike-tyrant *Agriornis albicauda (andicola),* strange-tailed tyrant *Yetapa risoria,* gray-breasted flycatcher *Empidonax griseipectus,* Minas Gerais tyrannulet *Phylloscartes roquettei,* Kaempfer's tody-tyrant *Idioptilon kaempferi,* ash-breasted tit-tyrant *Anairetes alpinus,* Bananal tyrannulet *Serpophaga araguayae,* and the Peruvian plantcutter *Phytotoma raimondii.*

TYRANT FLYCATCHERS AND THEIR ALLIES

SCOTT M. LANYON

The suborder Tyranni is restricted to the Americas, where it is widely distributed in virtually all habitats, elevations, and latitudes. There is great variability in size, plumage coloration, and behavior within this group, and the extreme diversity found both within and between families makes it very difficult to generalize about them. In fact, scientists remain uncertain as to which family some species belong. These diverse birds are united in this suborder by a particular arrangement of the muscles of the syrinx or vocal organ.

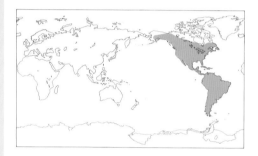

TYRANT FLYCATCHERS
The family Tyrannidae is found throughout North, Central, and South America. It includes not only the tyrant flycatchers but also birds commonly known as phoebes, elaenias, kingbirds, flatbills, and wood-peewees. The sexes look alike, generally with plumage that is a mixture of greens, browns, yellows, and white. In most species a strong pair-bond is formed for the duration of the breeding season, and both sexes help to raise the young. Of the nearly 400 species in 114 genera that have been identified so far, the greatest diversity is found in the New World tropics. Each species exploits the environment in a slightly different way, and so intricate is this division of resources that some avian communities in South America are known to contain in excess of 60 flycatcher species.

The difference in food preferences is one reason why multiple species can coexist. Obviously, fruit-eating and insect-eating flycatchers can live

▶ *The scissor-tailed flycatcher is a common and conspicuous bird of open country from the south-central United States to Panama. Noisy and pugnacious, it frequently harasses hawks and other birds much larger than itself.*

together without direct competition for food. However, the great majority of flycatchers, as their name suggests, eat insects. How can so many similar species eat the same type of food and still be found together? The answer is that each species employs a slightly different combination of prey size, habitat, vegetation type, foraging position, and capture technique. Scientists believe that it is this ability to finely divide the available resources, coupled with the fact that the New World tropics provide birds with rich and diverse resources to exploit, that gave rise to this largest of passerine families. (Their counterparts in the Old World are the flycatchers in the family Muscicapidae.)

MANAKINS

Manakins are small, brightly colored fruit-eaters with short tails and short, wide bills. There are 57 species (placed in 19 genera in the family Pipridae) confined to lowland forests of South and Central America. The females look remarkably similar between species, most with green or olive plumage. In contrast the males are generally brightly colored, usually with bright blues, reds, or yellows on a black background.

This high degree of color difference between the sexes is indicative of a complex social behavior. Male manakins congregate at traditional places in the forest, known as "leks", to attract mates. In some species, males clear dance arenas on the forest floor; in others, males prepare dance perches

Luiz Claudio Marigo

▲ *In a bizarre courtship display the male wire-tailed manakin (left) rapidly brushes the female's throat with the wire-like tips of his tail feathers. The great kiskadee (center) is one of the more conspicuous tyrant flycatchers of tropical America, often perching on roadside fenceposts and telephone wires. Only the male Guiana cock-of-the-rock (right) is brightly colored; the female is plain, dowdy, and gray. The common name hints at two features: males congregate at leks to display like cocks; and females come to the lek to mate, then retire alone to build their nests in rock crevices.*

◀ *Manakins are notable for their extraordinary group displays and complex courtship behavior, but many species remain unstudied. This species, the helmeted manakin, is a little-known bird that inhabits the tablelands of central and southern Brazil.*

▶ *The plantcutters, a family of only three species of exclusively South American distribution, are among the very few purely vegetarian passerines. They inhabit open country and closely resemble finches in general appearance and behavior. This is the rufous-tailed plantcutter.*

J. Dunning/Vireo

by stripping branches of leaves. Females are attracted to these leks by the calls of the males. When a female arrives the resident males perform their elaborate courtship displays which highlight their bright plumage. The displays are stereotyped within a species and can include vocalizations, wing snaps, short flights, hops, side-steps, and rapid tail movements. If the female is receptive, the pair will mate at the conclusion of the display. The display and mating constitute the extent of the pair-bond in manakins; the female, having previously built a nest, will lay eggs, incubate, and feed the young without further assistance from the male. In one genus, *Chiroxiphia,* the reproductive display requires cooperation between several males, only one of whom will have the opportunity to mate with the female.

COTINGAS

Members of the family Cotingidae (61 species in 25 genera) are fruit-eating birds, generally weighing 50 to 100 grams (1¾ to 3½ ounces). They are restricted to Central and South America, occur in low densities, and usually frequent the upper portions of forest canopies. These last two observations lead directly to the remaining generalization, which is that cotingas are very poorly known. In some genera (for example, *Lipaugus*) male and female plumage is similar to that of flycatchers; the sexes look alike, and have drab green and brown plumage. In most other genera, however, the males are brightly colored and only the females possess the more cryptic coloration. Complex social behavior appears to be common within this family, but only the cocks-of-the-rock (genus *Rupicola*) have been well studied. There is even a suggestion that, like the manakin genus *Chiroxiphia,* the reproductive displays of

some cotinga species require cooperation between males.

When birds specialize on fruit as a food source, as is true for the cotingas, there is ample opportunity for plant and bird to become strongly interdependent. For example, the white-cheeked cotinga *Ampelion stresemanni* of the Peruvian Andes feeds almost exclusively on mistletoe fruit. After a meal it regurgitates the seeds onto branches, where in time the seeds will germinate and grow. Clearly the mistletoe is dependent on the cotinga for seed dispersal because there is no other fruit-eater at this elevation. Therefore, without the cotinga the seeds would simply fall to the ground, where they cannot survive.

SHARPBILL

Named for its distinctive bill shape, the sharpbill *Oxyruncus cristatus* is a canopy-dwelling fruit-eating bird about which relatively little is known. Although its range is probably between Costa Rica and southern Brazil, it has been found at relatively few localities. Attempts to determine whether sharpbills are genetically more closely related to flycatchers or to cotingas have been equivocal. Scientists can only conclude that this species is a survivor of a very old lineage within the Tyranni.

PLANTCUTTERS

The plantcutters, three species in the genus *Phytotoma,* are confined to dry habitats in southern South America. These species have the distinction of being the only passerine birds known to depend on leaves and fleshy stems for the bulk of their diet. They have serrations on the edge of the bill that enable them to cut leaves and stems into pieces small enough to eat. Biochemical studies indicate that they are most closely related to the cotingas.

LYREBIRDS AND SCRUB-BIRDS

G.T. SMITH

Lyrebirds and scrub-birds are surviving members of an ancient and diverse radiation of Australian songbirds, so while quite distinct, the two families are each other's closest relative. All species are terrestrial, have weakly developed powers of flight, and live in dense vegetation. Despite the difference in size they have a number of ecological similarities because of common adaptations to a similar habitat.

Order Passeriformes
Suborder Oscines
Family Menuridae
1 genus, 2 species
Family Atrichornithidae
1 genus, 2 species

SIZE
Smallest Rufous scrub-bird *Atrichornis rufescens*, total length 16 to 18 centimeters (6½ to 7 inches); weight 30 grams (1 ounce).
Largest Superb lyrebird *Menura novaehollandiae*, total length 80 to 100 centimeters (31½ to 39½ inches), tail 50 to 60 centimeters (20 to 23½ inches), weight up to 1.2 kilograms (2½ pounds).

CONSERVATION WATCH
The rufous scrub-bird *Atrichornis rufescens* and noisy scrub-bird *A. clamosus* are listed in the ICBP checklist of threatened birds.

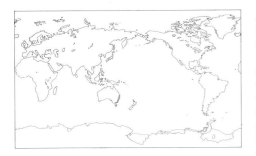

LYREBIRDS

The superb lyrebird *Menura novaehollandiae* is found in a narrow belt of eastern Australia from southeastern Queensland to southern Victoria. It was introduced into Tasmania in 1934. Its habitat is wet eucalypt forest and temperate rainforest.

Albert's lyrebird *M. alberti* is restricted to a small belt of subtropical rainforest in southern Queensland and northern New South Wales.

Lyrebirds are pheasant-sized, with brown to rufous plumage, long powerful legs, short rounded wings, and a long tail with modified feathers; the outer two feathers in the superb lyrebird are shaped like a Greek lyre, hence the name. They are elusive and difficult to observe in their dense habitat, where they are fast, agile runners when danger appears. They rarely perch in trees except to roost, ascending by jumping from branch to branch, and descending in the morning by gliding. The diet is mainly invertebrates which they expose by digging, ripping apart rotten logs, or turning over stones with their powerful feet.

Their most conspicuous characteristic is their loud territorial song, 80 percent of which may be

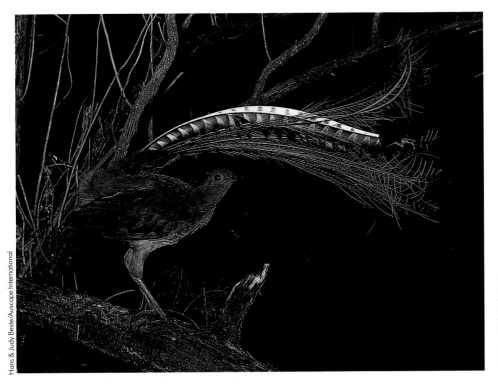

◀ Largest of all songbirds (passerines), the superb lyrebird feeds on insects and other invertebrates living on the forest floor, searching the leaf litter for them with methodical, raking sweeps of its large clawed feet. Smaller birds such as yellow-throated scrubwrens frequently attend it in a "crumbs from the lord's table" relationship, feeding on the smaller insects overlooked or ignored by the larger bird but disturbed by its raking.

Hans & Judy Beste/Auscape International

► *The noisy scrub-bird is aptly named: notorious among bird-watchers for being so elusive and so agile in dense cover that it is all but impossible even to glimpse, let alone observe, it is also notable for the almost ear-splitting intensity of its calls.*

Graeme Chapman

▼ *Like the noisy scrub-bird in behavior and general appearance, the rufous scrub-bird can barely fly. It lives close to the ground in dense cover, slipping mouse-like through the leaf-litter. In display the male cocks his tail, droops his wings, fluffs his chest feathers, and delivers his extraordinarily penetrating song. So inflexible are the bird's habitat requirements and so fragmented its distribution that its total numbers may not exceed 1,000, and its status is extremely precarious.*

Glen Threlfo/Auscape International

mimicry of other birds—occasionally they may mimic barking dogs. The adult male superb lyrebird establishes a territory of 2.5 to 3.5 hectares (6 to 9 acres) when sexually mature; the female's nesting territory may be within or overlapping the male territories. Males defend their territories, especially during the winter breeding season, by chasing intruders, singing, or displaying on earth mounds (well-concealed platforms of vines and fallen branches for Albert's lyrebird), which they have constructed throughout their territories. The displays are spectacular; the tail is fanned and thrown forward over the head and vibrated while the bird dances and sings.

The male mates with a number of females attracted to his displays. The female builds a domed nest, usually less than 2 meters (6½ feet)

above ground, incubates the single egg for 47 days, and feeds the nestling until it leaves the nest when about 50 days old. The young bird stays with its mother for about eight months after fledging. After the breeding season the birds are more mobile and wander in small groups through the forest.

The above is based on observations of the superb lyrebird. The little that is known about Albert's lyrebird suggests that there are few major differences in overall biology between the two species.

SCRUB-BIRDS

The noisy scrub-bird *Atrichornis clamosus* has survived in only one small locality on the south coast of Western Australia—plus a small population successfully translocated to another area nearby—and the main habitat is the boundary between swamp and forest and in wet gullies. The rufous scrub-bird *A. rufescens* has a discontinuous distribution on the east coast of Australia around the Queensland–New South Wales border, in wet temperate forest and subtropical rainforest where their diet is mainly invertebrates.

The habitat of both species is dense, so scrub-birds are rarely seen. They are small, solidly built, brown birds with long legs and tail, and short rounded wings; they rarely fly, and then only for a few meters. Their most conspicuous behavior is the loud territorial song of the males (the females give only alarm notes). The rufous scrub-bird male also uses mimicry, and the noisy scrub-bird has another quieter song which uses modified segments of other birds' songs. Noisy scrub-bird males have territories of 5 to 10 hectares (12½ to 25 acres) within which the females have nesting territories. The nesting behavior is similar to that of lyrebirds, although of shorter duration, and the rufous scrub-bird breeds in spring/summer. Both species maintain their territories throughout the year.

LARKS AND WAGTAILS

P.A. CLANCEY

Order Passeriformes
Suborder Oscines
Family Alaudidae
c. 15 genera, *c.* 79 species
Family Motacillidae
c. 5 genera, *c.* 60 species

SIZE
Larks (Alaudidae) total
length 11 to 19 centimeters
(4⅓ to 7½ inches); weight 13
to 45 grams (½ to
1½ ounces).
Wagtails (Motacillidae) total
length 14 to 17 centimeters
(5½ to 6⅔ inches); weight 13
to 32 grams (½ to 1 ounce).

CONSERVATION WATCH
The following species are
listed in the ICBP checklist of
threatened birds: Ash's lark
Mirafra ashi, Degodi lark *M.
degodiensis*, Somali long-
clawed lark *Heteromirafra
archeri*, South African long-
clawed lark *H. ruddi*, Sidamo
long-clawed lark *H.
sidamoensis*, Raso lark *Alauda
razae*, yellow-breasted pipit
Hemimacronyx chloris, Botha's
lark *Spizocorys fringillaris*,
Chaco pipit *Anthus chacoensis*,
Sokoke pipit *A. sokokensis*,
and ochre-breasted pipit *A.
nattereri*.

T hese two families, the Alaudidae and the Motacillidae, are small ground-dwelling birds, represented worldwide except in extreme latitudes and oceanic islands. Larks are of course known for the beautiful song of many species. Some of the pipits (which are in the same family as wagtails) also have a lovely musical call in flight and superficially resemble larks.

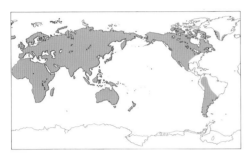

LARKS
Not all larks have the vocal powers of the renowned skylark *Alauda arvensis* of Eurasia, though many others sing well during the course of their aerial song-flights. Some sing from tree-stumps or posts or even anthills, and others produce clapping or fluttering sounds through wing-action during their nuptial displays. Because of its popularity, the skylark has been introduced to Vancouver Island on Canada's Pacific coast (it has spread to the nearby San Juan Islands of Washington State), New Zealand, and Australia (which already had one other species, the singing bushlark *Mirafra javanica*).

The lark family (Alaudidae) is centered mainly on temperate regions of the Old World in generally open habitats. It comprises about 79 species, with the greatest concentration in Africa. The most numerous is the shore or horned lark *Eremophila alpestris*, which has the widest distribution and occupies a great variety of habitats from Arctic tundra to temperate grasslands and even desert. Other species are usually more particular about their choice of terrain.

◄ *A pair of skylarks raise up to four broods per season. This bird may be feeding an early brood: as in many well-studied birds, the average clutch size has been found to vary not only geographically but also through the season, rising from just over three in early spring to four in summer.*

Richard T. Mills

▶ *Slender and graceful in build, wagtails are named for their habit of persistently waving their tails up and down. The gray wagtail is seldom found away from water, especially rushing mountain streams and swiftly flowing rivers.*

Laurie Campbell/NHPA

Typical larks are generally streaked brown over the upper parts, wings and tail, and white or buff on the underside, the breast usually streaked with dark brown. Some have a small crest on the head, and some show white over the lateral tail. In most, the bill is slender and slightly decurved, but in others it is robust. The legs and toes are long; in species that inhabit grassland the claw of the hind toe is extended. All walk rather than hop.

Larks feed on insects and other invertebrates as well as seeds and grain. They build simple cup-shaped nests on the ground; these are sometimes completely exposed and sometimes sheltered at the base of a tuft of grass or under a low bush. In desert a partial canopy may be added to shield the incubating female from the heat of the sun. Many larks exhibit plumage colors that match the color of the soil on which they breed, resulting in such forms within a species being classified as different subspecies or races. The horned lark, for example, has many differently colored forms—breeding as it

does in places as varied as Franz Josef Land and Novaya Zemlya in northern Eurasia, southern Mexico, and the savannas of Colombia.

Finchlarks
The finchlarks (genus *Eremopterix*) are small, the sexes differ markedly from one another, and their habitat is mainly semi-desert country. One species is restricted to India, Pakistan, and Sri Lanka; all others live in Africa. They are more variegated in their color patterns than true larks and, as their vernacular name implies, have short conical finch-like bills. Like other larks they are gregarious when not breeding, and they are nomadic when food becomes seasonally scarce in their severe habitat.

WAGTAILS AND PIPITS
The 60 or so species in the family Motacillidae are small and of slender build, some of them longish-tailed. They are mainly ground-nesters and almost entirely insectivorous.

Wagtails

Typical wagtails are characteristic birds near running water on riverbanks and in moist grassland. Four species are resident in southern and eastern Africa and Madagascar (they do not migrate), whereas the wagtails that breed in northern countries are highly migratory. In these northern forms a breeding plumage is assumed by the adult males, so although wagtails are generally classified as 11 species, there is extensive racial variation—the seasonal change of dress being most markedly illustrated in the yellow wagtail *Motacilla flava,* of which 17 subspecies are recognized. The yellow wagtail is the only one that breeds in North America, in the northwest of Canada and Alaska. This species and the yellow-headed wagtail *M. citreola* are inhabitants of Eurasian steppes and pasture where they feed among grazing cattle. Of the other *Motacilla* species, four are white or pied; the purely African *M. capensis* and its Madagascan counterpart *M. flaviventris* are plainer, and two are long-tailed and choose purely riverine habitats (*M. cinerea* and *M. clara*).

An atypical species, the eastern Eurasian forest wagtail *Dendronanthus indicus,* is restricted to forest on its breeding grounds.

Pipits

The longclaws (genus *Macronyx*) represent the largest and most colorful of the pipits. All seven species live in Africa, centered on the eastern and southern savannas. The largest are two yellow-throated species, *M. croceus* and *M. fuelleborni,* and the orange-throated Cape longclaw *M. capensis;* the pink-throated *M. ameliae* and *M. grimwoodi* are somewhat smaller; and the remaining two are small

and yellow-throated, *M. flavicollis* in the Ethiopian highlands, and *M. aurantiigula* in East Africa. These birds demonstrate an interesting evolutionary convergence with the meadowlarks found in the Americas.

Linking the decorative longclaws of Africa and the dull-colored pipits of the genus *Anthus* are, in Africa, the golden pipit *Tmetothylacus tenellus,* plus a couple of anomalous forms, the yellow-breasted pipit *Hemimacronyx chloris* in the Drakensberg Mountains of southeastern Africa, and the Kenyan yellow-breasted pipit *H. sharpei* (also known as Sharpe's longclaw) found in the highland grasslands of Kenya.

The *Anthus* pipits are lark-like birds of open country and montane environments, generally brownish above, plain or moderately streaked, light buff on the underside, with the breast and sides usually streaked with dark brown. In many, the outer tail-feathers are broadly marked with white, but because the differences between species are subtle and the plumage is altered through bleaching and wear, field determination can be extremely difficult. Northern species of pipits are highly migratory, and even in tropical species there is much post-breeding movement associated with local seasonal declines in the availability of food. The *Anthus* pipits probably originated in the temperate regions of the Old World, from which they radiated extensively: 14 species are seen as basically Palaearctic (Eurasian), 13 endemic to Africa, four to Indo-Malaysia and Australasia, eight to Central and South America, and two to North America. Why the family should be so poorly represented in the North American continent is currently inexplicable.

Richard T. Mills

◄ *A European meadow pipit. Pipits are very ordinary in appearance but they are among the most widespread of all songbirds. Though the group reaches its greatest diversity in Eurasia, one form or another occurs in open country of all kinds almost everywhere except Antarctica; one species even inhabits the remote and inhospitable island of South Georgia.*

SWALLOWS

P.A. CLANCEY

Order Passeriformes
Suborder Oscines
Family Hirundinidae
c. 20 genera, *c.* 82 species

SIZE
Total length 12 to 25
centimeters (4¾ to 10
inches); weight 10 to 45
grams (½ to 1½ ounces).

CONSERVATION WATCH
The following species are
listed in the ICBP checklist of
threatened birds: white-eyed
river martin *Pseudochelidon
sirintarae*, Bahama swallow
Tachycineta cyaneoviridis,
white-tailed swallow *Hirundo
megaensis*, and Red Sea cliff
swallow *H. perdita*.

T he family Hirundinidae includes about 82 swallows and martins which are aerial-feeding insect-eaters distributed throughout the world's temperate and tropical zones (except some islands), and two species of river martins.

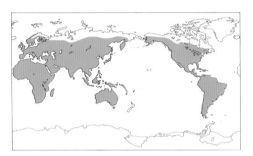

SWALLOWS AND MARTINS
The plumage of typical swallows (genus *Hirundo*) is glossy blue-black on the upper parts, dark on the wings and tail, the latter spotted subterminally with white, and the outer tail-quills in adults extended into narrow filaments; many species have patches of red-brown over the forehead or fore-throat. Others may have the tail squared and lack the streamers of the *Hirundo* species; the top of the head and lower back are tawny or red-brown and the underside is streaked (as in *Cecropis* and *Petrochelidon* species). Crag martins, which are nowadays classified in the genus *Hirundo* (formerly *Ptyonoprogne*), are dull-colored birds of dry and even desert regions in the Old World. The house martin *Delichon urbica,* and its allies the Asiatic house martin *D. dasypus* and the Nepal house martin *D. nipalensis* distributed across Europe and Asia, are glossy blue-black above and have short legs and toes covered with feathers as a protection

▶ *The barn swallow usually builds its nest in some human-built structure such as a bridge, shed, or barn, but it shows a marked preference for buildings in which cattle or other domestic stock are kept. Very occasionally it nests in ancestral sites such as crevices in cliffs. It breeds across much of the Northern Hemisphere, wintering in South America, Africa, India, and Southeast Asia.*

Wisniewski/Zefa

Richard T. Mills

◀ *Sand martins are strongly gregarious and nest in colonies, excavating holes in sandy river banks and quarries. They show strong fidelity to the site, returning every season but usually drilling a fresh burrow each time. The species is widespread across North America and Eurasia, but the European population suffered several abrupt declines during the period 1960–1990: one colony in Scotland collapsed from 900 pairs in 1982 to fewer than 200 pairs two years later. This decline has been linked with a concurrent series of devastating droughts affecting the region immediately south of the Sahara, from Eritrea west through the Sudan to the Sahel, where most of the birds spend the winter.*

against the low temperatures they encounter. Other martins are robust, glossy American species of the genus *Progne,* and the sand martins (genus *Riparia*) which are small and dull-colored.

These birds feed on insects taken during flight, and some may be found consorting when insects are temporarily locally abundant. They may also be seen alongside swifts when insect swarms are present near to the ground before a storm. Mainly silent birds, they give voice to twittering and short warbling songs delivered while on the wing or perched on bare twig. Most are strong migrants, especially the species of *Hirundo, Delichon, Riparia* and others, some of which move north to breed at high latitudes. Even tropical species have their seasonal movements—for example, the Malagasy Mascarene martin *Phedina borbonica* ranges to eastern Africa after breeding on Madagascar. In the Afrotropics the roughwings (genus *Psalidoprocne*) are usually quite sedentary.

Most swallows build solid nests in the shape of half-bowls, saucers and even retorts, fashioned out of mud pellets and straw, the eggs resting in a cup of fine grasses, hair, and curly feathers. Species of the genus *Riparia* nest in colonies, tunneling into vertical sandy banks and creating a nest of grasses and feathers at the end of the tunnel.

RIVER MARTINS
Both species of river martin are red-billed, short-tailed aberrant swallows. A native of the Zaire river system of Africa, the African river martin *Pseudochelidon eurystomina* is known to regulate its breeding cycle to the drop in the level of a major river so that it can breed (in colonies) in burrows on the exposed sandy bars. The white-eyed river martin *P. sirintarae* was discovered in Thailand and named only in 1968; its breeding grounds are still unknown but are probably the middle reaches of the Mekong River.

CUCKOOSHRIKES

P.A. CLANCEY

Order Passeriformes
Suborder Oscines
Family Campephagidae
9 genera, *c.* 72 species

SIZE
Smallest West African wattled cuckooshrike *Campephaga lobata,* 18 centimeters (7 inches); weight 20 grams (7/10 ounce).
Largest Ground cuckooshrike *Pteropodocys maxima,* 28 centimeters (11 inches); weight 111 grams (4 ounces).

CONSERVATION WATCH
The following species are listed in the ICBP checklist of threatened birds: slaty cuckooshrike *Coracina schistacea,* Mauritius cuckooshrike *C. typica,* Réunion cuckooshrike *C. newtoni,* black cuckooshrike *C. coerulescens,* Sula cuckooshrike *C. sula,* white-winged cuckooshrike *C. ostenta,* and West African wattled cuckooshrike *Campephaga lobata.*

C uckooshrikes are so-called because of their bustle of hard-shafted yet soft and loosely attached rump-feathers which is similar to that of Old World cuckoos. The family, Campephagidae, is confined to the tropics of Africa south of the Sahara and also from Afghanistan and the Himalayas, east across Asia to the Japanese islands, and south to Southeast Asia, Australia, and the Pacific islands.

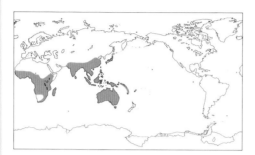

GRAY TREE-DWELLERS
Cuckooshrikes in the genus *Coracina* are moderately large, gray or blackish and white, and generally barred. Male and female are similar in appearance. Their bill is relatively heavy, broadly based, and hook-tipped, and they have short legs. The 43 species of this genus and its immediate allies are centered numerically on Southeast Asia and Australia, and with some species having a

remarkable variety of forms—no less than 33 subspecies of the cicadabird *C. tenuirostris* are recognized. In the wholly African genus *Campephaga,* comprising five species, the adults are markedly different: the males are an attractive glossy blue-black, with some exhibiting bright patches of yellow, orange or deep red on the wings; the females are dark olive-brown, barred and streaked in yellow, yet whitish below.

Cuckooshrikes are of secretive disposition, hiding among screening foliage as they forage for insects and fruit, but some *Coracina* species are more conspicuous and during the non-breeding season may be seen in small parties. Nests are usually constructed high up in trees and skillfully blended into the moss and lichens of tree limbs to make detection difficult.

The trillers (genus *Lalage*) reach Australia and the Pacific islands, but the 13 species are mostly distributed on the mainland and islands of Southeast Asia. Their plumage is blackish-gray and white, with some tinged a rust color.

The woodshrikes (genera *Hemipus* and *Tephrodornis*), close relatives of the trillers, are four robust species, gray, black, and white in color, or else pied, and dwell in the forests and woodlands of Southeast Asia and Indonesia.

The 10 species of minivets (genus *Pericrocotus*) are restricted to the east of the family's range, extending from the Amur River near the USSR–China border and the islands of Japan, east to Afghanistan, India, and south to the islands of Indonesia. Males of many species are strikingly colored in bright red, orange, and yellow, although some are by contrast dull-colored, and all females are less colorful than the males. The biology of this group appears to differ little from that of the cuckooshrikes.

◄ The barred or yellow-eyed cuckooshrike inhabits eastern Australia and the New Guinea region. It is a nomadic rainforest dweller that feeds mainly on fruit.

BULBULS AND LEAFBIRDS

P.A. CLANCEY

Order Passeriformes
Suborder Oscines
Family Pycnonotidae
15 genera, *c.* 120 species
Family Irenidae
3 genera, 14 species

SIZE
Bulbuls (Pycnonotidae)
Smallest Slender bulbul
Phyllastrephus debilis, total
length 14 centimeters
(5 inches); weight 13 to
16.5 grams (½ ounce).
Largest Yellow-spotted
nicator *Nicator gularis,* total
length 23 centimeters
(9 inches); weight 63 grams
(2 ounces).
**Fairy-bluebirds, leafbirds,
and ioras (Irenidae)**
Smallest
Common iora *Aegithina tiphia*,
total length 12 centimeters
(4¾ inches); weight 10 grams
(⅖ ounce).
Largest
Fairy-bluebird *Irena puella*,
total length 27 centimeters
(10½ inches); weight
70 grams (2½ ounces).

CONSERVATION WATCH
The following species are
listed in the ICBP checklist of
threatened birds: wattled
bulbul *Pycnonotus
nieuwenhuisii*, Prigogine's
greenbul *Chlorocichla
prigoginei*, spot-winged
greenbul *Phyllastrephus
leucolepis*, Appert's greenbul
P. apperti, dusky greenbul *P.
tenebrosus*, gray-crowned
greenbul *P. cinereiceps*,
yellow-throated olive
greenbul *Criniger olivaceus*,
mottle-breasted bulbul
Hypsipetes siquijorensis, and
Mauritius black bulbul *H.
olivaceus*.

T hese birds are in the main solitary, but readily congregate when food is freely available. The colorful and attractive leafbirds of the family Irenidae occur alongside bulbuls in forested areas of southern Asia.

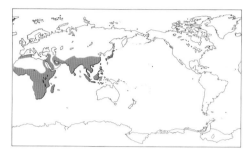

BULBULS

Bulbuls are a mainly tropical Old World family, Pycnonotidae, comprising 15 genera with about 120 species, which are distributed in Africa and from the Middle East across to Japan and south to Indonesia. Significantly they do not reach New Guinea or Australia—although the red-whiskered bulbul *Pycnonotus jocosus* has been introduced (successfully) to Sydney and Melbourne, and also to Florida in the USA. Bulbuls are small to medium-sized, and somberly colored in olive and brown, often whitish or yellow on the underside, with some exhibiting distinctive yellow or red undertail coverts. Many are crested and have hairlike filo-plumes over the back of the head. The bill is relatively robust, often notched towards the tip, and the legs and toes are strong.

Most bulbuls are insect- and fruit-eaters. When certain favored trees are bearing fruit, some species congregate in numbers, but otherwise they tend to be solitary and keep much to screening vegetation, especially in the canopy of high forest and woodlands. However, some exploit the understory and are partly terrestrial—especially so in the moderately gregarious species of *Phyllastrephus* in the African tropics. Several make their presence

◄ *The red-vented bulbul (far left) is a common inhabitant of parks and gardens in Southeast Asia. Its calls are cheerful but undistinguished, whereas the golden-fronted leafbird (near left) is notable for its flawless mimicry of the calls of other birds.*

Morten Strange

these remains to be resolved. Among these African bulbuls significant deviations from the norm are presented by the golden bulbul *Calyptocichla serina* and the honeyguide bulbuls *Baeopogon indicator* and *B. clamans*: the honeyguide bulbuls mimic the appearance of the true honeyguides, larger birds of the genus *Indicator,* family Indicatoridae, which are seemingly distasteful to predators.

The main concentration of species is in the equatorial rainforests of Africa and similar vegetation in Southeast Asia, the Philippines, and the Indonesian islands. The Asian genus *Hypsipetes* has an extensive range, several species reaching islands in the western Indian Ocean such as the Seychelles, Mauritius, Reunion, Madagascar, and even Comoros, but failing to gain a foothold on the African continent.

FAIRY-BLUEBIRDS, LEAFBIRDS, AND IORAS
An essentially Indo-Malaysian group of colorful arboreal birds, the family Irenidae extends from the Himalayas, the Indian peninsula and Sri Lanka eastwards to southern China, and south to the Philippines and Indonesia. Most species are largely sedentary, but some are on record as undertaking seasonal post-breeding migratory movements.

Fairy-bluebirds
The two fairy-bluebirds *Irena puella* and *I. cyanogaster* are the largest members of the family Irenidae. They are striking birds, the males shining ultramarine blue over the upper parts and black below, the females a duller, greener blue, and with black lores (the area extending from the eyes to the bill-base); all have red eyes. The bill is relatively stout with the culmen (the dorsal ridge from tip to forehead) arched, and the upper mandible notched short of the tip. Fairy-bluebirds feed on fruits from forest trees and shrubs, as well as flower nectar and small invertebrates, foraging in parties of up to eight individuals in the canopy of the monsoon forest. They are seemingly less aggressive towards competitors than their smaller relatives, the leafbirds, and are frequently seen consorting with other fruit-eating birds attracted to trees laden with fruit. Fairy-bluebirds nest in the upper story of forest trees, and their calls have been described as a "twing twing". *Irena cyanogaster* is confined to the Philippines, while the range of *I. puella* extends from there to northwestern India.

Leafbirds
The eight species of leafbirds (genus *Chloropsis*) are largely green; the males have a blue-black band on the breast, marked with vivid blue (as in the blue-whiskered leafbird *C. cyanopogon*); the forehead is green, yellow, or red; some are orange below, and in the blue-winged leafbird *C. cochinchinensis* the wings are blue. The females are duller than the males and lack the dark breast-band. They all have a more slender bill than the bluebirds, which is

▲ *Displaying the perky self-confident air so characteristic of its family, the yellow-vented bulbul is abundant in lowland forests and clearings throughout much of Southeast Asia. It congregates at night to roost in flocks.*

known in gardens in urbanized areas.

Bulbuls build loosely constructed nests of local plant materials, placing them where they will be concealed in trees, bushes, and creepers. Vocalization generally consists of ringing calls, often in answer to one another from different points of the forest; and when breeding the birds make short warbling and sometimes trilling songs—for example, the somber bulbul *Andropadus importunus* of southern and eastern Africa.

Most of the species are sedentary, but a few from the high northern latitudes are recorded as undertaking post-breeding migratory movements. The most widely distributed genus in the family is *Pycnonotus,* crested birds with colored undertail coverts; the common bulbul *P. barbatus* reaches northwest Africa and the eastern shores of the Mediterranean Sea.

In the African tropics the forest-dwelling species of *Andropadus* and its immediate allies are confusingly similar, and the validity of a couple of similarly notched on the upper mandible.

Leafbirds inhabit forests of various types, even mangroves, as well as gardens and parks. Despite their short, rounded wings, their flight is both sustained and swift. Field researchers describe the birds as highly pugnacious, driving away competitors as they forage for fruits, invertebrates, and flower nectar. Keeping as they do to the screening cover of the canopy and densely leaved trees, they are often difficult to see because of their dominant color of green, although while they are foraging, leafbirds are as lively and dexterous as tits of the family Paridae, hanging upside down and assuming various acrobatic postures as they search for elusive prey and food items.

Nests are described as a loose cup, semi-pensile in form, built of local plant materials felted together, and the eggs are pale buff, marked with spots and blotches of red-brown.

Ioras

Ioras (four species in the genus *Aegithina*) are distributed from northwestern India to Borneo and the Philippines. They are the size of small sparrows, predominantly green, with the wings starkly black crossed by two sharply-defined white bars. As in other members of the family, the upper mandible of their bill is notched towards the tip; but unlike the bills of *Irena* and *Chloropsis* the culmen is relatively straight and not arched or decurved. The great iora *A. lafresnayei* and green iora *A. viridissima* have an interesting mannerism: they fluff up their long lightly-colored flank feathers over their back to give the impression that the rump is white, which it is not. Ioras are fruit-eaters, but will also take insects, spiders, and nectar. They occur in pairs in forest and woodland environments, and their breeding details are similar to the leafbirds.

◀ *The lesser or Asian fairy-bluebird is a noisy, active, gregarious bird of the treetops that feeds largely on fruit. It is common in lowland evergreen forests throughout Southeast Asia, from India to Java and the Philippines.*

D. Avon/Ardea London Ltd

SHRIKES AND VANGAS

Order Passeriformes
Suborder Oscines
Family Laniidae
c. 11 genera, *c.* 78 species
Family Vangidae
c. 12 genera, *c.* 14 species

SIZE
Smallest
Coral-billed nuthatch
Hypositta corallirostris, total
length 12 centimeters
(4¾ inches); weight 10 grams
(½ ounce).
Largest
Long-tailed shrike *Corvinella melanoleuca*, total length 50 centimeters (20 inches); weight 87 grams (3 ounces).

CONSERVATION WATCH
The following species are listed in the ICBP checklist of threatened birds: Gabela helmet shrike *Prionops gabela,* Mount Kupe bush shrike *Malaconotus kupeensis,* green-breasted bush shrike *M. gladiator,* Monteiro's bush shrike *M. monteiri,* Uluguru bush shrike *M. alius,* São Tome fiscal shrike *Lanius newtoni,* Van Dam's vanga *Xenopirostris damii,* and Pollen's vanga *X. polleni.*

SHRIKES AND VANGAS

P.A. CLANCEY

The wide-ranging, composite family Laniidae (shrikes) is closely related to the small family Vangidae, confined to Madagascar and the Comoro Islands. Recent research indicates that vangas are descended from an ancestral form of helmet shrike from mainland Africa.

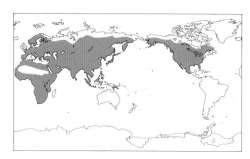

SHRIKES
This family of medium-sized perching birds comprises three distinct groups: the helmet shrikes and the white-headed shrikes (subfamily Prionopinae, nine species), in tropical Africa; the bush shrikes (Malaconotinae, about 44 species), also in tropical Africa, although the black-crowned tchagra *Tchagra senegala* ranges to northwestern Africa; and the "true", predatory shrikes (Laniinae, about 25 species), extensively distributed over much of Africa, Europe, Asia, and North America, but absent from South America, Australasia, and Madagascar.

True shrikes
The true shrikes are patterned in gray, chestnut, black and white, and many of the African species are pied. All have a robust, hooked and notched bill; and except perhaps for the two long-tailed African species (*Corvinella corvina* and *C. melanoleuca*) they are territorial and of solitary disposition. From a vantage perch they watch for terrestrial prey, which is seized on the ground, and many are renowned for their habit of impaling surplus food items on the thorns of bushes or barbed wire fencing. The sexes are well differentiated, and their calls are strident and include snatches from those of other birds. Nests are simple cups of twigs and grasses, lined with rootlets and hair, and these are placed in bushes and trees at no great height from the ground. The eggs are attractively marked.

True shrikes are characteristic of open habitats, preferring steppe—even when quite desertic—and lightly-treed savanna, and virtually all of the birds that breed in the far north (*Lanius* species) migrate to southern latitudes to avoid the winter.

Bush shrikes
The bush shrikes are generally sedentary, mostly in high evergreen forest or savanna woodland in Africa. These birds glean insects from tree limbs and foliage, but some species such as the bokmakierie *Telephorus zeylonus* of southern Africa forage extensively on the ground. Most are attractively colored on the underside, especially the tree-dwelling *Malaconotus* and *Chlorophoneus* species which are largely green on the back, wings and tail, but yellow, orange or even red below—this variation in abdomen color among the *Chlorophoneus* species is now understood to be mimicry of the color patterns of the larger, more powerful and feared *Malaconotus* bush shrikes alongside which they occur, thus gaining advantage for foraging groups of the smaller birds. The nests of all members of this subfamily are constructed in trees or among creepers, and eggs are spotted.

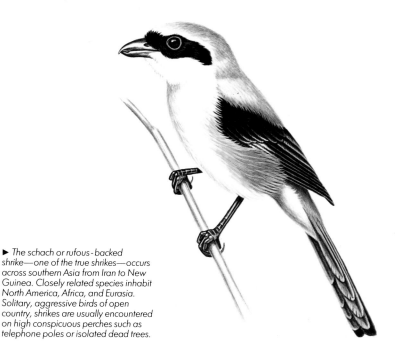

► *The schach or rufous-backed shrike—one of the true shrikes—occurs across southern Asia from Iran to New Guinea. Closely related species inhabit North America, Africa, and Eurasia. Solitary, aggressive birds of open country, shrikes are usually encountered on high conspicuous perches such as telephone poles or isolated dead trees.*

◄ A white helmet shrike incubating. Helmet shrikes differ from other shrikes in a number of ways, including their gregarious behavior and the rather deep, narrow, and tightly-woven structure of their nests. Exclusively African and totalling about 9 species, helmet shrikes are noisy, conspicuous birds that mainly inhabit arid scrublands.

Peter Johnson/NHPA

Helmet shrikes

Helmet shrikes (genus *Prionops*) are characterized by feathering on the crown which forms a brush-fronted "helmet". Moderately gregarious, they roam through the woodland in parties feeding in the trees. The white-headed forms (genus *Eurocephalus*) are also gregarious when on the move or at roost, but they feed independently, behaving much like the true shrikes. Both types construct beautifully fashioned and compacted nests, and among the helmet shrikes some uncommitted adults have been observed helping to rear the young of others of their family group.

VANGAS

The 14 species of vangas are all small, tree-dwelling perching birds found only on islands in the western Indian Ocean. The attractive blue vanga *Cyanolanius madagascarinus,* whose range extends to the Comoro Islands, is the only species to be found outside the main island of Madagascar. A species of doubtful attribution, the kinkimavo *Tylas eduouardi* of Madagascar, long considered to be an aberrant bulbul, is now seen as a member of the Vangidae.

The family exhibits a wealth of bill forms. The sickle-billed vanga *Falculea palliata* of the south-east has a long, scimitar-shaped bill; the three species of the genus *Xenopirostris* have a lower

O. Langrand/Bruce Coleman Ltd

◄ Confined to the Madagascan region, the 14 species of vangas have radiated into a variety of different habitats and ways of life in a manner reminiscent of the finches of the Galapagos or the Hawaiian honeycreepers. This is the rufous vanga.

mandible that is steeply upswept at the end; and the helmet bird *Euryceros prevostii* has a heavy bill which extends far back onto the fore-crown.

While generally viewed as closely related to the Laniidae, vangas differ widely from the true shrikes in their field behavior, and their feeding strategies most closely resemble those of the helmet shrikes. They prefer forest and woodland canopy, gleaning insect and lower vertebrate prey from the foliage, twigs, and limbs of trees. Many are gregarious, others relatively solitary. Nests are constructed in trees and among creepers, and the eggs are marked with dark spots, but very little is known about their breeding activity.

Order Passeriformes
Suborder Oscines
Family Bombycillidae
1 genus, 3 species
Family Ptilogonatidae
3 genera, 4 species
Family Hypocoliidae
1 genus, 1 species
Family Dulidae
1 genus, 1 species

Left to right: waxwing, silky flycatcher, hypocolius, palm chat.

Size

Waxwings (Bombycillidae)
total length 14 to
21 centimeters (5½ to
8¼ inches); weight 25 to
69 grams (⅕ to 2½ ounces).
**Silky flycatchers
(Ptilogonatidae)** total length
19 to 24 centimeters (7½ to
9½ inches).
Hypocolius (Hypocoliidae)
total length 17 centimeters
(6¾ inches).
Palm chat (Dulidae) total
length 12 to 17 centimeters
(4¾ to 6¾ inches).

Conservation Watch

These species do not appear
to be threatened.

WAXWINGS AND THEIR ALLIES

H. ELLIOTT McCLURE

Waxwings and their allies have been reclassified into 4 separate families: the waxwings (family Bombycillidae), the silky flycatchers (Ptilogonatidae), the hypocolius (Hypocoliidae), and the palm chat (Dulidae). They are fruit-eating birds of the Northern Hemisphere, and all except the palm chat have silky plumage.

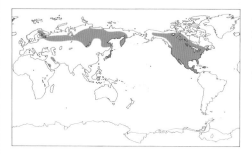

WAXWINGS

The name "waxwing" refers to a red waxlike droplet that forms at the tip of each secondary wing feather, a decoration with no known use. There are three similar species: the Bohemian waxwing *Bombycilla garrulus* in North America, northern Europe and Siberia; the cedar waxwing *B. cedrorum* in North America; and a small population of the Japanese waxwing, *B. japonica,* limited to north-east Asia. Waxwings are fawn-colored birds and are so sleek and trim, with a high crest, that they look carved from wood. Except when nesting they are highly gregarious, and when perched on a limb or wire stand almost touching.

Waxwings prefer northern cedar or evergreen forests for breeding, and their nest is an open cup usually placed high in a tree. The female is fed by the male while she is incubating. Upon hatching the nestlings are fed almost exclusively on insects

J. Kenning

▶ *Cedar waxwings. In all three waxwing species, males resemble females in plumage, but juveniles are dull brown and heavily streaked.*

until they fledge at 14 to 16 days. All three species feed heavily on cedar berries, and after the pulp is digested the seeds are passed, making this habit a very important element in reforestation. When nesting is completed waxwings gather into large flocks and move south erratically.

SILKY FLYCATCHERS

This is a New World family, almost restricted to Mexico and southwestern USA. Four species are recognized (in 3 genera), the best-known being the phainopepla *Phainopepla nitens,* a shiny black bird with white wing-patches and crimson eyes, which differs markedly from waxwings in habits, flight and actions. It migrates from dry areas in Mexico to the USA as far north as Sacramento, to nest as single pairs. The nest is built by the male high in a deciduous or evergreen tree, and he also does most of the incubation; both parents feed the nestlings. The pair may seek another territory in a different habitat for their second brood. The period of residency in the USA is very short, from February or March into July—although some birds may overwinter there—then they move as far south as Panama during fall migration. Their most recognizable call is a plaintive high-pitched note given when disturbed by predators or humans.

HYPOCOLIUS

Hypocolius ampelinus, a pale gray bird with black-tipped tail feathers, is limited to the Tigris–Euphrates valley and surrounding areas of Asia Minor. It travels about the scrub country in small flocks feeding almost entirely on small fruits. Male and female build an open cup nest in a pine tree or shrub.

PALM CHAT

The palm chat *Dulus dominicus,* of Haiti and the Dominican Republic in the Caribbean, is a gregarious species which builds communal nests high in palm trees. The entire structure is up to 1 meter (3¼ feet) in diameter, loosely woven to protect five or more compactly woven individual nests, each with a separate tunnel to the outside. Incubation and the nestling period are about two weeks, and the noisy chattering birds use the nest structure when resting and roosting.

MOCKINGBIRDS AND ACCENTORS

H. ELLIOTT McCLURE

Order Passeriformes
Suborder Oscines
Family Mimidae
12 genera, 32 species
Family Prunellidae
1 genus, 12 species

SIZE
Mockingbirds (Mimidae)
total length 20 to 33
centimeters (7¾ to 13
inches); weight 36 to 56
grams (1¼ to 2 ounces).
Accentors (Prunellidae)
total length 14 to 18
centimeters (5½ to 7 inches);
weight 25 to 35 grams (⅘ to
1⅓ ounces).

CONSERVATION WATCH
The Socorro mockingbird
Mimodes graysoni and white-
breasted thrasher
Ramphocinclus brachyurus are
listed in the ICBP checklist of
threatened birds.

Mockingbirds and their allies are thrush-like birds of the New World. Accentors are smaller, similar to sparrows, but with a more slender bill; they live in the Old World, from Europe to Japan.

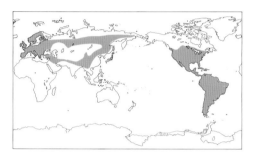

MOCKINGBIRDS AND THEIR ALLIES
The family Mimidae includes 32 species of mockingbirds, catbirds, mocking-thrush, trembler, and thrashers, about equally divided between North and South America. They are sturdy-legged terrestrial or low-vegetation birds, with strong downcurved beaks, short wings, and long tails. Only the blue-and-white mockingbird *Melanotis*

hypoleucos of South America and the black catbird *Melanoptila glabirostris* of Mexico vary from the usual gray or brown coloration of the family, but many have lovely patterns to distinguish them.

The family is noted for its beautiful singers and mimics. It is tempting to refer to the northern mockingbird *Mimus polyglottus*—which may imitate as many as 20 local bird species—as the most talented, but this may be simply a reflection of the fact that it is an urban species, studied and admired by many observers, whereas those species in less populated areas are not as well known.

All are strongly territorial, especially the mockingbirds which will aggressively attack transgressors whether mammal or bird, and many maintain territories the year round. The North American species live in forest edge, open country, and desert habitats. The successful urbanization of the northern mockingbird, gray catbird *Dumatella carolinensis,* and brown thrasher *Toxostoma rufum* is

Joe McDonald/Tom Stack & Associates

◄Named for the skill with which it mimics the calls of other birds, the northern mockingbird is a common and hardy suburban bird over much of North America. It is strongly territorial, and may lay claim to an ornamental berry bush or similar food resource through the winter, defending it from all comers. An active and fearless bird, it often teases dogs and cats.

▶ Common and widespread, the brown thrasher resembles the northern mockingbird in many respects, but it is much more strongly migratory, very shy and retiring, and seldom mimics other birds. It spends much of its time on the ground.

Don & Esther Phillips/Tom Stack & Associates

▼ Few European birds are so common yet so inconspicuous as the dunnock or hedge sparrow. It spends most of its time on the ground, quietly shuffling and creeping along in a highly distinctive manner, persistently twitching its wingtips. It nests close to the ground, often in ivy or heaps of garden refuse; the female builds the nest and incubates alone, but both parents cooperate in rearing the young.

Stephen Dalton/NHPA

partly the result of open tree and shrub plantings found in urban areas. These plantings simulate the forest edge conditions preferred by these species and provides them with insects, berries, and fruits. Most build open-cupped bulky nests of twigs, lined with grasses and fibers, in low vegetation. Incubation, mainly by the female but with the male's help, is about two weeks, and the young are fledged at about two weeks. Several species breed a second time each year, and the California thrasher *T. redivivum* may nest as late as fall. Most species forage on the ground, taking terrestrial invertebrates, which they find using their downcurved bill to dig in the soil or search under surface debris. They also eat small fruits in season.

The northern species are migratory to varying degrees, some moving only a few kilometers north or south, others going as far as Central America.

Like insular species elsewhere, those of the Caribbean are endangered through the introduction of small predators and the destruction of environments by humans. The white-breasted thrasher *Ramphocinclus brachyurus* of Martinique and Saint Lucia and the trembler *Cinclocerthia ruficauda* (noted for its habit of shivering and trembling) of the smaller islands of the West Indies are both now rarely seen, but two other species, the scaly-breasted thrasher *Margarops fuscus* of the Lesser Antilles and the pearly-eyed thrasher *M. fascatus,* which ranges from the Bahamas south through many islands, are still quite abundant. These island inhabitants are at risk, however, as they lay only two or three eggs.

ACCENTORS

Limited to Eurasia, 12 species of accentors (family Prunellidae, genus *Prunella*) occupy numerous high-latitude and high-altitude habitats, preferring brushy areas. Most move altitudinally with the seasons, and some are migratory. They are sparrow-sized birds with a thrush-like bill and round wings with 10 primary feathers. Not strong fliers, they forage on the ground or in low shrubbery, taking insects during warm weather, but seeds and berries in winter, having a crop and gizzard capable of handling this harsh food. The nest, placed on the ground in a crevice or among rocks, is neatly woven and the cup insulated by feathers. The three or four greenish-blue eggs are incubated by the female for about 15 days, and the young are quickly fledged, sometimes before they can fly. Usually there are two broods. The dunnock or European hedge sparrow *Prunella modularis* is well known as an occupant of lower scrub country and moorlands in Europe.

DIPPERS AND THRUSHES

C. PERRINS

Order Passeriformes
Suborder Oscines
Family Cinclidae
1 genus, 5 species
Family Turdidae
48 genera, c. 304 species

SIZE
Dippers (Cinclidae) total length 17 to 20 centimeters (7 to 8 inches).
Thrushes and allies (Turdidae) total length 11 to 33 centimeters (4½ to 13 inches).

CONSERVATION WATCH
There are 37 species listed in the ICBP checklist of threatened birds. They include the following: rufous-throated dipper *Cinclus schulzii*, rusty-bellied shortwing *Brachypteryx hyperythra*, east coast akalat *Sheppardia gunningi*, dappled mountain robin *Modulatrix orostruthus*, black shama *Copsychus cebuensis*, Javan cochoa *Cochoa azurea*, olomao *Myadestes lanaiensis*, Stoliczka's bushchat *Saxicola macrorhyncha*, Fuerteventura stonechat *S. dacotiae*, Benson's rockthrush *Monticola bensoni*, Everett's thrush *Zoothera everetti*, Taita thrush *Turdus helleri*, Yemen thrush *T. menachensis*, and gray-sided thrush *T. feae*.

T he five species of dippers making up the family Cinclidae are the only truly aquatic passerine birds. The family Turdidae—thrushes—is an important and widespread family with representatives in virtually every area of the world except the Arctic, Antarctic, and some oceanic islands.

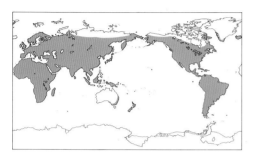

DIPPERS

Dippers (all five species are in the genus *Cinclus*) inhabit clear, swiftly-flowing streams and therefore tend to be found in hilly or mountainous regions throughout much of Europe, Asia, western North America and the northern half of the Andes in South America, wherever such streams occur. Because they do not migrate long distances, however, they do not occupy the streams and rivers of the high Arctic, which might be suitable for them in summer but would freeze over in winter.

Dippers feed largely on the larval forms of aquatic insects such as stone-flies and mayflies, though they take a few small fish as well. They dive into the water, where they either swim or walk along the bottom; when they walk, they usually do so in an upstream direction, holding onto stones with their powerful feet and using the force of the current to keep them on the bottom. They build a bulky, domed nest of mosses near the water, and the female lays and incubates three to six pure white eggs. Both parents raise the young, which take about three weeks to fledge.

◀ *Seldom encountered away from swiftly flowing water, dippers feed underwater on aquatic insects, using their wings to maneuver and bobbing up like corks when surfacing. Even in flight they follow streams, whirring low over the water on rapidly vibrating wings. This is the white-breasted dipper, which is common across Europe and much of temperate Asia.*

Richard T. Mills

405

Tom & Pam Gardner

THRUSHES

The family Turdidae is commonly divided into two subfamilies: Turdinae, comprising about 114 species, of which some 64 are classified in the genus *Turdus*; and Saxicolinae, with about 190 species.

Subfamily Turdinae

Turdinae includes the typical thrushes, such as the American robin *Turdus migratorius*, the Eurasian blackbird *T. merula*, the olive thrush *T. olivaceus* of southern Africa, and the white-necked thrush *T. albicollis* of South America. They are about 23 centimeters (9 inches) in length, predominantly quietly colored, mostly in brown with some in gray or black, and many are spotted underneath. In many species the sexes are similar in appearance, but in some the males are rather more brightly colored. Many have fine songs, audible at a considerable distance.

This subfamily has an almost worldwide distribution, although there are few in Australia and none in New Zealand, apart from introduced species. Most live in wooded areas, but they are commonly seen feeding both in trees and on the ground. Most species feed on a wide variety of fruits and animal prey, especially insects and worms; those living in colder climates may vary their diet seasonally, taking fruit in the fall, worms and snails in the winter, and insects during the summer..

The species that inhabit warmer climates are usually resident there throughout the year, whereas those that

▲ *The blackbird was once confined to dense forest, but its habitat requirements have relaxed over the past 150 years or so until now it is one of the commonest of birds in European parks and gardens, where it often reaches higher population densities than in its ancestral habitat.*

▶ *The robin in Britain is a tame, familiar bird of parks and gardens, but it is much more shy and retiring in disposition elsewhere in its range. The red breast is a badge used in territory defense rather than in courtship. Both sexes wear it, and defend territories all the year round—separately in winter, jointly in summer.*

Richard T. Mills

breed at higher latitudes migrate quite long distances between their breeding grounds and non-breeding quarters; for example, the redwing *T. iliacus* may breed in northern Scandinavia or Russia and winter in Ireland or Italy. Many of the New World species migrate much further; Swainson's thrush *Catharus ustulatus* breeds in the United States and then flies as far south as northwestern Argentina, while the gray-cheeked thrush *C. minimus*, which breeds largely in northern Canada, migrates as far south as Brazil.

The large majority of species build rather standard, cup-shaped nests, some lining them with mud. They lay two to five eggs, mostly bluish or greenish, and speckled with browns and blacks. They may raise two or more broods in a year. Both parents help to raise the young. The fieldfare *Turdus pilaris*, which breeds in northern Europe, is unusual in that it often nests in colonies. When these large, bold thrushes are threatened by a predator they dive-bomb it, defecating over it as they do so, leaving the would-be predator to withdraw in a sticky mess!

Subfamily Saxicolinae
The members of the other subfamily, the Saxicolinae, are mostly smaller birds—such as the chats, wheatears and robins of the Old World, and the bluebirds of the Americas—averaging 15 centimeters (6 inches) in length. Many are more brightly colored, some having quite vivid reds and oranges, and the bluebirds being predominantly blue. Many of the wheatears are strikingly patterned in blacks, grays, and whites. Males of many species generally have brighter plumage than

their mates. The Himalayan forktails (genus *Enicurus*) have very graduated, deeply-forked tails of black feathers tipped with white. Many species, such as the nightingale *Luscinia megarhynchos*, are fine songsters with a rich range of notes.

This subfamily is found in a wide range of habitats, from forest and thick scrub (some redstarts, robins, nightingales), at the edges of swiftly flowing streams (white-capped redstart *Chaimarrornis leucocephalus*), to desert (some of the African chats, some wheatears). They nest in a wider range of sites than the true thrushes, for although some build normal cup-shaped nests in a bush, many make their nest in a hole in a tree or a cavity under a rock. Many are primarily insect-eaters, others take fruit in season. Many, especially those that are primarily insect-eaters, migrate long distances to warmer winter quarters; for example, the wheatear *Oenanthe oenanthe* breeds in the Arctic areas of the Old World (from Greenland across to eastern Siberia and even into Alaska), yet all the birds migrate to Africa to avoid the winter. From Greenland they may face a non-stop flight over water of some 3,200 kilometers (2,000 miles).

Also included in this subfamily are several larger species, such as the rock thrushes (genus *Monticola*), and the grandala *Grandala coelicolor*, a bird of the high Himalayas. The male grandala's body is bright blue, whereas the female is dull brown, speckled with white on the head and wings. Grandalas spend most of the year in flocks, sometimes several hundred strong, feeding on the ground. They nest on rocky ledges at an altitude of about 4,000 meters (13,000 feet).

▲ Two forest thrushes of eastern Asia: White's thrush (left), and a white-rumped shama (right). The shama's song rivals that of the nightingale in richness and versatility, but White's thrush is a comparatively quiet and inconspicuous inhabitant of the forest floor.

Order Passeriformes
Suborder Oscines
Family Timaliidae
53 genera, 276 species
Family Troglodytidae
14 genera, *c.* 60 species

SIZE
**Babblers and allies
(Timaliidae)** total length
9 to 36 centimeters (3½ to
14 inches).
Wrens (Troglodytidae) total
length 10 to 22 centimeters
(4 to 8⅔ inches).
CONSERVATION WATCH
There are 34 species listed in
the ICBP checklist of
threatened birds. They
include the following: Zapata
wren *Ferminia cerverai,*
Clarion wren *Troglodytes
tanneri,* short-tailed scimitar-
babbler *Jabouilleia danjoui,*
rufous-throated wren-babbler
Spelaeornis caudatus, tawny-
breasted wren-babbler *S.
longicaudatus,* striped babbler
Stachyris grammiceps, snowy-
throated babbler *S. oglei,*
Hinde's pied babbler
Turdoides hindei, and gold-
fronted fulvetta *Alcippe
variegaticeps.*

BABBLERS AND WRENS

C. PERRINS

The Timaliidae is an Old World family comprising three subfamilies found throughout warmer areas. The wrens, Troglodytidae, are a New World family. A single exception is the winter wren *Troglodytes troglodytes,* whose ancestors presumably crossed the Bering Strait and eventually settled in most of Asia and Europe, where it is known simply as the wren.

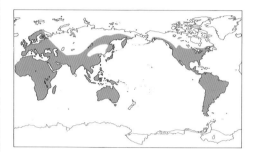

BABBLERS

The Timaliidae can be divided into three subfamilies: the babblers, Timaliinae, with about 255 species in 49 genera; the parrotbills, Paradoxornithinae, containing

19 species in 3 genera; and the bald crows, Picathartinae, with just two species in a single genus.

Babblers

Babblers are widespread throughout the warmer areas of the Old World. Of the four species (genus *Pomatostomus*) in Australia, Hall's babbler *P. halli* was discovered as recently as 1963 in acacia scrub in Queensland's dry interior. In parts of India and Southeast Asia, babblers form a very important part of the bird communities; for example, about one fifth of all bird species in Nepal are babblers. Only one related bird occurs in America: the wrentit *Chamaea fasciata,* which is sedentary on the west coast from Baja California to the Columbia River, Oregon's northern boundary.

▶ *The white-browed babbler lives in
arid Australian scrublands in permanent
groups of a dozen or so, feeding mostly
on the ground. Each group typically
consists of a dominant pair and several
of their offspring, which assist in raising
subsequent broods of young.*

Tom & Pam Gardner

Morten Strange

A typical babbler would be a rather nondescript brown bird about the size of a small thrush. However, they range in size from about 9 centimeters (3½ inches) for the pygmy wren-babbler *Pnoepyga pusilla*, a bird with a very short tail, to 36 centimeters (14 inches) for species of laughing-thrushes. While most are predominantly brown, a few are brightly colored with reds, blues, and yellows. The pied babbler *Turdoides bicolor* of South Africa is pure white with black wings and tail. There is very little difference in plumage between the sexes. Most have multipurpose, thrush-like or warbler-like bills, but those in the genera *Pomatostomus* and *Pomatorhinus* have longer, more decurved bills, while the slender-billed scimitar-babbler *Xiphirhynchus superciliaris*, which lives in the mountains of northeast India through Burma, is well named for its long, distinctive, decurved bill in the shape of a scimitar.

Babblers inhabit wooded country, many living in thick scrub where they are difficult to locate except by their noisy calls. A few are more specialized to particular habitats, such as the Iraq babbler *Turdoides altirostris* which lives mainly in reed-beds and other swampy areas, and the Arabian babbler *T. squamiceps*, one of the larger babblers, which lives in scrub along the edges of wadies in desert regions. The majority of

species are insect-eaters, hunting for insects or other small invertebrates among the foliage of trees or on the ground. The larger species also take small vertebrates such as lizards. Many will also take berries, and some feed on nectar when it is available.

They tend to be sedentary, defending their territories year-round. Few have been studied in detail,

▶ *A popular cage-bird, the red-billed leiothrix is widely known to aviculturists as the Pekin robin. In the wild it lives in small groups that forage on the ground in dense forests from northern India to China.*

▲ *Usually encountered in loose parties of a dozen or so, the blue-winged minla often joins other species of babblers and other birds in flocks wandering through the canopy in mountain forests of India and Southeast Asia. The bluish flight feathers are distinctive but very difficult to see.*

▶ Winter wren nests are globular structures of moss, twigs, and grass, stuffed into any available cavity and often well hidden. In the breeding season the male winter wren builds a number of nests within his territory. He may pair with several females; once mated, each female chooses one of his nests and raises her brood of young in it, largely unaided by the male.

but the following notes seem typical of many: they live in small parties of up to a dozen birds; they are highly social, and the birds remain together almost all the time; in some species they roost together, sitting tightly-packed, shoulder-to-shoulder along a branch. The group jointly defends the territory, which the birds do very noisily with a great variety of calls. They breed communally, the dominant pair building a nest of twigs in a tree or dense bush; the remainder of the group help to defend the pair's nest and raise the young. Young males stay within their own group and breed either by inheriting the territory when their father dies or, if the group becomes large enough, by "budding-off" with some of the other younger birds into a new territory taken from a neighboring, smaller group. The young females disperse to other groups nearby, presumably to avoid inbreeding.

Many species are widespread and common, some making use of man-made habitats, but others are a cause for concern. Admittedly, their habit of living in dense bush may mean that some have been overlooked and are more common than is believed, but nevertheless, several are known only from small areas of forest, and the rapid removal of this habitat means they are almost certainly threatened. At least five species are restricted to small areas in the Philippines and are considered endangered or threatened; for example, the striped babbler *Stachyris grammiceps* is found only on the slopes of mountains on northern Luzon.

Parrotbills

Parrotbills occur in northern India and Southeast Asia, except for the bearded tit or bearded reedling *Panurus biarmicus* which inhabits reed-beds from Central Asia westwards into Europe. They are small, brownish birds, ranging in size from 10 to 28 centimeters (4 to 11 inches). Most have stubby bills, but that of the spot-breasted parrotbill *Paradoxornis guttaticollis* is particularly deep and rather parrot-like. They generally live in thick scrub, many of them in bamboo thickets.

Bald crows

The two species of bald crow, or rockfowl, are extraordinary birds whose relationship to this and other families has long been debated. Reminiscent of extremely long-legged thrushes, they are bald-headed —hence one of their names—the skin being bright yellow in the white-necked baldcrow *Picathartes gymnocephalus*, and bright blue and pink in the gray-necked bald crow *P. oreas*. They inhabit dense forests in West Africa, where they nest in large caves or on deeply shaded cliffs, building a nest of mud on the rock-face. They tend to nest in groups, perhaps because of the specialized nature of their nest sites. The female lays one or two eggs, and both parents bring insects and worms to their young. Although they are not thought to be in immediate danger, their special habitat requirements make them vulnerable.

WRENS

The majority of wren species occur in South and Central America. Just nine species breed in North America, and while several of these migrate south to milder climates for the winter, almost all other species are sedentary.

Most are small or smallish species, the largest being the cactus wren *Campylorhynchus brunneicapillus*, about 22 centimeters (8⅔ inches) in length. They have short, rounded wings and are not strong fliers. All are basically grayish or brownish in color, many heavily streaked with black, and some have white eye-stripes or white throats. Many wrens have powerful voices, and some such as the flutist wren *Microcerculus ustulatus* and the musician wren *Cyphorinus aradus* are highly musical. Some sing antiphonally—a couple of birds giving responses, alternately, to each other. All species build domed nests. Although most of the tropical species are thought to be monogamous, several of the North American species and the winter wren are polygamous, the males building a succession of nests in the hope of attracting a new mate to each. The cactus wren lives in small groups in which juveniles of previous broods help their parents to raise the young of the current brood.

Apolinar's wren *Cistothorus apolinari* is restricted to waterside vegetation in a small area of the eastern Andes in Colombia. The Zapata wren *Ferminia cerverai* is found only in the Zapata Swamp on Cuba, and numbers have been greatly reduced by habitat loss, especially burning.

▼ Aptly named, the cactus wren often builds its nest amid the formidable spines of the chola cactus. Largest of the wrens, it occurs from southern Mexico north to the southwestern USA.

Jeff Foott/Auscape International

Order Passeriformes
Suborder Oscines
Family Sylviidae
c. 66 genera, c. 361 species
Family Muscicapidae
c. 28 genera, c. 179 species

SIZE
Warblers (Sylviidae) total
length 9 to 16 centimeters
(3½ to 6¼ inches).
**Old World flycatchers
(Muscicapidae)** total length
10 to 21 centimeters
(4 to 8¼ inches).

CONSERVATION WATCH
There are 47 species listed in
the ICBP checklist of
threatened birds. They
include the following: Cuban
gnatcatcher *Polioptila
lembeyei*, Grauer's swamp
warbler *Bradypterus graueri*,
aquatic warbler *Acrocephalus
paludicola*, Rodrigues warbler
A. rodericanus, Seychelles
warbler *A. sechellensis*, white-
winged apalis *Apalis chariessa*,
São Tome short-tail
Amaurocichla bocagii, bristled
grass-warbler *Chaetornis
striatus*, Japanese marsh
warbler *Megalurus pryeri*,
Nimba flycatcher *Melaenornis
annamarulae*, Sumba
flycatcher *Ficedula harterti*,
blue-breasted flycatcher
Cyornis herioti, Chapin's
flycatcher *Muscicapa lendu*,
Gabon batis *Batis minima*, and
banded wattle-eye *Platysteira
laticincta*. The Aldabra
warbler *Nesillas aldabranus* is
possibly extinct.

▶ *The golden-crowned kinglet. Kinglets
belong to a small genus of tiny, plump
warblers characterized by a patch of
vivid red or yellow on the crown. They
are widespread in the Northern
Hemisphere, and most live in coniferous
forests.*

WARBLERS AND FLYCATCHERS

C. PERRINS

Most of the birds in the family Sylviidae are classified as Old World warblers of the subfamily Sylviinae, which comprises about 349 species in 63 genera. The Muscicapidae, a large, wholly Old World family, is quite unrelated to the Tyrannidae, or New World flycatchers (see page 386). It is divided into two subfamilies: the Muscicapinae, which comprises 153 species in 24 genera; and the Platysteirinae, with 26 species in 4 genera.

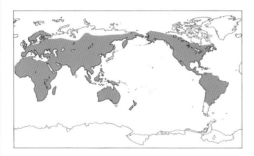

WARBLERS

Very few members of the family Sylviidae have succeeded in penetrating the New World. For example, the Arctic warbler *Phylloscopus borealis* breeds in forests across Siberia and extends just into western Alaska; all the birds migrate to Southeast Asia each winter. Only two species, the golden-crowned kinglet *Regulus satrapa* and the ruby-crowned kinglet *R. calendula*, are widespread in North America.

Within the family Sylviidae is a second, much smaller subfamily, the Polioptilinae or gnatcatchers, containing 12 species in 3 genera. These are restricted to the Americas and do not occur in the Old World. Gnatcatchers are all very small birds, 10 to 12 centimeters (4 to 4¾ inches) long, including their long tails. Like the other warblers, they are insect-eaters. Most occur in South or Central America, but three reach as far north as the USA—the only species that is widespread there, the blue-gray gnatcatcher *Polioptila caerulea*, migrates south for the winter. One South American species, the long-billed gnatwren *Ramphocaenus melanurus*, is 12.5 centimeters (5 inches) long, but this includes its tail measuring 4.5 centimeters (1¾ inches) and an extraordinarily long bill of about 3 centimeters (1¼ inches).

The subfamily Sylviinae is dominated by six genera: *Acrocephalus* (28 species), which is widespread throughout the Old World, especially in reed-beds; *Sylvia* (20 species), most common in Europe and North Africa; *Phylloscopus* (40 species), which forms an important component of forest birdlife, especially in some of the Himalayan region; *Cisticola* (41 species) and *Prinia* (26 species), prevalent in Africa; and *Apalis*

(20 species), exclusive to Africa. Birds in this subfamily have reached almost all of the Old World, including fairly remote islands in the Pacific. In spite of being so widespread, only eight species occur in Australia, including grassbirds and the spinifexbird and only one, the fernbird *Bowdleria punctata*, in New Zealand. Basically they are birds of woodland or scrub, but some, including several members of the genus *Cisticola* and the spectacled warbler *Sylvia conspicillata*, live in very sparse, low vegetation. Many breed in wooded areas at high latitudes or high altitudes, but migrate to warmer climates for the winter, undertaking very long migrations for such small birds. For example, the willow warbler *Phylloscopus trochilus*, which breeds in eastern Siberia, migrates to Central Africa to avoid the winter, making a round trip of some 25,000 kilometers (15,500 miles).

Most warblers are small birds, less than 15 centimeters (6 inches) in length, including the quite long tails of some species. An exception is the grassbird *Sphenoeacus afer* of South Africa, which

Tom & Pam Gardener

reaches about 23 centimeters (9 inches), although again this includes a long tail. The majority are dull in color, predominantly green or brown, although some are quite heavily streaked with black. In these the sexes are generally similar in appearance. An exception is the genus *Sylvia*; the males of many species are quite brightly colored, with orange or reddish underparts or black and gray patterning; the females tend to be duller. Some, such as the largely African *Apalis* and *Prinia*, have long, strongly gradated tails. In contrast, the crombecs (genus *Sylvietta*) of Africa have almost no tail at all; they climb about on the trunks and branches of trees in a way similar to the nuthatches (family *Sittidae*). The main food of almost all species is insects, and this is probably the main reason why so many of those that breed at high altitude or in northern latitudes migrate to warmer areas for the winter. To match this diet, the bill is small and finely pointed. Many warblers also take small fruits and berries when available, and a few take nectar or tiny seeds.

What they lack in appearance, many of the warblers more than make up for in song, being fine, strong singers. Many indulge in song-flights, soaring up and

"parachuting" down in a striking fashion. Many of the cisticolas, while perhaps not qualifying as fine singers, have elaborate flight displays; indeed many of these are difficult to see in the field, and the species are most easily separated on the basis of their calls.

Most Old World warblers are monogamous, but in some the males are regularly polygamous. For example, the male Cetti's warbler *Cettia cetti* may have as many as five or more females breeding in his territory. The majority build simple cup-nests, often very neatly woven, in thick vegetation. Some build domes or purse-like nests, and the tailorbirds (genus *Orthotomus*) of India and Southeast Asia are renowned for taking two or more large leaves and stitching them together with small fibers or cobwebs threaded through small perforations made in the edges of the leaves; the nest is then built within the leaves. The normal clutch is two to six eggs, depending on the species. The young are raised by both parents (except in polygamous species) and take about two weeks to reach the flying stage, although in some species the young may scatter from the nest before that.

Because many species occur on small islands, their

▲ *Africa is the home of a bewildering range of species of cisticolas, but several forms also occur in Asia and two extend to Australia. Most inhabit grasslands of various kinds. This is the golden-headed cisticola of India, Southeast Asia, and Australia, where it is often known as the tailorbird for its habit of stitching leaves to its nest, aiding in its concealment.*

413

▲ *A sedge warbler in full song in a German meadow. Although warblers are so plain in plumage that many species are very difficult to identify, they often have loud, rich, and varied songs. Birdsong may be uttered to announce territory – keeping trespassers away – or to attract females, or both; some species have separate songs for each purpose.*

population sizes are often very small and so the species are extremely vulnerable, especially to habitat destruction. For example, in the western Indian Ocean the Aldabra warbler *Nesillas aldabranus*, which was discovered in 1967, inhabits only a tiny area of Aldabra Island; there may be fewer than ten individuals. Slightly less endangered is the Seychelles warbler *Acrocephalus sechellensis*; in 1967 there were probably only 20 to 30 individuals of this species, all on Cousin Island. The island was purchased by the International Council for Bird Preservation (ICBP), and the bird's habitat increased. The numbers have built up well, and some have been distributed to other islands in the Seychelles.

OLD WORLD FLYCATCHERS
Members of the subfamily Muscicapinae occur almost everywhere except in treeless areas, avoiding the

center of deserts, high latitudes, and high altitudes. They are primarily birds of wooded areas, from dense forest to very open woodland, almost anywhere they can find an available perch from which to hawk for insects. The orange-gorgetted flycatcher *Muscicapa strophiata* nests at 4,000 meters (13,000 feet) in the Himalayas, where the forest is at its altitudinal limits, but all depend on woodland or shrubs of some sort.

Species that breed in the northern latitudes migrate south to avoid the cold winter, the spotted flycatcher *Muscicapa striata* from Europe going as far as South Africa. The red-breasted flycatcher *M. parva* is unusual in that those from the western end of the range, in Europe, migrate southeastwards to winter in India and Southeast Asia. Those living in warmer areas of the world mostly remain resident throughout the year.

Flycatchers are small birds, about 10 centimeters (4 inches) long, although some have long tails which make their total length much greater than this. They vary markedly in color. Many are rather dull brown birds, but in others the males are more striking—black and white such as the collared flycatcher *Ficedula collaris*, or blue as in the Hainan blue flycatcher *Cyornis hainana*, or other colors. In the vanga flycatcher *Bias musicus* the male is largely black with a white belly and the female is rich chestnut above, with a black head.

As their name suggests, these birds feed mainly on insects. Many literally catch flies, sitting conspicuously on perches and darting out to snap up passing insects. Others forage more among the foliage and take mainly perched insects or caterpillars. In cold weather, when flying insects are scarce, spotted flycatchers may even bring wood-lice to their young. Most have rather broad, flattened bills for catching their flying prey, but some have finer bills and take many of their prey from the ground.

The flycatchers breed in a variety of sites. Many make simple cup-shaped nests in trees or on ledges on cliffs or buildings, whereas others such as the pied flycatcher *Ficedula hypoleuca* nest in holes in trees. Hole-nesting species lay larger clutches than those of open-nesting species, up to eight eggs as opposed to two to five. The pied flycatcher, which winters in Africa and breeds in Europe, is sometimes bigamous: the male displays to a female when she arrives on the breeding grounds and mates with her, but when the female is incubating eggs he may set up another territory nearby and try to attract a second female. The male usually feeds the young in the nest of the first female, so it is the second one that loses out; she gets no help in the rearing of the young, and frequently some die.

A number of species live on islands in Southeast Asia and are poorly known, but possibly endangered. For example, the white-throated jungle-flycatcher *Rhinomyias albigularis* occurs only on the islands of Negros and Guimaras in the Philippines, and the few remaining patches of forest where it lives are being cleared. Many of the inhabitants of West African forests are threatened by continued heavy logging; among these is the Nimba flycatcher *Melaenornis annamarulae*, an all-black flycatcher from the foothills of Mount Nimba.

The other subfamily in the Muscicapidae, the Platysteirinae, is found only in Africa. The two main genera are the puff-backs, genus *Batis*, and the wattle-eyes, genus *Platysteira*. The latter are so-called because of striking wattles above the eye; these are often red, but in Blisset's wattle-eye *P. blissetti* of West Africa they are green or blue depending on the race. The birds are found in a range of wooded habitats from thick forests to open woodland. They make open cup-shaped nests placed in a fork among branches or on a larger branch and lay two eggs.

Most members of this subfamily seem not to be endangered, but the banded wattle-eye *P. laticincta* occurs only in the Bamenda Highlands of West Cameroon, another area of West African forest that is suffering from habitat destruction.

◄ The rufous-bellied niltava is common in dense forests from the western Himalayas to Burma and Malaysia. Highland populations migrate to lower elevations in winter. Sparrow-sized and inconspicuous despite its brilliant plumage, it resembles other flycatchers in behavior, hawking insects from low perches in dense cover.

▼ The forests of India and Southeast Asia are the home of the ferruginous flycatcher. It is an inconspicuous bird most often encountered in the lower branches of the forest canopy, hawking for insects in brief sallies.

Morten Strange

Order Passeriformes
Suborder Oscines
Family Maluridae
6 genera, 26 species
Family Acanthizidae
11 genera, 65 species
Family Ephthianuridae
2 genera, 5 species

SIZE
Smallest Weebill *Smicrornis brevirostris*, total length 8 to 9 centimeters (3½ inches); weight 5.5 grams (⅕ ounce).
Largest Rufous bristlebird *Dasyornis broadbenti*, total length 27 centimeters (10½ inches); weight 40 grams (1⅖ ounces).

CONSERVATION WATCH
The following species are listed in the ICBP checklist of threatened birds: purple-crowned fairy-wren *Malurus coronatus*, thick-billed grass-wren *Amytornis textilis*, gray grass-wren *A. barbatus*, Carpentarian grass-wren *A. dorotheae*, mallee emu-wren *Stipiturus mallee*, eastern bristlebird *Dasyornis brachypterus*, western bristlebird *D. longirostris*, Biak gerygone *Gerygone hypoxantha*, and chestnut-breasted whiteface *Aphelocephala pectoralis*.

FAIRY-WRENS AND THEIR ALLIES

STEPHEN GARNETT

Australasia—Australia, New Zealand, New Guinea and nearby islands—has a distinctive bird fauna developed during many millions of years of isolation. Although many species derive their common names from European birds with a similar appearance, like the wrens described in this chapter, they are in fact no relation.

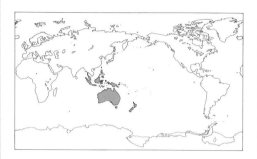

FAIRY-WRENS

The main similarities between the fairy-wrens, grass-wrens, wren-warblers and emu-wrens of Australia and New Guinea (family Maluridae) and the northern wrens is that they are all small and all cock their tails. Otherwise the two groups are quite different. Male fairy-wrens (genus *Malurus*) are among the jewels of the Australian bush, tiny creatures arrayed in stunning combinations of turquoise, red, black, and white. Their mates, and most other Australasian wrens, are more somber. Although the emu-wrens (genus *Stipiturus*), which derive their perverse name from sparse emu-like tail feathers, do have blue bibs, and some of the secretive grass-wrens (genus *Amytornis*) have bold black and chestnut colors, the overall design is for camouflage rather than display.

The family includes 26 species in six genera. All are insectivores, and most forage on the ground or among the underbrush. The wren-warblers of New

► *Alert for danger, a variegated fairy-wren fetches food for its young. As in most fairy-wrens, adult males are clad in glittering blue, females in mousy brown. The variegated fairy-wren is by far the most wide-ranging species, occupying a variety of habitats across Australia.*

Tom & Pam Gardner

Guinea and some fairy-wrens occupy rainforest, where a few species venture into the canopy, but most members of the family are found in grassland or the understory of woodland. The Eyrean grass-wren *Amytornis goyderi,* confined to the sandhills in the driest of Australia's deserts, and several other grass-wrens and emu-wrens use tussocks of spiny grass for protection.

Most members of the family are sedentary and build domed nests in dense vegetation. Like so many Australian birds, the young from one brood frequently remain with their parents to help raise later offspring. Detailed studies of species such as the superb fairy-wren *Malurus cyaneus*, a common bird even in suburban gardens in southeastern Australia, have shown that pairs with helpers are able to rear more young than those without.

AUSTRALASIAN WARBLERS

The Australasian warblers (family Acanthizidae) are a diverse group of inconspicuous small to medium-sized songbirds with slender bills and relatively long legs. The plumage is typically olive, gray or brown, but some thornbills (genus *Acanthiza*) have flashes of yellow or chestnut at the base of the tail, and many gerygones (genus *Gerygone*) have yellow underparts.

Most of the species in this group are found only in Australia, although gerygones, small birds with far-carrying tinkling calls, have spread to New Zealand, many Pacific islands and much of Southeast Asia. Gerygones are also the only migratory members of the group, most of the others remaining within a territory all year round. Typically the thornbills, whitefaces (genus *Aphelocephala*), and scrubwrens (*Sericornis*) are terrestrial, finding insects and some seeds on the ground or low in the underbrush. A few thornbill species, the weebill *Smicrornis brevirostris* and the gerygones forage in the treetops, while the origma *Origma solitaria* is confined to sandstone outcrops where it builds its domed nest in caves.

Many species breed communally, the younger members of family parties helping to find food for the nestlings and to defend the territory. Studies of marked birds have shown that adults often live more than ten years, with one striated thornbill *Acanthiza lineata* surviving at least 17. Because so many birds survive the winter there is less extra food available in the spring for breeding individuals; this may explain why the Australasian warblers usually lay only one or two eggs in a clutch, whereas in harsher climates similar-sized species often have clutches of a dozen or more.

Most forests contain birds that glean food from the bark of trees. In New Zealand this role is filled by a distinctive group, the Mohouinae (usually classified as a subfamily of Acanthizidae), consisting of the New Zealand creeper *Finschia novaeseelandiae*, the whitehead *Mohoua albicilla*, and the yellowhead *M. ochrocephala*. These are

small birds with spines at the end of their tails. Though they also take food from the leaves, much of their time is spent searching fissures in the bark for insects, sometimes supported by their tails and often hanging upside down. All three form flocks in the winter, and the yellowhead and the whitehead breed communally. The nests of these species are domed; that of the whitehead is invariably inside a tree hollow.

AUSTRALIAN CHATS

Male Australian chats (family Ephthianuridae) are boldly marked in red, orange, yellow, or black and white—colours that stand out in the swamps or arid open shrublands they inhabit—whereas the females are generally more muted versions of their mates. In only the gibberbird *Ashbyia lovensis* are the sexes a similar drab yellow. These five small species have the long legs typical of birds that spend much of their time on the ground. Though all the chats have tongues tipped with a brush, which elsewhere is usually associated with nectar-feeding, the main food of Australian chats appears to be insects. They are sociable birds, often occurring and sometimes nesting in loose flocks. The crimson chat *Epthianura tricolor* and orange chat *E. aurifrons* undertake extensive nomadic movements, sometimes erupting from central Australia to areas nearer the coast in dry years.

▲ *Two colorful Australians: the orange chat (top) and the splendid fairy-wren. Both are birds of arid interior scrublands, and in both cases females are considerably duller than the males portrayed here.*

Order Passeriformes
Suborder Oscines
Family Orthonychidae
8 genera, 21 species

SIZE
Smallest Nullarbor quail-thrush *Cinclosoma alisteri*, total length 17 centimeters (6⅗ inches); weight 20 grams (⁷⁄₁₀ ounce).
Largest Eastern whipbird *Psophodes olivaceus*, total length 30 centimeters (11⅘ inches); weight 80 grams (2⅘ ounces).

CONSERVATION WATCH
One subspecies of the northern scrub-robin *Drymodes siperciliaris colcloughi* is extinct and some species may be threatened by deforestation and habitat destruction.

▶ *Wedgebills live in loose communities in arid scrublands. They feed on the ground but also spend much time on conspicuous perches, delivering their distinctive calls.*

▼ *The logrunner inhabits rainforests in New Guinea and southeastern Australia. Almost entirely terrestrial, it sifts through leaf litter for insects with wide sideways sweeps of its powerful feet.*

LOGRUNNERS AND THEIR ALLIES

STEPHEN GARNETT

The affinities between the eight genera in the family Orthonychidae are still being examined by scientists, and field guides and other reference books give a variety of common names for them. All but one species live in Australia and New Guinea.

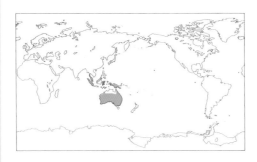

R. Drummond

DISTINCTIVE VOICES
In the classification system adopted for this book, the family includes 21 species, generally known as: logrunner and chowchilla (genus *Orthonyx*), whipbirds and wedgebills (*Psophodes*), quail-thrushes (*Cinclosoma*), scrub-robins (*Drymodes*), and in New Guinea the rail-babblers and jewel-babblers (genera *Androphobus, Eupetes, Infrita,* and *Melampitta*). One species occurs in Southeast Asia: the Malaysian rail-babbler *Eupetes macrocerus*.

The logrunner *Orthonyx temminckii* and the chowchilla *O. spaldingii* are birds of the forest floor, solid and medium-sized with stout legs for scratching in the leaf litter while they rest on strong spines extending from the ends of their tail feathers. Although hard to see in their dense habitat, they keep in touch with each other using calls of exceptional volume. Other members of the family have the same habit. The eastern whipbird *Psophodes olivaceus*, a dark olive bird of dense forest, gets its name from the whip-crack call of the male, while the chiming wedgebill *P. occidentalis* and chirruping wedgebill *P. cristatus,* identical but for their song, have ringing calls that carry a long way across the arid shrubby plains they inhabit. Typically these calls are ventriloqual. For instance, despite the strength of scrub-robin calls the birds are very difficult for a predator, or a birdwatcher, to find. Most members of the family are colored cryptically, although the complex combinations of black, chestnut, and white in the quail-thrushes of southern and inland Australia are very striking. The one really colorful species is the blue jewel-babbler *Eupetes caerulescens* of the New Guinean rainforests, the male being a vivid blue and white.

Nearly all species obtain their food on or near the ground. Most take insects, but the chowchilla, at least while nesting, specializes in leeches, sometimes carrying ten or more of these slippery, squirming prey at the one time. Different species occur in the full range of wooded habitats from Australia to Southeast Asia but reach their greatest diversity in the rainforests where the leaf litter is deep and rich. All of them appear to be sedentary, with family parties defending the same territories year-round. Most species appear to be monogamous; however, at least some, such as the logrunners, breed communally and several non-breeding individuals help to rear the brood.

MONARCHS AND
THEIR ALLIES

STEPHEN GARNETT

Order Passeriformes
Suborder Oscines
Family Rhipiduridae
1 genus, 40 species
Family Monarchidae
17 genera, 97 species
Family Petroicidae
11 genera, 42 species
Family Pachycephalidae
10 genera, 45 species

SIZE
Smallest Rose robin *Petroica rosea*, total length 10 centimeters (⅖ inch); weight 10 grams (⅖ ounce). **Largest** Asian paradise flycatcher *Terpsiphone paradisi*, total length 20 centimeters (8 inches); weight 20 grams (⁷⁄₁₀ ounce).

A lthough the four families of flycatchers and insect-eating birds in this chapter are more diverse in Australia, they also occur in New Guinea, New Zealand, and other islands in the Pacific, with a few farther afield.

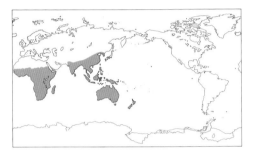

FANTAILS

Casual observation would suggest that the gray fantail *Rhipidura fuliginosa* spends more energy catching food than it actually gains from eating it. Its long tail is constantly being fanned or waved from side to side, and the little bird itself is forever pirouetting through the air in pursuit of invisible prey. Yet such behavior must be successful, because all 40 species of fantail (genus *Rhipidura*) from India to New Zealand are a similar shape and size, and most appear hyperactive. The yellow-

bellied fantail *R. hypoxantha* of India is sometimes placed in a separate genus, but the remaining fantails are all combinations of black, gray, white, or rufous, with a tail longer than their body and a short broad bill surrounded by spines called rictal bristles which are thought to increase the catching area of their mouth.

Most fantails forage in or from the foliage at all levels in the woodlands or forest they inhabit. Some are found only in mangroves. In rainforests there is a great diversity, because several species can coexist by feeding at different levels—the canopy, the middle stories, and the ground level. The Australian willie wagtail *R. leucophrys* often feeds on the ground out in the open, for which its long legs are an adaptation.

Those that live in colder parts of Australia or in the Himalayas migrate to warmer climates in the winter, the rufous fantail *R. rufifrons* from southern Australia traveling as far north as New Guinea. Fantails nest in the spring, vigorously defending territories and often rearing several broods in delicate cup-nests. The willie wagtail, one of the

CONSERVATION WATCH
The following species are listed in the ICBP checklist of threatened birds: Seychelles paradise flycatcher *Terpsiphone corvina*, cerulean paradise flycatcher *Eutrichomyias rowleyi*, short-crested monarch *Hypothymis helenae*, celestial monarch *H. coelestis*, Rarotonga monarch *Pomarea dimidiata*, Marquesas monarch *P. mendozae*, iphis monarch *P. iphis*, Fatu Iva monarch *P. whitneyi*, Rennell shrikebill *Clytorhynchus hamlini*, Truk monarch *Metabolus rugensis*, black-chinned monarch *Monarcha boanensis*, white-tipped monarch *M. everetti*, Biak monarch *M. brehmii*, Biak black flycatcher *M. atra*, Malaita fantail *Rhipidura malaitae*, red-lored whistler *Pachycephala rufogularis*, and Vogelkop whistler *P. meyeri*. The Tahiti monarch *Pomarea nigra* and Guam flycatcher *Myiagra freycineti* are close to extinction.

Tom & Pam Gardner

◄ *A rufous fantail incubating in an Australian rainforest. Fantail nests are typically slung in low thin horizontal forks, sheltered by foliage.*

▲ Three representatives of a cluster of related families that reaches its greatest diversity in Australia and New Guinea: the red-capped robin (left) inhabits dry scrublands; the spectacled monarch (center) is a bird of tropical rainforests; and the shrike-tit (right) inhabits eucalypt woodlands.

▼ The willie wagtail has markedly longer legs than other fantails, reflecting the fact that it is the only species to inhabit open country and feed largely on the ground.

Graeme Chapman

better-studied species, threatens rivals by displaying white eyebrows which in defeated contestants are reduced to a mere sliver.

MONARCH FLYCATCHERS

The family Monarchidae comprises more than 90 species in some 17 genera. They are small songbirds with a predilection for flicking their tails and occasionally raising incipient crests, which give many of the species a characteristically steep forehead. Most have long tails, though the Hawaiian elepaio *Chasiempis sandwichensis* has a cocky little tail that it generally keeps erect. Rufous, black, white, blue, or gray, often with iridescent highlights, are the most common plumage colors, but spectacular exceptions include the black and yellow *Machaerirhynchus* boatbills and the paradise flycatchers (genus *Tersiphone*) in which the males have very striking, long tails which may exceed 30 centimeters (11¾ inches) in length. Often the males are more brightly colored than the females, but in many species the sexes are similar.

Most members of the family are forest birds, but there are representatives present in most wooded habitats in Africa south of the Sahara, Southeast Asia, Australia, and islands in the Pacific. The greatest diversity is in New Guinea, where species sally for insects from the very tops of rainforest trees, others feed in the canopy, and many more live in lower parts of the forest. In the frilled monarch *Arses telescophthalmus* there is even a difference in foraging behavior between the sexes, with males searching tree trunks and females sallying after insects in flight. Monarchs often forage in mixed-species parties of birds, which sometimes contain four or five different monarch species moving through the forest together, the insects disturbed by one bird being caught by another.

Most tropical monarchs are probably resident within a territory, but those breeding in temperate regions—such as some populations of the Asiatic paradise flycatcher *Terpsiphone paradisi,* the satin flycatcher *Myiagra cyanoleuca,* and the black-faced monarch *Monarcha melanopsis* of southeastern Australia—migrate to the tropics in the winter. In temperate areas, breeding is restricted to spring, but it is extended in the tropics. Their nests consist of delicate cups sometimes elaborately decorated with lichen.

AUSTRALASIAN ROBINS

The term "robin" has been applied to birds with red breasts all over the world, but the flame, scarlet, rose, and pink robins of Australasia are related to neither their European nor American namesakes. Even more incongruous is the use of the name for other dumpy little flycatchers in the same group; yellow, dusky, and black robins have no red on them at all. Nevertheless they do share many features of the European robin *Erithacus rubecula* in shape, size, and endearing nature.

The Australasian robins are currently classified in the family Petroicidae, comprising 42 species in 11 genera. They are mostly birds of woodland and forest where, rather than search actively, they like to sit and wait for prey to reveal itself. Most catch insects on the ground, although the mangrove robin *Eopsaltria pulverulenta* will take shrimps and small crabs. A few species are more energetic—the lemon-breasted flycatcher *Microeca flavigaster* will glean food from foliage and sally after flying insects, and the Jacky Winter M. *leucophaea* sometimes hovers.

Most members of the family are sedentary, but some of the robins in southern Australia migrate north from Tasmania or out of the mountains during winter. When breeding, the flame robin *Petroica phoenica* and the scarlet robin *P. multicolor* exclude each other from their individual territories; when flame robins migrate, the territories of the sedentary scarlet robins expand. Breeding in temperate areas is in spring, and the nests are usually built with bark, moss, and lichen placed in forks of trees.

The majority of species are common but some are restricted to islands and have become seriously endangered. In New Zealand the Chatham Island robin *Petroica traversi* was reduced to seven birds by 1976 when the population was moved from Little Mangere Island, where the habitat was thought to be highly degraded, to Mangere Island, where the habitat was thought to be better; two of the seven died, but with the help of an innovative fostering program using the closely related Chatham Island tit *P. macrocephala chathamensis*

and transfers to a third island, South East Island, the population was 119 by the end of the 1989/90 breeding season.

WHISTLERS, SHRIKE-THRUSHES, AND THEIR ALLIES

Ten genera with 45 species are included in this family (Pachycephalidae), which is centered on New Guinea and Australia. The whistlers, shrike-thrushes, and their relatives are stout flycatchers with sturdy bills, and many have lovely calls. Most are rufous, brown or gray, but among them are some yellow, black and white birds—the shrike-tit *Falcunculus frontatus* and the male golden whistler *Pachycephala pectoralis.* A few, such as the shrike-tit, the crested bellbird *Oreoica gutturalis,* and the crested pitohui *Pitohui cristatus* have crests, but most are unadorned. Most species feed on insects gathered from leaves and branches of the forest or from among the leaf litter, with little of the frenetic activity characteristic of fantails. There is some variation in diet: shrike-tits use their powerful bill to search under bark; the white-breasted whistler *Pachycephala lanioides,* an Australian mangrove species, eats fiddler crabs; and the mottled whistler *Rhagologus leucostigma* of New Guinea eats fruit.

Banding studies suggest that the tropical members of the family are largely sedentary, so pairs occupy a territory all year round. The rufous whistler *Pachycephala rufiventris,* however, migrates north after breeding in southeastern Australia, and individuals of the golden whistler certainly disperse over enormous distances. The golden whistler shows more geographical variation than any other bird species, with more than 70 races having been described, mostly from isolated populations on the Pacific islands.

Hans & Judy Beste/Auscape International

▲ *A frilled monarch on its nest. This monarch inhabits rainforests of New Guinea and Cape York Peninsula, Australia.*

▼ *The white-browed robin is a sedentary species that favors vine scrub and palm thickets along streams in tropical Australia.*

Tom & Pam Gardner

TITS

C. PERRINS

Order Passeriformes
Suborder Oscines
Family Aegithalidae
3 genera, 7 species
Family Remizidae
4 genera, 10 species
Family Paridae
3 genera, 46 species

SIZE
**Long-tailed tits
(Aegithalidae)** total length
10 to 12 centimeters
(4 to 4¾ inches).
Penduline tits (Remizidae)
total length 8 to 11
centimeters
(3⅕ to 4¼ inches).
**Tits, chickadees, and
titmice (Paridae)** total length
10 to 22 centimeters
(4 to 8⅗ inches).

CONSERVATION WATCH
The white-winged tit *Parus
nuchalis* and the yellow tit
P. holsti are listed in the ICBP
checklist of threatened birds.

▼ *The tail accounts for just over half of
the long-tailed tit's total length of 14
centimeters (5 inches).*

The three families commonly grouped under this heading—the Aegithalidae, the Remizidae, and the Paridae — are widespread, and consist for the most part of small forest-dwelling birds. Some are familiar urban dwellers; the great, blue, and coal tits are among the most intensively studied species in the world.

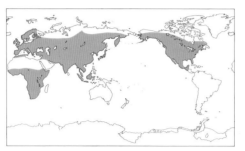

LONG-TAILED TITS
The seven species of long-tailed tits, family Aegithalidae, include the bushtit *Psaltriparus minimus* of western North America and Mexico, and the long-tailed tit *Aegithalos caudatus* which has a very wide range from Britain across Europe and Asia to Japan. The other five species live in the Himalayas and associated mountain ranges. All are tiny, weighing only 6 to 8 grams (¼ ounce) and are 10 to 12 centimeters (4 to 4¾ inches) long, including their long tail. They are mostly rather dull brownish or grayish birds, although the long-tailed tit has bright pink in its plumage and some races have a white head. They have small bills, and feed on small insects and tiny seeds. All species make a beautifully constructed, domed nest, woven from cobwebs and lichens and lined with feathers. Nests are well hidden in thick foliage.

PENDULINE TITS
The penduline tits, family Remizidae, comprise ten species, seven of them (genus *Anthoscopus*) confined to Africa, where they inhabit woodlands. Of the others, the penduline tit *Remiz pendulinus* is widespread across Europe and Asia at low latitudes, where it lives in marshes and along riverbanks; in areas where the winter is particularly severe, it migrates southward. The verdin *Auriparus flaviceps* occurs in the southwestern USA and northern Mexico in dense desert scrub. The tenth species, the fire-capped tit *Cephalopyrus flammiceps*, lives in high-altitude forest in the Himalayas. Most species are largely sedentary. They have small bills and feed on insects and small seeds; the penduline tit takes many seeds from reed mace.

They are mostly rather dull in plumage, although the male penduline tit is quite smartly plumaged, and the male fire-capped tit has, as its name suggests, a bright orange head and throat. They build intricately woven purse-like nests, often suspended from the outer twigs of a bush. The Cape penduline tit *Anthoscopus minutus* is noted for building a nest entrance whose lips can be shut when the bird is in or away from the nest, and to make it even harder for a potential predator, there is a false hole beneath the entrance which is a dead end.

TITS, CHICKADEES, AND TITMICE
The family Paridae comprises 46 species of tits, chickadees, and titmice, all but two in the genus *Parus*. The exceptions are the yellow-browed tit *Sylviparus modestus*, a dull greenish bird with a tiny patch of yellow above the eye, which occurs throughout the Himalayas, and the sultan tit *Melanochlora sultanea*, a black bird 22 centimeters (8½ inches) long, with a striking yellow crest and and yellow underparts which occurs in low-altitude forest from Nepal to the island of Sumatra in Indonesia. Few exceed 13 centimeters (5 inches) in length. They tend to be brown, gray or green above, and paler or yellow underneath. The majority have black caps, some of them with crests, and white cheeks.

The azure tit *Parus cyanus* and blue tit *P. caeruleus* are predominantly blue above.

Species of *Parus* occur throughout most of Europe, Asia, Africa, and North America down into Mexico, but not in the rest of Central America, nor in South America or Australasia. Most species are largely sedentary, although the most northerly populations of species such as the black-capped chickadee *P. atricapillus* in Canada and the great tit *P. major* in northern Europe may move considerable distances to areas with milder winters. They are primarily forest birds. They feed on a wide range of insect and seeds, although all bring insect food to their nestlings; many are seed-eaters through the colder parts of the year. They have short, straight bills; in many species these are slightly stubby and capable of hammering open

small nuts. Those species that live in conifer forests have finer bills, probably associated with their habit of probing into clusters of needles.

All species nest in holes, usually in trees, although some will nest in holes in the ground or among piles of rocks. Some species depend on holes left by natural causes or excavated by woodpeckers, whereas others excavate their own in rotten wood. Although the clutches may be as small as three eggs in tropical species, some of the tits in temperate climates lay large clutches. The blue tit lays the largest clutch of any bird that raises its young in the nest; in oak woodland in central Europe, the average number is 11 eggs, but clutches of 19 are recorded occasionally. Being hole-nesters, some species readily accept nest-boxes and as a result have been intensively studied.

▲ *Wing-feathers spread for braking and feet thrown forward for landing, a blue tit approaches its cavity nest with food for its young. Active, hardy, enterprising, and entertaining, the blue tit is a popular visitor at garden bird-feeders across Europe.*

NUTHATCHES AND TREECREEPERS

C. PERRINS

Order Passeriformes
Suborder Oscines
Family Sittidae
3 genera, 26 species
Family Certhiidae
2 genera, 6 species
Family Rhabdornithidae
1 genus, 2 species
Family Climacteridae
1 genus, 7 species

SIZE
Nuthatches and sitellas (Sittidae) total length 10 to 18 centimeters (4 to 7 inches).

Treecreepers (Certhiidae, Rhabdornithidae, and Climacteridae) total length 12 to 18 centimeters (4¾ to 7 inches).

CONSERVATION WATCH
The following species are listed in the ICBP checklist of threatened birds: white-browed nuthatch *Sitta victoriae*, Algerian nuthatch *S. ledanti*, Yunnan nuthatch *S. yunnanensis*, yellow-billed nuthatch *S. solangiae*, giant nuthatch *S. magna*, beautiful nuthatch *S. formosa* and long-billed creeper *Rhabdornis grandis*.

▼ *A Eurasian treecreeper flies from its nest. Built by both parents, treecreeper nests are usually hidden behind loose bark on a tree-trunk.*

The family Sittidae is divided into two groups: the Eurasian nuthatches (subfamily Sittinae) and the Australasian sitellas (subfamily Neosittinae). Of the 22 species of nuthatches (genus *Sitta*), most occur in Eurasia, and there are four species in North America. The wallcreeper *Tichodroma muraria* is usually considered to be an aberrant member of this family. The other three families in this group—the Certhiidae, the Climacteridae, and the Rhabdornithidae—are at least superficially similar to each other and are called creepers.

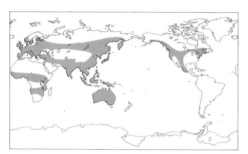

NUTHATCHES AND SITELLAS

Typical nuthatches are stocky small birds, with longish straight bills. The plumage is typically blue-gray above, and white, pale gray or reddish brown below; some have a dark cap, and some have a dark streak through the eye area. They are unique in having the ability to run up and down tree trunks, which they do while foraging for insects hidden in the bark. Outside the breeding season many species also take seeds, some specializing on these in winter. Most nest in holes in trees or rocks, often reducing a large entrance to one that they can only just squeeze through, by walling it with mud, dung, and other sticky substances which dry hard.

The three species of sitellas occur in Australia and New Guinea. Although they seem similar to nuthatches in the way in which they clamber about in trees, they may not be closely related. They do not nest in holes, but build a beautifully woven cup-shaped nest in the fork of a tree, and camouflage it with pieces of bark attached to the outside. In Australian woodlands, the varied sitella *Daphoenositta chrysoptera* has half a dozen different forms, the colors and patterns varying according to the geographical location; they are now generally recognized as races within the one species.

The wallcreeper inhabits mountain ranges in Europe and Asia, from the Pyrenees to the Himalayas. It nests at high altitudes adjacent to running water, but descends to lower altitudes for the winter. It has bright red in the wings and a long, decurved bill with which it feeds on insects.

TREECREEPERS

The most widespread family is the Certhiidae, the Holarctic treecreepers, which includes the treecreeper *Certhia familiaris* of Eurasia and the brown creeper *C. americana* of North America; three other species are distributed in the Himalayas and associated mountain ranges. These small brownish birds use their stiffened tail feathers as props for supporting themselves as they climb up trees. The spotted creeper *Salpornis spilonotus* of Africa and India is sometimes placed in a separate subfamily, Salpornithinae.

The Australian creepers (family Climacteridae) comprise seven species, six of them occurring only in Australia and the seventh in New Guinea. The Philippine creepers (family Rhabdornithidae) comprise only two species, both of which are confined to the Philippines.

Stephen Dalton/NHPA

HONEYEATERS AND THEIR ALLIES

TERENCE LINDSEY

F lowerpeckers, sunbirds, white-eyes, and honeyeaters are families of small tree-dwelling birds distributed in Africa, tropical Asia, Australasia, and islands in the southwestern Pacific. The diet of many species includes nectar ("honey") taken from flowers.

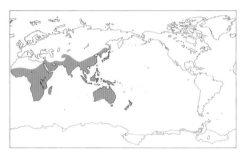

FLOWERPECKERS

The flowerpeckers constitute a family, the Dicaeidae, of about 55 species widespread in southern Asia and Australasia. The family is rather uniform, and at least 40 species are grouped in only two widespread genera (*Prionochilus* and *Dicaeum*). Most flowerpeckers are dumpy little birds with long pointed wings, stubby tails, and short conical bills. The tongue has a characteristic structure, short and deeply cleft, with the edges curled upwards to form twin partial tubes, thought to be an adaptation for gathering nectar from flowers. Many species have dull plumages, but the males of many others have bright colors and patterns, often involving areas of crimson and glossy blue-black. Almost all are strictly arboreal and feed largely on nectar, fruits, and insects; they inhabit a wide range of environments, from dense rainforest and highland moss forest to arid savanna woodland.

Order Passeriformes
Suborder Oscines
Family Dicaeidae
6 genera, 55 species
Family Nectariniidae
5 genera, 116 species
Family Zosteropidae
10 genera, 85 species
Family Meliphagidae
42 genera, 167 species

Left to right: flowerpecker, sunbird, white-eye, honeyeater.

SIZE
Flowerpeckers (Dicaeidae)
total length 7 to 19 centimeters (2¾ to 7½ inches).
Sunbirds (Nectariniidae)
total length 9 to 21 centimeters (3½ to 8¼ inches).
White-eyes (Zosteropidae)
total length 10 to 14 centimeters (4 to 5½ inches).
Honeyeaters (Meliphagidae)
total length 7 to 50 centimeters (2¾ to 19¾ inches).

CONSERVATION WATCH
There are 41 species listed in the ICBP checklist of threatened birds. They include the following: brown-backed flowerpecker *Dicaeum everetti*, forty-spotted pardalote *Pardalotus quadragintus*, Rockefeller's sunbird *N. rockefelleri*, Gizo white-eye *Zosterops luteirostris*, white-breasted white-eye *Z. albogularis*, Mauritius olive white-eye *Z. chloronothus*, great Ponape white-eye *Rukia longirostra*, rufous-throated white-eye *Madanga ruficollis*, crimson-hooded honeyeater *Myzomela kuehni*, Bonin Islands honeyeater *Apalopteron familiare*, stitchbird *Notiomystis cincta*, Kauai o'o *Moho braccatus*, and bishop's o'o *M. bishopi*.

Morten Strange

◄ *The orange-bellied flowerpecker is widespread in lowland forests of Southeast Asia.*

Morten Strange/Flying Colours

▲ *Vivacious and intensely active like most sunbirds, the crimson sunbird inhabits India and Southeast Asia. It often hovers to extract nectar directly from small blossoms, but with large flowers such as hibiscus and cannas it usually gains access by piercing the corolla near the base. The call has been likened to a pair of scissors being snapped open and shut.*

This family is of particular interest because of the almost symbiotic relationship that exists between certain species and the plants (mainly mistletoes) upon which they feed. A good example is the mistletoebird *Dicaeum hirundinaceum* of Australia, which feeds almost entirely on the berries of various species of mistletoes. Its gut is modified to form little more than a simple tube, from which the muscular stomach diverts as a blind sac reached through a sphincter. This arrangement allows berries to bypass the stomach entirely; the fruit pulp is absorbed and digested in passage, but the central seed travels through undamaged, to be voided intact but with a sticky coating. The seed sticks to the bark on the branches of trees, germinates, and in due course forms a new plant.

The relationship thus provides the bird with a reliable source of food, and the plant with an effective means of dissemination. So intimate is this partnership that one is seldom found where the other does not also occur.

Most flowerpeckers build pendent globular nests of closely-matted vegetable fiber, placed in shrubs and saplings, but the pardalotes (genus *Pardalotus*), whose four or five species are found only in Australia, form a conspicuous exception in that they nest in tunnels in the ground, or in tree cavities. Pardalotes also differ from flowerpeckers in their extremely intricate plumage patterns; most species have a bright yellow or red rump, and a black crown which is minutely spotted or streaked white. Pardalotes also exhibit an intimate relationship with their food supply: they feed almost entirely on the larvae (and an exudate, known as lerp, produced by these larvae) of leaf-eating insects found on the foliage of eucalyptus trees. Some researchers feel that pardalotes should be afforded family rank, "Pardalotidae".

SUNBIRDS

Strongly associated with flowers wherever they occur, sunbirds constitute a family of about 116 species of small, vivacious, extremely active, colorful birds found throughout much of the Old World tropics. The group is best represented in Africa, where most habitats support at least one or two species, but one sunbird extends as far east as Australia. Most species belong to a single genus, *Nectarinia*.

Characteristic features include a long, slender, decurved bill with fine serrations along the margins of both mandibles, and a tubular, deeply cleft tongue. A few species are dull, but more typically the male is brilliantly colored, often glittering vivid green or blue above, and red or yellow below. Many species have long pointed extensions on the central tail feathers, and some have brightly colored tufts of feathers on the sides of the breast.

Sunbirds feed on small insects and nectar, and in Africa at least it seems probable that many flowers are dependent on sunbirds for pollination. The birds sometimes congregate at flowering trees or shrubs, but they are not truly gregarious—males in particular often vigorously defend favored flowers against all other sunbirds. Their calls are usually abrupt and metallic, but many species have rapid high-pitched twittering songs. Sunbirds breed as monogamous pairs and build elaborate pendent domed nests; the male does not help in nest-building or incubation but usually assists in rearing the young.

Half a dozen species inhabiting the rainforests of Southeast Asia, especially in Malaysia, are very distinct from other sunbirds: their plumage is mainly dingy brown, they have long, strong, downcurved bills, they build cup-shaped nests, and both sexes incubate. Usually called spider-hunters, these species are grouped in the genus *Arachnothera*.

WHITE-EYES

About 85 species of white-eyes occur in Africa and across southern Asia to Australasia, extending north to Japan, and east to the islands of Samoa. They occupy almost every kind of wooded habitat and are often common in suburban parks and gardens. The great majority form a single widespread genus, *Zosterops,* whose members are so uniform in appearance and structure that species limits are extremely difficult to determine (*Zosterops* is among the largest of all avian genera). Most species are about 10 to 12 centimeters (4 to 4¾ inches) long, unremarkable in

proportions, with rather slender pointed bills. The plumage is usually green above, pale gray or yellow below, with a distinct ring of tiny, dense, pure-white feathers around each eye. The tongue is brush-tipped, and the birds feed on nectar, fruit, and small insects. White-eyes are also characterized by an almost complete lack of any sexual, age, or seasonal variation in plumage.

Most are gregarious, living in wandering groups from which pairs periodically drop out in order to breed, returning to the flock when done. It is known that several species mate for life, and this may be typical of the group as a whole. These characteristics of sociability and a strongly nomadic tendency confer great powers of dispersal, and white-eyes are successful colonizers of remote oceanic islands. Most islands in the Indian Ocean and western Pacific are occupied by one or another species, and some have two. Sometimes these share the same ancestral form, a result that may occur when the interval between arrivals is sufficiently long to render the original colonists incapable of interbreeding with the new arrivals. Remarkably, two islands in the southwestern Pacific (Norfolk Island, and Lifou Island near New Caledonia) are the site of triple invasions of this kind: on both islands there are three distinct species, all originating from the same ancestral stock.

Graeme Chapman

▲ The yellow white-eye is a mangrove-inhabiting bird of the coasts of tropical Australia.

▼ The regal sunbird (top) is restricted to high mountain ranges in central Africa. The spotted pardalote (bottom) inhabits eucalypt woodlands in southern Australia.

▲ Two Australian honeyeaters: the eastern spinebill (above) is abundant in thickets of flowering shrubs along the east coast and in Tasmania, and the regent honeyeater (right) is an uncommon and perhaps declining nomad that inhabits dry eucalypt woodlands on the western slopes of the Great Dividing Range.

▼ Coastal heaths of southern Australia are the home of this species, the New Holland honeyeater.

HONEYEATERS

The family Meliphagidae comprises about 167 species of honeyeaters, friarbirds, spinebills, miners, and wattlebirds, all largely restricted to the Australasian region. New Zealand, New Caledonia, and other islands in the southwestern Pacific each have a few species, but most family members inhabit New Guinea (about 63 species) and

Jean-Paul Ferrero/Auscape International

Australia (about 68 species). A prominent characteristic of the family is the unique structure of the tongue, which is deeply cleft and delicately fringed at the tip so that it forms four parallel brushes, an adaptation to nectar-feeding. Otherwise honeyeaters vary very greatly in size, structure, and general appearance. All are chiefly arboreal and normally gregarious, but there are very few features common to all.

Some species inhabit rainforest, and these tend to feed mainly on fruit. Many others rely primarily on nectar; and one large genus (*Myzomela*, with 24 species) shows some striking similarities to sunbirds (family Nectariniidae), being very small, vivacious, pugnacious, brightly colored (in the males), and strongly associated with flowering trees and shrubs. But many others rely heavily on insects, and these tend to vary more widely in habitat, general appearance, and behavior.

The honeyeater family forms one of the most prominent elements of the Australian songbird fauna, and it is there that the group reaches its greatest diversity, inhabiting virtually all habitats from highland rainforest to coastal heath and arid scrub in the interior. A birdwatcher taking a casual stroll at any time in any Australian habitat can hardly fail to see many individuals of several species of honeyeaters. Some, such as the New Holland honeyeater *Phylidonyris novaehollandiae,* its close relatives, and the spinebills (genus *Acanthorhynchus*), inhabit coastal heaths and flowering shrubs, where they are colorful and conspicuous in appearance and behavior. Insect-eating forms in woodland and forest are often duller in plumage and less conspicuous in behavior but no less numerous; some glean insects from foliage, while others probe the bark of trees for food. Some species are very specialized in diet and habitat, whereas others are generalists, feeding almost indiscriminately on fruit, nectar and insects in a range of habitats.

Honeyeaters are also notable in the range of social behavior they exhibit. Some are gregarious only in the limited sense that they tend to congregate wherever food is abundant, but many others live in permanent structured communities. In the miners (genus *Manorina*), this communalism reaches extraordinary sophistication. The situation has been best studied in the noisy miner *Manorina melanocephala,* which usually lives in loose communities of several hundred individuals occupying permanent territories many acres in extent; the communal territory is vigorously defended by all members against all avian trespassers (itself a feature virtually unique among birds), but each individual is at the same time a member of a smaller subgroup within the larger community. These subgroups are themselves distinct communities, each consisting of an adult female with several male consorts, which constitute the basic breeding units within the community.

VIREOS

KENNETH C. PARKES

Order Passeriformes
Suborder Oscines
Family Vireonidae
4 genera, 43 species

SIZE
Total length 10 to 16
centimeters (4 to 6¼ inches);
weight 8 to 40 grams
(³⁄₁₀ ounce to 1½ ounces).

CONSERVATION WATCH
The black-capped vireo *Vireo
atricapillus* is listed in the
ICBP checklist of threatened
birds.

V ireos belong to a strictly New World family, Vireonidae, whose relationships to other songbirds are uncertain. The family was once considered to belong to the large group of "New World nine-primaried oscines"—songbirds in which the tenth (outermost) primary feather of the wing is reduced or absent—but recent research places the vireos well outside this assemblage. Their true nearest relatives are as yet unknown.

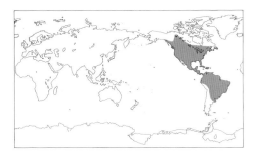

SMALL GRAY-GREEN SONGBIRDS

The 25 typical vireos (genus *Vireo*) are widely distributed in North, Central, and South America. They are small, and most are inconspicuously colored in combinations of dull green, brown, yellow, gray, and white. Of the 11 species that breed as far north as the USA, all but one (Hutton's vireo *Vireo huttoni*, of the western states south to Mexico and Guatemala) are strongly migratory. The most widely distributed species, the red-eyed vireo *Vireo olivaceus*, avoids winter by migrating as far south as Amazonian Brazil, where it shares the equatorial forests with resident vireos. Several months later, others from southern South America

migrate northward to the same equatorial forests in the Southern Hemisphere's autumn.

The smallest members of the family are the greenlets (13 species, in the genus *Hylophilus*), which live in forest or scrub, only in tropical regions of Central and South America. Peppershrikes (two species, in the genus *Cyclarhis*) and shrike-vireos (three species, in the genus *Vireolanius*) are quite distinctive, and are classified in their own subfamilies. These are the largest vireos, and are found from Mexico south through tropical South America. The peppershrikes have heavy shrike-like bills. The bills of the shrike-vireos are not as heavy, but this subfamily includes the most brightly colored members of the family.

Vireos are found in a variety of habitats. Some prefer the forest canopy, while others inhabit dense tangled undergrowth, forest edges, or mangroves. They are compulsive singers, often heard during the heat of the day when most songbirds are silent. Vireos are primarily insect-eaters, but many add fruits to their diet. Their nests, whether located high in trees or in low shrubs, are almost always woven cups, suspended by their rims from forked branches. Their eggs are white with small spots of various colors.

John S. Dunning/Ardea Photographics

◄ *Striking in appearance but inconspicuous in behavior, the slaty-capped shrike-vireo inhabits tropical lowland forests of South America, favoring the vicinity of streams. It occupies the middle and upper levels of foliage, and frequently joins wandering mixed-species foraging parties of other birds.*

Order Passeriformes
Suborder Oscines
Family Emberizidae
134 genera, 560 species

SIZE
One of the smallest Ruddy-breasted seedeater *Sporophila minuta*, total length 9 centimeters (3½ inches); weight 8 grams (⅓ ounce).
Largest White-capped tanager *Sericossypha albocristata*, total length 24 centimeters (9½ inches); weight 114 grams (4 ounces).

CONSERVATION WATCH
There are 49 species listed in the ICBP checklist of threatened birds. They include the following: Sierra Madre sparrow *Xenospiza baileyi*, grosbeak bunting *Nesospiza wilkinsi*, marsh seedeater *Sporophila palustris*, Floreana tree-finch *Camarhynchus pauper*, yellow cardinal *Gubernatrix cristata*, black-cheeked ant-tanager *Habia atrimaxillaris*, sooty ant-tanager *H. gutturalis*, black-backed tanager *Tangara peruviana*, white-bellied dacnis *Dacnis albiventris*, and scarlet-breasted dacnis *D. berlepschi*.

BUNTINGS AND TANAGERS

LUIS F. BAPTISTA

The family Emberizidae comprises five subfamilies, which formerly were treated as separate families: Emberizinae, Catamblyrhynchinae, Cardinalinae, Thraupinae, and Tersininae. These names are used in the text below because the common names, "buntings", "sparrows", and "finches", apply to birds in different groups in the New World and Old World. All members of the family have wings with nine primary feathers, although a vestigial tenth primary may be present.

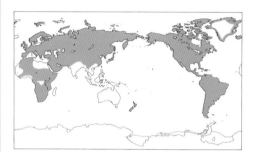

EMBERIZINAE

The Emberizinae includes 279 species in 65 genera, which are thought to have originated in the New World and then dispersed to the Old World in several separate colonizations. The subfamily includes the "sparrows" and various "finches" of the New World, and the "buntings" of the Old World. These birds are small, averaging about 15 centimeters (6 inches) in length, and all have short, conical, attenuated finch-shaped bills reflecting

► *A male yellowhammer at its nest. The female alone incubates the eggs, but both parents cooperate in rearing the young. A typical Old World bunting, the yellowhammer is common in open, scrubby habitats across Europe and western Asia. Strongly gregarious when not breeding, it often joins flocks of other buntings and finches in winter.*

Brian Bevan/Ardea London Ltd

Wardene Weisser/Ardea Photographics

Wayne Lankinen/DRK Photo

▲ The white-crowned sparrow inhabits grassy clearings and open woodland across northern Canada and alpine meadows in the Rocky Mountains.

their seed-eating habits. They are mostly terrestrial but may occupy a variety of habitats, including grassland, brushy areas, forest edge, and marsh.

In the Old World there are 42 species, about 28 of which breed in Europe. These include the typical buntings of the genus *Emberiza* which are Eurasian and Afrotropical in distribution, and the Lapland longspur *Calcarius lapponicus* and snow bunting *Plectrophenax nivalis* which breed in Arctic America and Siberia. The latter species breeds in northern Greenland, farther north than any other land bird. Most species tend to have brown streaked body plumage. The pointed crest and glossy blue-black plumage of the male crested bunting *Melophus lathami* and the dark-blue plumage of the Chinese blue bunting *Latoucheornis siemsseni* are unusual in the group.

In the New World there are 234 species in about 60 genera. The sparrows (especially in the genera *Zonotrichia* and *Ammodramus*) are mostly brown and streaked. But many species are very attractively colored—for example, the rufous and green of some towhees (genus *Pipilo*) and brush-finches (genus *Atlapetes*), and the bright orange-yellow of the saffron finches (genus *Sicalis*) of tropical Central and South America.

Emberizines are generally seed-eaters but often switch to a diet of insects when feeding their young. In a few species of juncos (genus *Junco*) and crowned-sparrows (genus *Zonotrichia*) of the New World, bill lengths have been shown to change with the seasons, being shorter when feeding on seed. Some of the New World genera (for example, *Melospiza, Pipilo, Zonotrichia*) feed on the ground with a "double kick"—they jump back with both feet at once, raking the soil to reveal a tasty morsel. None of their Old World cousins feeds in this manner. Birds in the grasslands of tropical Central and South America (for example *Sporophila, Tiaris, Geospiza*) may clamp food items onto a perch with their feet and then pull off pieces with their bill.

Most emberizines build a cup-shaped nest of grasses, roots and other plant fibers. However, members of several genera in tropical America (*Tiaris, Loxigilla, Geospiza, Melanospiza*) build domed nests; and the Cuban grassquit *Tiaris canora* often constructs a long tubular entrance to its nest. Most species tend to be monogamous or occasionally bigamous. Biochemical studies of blood proteins have revealed that in the white-crowned sparrow *Zonotrichia leucophrys,* which was thought to be monogamous, females engage in "extramarital affairs" so that some 40 percent of their offspring are "illegitimate".

With a few exceptions—for example, the song sparrow *Melospiza melodia* and the five-striped sparrow *Aimophila quinquestriata*—most species tend to possess small song repertoires consisting of short simple utterances. A few species are well known for their regional song dialects—for example, the ortolan bunting *Emberiza hortulana* in the Old World, the white-crowned sparrow in

▲ The red-legged honeycreeper (top) is widespread from Mexico to central Brazil. It is a common, vivacious, and gregarious forest bird that feeds on nectar, fruit, and small insects. The superb tanager (right) is confined to the forests of Brazil. The painted bunting (bottom) breeds in thickets and weedy tangles across the southeastern United States from Texas to North Carolina; it winters in Central America. In the superb tanager the sexes are similar, but in the other two species the female is very much duller than the male.

North America, the rufous-crowned sparrow *Zonotrichia capensis* of tropical America, and the 14 Galapagos finches (genera *Geospiza, Cactospiza, Platyspiza, Camarhynchus, Certhidea, Pinaroloxias*) made famous by Charles Darwin. The *Melospiza* and *Zonotrichia* sparrows have been used in extensive studies of song-learning, which have helped scientists to understand the relative roles of learning versus inheritance in vocal traditions.

CATAMBLYRHYNCHINAE

The plush-capped finch *Catamblyrhynchus diademata* of the Andes of South America is a finch-like bird about 15 centimeters (6 inches) long, and its stiff erect golden-brown crown feathers give it its name. This species favors bamboo groves, where it forages on insects and vegetable matter; its bill is similar in shape and structure to some of the Old World parrot-billed babblers (genus *Paradoxornis*), which also are bamboo specialists. Little is known of the plush-capped finch's life history. In some classification systems it is placed with the tanagers in the subfamily Thraupinae.

CARDINALINAE

This group consists of 39 species in 9 genera, and is distributed throughout the Americas. Some, notably the "buntings" (genus *Passerina*), are very colorful. The males may have much brighter plumage than the females—for example, the

cardinals (genus *Cardinalis*), the buntings (*Passerina*) and some of the grosbeaks (*Pheucticus*); but in the 12 *Saltator* species of Central America, the West Indies, and South America, both sexes are similar in color, being mostly green and brown.

All cardinalines are thick-billed finches, and they differ in structure of palate, tongue and jaw musculature from the more slender-billed emberizines, described earlier. The former are adapted to crush seeds, whereas the latter tend to be seed-peelers, removing the husks and swallowing the kernels whole. The diet of cardinalines consists of a mixture of fruit, grain, and insects; the saltators tend to feed on berries, fruit, and insect larvae.

All species build cup-shaped nests placed in trees or bushes. Males of the black-headed grosbeak *Pheucticus melanocephalus* and rose-breasted grosbeak *P. ludovicianus* assist the female in incubation and have the peculiar behavior of singing while sitting on the eggs. The cardinal *Cardinalis cardinalis* is well known for its regional song dialects.

Some of the species breeding in North America, such as the dickcissal *Spiza americana* and the indigo bunting *Passerina cyanea,* winter in Central America; and the latter has been the subject of fascinating studies on various aspects of migration, especially orientation using the stars and the Earth's magnetic field.

THRAUPINAE

The tanagers, about 240 species in 58 genera, are entirely New World in distribution. Four species (genus *Piranga*) breed in the USA and migrate to the tropics in the fall; some 163 species are confined to South America. The Andes is the center of radiation for the group. Most (about 149 species) are forest-dwellers, whereas others (about 54 species) tend to prefer semi-open areas; the rest do not show obvious habitat preferences.

Although a few species are drab and secretive, the tanagers include some of the most colorful of all birds, especially in the genus *Tangara*. Most are primarily fruit-eaters but they also take insects. Some such as the honeycreepers (genera *Cyanerpes* and *Dacnis*) and flower-piercers (*Diglossa*) are nectar specialists and have long thin delicate bills. A few (for example, the genera *Lanio* and *Habia*) are insect specialists, and have heavy bills equipped with notches to better grasp insects. Some tanagers are noted for their behavior of following troops of army ants that stir up insects in their wake, and others are known to participate in mixed-species flocks. Some species have attractive songs. Upon seeing a potential predator, the thick-billed euphonia *Euphonia laniirostris* imitates the alarm calls of other songbirds, for example the variable seedeater *Sporophila aurita*. This attracts the latter species to come and harrass ("mob") the predator, leaving the euphonia to go about its

business and avoid a dangerous situation.

Most tanagers build cup-shaped nests placed among moss or dead leaves. Others, such as the *Euphonia* and *Chlorophonia* species, build globular nests. The female does most of the nest-building and incubation. Some males may feed the incubating female, and both sexes feed the young. Incubation is shorter (10 to 13 days) for the open cup-nesters, which are subject to higher predation rates, and longer (18 to 24 days) for the *Euphonia* species, which build domed camouflaged nests. Breeding pairs and their broods form small post-breeding flocks of up to a dozen individuals. Five species are known to have helpers at the nest.

TERSININAE

The swallow tanager *Tersina viridis,* which is merged with the thraupine tanagers in some classification systems, occupies a range from Panama to the northern parts of South America and Trinidad. It is a bird of montane areas and is migratory in at least part of its range. It is about 15 centimeters (6 inches) long, with longer wings and shorter legs than the thraupine tanagers, and is endowed with a wide, flattened bill, slightly hooked at the tip, and a distinct palate. The bill is adapted for digging and for capturing insects on the wing in the manner of swallows (hence its name). It also eats fruit which it plucks from trees; large pulpy fruit is preferred to small berries, which are rarely taken.

Courtship consists of a unique curtsey display

which involves flattening of the head feathers, lowering and quivering of wings, and bowing of male and female to each other. This display is also used in aggressive encounters, notably with territorial neighbors. Song is delivered only by the male, mostly during the nest-building phase. A cup-shaped nest is placed in a horizontal burrow excavated in a bank or in a hole in a wall. The swallow tanagers may also take over abandoned burrows. The female alone incubates the three eggs, but the male may assist in feeding the young.

▲ *The paradise tanager is a bird of the treetops in South American tropical forests.*

▼ *The Brazilian tanager is confined—as its name suggests—to Brazil, where it inhabits forest edges and clearings.*

WOOD WARBLERS AND ICTERIDS

KENNETH C. PARKES

Order Passeriformes
Suborder Oscines
Family Parulidae
22 genera, 114 species
Family Icteridae
25 genera, 92 species

SIZE

Wood warblers (Parulidae)
Smallest Lucy's warbler, *Vermivora luciae*, total length 10 centimeters (4 inches); weight 6 grams (⅕ ounce).
Largest Yellow-breasted chat *Icteria virens*, total length 19 centimeters (7½ inches); weight 33 grams (1 ounce).

Icterids (Icteridae)
Smallest Orchard oriole *Icterus spurius*, total length 15 centimeters (6 inches); weight 18 grams (⅗ ounce).
Largest Olive oropendola *Psarocolins bifasciatus*, total length 52 centimeters (20½ inches); weight 445 grams (15⅗ ounces).

CONSERVATION WATCH
There are 25 species listed in the ICBP checklist of threatened birds. They include the following: Bachman's warbler *Vermivora bachmanii*, golden-cheeked warbler *Dendroica chrysoparia*, Kirtland's warbler *D. kirtlandii*, whistling warbler *Catharopeza bishopi*, Altamira yellowthroat *Geothlypis flavovelata*, Semper's warbler *Leucopeza semperi*, Paria redstart *Myioborus pariae*, gray-throated warbler *Basileuterus cinereicollis*, white-winged ground-warbler *Xenoligea montana*, pearly-breasted conebill *Conirostrum margaritae*, Tamarugo conebill *C. tamarugense*, selva cacique *Cacicus koepckeae*, Martinique oriole *Icterus bonana*, saffron-cowled blackbird *Xanthopsar flavus*, pampas meadowlark *Sturnella defilippi*, and red-bellied grackle *Hypopyrrhus pyrohypogaster*.

These two families inhabit the Americas. The birds of the wood warbler family, Parulidae, are commonly known as warblers, redstarts, waterthrushes, ovenbirds (quite distinct from the birds belonging to the family Furnariidae of the same name) and yellowthroats. The most morphologically and ecologically diverse group of the New World songbirds is the family Icteridae, for which there is no really good comprehensive English name. Often called "American blackbirds", relatively few of its approximately 92 species are wholly or predominantly black.

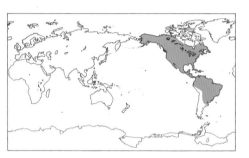

WOOD WARBLERS

Wood warblers reach their greatest diversity in North America, Central America, and the West Indies; about 13 species and many additional subspecies are actually confined to islands in the West Indies. However, members of the family occupy virtually every conceivable habitat south of the tundra. Most nest in trees, shrubs, or vines; those species that nest on the ground tend to do so in wooded habitats rather than open country. Several species are typical of the belt of coniferous forest stretching across Canada and the northern USA, extending south in the mountains into Mexico. Others prefer deciduous and mixed forests. The yellowthroats (genus *Geothlypis*) are unusual in nesting in marshes and upland areas that may lack any woody vegetation. Habitat requirements for some species are quite restrictive: some populations of the most widely distributed species, the yellow warbler *Dendroica petechia*, are confined to coastal mangroves. The buff-rumped warbler *Phaeothlypis fulvicauda* in tropical climates and the Louisiana waterthrush *Seiurus motacilla* in temperate eastern USA seldom nest far from running water.

Although primarily insect-eaters, wood warblers exhibit many foraging methods. Some are excellent flycatchers and have evolved the flatter bills surrounded by the longer hair-like rictal bristles typical of birds that forage aerially. Narrow, thin bills exemplify the many species that pick small insects or their eggs from leaves or twigs. One species, the black-and-white warbler *Mniotilta varia* of eastern North America, has creeper-like habits—climbing on trunks and limbs of trees to search crevices in the bark for small insects and eggs—and has evolved a bill, toes, and claws longer than those of its nearest relatives.

Most of the wood warblers are brightly colored and patterned. Among the North American species, males in the breeding season are often more brilliantly colored than females, but both sexes usually molt into a less-conspicuous plumage before the autumn migration, similar to the plumage of their young; in fall, many conspicuous details of the breeding plumage are lost, so these species present notorious identification problems for birdwatchers and are known as "confusing fall warblers". Among the few dull-plumaged northern species in which the sexes are alike year-round are the waterthrushes and ovenbird (all in the genus *Seiurus*), the worm-eating warbler *Helmitheros vermivora*, and Swainson's warbler *Limnothlypis swainsonii*.

In most of the South American species the sexes are alike in color and do not have obvious seasonal plumage changes. Most of these belong to the large tropical genera *Myioborus* and *Basileuterus*, which barely reach the southwestern USA.

ICTERIDS, OR AMERICAN BLACKBIRDS

The family Icteridae is divided into three distinctive subfamilies. Oropendolas (genus *Psarocolius*), caciques (four genera) and American orioles (genus *Icterus*, including the troupial *I. icterus*) constitute the subfamily Icterinae; most are tree-dwelling, and all (except one cacique) build woven pendulous nests. The subfamily Agelaiinae includes the "blackbird" members of the family as well as meadowlarks (genus *Sturnella*), which are terrestrial birds of grasslands, grackles (genus *Quiscalus*), cowbirds (genera *Molothrus* and *Scaphidura*), and several small genera endemic to South America. Members of this subfamily differ from virtually all other songbirds by molting their tail feathers in a centripetal sequence, that is from the outermost to the innermost. The bobolink *Dolichonyx oryzivorus* is the sole member of the subfamily Dolichonychinae. It breeds in the grainfields and marshlands of Canada and the northern USA, and migrates to Brazil and Argentina. It is the most bunting-like in appearance, and has two complete molts annually (very rare in birds in general).

Although the family is predominantly tropical, it has numerous representatives in both the northern and southern temperate regions. The red-winged blackbird *Agelaius phoeniceus,* one of the most abundant of North American birds, breeds as far north as Alaska—as does the rusty blackbird *Euphagus carolinus.* At the other end of the world, the long-tailed meadowlark *Sturnella loyca* breeds as far south as Tierra del Fuego and the Falkland Islands.

Except for some of the oropendolas and caciques, relatively few members of this family are true forest birds. Many inhabit swamps, marshes, and savannas, and many of the orioles prefer open woodlands and arid scrub. Some grackles have become highly urbanized; the common grackle *Quiscalus quiscula* is abundant in city parks in much of the USA and Canada, and large noisy roosts of the great-tailed grackle *Q. mexicanus* are

typical of towns and villages throughout Mexico and Central America. Size difference between males and females is especially conspicuous in this family, particularly in the larger species—in the great-tailed grackle the male may weigh as much as 60 percent more than the female. In fact, this family has a greater range of sizes among species than any other family in the order Passeriformes.

Nesting habits are highly diverse. Some of the oropendolas, caciques, and marsh-inhabiting blackbirds choose to nest in dense colonies. Only one of the six species of cowbird (genera *Molothrus* and *Scaphidura*) builds its own nest; all of the others are parasitic, depositing their eggs in the nests of other birds. Many host-bird species have evolved defenses against this parasitism, but when cowbirds move into an area they haven't previously inhabited, the new host species, especially if already rare, may be severely affected. This happened when the brown-headed cowbird *Molothrus ater* invaded the range of Kirtland's warbler *Dendroica kirtlandii* in the state of Michigan and the black-capped vireo *Vireo atricapillus* in Oklahoma, and when the shiny cowbird *M. bonariensis* spread into the limited area occupied by the yellow-shouldered blackbird *Agelaius xanthomus* in Puerto Rico.

▲ *A male bobolink in its summer home; the female is dull, brownish, and streaked. This is a bird of lush grassland, including hay meadows and clover fields, across temperate North America. It winters in South America. Its rich, bubbling song—often uttered in flight—has a distinctive banjo-like timbre.*

◄ *One of the loveliest of wood warblers, the prothonotary warbler is seen to best advantage in its preferred breeding habitat, the swamps and flooded forests of the southeastern United States, where its orange-yellow head and breast glow against the gloom of the cypress trees. It spends most of its time near the ground and, like most wood warblers, it winters in Central and South America.*

FINCHES

ANTHONY H. BLEDSOE
AND ROBERT B. PAYNE

Order Passeriformes
Suborder Oscines
Family Fringillidae
20 genera, 122 species
Family Drepanididae
17 genera, 29 species
Family Estrildidae
26 genera, 123 species
Family Ploceidae
15 genera, 126 species
Family Passeridae
3 genera, 33 species

SIZE
Smallest Orange-bellied
waxbill *Amandava subflava*,
total length 9 centimeters
(3½ inches); weight 7 grams
(¼ ounce).
Largest Long-tailed widow
Euplectes progne, total length
(including tail)
60 centimeters (24 inches);
weight 42 grams
(1⅗ ounces).

CONSERVATION WATCH
There are 44 species listed in
the ICBP checklist of
threatened birds. They
include the following:
Molokai creeper *Paroreomyza
flammea*, Kauai akialoa
Hemignathus procerus, Oahu
creeper *Paroreomyza
maculata*, nukupuu
Hemignathus lucidus, ou
Psittirostra psittacea, poo-uli
Melamprosops phaeosoma,
Maui parrotbill *Pseudonestor
xanthophrys*, blue chaffinch
Fringilla teydea, São Tomé
grosbeak *Neospiza concolor*,
red siskin *Carduelis cucullata*,
Bates's weaver *Ploceus batesi*,
black-chinned weaver *P.
nigrimentum*, Tanzanian
mountain weaver *P. nicolli*,
Ibadan malimbe *Malimbus
ibadanensis*, Gola malimbe *M.
ballmanni*, and Rodrigues fody
Foudias flavicans. The yellow-
throated serin *Serinus
flavigula*, and golden-naped
weaver *Ploceus aureonucha* are
possibly extinct.

The word "finch" commonly describes several groups of seed-eating birds, divided into five families. Although typical finches have a cone-shaped bill, not all members of these five families are seed-eaters and indeed some of them have very different bill shapes.

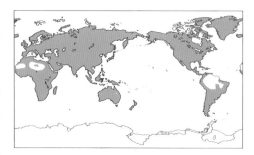

WINGS WITH NINE PRIMARIES
The chaffinches and cardueline finches (family Fringillidae) and the Hawaiian honeycreepers (family Drepanididae) have only nine large primary feathers in each wing (the tenth primary is vestigial) and are therefore grouped with the mainly American assemblage of songbirds called "New World nine-primaried oscines". The geographic name for this assemblage is hardly appropriate, because the center of diversity in the family Fringillidae is outside the New World, and the Hawaiian honeycreepers occur only in the Hawaiian islands. Nonetheless, anatomical and genetic studies indicate that these two families are close relatives of the Old World buntings and of the other members of the nine-primaried assemblage—the strictly New World tanagers, wood warblers, cardinal grosbeaks, and icterine blackbirds and orioles.

Fringillid finches and Hawaiian honeycreepers are each other's closest relatives, a relationship that belies their marked differences in appearance, distribution, and behavior. All fringillid finches have conical bills, and most inhabit temperate climates, are migratory, and eat large quantities of seeds and buds. In contrast, the Hawaiian honeycreepers are restricted in distribution, exhibit an incredible range of bill sizes and shapes, inhabit tropical and semi-tropical forests, and many eat nectar from flowers.

▶ *The crossed mandibles of the crossbill are used to lever seeds from the cones of pine and spruce trees, upon which the birds feed almost exclusively.*

The only obvious external similarity linking Hawaiian honeycreepers with the cardueline subfamily of the Fringillidae is the possession of red or yellow in the plumage at some stage of the life cycle. However, similarities in muscle patterns, protein structure, and DNA sequences do unite these two apparently disparate families.

FRINGILLID FINCHES
The family Fringillidae is made up of two subfamilies: chaffinches and cardueline finches.

Chaffinches
The chaffinches (subfamily Fringillinae) are the only fringillid finches that lack red or yellow in their plumage. There are three species. The common chaffinch *Fringilla coelebs* and the brambling *F. montifringilla* are primarily Eurasian, inhabiting open woods, gardens, and farmlands; they have boldly patterned males, but the females are duller. The blue chaffinch *F. teydea* is confined to the Canary Islands, where it inhabits mountain pine forests; the males are evenly colored blue. Chaffinches feed on seeds, buds, fruits, and insects.

The common chaffinch has a distinctive place in the history of song research in ornithology. It was the first species in which the interplay between song, innate vocal drives and learning was studied extensively. By playing recordings of songs at specific times during the growth of chaffinches, ornithologists discovered that a learning period early in life lasts roughly a year and is followed by a silent period. After the silent period, the birds engage in a period of practice, during which their vocalizations crystallize into a song. The stages of song development are closely linked to levels of endocrine hormones.

Cardueline finches
The 119 species of cardueline finches (subfamily Carduelinae) are distributed widely in the Americas, Africa, and Eurasia, with the greatest diversity in the Himalayan region. Most species occur in subarctic, temperate, or desert regions; the few that occur near the Equator inhabit mountainous regions with temperate climates. They do not breed natively in Madagascar, the Indian subcontinent south of the Himalayas, or Australasia, although humans have introduced them to many non-native regions and islands such as Bermuda and New Zealand. The genera in this subfamily include *Serinus* (commonly known as canaries and seed-eaters), *Carduelis* (greenfinches, goldfinches, redpoll, siskins), *Acanthis* (redpolls, twite, linnet), *Leucosticte* (rosy finches), *Carpodacus* (rosefinches), *Pinicola* (pine grosbeaks), *Loxia* (crossbills), *Pyrrhula* (bullfinches), *Coccothraustes* (hawfinch, evening grosbeak), and 11 others.

Most Northern Hemisphere cardueline finches are migratory, and many species exhibit nomadic behavior. Certain nomadic species such as the

S.R. Cannings

▲ A male American goldfinch tests the water. Bathing, in either water or dust, is a vital part of routine feather care and maintenance for birds. This goldfinch is one of many that use shallow, secluded pools.

◄ The goldfinch (top left) and the chaffinch (lower left) are two of the commonest and most widely-known finches of European fields and gardens.

crossbills will breed at any time of the year, as long as food is abundant. This nomadic tendency may in part explain the ability of the group to colonize distant oceanic islands—in prehistoric times a cardueline-like colonist gave rise to the range of Hawaiian honeycreepers.

Cardueline finches eat primarily seeds, buds, and fruits. Insects do not make up a substantial proportion of the diet, even during the nesting season. This is distinctive, because most grain- and fruit-eating songbirds switch over to insects during nesting. The carduelines, in contrast, feed their nestlings a mix of predigested plant material, including seeds and buds. They typically nest well off the ground in shrubs and trees, although a few species have developed close commensal relationships with humans (for example, the house finch *Carpodacus mexicanus* of North America) and may nest on buildings. An unusual aspect is that cardueline young defecate in the nest, so that the fecal material builds up on the walls and the rim; in other birds, the young either eject feces out of the nest or package fecal material in a sac that is removed by the adults. Although canaries are the most renowned cardueline singers, many other species produce beautiful songs. Domesticated canaries are descendants of the canary *Serinus canaria*, a species endemic to the Azores, Canary,

and Madeira islands. Brought to Europe as a cage-bird in the sixteenth century, the species has subsequently been bred with other carduelines with pleasant songs, such as the bullfinches; the resulting hybrids often have exceptional singing abilities. Hybrids between different species of carduelines are rare in nature, yet in captivity many hybrids have been produced, including several from genetically distant parental species.

THE HAWAIIAN HONEYCREEPERS
Several Hawaiian honeycreepers are quite similar in appearance to the fringillid finches, probably because of the close evolutionary relationship between the two families. In the family Drepanididae there are 29 extant or recently extinct species in 17 genera.

Finch-like Hawaiian honeycreepers
The finch-like honeycreepers (subfamily Psittirostrinae) consist of nine modern species; of these, only five are extant, and even they survive in but small numbers. Each of the finch-like Hawaiian honeycreepers has a conical bill; in one species, the Maui parrotbill *Pseudonestor xanthophrys*, the bill is modified into a structure remarkably like the bill of a small parrot.

Thirteen more species (in four genera) are sometimes classified in a separate subfamily (Hemignathinae); of these, two species are extinct, two are probably extinct, and the remaining nine persist in very small numbers or in greatly reduced distributions. These birds resemble the finch-like honeycreepers but have less conelike bills, which are either relatively straight, as in the Kauai creeper *Oreomystis bairdi,* or have the upper portion of the bill greatly elongated and curved, as in the common amakihi *Hemignathus virens*.

All finch-like Hawaiian honeycreepers eat seeds, buds, and insects; and two of them, the Laysan finch *Telespiza cantans* and the Nihoa finch *T. ulima,* also include the eggs of seabirds in their diets. The calls and songs are poorly known or inadequately studied, but are variously described as canary-like or linnet-like. Of the species for which there is nesting information, nests are formed of grass, stems, and rootlets; the Laysan finch places it up to 10 centimeters (4 inches) above the ground in grass tussocks, the Nihoa finch in rocky outcroppings of cliffs or crevasses, and the palila *Loxioides bailleui* in trees. The breeding season of the honeycreepers on the leeward islands—the Laysan and Nihoa finches—begins about January and appears to end by mid-July, whereas elsewhere the species nest in May–July, as far as is known.

Nectar-feeding Hawaiian honeycreepers
The other subfamily of Hawaiian honeycreepers (Drepanidinae) consists of the mamo, the iiwi, and their allies, seven species in all; of these, only four

▼ *At least eight species of Hawaiian honeycreepers are now extinct, but a few, like this apapane, remain common. Like others in the subfamily Drepanidinae, the sexes are almost identical in appearance.*

P. La Tourrette/Vireo

P. La Tourrette/Vireo

are extant and even they persist in but small numbers or restricted distributions. The members of this group eat nectar, giving rise to the vernacular name "Hawaiian honeycreeper". However, as with "finch", the term "honeycreeper" is meant to convey a lifestyle and associated feeding habits rather than a connection to other birds called honeycreepers, which have evolved independently as nectar-feeders (for example, the *Cyanerpes* honeycreepers of tropical America, in the family Emberizidae). The nectar-feeding Hawaiian honeycreepers nest well above ground, and eat a variety of foods. Several species specialize on flowers and have tongues that are brush-like at their tips, an adaptation for soaking up nectar.

ANTHONY H. BLEDSOE

WINGS WITH TEN PRIMARIES
The three families described below—Estrildidae, Ploceidae, and Passeridae—are small finches of the Old World. Typically they have a cone-shaped bill,

short wings with 10 primaries, and rear their brood in covered nests. They display a remarkable variety of group living, mating systems, and parental care.

ESTRILDID FINCHES
The estrildid finches (family Estrildidae) include five subfamilies: waxbills Estrildinae, grassfinches Poephilinae, mannikins Lonchurinae, and parrot-finches Erythrurinae. They are small birds 9 to 15 centimeters (3½ to 6 inches) long, which feed on small grass seeds, mainly in tropical areas. Most live in Africa, where they are common in grassy areas; several are familiar around villages; and a few feed and nest in forests and swamps. Their cheerful air, bobbing courtship displays, and diet of seeds make them attractive as cage-birds, so the behavior of many species is better known from watching them in captivity than in the field. Males and females are similar in size, and in some species they have similar plumage. The nests are thatched with grass, and often lined with feathers.

Waxbill pairs stay together in a close social

▲ *Unusually among birds, many of the Hawaiian honeycreepers, like this iiwi, have kept the original names bestowed upon them centuries ago by the native Hawaiians. Some of these birds played a prominent role in Hawaiian cultural and ceremonial life.*

▲ *Three colorful grassfinches and weavers: the Gouldian finch (left) is confined to tropical northern Australia, where its population is steadily declining; the melba finch (center) and the red bishop (right) inhabit African thornveldt and reedbeds respectively.*

bond, perch together in physical contact, preen each other, and rear their young together. One species is the small, red-billed firefinch *Lagonosticta senegala,* which builds its nest of grass and the feathers of domestic or guinea fowl, often in the thatched roofs of African villages, where they are called "animated plums". Adults usually live less than a year and maintain their populations by repeated breeding, rearing several broods in a season. They breed when the rains produce a fresh crop of grass seeds; the adults eat seeds after they fall to the ground, sometimes raiding the raised cones of harvester-termite holes for drying seeds.

The waxbills' nestlings have colorful markings inside the mouth. In the red-billed firefinch the juvenile mouth marking is bright orange-yellow with black spots on the palate and two shades of blue in the light-reflecting tubercles at the corners of the mouth. The bright gape guides the feeding behavior of the parent to the begging young. Other kinds of firefinches—indeed, most other estrildids—each have their own mouth pattern. The patterns may be important to the parents in recognizing their young. The mouth colors of the young disappear when the birds are grown.

Waxbills of the genus *Estrilda* take seeds from the standing stems of grasses, and they nest early in the rains. The melba finch *Pytilia melba* uses its sharp bill to dig into a cover of "runways" on the ground and tree trunks in the dry season so that it can feed on the termites below. One species of

seed-eater, *Pyrenestes ostrinus,* has thick- and thin-billed forms which live together and interbreed in swampy areas of Central Africa. They feed on the seeds of sedges and take the same diet when food is abundant, but in seasons of scarcity the thick-billed form specializes on hard-seeded sedges while the thin-billed form takes other foods.

Mannikins have thicker bills and different courtship displays. The white-rumped munia *Lonchura striata* feeds in irrigated rice fields, eating the rice and the protein-rich filamentous green algae that grows there. Some populations in Malaysia breed twice in a year, each time corresponding with a crop of rice and a bloom of algae, so their breeding seasons are determined by manmade cycles of rice cultivation. Society finches are a domesticated strain of the munia developed centuries ago in China; they are used as foster parents for other estrildid finches and whydahs in captivity, incubating the eggs and rearing the young by regurgitation—although, as the young beg with their heads twisted upside down, society finches are limited as foster parents to birds with this feeding behavior.

In the grassfinch subfamily the best-known species is the zebra finch *Taeniopygia guttata,* which is widespread in dry regions of Australia. It is highly nomadic and breeds in loose colonies after rain. Parrot-finches, 12 species in the genus *Erythrura,* live on tropical islands of the western Pacific, and two species have reached Australia.

They too have bright and unusual colors inside the mouths of the young.

WEAVERS, OR PLOCEID FINCHES
The Ploceidae is a largely African family of 126 species, with a few of those species in Asia and islands of the Indian Ocean. The family is subdivided into buffalo-weavers, sparrow-weavers, true weavers which include bishops and widow finches, and whydahs.

Buffalo-weavers
Buffalo-weavers (subfamily Bubalornithinae) differ from other ploceid finches in that the outer (tenth) primary feather of each wing is large. They are noisy birds which live in flocks and colonies in African savannas (and villages) year-round. One species builds communal nests of thorns and sticks, and the other builds nests of thatched grass. The male black buffalo-weaver *Bubalornis albirostris* is unique among songbirds in having a phallus and in growing a protuberance on the bill in the breeding season.

Sparrow-weavers
The African sparrow-weavers and social weavers (subfamily Plocepasserinae) insert the ends of grass or twigs into a mass of thatch and are "thatchers" rather than "weavers". Several are cooperative breeders. Cooperative colonies of white-browed sparrow-weavers *Plocepasser mahali* last over many generations. The young often remain with their parents and aid them in breeding, or they may move from their natal colony and make a place for themselves in another colony by becoming helpers of the breeding pairs. Sociable weavers *Philetarius socius* build a huge structure of sticks (weighing up to 1000 kilograms, or almost 1 ton) with a common roof and separate nest holes below, each occupied by a pair of breeding birds; the compound nest protects the birds from the extremes of temperature in the Kalahari desert as well as from predators.

Weavers, bishops, and widow finches
Members of the Ploceinae subfamily vary in their behavior. Some are forest-dwelling birds, monogamous, and eat insects. However, most of the 100 species live in grasslands or marshes, are polygynous (one male mating with more than one female), and eat grass seeds, though they may feed insects to their young. Colonies of hundreds of weaver nests are conspicuous in the African landscape. A male builds several nests, attracting several females. He first builds a swing to perch upon, then extends the swing into a ring, and adds a covered basket, keeping the ring as the entrance. He pushes and pulls the nest material into loops and knots around other blades of grass—a process similar to basket-weaving. Some species build a ball-like nest with an entrance hole in one side.

Others finish the nest with a long entrance tube of woven grass. Young males build practice nests which usually fall apart. In the village weaver *Ploceus cucullatus,* males raid each other's nests for green grass. Females are attracted to the male as he hangs upside-down while clinging to the nest, waving his wings in flashes of yellow and black; the female inspects the nest, and if it is to her liking she lays and rears a brood by herself. Females prefer to nest with males in larger colonies. Several weavers nest in mixed-species colonies, where they benefit as in single-species colonies, by seeing other birds returning from new sources of food and by group defence against predators.

The red-billed quelea *Quelea quelea* is a locust-like scourge of cultivated grains. They breed in colonies, a single acacia bush bearing hundreds or thousands of nests, in thousands of bushes. The time when conditions are right for breeding, with seeds in the soft, milky stage suitable for feeding to their young, is short. They breed rapidly: the male builds the nest in two or three days, the female lays three eggs and incubates 10 days, and the young fledge by 10 days and are quickly independent. Flocks follow the seasonal rains over Africa. A bird may breed repeatedly, first where rains fall early, then again hundreds of kilometers away where rains fall a few weeks later.

Among bishops and widow-birds, sexual dimorphism in size and plumage is extreme; males are larger and brightly colored, whereas females and non-breeding males are streaky brown and blend with their grassy habitat. Bishops are named for the hood of red or yellow in the males' breeding

▲ *The male baya weaver builds a nest then coaxes a female to accept it with vigorous wing-flapping displays. As soon as one accepts, moves in and begins a brood, the male proceeds to build another nest close by, repeating the process until he has three or four mates and families.*

▼ *The intensely gregarious zebra finch inhabits desert and grassland across Australia.*

Peter Johnson/NHPA

waxbills, to which the whydahs are related on the basis of feather arrangement and biochemical genetics, and most whydahs lay in the nests of a single kind of waxbill. One of the whydah species, the village indigobird *Vidua chalybeata,* is a brood-parasite of the red-billed firefinch. The female indigobird is attracted to males that have songs like her foster father's; the song brings her together with a male of her own species, and with a foster pair whose nest she will lay in. A male has a variety of mimicry songs, learnt not only from foster parents, but also from older male indigobirds in his neighborhood. Males visit each other's call-sites and share the same songs, both mimicry songs and the non-mimetic songs. One male in a neighborhood is the favorite of most females. When he changes his songs, other males copy him.

The young whydahs have the same juvenile mouth markings and begging calls as the nestlings of their foster species, so nestling village indigobirds show the mouth colors of the nestling red-billed firefinches, and the indigobird *V. raricola* has a pink palate and red and light blue tubercles like those of the black-bellied firefinch *Lagonosticta rara.* Juvenile plumage in some whydahs resembles the foster young, especially the unmarked gray of the paradise whydah *Vidua paradisaea* like the melba finch, and the golden plumage of the straw-tailed whydah *V. fischeri* like the purple grenadier *Granatina ianthinogaster.*

SPARROWS

Old World sparrows (family Passeridae) build a covered nest of thatched grass with a side entrance. They differ from other seed-eating birds by having a vestigial dorsal outer primary feather and an extra bone in the tongue. Several of the 33 species have lived in close association with humans for centuries. One is a highly successful immigrant in areas far from its original range: the house sparrow *Passer domesticus* probably spread from the Middle East along with the movement of agriculture into Europe about 7,000 years ago, then as Europeans colonized other parts of the world in the nineteenth century it went with them. In North America, distinct populations of house sparrows evolved within less than a century and now differ from the founding forms in size and color—those in the northern Great Plains are large, and those in dry areas of the southwest are pale. Different *Passer* species often take over the nests of other birds, like pirates, driving away the nest-builders. The chestnut sparrow *P. eminibey* sometimes is a pirate; other times it builds its own nest.

Two other groups of sparrows are rock sparrows (genus *Petronia*) which nest in holes, and snow finches (genus *Montifringilla*) which feed on insects and seeds blown onto high snowfields and nest in rocks above the treeline in Europe and Asia.

▲ *A long-tailed whydah. These birds use their long tails to attract females in impressive display flights over their grassland territories—the longer the tail, the more successful the display.*

plumage. Some long-tailed species are called "widows" (after the train of tail, as in a widow's veil of mourning), and a few are whydahs (after a Portuguese word with the same meaning). In the long-tailed widow *Euplectes progne,* the male displays with a slow, flapping flight over his grassland territory, tail dangling behind and looking good to females—males in Kenya whose tails were experimentally lengthened by adding parts from other males were able to attract more females to nest. The African cuckoo-weaver *Anomalospiza imberbis* is a brood-parasite; it lays its eggs in the nests of grass-warblers, which rear the alien young at the expense of their own brood.

Whydahs, or viduine finches

Whydahs lack a social family life. These brood-parasites (14 species, all in Africa) lay their eggs in the nest of a foster species. The foster parents incubate the eggs and rear the young whydahs along with their own brood. The foster group are

◄ *Native to Southeast Asia, the Java sparrow is widely known as a popular cagebird.*

ROBERT B. PAYNE

STARLINGS AND THEIR ALLIES

TERENCE LINDSEY

Order Passeriformes
Suborder Oscines
Family Sturnidae
27 genera, 113 species
Family Oriolidae
2 genera, 25 species
Family Dicruridae
2 genera, 20 species

Left to right: starling, oriole, drongo.

SIZE
Starlings (Sturnidae) total length 18 to 45 centimeters (7 to 17¾ inches); weight 30 to 105 grams (1 to 3¾ ounces).
Orioles (Oriolidae) total length 18 to 31 centimeters (7 to 12¼ inches); weight 45 to 120 grams (1½ to 4¼ ounces).
Drongos (Dicruridae) total length 18 to 38 centimeters (7 to 15 inches); weight 45 to 90 grams (1½ to 3 ounces).

S tarlings and mynahs (family Sturnidae), orioles and figbirds (Oriolidae), and drongos (Dicruridae) are Old World birds, with strongest representation in hotter climates. A number of species have glossy black plumage, but many are boldy patterned and brilliantly colored. Most are arboreal, but there are some terrestrial species.

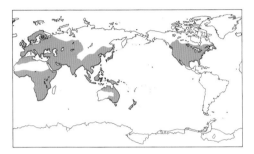

STARLINGS

Almost throughout the English-speaking world the common starling *Sturnus vulgaris* is among the commonest of garden birds. It is a stumpy, rather short-tailed bird with a confident strut and alert, pugnacious manner; its plumage is black, glossed with green and purple, and spangled (in the non-breeding season) with pale brown. Its song, an extraordinary jumble of squeaks, rattles, whistles, and other apparently random sounds, is uttered freely from exposed perches on telephone wires or television aerials. It nests in an untidy jumble of grass and litter stuffed into any available cavity in a tree or a building, and it gathers in large noisy flocks to roost at night.

This bird is in fact a fairly typical member of the Sturnidae, a family of about 113 species found throughout the Old World. The common starling originated in Europe and owes its present more extensive distribution to deliberate introductions by humans, especially during the late nineteenth century when birds were released in North America, South Africa, Australia, New Zealand, and elsewhere, and subsequently prospered.

Some species of starlings are arboreal, fruit-eating birds of jungle and rainforest, but most are insect-eaters inhabiting open country, spending much of their time on the ground. Most species are gregarious, many roost in flocks, and some nest in dense colonies.

There are a number of highly specialized forms with very restricted distributions, but there are also several larger, more significant groups. The genus *Aplonis* comprises about 20 arboreal fruit-eating species found throughout Polynesia and the New Guinea region. About 35 African species, mainly in the genera *Onychognathus, Lamprotornis,* and *Spreo,* include some strikingly beautiful birds with glossy, iridescent plumage, often featuring patches of vivid violet, green, orange, and blue. India is the home of the mynahs (genus *Acridotheres*). Mynahs are mainly dull brown, with patches of naked yellow skin on the head and bold white flashes in the wing. One species, the Indian mynah *Acridotheres tristis,* has been introduced to southern Africa, Australia, Polynesia, and elsewhere; it is a vigorous and aggressive species, often vying with the common starling for domination of the urban environment.

The family also includes the two tickbirds or oxpeckers (genus *Buphagus*), which spend much of their time riding around on buffaloes, rhinos, and other large mammals of the African plains. Researchers are still undecided as to whether the two species of sugarbird (genus *Promerops*) of South Africa are starlings or honeyeaters.

ORIOLES

The typical calls of most orioles have a distinctive timbre—a rich, liquid, bubbling quality that is difficult to describe but easily recognized—and these calls are a characteristic sound of forest and woodland almost throughout the Old World, from Europe to Japan and Australia to southern Africa. The birds themselves are often difficult to observe because they spend most of their lives in dense foliage high in the treetops.

The family contains about 25 species in two genera: *Oriolus* (orioles) and *Sphecotheres* (figbirds). Typical orioles are rather sturdy in build but have long pointed wings; the bill is moderately long, moderately slender, and slightly decurved; the eyes are usually red. The males of many species are boldly patterned in black and bright yellow (or crimson, maroon or chestnut in a few species), but some are dull green, and streaked below; females of most species are also dull green and streaked. Most orioles live alone or in pairs. Though they call persistently, they are unobtrusive and rather deliberate in their general behavior. A few species migrate, but most are sedentary. Orioles eat insects and fruit.

One species in this family differs markedly from other orioles in several respects: the figbird

CONSERVATION WATCH
The following species are listed in the ICBP checklist of threatened birds: Santo mountain starling *Aplonis santovestris,* Ponape mountain starling *A. pelzelni,* Rarotonga starling *A. cinerascens,* Abbott's starling *Cinnyricinclus femoralis,* Bali starling *Leucopsar rothschildi,* helmeted mynah *Basilornis galeatus,* bare-eyed mynah *Streptocitta albertinae,* Isabella oriole *Oriolus isabellae,* silver oriole *O. mellianus,* Grand Comoro drongo *Dicrurus fuscipennis,* and Mayotte drongo *D. waldeni.*

◄▲ *The greater raquet-tailed drongo (far left) inhabits India and Southeast Asia; the spatulate, twisted tips to the tail feathers cause a distinctive humming noise in flight. The superb glossy starling (above) is a familiar sight to safari tourists in East Africa, commonly visiting campsites and picnic grounds. The figbird (near left) is common and widespread in Indonesia, New Guinea, and northern and eastern Australia.*

Sphecotheres viridis is gregarious, noisy, active, and quarrelsome. This bird inhabits Indonesia, New Guinea, and northern and eastern Australia, and it is often common in suburban parks and gardens. Its diet consists almost entirely of fruit, especially figs, and a flock can almost strip a tree.

Birds called orioles also occur in North and South America; these species are strikingly similar to Old World orioles in several respects—they are strongly arboreal, have loud ringing songs, build hammock-shaped nests, and the males are boldly patterned in black, yellow, or orange—but they are not related.

DRONGOS

About 20 species of birds make up the family Dicruridae. The word "drongo" derives from the vernacular name used for the local species by indigenous peoples of northern Madagascar, and it has passed into general use in English for all species. Drongos are bold, aggressive, medium-sized songbirds with relatively short legs, long,

forked tails, stout bills, an alert upright stance, and flamboyant behavior. The sexes are similar, and the plumage of almost all species is entirely glossy black. Many species have crests, and in several the tail feathers are enormously elongated or otherwise modified into an ornamental structure. Typical drongo calls include grating metallic notes, modulated whistles, and harsh scolding notes; few have complex or attractive songs. Drongos occur almost throughout the Old World tropics, from Africa to Australia, but the family is best represented in southern Asia. They inhabit wooded country of all kinds, from dense jungle to savanna, but a characteristic environment is a forest clearing or regrowth at the edge of rainforest. Most species feed primarily on large flying insects captured in flight during sorties from a perch in a tall dead tree or a similar vantage point. Nectar is an important food for some species, and most drongos also capture lizards, small birds, and mammals at least occasionally. They sometimes rob other birds of food in flight.

NEW ZEALAND WATTLEBIRDS

TERENCE LINDSEY

Order Passeriformes
Suborder Oscines
Family Callaeidae
3 genera, 3 species

SIZE
Smallest Saddleback
Creadion carunculatus,
total length 26 centimeters
(10 inches).
Largest Kokako *Callaeas
cinerea,* total length 38
centimeters (15 inches).

CONSERVATION WATCH
The saddleback and kokako
are listed in the ICBP
checklist of threatened birds.

C onfined to New Zealand, the Callaeidae show some similarities with Australian currawongs and apostlebirds, but their ancestry remains uncertain. They may be descendants of an early crow-like stock.

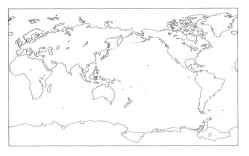

HUIA

The huia *Heteralocha acutirostris* was remarkable for the sexual difference in its bill structure: the female had a very long, slender, and deeply curved bill, while the male's bill was only moderately long and more or less conical. (Among birds, males and females often differ dramatically in plumage and may differ in size, but marked sexual differences in body structure are very rare.) First described in 1835, this bird was known only from dense forests on the North Island of New Zealand, but it has not

been reliably reported since 1907 and presumably is extinct.

THE KOKAKO AND THE SADDLEBACK

The huia has two surviving relatives. The kokako *Callaeas cinerea* and the saddleback *Creadion carunculatus,* which together make up the family Callaeidae, are a very distinct group restricted to New Zealand. They are usually (but tentatively) considered to be most closely related to starlings, bowerbirds, Australian apostlebirds, and mudlarks. All three species are (or were, in the case of the huia) medium-sized songbirds of dense forests, with very reduced powers of flight, strong legs, rounded wings, and conspicuous naked wattles at the gape of the bill. They feed on insects, especially insect larvae. They are sedentary, long-lived, mate for life, and maintain permanent territories. Even the two surviving members are seriously threatened: they are extremely vulnerable to predation from introduced stoats, weasels, and rats, and to degradation of their environment by the introduced Australian possums.

▲ *The huia was remarkable for the difference in bill shape between the sexes (the female's was the longer). Reluctant to fly, it had no defenses against the twin threats of introduced predators and forest destruction, and disappeared by about 1907.*

Brian Chudleigh

◄ *Formerly widespread on the main islands of New Zealand, the saddleback now survives only on a few small offshore islands.*

MAGPIE-LARKS AND THEIR ALLIES

Order Passeriformes
Suborder Oscines
Family Grallinidae
3 genera, 4 species
Family Corcoracidae
2 genera, 2 species
Family Artamidae
1 genus, 10 species
Family Cracticidae
3 genera, 10 species

Size
Smallest Little wood-swallow
Artamus minor, total length
12 centimeters (4¾ inches);
weight 15 grams (½ ounce).
Largest Gray currawong
Strepera versicolor, total length
50 centimeters (19¾ inches);
weight 350 grams
(12½ ounces).

Conservation Watch
These species do not appear
to be threatened.

IAN ROWLEY

A feature of the Australian landscape is the presence of black, or black and white, sociable birds calling to each other. In city parks they could be currawongs or the local magpies. In farmlands they may be any one of a dozen species, for the families in this chapter are common across the whole of Australia.

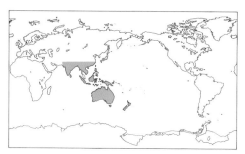

families of birds use mud to build nests, and it is now recognized that these four birds belong to two quite different families: the Grallinidae (magpie-lark and torrent-lark) and the Corcoracidae (white-winged chough and apostlebird).

Magpie-lark and torrent-lark
These are thrush-sized birds with glossy black and white plumage, and clear differences between the male and female. The magpie-lark *Grallina cyanoleuca* is common throughout Australia and southern New Guinea in most habitats except forests. Both parents build the nest, incubate the eggs, and brood and feed the nestlings. The young stay with their parents for several weeks after they fledge but join flocks of other young birds in late summer. Pairs tend to remain together throughout the year on the same territory, maintaining contact with a characteristic piping duet that has given rise to their other common name of "pee-wee". The torrent-lark *G. bruijni* is less well known, as it inhabits the margins and boulders of fast-flowing streams in the mountains of New Guinea.

White-winged chough and apostlebird
These two species are confined to the shrublands and woodlands of eastern Australia, where they forage on the ground. The white-winged chough *Corcorax melanorhamphus* rakes the litter and probes the ground with its long curved bill in search of insects and their larvae, while the apostlebird *Struthidea cinerea,* with its broad finch-like bill, is mainly a seed-eater. Both species are very sociable and are usually encountered in groups of five to twelve, consisting of the parent birds and their offspring from the previous two or three breeding seasons. It takes several years to acquire the necessary foraging skills and experience to be able to raise a family, and even then pairs are unlikely to be successful unless they have helpers to share the task of provisioning the young. Helpers also share in nest-building, incubation, brooding, and tending fledglings, so breeding is very much a cooperative affair.

MUD-NEST BUILDERS
The four species grouped under this heading share one characteristic: they all build nests of mud on horizontal branches well off the ground. But many

▼ *Strongly gregarious, the masked wood-swallow is a widespread nomad of the arid interior of Australia.*

R. Drummond

WOOD-SWALLOWS

The ten members of the family Artamidae are all in the same genus, *Artamus*. They have blue-gray bills with a black tip, short legs, and long strong wings shaped so that their flight silhouette appears very similar to that of the common starling. The family is largely Australasian in distribution and they forage in all types of habitat if there are enough insects to make it worthwhile. Although they are primarily insect-eaters, their brush-tongues enable them to lap nectar and pollen when available. Most of the family remain in the same area year-round, but at least three species are widely nomadic and exploit tropical and temperate environments at different times of the year, in large flocks that sometimes contain more than one species. Several species occur throughout islands to the north of Australia, reaching Fiji to the east, and Southeast Asia and India.

When not trawling for aerial insects these very sociable birds frequently perch side by side on conspicuous branches (or telegraph poles) and preen each other; at night they tend to roost together in a swarm clustered on the trunk of a tree or in a hollow. Wood-swallow nests are usually frail structures of fine twigs, and birdwatchers can often see the eggs through the bottom of the nest. In arid regions, nests may be built within six days of rain falling and eggs laid within 12 days, which is considerably shorter than the normal nest building period. Both parents build the nest, incubate the eggs, and feed the young, sometimes assisted by an extra helper.

AUSTRALIAN MAGPIES, OR BELL-MAGPIES

The Cracticidae is an endemic Australasian family comprising about 10 species, plus different races or subspecies, but bearing no relationship to magpies of the family Corvidae (see page 232) despite their shared common name. They all tend to be black, white, and gray, stockily built, with strong bills usually colored blue-gray with a black tip, and are usually fine singers. The three genera are quite distinct: the *Cracticus* butcher-birds are a little larger than a thrush and are very similar to the shrikes of other continents; the *Gymnorhina* magpies are about twice their size; and the *Strepera* currawongs are the largest, reaching about 50 centimetres (19¾ inches) in length.

Butcher-birds and magpies are chiefly found in savanna woodlands and shrublands, whereas the currawongs are mainly birds of the forests. Four species of butcher-birds and a magpie are found in New Guinea but no currawongs. The magpie *G. tibicen* has been introduced to New Zealand, which had no native members of the family. Apart from these examples and the very doubtful inclusion of the Bornean bristlebird *Pityriasis gymnocephala,* of which virtually nothing is known, the family is confined to Australia. Most parts of the country have a couple of species, and they are among the commonest local birds. Magpies and butcher-birds tend to be resident year-round, often in groups larger than the simple pair because the young stay on long after they become independent. Currawongs tend to wander nomadically through the forests in large flocks after breeding is finished; during the winter they are common sights in many towns in eastern Australia, where their "currawong" call is well known.

All species are basically insect-eaters, but when such food is scarce they will eat seeds and carrion. Butcher-birds generally take their prey in the air or on the ground after flying a swift sortie from a convenient perch. Magpies usually forage as they walk over open ground, probing soft areas of soil and turning over sticks, cow-pats, and other likely hiding places for insects. Currawongs, with the largest, dagger-shaped bill, concentrate on larger prey and have been seen plundering the nests of other birds and even killing small passerines. All members of this family build bulky stick-nests, although the female does most of the work and the incubating; however, the male and sometimes other group members help to feed the nestlings and fledglings.

Tom & Pam Gardner

▲ A bold, alert, and aggressive bird, the black currawong is a common panhandler at forest picnic grounds and campsites in the highlands of Tasmania. In winter many form flocks and descend to coastal lowlands. The sexes are similar.

Order Passeriformes
Suborder Oscines
Family Ptilonorhynchidae
8 genera, 19 species
Family Paradisaeidae
16 genera, 42 species

SIZE

Bowerbirds

Smallest Golden bowerbird *Prionodura newtoniana,* total length 22 centimeters (8¾ inches); weight 70 grams (2½ ounces).

Largest Great bowerbird *Chlamydera nuchalis,* total length 40 centimeters (15¾ inches); weight 230 grams (8 ounces).

Birds of Paradise

Smallest King bird of paradise *Cicinnurus regius,* total length 25 centimeters (9⅘ inches); weight 50 grams (1⅘ ounces).

Largest Black sicklebill *Epimachus fastuosus,* total length 110 centimeters (43⅓ inches); weight 320 grams (11⅓ ounces).

CONSERVATION WATCH

The following species are listed in the ICBP checklist of threatened birds: golden-fronted bowerbird *Amblyornis flavifrons,* Adelbert bowerbird *Sericulus bakeri,* long-tailed paradigalla *Paradigalla carunculata,* black sicklebill *Epimachus fastuosus,* ribbon-tailed astrapia *Astrapia mayeri,* Wahnes's parotia *Parotia wahnesi,* and Goldie's bird of paradise *Paradisaea decora.*

BOWERBIRDS AND BIRDS OF PARADISE

CLIFFORD B. FRITH

These two groups, found only in New Guinea, Australia, and the Moluccas, include some of the most behaviorally complex and bizarrely plumaged birds. Bower-building and decorating by male bowerbirds, and the fantastically ornate plumages and associated displays of male birds of paradise, have long attracted the attention of ornithologists all over the world. The two groups were once considered closely related, forming one family, but recent genetic studies have shown that they are quite separate.

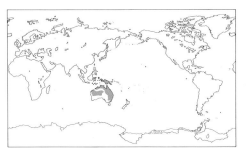

BOWERBIRDS

Bowerbirds (family Ptilonorhynchidae) are closely related to Australian lyrebirds, scrub-birds, and, surprisingly, treecreepers. They are stout, heavy-billed, strong-footed birds. Nine species are peculiar to New

Guinea, where the group presumably originated, eight are found only in Australia, and two occur in both areas. The three species that are called catbirds, because of their cat-like call, are monogamous, the sexes forming pairs to defend a territory and care for offspring. All 16 other species of bowerbirds are polygynous; the females build nests, incubate eggs, and raise young alone, and are appropriately drably colored in browns, grays, or greens. Both sexes in most of the polygynous species mimic other bird calls, especially those of predatory birds, and other sounds.

Bowerbirds live in rainforest or other wet forests, except the five avenue-bower-building *Chlamydera* species, which inhabit dry woodlands, grasslands, and, in three cases, the arid Australian interior.

Bowerbirds eat mostly fruit, but nestlings are fed differing proportions of animals, usually insects;

► *A male satin bowerbird rearranges his treasures, displaying his species' strong bias toward blue. As in other bowerbirds, the display area is decorated with a variety of small objects, often including plastic drinking straws and similar picnic litter.*

Leo Meier/Weldon Trannies

Frithfoto

catbird nestlings are frequently fed other birds' nestlings. Some species also eat flowers, flower and leaf buds, fresh leaves, and succulent vine stems.

Nests are bulky cups of leaves, fern fronds, and vine tendrils on an untidy stick foundation built in a shrub, tree fork, or vine tangle, except for the golden bowerbird *Prionodura newtoniana*, which builds its nest in a tree crevice. One or two eggs, sometimes three, are laid; incubation is from 19 to 24 days.

Some researchers favor including the enigmatic and probably extinct pio pio *Turnagra capensis* of New Zealand with the bowerbirds.

BIRDS OF PARADISE
Generally stout-billed or long-billed and strong-footed, birds of paradise are crow-like in general appearance and size; genetically the family Paradisaeidae is closest to crows and their allies. The group is divided by anatomical characteristics into the subfamilies Paradiseinae, comprising 38 typical birds of paradise, and Cnemophilinae, consisting of three little-known species—Loria's *Loria loriae*, yellow-breasted *Loboparadisea sericea*, and crested *Cnemophilus macgregorii* birds of paradise. They live in rainforest, moss forest, or swamp forest and may also frequent nearby gardens. Fruit dominates their diet, almost exclusively so in some manucodes and the crested bird of paradise, although sicklebills and riflebirds may eat more arthropods than fruits.

Most of them live in mountainous New Guinea and immediately adjacent islands. The exceptions are the paradise crow *Lycocorax pyrrhopterus* and Wallace's standardwing *Semioptera wallacii* on the Moluccan islands, and the paradise riflebird *Ptiloris paradiseus* and Victoria riflebird *P. victoriae* in eastern Australia. The magnificent riflebird *P. magnificus* and trumpet manucode *Manucodia keraudrenii* extend their New Guinea ranges to include the rainforests of Cape York, in northern Australia.

Of the typical birds of paradise in which the sexes are almost identical (monomorphic), several species have been studied and found to be monogamous—the male and female sharing nesting duties. As a result the monomorphic paradise crow and two paradigallas have also been considered monogamous. However, I recently studied a nesting short-tailed paradigalla *Paradigalla brevicauda* and observed only a single parent, presumably the female. Of the remaining 30 typical birds of paradise which are sexually dimorphic (sexes unlike), almost half are known to be polygynous—one male fertilizing several females, and the females raising their young alone and unaided. The remainder are assumed to be likewise.

Nests are bulky cups of leaves, ferns, orchid stems, and vine tendrils placed in a tree fork. Few include sticks, except that of the crested bird of paradise which is a domed mossy nest atop a few sticks. This species was considered to be a bowerbird, and to be monogamous, but I recently studied it nesting and confirmed that it is a bird of paradise, with a lone (presumably female) nesting parent. The king bird of paradise *Cincinnurus regius* is unusual in that it nests in a tree hollow. Birds of paradise lay one to three eggs, beautifully marked and colored, and these take between 16 and 22 days to hatch. Nestlings leave the nest when they are 16 to 30 days old.

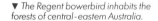
▲ *The emperor bird of paradise is restricted to the mountains of the Huon Peninsula, Papua New Guinea.*

▼ *The Regent bowerbird inhabits the forests of central-eastern Australia.*

Tom & Pam Gardner

Hans & Judy Beste/Auscape International

▶ A male Victoria riflebird at the climax of his extraordinary display. Preferred display perches are usually broken-off tree stubs several meters from the ground in dense rainforest.

D. Parer & E. Parer-Cook/Auscape International

▲ Most members of the genus Paradisaea, like this Raggiana's bird of paradise, congregate to display in leks: a tree full of these birds all in simultaneous display, quivering their long lacy flank plumes and calling hysterically, is one of the most spectacular sights among birds.

SPECTACULAR MATING RITUALS

Birds of paradise are well known for the males' spectacularly ornate plumage and fantastic courtship displays. Males of the various polygynous species court females in differing ways, but the best known are species in which males congregate to display in groups, known as leks. These include all but one of the plumed or "true" birds of paradise of the *Paradisaea* genus. Males of these species call from and hop about lek perches and spread raised plumes while posturing and performing dances in unison. Recent studies of raggiana bird of paradise *P. raggiana* leks revealed that most visiting females mate with the same individual male, and it is assumed that this male is the dominant and most fit on the lek. In this way, females optimize the choice

of male genes for their offspring.

Lekking behavior is also known in Wallace's standardwing, and in Stephanie's astrapia *Astrapia stephaniae* of New Guinea.

Males of some other conspicuously sexually dimorphic (sexes unalike) species do not congregate, but defend an exclusive territory and display alone; for example, in the superb bird of paradise *Lophorina superba* and the riflebirds, a solitary male displays on a tree stump, fallen tree trunk, or perch. Male parotias (genus *Parotia*) perform elaborate dances on cleared ground courts, which may form dispersed leks.

In the polygynous species of bowerbirds, the male constructs and decorates a bower of grasses, ferns, orchids, or sticks, and there he calls and displays to attract and impress females. In some adult male bowerbirds, plumage is spectacularly colorful, but in others the plumage is uniformly dull like that of the female; males with colorful plumage build simple bowers, whereas unadorned males build complex ones. This clearly indicates that bowers represent a transfer of sexual attraction from the male's appearance to a structure and its decorations. Males go so far as to steal decorations from one another's bowers. Studies of the satin bowerbird *Ptilonorhynchus violaceus* reveal that females choose to mate with males at larger, better-decorated bowers.

There are four bower types: court, mat, maypole, and avenue. Only the tooth-billed bowerbird *Scenopoeetes dentirostris* makes a "court", by clearing a patch of forest floor and decorating it with upturned fresh leaves. The only mat bower is that of the rare Archbold's bowerbird *Archboldia papuensis*, a large black bird of highland New Guinea, which accumulates a mat of fern fronds on the forest floor and decorates it with snail shells, beetle wing cases, fungus, charcoal, and other items, and drapes orchid stems on the perches above. Maypole bowers—stick structures built around one or several sapling stems—are constructed by the four New Guinea gardener bowerbirds (genus *Amblyornis*) and the golden bowerbird of northeast Australian upland rainforests. The remaining eight species construct avenue bowers—two parallel vertical stick-walls standing in the ground—in simple or complex forms. Satin bowerbird bowers have been known to exist at one site for 50 years.

▲ Smallest and most vivid of its family, the king bird of paradise is very common, very noisy, but very difficult to observe in the dense thickets it prefers. Though each male displays independently there are usually other males not too far away.

◄ A male great bowerbird coaxes a female to his display area. This particular bower is unusual in being roofed; most are left open at the top.

CROWS AND JAYS

IAN ROWLEY

Order Passeriformes
Suborder Oscines
Family Corvidae
c. 20 genera, *c.* 130 species

SIZE
Smallest Hume's ground jay *Pseudopodoces humilis,* total length 20 centimeters (8 inches).
Largest Raven *Corvus corax,* total length 66 centimeters (26 inches).

CONSERVATION WATCH
There are 12 species listed in the ICBP checklist of threatened birds. They include the following: beautiful jay *Cyanolyca pulchra,* azure jay *Cyanocorax caeruleus,* Sri Lanka magpie *Urocissa ornata,* hooded treepie *Crypsirina cucullata,* Ethiopian bush-crow *Zavattariornis stresemanni,* Marianas crow *C. kubaryi,* and Hawaiian crow *C. hawaiiensis.*

▼ *An Australian raven utters its harsh unmusical call.*

The crow family Corvidae is a very successful branch of passerines, which probably originated in Australia when it was isolated from the Asian landmass. Only when the continents drifted closer together, about 20 to 30 million years ago, could landbirds cross the water and an exchange of species take place. Once the original corvids reached Asia, extensive evolution seems to have occurred, resulting in the many different forms that have spread to all parts of the world. Approximately 110 species may be broadly grouped into crows (including jackdaws, rooks and ravens), choughs, nutcrackers, typical magpies, blue-green magpies, treepies, typical jays, American jays, gray jays, and ground jays.

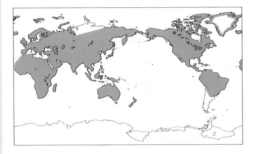

LONG-LEGGED FORAGERS

Most corvids are medium to large in size, have nostrils covered with bristles, and relatively long legs with characteristic scaling. Their color varies from the somber black of the raven through to the brilliant reds and greens of the Asian magpies. Both parents build the nest and feed the young, but only the female incubates the eggs and broods the nestlings; during that time she is fed by the male. Many corvids hide surplus food in caches which they relocate and use long afterwards.

Members of the family are mainly found in forests, open woodland, scrubland, and plains. Their long legs enable them to walk and hop over the ground quickly, foraging as they go, probing with their strong bills to search for insects, tear meat from carcasses, harvest berries, or pick up fallen seeds. Many of the smaller species forage extensively in the forest canopy and some specialize in harvesting particular foods such as nuts and pine seeds, which they may cache. Many corvids do not breed until they are at least two years old, and in the meantime the immature birds form large flocks that feed, fly, and roost together, exploiting the food available over a wide area, sometimes becoming pests. In many species, a breeding pair tends to be resident in one area or territory with the same partner for as long as they both survive, and they defend that area against intrusion by others of the same species. Other species, such as the rook *Corvus frugilegus,* jackdaw *C. monedula,* and pinyon jay *Gymnorhinus cyanocephala,* prefer to nest colonially. A few species, such as the Florida scrub jay *Aphelocoma coerulescens* and the Mexican jay *A. ultramarina,* have adopted a cooperative way of life and live in social groups consisting of the progeny from previous years, which stay in the family long after they have reached sexual maturity, helping the breeders to raise the current crop of young.

CORVID GROUPS

Crows, jackdaws, rooks, and ravens (40 species in the genus *Corvus*) vary in size from 33 to 66 centimetres (13 to 26 inches) and are basically black, although a few species show areas of white, gray, or brown. This genus has representatives in every continent except South America, with several living in close proximity to humans; for example, I have seen the hooded crow *C. corone cornix* fighting its reflection in the gilded domes of the Kremlin and the house crow *C. splendens* peering into bicycle baskets in Mombasa.

C.A. Henley

The two black choughs (genus *Pyrrhocorax*) closely resemble the true crows but have finer bills, one scarlet and one yellow. Both live in rocky environments in Eurasia, either on coastal cliffs or in mountains where they live above the treeline for most of the year, eating mainly insects and berries.

In contrast the two nutcrackers (genus *Nucifraga*) are usually permanent residents of conifer-clad mountains, one in America and the other in Europe and Asia; they are remarkable for the extent of their dependence on cached stores of pine seeds to last them through hard winters. The distinct, but very similar gray jays (genus *Perisoreus*) have much the same distribution to that of the nutcrackers; they too are permanent residents of coniferous forests, one species in North America and the other in Eurasia and Asia.

The true magpie *Pica pica* is common throughout most of North America, Europe, and Asia; it is easily identified by its black and white plumage, long tail, and large family groups. Magpies like thick shrubby cover for their roofed nests, but they forage for insects and seeds in the open, especially in agricultural lands and, more recently, suburbia. They occasionally rob other birds' nests of eggs and young. In California a second species, *P. nuttalli,* is clearly recognized by its yellow bill. The azure-winged magpie *Cyanopica cyana* of Spain and eastern Asia is a near relative with a similar lifestyle.

The blue, green, and Whitehead's magpies (genus *Cissa*) are about the same size as true magpies but have even longer tails. The blue magpies are birds of thickets and woodlands occurring throughout the Indian subcontinent, and from Southeast Asia to Taiwan and to the larger Indonesian islands of Java and Borneo. In this region the green and chestnut-red species of this genus tend to live in jungle and forest.

The 10 treepies (genus *Crypsirina*) share much the same distribution as the *Cissa* magpies but are more plainly colored in shades of gray, black, and white; they have black feet, strongly curved black bills, and long tails. They spend most of their time in trees, frequently in mixed-species flocks of foraging birds, especially drongos.

True jays (three species in the genus *Garrulus*) have unusual blue and black barred feathers, and are confined to Europe and Asia, where they eat a wide range of foods. They are especially fond of acorns, which they may cache. The blue or American jays are very varied and many have strikingly beautiful plumage. Although the 30 or so species are placed in 7 different genera, they are thought to have originated from the same stock and occur throughout the Americas from Canada to Argentina, in scrub, woodland, and forest. Most feed mainly on insects and fruits, but some have become extremely specialized. For example, wherever the piñon pine grows in western North America it provides the pinyon jay with a staple

Wilhelm Möller/Ardea Photographics

John Cancalosi/Auscape International

▲ Both sexes of the Eurasian jay cooperate in raising their brood of young. Normally noisy and demonstrative birds, jays are extremely quiet and inconspicuous when nesting.

◄ Noisy, lively, and inquisitive, the green jay often visits farmyards, ranches, and suburban gardens of the more arid parts of Central America. It favors bushy thickets and streamside vegetation.

diet of seeds; the birds cache the seeds to provide a reserve for the breeding season.

Ground jays are a distinct group of corvids that inhabit the high deserts of Central Asia. Four species (genus *Podoces*) are much the same size as the other jays, but the fifth, Hume's ground jay *Pseudopodoces humilis,* is the size of a sparrow.

Finally, three problematical birds are currently included in the Corvidae for want of a better place to put them. These are the Ethiopian bush-crow *Zavattariornis stresemanni;* the piapiac *Ptilostomus afer* of West and Central Africa, and the crested jay *Platylophus galericulatus* of Malaysia and Indonesia.

PART THREE
REPTILES &

AMPHIBIANS

INTRODUCING REPTILES AND AMPHIBIANS

HAROLD G. COGGER

Scientists divide the animal kingdom into several major groups for classification purposes. By far the largest group is the invertebrates: it contains about 95 percent of the millions of known species of animals, including sponges, mollusks, crustaceans, and insects. Reptiles and amphibians are vertebrates—they belong to a group of animals characterized by having a bony backbone, a flexible but strong support column of articulated sections (vertebrae) to which the other body structures are attached. Vertebrates include not only amphibians (class Amphibia) and reptiles (class Reptilia), but also fishes (about three classes), birds (class Aves), and mammals (class Mammalia). The differences between reptiles and amphibians are generally more obvious than their similarities, although their joint study under the name "herpetology" (from the Greek *herpo*, to creep or crawl) is a scientific tradition dating back nearly two centuries.

DIVERSE FORMS

The fossil record shows that both amphibians and reptiles were once much more diverse in size and structure than they are today (dinosaurs being the best known examples), and so there is a tendency to regard today's forms as a mere shadow of their former glory, having been displaced in dominance by the warm-blooded birds and mammals. However, the amphibians, with 4,250 living species, together with the 7,000 living reptiles, jointly outnumber the birds (9,000 species) and mammals (about 4,250 species). They are in no way more primitive than birds and mammals. Their behavior and physiology are just as complex and equally well adapted to the astonishingly varied environments they inhabit.

BODY TEMPERATURE AND METABOLISM

One significant factor in the distribution and behavior of modern reptiles and amphibians is their inability to produce sufficient internal metabolic heat to maintain a constant body temperature, as do almost all birds and mammals. For this reason they are often termed "cold-blooded", or "ectotherms". Although some reptiles are able to generate enough metabolic heat to raise their temperature for a specific purpose for limited periods (for example, female pythons brooding their eggs), none can constantly maintain a temperature much higher than that of their surroundings without an external heat source. Their dependence on external sources of heat to reach and maintain an "active" body

▲ Reptiles vary enormously in size and structure, reflecting adaptations suited to almost every conceivable habitat throughout the tropical and temperate regions of the world. Because of their somewhat greater dependence on moisture, amphibians tend to be rather less diverse in size and shape. In both groups many species are unremarkable in color but certain poisonous snakes and tropical frogs rival birds in their bold patterns and gaudy hues.

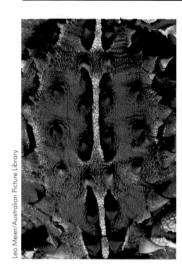

Leo Meier/Australian Picture Library

▲▼ *(Above) Covered with a water-impermeable barrier of dry, horny scales, like the skin of this Australian thorny devil* Moloch horridus *in close-up, reptiles can in general thrive in much drier habitats than amphibians can tolerate. (Below) In contrast to reptiles, many amphibians can extract some or all of their oxygen requirements through their skin. This is Darwin's frog* Rhinoderma darwinii.

temperature greatly limits their ability to live and breed in colder climates, with the consequence that they are most abundant in tropical and warm temperate regions.

While being cold-blooded can have disadvantages, such as being more vulnerable to predators at low environmental temperatures, it also has a number of advantages. Amphibians and reptiles can simply "shut down" when conditions are unsuitable—for example, if it is too cold or food is scarce—and therefore do not have to use up large amounts of stored energy to keep themselves warm.

SKIN AND SCALES

The skin of most amphibians is thin and highly glandular, and needs to be kept relatively moist to function effectively. It is also highly permeable, allowing water to be absorbed or lost, and is often involved in gas exchange as an adjunct to normal respiration through the lungs. This means that most amphibians are restricted to moist or humid environments.

The skin of reptiles differs from that of amphibians in that the outer layer is thickened to form scales, composed of keratin; some areas are further thickened to form tubercles and crests. In many reptiles (and a few amphibians) small bones develop in the skin just below the

surface. These dermal bones, or osteoderms, add strength and protection to the skin and make it even more impervious to water loss. Skin color and color changes are largely determined by pigment cells just below the outer layers of skin.

BODY STRUCTURE

From the muscular fins of their fish ancestors, the first amphibians evolved four limbs for moving about on land. Subsequent amphibians and reptiles share this basic tetrapod (four-limbed) vertebrate skeletal structure.

This basic skeletal structure has become highly modified in some amphibians and reptiles. An extreme adaptation is found in turtles, in which the greater part of the vertebral column, the ribs, and the limb girdles, have become fused with dermal bones to form a protective shell. In frogs, the development of an efficient jumping mechanism has called for greater rigidity of the vertebral column and a solid structure to support the enormous mass of hind limb muscles. This has been achieved by a great reduction in the number of vertebrae and the incorporation of several vertebrae into a single long supporting bone called the urostyle.

Modern amphibians generally have simple pedicellate teeth, which are used to grasp the invertebrates on which they feed, but in many species the teeth are reduced or absent. Teeth in reptiles vary greatly in form and structure, from simple blunt teeth used to grasp and partially crush prey, to those with broad grinding surfaces or sharp shearing edges. Perhaps the most specialized are the fangs of snakes that have evolved grooves or hollows, making them efficient venom-delivering hypodermic needles.

In most amphibians and reptiles the tongue is an important adjunct to the teeth and jaws in catching prey. It is usually very muscular, flexible, and extrusible, and is often sticky with mucus to adhere to prey such as insects. It is also used to manipulate food within the mouth. In some groups of reptiles, such as monitor lizards and snakes, the tongue has become a specialized sense organ which senses and locates prey, but otherwise is not directly involved in the capturing of prey.

INTERNAL ORGANS

The soft body structures of adult amphibians and reptiles are similar to those of other vertebrate animals. A trachea (windpipe) leads from the glottis to paired lungs, although the lungs may be reduced or absent in some aquatic species that can absorb almost all the oxygen they need through their skin. In most snakes the left lung has been lost during the course of evolution; and in many sea snakes the lung now extends forward along the length of the trachea and is involved in

ZEFA/Ziesler

regulating the snake's buoyancy when diving. The circulatory system is complex and efficient, but, with the exception of the crocodiles, the heart of amphibians and reptiles has only three chambers—two auricles or atria and a partially divided ventricle—although the flow is arranged to minimize mixing of oxygenated (arterial) and de-oxygenated (venous) blood in the single ventricle. In crocodiles, as in birds and mammals, the ventricle is completely divided by a septum, which prevents any mixing of arterial and venous blood and is much more efficient than the system found in other reptiles and in amphibians. While many different food specialists have developed among the amphibians and reptiles, once past the mouth the food is digested in a fairly standardized vertebrate digestive system. An important feature of both amphibians and reptiles—in which they differ from most mammals—is the cloaca, a large chamber just inside the vent, into which the gut and the ducts of urinary and sexual organs all enter.

For the majority of amphibians, living as they do in humid environments, water conservation is not a major problem. Consequently their nitrogenous waste products are converted into urea in the liver and then excreted by the kidneys and leave the body in a fluid urine. The same method is used by reptiles that live in moist or aquatic environments, but for reptiles in very dry environments such as deserts, where free water is scarce or rare, body water is too precious to be lost as urine. Consequently in the majority of reptiles the liver converts nitrogenous waste into uric acid which, after excretion by the kidneys, passes to the cloaca where water is resorbed and solid uric acid crystals deposited; these leave the · body as a moist paste, using only a fraction of the water that would be needed to excrete urea as urine.

THE SENSES

For most amphibians and reptiles three senses are critical to survival: sight, hearing, and olfaction (smell). The relative importance of each is strongly correlated with the lifestyles of any particular species.

In frogs in which the males attract females by calling to them, hearing clearly needs to be acute and discriminating if the calls of individual males of one species are to be distinguished from the background noise of other calling males or other species. The males of many lizards use brilliant colors to attract mates and to warn other males away from their territories. Clearly, acute vision which recognizes those colors is essential to such species.

The eyes of amphibians and reptiles are similar in general features to those of other land vertebrates, with characteristic cornea, iris, lens, and retina. In most amphibians there is an

Anthony Bannister/NHPA

immovable upper eyelid and a movable lower eyelid, the upper part of which (the nictitating membrane) is usually transparent. In reptiles the lower eyelid is also usually movable and at least partly scaly, but in many lizards, and in all snakes, the lower and upper eyelid have become fused and the eye is covered by a large, fixed, transparent disk—the spectacle—which protects the eye from damage. The spectacle becomes scratched and dirty over time, and is shed with the skin at each sloughing cycle.

The ears are also typical of other land vertebrates. Sounds are received on a membrane on each side of the head—the tympanum—and transmitted by one or more fine stapedial bones to the inner ear. However, many amphibians and reptiles have limited reception of airborne sounds, as the tympanum is absent or covered by skin, and the inner bony structures are often modified to receive vibrations through other body tissues.

Olfactory senses (smell and taste) are also moderately to well-developed in most amphibians and reptiles. The most important olfactory organ in both amphibians and reptiles is a vomeronasal organ called Jacobson's organ, lying in the roof of the mouth. In amphibians and many reptiles it "tastes" the food in the mouth. In other reptiles, especially monitor lizards and snakes, the protruding tongue "tests" its environment by

▲ *Mammals and birds shed and replace their skins constantly, cell by cell, but reptiles typically do so all at once, like this sloughing Bibron's gecko Pachydactylus bibroni of Africa.*

▼ *Containing both rods and cones, the eye of a crocodile is thought to be capable of color vision.*

David P. Maitland/AUSCAPE International

collecting minute particles which are passed for identification to Jacobson's organ by the tip(s) of the tongue. In this way such reptiles can follow the trail even of moving prey.

REPRODUCTION AND LIFE CYCLES

The life cycle of an amphibian is fundamentally different from that of a reptile. Reptiles produce eggs within which the embryo develops and hatches (or is born, in the case of live-bearing species), generally as a miniature replica of the adult. Amphibians are characterized by a two-stage life cycle in which the eggs hatch into aquatic, gill-bearing larvae (for example, tadpoles) which eventually metamorphose into air-breathing, mostly terrestrial adults. Occasionally the entire tadpole stage takes place within the egg or within the body of the mother.

Almost all amphibians produce eggs, varying in number from a single egg to many thousands in a clutch, which are deposited in water or in humid sites on land. They are fertilized externally or internally. Most male amphibians lack any form of intromittent organ with which to introduce sperm inside the female, but some are able to protrude part of their cloaca into that of the female. In most frogs, the male simply expels his seminal fluid over the eggs as they are extruded from the female's body. In salamanders, the male deposits a package of sperm (the spermatophore), which is picked up by the cloacal lips of the female and is held in her cloaca until needed for fertilization.

The egg of a reptile results from internal fertilization. With the exception of the tuatara, in which sperm are exchanged between male and female through cloacal contact, reptiles have well-developed intromittent organs to deliver the seminal fluid. In lizards and snakes these organs are paired, so that each male has two functioning penises. Turtles and crocodilians have only a single penis.

The reptile egg is substantially more complex than that of an amphibian, and much better adapted to survival on land. In many reptiles, the egg does not develop a shell, and development occurs within the mother's body; the embryonic membranes are retained and in some cases form a primitive placental connection with the mother.

REPTILE AND AMPHIBIAN EGGS

The eggs of amphibians do not have a shell or any other protective membrane and are therefore very vulnerable to desiccation. Each fertilized egg is surrounded by a dense protective jelly, and contains yolk to nourish the developing young until it hatches; the embryo's waste products simply permeate out through the surrounding jelly.

Most modern reptiles lay eggs, but they differ significantly from those of amphibians. Typically, the developing embryo is cushioned by a series of fluid-filled sacs within a protective calcareous shell. The shell physically protects the embryo and although permeable to gases, and somewhat permeable to water, greatly reduces water loss, allowing the egg to develop safely in relatively dry locations on land. Inside the shell the yolk provides food for the developing embryo, and three embryonic membranes carry out special functions. The allantois is a sac in which the embryo's waste products are safely stored. It is partly fused with another sac, the chorion, which lies against the inner surface of the shell, and both contain blood vessels, which assist in exchanging oxygen needed by the embryo through the permeable egg shell. The third membrane, the amnion, forms a fluid-filled sac around the embryo, which cushions it from external disturbance and prevents dehydration.

This kind of egg, unlike that of an amphibian, is essentially a closed system, in which all of the embryo's basic needs are met. It is known as a cleidoic egg.

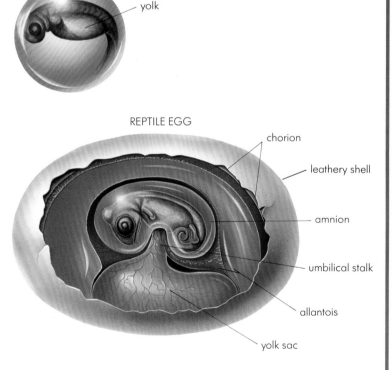

AMPHIBIAN EGG

yolk

REPTILE EGG

chorion

leathery shell

amnion

umbilical stalk

allantois

yolk sac

▲ A typical amphibian egg (top) and a reptilian egg, showing membranes and leathery shell (bottom).

CLASSIFYING REPTILES AND AMPHIBIANS

JAY M. SAVAGE

Biologists have discovered, described, and given names to about 1.5 million of the many millions of species of plants, animals, and microorganisms that exist at present. They have also named many species that formerly lived on Earth but are now extinct. The role of classification is to provide a unique name for each of these organisms and put them in a hierarchy of increasingly inclusive taxa (groups) based on their evolutionary relationships, so that they can be unequivocally recognized and associated with other species having a common ancestry.

SYSTEMATICS AND TAXONOMY

The science of discovering the diversity and evolutionary relationships of organisms is called systematics, and aspects of systematics involved with naming and classifying organisms form the subdiscipline of taxonomy. The basic taxon (group) is the species, and ideally all higher or more inclusive taxa comprise an ancestral species and all its descendants. Determination of this relationship is based on all members of a group sharing one or more derived features (evolutionary novelties). For example, all frogs and toads have two of the ankle bones greatly elongated to form a fourth segment in the hind leg (all other limbed amphibians and reptiles have three segments in the hind leg). This condition provides greater leverage for leaping and serves to distinguish the group, the order Anura, in which frogs and toads are placed.

NAMES AND CLASSIFICATION

Scientists use Latinized names for groups of organisms in order to provide universality and stability and to avoid the necessity of translating the name into many different languages. No matter what one's native language, the Latinized scientific name is immediately associated with the same group of organisms.

The fundamental category in all classifications is the species. Species are populations of organisms that share one or more similarities not found in related species. Species form closed genetic systems, although some closely related species may occasionally hybridize. The name of a species is composed of two words: the first is the name of its genus (the generic name) and indicates its closest relationships; the second name (the specific name) is uniquely used for that species within a genus. The generic name always

▼ A red-eyed leaf-frog Agalychnis callidryas clasps a heliconia in a Costa Rican rainforest. One obvious external feature sets frogs and toads apart from all other amphibians: their hind limb consists of four hinged elements, not three.

J. Cancalosi/AUSCAPE International

▲ *Even such intangible areas as vocalizations and behavior can yield features of value in classifying animals, and in principle there is no reason why any one set of features is inherently any more valid than another. But for obvious, practical reasons those features involving external appearance and structure (morphology) are most useful in the widest range of circumstances. For example, careful inspection of the scales on a snake's head, like this carpet python Morelia spilotes, reveals patterns that vary from one group to another, and are useful in distinguishing them.*

the nerve cord and extends the length of the body; and gill arches and gill pouches (which are present only embryonically in reptiles, birds, and mammals). In other chordates, including larval amphibians, the gill arches support functional gills and the gill pouches in the throat region open to the outside.

Three subphyla are recognized within the Chordata; amphibians and reptiles are located in the subphylum Craniata. Animals belonging to this group have chambered hearts, semicircular canals in the inner ear, skeletal gill supports, and a protective cartilaginous or bony cranium around the brain. Most members of this group also have segmentally arranged skeletal elements (vertebrae) surrounding and protecting the spinal cord.

The next division in the hierarchy is the class. Amphibians and reptiles are placed in different classes, the Amphibia and Reptilia, respectively. Classes are further divided into orders; there are three living orders of amphibians and four of reptiles. Orders are divided into families, families into genera, and each genus comprises one or more species. The classification adopted in this book also makes use of the suborder, which is a category more inclusive than a family but less inclusive than an order.

These hierarchical categories form a set of taxa defined by shared similarities. This means that species within a genus are more similar to one another than to species in other genera within the same family; genera in the same family are more similar to one another than to genera in other families; and so on. It also means that, progressing up the hierarchy, fewer and fewer features are shared by the species included in an order, a class, a phylum, or a kingdom. The hierarchy thus reflects in reverse the branching pattern of the evolution of life, because the decrease in similarity correlates with time of divergence—that is, the greater the similarity the more recent the divergence.

begins with a capital letter, and the specific name is always written in the lower case, and both names are always printed in *italics*; for example, the name of the Gaboon viper of tropical Africa is written *Bitis gabonica*. Species names are frequently based on a distinctive feature, or may refer to the species' place of origin, or may honor a person by using a Latinized form of their name.

In some instances when populations of any one species are separated geographically and have relatively consistent differences, they may be recognized as subspecies and then have a third name: for example, *Lampropeltis triangulum triangulum* for the eastern milk snake and *Lampropeltis triangulum gentilis* for the plains milk snake of North America.

Scientists also use a standard set of terms for each level in the hierarchical arrangement of taxa and a single Latin name (always capitalized) for each different taxon. The most inclusive category to which amphibians and reptiles belong is the kingdom Animalia. This taxon contains all organisms that are multicellular, obtain their energy by ingesting food from the surroundings, and usually develop from the coming together of two cells, a large egg and a small sperm.

The kingdom Animalia is composed of 35 phyla, each phylum representing a distinctive major biological organization and lifestyle. Amphibians and reptiles belong to the phylum Chordata, which consists of animals that have a single dorsal nerve cord; a cartilaginous notochord (at least embryonically) that lies below

VERNACULAR AND COINED NAMES

Amphibians and reptiles are known by vernacular or common names that vary from language to language, country to country, and even within the same country. In some cases, the same name may be used for a number of similar appearing species. In others, one name is used for entirely different creatures; for example, in different parts of the United States the name "gopher" is applied to a turtle, a ground squirrel, and a subterranean rodent called a pocket gopher. Because of the potential for confusion these names are not used in formal, scientific identification. Species names that had never been used by local people for amphibians and reptiles have in recent years been coined for use in popular works or field guides, especially in English. These names, alas, lack even the dignity of being truly vernacular, as they are inventions of their authors, duplications of scientific names, and usually uninformative when read by someone who has a different native language.

SPECIES

There are about 4,250 species of living amphibians and nearly 7,000 species of living reptiles. Although some of the larger and more spectacular representatives—such as the "hairy frog", the leatherback turtle, and the Komodo dragon—are very distinctive, many closely related forms are characterized by subtle differences. External form and structure, size and body proportions, and coloration are usually the basis for separating allied species, with differences in the number, shape, and structure of scales being emphasized in reptiles. Recent studies of chromosomes have been valuable in distinguishing between cryptic species—those whose external features differ only slightly from their closest relatives.

Because many kinds of amphibians and reptiles are small and secretive, occur very locally, or are inhabitants of unexplored regions of the Earth (especially tropical rainforests), previously undescribed species are discovered each year. During the past decade about 40 new species of amphibians (mostly frogs and toads) and 25 species of reptiles (mostly lizards) have been described each year. Sometimes, as our knowledge of a particular group increases, forms previously thought to be distinctive are shown to belong to a single species. These factors, plus the cases in which conflicting evidence (or its interpretation) leads different scientists to disagree on the status of named forms, reflect the differing numbers of species recognized by one or another authority. It is safe to say, however, that new species of amphibians and reptiles will continue to be discovered at these rates well into the future. It should be noted that while amphibians and reptiles are often regarded as minor remnants of an earlier evolutionary radiation, the number of species of living amphibians is similar to that of mammals, and there are many more species of living reptiles than of mammals.

GENERA

A genus contains one or more species that share at least one derived, or advanced, feature. The classification of amphibians into different genera is usually based on internal features, especially those of the skeleton, because their smooth bodies possess relatively few significant external differences. Skeletal morphology is also used in reptiles, but scalation (especially the scales on the head) frequently provides a basis for generic groupings. Occasionally a single, very distinctive living species may be classified in a genus of its own because of one or more unique features; for example, most American and Eurasian treefrogs are placed in the genus *Hyla*, but the spiny-headed or crowned treefrog is placed in a separate genus, *Anotheca*, as *Anotheca spinosa*. Frequently, however, genera originally thought to contain only a single species have newly discovered living or fossil species added to them as the result of additional research. Currently we recognize about 410 genera of living amphibians and 898 genera of living reptiles.

FAMILIES

Genera are clustered into different families on the basis of one or more shared features of internal morphology. Frequently it is difficult for the layperson to readily identify which family a species belongs to, from only its external features; even a specialist may have a problem with family placement of unfamiliar amphibians, for external resemblances are deceiving. A few families of amphibians have only a single very distinctive genus containing one or a few species. In theory such families should be regarded as sole relicts of a larger group of species, the majority of which have become extinct.

All family names end in the suffix "-idae", and because many groups of amphibians and reptiles lack vernacular names the family name is often rendered into English as a word ending in "-id"; for example, the lizard family Agamidae may be called agamids. Families may be divided into two or more subfamilies when there are marked differences among the included genera. Amphibians are placed into 37 families represented by living species, and reptiles are placed into 50 families.

▼ A species is essentially a closed genetic system: interbreeding takes place between individuals within the system but not outside it. Usually such systems have close neighbors, forming clusters called genera. But in some cases most of the species forming the cluster die out, leaving perhaps only a single survivor at the present time. This probably happened in the case of the crowned frog Anotheca spinosa of Central America, a species that has no close relatives anywhere.

Michael Fogden/Oxford Scientific Films

▲ *Now confined to a few tiny islands in New Zealand, the two species of tuatara are the sole surviving remnants of an entire order of reptiles that had their heyday some 200 million years ago, but died out one by one during the intervening eons.*

ORDERS

Orders are clusters of families that share one or more derived features and are obviously separated from other such groups. Most species of amphibians may be immediately recognizable as a salamander or newt (order Caudata), a limbless caecilian (order Gymnophiona), or a long-limbed, tailless frog or toad (order Anura). Similarly, it is difficult to confuse a turtle (order Testudinata), a typical lizard, amphisbaenian, or snake (order Squamata), or a crocodilian (order Crocodilia)

with one another. The sole living representatives of the order Rhynchocephalia, the tuatara of New Zealand, superficially resemble lizards but are unlike any lizard in obvious details of scalation and numerous internal features. Tuatara are the only contemporary members of their order, but a number of extinct rhynchocephalians are known from the Mesozoic, 245 to 65 million years ago.

There are also many extinct orders (and their families, genera, and thousands of extinct species) of amphibians and reptiles that have no representatives alive today.

HIGHER CLASSIFICATION

Scientists agree that modern amphibians are related to one another and in turn are related to a cluster of ancient amphibians that date back to the upper Devonian, about 380 million years ago, but became extinct by the end of the Triassic, 200 million years ago. The difficulty of interpreting these relationships and the status of other fossil "amphibians"—including the lineage most closely allied to early reptiles—leaves the higher classification of all so-called "amphibians" in a state of flux. It therefore seems best not to use the conflicting schemes of classification at this time.

Reptiles, on the other hand, form three distinct subclasses, based primarily on differences in skull and jaw characteristics. The subclass Anapsida includes the turtles and a number of extinct lines.

THE WARTY NEWT: AN EXAMPLE OF CLASSIFICATION

The concepts of zoological classification and nomenclature may be illustrated by using the warty newt *Triturus cristatus* of western Eurasia as an example. The genus *Triturus* contains 12 species of newts all from this region. *Triturus* is grouped with a series of 14 other genera found in the temperate zone of North America and eastern Eurasia and the temperate and subtropical zones of eastern Asia, into the family Salamandridae. Together with a number of other families of amphibians, in which the adults have limbs and a tail and their aquatic larvae have true teeth, the Salamandridae belong to the order Caudata. This order in turn belongs to the class Amphibia. The Amphibia are included with the class Reptilia, class Mammalia and several classes of aquatic jawed and jawless fishes, in the subphylum Craniata which, together with tunicates (subphylum Urochordata) and lancelets (subphylum Cephalochordata) constitute the phylum Chordata. The Chordata are one of 35 phyla of multicellular, heterotrophic organisms that form the kingdom Animalia.

Kingdom: Animalia
Phylum: Chordata
Subphylum: Craniata
Class: Amphibia

Order: Caudata
Family: Salamandridae
Genus: *Triturus*
Species: *Triturus cristatus*

▲ *The warty, or crested newt Triturus cristatus of Europe.*

The subclass Lepidosauromorpha contains a diversity of fossil forms (for example, ichthyosaurs and plesiosaurs), plus tuatara, lizards, and snakes. All other reptiles are placed in the subclass Archosauromorpha, which includes the extinct dinosaurs, pterosaurs, and phytosaurs, the crocodilians, and technically the birds, although birds are traditionally placed in a separate class.

CHANGES IN CLASSIFICATION

Classifications are modified to reflect newly understood evolutionary relationships. In this way the information content of the classification system increases. The recent integration of chromosomal and molecular data into evolutionary studies, and better ways of evaluating existing classifications, demonstrate this process. As a result, classifications evolve, and they must be understood as impermanent statements of current knowledge. The present framework that recognizes three orders of living amphibians and four orders of living reptiles seems firmly established, but some revision in the number and positioning of families, especially for those with larger numbers of species, is to be expected as relationships become clearer.

LIVING ORDERS, SUBORDERS, AND FAMILIES OF AMPHIBIANS AND REPTILES

Because the classification of amphibians and reptiles is undergoing constant revision, there is no definitive classification list available. The following represents a consensus of the academics who have contributed to this publication.

Some families of amphibians and reptiles have vernacular English names that are widely used. Other families have no equivalent English common name, and it seems best to use an anglicized version of their scientific name where such names are needed, rather than a coined one. For example, salamanders of the family Hynobiidae are called hynobiids.

CLASS AMPHIBIA

ORDER
CAUDATA — SALAMANDERS AND NEWTS

Suborder Cryptobranchoidea
Cryptobranchidae — Cryptobranchids
Hynobiidae — Hynobiids

Suborder Salamandroidea
Proteidae — Mudpuppies, waterdogs, and the olm
Dicamptodontidae — Dicamptodontids
Amphiumidae — Amphiumas (congo eels)
Salamandridae — Salamandrids
Ambystomatidae — Mole salamanders
Plethodontidae — Lungless salamanders
Sirenidae — Sirens

ORDER
GYMNOPHIONA — CAECILIANS
Rhinatrematidae — Rhinatrematids
Ichthyophiidae — Ichthyophiids
Uraeotyphlidae — Uraeotyphlids
Scolecomorphidae — Scolecomorphids
Caeciliidae — Caecilids
Typhlonectidae — Typhlonectids

ORDER
ANURA — FROGS AND TOADS

Suborder Archaeobatrachia
Leiopelmatidae — The tailed frog and New Zealand frogs
Discoglossidae — Discoglossid frogs

Suborder Pipoidea
Pipidae — Pipas and "clawed" frogs
Rhinophrynidae — Cone-nosed frogs

Suborder Pelobatoidea
Pelobatidae — Spadefoots and megophryine frogs
Pelodytidae — Parsley frogs

Suborder Neobatrachia
Sooglossidae — Seychelles frogs
Myobatrachidae — Australasian frogs
Heleophrynidae — Ghost frogs
Leptodactylidae — Neotropical frogs
Bufonidae — Toads and harlequin frogs
Allophrynidae — The allophrynid frog
Brachycephalidae — Saddleback frogs
Rhinodermatidae — Darwin's frogs
Pseudidae — Natator frogs
Hylidae — Hylid treefrogs
Centrolenidae — Glass frogs
Ranidae — Ranid frogs
Hemiotidae — Shovel-nosed frogs
Hyperoliidae — Reed and lily frogs
Rhacophoridae — Rhacophorid treefrogs
Arthroleptidae — Squeakers
Dendrobatidae — Poison frogs
Microhylidae — Microhylids

CLASS REPTILIA

SUBCLASS ANAPSIDA

ORDER
TESTUDINATA — TURTLES, TERRAPINS, AND TORTOISES
Suborder Pleurodira — SIDE-NECKED TURTLES
Chelidae — Snake-neck turtles
Pelomedusidae — Helmeted side-neck turtles

Suborder Cryptodira — HIDDEN-NECKED TURTLES
Chelydridae — Alligator (snapping) turtles
Cheloniidae — Hardback sea turtles
Dermochelyidae — Leathery turtle
Trionychidae — Holarctic and paleotropical softshell turtles
Carettochelyidae — Papuan softshell turtle
Kinosternidae — Mud and musk turtles
Dermatemydidae — Central American river turtle
Emydidae — New World pond turtles and terrapins
Testudinidae — Land tortoises
Bataguridae — Old World pond turtles

SUBCLASS LEPIDOSAUROMORPHA

ORDER
RHYNCHOCEPHALIA — TUATARA
Sphenodontidae — Tuatara

ORDER
SQUAMATA — SQUAMATES

Suborder Sauria — LIZARDS
Iguanidae — Iguanids
Agamidae — Agamids
Chamaeleontidae — Chameleons
Gekkonidae — Geckos
Pygopodidae — Flap-foots
Xantusiidae — Night lizards
Lacertidae — Lacertids
Teiidae — Macroteiids
Gymnopthalmidae — Microteiids
Scincidae — Skinks
Cordylidae — Girdle-tailed lizards
Dibamidae — Dibamids
Anguidae — Anguids
Xenosauridae — Knob-scaled lizards
Helodermatidae — Beaded lizards
Lanthanotidae — Earless monitor lizard
Varanidae — Monitor lizards

Suborder Amphisbaenia — AMPHISBAENIANS
Bipedidae — Ajolotes
Amphisbaenidae — Worm-lizards
Trogonophidae — Desert ringed lizards
Rhineuridae — Rhineurid

Suborder Serpentes — SNAKES
Anomalepididae — Blind wormsnakes
Typhlopidae — Blind snakes
Leptotyphlopidae — Thread snakes
Aniliidae — Pipe snakes
Uropeltidae — Shield-tailed snakes
Boidae — Boas, pythons, and wood snakes
Acrochordidae — File snakes
Colubridae — Harmless and rear-fanged snakes
Elapidae — Cobras, kraits, coral snakes, and sea snakes
Viperidae — Adders and vipers

SUBCLASS ARCHOSAUROMORPHA

ORDER
CROCODILIA — CROCODILIANS
Alligatoridae — Alligators and caimans
Crocodylidae — Crocodiles
Gavialidae — Gharials

REPTILES AND AMPHIBIANS THROUGH THE AGES

OLIVIER RIEPPEL

The first backboned animals (vertebrates) evolved in the sea at the end of the Cambrian period about 500 million years ago. During the next 150 million years they continued to evolve in the world's oceans into amazingly diverse groups of fishes. But to colonize the land these aquatic vertebrates had to meet and overcome the tremendous challenges involved in exchanging an aquatic existence for a terrestrial one. These included breathing atmospheric rather than dissolved oxygen, abandoning the buoyancy of water, and modifying the body structure to meet the high gravitational forces encountered on land.

FROM WATER TO LAND

The first vertebrates to make the transition from water to land were the amphibians. In the late Devonian period about 360 million years ago, they became the first tetrapods—that is, the first backboned animals with four articulated legs—from which all later vertebrates (reptiles, birds, and mammals) evolved. Indeed, once the amphibians had evolved the basic structures needed to colonize the land, the evolutionary opportunities for vertebrates were immense. Only about 50 million years after the first amphibians appeared, a relatively short time in geological terms, the first reptiles evolved

from an amphibian ancestor, and by the Triassic period about 240 million years ago they had become the dominant land-dwelling animals. This was the beginning of the era of the mighty dinosaurs.

FROM FISHES TO AMPHIBIANS

The origin of early tetrapods is still a matter of scientific debate. Among the animals alive today, the closest relatives to the tetrapod ancestor are probably the lungfishes. They too have internal nostril openings (choanae), allowing air to be taken in while the mouth is closed, as well as lungs and a functionally divided heart.

▼ Eusthenopteron was a lobe-finned fish of the Devonian period, close to, if not directly on, the evolutionary line that led to the amphibians. It had lungs as well as gills and could breathe air during excursions over land when its pond dried up. Its paired fins, supported by muscular lobes with bony elements, were the precursors of the tetrapod limb bones.

Another living fish sharing similarities with tetrapods is the famous coelacanth, which was discovered in the seas near South Africa as recently as 1938, bearing testimony to the survival of a group thought to have died out 65 million years ago. The coelacanth's paired fins make the kind of movements that we assume tetrapod ancestors made, although its anatomical details are not closely related to those of tetrapods. Fossils of lobe-finned fishes of the Devonian period show closer similarities, both in skull structure and in the structure of the pectoral fins.

EARLY AMPHIBIANS

The earliest known amphibians, *Ichthyostega* and *Acanthostega*, whose fossils were found in Greenland, lived about 360 million years ago. They belong to a group known as labyrinthodonts, named for the labyrinthine infoldings of the pulp cavity of their teeth. These early forms are interesting because they retain some fish-like characteristics, such as a fish-like tail and a rudiment of the bony gill cover in the cheek region of the skull. In fish the bone supporting the bony gill cover plays a crucial role in the mechanism of gill ventilation, whereas during the course of tetrapod evolution this bone was eventually reduced to a slender rod (the stapes) capable of transmitting sound waves picked up by a tympanic membrane (the eardrum) to the pressure-sensitive inner ear.

In fish the pectoral or shoulder girdle forms the back end of the skull. In tetrapods such as amphibians and reptiles, this girdle is not attached to the skull but is separated from it by a neck which allows the skull to move more freely. The pectoral and pelvic girdles anchor and support the limbs, and the vertebral column becomes suspended between the pectoral and pelvic girdles like a bridge, requiring not only modification of soft tissues but also the evolution

of special joints between the vertebrae to permit the backbone to bend sideways. Simultaneous breathing and locomotion may not have been possible for the early amphibians during their excursions on land, because the rib movements needed to ventilate their lungs may have conflicted with the muscle contraction required to undulate the vertebral column from side to side.

"REPTILIOMORPH" AMPHIBIANS

One very primitive tetrapod from the early Carboniferous period about 340 million years ago was *Crassigyrinus*, whose fossils were found in Scotland. From its rather weak limbs, we assume it had aquatic habits. It too retained a rudiment of the original gill cover and had skull proportions similar to those of lobe-finned fishes. But the structure of the skull indicates that *Crassigyrinus* may be related to the Anthracosauria, one of the major divisions of labyrinthodont amphibians.

The anthracosaurs are now considered to belong to the "reptiliomorph" group, including the ancestor of reptiles, birds, and mammals—as opposed to the "batrachomorph" group, which included the ancestor of modern amphibians.

The reptiliomorphs were predominantly aquatic amphibians and retained traces of a lateral line canal system in the skull bones (used to detect changes in water pressure or water currents) and had an elongated body with short limbs adapted for eel-like locomotion. A few had well-developed limbs, such as *Proterogyrinus*, a predominantly terrestrial animal up to 60 centimeters (24 inches) long.

During the Permian period 285 to 245 million years ago, the evolutionary trend in some amphibians seems to have been toward adaptations for terrestrial life. Fossils have been found in "red beds" (red sandstone) in locations that in the Permian were dry coastal plains. *Seymouria*, from the Permian red beds of Texas,

▲ Ichthyostega, *the earliest known amphibian, had a skull similar to that of its fish ancestors, but also had strong limbs and a sturdy ribcage, indicating its terrestrial habits. However, it probably never ventured far from water because it retained a bone-supported fin on its tail and lateral line sensory canals in its skull.*

was about 60 centimeters (24 inches) long, rather massively built, and would have lived mainly on land; it was a reptiliomorph in a number of features, most notably the skull and the vertebral column. *Diadectes*, the earliest tetrapod that can be positively identified as a plant-eater, was up to 3 meters (10 feet) long. The fact that this reptiliomorph amphibian was for a while classified as a reptile highlights the difficulty scientists have in separating reptiles from amphibians on their skeletons alone.

"BATRACHOMORPH" AMPHIBIANS

Among the earliest labyrinthodonts were two representatives of the "batrachomorph" amphibians (Greek *batrakhos* means "frog"), which lived during the early Carboniferous period about 360 million years ago. They are classified in the order Temnospondyli. *Caerorhachis* was predominantly terrestrial, and unlike the reptiliomorphs (and all modern reptiles and amphibians) it did not have the "otic notch", a concavity in the skull above the cheek region which accommodates the tympanic membrane. *Greererpeton*, an aquatic animal, also had no sign of this otic notch; in its skull the stapes was a massive element (similar to that in fossils of lobe-finned fishes) supporting the braincase. The development of a middle ear structure occurred only later in batrachomorph evolution, among the Temnospondyli, apparently independent from the reptiliomorph group. This suggests that the middle ear of modern amphibians has quite a different origin from the middle ear of reptiles and mammals.

Dendrerpeton from the late Carboniferous period of North America, about 290 million years ago, was one of the earliest representatives of the Temnospondyli with an otic notch and predominantly terrestrial habits. In later representatives of this group, such as the larger, massively built *Eryops* and the heavily armored *Cacops*, both found in North America, the skull was still relatively high and narrow, with the eyes on the sides but quite high; the vertebrae and the ribcage were well developed; and strong girdles and limbs supported the animals on their predatory excursions. During the Triassic period when seas inundated the northern continents, many amphibians grew to astonishing size, the body was elongated, and the limbs reduced; in *Metoposaurus* and *Cyclotosaurus* the skull was large (up to 1 meter, or 3¼ feet long) and flat, with the eyes on the top. Among other representatives of the Temnospondyli were the Branchiosauridae, a diverse family whose fossils are abundant in Europe, including not only the aquatic larvae of predominantly terrestrial adults, but also creatures that reached sexual maturity as an aquatic larva, like the axolotl today (order Caudata).

It is generally accepted, on the basis of a

number of characteristics (most importantly the structure of the inner ear and the pedicellate teeth), that all modern amphibians have a common ancestral origin within Temnospondyli.

MODERN AMPHIBIANS

The class Amphibia, subclass Lissamphibia, is represented by the three living orders of amphibians: the Anura (frogs and toads), the Caudata (salamanders and newts), and the Gymnophiona (caecilians).

Perhaps the ancestors of today's amphibians belonged to one particular branch of the Temnospondyli. *Doleserpeton* was a small terrestrial amphibian of the early Permian period about 285 million years ago. Fossil bones found in Oklahoma, United States, share with modern amphibians many common features in vertebral structure. *Doleserpeton* also had teeth of a type known as pedicellate, which occur in today's amphibians, in which the tooth is divided horizontally by unmineralized tissue—quite unlike the labyrinthodont infolding described earlier. Most fossil skulls of *Doleserpeton* are smaller than 12 millimeters (½ inch) in length, so it is possible that they are the skulls of immature animals, and the peculiar tooth structure may be a juvenile feature.

The fossil record of lissamphibians is rather poor, frogs being the notable exception. *Triadobatrachus* from the early Triassic of Madagascar, 245 million years ago, resembled modern frogs in a number of features such as the skull and the elongated hind limbs.

The earliest "true" salamander fossil was found in Russia and lived during the late Jurassic, about 150 million years ago. The earliest caecilian fossil is from the late Cretaceous, more than 65 million years ago, but only a vertebra was found. However, evidence suggests that the caecilian group is as old as the break-up of the Gondwana landmass into the southern continents, which occurred some 200 million years ago.

THE FIRST REPTILES

The oldest supposed reptile fossil comes from the early Carboniferous of Scotland, about 340 million years ago. By 315 million years ago the Captorhinomorpha had evolved; this was a group of rather small and agile lizard-like animals known from fossils found in tree stumps of the giant lycopod *Sigillaria* tree of the late Carboniferous, as well as in subsequent Permian sediments. Skull proportions had changed, so that these predators no longer had the powerful snapping bite characteristic of labyrinthodont amphibians, but had a hard crushing bite suited to deal with the chitinous armor of their insect prey. The Captorhinomorpha represent a basal group of reptiles from which it is believed all other reptiles evolved.

Until recently, classifications of reptilian diversity have been based mainly on the structure of the skull, with up to five subclasses being recognized. The oldest of these subclasses, which contained the captorhinomorphs and so gave rise to all other reptiles, is the Anapsida. In this group the bony covering of the upper cheek region of the skull is complete, although in turtles (the only surviving anapsids) this region of the skull may have openings around the edges to accommodate various muscles. Their fossil record reaches back into the late Triassic with *Proganochelys*, which differed from modern turtles only in having teeth on the palate and a few other features. Turtle diversity increased in the Jurassic, but has subsequently declined. Despite their apparent ponderousness, turtles have survived on Earth with surprisingly little change in their basic structure for nearly 200 million years.

The Synapsida is a group that appeared in the late Carboniferous, about 300 million years ago, and is characterized by a single low opening in the cheek region of the skull. The earliest synapsids were the pelycosaurs, a diverse group which contained both swift-moving carnivorous reptiles and large, heavily-built plant-eating reptiles. Evolving from pelycosaurs in the mid-Permian period about 250 million years ago, and surviving for nearly 50 million years, were the therapsid or mammal-like reptiles. Formerly believed to have descended from captorhinomorph ancestors, the Synapsida and their descendants, the mammals, are now recognized as a separate evolutionary lineage.

The Diapsida is a group characterized by one (upper) or two (upper and lower) openings in the cheek region of the skull. The earliest example is *Petrolacosaurus* from the late Carboniferous of Kansas. The Diapsida split into two major lineages, the Archosauromorpha (the "ruling reptiles"), and the Lepidosauromorpha, which include the squamates (lizards, amphisbaenians, and snakes) and the tuatara.

The Archosauromorpha is the most diverse group of reptilian evolution. Early representatives were the plant-eating rhynchosaurs which, together with mammal-like reptiles (synapsids), were the dominant animals on land for perhaps 60 million years or more—until the rise of those best known archosaurs, the dinosaurs and their relatives, with their improved locomotor abilities and perhaps more sophisticated temperature-regulating physiology. Along coastal areas lived the prolacertiforms, some highly agile, predominantly terrestrial predators of about 80 centimeters (31 inches) total length, such as *Macrocnemus*. Perhaps the oddest of all was *Tanystropheus*, which lived in the sea preying on fish and cephalopods (squid, octopus, etc.), reaching out at them with its grotesquely elongated neck, as long as body and tail together. Its total length was 6 meters (almost 20 feet).

THE AGE OF DINOSAURS

Dinosaurs ruled the Earth for 140 million years, throughout the Jurassic and Cretaceous periods. This reign came to an abrupt end when, some 65 million years ago, a catastrophic event wiped out the dinosaurs and all other large reptiles, except the crocodiles and turtles.

The early dinosaurs split into two groups: the Saurischia (with a reptile-like pelvis), and the

▼ *The discovery of Deinonychus, a saurischian, helped to shed the myth of the clumsy, sluggish, tail-dragging dinosaur. Grabbing its prey with long forelimbs, this active predator disembowelled its prey with scythe-like claws on its hind limbs, a feat requiring great speed and agility. Its tail was stiffened with bony rods, enabling it to be held well off the ground for balance.*

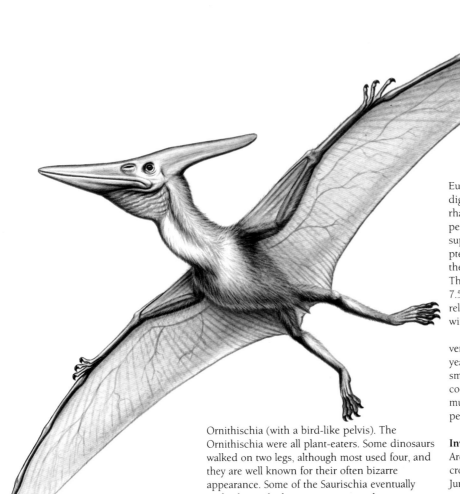

▲ With hollow bones, a reinforced spine, "warm-blooded" metabolism, and an insulative layer of hair, the pterosaurs are now believed to have been more efficient fliers than was first thought. Pteranodon soared far out to sea, feeding on fish caught in a pelican-like pouch. Its narrow wing membranes, supported by elongated fingers, spanned up to 7.5 meters (24½ feet).

Ornithischia (with a bird-like pelvis). The Ornithischia were all plant-eaters. Some dinosaurs walked on two legs, although most used four, and they are well known for their often bizarre appearance. Some of the Saurischia eventually evolved into the largest meat-eaters that ever roamed the continents: the Carnosauria, such as *Allosaurus* and *Tyrannosaurus*. Others became the largest plant-eaters, such as *Diplodocus*, which reached a length of 30 meters (almost 100 feet), and *Brachiosaurus*, whose body, 20 meters (65 feet) long, stood 12 meters (40 feet) high.

Theropods, a group of saurischians, walked on two legs. They were agile predators, including not only one of the smallest dinosaurs known, *Compsognathus*, with an adult size comparable to that of a chicken, but also larger fast-moving animals. Some may have had quite complex brains, and nervous control of their physiology (including internal temperature regulation).

Taking to the air
Birds are thought to have descended from advanced theropods, and one of the links is *Archaeopteryx* which lived during the late Jurassic, 150 million years ago. Birds can therefore be viewed as the last surviving dinosaurs! Birds, however, are not the first vertebrates to have acquired the power of flight. The Pterosauria, first known from late Triassic fossils found in the

European Alps, evolved a much-enlarged fourth digit which supported a wing membrane. The rhamphorhynchoids of the subsequent Jurassic period remained rather small and may superficially have resembled modern bats, but the pterodactyloids of the Cretaceous period included the largest animals that ever took to active flight. The best known is *Pteranodon* with a wingspan of 7.5 meters (24½ feet), and there is evidence that related forms such as *Quetzalcoatlus* had a wingspan of 11 to 12 meters (36 to 39 feet).

But even the pterosaurs were not the first vertebrates invading the air. Perhaps 100 million years earlier, during the late Permian, there was a small group of lizard-like diapsid reptiles, the coelurosauravids, whose greatly elongated ribs must have supported a wing membrane permitting gliding flight from tree to tree.

Invasion of the seas
Archosaurs also invaded the sea. Fossil crocodilians such as the thalattosuchians of the Jurassic period had greatly elongated jaws resembling those of the modern gharial and were well suited to capturing fish. Some forms such as *Geosaurus* and *Metriorhynchus* had limbs transformed into paddles and a tail-bend which may have supported some fin-like structure. We do not know when the crocodiles' ancestors branched off from other archosaurs in the evolutionary tree, but they first appeared in the middle Triassic as terrestrial animals.

The shallow seas accommodated yet another group of reptiles, the Euryapsida, characterized by a single high opening in the cheek region of the skull. This group includes the placodonts, whose name refers to the large crushing tooth-plates on the palate and the hind parts of the jaws. Their front teeth were long and chisel-shaped, suited to pick up hard-shelled mollusks from the sea bed. Some placodonts developed body armor which superficially resembled the shell of a turtle. Also in this group were the small pachypleurosaurs, the larger nothosaurs (growing up to almost 4 meters, or 13 feet), and the plesiosaurs. Pachypleurosaurs and nothosaurs moved through the water by undulating, like an eel, whereas plesiosaurs had large limbs (paddles) which, supported by strong girdles and a rigid trunk, functioned as hydrofoils much as in modern sea turtles.

Ichthyosaurs were the reptiles most perfectly adapted to living in water and resembled dolphins in their external appearance. The relatively small mixosaurs, rarely exceeding 1 meter (3¼ feet) in length, coexisted in great numbers with pachypleurosaurs in inshore areas, but some other species grew to a very large size. *Shonisaurus*, of the late Triassic, reached 15 meters (50 feet) and roamed the open sea.

THE HISTORY OF SQUAMATES

Lizards, amphisbaenians and snakes, which make up the great majority of modern reptiles, are squamates. Squamates are members of the subclass Lepidosauria. The origin of squamates must date back to at least the late Triassic, although the oldest fossils found are from the late Jurassic. By the end of the Cretaceous, there were squamates that we can classify as representatives of living families. However, a study restricted to France showed that modern species of reptiles, like those of amphibians, did not appear until the middle Miocene, about 20 million years ago. Because the sea is a barrier for most squamates, the geographical distribution of their various subgroups more or less reflects the pattern of plate tectonics—the slow drift of continental landmasses across the globe at the rate of a few centimeters a year.

By the Devonian period about 400 million years ago, the world's landmasses had drifted together to form a single supercontinent: Pangea. This was the world into which the first amphibians and reptiles evolved. In the Jurassic period, some 200 million years later, Pangea began to split into two· large supercontinents: a northern one called Laurasia, which included most of modern-day Europe, Asia, and North America; and a southern one called Gondwana, which included modern-day South America, Antarctica, Australia, New Zealand, Africa, and India. Accordingly, most squamate subgroups can be understood as either gondwanan or laurasian in origin.

There was a short period in the late Cretaceous when intermittent land connections allowed animals to cross between North and South America, as they can today, but after the Cretaceous period North and South America remained separated until recent times. Iguanid lizards are an example of a group of southern origin which made its way from South into North America during a Cretaceous connection and, from there, dispersed overland to Europe, where the fossil iguanid *Geiseltaliellus* has been found.

The Atlantic Ocean separated North America from Europe during the early Eocene, but at the beginning of that epoch, 57 million years ago, the animals of the two continents still showed close affinities. (During that time there was a sea between Europe and Asia.) After the complete opening up of the North Atlantic, the European reptiles and amphibians evolved into types restricted to Europe, quite separate from those of North America, examples being the anguid lizards and an extinct group of varanid lizards known as necrosaurids.

Squamate fossils in Australia generally date back to about 30 million years, but the oldest of them are all representatives of typical gondwanan groups such as varanid lizards. Of particular interest is the giant varanid *Megalania* from the Pleistocene (2 million to 10,000 years ago), which reached a body size of 7 meters (23 feet) and weighed almost 600 kilograms (about half a ton).

The disappearance of *Megalania* and many other large species of reptiles such as the 8 meter (24 feet) Australian python *Montypythonoides* (named *after* the comedy series on television), coincides with the extinction of many large mammals toward the end of the Pleistocene, during the period from about 100,000 to 10,000 years ago. The causes of this "megafauna" extinction remain controversial. Some scientists favor climatic changes, whereas others believe that excessive hunting by humans had a major impact on the number of species. The evidence remains inconclusive.

▼ *Ichthyosaurs, such as Ichthyosaurus, were the most aquatic of all reptiles, having a streamlined body and shark-like fins and tail. Well-preserved fossils have indicated that ichthyosaurs were live-bearers, and the young were born tail-first, as they are in modern dolphins and whales.*

HABITATS AND ADAPTATIONS

WILLIAM E. DUELLMAN
AND HAROLD HEATWOLE

Amphibians and reptiles occur throughout much of the world, even in the Himalayas and the Andes above 4,500 meters (15,000 feet). They are present on all continents except Antarctica and some reptiles can even be found on tiny, remote islands. They live in rainforests, woodlands, savannas, grasslands, deserts, and scrub; they occur on the ground, underground, in rock crevices and under debris, up trees, and in marshes, swamps, lakes, streams, ponds, and the sea. Some even take to the air briefly.

WHERE AMPHIBIANS LIVE

Only Antarctica, extreme northern Europe, Asia, and North America, and most oceanic islands do not have native amphibians. Caecilians are pantropical, but are absent from Madagascar, New Guinea, and Australia. Salamanders are widely distributed in temperate regions of the Northern Hemisphere, although one family, the Plethodontidae, is most diverse in tropical Central and northern South America. In contrast, frogs are found nearly worldwide, with the greatest diversity of species in the tropics—the same number of species (81) occurs at a single locality in the upper Amazon Basin in Ecuador, as in all of the United States.

Relatively few species of amphibians are aquatic (water-dwelling) as adults, although most of them have aquatic larval stages (for example, tadpoles in frogs). Most salamanders and caecilians, and about half of the species of frogs, are terrestrial (ground-dwelling). Some of these, especially the caecilians, spend most of their time below ground. A few salamanders and many frogs are arboreal (tree-dwelling).

▶ In a Costa Rican rainforest, a puddle frog Smilisca phaeota dozes securely in its daytime roost in an unfurling heliconia leaf. In temperate regions many frogs are aquatic, but tropical rainforests offer a much wider range of potential niches for frogs, where many species are tree-dwelling.

Michael and Patricia Fogden

Reg Morrison/AUSCAPE International

◄ Despite their dependence on water for at least part of their life cycles, a number of frogs have successfully adapted to deserts and other arid environments. One of the most bizarre is the water-holding frog Cyclorana platycephalus of the Australian interior. During drought it buries itself underground, shedding its loose, baggy skin and lining it with mucus; the skin dries and hardens to form a waterproof barrier, within which the frog lies dormant until the rains come again.

R.W. VanDevender

▲ Caecilians are amphibians that have evolved to suit an almost entirely subterranean life style. Adaptations include a heavily ossified (bony) skull, reduced eyes, and lack of any trace of a pelvic girdle. This is Oscaecilia ochrocephala of Central America.

AQUATIC AMPHIBIANS

Four families of salamanders in North America (Proteidae, Cryptobranchidae, Amphiumidae, and Sirenidae) are strictly aquatic, as are two cryptobranchids in China and Japan (the giant salamanders), and one proteid in southeastern Europe (the olm). These, and some North American plethodontid salamanders that live in subterranean waters, are neotenic; that is, they never complete metamorphosis from the larval stage; they include some of the largest salamanders, which breathe by means of gills and retain larval tail fins. The sirens (family Sirenidae) and congo eels (family Amphiumidae) are limbless, or nearly so, and swim in a serpentine manner. In contrast, the mudpuppies and olm (Proteidae) and the giant salamanders and hellbenders (Cryptobranchidae) have depressed bodies and normal limbs; they are more sedentary and crawl about on the bottoms of lakes or rivers.

Most of the so-called pond frogs of the genus *Rana* have powerful hind limbs and fully webbed feet; they are well adapted for leaping from land into water, where they are excellent swimmers. Practically all frogs are capable of swimming, but members of two families, in particular, are highly specialized swimmers. The paradox frog and its allies (family Pseudidae) in South America have extremely powerful hind limbs, huge fully webbed feet, and dorsally protruding eyes; these frogs are active near the surface of ponds. The clawed frogs of Africa and the Surinam toad and its relatives in South America (family Pipidae) also have fully webbed feet, but they have depressed bodies and relatively small eyes. They rest and feed on the bottoms of ponds and come to the surface

periodically to gulp air. If ponds dry up, these frogs burrow into the mud and remain there until the ponds fill with water again.

One family of caecilians (Typhlonectidae) in South America is strictly aquatic. These eel-like amphibians live in lakes and rivers.

BURROWING AMPHIBIANS

While a few salamanders are known to burrow in mud or soft soils, the caecilians are the most adept burrowing amphibians. These slender, worm-like, limbless animals have muscular bodies and compact skulls. Their eyes are covered by skin or bone, so they are blind; but with a protrusible, sensory tentacle lying in a groove or cavity on the head, caecilians are capable of detecting environmental cues and prey (mostly earthworms) by chemosensory means. Caecilians propel themselves through the soil by concertina locomotion, in which part of the body remains in static contact with the earth, and the adjacent parts of the body are pushed or pulled forward at the same time.

Frogs of diverse groups have a large, keratinized, spade-like tubercle on the inner edge of each hind foot. Spadefoot toads (family Pelobatidae) in North America, rain frogs (Microhylidae) in Africa, sand frogs (Ranidae) in Africa, the Central American burrowing frog (Rhinophrynidae), and the burrowing frog and the crucifix toad (Myobatrachidae) in Australia, all use lateral scooping motions of the feet to bury themselves quickly in soft dirt or mud. A few frogs burrow head-first and have calloused pads on their snouts. African shovel-nosed frogs (Hemiotidae) have pointed snouts and burrow by

These frogs walk about in trees by grasping branches. A few treefrogs (some *Agalychnis* and *Hyla* species in Central America and some *Rhacophorus* species in southeastern Asia) not only have huge, fully webbed hands and feet but also fringes of skin along the limbs. These modifications combine to provide an extensive surface area when the limbs are partially extended and the fingers and toes are spread apart; after leaping from high perches these frogs can glide considerable distances.

Some frogs have other adaptations for life in the trees. Some of the casque-headed treefrogs in the American tropics have the skin co-ossified with the underlying bones of the head and have a greatly reduced blood supply to this skin. During dry seasons, to reduce water loss, these frogs back into holes in trees or the "cup" of bromeliad plants containing small amounts of water; they flex their heads at right angles to the body and block the holes with their heads.

Many kinds of treefrogs mate and deposit their eggs on vegetation above water, into which the hatching tadpoles later drop. Others adhere their eggs to the walls of water-filled cavities in trees or deposit eggs in arboreal bromeliads. In this way, adults do not have to leave the trees to reproduce. The ultimate adaptation for arboreal reproduction occurs in some of the South American marsupial frogs (Hylidae), in which the fertilized eggs develop into miniature froglets in a pouch on the mother's back. Thus, it is possible for generations to pass without any individuals leaving the trees.

WILLIAM E. DUELLMAN

Michael and Patricia Fogden

▲ *In general, amphibians require water for their early development. This presents some special challenges for those frogs that live as adults in trees but must somehow ensure the presence of water for their tadpoles. This challenge has been met in a variety of extraordinary ways; the female pygmy marsupial frog* Flectonotus pygmaeus *of South American highland rainforests, for example, carries her eggs in a moist pouch on her back.*

vertical motions of their heads. The Australian turtle frog and the sandhill frog (Myobatrachidae) have blunt snouts and dig with their forefeet.

Frogs may bury themselves to seek daytime retreats or more importantly for long periods of aestivation (in a torpid condition) during dry seasons, which may last for many months. During these times, frogs are subject to desiccation, so before aestivating they fill their urinary bladder with water. Some frogs, such as the African bullfrog (Ranidae), the water-holding frog (Hylidae) in Australia, and horned frogs and their allies (Leptodactylidae) in South America, reduce the risk of desiccation even further by making a cocoon. Once they are deep in their burrows the frogs shed the outer layer of their skin and secrete mucus, which together with the shed skin hardens to form a cocoon around the frog.

ARBOREAL AMPHIBIANS
The terminal segment of each finger and toe of many treefrogs (Hylidae, Centrolenidae, Hyperoliidae, Rhacophoridae, and a few Microhylidae) is expanded into a specialized toepad, by means of which these frogs are able to adhere to vertical surfaces. Many treefrogs also have slender bodies and long limbs, modifications that allow them to leap from one leaf or branch to another and to hold onto their perches. In one group of hylids (*Phyllomedusa* species) in the American tropics, the innermost fingers and toes are elongated and opposable to the outer digits.

WHERE REPTILES LIVE
Like amphibians, reptiles are ectotherms; that is, they do not produce very much of their own body heat and must rely on the external environment for it. For this reason they are sensitive to temperature and the number of species decreases toward higher latitudes and elevations, until eventually they drop out altogether. Nevertheless, one hardy species of lizard and one species of snake occur above the Arctic Circle in Scandinavia, and on some mountains lizards skirt around snow banks in their daily activities. The tuatara has been known to chase and catch seabirds on a night when air temperature was 7°C (45°F) and there was a driving rain and a wind of 50 knots! Such examples are unusual, however, and there are two kinds of places where reptiles really excel. They are abundantly tropical, and they form a conspicuous part of the desert fauna.

Turtles and crocodilians are mostly aquatic, whereas lizards and snakes are mainly terrestrial or arboreal. There are interesting exceptions, however: some tortoises not only live away from water, but they do so in desert regions, and some sea snakes live a totally aquatic existence.

TERRESTRIAL REPTILES

Because living on land is so familiar to humans, adaptations to other habitats often are considered "special" and terrestrial adaptations merely as the norm. However, terrestrial life imposes unique conditions that require modification of an animal's form and function just as much as does life in water or in trees. The skeleton needs to be strong enough to support body weight without the buoyant support offered by water. There needs to be a means of renewing the air over the respiratory surface (breathing) and a means of propelling the animal along the ground. Ability to see at a distance is important. Lungs, limbs, and eyes seem standard equipment to humans because we have them and we live on land, and because most of the animals we see have similar structures. In a total biological context, however, they are remarkable adaptations.

The limbs of reptiles are highly adapted to the kind of environment they occupy. Terrestrial lizards that run rapidly usually have relatively long legs with well-developed toes and claws that aid in gaining purchase on the ground. Some lizards with exceptionally long legs rear up and run on the hind legs only, thereby increasing their speed. These species have an exceptionally long tail, which serves as a counterbalance during such bipedal running.

Terrestrial snakes, entirely lacking limbs, have a different suite of adaptations. They are able to move across the ground by bending the body and pushing backward against irregularities of the surface. The large transverse scales on the underside are firmly attached in front, but have a free edge behind, where they overlap the scale following. These free edges catch against the ground and prevent backward slippage. (In sea snakes, which no longer need to have a grip on the ground, the belly scales are reduced in size, and in some cases are the same size and shape as the dorsal ones.) Many terrestrial snakes are long and narrow and have relatively long tails, attributes that are associated with rapid movement over the ground.

ARBOREAL REPTILES

Living in trees requires being able to grip trunks and branches to avoid falling, and being able to get from one branch to another. Many climbing lizards have sharp claws that can dig into bark and assist in climbing. Others, such as geckos, have expanded pads of lamellae on the toes, which allow them to grip almost smooth surfaces. Their effectiveness is obvious when you see a gecko run upside-down on the ceiling! Other arboreal lizards, such as chameleons, have toes that are opposable (like the thumb and fingers of a human) and can grasp twigs. Chameleons also have a prehensile tail; they can curl it around a branch and hold on, monkey-style.

The limblessness of snakes requires that they have different adaptations. Many arboreal snakes are amazing climbers, some even able to go up a vertical tree trunk without coiling around it; they merely utilize crevices in the bark for purchase. Some arboreal snakes, called vine snakes, look very much like vines. They are long and thin and drape over branches. This is an effective camouflage, but their length also assists in bridging gaps when moving among branches. Some arboreal snakes have a triangular shape in cross-section that gives greater strength and rigidity to a body extended, unsupported, across open areas.

Coping with a three-dimensional environment requires ability to judge distances. Many arboreal lizards and snakes have the eyes directed forward in such a way that both eyes can focus in front and achieve stereovision.

Perhaps the most unusual adaptation of arboreal reptiles is the ability to glide. A genus of Asian lizards (*Draco*) have the ribs extended from the sides of the body and covered with a flap of skin to form "wings". If molested these "flying lizards" will escape out of their tree and glide long distances, either to the ground or to another tree trunk. There is also an Asian "flying" snake (*Chrysopelea*). It leaps from trees and by flattening its body can glide and break its fall. But there are no modern reptiles that match the soaring of the extinct pterodactyls.

▼ *Tree-dwelling snakes tend to be long and thin, like this plain tree-snake* Imantodes inornata *hunting among the twigs and foliage of a Heisteria tree in Costa Rica. A slender snake is better able than a fat one to span the gaps between branches, and many species are also triangular in cross-section, to combine to best effect the contrary advantages of suppleness and rigidity.*

Michael Fogden

SUBTERRANEAN REPTILES

Some of the special modifications of burrowing reptiles are not true adaptations, but rather loss of structures that no longer have a biological function. For example, in a dark burrow where the head is in direct contact with the soil, eyes serve no useful purpose, and some burrowing snakes and lizards have only rudimentary eyes. Limblessness also is a consequence of a burrowing existence. Although limbs can be useful in digging, and many terrestrial reptiles use them for digging burrows for shelter, they increase friction and require a larger burrow for animals that push through the soil, rather than dig. Accordingly, most truly subterranean reptiles have lost such encumbrances or have them greatly reduced in size.

By contrast, other features *are* adaptations to a burrowing life. Many burrowing lizards and snakes have the skull bones fused into a solid, compact structure that can support the head as a battering ram. Blindsnakes (family Typhlopidae) have a sharp point on the tail, which serves as an anchor when pushing the smooth, highly polished body through the soil. Amphisbaenians employ another tactic: they have grooves around the body that provide traction, and they "inch" forward in their burrows.

The limblessness of snakes is probably a heritage from their burrowing ancestors. It is believed that snakes arose from burrowing lizards that had become limbless and nearly blind. The snake eye lacks structures present in most vertebrates and appears to have been redeveloped from a rudimentary lizard's eye. Snakes have never regained limbs, and as indicated above, this condition has channeled their adaptation to arboreal and terrestrial habitats along different pathways from those taken by lizards.

AQUATIC REPTILES

The terrestrial ancestry of modern reptiles has imposed limitations upon their life in water. Being egg-layers, females of most species must come to land to breed; only some of the marine snakes give birth to live young in the water and never voluntarily emerge onto land. Air-breathing is another limitation. All reptiles must come to the surface to breathe periodically, yet some have been able to prolong their diving time to a remarkable degree. Sea snakes can submerge for up to two hours, partly because they can absorb oxygen through their skins to a degree that land snakes cannot. They can dive to 100 meters (330 feet) and not suffer the bends, again perhaps because of skin permeability to gases; the excess nitrogen absorbed into the blood under pressure may pass through the skin into the sea and not build up enough to cause the bends. To keep air from leaking out, sea snakes have valves that close off the nostrils during dives, and the mouth is tight-fitting.

The nostrils and eyes of crocodilians and aquatic snakes tend to be located further onto the upper surface than in other reptiles. This allows an animal at the surface to ride low in the water, with fewer buoyancy problems, but at the same time be able to breathe and see.

Locomotion under water is completely different from moving on land, as pressure needs to be applied to a surrounding medium, not against a surface. Reptiles have adapted to this requirement in two ways. Those that have limbs have developed webbed feet or, in the case of sea turtles, flippers; and sea snakes have evolved a flattened paddle-shaped tail. Crocodilians and some semi-aquatic lizards also have the tail flattened, thereby increasing its surface area and propulsive thrust.

The sea poses problems in addition to those encountered by reptiles in fresh water. Reptilian kidneys cannot cope with high salinities, and life in the sea is possible by virtue of special salt-excreting glands. Sea snakes, sea kraits, and file snakes have one of their salivary glands modified as a salt-excreting gland. It is located beneath the tongue, and brine is excreted into the tongue sheath. When the snakes protrude their tongues the brine is pushed out into the sea. Another group of snakes inhabiting salty water, the homalopsines, have a similar gland, but located in the front of the roof of the mouth. Sea turtles have a modified tear gland that excretes brine from the eye, and the saltwater crocodile has small salt glands scattered over the surface of its tongue. The marine iguana, a large lizard from the Galapagos Islands, which dives into the sea and

▼ Reptiles as well as amphibians have many representatives that live almost entirely underground. Some burrowing snakes such as this Central American threadsnake Leptotyphlops albifrons feed mainly on ants and termites. Their sensitive sense of smell enables them to track foraging ants or termites back to the nest, where they are protected during their raids apparently by their body secretions of powerfully repellant chemicals.

Gerald Thompson/Oxford Scientific Films

feeds on marine algae on the bottom, has a salt gland in its nasal passage. The excreted brine is sneezed into the air when the lizard is on land.

Few animals are adapted to move over the surface of the water. One such rarity is a reptile, the basilisk, or Jesus lizard, from Central America. It has flaps of skin on the sides of the toes of the hind feet. These are folded up when it walks on land. If frightened it skitters bipedally out over the surface of a stream or pond; the flaps open up and provide sufficient additional surface area for it to be able to run across the top of the water, as long as it maintains momentum. If it stops running it sinks and has to swim.

REPTILES ON ISLANDS

Remote oceanic islands usually have very few species of plants and animals, mainly those capable of surviving long voyages. Certain reptiles, notably geckos, are usually well represented, because several adaptations contribute to their success in dispersing to islands and their establishment once there.

Many lizards live in and under driftwood on beaches. Logs are washed out to sea and eventually drift to a distant shore, carrying with them lizards or their eggs ensconced in crevices. Some geckos have salt-tolerant eggs, which are sticky when laid but after drying adhere tightly to

the sides of cracks and crevices.

A major problem colonizing animals face is establishing a population once a remote island is reached. Dispersal by rafting is a relatively rare phenomenon, and a second raft may not arrive at an island during the life-span of an individual reptile. Thus, most arriving species would die out with the death of the colonizing animal, unless such a miniature ark carried at least one individual of each sex. But parthenogenetic species (those in which the females can lay fertile eggs without being inseminated by a male) do not have such restrictions. It is probably significant that some of the geckos characteristic of remote islands are parthenogenetic, the males being either scarce or completely absent. In such species only one individual or egg need arrive at an island for a colony to become established. A widely distributed Pacific blindsnake is also parthenogenetic.

Perhaps the land reptiles with the record for overwater dispersal are the iguanas of Fiji and neighboring islands in the Pacific. Their only relatives are in South and Central America. Clearly, their ancestors rafted across the Pacific Ocean and, once isolated, evolved into the species of today.

HAROLD HEATWOLE

▲ *A combination of momentum, rapid leg movement, and fold-away flaps along each side of the toes enables the basilisk lizards of Central and South America, sometimes called Jesus lizards (genus Basiliscus), to run over the surface of water for some considerable distance.*

REPTILE AND AMPHIBIAN BEHAVIOR

WILLIAM E. DUELLMAN
AND CHARLES C. CARPENTER

Reptiles and amphibians are the only land vertebrates unable to control their body temperature by internal, physiological means. By and large, this means that unless they regulate their body temperature behaviorally, it will simply rise and fall with the temperature of their immediate surroundings. This places enormous constraints on every aspect of their life. Hatching, growing, feeding, reproducing, escaping from predators, and even just being active depends on attaining and maintaining a body temperature that allows them to function normally. Consequently much of the behavior of reptiles and amphibians is aimed at temperature regulation, and this requirement underlies much of the complex and often bizarre behavior of individual species.

AMPHIBIAN BEHAVIOR

Because amphibians have permeable skin and, as a rule, are unable to regulate their body temperature by physiological means, many aspects of their behavior are dictated by the environment. Most amphibians are nocturnal and are active only when environmental conditions are sufficiently moist to prevent their bodies losing too much water by evaporation.

However, some amphibians exhibit water-conserving behavior and even behavior that regulates their body temperature. For example, diurnal (day-active) frogs living in cool climates often bask in the sun and thereby raise their body temperature before foraging for food. Some of them periodically enter water to acquire moisture and to lower their temperature. The North American bullfrog regulates its temperature by changing its position in relation to the rays of the sun. Other postural changes, evaporative cooling by mucous secretions, and periodic rewetting of the skin, all serve to reduce or stabilize the bullfrog's body temperature. Conversely, tiger salamanders, living in the Rocky Mountains of North America, and tadpoles in many parts of the world increase their body temperature by moving into shallow water on sunny days. However, almost nothing is known about the behavior of the secretive, burrowing caecilians.

FEEDING BEHAVIOR

Most amphibians simply sit and wait for prey to come within reach. A few heavy-bodied frogs, such as South American horned frogs, lure agile prey by waving the long toes on their hind feet. Other frogs, especially the small poison frogs in tropical America, actively forage for small insects

by day and find their prey by sight. A Central American burrowing frog (genus *Rhinophrynus*) apparently locates subterranean termite tunnels by their odor, then breaks open the tunnel, inserts its nose, and laps up the termites.

DEFENSIVE BEHAVIOR

Faced with potential predators, many amphibians feign death, present an enlarged image or the least-palatable part of the body to the predator, confuse the predator by changing the characteristic shape of the body, or change their color as a warning. In some cases they may even attack the predator. Salamanders that have concentrations of poison glands on their tail may lash it at predators; others have many poison glands on the back of the head and butt their head at the predator. Some Asiatic salamandrids are capable of protruding the tips of their ribs through the poison glands on their flanks, greatly increasing the chances of a predator encountering unpalatable secretions. Many salamanders exhibit caudal autotomy—that is, they can break off a portion of their tail—so if a predator grasps the tail, it will break off and continue to wriggle, thereby distracting the assailant while the salamander escapes.

Feigning death is widespread among frogs: some fold the limbs tightly against the body and lie motionless on their backs; others assume a rigid posture with the limbs outstretched. Toads commonly inflate their lungs, thereby puffing up the body and presenting a larger image to the predator, at the same time lifting their body off the ground and sometimes tilting it toward the predator. Some South American leptodactylid frogs (*Physalaemus* and *Pleurodema*) have large glands in the groin resembling brightly colored

478

"eyespots", so that when a frog lowers its head and lifts its pelvic region, it looks like the head of a larger animal and presumably frightens off the would-be predator. When faced with danger, some African and South American treefrogs open their mouth (often displaying a brightly colored tongue) in an apparently threatening pose. Some large frogs actually leap and snap at predators. Many kinds of frogs emit a loud scream when disturbed or grasped by a predator; the noise may serve not only to frighten the predator but also warn other frogs of danger.

AMPHIBIAN REPRODUCTIVE BEHAVIOR

The most obvious reproductive behavior of amphibians is the sound made by frogs, especially the mating call made by males. Each species has a distinctive call, which is recognized by other individuals of the same species. Many frogs that live in seasonal environments have short breeding periods during which the males move to temporary ponds and vocalize; these calls usually are acoustically simple and serve only to attract females. Frogs that are active for most of the year in tropical and subtropical regions commonly have more complex calls with both courtship and territorial components—a male advertises his territory acoustically to other males and also attracts females. Not all species of frogs call.

Salamanders are not known to use vocalizations in their sexual behavior, although some can produce barking or squeaking sounds. At least some terrestrial species mark their territories with scent trails, which are recognized by members of the same species. Each species of European newt (genus *Triturus*) has a unique, complex courtship behavior, in which the male must elicit a response from the female before he can initiate the next phase of courtship. Courtship takes place in water, and during the mating season males develop bright colors and elaborate tail fins. The male positions himself in front of the female and initiates a series of tail movements, stimulating her visually, by touch, and by smell. New World lungless salamanders engage in a tail-straddling walk in the water or on land. The male initiates

▼ *In a defensive threat display against a Central American toad-eater snake* Xenodon rabdocephalus, *a cane toad* Bufo marinus *inflates its lungs, lifts its body clear of the ground, and tilts toward its attacker. If further provoked, the toad can eject a spray of poisonous secretion from glands on its head.*

Michael and Patricia Fogden

Michael and Patricia Fogden

The only frogs that have elaborate courtship rituals are some of the aquatic tongueless frogs. In the South American genus *Pipa*, the male grasps the female around the waist while she performs a series of somersaults during which eggs are expelled; the male then fertilizes the eggs and sweeps them onto her back with his feet, where they implant and develop.

CARE OF EGGS AND TADPOLES

Most amphibians that deposit their eggs in water immediately abandon them. In species that deposit their eggs on land, however, it is usual for a parent to stay with the eggs, preventing marauders from stealing them. Many frogs take care of their young in more active ways. For example, females of the South American marsupial frogs transport eggs on their backs or in a dorsal pouch. Males of Darwin's frog in Chile pick up hatchling tadpoles from egg clutches on the ground and carry them in a vocal sac until they complete their development. Males of the hippocket frog in Australia pick up tadpoles and transport them in pockets in their flanks. Females of the Australian gastric-brooding frog swallow fertilized eggs; the eggs hatch, and the tadpoles complete their development in the stomach, which has to turn off its digestive functions during brooding.

Observers have seen the parents of some South American pond frogs (genus *Leptodactylus*) protect their free-living tadpoles; the tadpoles move in schools, and the female remains with them. Females have even been seen leaping at potential predators of the tadpoles. Perhaps the ultimate type of parental care in amphibians is the feeding of free-living tadpoles by the mother. In some tropical American poison frogs, the mother transports individual tadpoles to bromeliad plants which hold water; she periodically returns to each bromeliad and deposits unfertilized eggs to provide nourishment for each tadpole. Similarly, a Jamaican treefrog will deposit her fertilized eggs in water cupped by a bromeliad, and she subsequently provides other batches of unfertilized eggs for food.

WILLIAM E. DUELLMAN

▲ *The female strawberry poison frog Dendrobates pumilio of Central America transports her tadpoles, one by one, to tiny pools formed by rainwater in bromeliad plants. She will return later to lay unfertilized eggs in the water as food for her young.*

▶ *The female Surinam toad Pipa pipa carries her developing young about with her on her back, snuggled tightly into specially formed pockets in her skin.*

courtship by rubbing the female's head with his chin, secreting chemicals from glands on the chin which stimulate her. He then moves forward and the female straddles his tail, pressing her chin to the base of his tail. If the female does not follow closely, the male may swing around and violently slap her head with his chin. Courtship terminates successfully when the female follows the male until he deposits a gelatinous capsule with a cap of sperm (spermatophore), which she then picks up with the lips of her cloaca.

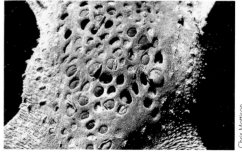

Chris Mattison

REPTILE BEHAVIOR

Reptiles have evolved into an impressive spectrum of shapes and sizes, which include shell-covered turtles and tortoises, long sinuous snakes, swift-moving lizards, and heavy-bodied crocodiles. They have therefore developed a wide variety of strategies to help them survive and reproduce. Most reptile species lay eggs, but some lizards and snakes give birth to living young.

Species recognition is critical. Individuals need to communicate with others of their own species so that they can get together to reproduce. This communication can be visual (as in displaying iguanid lizards), olfactory, including chemical substances called pheromones (as in most snakes), or vocal (as in lizards such as geckos, and in crocodiles and alligators). The signals provide not only information for species-identity but also the gender and reproductive readiness of an individual. Males in some species advertise their territory and perhaps their social dominance over other males using displays that are typical of their species.

TERRITORIALITY

In most families of lizards, males declare their territories by performing ritualized displays. Males in the Iguanidae and Agamidae families posture in ways that emphasize sexual differences in color and pattern, asserting to other males the ownership of a territory. These postural displays are reinforced by ritualized movements which, depending on the species, may be "push-ups" or bobbing of the head. Male marine iguanas may butt their heads together when one tries to claim the territory of another. In many monitor lizards, the males engage in combat by grasping each other with their front legs and, while standing high, try to push each other down. In some types of lizards a superior male asserts his dominance over other males with his displays, and the subordinate male assumes a submissive posture. Geckos, usually active at night, may declare this information with vocal signals. The territory provides an area for the adult male lizard to hunt for food and find a female.

Very little territoriality has been observed among snakes. However, in some families, such as the vipers and pit vipers, elaborate combat rituals take place between males—these may serve to establish dominance and the right to mate with a female. In the rattlesnakes' combat ritual, each male raises his head and body high, then twists against his opponent and attempts to topple him to the ground. In some tortoises, two males may butt their shells together in tests of strength to establish superiority.

COURTSHIP AND MATING OF REPTILES

During courtship the male performs certain behaviors, often ritualized and aided by pheromones, to attract the female and induce her

◄ A male banded anole Anolis insignis of Central America advertises his ownership of a territory by conspicuously displaying a colorful, sail-like dewlap on his throat.

Michael and Patricia Fogden

▲ *Plodding across a beach in Queensland, Australia, a female loggerhead turtle Caretta caretta returns to the sea at dawn after egg-laying at night. These and other marine turtles migrate vast distances across open ocean to reach traditional nesting sites, usually the same beach at which they themselves were hatched.*

to become sufficiently passive for copulation to take place. During the courtship of snakes the male usually crawls repeatedly over the female and orients his head to hers, at the same time bringing his tail into position next to hers to allow copulation to take place. In some species the male may hold the female by biting her head or neck region. In many lizards the male gets a biting grip on the female to hold her while he brings the base of his tail into position for mating. The male slider turtle courts his female by swimming backwards in front of her, stroking her face with his long fingernails. The male gopher tortoise circles his female on land, perhaps butting her shell, to induce her to become passive so that he can mount her from the rear. In North American box turtles, which have movable lower shells, the female may clamp her lower shell shut on the hind feet of the male, thus pinioning him as he falls on his back to mate; this position seems to be necessary because of the high-domed shells of these turtles.

REPTILES' PARENTAL CARE

In the typical nesting behavior of many reptiles, after mating the female digs a burrow, deposits her eggs, and leaves the eggs to hatch on their own. There is no protection for the young when they hatch, so many die and only a few survive to adulthood. In some species of snakes and lizards the female retains the eggs inside her body until the young are ready to be born; she gives birth to live young, which are on their own after emerging from their fetal membranes.

Some female North American skinks stay with the eggs and brood them until hatching, even retrieving any eggs if a predator tries to make off with them. The females of some pythons coil around their eggs and incubate them; by contracting the muscles of her body, the mother can produce metabolic heat when needed to regulate the temperature of the eggs. Perhaps the most obvious nesting behavior is the mound building of the American alligator. The female then remains near the nest to protect it from intruders, and she may assist the young in hatching and escaping from the nest.

MIGRATION

Most reptiles do not migrate. For those that do, it is usually to reach a favorable nesting site or more favorable feeding sites. Prairie rattlesnakes may migrate up to 5 kilometers (3 miles) from their hibernation dens in the spring and then return to the same den in the fall. Marine iguanas of the Galapagos Islands migrate from the lava reefs to nesting beaches to lay their eggs, and then return to the reefs 100 meters (300 feet) away. To reach their nesting beaches, sea turtles may migrate 2,000 kilometers (1,240 miles). This involves sophisticated guidance mechanisms, which we do not yet understand. Certain populations of the green sea turtle migrate from the coast of Brazil to nesting beaches on Ascension Island in the mid-Atlantic Ocean, a distance of 5,000 kilometers (3,000 miles).

REPTILIAN DEFENSE

The most common type of defense in reptiles is biting. This is highly developed in many families of snakes, especially the true vipers, pit vipers, and elapid snakes (cobras and their relatives), which all produce venom. The only lizards that produce venom are the beaded lizards of the Americas.

The body structure of reptiles, and their habitat, affects the way they express aggressive behavior. Many lizards flatten their bodies and raise them high to appear larger, thus deterring a would-be predator. The bearded dragons and frilled lizard of Australia have elaborate throat fans, which they present to a predator, greatly enlarging their apparent size. Many snakes inflate or enlarge their neck in a threatening manner—as exemplified in some of the cobras. Many snakes and lizards threaten a predator by gaping their mouth, and some snakes flash a bright color on the underside of the tail. The North American hognose snakes respond to the threat of a predator by a sequence of unusual antics, first posturing as if to strike, then rolling over onto their back to simulate death, gaping the mouth, defecating, and becoming limp. If someone turns the snake over onto its belly, it immediately turns over again, belly up, as if to "play dead".

Cryptic coloration (camouflage) is common in reptiles, some of them blending with the habitat when they are motionless, while others have color patterns that produce a confusing appearance when in motion. Chameleons are freely able to change their color or color pattern as they move among different types of vegetation.

CAPTURING FOOD

The primary function of venom in a poisonous snake is to immobilize prey so that it can be devoured. Non-venomous snakes use other ways of obtaining food, such as coiling around prey to constrict it, as in the boas and rat snakes. Some reptiles actively pursue their prey, while others have a strategy of "sit and wait" until prey comes close. The death adder of Australia may lie hidden in the leaf-litter with its brightly colored tail exposed and intermittently waved from side to side, as an insect-like lure to attract prey. The North American alligator snapping turtle lies on the bottom of a river with its mouth open, wiggling a small fleshy "worm-like" protuberance on its tongue, to lure an unwary fish.

CHARLES C. CARPENTER

▲ Many snakes, like this African twig snake Theletornis capensis, respond to threat by inflating the throat to exaggerate the apparent size of the head.

Michael Fogden/Oxford Scientific Films

Jack Dermid/Oxford Scientific Films

◀ A range of snakes and lizards "play dead" when threatened. An alarmed eastern hognose snake Heterodon platyrhinos of North America will roll over onto its back, defecate, open its mouth, and go limp. The response is so rigidly programed that, if rolled over onto its stomach, the snake will promptly "come to life" again, just long enough to roll over on its back and repeat the performance.

ENDANGERED SPECIES

BRIAN GROOMBRIDGE

There have always been endangered species. Over periods of time measured in tens of millions of years, different lineages of species have evolved, expanded in numbers and importance, and then disappeared from the fossil record. By analyzing the different types of animals and plants fossilized in sedimentary layers from successive geological periods, we know that unrelated clusters of species have disappeared one after the other, during the course of millions of years, and probably for a variety of reasons. This happened to many amphibian groups during the Permian and Triassic periods (approximately 285 to 200 million years ago). But there have been times when a number of distinct groups completely disappeared in one relatively brief period, and perhaps with some common cause—as in the case of many reptile groups at the end of the Cretaceous, about 65 million years ago. It was during this period that all the dinosaurs then remaining became extinct.

► A land iguana Conolophus subcristatus *nibbles at a cactus plant. Confined to the Galapagos Islands, this species has become seriously endangered because of predation from dogs and cats released on the islands. It lives in burrows in sandy ground and feeds on plants and insects.*

K.H. Switak/NHPA

A NATURAL COURSE OF EVENTS?

Evolutionary theory predicts that species are likely to become extinct as one species out-competes another for some limited set of resources, such as the right type of food, or as a species becomes unable to adapt to a changing environment. In addition, over evolutionary time the genetic identity of species will change as populations evolve into what might eventually be a new and distinct species, or as some geographic barrier forces one species to split into two separate populations, each of which can embark on a new evolutionary career.

If both the evidence of the fossil record and evolutionary theory indicate that the decline and extinction of species is a natural course of events—that is, there have always been "endangered species"—why is there so much concern now about declining populations and contemporary extinctions?

The answer is that the rates of endangerment and extinction around the world are feared by many to be higher than ever before. In gathering up a disproportionate share of the Earth's resources and becoming the most successful single species the world has ever seen, humankind has modified virtually every component of the biosphere—from the deep ocean floor to the atmosphere itself. On a global scale, the rate and magnitude of these changes have increased sharply during the past few human generations, and continue to do so, as some people strive to meet the most basic needs of existence while others pursue a life style of excessive consumption. We have now observed the depletion and fragmentation of formerly widespread wild species, the diminution of species-rich habitats, and are able to compile lists of species seen 50 or 100 years ago but now apparently gone forever.

TWENTIETH-CENTURY EXTINCTIONS

No one should doubt that many species and their habitats face acute problems. We can predict that many sensitive species with small or now-fragmented populations will probably be lost. Yet it should be stressed that we actually know of few species that have been lost in our lifetime. The number of vertebrate species (those with a backbone) reliably known to have become extinct in the twentieth century is about 110. In purely numerical terms, this is of little significance compared with the 42,000 living vertebrate species known to science. The relative proportions are even more unbalanced among invertebrates and plants, but this has little meaning because the scientific inventory of these groups is far from complete, and evidence of extinction is very difficult to obtain.

Of course, it is impossible to know of all contemporary extinctions, and a significant number of species have certainly disappeared without ever having been collected, described, and named by science. This must be particularly true among invertebrates and plants, especially those inhabiting inaccessible areas.

FOUND JUST IN TIME

Some new species are named as a result of a taxonomist recognizing that a previously known single species is actually two, perhaps following new evidence of anatomical, biochemical, or behavioral differences between them. Other new species really are "new", in the sense of never having been recorded by science before. Sometimes these are discovered in circumstances that reveal how difficult it is for us to monitor the existence or status of such species.

For example, in 1981 the three-year-old son of a herpetologist was left at the roadside near a small area of forest in the Palni Hills of South India; he was to play at snake hunting while his elders got on with the real hunting nearby. The boy idly picked some cement out of a stone wall, and a small snake fell at his feet; the snake turned out to be a species new to science, eventually named *Oligodon nikhili*. Had that patch of forest been chopped down like the rest of the original forest cover, the species would almost certainly have disappeared even before its discovery.

Similarly, the anguid lizard *Diploglossus anelpistus* was described after only four specimens were found during the clearance of a forest area in the Dominican Republic. The species has never been seen for certain anywhere else on the island, and the forest area in which it was discovered has since been totally cleared. This species may now be extinct, although there are indications that it is nocturnal and semi-burrowing, so there is a possibility that additional populations remain

M.J. Tyler

▲ The gastric-brooding frog Rheobatrachus silus *inhabits mountain forest streams in eastern Australia. When its extraordinary breeding habits (the female incubates her young in her stomach and "gives birth" to them through her mouth) were first unraveled during the 1970s, it was reasonably common, yet during the 1980s it disappeared, for unknown reasons. Several subsequent thorough searches have failed to find it.*

▲ *Two butterflies imbibe eye and nasal secretions from basking yellow-spotted Amazon River turtles* Podocnemis unifilis *in Manu National Park, Peru. Inhabiting large rivers and wetlands in South America, several species of the genus* Podocnemis *have breeding habits somewhat similar to those of marine turtles, coming ashore to nest communally. Some have been brought to the brink of extinction by human exploitation of their eggs for food.*

imminent danger of extinction unless the causal factors are removed; these, together with species in less acute danger, are referred to as "Threatened Species". The terms are often used in somewhat different senses outside the IUCN system. Since the 1960s many regional, national, and provincial "Red Books" have appeared, produced by official bodies and by non-governmental organizations.

Despite its immense success, the Red Book approach does have limitations. Firstly, the biological assessment and monitoring that should precede inclusion of species on any "Threatened Species" list is most feasible in countries with high levels of scientific expertise, financial resources, and public concern. This inevitably leads to geographic imbalance in coverage: in the amphibian and reptile Red Data Book, for example, the large number of species from North America and the small number from tropical countries is more likely to reflect the density of concerned observers than the number of threatened species. There is also considerable taxonomic imbalance, as some groups of animals are studied by a number of active specialists. Thus the conservation status of all species of turtles and tortoises, for example, has been assessed to some extent, and largely as a result of this interest there are many more turtles and tortoises in the IUCN Red List than there are snakes or salamanders. This is linked to the problem of size and visibility: far more is known about large species, such as crocodiles, turtles, and large iguanid lizards, than about small lizards, snakes, and salamanders, many of which are also highly secretive and seasonal in appearance. Furthermore, it is actually impossible to monitor the status of individual species in species-rich faunas, such as those of tropical rainforests.

"FLAGSHIP SPECIES"
Although attention is swinging away from individual threatened species, toward habitats and ecosystems generally, the great effort still made to protect individual species is not necessarily misdirected. Species such as the tiger and the giant panda are not only important in their own right, but they also serve as "flagship species"—that is, conservation and management of the extensive areas of habitat needed to support viable populations of these large animals will simultaneously improve the prospects of countless smaller animals and also plants.

There is no exact amphibian or reptile analog of the tiger, although protected areas set up in India for the gharial *Gavialis gangeticus* and the Indo-Pacific crocodile *Crocodylus porosus* (both threatened species) do include populations of other reptiles of conservation concern, notably freshwater turtles and the water monitor *Varanus salvator*. In general, conservation of amphibians

hidden at other sites. The point is that there must be many cases where chance discoveries, such as the two mentioned here, simply do not happen, and therefore species are lost before their existence is known to science.

THE NEED FOR INFORMATION
There is still hope. Some of the trends mentioned can be slowed, some may be reversible, and our information base can be expanded. Armed with information and political strength, it may be possible for us to relieve and sometimes remove many of the pressures that reduce the world's biological diversity.

Information is a key requirement. The "Red Data Book" concept, originated during the 1960s by the late Sir Peter Scott, has been an important means of stimulating public involvement in conservation action. The intention was to catalog species thought to be, in varying degrees, in danger of extinction. Brief notes on distribution, threats, and conservation needs were included. A system of categories was devised to reflect the supposed imminence of extinction, and this has been only slightly modified over the years.

In the Red Data Book system of categories developed by IUCN (the International Union for Conservation of Nature and Natural Resources, now known as the World Conservation Union), "Endangered" has a relatively precise meaning and refers to species considered to be in and reptiles depends either on their incidental presence in protected areas which have been set up for other reasons or on individual-species action. The latter is sometimes the most appropriate course of action, because among the most-threatened species are crocodilians and marine turtles whose present predicament is the result of human exploitation—also an individual species-action.

STEPS TOWARD EXTINCTION

Although there is justifiable concern about the extinction of species, and perhaps a curious fascination about its finality, action can only be directed at species before they reach this stage. Some species may be beyond remedial action; these have been termed "proto-extinct". Others, while stressed to a greater or lesser extent, have the potential to respond to management interventions, and conservation action should therefore be directed at them rather than those probably beyond help. Information on the biology of the extinction process should assist in distinguishing one group from another.

A species can persist only when reproduction continues to be successful—when births are sufficient to offset deaths. This process can be disrupted in various ways. The direct exploitation of those individuals most reproductively active in a population will obviously reduce the birth rate. If carried on for a significant length of time—depending on the age when they first breed, and the age when they cease to breed, among other factors—the entire population may collapse. Large-scale disruption of habitats can reduce the availability of shelter and food, and therefore hinder reproduction just as effectively.

It has only recently become clear to what extent random events can affect populations, especially when populations have become numerically small and fragmented because of habitat loss or exploitation. Small populations are more likely than large ones to lose the genetic variability needed to meet new environmental challenges. Small and isolated populations are more likely to be eradicated by disease or natural disaster than large ones. In an extreme case, reproduction might cease because individuals of opposite sexes simply do not meet or have lost the social context in which migration or some other key breeding behavior usually occurs. Where fragmented populations are separated by large areas of now-unsuitable habitat, local populations that become critically depleted cannot be replenished by distant populations.

Species with these characteristics can in effect be doomed to extinction although individual members are still living. Among sea turtles, for example, virtually all the females that emerge to nest on a particular beach might be slaughtered, or all their eggs collected. There are beaches where this has occurred year after year. Because females are potentially long-lived, they still appear on the nest-beach every year, even though no young have entered the water for many years. If the period of intense exploitation exceeds the reproductive span of the females, the population will eventually disappear as the ageing females fail to be replaced by their progeny.

A prime example is one of the island populations of the Galapagos giant tortoise *Chelonoidis elephantopus*—the subspecies *C. e. abingdoni* in particular. This population now consists of a single male, found on Pinta Island in 1971 and held at the Charles Darwin Research Station since 1972. The tortoises on Pinta were heavily exploited by whalers and fishermen, and the vegetation severely degraded by the grazing of goats, introduced in 1958. Chance events doubtless played a major part in the reduction of this heavily stressed population to a single individual. While this concerns what is usually treated as a subspecies, rather than a full species, it is probably a good model of what has happened to extinct species.

▼ Inhabitants of high mountain cloud forest in Central America, golden toads Bufo periglenes congregate in the breeding season to mate at small pools and streams. During the 1980s, this species (as well as many other frogs of highland forest around the world) suffered a catastrophic and largely mysterious decline, in many cases to apparently total extinction.

Michael and Patricia Fogden

▶ Feeding largely on lizards and frogs, the broad-headed snake Hoplocephalus bungaroides is confined to forests extending only a hundred miles or so around Sydney, Australia, where it is threatened by urban sprawl and related development.

J.C. Wombey/AUSCAPE International

THE EXTENT OF THE PROBLEM

It is impossible to assess just how many species are threatened at a global level. We can only deal with species that we actually know about, and we can only enumerate those that have been formally designated as threatened and listed in some organized format. The IUCN Red List is the most wide-ranging compilation of such species. Numbers of amphibians and reptiles included in the 1990 edition are given below, together with the number of each categorized as "Endangered". This table refers to full species only; it does not include the small number of geographic subspecies also in the IUCN Red List.

HABITAT LOSS

Habitat loss is the most common and pervasive cause of population decline and fragmentation; usually it is a complex of interacting factors rather than one simple one. Whatever immediate causes are involved, they are often the result of more distant social and economic factors operating at the national or global level.

Sometimes things are simple. The Israel painted frog Discoglossus nigriventer was described a few decades ago from specimens collected at Lake Huleh on the Israel–Syria border. Soon afterwards the lake was drained to allow agricultural development, and the species has never been

	NUMBER OF SPECIES IN IUCN RED LIST OF THREATENED SPECIES	THREATENED SPECIES THAT ARE "ENDANGERED"		NUMBER OF SPECIES IN IUCN RED LIST OF THREATENED SPECIES	THREATENED SPECIES THAT ARE "ENDANGERED"
Reptiles			**Amphibians**		
Turtles and tortoises	79	11	Frogs and toads	32	5
Crocodiles and alligators	15	11	Newts and salamanders	25	2
Tuatara	1	–			
Lizards	42	8			
Snakes	32	6			
TOTAL	**169**	**36**	TOTAL	**57**	**7**

seen since despite searches being made. It is presumed extinct.

When species are more widespread than the Israel painted frog appears to have been, disruption of habitat will more typically result in loss or depletion of local populations. This is the case with many amphibian species. In Europe, for example, countless small ponds and swamps have been drained to meet the demands of intensive agriculture, and countless local amphibian populations have disappeared along with them, simply as a result of the loss of their essential breeding habitat.

Less frequently, human activity provides new habitat for wild species and perhaps allows larger populations to persist than would otherwise have been the case. The rare and localized populations of the viperid snake *Vipera ursinii* in the mountains of southern Europe, for example, favor short grassland with abundant cover of juniper bushes—a habitat created partly by grazing sheep, under the traditional low-intensity system of subsistence farming. Now there tend to be fewer sheep, and the habitat is changing to the detriment of the snake populations.

More typically, human activity involves a piece-by-piece expansion into relatively unmodified habitats. Better roads lead to easier access, which leads to more people, more hunting or

disturbance, more construction, more forest clearance, more agricultural development, and so on. Narrowly distributed specialist-species are likely to be wiped out; wider ranging generalist-species will persist, but often in fragmented populations more susceptible to chance events, leading slowly to further population loss.

EXPLOITATION AND INTERFERENCE

Direct exploitation by humans is a cause of decline in relatively few animal species, but where it does occur its effects can be even more rapid and drastic than habitat modification. The traditional diet of many indigenous people often included several species of reptiles and their eggs, and sometimes the larger species of frogs. Although it was probably sporadic and moderate exploitation, it is difficult to establish how badly it affected wild species in prehistoric times. Today it is almost as difficult to distinguish between "subsistence" and "commercial" utilization. The latter, and international trade in particular, has caused severe decline in several species.

Frogs were collected commercially in the United States at the end of the nineteenth century. They are still collected in parts of South America and Europe, and the trade in frog legs collected in Southeast Asia and sent to Europe and elsewhere has now reached a high volume. Toad skin is

▼ *Amerindian youths with a captured anaconda (genus Eunectes) in Brazil. Anacondas are heavily persecuted by local human populations, and must be regarded as threatened. A few python species may grow longer, but anacondas are the largest and heaviest of snakes, sometimes exceeding 10 meters (about 33 feet) in length and 250 kilograms (550 pounds) in weight.*

Michael and Patricia Fogden

▲ *The American crocodile Crocodylus acutus, here represented by a juvenile, remains fairly widespread over most of its former range, but its numbers have been severely reduced by hunting.*

▼ *A Pacific ridley Lepidochelys olivacea on a nursery beach in Santa Rosa National Park, Costa Rica. Scattered broken eggshells symbolize the fact that both this and the related Atlantic ridley are among the most critically endangered of the world's marine turtles.*

Mills Tandy/Oxford Scientific Films

processed into leather on a small scale.

Most species of crocodilians have been adversely affected by excess hunting for the skin trade, although the smallest species and those with too many bony plates in the skin have been less exploited. The worst period was during colonial times in the first half of the twentieth century, particularly as firearms became readily available during and after the Second World War. It became possible for hunters to cruise along waterways, detect the animals by torchlight (when the eyes give a distinctive glow in reflected light), and kill each one with a well-placed rifle shot. Species that frequented beds of dense aquatic vegetation were less affected, but entire breeding populations of less-secretive species were eradicated. The black caiman *Melanosuchus niger* of South America, the Nile crocodile *Crocodylus niloticus* in Africa, and the saltwater crocodile *C. porosus* in Asia, have all been major targets. Simple persecution or target practice also accounted for many crocodilians. Hunting, malicious persecution, and habitat loss because of drainage or dam construction, provide a powerful combination of threats. The precarious status of the Chinese alligator *Alligator sinensis* appears to be a result of such a combination.

Sea turtles have also been the victims of heavy trade pressures, with eggs, meat, shell, or skin variously taken as the main commodities. Nesting populations are especially vulnerable to excess hunting; it is theoretically possible for dedicated hunters to collect every female turtle that emerges during the nesting season. In practice, bad weather or sea conditions, or simple logistics, sometimes make this impossible—thereby

ensuring the continued survival of some populations that would otherwise become extinct. Nevertheless, many nesting populations of olive ridley *Lepidochelys olivacea*, the green turtle *Chelonia mydas*, and the hawksbill *Eretmochelys imbricata* have been irretrievably lost. Because females typically nest on the beach where they first nested and laid their clutch, it is almost impossible for lost nesting populations to be replaced by members of another population.

Larger lizards, including monitors (genus *Varanus*) in Africa and Asia, and tegus (genus *Tupinambis*) in South America, and also many snakes, such as boas, pythons, cobras, and vipers, are intensely exploited for the skin trade. Many amphibians and reptiles are in demand for the live animal trade. It is difficult to measure the effects of this activity on wild species, but there is little doubt that local populations can be adversely affected.

The accidental or intentional introduction of species into habitats where they were formerly absent has been a major problem, especially on small islands. The mongoose in the West Indies; goats and rabbits in the Mascarenes; cats, dogs, and pigs almost everywhere else; all have had detrimental impacts on indigenous species of animals. The initial effect might be caused by direct predation, as with the mongoose preying on lizards of the genus *Cyclura* and snakes of the genus *Alsophis* in the West Indies. Or the initial effect might be on habitats, as with goats and rabbits on Round Island north of Mauritius, where much of the original native plant-life has been destroyed, and at least one of the endemic reptiles has disappeared.

◄ *Komodo dragons Varanus komodoensis mating. Largest of all lizards, komodo dragons are confined to a few small islands in Indonesia. Although threatened, their notoriety as potential man-eaters and consequent high status as a tourist attraction offers some hope that their population levels might be sustained by careful management.*

CONSERVATION MEASURES

Exploitation tends to be a direct problem, although one with complex socio-political origins. For this it is theoretically possible to find direct solutions. One option is legislation prohibiting the exploitation, and the creation of national laws intended to meet the requirements of international conventions can also be effective. The Convention on International Trade in Endangered Species of Wild Fauna and Flora (CITES) is the most wide-ranging of these, and it has had significant success in reducing or preventing international commerce of threatened species listed in its Appendices. Crocodilians, sea turtles, and many other reptiles and amphibians are prominent in these lists.

For some species, notably certain crocodilians, trade in animals reared in captivity from eggs or young collected in the wild is allowed. This can let economic benefits flow to local people, and also provides a reason for protecting the habitat of wild populations needed to replenish farmed stock. This option is not feasible for species with lower-value products or those more difficult to raise in captivity.

Habitat conservation is a more complex problem. But in a growing number of countries, development projects such as roads and dams cannot legitimately be started before an investigation is made into the probable impact on wild species and habitats. Local authorities often have the power to protect small areas of habitat, such as farm ponds. In general, "Protected Area" designation is a powerful conservation tool, particularly when a nation has the resources to implement land-use policies within its protected areas. This remains the best starting point for conservation of large areas where habitats (intact or man-modified) and their component species require management.

▼ *The crested iguana Brachylophus vitiensis of Fiji. Because of low population size and small recruitment potential, reptile species inhabiting oceanic islands generally have an even more precarious status than those species inhabiting continental land masses.*

▼ *Most caecilians are so obscure and
little-known that they lack English
names. This is* Dermophis mexicanus
of Central America.

CAECILIANS

RONALD A. NUSSBAUM

Caecilians are worm-like, mostly secretive, burrowing amphibians confined to the tropical and subtropical regions of the world. Although their name (pronounced "see-sil-e-an") means "blind", most caecilians have small eyes, which in some species are hidden beneath the bones of the skull. The order Gymnophiona is the least diverse and most poorly known of the three orders of amphibians.

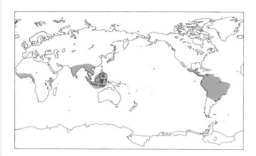

BUILT FOR BURROWING

Caecilians are the only living amphibians that are completely legless. They are capable of moving snake-like across surfaces when forced to do so, but normally they live underground and move through pre-formed tunnels or create new tunnels by pushing their head through loose mud or moist soil. Caecilians have many adaptations for burrowing. The skull is powerfully constructed with a pointed snout and an underslung lower jaw or recessed mouth, features that allow the head to be used as a ram. The eyes are reduced in size and importance, as there is no light in their underground world.

Caecilians have a unique pair of sensory organs called tentacles, one emerging from a groove or cavity on each side of the snout between the eye and the nostril. The tentacles, which are probably organs of taste and/or smell, are admirably suited for sensing the environment of tunnels. Another peculiar feature of caecilians is the manner in which they close their jaws; instead of having a single set of jaw-closing muscles as in all other terrestrial vertebrates (land-dwelling animals with backbones), caecilians have a dual mechanism that consists of two sets of jaw-closing muscles. Even this mechanism may be considered an adaptation for burrowing. Another interesting feature of caecilians is the presence of numerous skin folds, or rings, called annuli, which partially or completely encircle the body, although it is not

J. Campbell

Povel German

yet known if the annuli are directly related to locomotion. Many species also have small fish-like scales hidden under the skin folds.

The body muscles of caecilians are arranged in such a manner that the body can act like a rod moving within a tube. The "tube" is the skin and outer layer of body muscles. The "rod" is the head, the vertebral column, and the deep body muscles associated with the vertebral column. With the tube fixed in position in a tunnel, the rod can be pushed slightly forward through the soil to extend the tunnel. When the tip of the rod (the head) has reached its maximum forward progression, the tube is pulled forward in waves and fixed in position again so that the rod can again be pushed through the soil. However, in very loose mud, caecilians use a different kind of locomotion; they simply swim eel-like through the watery medium.

REPRODUCTION

Unlike most other amphibians, all caecilians have internal fertilization; the males have a distinctive protrusible copulatory organ, called a phallodeum, which serves to inseminate the females. Like all amphibians, caecilians have eggs and embryos similar to those of fish, and unlike those of

reptiles, birds, and mammals. The latter three groups have a pair of special membranes (amnion and chorion) that enclose the embryos. Fish, caecilians, and other amphibians lack these membranes, and the embryos typically develop within a gelatinous egg case. Some caecilian species lay eggs; the eggs hatch into larvae; the larvae later metamorphose into adults. Other species are viviparous: they give birth to living young, and there is no larval stage.

Females of egg-laying species remain with their eggs until they hatch. The function of this parental care is unknown, but presumably the female protects her embryos against predation and perhaps other sources of embryo mortality such as destructive molds and desiccation.

Viviparous species have a number of unusual adaptations. The embryos developing in the female's oviducts soon use up all of the yolk provided in the egg, and the embryos then begin feeding on a substance called "uterine milk", which is secreted by the oviducts. Embryos are provided with numerous tiny teeth, which have distinctive shapes and are shed soon after birth. The function of these embryonic teeth is not understood, but they may have something to do with intrauterine feeding. The gills of embryos

▲ *Most caecilians give birth to live young but some, like this sticky caecilian* Ichthyophis glutinosus *of southeast Asia, lay eggs in underground chambers. In these species the mother remains with her eggs until they hatch.*

▶ Ichthyophis kohtaoensis of southeast Asia. Most caecilians spend almost their entire lives underground, surfacing only in unusual circumstances. All appear to be carnivorous, feeding mainly on insect larvae and other invertebrates such as earthworms.

R. Altig

developing in the oviducts are enormously developed in some species and are thought to function in gas exchange between the tissues of the female and her embryo. This has not been proved, and other functions are certainly possible.

FOOD AND ENEMIES

The food and feeding habits of caecilians have not been carefully studied, but the larval caecilians that have been dissected by scientists had eaten a variety of immature insects, earthworms, and other invertebrates. Land-dwelling caecilians seem to feed primarily on earthworms. Beetles and other insects have also been found in their digestive tracts, and occasionally small frogs and lizards.

Snakes, especially coral snakes, seem to be the primary predators of caecilians. In the Seychelles islands, chickens, pigs, and the shrew-like tenrecs (introduced from Madagascar) at least occasionally eat caecilians. None of these predators is native to the Seychelles.

Like all amphibians, the skin of caecilians contains numerous poison glands, which presumably function to discourage predators. Many frogs and salamanders with highly toxic skin secretions are brightly colored, and experimental evidence indicates that the bright colors warn predators of the presence of the toxins and therefore help the amphibian to avoid injury from predator attack. Most caecilians have subdued colors, usually various shades of gray, but some (for example, members of the Rhinatrematidae and Ichthyophiidae families) have bright yellow lateral stripes. Although the function of the bright colors in these caecilians has not been studied, it seems likely that they too will prove to be warning colors.

PRIMITIVE CAECILIANS

Two families are considered to be relatively primitive, which means they are more like the common ancestor of all caecilians than are other caecilian families alive today. These are the Rhinatrematidae of northern South America and the Ichthyophiidae of India and Southeast Asia.

The rhinatrematids (two genera, nine species) of South America have retained the highest number of primitive characteristics: they have a "terminal" mouth (which means the mouth-opening is not recessed on the underside of the snout); their tentacles are in contact with the relatively large eyes; they have numerous skull bones; and they have a tail. These primitive caecilians also have the annuli (skin folds, or rings) subdivided into secondary and tertiary annuli which completely encircle the body, and numerous scales throughout the length of the body. The significance of secondary and tertiary annuli is still uncertain, and scientists know of their existence only by studying the embryonic and larval development of certain species. The specialized dual jaw-closing mechanism of caecilians is least developed in members of this family.

The life history and ecology of these seldom-seen caecilians is very poorly understood. However, we do know that they deposit eggs in soil cavities and that the eggs hatch into larvae with tiny external gills and gill slits. The larvae live in seepages and streams until they metamorphose into the adult form. The adults live in moist soil, leaf litter, and rotten logs.

The ichthyophiids (two genera, 36 species) of tropical Asia look very much like rhinatrematids, and until recently they were classified in the same family. They too have terminal mouths, numerous skull bones, a short tail, abundant scales, and primary annuli that are subdivided twice. Relatively advanced traits of the Ichthyophiidae include the position of the tentacle, which is forward of the eye, and a well-developed dual jaw-closing mechanism. The life history of ichthyophiids is similar to that of rhinatrematids.

Although few clutches of eggs have been found, all of these were attended by a female.

TRANSITIONAL CAECILIANS

A small group of caecilians confined to India is placed in the family Uraeotyphlidae (one genus, four species). Uraeotyphlids are in some ways intermediate between the primitive and advanced

▲▼ *Primitive caecilians: a species of Epicrionops (above) and Ichthyophis kohtaoensis (below). Unique to caecilians, the tiny, inconspicuous sensory tentacles before the eye are just barely visible.*

Ronald A. Nussbaum

caecilians. Primitive traits include the presence of many bones in the skull, a short tail, and numerous scales. Advanced features include a mouth-opening recessed below the snout, tentacles very far forward of the eyes, and the condition of the primary annuli—these skin rings are subdivided only once (instead of twice as in the primitive families), and some of the annuli at the forward end of the body do not completely encircle the body. The dual jaw-closing mechanism is well developed. Although very little is known about the life history of uraeotyphlids, the few observations suggest their way of life is similar to that of ichthyophiids and rhinatrematids, but with a shorter larval period.

ADVANCED CAECILIANS

The remaining three families of caecilians— the Scolecomorphidae, the Caeciliidae, and the Typhlonectidae—may be viewed as relatively advanced groups with specialized form, structure, and life history. Scolecomorphids and caeciliids live on land and are well adapted for burrowing in moist soil. Typhlonectids are aquatic or semi-aquatic. These three families share a number of advanced characteristics, such as a reduction in the number of skull bones, a recessed mouth, absence of tertiary annuli, and reduced number of scales, and they are tailless. Even so, the evolutionary relationships of these advanced families to each other and to the more primitive families are still not fully understood.

Scolecomorphids (two genera, five species) occur in equatorial East and West Africa. These are among the most bizarre of caecilians. They have large tentacles placed far forward on the underside of the snout in front of the mouth. The vestigial eyes are attached to the base of the tentacles. In resting position, the eyes are under the skull bones, but when the tentacles are protruded the eyes are carried outside the skull along with the tentacles. Scolecomorphids are the only caecilians that lack stapes, the tiny bones that conduct sound vibrations from the environment to the inner ear. Other advanced features of this family include the loss of secondary annuli (only primary annuli that do not completely encircle the body are present); and, as is usually the case with caecilians that have lost their secondary and tertiary annuli, scolecomorphids do not have scales. The body ends in a small, blunt "terminal shield", which lacks annular grooves. The East African species (in the genus *Scolecomorphus*) are viviparous; that is, they give birth to living young. The eggs are fertilized internally and are retained in the oviducts of the female, where embryonic development takes place. The aquatic larval stage has been lost, and the young are born as fully terrestrial juveniles. Nothing is known about the life history of the West African species of the genus *Crotaphatrema*.

▶ Siphonops annulatus of *South America. Caecilians are confined to tropical and subtropical areas around the world. With their minute eyes and underslung mouths, it is often difficult at a casual glance to determine which end is which. Caecilians can be distinguished from all other amphibians by the ringlike folds or segments, called annuli, in their skins.*

Chris Mattison

Ronald A. Nussbaum

The Caeciliidae (23 genera, 88 species) is the most diverse and most geographically widespread of the caecilian families. Species of this family are found in tropical Central and South America, equatorial Africa, the Seychelles archipelago in the Indian Ocean, and India. Caeciliids are recognized largely by the condition of the skull, which consists of a characteristic number and arrangement of bones. The skull bones are relatively few, and they are strongly joined to transform the skull into a solid ram used in burrowing. The mouth of all caeciliids is recessed, with the exception of *Praslinia cooperi* of the Seychelles Islands, which has a terminal mouth. The position of the tentacle opening varies considerably among caeciliids; in some species, the opening is close to the eye; in others, it is closer to the nostril. Only in the enigmatic *P. cooperi* is the tentacle in the primitive position adjacent to and in contact with the eye.

Caeciliids have a variable number of their primary annuli subdivided into secondary annuli (these secondary annuli always occur at the tail end of the body); or they have only primary annuli. Similarly, caeciliids have a variable number of scales in the skin folds; and species that have no secondary annuli generally also have no scales. None of the caeciliids has a tail; in some, the body ends in a blunt terminal shield.

Caeciliids have a variable life history. A few species such as *Praslinia cooperi* deposit eggs on land that hatch into water-dwelling larvae with a prolonged larval period. Others, such as *Grandisonia larvata* of the Seychelles Islands, have a very brief larval period. Still others, for example, *Afrocaecilia taitana* of East Africa, lay eggs on land and the eggs hatch directly into terrestrial juveniles without an aquatic larval stage. And some, like *Schistometopum thomense*, give birth to living young. Larval caeciliids are found in seepages and streams. The metamorphosed juveniles and adults are terrestrial burrowers. One species, *Hypogeophis rostratus* of the Seychelles, is found in streams, especially at night, as well as on land.

The family Typhlonectidae (four genera, 12 species) is found throughout much of tropical and subtropical South America. Relatively primitive typhlonectids are semi-aquatic, whereas the more advanced species are aquatic. The skull of typhlonectids is basically like that of the burrowing caeciliids, with relatively few bones that are solidly conjoined and with a recessed mouth. Even the most aquatic typhlonectid species are adept at burrowing in soft mud and gravel in the bottoms and edges of aquatic

▲ Viviparous caecilians, like Schistometopum thomense, *give birth to small but fully formed versions of themselves. Many caecilians are brightly colored or boldly patterned, perhaps as a warning to predators of their poisonous skin secretions.*

Ronald A. Nussbaum

▲ Typhlonectes natans of South America. Members of the family Typhlonectidae differ from other caecilians in being aquatic or semi-aquatic rather than subterranean burrowers. Lacking tails or scales, they have laterally compressed bodies and a pronounced dorsal fin running the length of the back to facilitate movement through the water.

habitats. Typhlonectids have primary annuli only, which are sometimes wrinkled, giving the appearance of being divided into secondary annuli, and they have no scales. They are tailless, with a terminal shield that lacks rings. The aquatic species have a skin fold or fin that runs along the back from the end of the body toward the head; the body is compressed laterally, which together with the fin increases swimming ability. The tentacular opening of typhlonectids is very small and variously placed between the eye and nostril, but it is never in contact with the well-developed eye. Unlike other caecilians, typhlonectids apparently do not protrude their tentacles. One recently discovered typhlonectid is peculiar in that it has no lungs (all other caecilians have lungs), and the passages between the nostrils and the mouth cavity are sealed. Typhlonectids are viviparous, producing young that are fully metamorphosed at birth.

EVOLUTIONARY HISTORY

The origin of caecilians is unknown. Some scientists believe they are derived from microsaurs, a group of salamander-like amphibians that became extinct during the Permian period, about 250 million years ago. The fossil record of caecilians is very poor—in fact only two discoveries have been reported and it is difficult to draw any conclusions from these specimens. The first consists of a single vertebra from Paleocene deposits in Brazil; this 60-million-year-old fossil has been classified in the family Caeciliidae, suggesting that there has been very little change in this family over long periods of time.

The second discovery, which has not yet been fully reported, consists of several fossils from Lower Jurassic deposits in northeastern Arizona. These fossils, which are at least 170 million years old, are notable for having small but well-developed legs.

IMPROVEMENTS FOR BURROWING

A typical vertebrate has a single set of jaw-closing muscles (one on each side of the head), which act by pulling up on the lower jaw from a point in front of the jaw hinge. Caecilians have this set of muscles too—labelled "adductor mandibulae" in the drawings. But they are unique in having a second set of jaw-closing muscles—labelled "interhyoideus posterior" in the drawings— which assist in jaw closure by pulling back and down on a special extension projecting behind the hinge of the lower jaw, in much the same way as pulling down on one end of a seesaw causes the other end to swing up. This set of muscles is present in other vertebrates also, but is not involved in jaw closure.

Why have these muscles been pressed in to service as jaw-closers in caecilians? The answer seems to lie in the evolution of burrowing efficiency. Relatively primitive caecilians, such as the rhinatrematid *Epicrionops petersi*, are relatively inefficient burrowers, which is reflected in both their behavior and their anatomy. Their mouth is at the terminal point of the snout, and they have a relatively weak skull, characters that would not be expected in a highly efficient burrower. In *E. petersi*, the adductor mandibulae muscles are large and dominate the jaw-closing system; they project up through openings in the skull, called temporal fossae, to attach on the top and side of the skull. The novel set of jaw-closing muscles— labelled "interhyoideus posterior"—while present and functional in these caecilians, is relatively small.

In somewhat more specialized burrowers such as *Ichthyophis glutinosus*, the temporal fossae are closed, making the skull more rigid. Closure of the temporal fossae restricts the size of the ancestral set of jaw-closing muscles by confining them under the roof of the skull. In *I. glutinosus*, the novel set of jaw-closing muscles has become dominant.

This evolutionary progression continues in forms such as *Microcaecilia rabei* and *Crotaphatrema lamottei*, in which the novel component of jaw closure is even more dominant. The highly specialized burrowing efficiency of these two species is indicated by the reduced number of skull bones, the rigidity of the skull, and the recessed or subterminal mouth.

FAMILY RHINATREMATIDAE
Epicrionops petersi

interhyoideus posterior

depressor mandibulae

adductor mandibulae

nostril

eye

opening for tentacle

FAMILY ICHTHYOPHIIDAE
Ichthyophis glutinosus

▲▼ *Caecilians are notable for the musculature of their jaws, incorporating a feature found in no other vertebrate. In most vertebrates, as well as some relatively primitive caecilians such as* Epicrionops petersi *and* Ichthyophis glutinosus *(above), jaw closure is primarily achieved by a set of muscles known as the adductor mandibulae. But in other caecilians, such as* Microcaecilia rabei *and* Crotaphatrema lamottei *(below, left), the jaw-closing function is dominated by the use of an entirely separate set of muscles, the interhyoideus posterior. The depressor mandibulae muscles are involved in jaw-opening.*

FAMILY CAECILIIDAE
Microcaecilia rabei

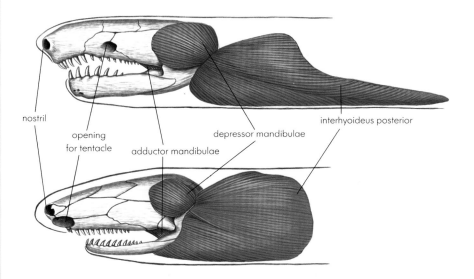

nostril

opening for tentacle

adductor mandibulae

depressor mandibulae

interhyoideus posterior

FAMILY SCOLECOMORPHIDAE
Crotaphatrema lamottei

Size
Smallest Lungless
salamanders of the genus
Thorius, in southern Mexico,
may reach maturity when
only 27 millimeters (1 inch)
long, head-body length
14 millimeters (½ inch).
Largest Chinese giant
salamander *Andrias
davidianus*, total length
1.8 meters (almost 6 feet);
weight 65 kilograms
(143 pounds).

Conservation Watch
The following species and
subspecies, all from North
America, are listed as
endangered in the IUCN Red
Data Book of threatened
animals: Santa Cruz long-
toed salamander *Ambystoma
macrodactylum croceum*,
desert slender salamander
Bactrachoseps aridus, and
Texas blind salamander
Typhlomolge rathbuni.

▼ *Newts, like this crested newt* Triturus
cristatus, *are common in freshwater
streams and swamps of all kinds,
although more or less restricted to the
Northern Hemisphere. This male is
eating its sloughed-off skin.*

SALAMANDERS AND NEWTS

B. LANZA, S. VANNI, AND A. NISTRI

S alamanders have been the subject of countless myths and legends since remote times. The origin of the salamander name is an Arab–Persian word meaning "lives in fire". In fact, until a few centuries ago it was believed that the black and yellow fire salamander *Salamandra salamandra* of Europe could pass unscathed through flames, a belief still held in some areas. Such modern names as the English "fire salamander" and the German "*Feuersalamander*" come from this ancient legend. In recent times salamanders have become a favorite of terrarium and aquarium keepers around the world because of the ease with which they can be raised in captivity.

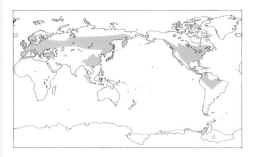

WHAT'S IN A NAME?

Salamanders belong to the order Caudata (from the Latin word *caudatus*, meaning "provided with a tail") and are also referred to as caudates; as the scientific name implies they retain a tail even after metamorphosis from the larval stage. In this they differ from other amphibians—caecilians have no tail or only a rudimentary one, and frogs and toads have a tail only in the larval stage.

"Salamander" is a broad term applicable to any member of the Caudata, whereas "newt" is more restrictive. Members of 10 genera *Cynops*, *Euproctus*, *Neurergus*, *Notophthalmus*, *Pachytriton*, *Paramesotriton*, *Pleurodeles*, *Taricha*, *Triturus*, and *Tylototriton*, which are mostly aquatic, at least during the breeding season, are commonly called newts; the others, whether amphibious or exclusively aquatic or exclusively terrestrial, retain the salamander name. "Newt" has a curious origin: it derives from the Anglo-Saxon word *efete* or *evete*, used to indicate newly metamorphosed newts, becoming *ewte* in Middle English, and then *newte*: an ewte = a newte.

Salamanders and newts are found almost exclusively in the Northern Hemisphere—in Europe, central and northern Asia, northwestern Africa, North America, and Mexico. In the Southern Hemisphere there are only a few members of the families Hynobiidae and Salamandridae in Southeast Asia, as well as the Chinese giant salamander, and species of lungless salamanders are found as far south as central Bolivia and southern Brazil. Caudates are found from sea level to about 4,500 meters (14,700 feet), wherever there is sufficient humidity during at least part of the year. Some specialized species dwell entirely in subterranean waters, others are tree and/or rock dwellers.

There are very few fossil remains testifying to the origin of this order. The earliest known true salamanders are three specimens dating back to the late Jurassic, about 150 million years ago. In 1726 the fossil remains of a giant salamander were curiously attributed by J.I. Scheuchzer to that of a human "witness to the Universal Flood", named *Homo diluvii testis*, and it was not until a century later that the remains were correctly identified and named *Andrias scheuchzeri*.

LIKE LIZARDS WITHOUT SCALES

Superficially salamanders resemble lizards but are immediately distinguished by their complete lack of scales. Even Linnaeus, the father of modern taxonomy, erroneously assigned some species of

salamanders to the lizard genus *Lacerta*. Salamanders also resemble lizards in usually having four limbs; but again, like some lizards, certain salamander species have only forelimbs.

The body is lizard-like, with the head more or less distinct from the trunk, except in permanently aquatic species, whose head is elongated and rather eel-like; in this case the limbs are short, as in the olm *Proteus anguinus* of Europe and the American *Amphiuma* species, which some people call "congo eels", or the hind limbs are entirely absent, as in American sirenids.

The tail is laterally compressed and often crested in the aquatic species, and rounded or only slightly compressed in the terrestrial ones. In some species, particularly tree-dwellers, the tail is also somewhat prehensile. When threatened by a predator, some lungless salamanders do what many lizards can do: voluntarily sever their tail (known as autotomy) as a decoy; the tail then regenerates. In most species the adults are 10 to 20 centimeters (4 to 8 inches) total length.

THE SKIN

Salamanders and newts continue to grow even after reaching sexual maturity, which is one of the reasons why the superficial horny layer of the skin is periodically shed. Depending on the species, the old skin comes off (as frequently as once a week in some cases) either in fragments or in one piece. This slough, or exuvia, is then usually devoured by the salamander.

▲ *Perhaps from their natural propensity to crawl from crevices in logs tossed onto a campfire, salamanders have had a persistent association with fire in folk legends. It was widely believed that they cannot be harmed by fire. Common over much of southern Europe, this is probably the species that originated such myths, the fire salamander* Salamandra salamandra.

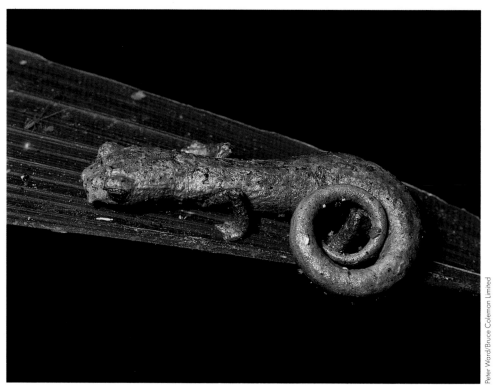

▶ *A salamander of the genus Bolitoglossa at night in a Costa Rican rainforest. Many newts and salamanders are nocturnal, but despite their obscurity and inconspicuousness, they are often very abundant. In some woods and forests, the total mass of these animals may outweigh that of all mammals and birds combined.*

Peter Ward/Bruce Coleman Limited

Unlike that of reptiles, the skin of salamanders contains three different types of glands distributed rather uniformly over the entire body surface: mucous, granular, and mixed. The secretion of the mucous glands, and the mucous part of the mixed glands, is homogenous and frothy, neutral or alkaline, and sticky. On land its primary function is to protect the skin from drying out, permitting respiratory exchanges which could not occur through a dry surface. In water, it helps to maintain the body's internal osmotic pressure (the salt and water balance in the body fluids) and simultaneously acts as a lubricant during swimming. The granular glands, and the granular part of the mixed glands, produce a granular secretion containing various types of poisons and often giving off a specific odor; they are located especially on the upper part of the head behind the eyes (paratoid glands), on the tail, and on the sides of the back. The mixed glands, as the name implies, produce both a mucus and a granular secretion and are located over almost the entire body. Some members of the families Plethodontidae, Ambystomatidae, and Salamandridae possess a fourth type known as the hedonic glands, which resemble the granular and mixed glands and are located in various zones. Their secretion, containing pheromones, plays an important role during courtship and mating; particularly noticeable is the chin gland of most plethodontid males.

In some species the skin is smooth and in others it is bumpy, or even wart-like at the poison gland outlets. The skin of amphibious species is smooth when the animals live in the water but can become roughish when they live on land. The skin is rich not only in glands but also in pigment cells. In many species the body color is brownish, yellowish, or grayish, sometimes with a barely contrasting pattern. Several have gaudy coloration on their back or underside throughout the year in both sexes (as in the fire salamander and various lungless salamanders), or else only in the males, particularly during the season of courtship. The courtship period is also the time when many male newts develop conspicuous skin folds over their body, tail, fingers, and toes. Species that live permanently underground lack pigment and thus are white or pink. The color of the larvae usually differs from that of the adults.

With the exception of the two species of the Asian genus *Onychodactylus*, whose toes have a horny sheath which turns into a little blackish claw at least during the mating season, salamanders and newts do not have nails.

THE SKELETON

Some parts of the skull begin as cartilage and then turn into bone during growth. Others, not derived from cartilaginous elements, are known as membranous bones. The shape, number, and arrangement of these two types of bones are important keys for classification.

The spine is generally divided into five regions: cervical, dorsal, sacral, sacro-caudal, and caudal. The cervical region consists of a single vertebra joined to the skull at four points (instead of two points in all other amphibians). The dorsal region usually has 13 to 20 vertebrae, although the minimum number is 11 (in the mole salamander *Ambystoma talpoideum*) and the maximum 63 (in *Siren* species). The sacral region consists of one vertebra; and the sacro-caudal from two to four. The caudal (tail) region varies from 20 vertebrae to more than 100 in some *Oedipina* species of Central and South America.

Those with an eel-shaped body, such as *Siren*, *Proteus*, and *Amphiuma*, have ribs corresponding with only the dorsal vertebrae closest to the head, but in other species the ribs occur along all or almost all the dorsal tract and sometimes also in the sacro-caudal region. In two types of newts, the sharp-ribbed newt *Pleurodeles waltl* and the alligator newt *Tylototriton andersoni*, the tips of the ribs can actually perforate the skin, increasing the likelihood that the newt's poison will enter the body of any predator.

The limbs have the basic structure common to all vertebrates; the front legs are linked to a thoracic girdle which does not articulate with the vertebral column, and the back legs to a pelvic girdle articulating with the single sacral vertebra. Most species have four fingers on each forelimb and five toes on each hind limb.

THE DIGESTIVE SYSTEM

Salamanders have no salivary glands. Those that spend all their life in water have what is called a "primary" type of tongue, a fleshy fold on the floor of the mouth with very little mobility because it has no intrinsic muscles; it is not used to capture prey. All other salamanders have a fairly mobile and well-developed tongue, which is used—when they hunt on land—to procure food. In some lungless salamanders it is mushroom-shaped and can be darted onto the prey, just as chameleons do. The teeth, usually small, are implanted on the margins of both the upper and lower jaws, as well as on the roof of the mouth; the sirenids have them only on the latter. The teeth of some lungless salamander males have a sexual function (discussed later). Salamander larvae also have true teeth, in contrast to those of frogs and toads which have only horny structures.

The esophagus is fairly short and leads into the stomach with no regional differentiation and then an almost-straight intestine with little distinction between the small and large components. The terminal part of the rectum enlarges to form a cloaca, which contains the outlets of the urinary and reproductive tracts; its opening is located underneath the base of the tail.

▼ *The skin of amphibians as diverse as toads and salamanders contains glands that secrete toxic substances as a defense against predators. The sharp-ribbed newt Pleurodeles waltl of eastern Asia adds an unusual component to this defense: its ribs are long and sharp, and if grabbed by a predator, the tips of the ribs penetrate through the skin and its associated poison glands and, in effect, inject their toxins into the soft tissues inside the mouth of the attacker.*

R. König/Jacana AUSCAPE International

Stephen Dalton/NHPA

▲ *Often kept as a pet, the Mexican axolotl* Ambystoma mexicanum *is the best-known of the newts and salamanders exhibiting neoteny. It may retain gills and other larval features throughout its life, even as a breeding adult.*

RESPIRATION AND BLOOD CIRCULATION

The lungs of salamanders (in species that have them) are sac-like. Usually they are identical in size, although in some species the right lung is slightly shorter, and in amphiumids the left lung is rudimentary. The lungs of the spectacled salamander *Salamandrina terdigitata* and a few other genera, such as *Euproctus*, are tiny. All plethodontids and some hynobids are lungless; respiratory exchanges occur only through the skin and through the mucous membranes of the mouth and throat. Respiration through the skin also plays an important role, at times essential, even in species with lungs, for instance when they hibernate underwater. The olm and other species that never lose their gills (known as perennibranchiates), as well as larvae, and neotenics (those species which occasionally retain larval features in the adult form—see page 505) can breathe either through their skin or through external gills which stick out in bright red tufts on the sides of the head. Only amphiumas have internal gills. The color of the gills is due to

the rich supply of blood, and respiration through the skin is enhanced by the marked vascularization of the skin.

The heart consists of a single ventricle in which the arterial and venous blood mix only partially. The left and right auricles are distinct in the lunged forms, and not completely separate in the perennibranchiate and lungless forms. The latter obviously do not have lung veins, and they have a smaller left auricle. Red blood cells are large (the eel-like amphiumas of the southeastern United States have the largest of all vertebrate animals), usually elliptical, and usually have a nucleus. According to the species, their number ranges from about 30,000 to 100,000 per cubic millimeter of blood; but as recently discovered in *Triturus* newts, the number of circulating red blood cells probably increases markedly also in other salamanders when environmental conditions prevent the blood from being sufficiently oxygenated.

The lymphatic system is fairly well developed. Salamanders have numerous lymph hearts under the scapula (shoulder-blades), at the root of the tail, and along the body flanks just under the skin. The lymph hearts are heart-like structures that improve the lymphatic circulation.

NERVOUS SYSTEM AND SENSE ORGANS
Salamanders have a relatively simple nervous system. The brain is rather small, but its front lobes are well developed, paralleling the marked development of the olfactory organs (detecting smell). In most species the eyes are large, with round pupils, but in permanently subterranean forms they are reduced in size and sometimes hidden under the skin. Species with well-developed eyes usually have upper and lower eyelids, plus a nictitating membrane known as the third eyelid.

There is no external ear, only a vestigial middle ear which does not have a cavity of its own. Vibrations are probably transmitted to the internal ear through the jaws or the bones connecting them to the skull, or else through the forelimbs and pectoral girdle. In the water, low-frequency vibrations are also perceived by lateral line organs (which also detect changes in water pressure and currents) found on the head and sides of the trunk in all larvae and in adults of many prevalently aquatic species. But only female spectacled salamanders, which enter the water to lay their eggs, develop lateral line organs that are active exclusively during this short period.

REPRODUCTION
Some salamander and newt species are exclusively aquatic, and some exclusively terrestrial. Others are clearly amphibious, and the females, or both sexes, return periodically to water

Jack Dermid/Bruce Coleman Limited

▲ Amphiumas, like this one-toed amphiuma Amphiuma pholeter, are entirely aquatic, although they sometimes emerge on land during heavy rain. Lateral line organs, visible here as pitted areas on the head, perceive low-frequency vibrations as well as changes in water pressure.

to reproduce. This usually occurs in the spring and can involve large migrations.

In the most primitive families, Cryptobranchidae and Hynobiidae, males fertilize the eggs outside the female's body. Fertilization is internal in other salamanders, although for sirenids observations still have to be confirmed. In species with internal fertilization the male deposits a gelatinous spermatophore (a capsule containing sperm) during mating, which can have a shape peculiar to the species or genus, with a

NEOTENY: A JUVENILE ALTERNATIVE

In many different groups of animals, including both amphibians and reptiles, some species achieve sexual maturity while still retaining many juvenile or larval characteristics. This phenomenon, known as neoteny, is common in amphibians, especially salamanders, where it may take a number of different forms. Although not universally accepted, one classification of these forms is that by DuBois (1979):

Total neoteny: some individuals remain permanently in a larval state and never reach sexual maturity. This occurs in some populations of the European olm;

Temporary neoteny: delayed metamorphosis, often accompanied by large larval size, in some species of frogs and salamanders;

Partial neoteny or paedogenesis: sexual maturity

and reproduction occurs in animals which are still in the larval stage of their life cycle. It occurs only in salamanders (among the amphibians) and can be of three types:

Obligatory paedogenesis: metamorphosis cannot be induced even with hormone treatment;

Quasi-obligatory paedogenesis: metamorphosis can be induced under certain environmental conditions, including hormone treatment. For example, the axolotl, a famous paedogenetic form of the Mexican mole salamander, will metamorphose if treated with the hormone thyroxin;

Facultative paedogenesis: the ability to achieve sexual maturity and reproduce for a long time before metamorphosis, as in some populations of the salamander genus *Triturus*.

▲ *The alpine newt* Triturus alpestris *lays its eggs in water. Protected by a gelatinous translucent capsule, the embryos develop within, in a manner not unlike that of typical frogs.*

females go to the water only to lay their larvae and eggs, respectively. Rituals linked to courtship, such as nuptial dances, are fairly complex and differ from species to species. It can involve, as in *Euproctus* newts, a sort of embrace which is obviously not accompanied by copulation, as salamanders have neither a penis nor any other form of intromittant organ.

The eggs do not have a protective shell but are normally surrounded by a gelatinous layer whose development and consistency differ from species to species. On land eggs can be laid in several different places as long as there is sufficient humidity and protection. In water they are attached to rocks or submerged logs or roots. Some tree-dwelling lungless salamanders lay their eggs in bromeliads, in the water cupped in the base of the leaves. Some species, such as the olm, have parental care, a task usually carried out by the female but also by the male in the giant salamanders, the mud salamander *Pseudotriton montanus* and the clouded salamander *Hynobius nebulosus*. The number of eggs depends on the species and the size of the individual, varying from five to six in some terrestrial salamanders to a maximum of about 5,000 in some *Ambystoma* mole salamanders which lay their eggs in water.

Several salamanders are ovoviviparous—instead of laying eggs the female gives birth to well-developed larvae or to young that almost perfectly resemble the adults. The fire salamander usually gives birth to well-developed larvae but can in some regions give birth to already meta-morphosed young. In some ovoviviparous species adelphophagy takes place within the maternal genital tract; this consists of the more-developed

whitish mass of sperm adhered to its apex. The spermatophore, or only the sperm cap, is sucked in by the cloacal lips of the female.

Ovulation and fertilization may either follow insemination or be delayed for several months, in some cases up to two and half years, as in the fire salamander. When fertilization is delayed, sperm are stored in special diverticula of the female genital tract and subsequently fertilize the eggs at the time of ovulation.

Rutting takes place on land in exclusively terrestrial species and in those such as the fire salamander and spectacled salamander, whose

▶ *Alpine newts in courtship. Caudates lack a penis or any similar organ, yet fertilization is usually internal. The male deposits his sperm in the form of a small packet, known as a spermatophore. Then, in an intricate courtship ritual, he maneuvers the female into a position to pick it up with the lips of her cloaca.*

embryos absorbing the eggs and smaller embryos, thus feeding on their siblings or "brothers" (*adelphós* is an ancient Greek word for brother). Many terrestrial lungless salamanders lay eggs that hatch already metamorphosed young.

The larvae usually have tail and dorsal membranes. In some species that reproduce in the water, before the limbs appear the larvae have a stick-like organ called the balancer or adhesive organ on the sides of the head in front of the gills, which serves to adhere the larvae to the substrate. After a period of time varying from a few days to several years according to the species, the more-or-less grown larvae undergo metamorphosis, losing their external gills and changing in other external and internal parts of the body. However, for environmental and/or genetic reasons, some or all members of a species can become sexually mature and capable of reproducing while maintaining such larval or juvenile morphological characters as external gills. The so-called perennibranchiate forms such as *Proteus* and *Necturus* never metamorphose, even if treated with a growth stimulant such as iodine or thyroid extract. Others do metamorphose spontaneously after a short time or when induced to do so when administered a hormonal treatment under experimental conditions. This is true for various species of *Triturus* newts and the axolotl *Ambystoma mexicanum*.

FOOD AND PREDATORS

Most salamanders are active at dusk and at night. When it is too cold or too dry many species take refuge under rotting vegetation or deeply buried rocks, in rock crevices or deep in the ground,

Hans Reinhard/Bruce Coleman Limited

▲ Several unusual features characterize the fire salamander's reproductive strategy. The female retains her fertilized eggs within her body until they hatch, and the emerging young may be either larvae (as in this case) or fully formed juveniles.

LIFECYCLE OF AN AMPHIBIOUS SALAMANDER

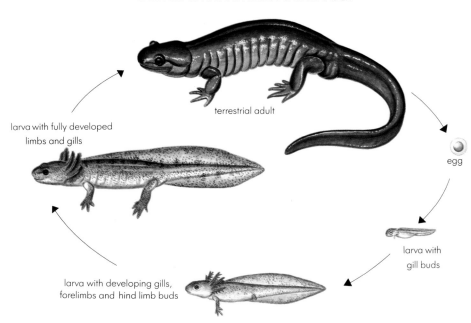

terrestrial adult

larva with fully developed limbs and gills

larva with developing gills, forelimbs and hind limb buds

egg

larva with gill buds

◄ Larval development in salamanders varies greatly between species. Illustrated is the basic amphibious lifecycle, in which much of the development occurs outside the egg. In many terrestrial salamanders, however, the young emerge from the eggs as miniature adults, with no aquatic larval stage.

becoming active only when the external environment is again suitable. They are all carnivorous and usually feed on small invertebrates such as insects, spiders, crustaceans, mollusks, and worms. The larger species also prey on small vertebrates. Cannibalism is not rare and may occur in both larvae and adults.

Unlike frogs and toads, salamanders do not usually make any sound as they have no larynx, or only a rudimental one, and no vocal cords. However, if disturbed or excited some species are capable of producing a weak squeak (the fire salamander, *Aneides*, *Dicamptodon*, and some species of newt), a faint yelp (sirenids) or a tiny shout (*Pleurodeles* newts).

Salamanders are preyed upon by other amphibians and by water tortoises, snakes, lizards, fish, birds, mammals, and large invertebrates. When in danger they defend themselves by secreting toxic or sticky substances, or by resorting to autotomy of the tail (as in some lungless salamanders) or by assuming a typical reflex posture which differs according to the species. One posture involves arching the body with the tail perpendicular to the body (or sometimes rolled up) to show off the gaudy ventral coloration that seems to warn the predator of a toxin in the skin. This particular posture is known as the unken reflex.

To humans the economic importance of salamanders is insignificant and is virtually limited to species of particular beauty and/or rarity, which are sought after as pets. Some species are used in biological research. The giant salamanders from Japan and China and the axolotl (the neotenic form of the mole salamanders) are appreciated as food, and the latter also as a supposed aphrodisiac. Others are used as live bait by fishermen, which has caused a great reduction in many populations of the lungless seal salamander *Desmognathus monticola* throughout the southeastern United States, and in some places have pushed them to the edge of extinction. The hynobid *Batrachuperus pinchonii* of eastern China is venerated as a divinity under the name "White Dragon", and the pool on Mount Omei where the species lives is the destination of pilgrimages, so much so that a sanctuary has been built nearby. This does not, however, prevent it from being dried and ground to a powder as a remedy for stomach disorders, a fate suffered also by its close relative, the Japanese clawed salamander *Onychodactylus japonicus*, which in traditional Japanese medicine has been used to rid the patient's body of worms.

Humans have had a direct impact on some populations through the capture of certain species for sale or study, and an even greater impact indirectly through the disruption or destruction of their habitat. Predation, especially on larvae, by fish stocked in pools for sport fishing has a very deleterious effect, but an even worse threat may come from acid rain.

THE MOST PRIMITIVE FAMILIES

The suborder Cryptobranchoidea includes the most primitive living salamanders, the only ones that have external fertilization. Their eggs are always deposited in two groups, each contained in a gelatinous sac. In the adults, two bones of the lower jaw (the angular and prearticular) are clearly separated from each other. In other respects the two families of this suborder, Cryptobranchidae and Hynobiidae, differ greatly in appearance.

► The hellbender Cryptobranchus alleganiensis *is entirely aquatic, living in mountain streams in the eastern United States. Much the largest of North American salamanders (up to 70 centimeters, or about 28 inches, in length), it is nevertheless only about half the size of its relative the Chinese giant salamander, largest of all the caudates. Baggy folds of skin along the flanks increase the total surface area of skin, improving oxygen transfer from the water, but hellbenders also have lungs and periodically come to the surface to gulp air.*

R.J. Erwin/NHPA

Jack Dermid/Oxford Scientific Films

Hellbenders and giant salamanders

The cryptobranchids (family Cryptobranchidae) always live in running water. They are corpulent and have large skin folds along the flanks, which increase the body surface and thus enhance the absorption of oxygen from the water. Their metamorphosis is incomplete, so the adults still have gill slits or grooves and no eyelids. All species have four fingers on each forelimb and five toes on each hind limb.

The hellbender *Cryptobranchus alleganiensis*, which can reach a total length of 75 centimeters (30 inches), inhabits central to northeastern United States. The genus *Andrias* is today represented by the Chinese giant salamander *A. davidianus*, growing to 180 centimeters (almost 6 feet), and the Japanese giant salamander *A. japonicus*, whose length never exceeds 150 centimeters (almost 5 feet). They feed on various invertebrates such as crustaceans but occasionally also eat small aquatic vertebrates. Each female lays up to 450 eggs, in paired rosary-like strings. The male releases his sperm on the eggs and seems to guard them until they hatch 10 to 12 weeks later.

Hynobids

In the family Hynobiidae none of the 35 species grows larger than 25 centimeters (10 inches) total length. They live in Asia, with the exception of the Siberian salamander *Salamandrella keyserlingi*, which has spread westward as far as European Russia. The family consists of nine genera: *Batrachuperus*, *Hynobius*, *Liuia*, *Onychodactylus*, *Pachyhynobius*, *Pachypalaminus*, *Ranodon*, *Salamandrella* and *Sinobius*. They all reproduce in water (for example streams, ponds, tarns) but

outside the breeding season are terrestrial. The female lays two spindle-like capsules, each containing 35 to 70 eggs, according to the species. Observers have often seen examples of parental care.

Some *Hynobius* and *Salamandrella* and all the *Batrachuperus* species have four-toed feet. Other types have five toes on each hind limb.

THE SEVEN ADVANCED FAMILIES

The suborder Salamandroidea, which includes the most advanced salamanders, differs from the primitive families in having the angular bone fused with the prearticular bone in the lower jaw, at least in the living species. With the probable exception of sirens, the sperm are taken into the female's body to fertilize the eggs.

There are seven families, some 50 genera, and a total of about 350 species. Members of this suborder occur everywhere salamanders exist, except Antarctica, Australasia, Oceania, and Africa south of the Sahara.

Sirens

The family Sirenidae has so many peculiar characters that some authors classify it in a separate suborder, Sirenoidea, or even a different order, Trachystomata. Being neotenic (see page 505), sirens have gills throughout their lifetime, and they lack eyelids; they have small eyes, no hind limbs and an eel-shaped body.

Sirens live in southeastern United States and adjacent regions of Mexico, spending their life in ponds and swamps with rich aquatic vegetation and a muddy bottom; like eels, they can cover short distances on land at night during rainy periods. The greater siren *Siren lacertina*, which

▲ Nearly one meter (3 ¼ feet) long, the greater siren *Siren lacertina* has a slender body resembling that of an eel; it has forelimbs but no hind limbs, and is equipped with lungs as well as gills. Sirens are exclusively North American.

▲ Two fully aquatic salamanders: the mudpuppy Necturus maculosus (top) and the dwarf siren Pseudobranchus striatus (above).

▼ The axolotl is normally muddy gray in color, but albinism, as in this pinkish individual, is not unusual.

Stephen Dalton/NHPA

grows to 95 centimeters (37 inches) total length, and the lesser siren *S. intermedia*, reaching 68 centimeters (27 inches) have three pairs of gill slits and four fingers on each hand. The dwarf siren *Pseudobranchus striatus*, which grows to 25 centimeters (10 inches), has only one pair of gill slits and three-fingered hands.

Little is known about their reproductive biology, but judging by their uro-genital apparatus, one may suppose that they have external fertilization; their eggs have been found isolated or in small clumps attached to submerged plants. Like the lungfishes, if their habitat dries they can survive for weeks or months embedded in the mud, enveloped by a kind of mucus cocoon with only the tip of the snout jutting out.

Waterdogs, the mudpuppy, and the olm

Proteus, the sea god who had the power of assuming whatever shape he pleased, inspired the Viennese naturalist Laurenti in 1768 to use the name *Proteus anguinus* for a curious amphibian discovered 24 years earlier near Ljubljana (now in Yugoslavia). Its discoverer, Baron J.W. Valvasor, had considered it to be the juvenile form of an animal destined to change into a dragon. The nobleman was actually not far from the truth, as the little animal, fortuitously swept above ground by the flood of a subterranean river, really looked like a larva with its large red gill tufts. This was not only the first true dweller of caves and underground passages to be discovered, but also the first member of the small but extremely interesting salamander family, Proteidae.

Proteus, also known as the olm, can grow to a total length of 33 centimeters (13 inches) and is found only in the subterranean waters of the western Balkan Peninsula and northeastern Italy. It is eel-like and usually a pale rose color because there is almost no skin pigmentation. The eyes are small and concealed under the skin. The forelimbs and hind limbs bear three and two digits, respectively. Larvae and juveniles, darker and with larger exposed eyes, are noteworthy for feeding on bacteria, protozoans, and organic matter contained in the slime.

The family includes five other species, all in genus *Necturus*, which inhabit central and eastern United States. Because of the popular misconception that they can bark, four are called waterdogs, and these do not exceed a total length of 28 centimeters (11 inches); the fifth is the mudpuppy *N. maculosus*, sometimes a little more than 40 centimeters (15¾ inches) long.

All members of the Proteidae family are neotenic and consequently never lose their gills; they have two gill slits. They also lack eyelids and upper jaws. The eggs, which are tended by the parents, are attached to submerged stones or logs.

Two families of mole salamanders

The mole salamanders spend their life almost entirely underground, emerging from their subterranean world only to reach the ponds or streams in which they reproduce. The breeding season is usually spring, but at least one species, the marbled salamander *Ambystoma opacum*, reproduces during fall. Fertilization, often preceded by a nuptial dance, is immediately followed by the laying of up to 200 eggs, generally in water but alternatively buried where rising water will flood the nest.

The family Ambystomatidae comprises about 30 species in the genus *Ambystoma* and four in *Rhyacosiredon*; they are found from southern Alaska and Canada throughout the United States and most of Mexico. Pacific mole salamanders are members of the family Dicamptodontidae, with three species in the genus *Dicamptodon* and one in the genus *Rhyacotriton*, all confined to northwestern United States.

These two families differ in that one (Dicamptodontidae) has a lacrymal bone, while members of the Ambystomatidae do not. They also differ in their habitat preferences during the breeding season: the former prefer mountain brooks, while the latter prefer still water such as lakes. Some species, such as the axolotl, are neotenic (see page 505).

Two *Ambystoma* species consist only of females. Unlike other sexually reproducing species, in which half the chromosomes come from the female and the other half from the male, these species possess cells with 42 chromosomes—

▼ *The Pacific giant salamander* Dicamptodon ensatus *(top) is restricted to western North America. Unlike most salamanders, which are silent, this salamander frequently emits low-pitched sounds when disturbed. With a maximum length of 40 centimeters (15 ¾ inches), the tiger salamander* Ambystoma tigrinum mavortium *(bottom) is one of the largest terrestrial salamanders, and adults often feed on small vertebrates. It is highly variable in color and pattern, with spotted varieties as well as banded forms.*

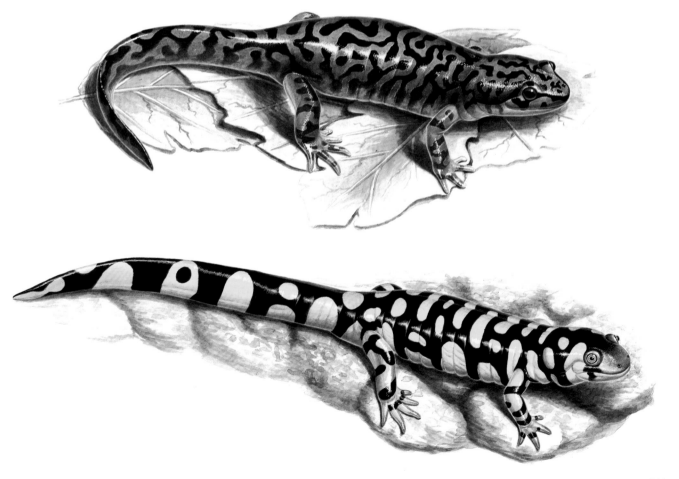

▶ *Like this European crested newt (opposite), the skin of many species of newts and salamanders carry bold patterns in striking colors, signalling to potential predators the warning of poison glands in the skin. These patterns are sometimes on the throat or belly, and are often exhibited in a gesture unique to caudates, the unken reflex. In this rigid posture, the head is thrown back, the spine arched and the tail held stiffly upright, the better to display the visual threat.*

three times the normal number, known as triploid number—and all of maternal origin. Moreover their eggs develop by gynogenesis—that is, they develop without the male's chromosomes entering the genetic complement of the new individual. The development of the eggs may be activated by sperm from a male of a related *Ambystoma* species (obviously one that has males); the individuals that develop from this type of reproduction are destined to be females.

All mole salamanders have a rather squat body, less than 35 centimeters (13 inches) in total length. Some have bright color patterns, which contrast with the dark color of the ground.

Amphiumas

If it were not for the presence of four tiny limbs, the amphiumas of the southeastern United States could easily be mistaken for eels, especially as they are neotenic and thus live mostly in water and, like eels, burrow in the mud; they can also move across wet ground. Adults have one pair of gill slits and inner gills, characters that increase their resemblance to eels. They have no eyelids or tongue. They lay their eggs under different kinds of shelters or on wet mud, in long strings each containing up to 150 eggs or more. The female remains coiled around them until they hatch, which takes about five months. On hatching the larvae must make their way to bodies of water, usually when it rains.

The family (Amphiumidae) includes three species: the one-toed amphiuma *Amphiuma pholeter*, at 30 centimeters (12 inches) total length; the two-toed amphiuma *A. means*, at 116 centimeters (46 inches); and the three-toed amphiuma *A. tridactylum*, at 106 centimeters (42 inches).

Salamandrids

The family Salamandridae includes some 60 species in 14 genera. Members of 10 genera more or less linked to water are commonly called "newts" (the scientific names are listed at the beginning of the chapter). The other four genera, predominantly or exclusively terrestrial, have the popular name "salamanders" in English, and are found in Europe, northwestern Africa, and southwestern Asia. The gold-striped salamander

Chioglossa lusitanica is endemic to Spain and Portugal, while the spectacled salamander is endemic to peninsular Italy. The genus *Mertensiella*, with two species in Turkey, some of the Greek islands, and the Caucasus, has males that bear a spur on the tail base. In the genus *Salamandra* there are two mountain-dwelling species, which give birth to perfectly metamorphosed young—the black, or black and yellow Alpine salamander *S. atra* from northern Albania to the western Alps, and Lanza's salamander *S. lanzai*, always black, endemic of the southwestern Alps—as well as the largest member of the family, the fire salamander *S. salamandra*. This magnificent black and yellow animal, sometimes longer than 30 centimeters (12 inches), has different subspecies throughout most of Europe extending to Iran and northwestern Africa.

Among the newts only three species of *Notophthalmus* and three of *Taricha* inhabit North America; the black spotted newt *Notophthalmus meridionalis*, the striped newt *N. perstriatus*, and the eastern newt *N. viridescens*, inhabit eastern North Amercia, while the larger (up to 22 centimeter or 8¾ inches) rough-skinned newt *Taricha granulosa*, red-bellied newt *T. rivularis* and Californian newt *T. torosa* may be found only in the marginal western United States. Two *Pleurodeles* species occur in northwestern Africa, with *P. waltl* also in Spain. All the other newts, about 45 species, live in Europe or Asia. The distribution of the genus *Euproctus*, sometimes referred to as brook salamanders, is interesting to biogeographers: while *E. asper* occurs in the Pyrenees mountains between France and Spain,

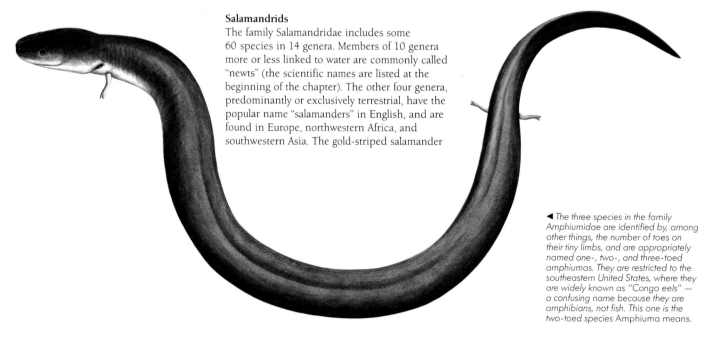

◀ *The three species in the family Amphiumidae are identified by, among other things, the number of toes on their tiny limbs, and are appropriately named one-, two-, and three-toed amphiumas. They are restricted to the southeastern United States, where they are widely known as "Congo eels" — a confusing name because they are amphibians, not fish. This one is the two-toed species Amphiuma means.*

▲ ▶ *Newts in the terrestrial phase, like this marbled newt* Triturus marmoratus *(above), take on the appearance of typical salamanders. The aquatic phase, as represented by this male alpine newt* Triturus alpestris *(right), is characterized by fin-like extensions on the tail and back, and sometimes the toes.*

E. montanus is endemic to Corsica and *E. platycephalus* to Sardinia, both islands of continental origin which detached from the south of France about 20 million years ago.

Only a few species are dull in color. Usually these salamanders and newts have lively and contrasting colors, some of them in gaudy patterns. Especially famous for the bright pattern of the breeding males are the 12 *Triturus* newts, and above all the marbled newt *T. marmoratus*, the alpine newt *T. alpestris*, and the extraordinary banded newt *T. vittatus*. The male banded newt develops a vertically shaped and incredibly high dorsal crest. With the exception of a few species that give birth to live young, the Salamandridae lay their eggs in water. In *Triturus* and some *Cynops* newts the female secures each egg in the cavity of a leaf, which she folds between her hind limbs, ensuring that the egg is protected and oxygenated. Neoteny occurs also in the newts.

Salamandridae differ from the other salamander families in several skeletal characters too complicated to be discussed here, but they always have four well-developed limbs, with four fingers and generally five toes; only the spectacled salamander and some individuals of the newt *Tylototriton andersoni* have four-toed feet. They range in size from 7 to 30 centimeters (2¾ to 12 inches) in total length.

Lungless salamanders

Plethodontidae, the largest family of the order Caudata, includes about 60 percent of the known living species—about 240 in some 30 genera— which are found in the Americas from Nova Scotia and extreme southeastern Alaska to central Bolivia and eastern Brazil. Only a few species are European, inhabiting Sardinia, the southeast of France, and Italy from the Maritime Alps to the central Apennines.

After metamorphosis they may be distinguished from all the other salamanders and newts by having a special structure, the nasolabial groove, which extends from the nostril vertically to the upper lip and enhances chemoreception (the sensory reception of chemical stimuli). They always have four limbs, four fingers, and five (rarely four) toes. Their coloration differs according to the species; some of the cave-dwelling forms resemble the olm in having a long flesh-colored body and scarlet gill tufts. The smallest are *Thorius* species, some of which reach only 2.7 centimeters (1 inch); the largest is *Pseudoeurycea bellii* at 32.5 centimeters (12½ inches).

As indicated by their common name, these salamanders are always lungless. They breathe through the mucous membrane in the mouth and throat and through the skin, both well supplied with many blood vessels. To keep their skin wet and thus able to absorb oxygen, these animals are linked even more to wet habitats than the lunged salamanders. They shelter in caves, crevices in rocks, spaces between roots and stones, or under logs, and will venture out only when it is humid enough and the temperature is mild. Like most other salamanders they do not tolerate heat well.

Some species, such as cave salamanders, genera *Hydromantoides* and *Speleomantes* (a European genus formerly known as *Geotriton* or *Hydromantes*), are mostly rock-dwelling; others, such as *Bolitoglossa*, are tree-dwellers. Some climbing salamanders (genus *Aneides*) and dusky salamanders (genus *Desmognathus*) like to climb on trees or rocks. The shovel-nosed salamander *Leurognathus marmoratus*, and the neotenic and cave-dwelling *Typhlomolge* and *Haideotriton* species are totally aquatic. The many-lined salamander *Stereochilus marginatus* and some *Eurycea*, *Desmognathus*, and *Gyrinophilus* species are mostly aquatic.

Each species of lungless salamander has a different type of nuptial dance. Males of species that have a chin gland also have enlarged teeth, with which they scarify the skin of their partner in order to "vaccinate" her with the aphrodisiac secretion of the chin gland.

The lungless salamanders are widespread, almost exclusively in America and probably originated there, although they also occur in Europe. According to some scientists, the ancestors of the European species (all belonging to *Speleomantes*, a genus closely related to the Californian *Hydromantoides*) reached western Europe by crossing the Bering land bridge and then became extinct all over Asia and most of Europe; other scientists, including the writers of this chapter, suppose that the ancestors of today's European species had colonized western Europe before its detachment from North America about 50 million years ago.

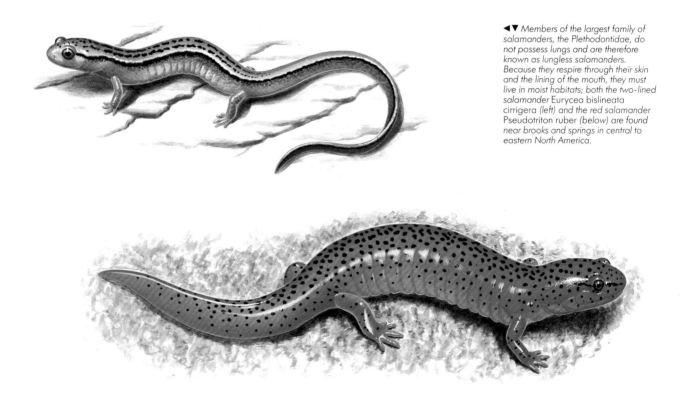

◄▼ Members of the largest family of salamanders, the Plethodontidae, do not possess lungs and are therefore known as lungless salamanders. Because they respire through their skin and the lining of the mouth, they must live in moist habitats; both the two-lined salamander Eurycea bislineata cirrigera (left) and the red salamander Pseudotriton ruber (below) are found near brooks and springs in central to eastern North America.

FROGS AND TOADS

R.G. ZWEIFEL

I n Europe, where there are relatively few species of the order Anura, the most conspicuous are smooth-skinned, long-legged frogs of the genus *Rana* and short-legged, warty toads of the genus *Bufo*. So it is only natural that every European language should have specific words for these two kinds. But where species are more diverse, this distinction breaks down, and it is neither possible nor necessary to force animals into one or other category. The name "frog" may properly be used for any member of the Anura, whereas "toad" is loosely applied to members of the genus *Bufo* as well as to other frogs of similar body form.

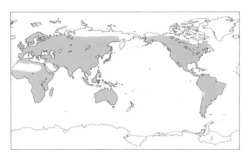

EASILY RECOGNIZED ANIMALS

A frog cannot be mistaken for any other animal. Early in their evolution, frogs acquired a body structure well suited to jumping: the ankle bones are elongated so that, with the femur and tibiafibula, they form a third major segment that gives the hind legs additional mechanical advantage in jumping. The short, rather inflexible vertebral column with no more than ten free vertebrae followed by a bony rod (the coccyx, representing fused tail vertebrae) is another adaptation to leaping. Given such limitations on its structure, a frog—whether it burrows in desert sand, swims in a high Andean lake, or lives in a rainforest tree—is eminently recognizable.

The diversity in body form displayed by frogs expresses their adaptation to particular ways of life. Almost identical solutions to problems of adaptation have evolved time and time again, in different evolutionary lines. Yet any one group of related species may include a diverse range of adaptive types. For example, the so-called "treefrog" family, Hylidae, includes mostly species with the tips of the fingers and toes enlarged into adhesive pads (an aid in climbing) and with the eyes positioned to provide binocular vision ahead, as well as visual fields below and above. But some hylid frogs of semi-arid regions are

▶ *Largest of the North American frogs, the bullfrog* Rana catesbeiana *is familiar and widely known by its deep, bellowing "jug-o-rum!" call. It is common and widespread, favoring marshes and wetlands of all kinds, especially cattail swamps.*

Steve M. Alden/Bruce Coleman Limited

Stephen Dalton/Oxford Scientific Films

adapted for terrestrial life, with eyes set higher on the head, narrow toe pads, and a tubercle on each hind foot that facilitates burrowing. Other hylids may lack any trace of toe pads and live in marshy fields and ditches. These various forms and structures are matched in several other families of frogs. Indeed, given an unfamiliar frog to identify, even an experienced specialist would first have to examine its internal anatomy to ascertain its family relationship, before going on to determine the genus and then the species.

DISTINGUISHING SPECIES
The number of known species of frogs (about 3,800) is approaching that of mammals (about 4,250). It is impossible to give an exact number because new species of frogs are continually being recognized and given scientific names. This is largely because of discoveries in poorly known regions, especially tropical rainforests. Also, in recent years means other than morphology (form

and structure) have been used to detect differences between species. Many species of frogs prove to be virtually identical in external appearance but can be distinguished by their calls (as indeed they distinguish each other), and with modern electronic techniques for recording and analyzing sounds it becomes possible for biologists to separate these so-called "sibling species". Biochemical techniques, too, make an important contribution by demonstrating genetic (DNA) differences between species whose morphology is virtually the same.

SIGHT, SMELL, AND SOUND
Frogs are primarily visually oriented animals. The eyes are generally large and protrude from their sockets, providing a broad visual field and compensating for lack of rotatory movement. The eyes can, however, be retracted into their sockets, whereupon they bulge against the roof of the mouth and assist in swallowing! The upper eyelid

▲ *Another North American, the leopard frog Rana pipiens is extremely well-adapted to cold climates, and its distribution extends northward to Hudson Bay and other parts of northern Canada.*

▶ *The red-eyed treefrog* Litoria chloris *is common along the coast and hinterland of eastern Australia, roughly from Bundaberg to Sydney. Like many other arboreal frogs, it is slender-bodied and has conspicuously disk-shaped pads on its fingers and toes.*

Hans and Judy Beste/AUSCAPE International

is a cover without independent movement. The lower eyelid can be moved and has a transparent upper portion, the nictitating membrane. The pupil is a horizontal oval in most frogs, but in others it is a vertical oval or may even be triangular or diamond-shaped. The iris commonly resembles the side of the head in color and may even be part of some aspect of facial pattern. Presumably this helps to conceal the eye, although some frogs have brightly colored eyes. There is no evidence that frogs have color vision.

A pineal organ (homologous with the "third eye" of reptiles) is found in some frogs and may be detected as a pale spot atop the head. The organ reacts to light, and experiments suggest that it is involved in sun-compass orientation and other aspects of behavior.

Frogs possess both the nasal olfactory organ and a separate vomeronasal organ that functions in a similar fashion. These two organs facilitate homing on breeding sites (presumably by recognition of chemical clues) and may serve in identification of prey.

Hearing is important to frogs, for finding mates and in territorial behavior. The main receptor is a membrane, the tympanum, which is stretched across an oval or round cartilaginous ring behind the eye. A rod of bone, the columella, transmits the vibrations of the tympanum to the inner ear, where sensory cells detect and sort them, passing the information on to the brain. This system is most effective in detecting higher-frequency airborne vibrations, whereas low frequencies pass into the body from the ground or whatever substrate the frog is resting on. The tympanum may be sexually dimorphic, being much larger in males. Frogs of some species have the tympanum concealed beneath skin, but others lack a tympanum completely.

ACTIVITY

Whether frogs are active—that is, engaged in feeding or breeding—depends largely on moisture, temperature, and time of day. Some species are principally nocturnal, others mostly diurnal (day-active), while others may be active day and night. Where moisture is adequate, activity may be limited to times when the temperature is above a level critical for the species concerned. For example, in North America the wood frog *Rana sylvatica* emerges from hibernation and breeds in waters barely above freezing, whereas other species of *Rana* in the same immediate area wait to breed until many weeks later because their eggs are less tolerant to low temperatures. In tropical and subtropical regions,

---

FROG CALLS

In a word-association test, upon hearing the word "frog" most people would think of "jump" or "croak." How and why do frogs make sounds? Female frogs are for the most part mute. The singing that we hear is done by males. In calling, a frog forces air from its lungs through the larynx, causing the vocal cords to vibrate and produce the sound. The sound is amplified and given a characteristic timbre by the vocal sac or sacs. (Frogs without vocal sacs have very soft voices or may not call at all.) These sacs are pouches of skin beneath the floor of the mouth or at the corners of the mouth, and have openings into the mouth cavity. When calling, a frog keeps its nostrils and mouth closed and uses muscles of the body wall and throat to shunt air back and forth between the mouth-sac cavity and the lungs. The frequency level of the call is largely determined by the frog's size but can be varied somewhat. Other characteristics of the call depend on the pattern of air flow, producing, for example, a long, drawn-out note or perhaps a series of short notes or even clicks. Most calling is done for advertisement: calls serve to attract females ready to mate, and to repel other males from a territory. Some species apparently make one sort of call to serve both purposes, whereas other species have a two-part call: one part of significance to other males, the other to the females. Others have a more varied repertoire, with even an escalating series of territorial calls. One poorly understood kind of call, the so-called "fright scream", is uttered with the mouth open when a frog is seized.

Frog calls are highly species-specific, as we might expect of a behavioral trait that serves to promote selection of an appropriate mate. Listening to a chorus with several species calling simultaneously, a person has no difficulty in distinguishing the different calls. The frogs, however, may hear much less of the sound around them, because their ears tend to be tuned to the frequency level of their own species' call—their hearing is not "jammed" by extraneous noise. Frogs of the same or even different species calling nearby may lessen the confusion by alternating their calls.

Scientists, as well as female frogs, make use of calls for distinguishing species of frogs. There are many instances where physical differences between species are very slight or ambiguous but where the calls provide sure identification of the males, at least.

C.A. Henley/AUSCAPE International

▲ The call of the bleating treefrog *Litoria dentata* is a penetrating, wavering bleating sound, hence the name. It is found in eastern Australia, especially in paperbark swamps.

FROGS WITH ALTERNATIVE LIFE HISTORIES

A surprising number of frog species—perhaps 30 percent or more—have a life history differing from the conventional one in which eggs laid in water hatch into tadpoles that grow and metamorphose. Many species omit the free-living tadpole stage and deposit a few relatively large eggs in a moist place out of the water, such as within a rotting log, in a burrow in the ground, or even in epiphytic plants that hold moisture. Usually these eggs are attended by one of the parents, often the male; development takes place within the egg membrane, and the hatchling is a metamorphosed froglet.

Two Chilean frogs, Darwin's frog *Rhinoderma darwini* and the similar *R. rufum*, lay their eggs on land. When the tadpoles hatch, the male picks them up in his mouth and they go into his vocal sac. In the first species the tadpoles develop through metamorphosis in the sac, whereas in the second species the frog carries them to water where they are set loose on their own.

Modifications of life history may involve morphological as well as behavioral adaptations of the parents. The marsupial treefrogs (exemplified by the genus *Gastrotheca* in tropical America) have a brood pouch on the female's back. As the female extrudes the eggs in the course of mating, the male uses his feet to guide them into her pouch. Depending on the species, the eggs may have direct development as described above (with baby frogs being "born" from the pouch) or tadpoles may develop there, later to be released into water to complete their development.

Species of live-bearing frogs occur in Puerto Rico (*Eleutherodactylus jaspari*) and West Africa (*Nectophrynoides* species). The eggs remain in the female's oviducts, where they undergo development through metamorphosis, before the froglets are born.

Michael and Patricia Fogden

Among the many deviations from the standard life history, that of the two species of gastric-brooding frogs of Australia (genus *Rheobatrachus*) is the most bizarre. In these totally aquatic species the female swallows the fertilized eggs, which then develop within her stomach; metamorphosed froglets eventually emerge through the mouth. The stomach's normal digestive mechanism (and, presumably, the urge to eat) is suppressed by a special chemical during this peculiar pregnancy.

▲ *Darwin's frog Rhinoderma darwinii of South America is notable for its unusual version of parental care. Males gather at clusters of hatching eggs and snap the tadpoles up as they emerge. They are deposited in the male's vocal sacs, where they remain until ready to emerge as small froglets.*

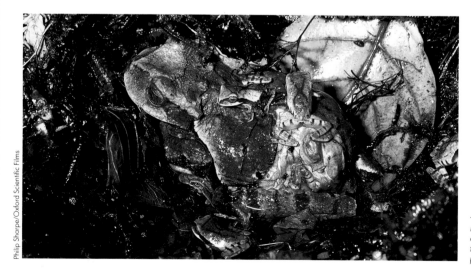

Philip Sharpe/Oxford Scientific Films

◀ *Another South American, the marsupial frog Gastrotheca ovifera has another parental care strategy in which the female "incubates" the eggs in a pouch on her back.*

where low temperatures are less likely to be important, active periods nevertheless may be controlled by seasonal changes in moisture. This is especially true for species that live away from permanent sources of water and depend on rainy seasons for both feeding and breeding activity.

BREEDING

With some specialized exceptions, frogs fertilize their eggs externally. The male approaches the female from behind and grasps her around the body with his arms. The hold may be at the waist (considered the primitive mode) but in most species is just behind the female's arms or (rarely) around the head. Males of many species have horny areas on their hands that help in gripping a slippery mate. The frogs maintain this posture, while the eggs are fertilized as they are extruded. When eggs make contact with water, the jelly membranes surrounding them swell, encasing each egg in a transparent sphere.

Eggs deposited in water are most commonly grouped in globular masses containing a few to hundreds of eggs. Some species that breed in still, warm water likely to be depleted of oxygen spread the eggs in a single layer on the surface, whereas those breeding in fast-flowing streams may attach the eggs singly to submerged rocks. Eggs deposited out of water may be in a variety of moist locations such as in tree holes, in plants growing on trees, underneath moist leaf-litter, or in holes in the ground. These are only a few examples of the many known sites and modes of egg deposition.

LARVAL LIFE

Frogs resemble other amphibians and differ from reptiles in that most species (75 to 80 percent of frogs) have a larval stage interposed between the egg and the mature body form. This period may be as brief as one week in species breeding in short-lived desert rainpools, or up to two years or more, but generally it averages a few weeks. A typical frog larva, or tadpole, lives in water and has a rather oval body with a strong finned tail, but no clear distinction between head and body. The mouth has a beak and rows of "teeth" made of a chitinous substance. Water taken in through

Michael Fogden/Oxford Scientific Films

▲ Centrolenella valerioi is typical of most of the arboreal glass frogs of Central and South America. A female lays her eggs on the underside of leaves overhanging running water, and the male then guards them until they hatch and fall, as tadpoles, into the water below.

Patrick Clement/Bruce Coleman Limited

◄ Fertilization is external in almost all frogs, and the male typically clasps the female in a tight embrace, known as amplexus. This embrace may be maintained for several days, perhaps as the least complicated means of ensuring he is on the spot when she is ready to lay her eggs. These are European common frogs Rana temporaria, which lay their eggs in gelatinous masses.

▲ *A golden mantella* Mantella aurantiaca *draws a bead on an unsuspecting fly. In most frogs and toads the tongue is attached at the front, free at the back, and can be flicked forward some distance and with considerable speed . . .*

▶ *Generally speaking, adult frogs are carnivores but their larvae are vegetarians. Nevertheless, there are some predatory tadpoles, and cannibalism is not unusual. Even vegetarian larvae, such as these meadow tree frog tadpoles* Hyla pseudopuma *may occasionally make a meal of a dead sibling.*

Michael and Patricia Fogden

the mouth passes over gills concealed within a chamber behind the mouth, before being expelled through the spiracle, a hole usually on the left side of the body. The gills not only serve for respiration, but also filter tiny food particles from the water, diverting them toward the stomach and supplementing the algae and detritus obtained by biting and scraping. Also concealed within the gill chamber are the animal's growing front legs. The rear legs develop externally where the tail and body meet. At the end of the larval stage the tadpole undergoes metamorphosis into the adult form. This transformation is more than simply growing limbs and resorbing the tail. Profound changes take place in both morphology and physiology. For example, the digestive tract shortens, becoming suited to a carnivorous diet

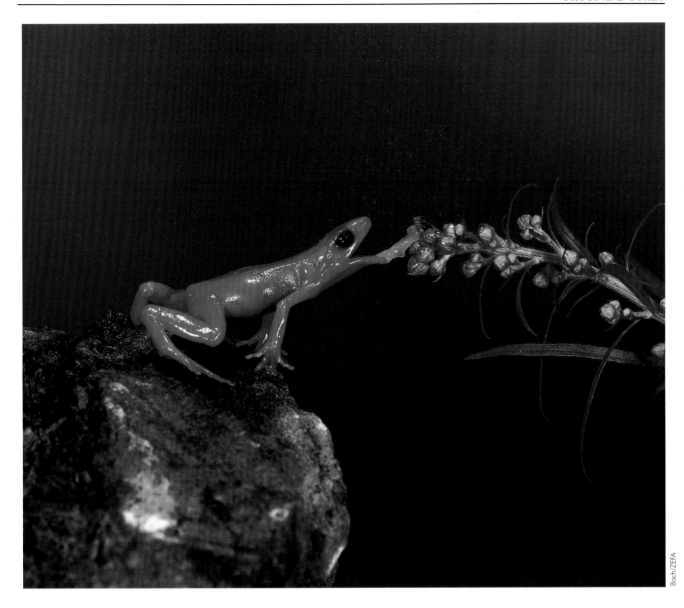

Bach/ZEFA

instead of a vegetarian diet; the gills disappear, and respiration is taken over by the lungs (tadpoles also may use their lungs long before metamorphosis is completed); the skeleton is extensively modified, and true teeth develop in the upper jaw.

FOOD AND ENEMIES

All frogs are carnivores, and many are generalized feeders that will take whatever small animals— vertebrate or invertebrate—their capacious mouths can accommodate. Relatively few frogs are large enough to eat other vertebrates, so most of them eat insects and other arthropods, and earthworms. But a large frog such as the North American bullfrog *Rana catesbeiana* can take birds and mice, small turtles and fish, and under

crowded conditions is a fearsome predator on smaller frogs of its own and other species. Tadpoles for the most part are vegetarians— filtering organisms from the water, scraping algae from stones, consuming bottom debris. Some species have predaceous tadpoles, however, which capture invertebrates or other tadpoles.

Frogs are prey to a host of enemies, ranging from tarantulas to humans. Many snakes live largely or entirely on frogs. Herons skewer frogs in shallow water, bats snatch them from branches over tropical pools, turtles ambush them from under water, big frogs eat smaller frogs, parasitic flies lay eggs on them, and leeches attack them externally and even enter their bodies. The larval stage, too, is a perilous time. Almost any meat-eater finds a tadpole a succulent morsel.

. . . snapping up prey with the aid of sticky secretions from glands in the tongue. Vision seems to be an important hunting sense, although there is no indication that frogs have color vision.

523

▶ One of a small group of primitive frogs restricted to New Zealand, Archey's frog Leiopelma archeyi lives on rough mountainsides. The eggs are laid on damp ground under logs or rocks, and the young are carried about on the male's back until they reach independence.

Frances Furlong/Bruce Coleman Limited

WHERE FROGS LIVE

Frogs are native to all the continents except Antarctica but are absent from almost all remote oceanic islands and from Greenland and other Arctic islands. Although widely distributed over the Earth, frogs are far more abundant and diverse in moist areas of the tropics, especially tropical rainforests, than farther north or south; regions closer to the poles are subject to extremes of weather, and dryness and cold are especially hostile to frog life.

The near-absence of native frogs from oceanic islands is rooted in the geological history of the islands, most of which are volcanoes or coral islands built on volcanic footings. There has simply been no way for animals so sensitive to desiccation and heat to reach them (frogs cannot live in salt water). Many of these islands may have suitable habitats; for example, several species of frogs have been introduced to the Hawaiian islands, where they prosper. Some islands—such as the Fiji group in the Pacific and the Seychelles group in the Indian Ocean—are unusual in having endemic frogs (species found nowhere else), presumably because these landmasses were formerly part of or close to continents but were moved by geological processes, carrying with them the ancestors of the present frog species.

ANCESTORS AND RELATIVES

All living amphibians—frogs (order Anura), salamanders and newts (order Caudata), and caecilians (order Gymnophiona)—are thought to be more closely related to each other (in the subclass Lissamphibia) than to other groups of amphibians now extinct but recognized in the fossil record. The relationships of the Lissamphibia are not well understood, but their common ancestry may be with the Dissorophoidea amphibians of the Palaeozoic era, more than 245 million years ago. Whatever their exact ancestry, today's amphibians lie on an ancient branch of the amphibian tree and are not on the evolutionary line that led to reptiles and, eventually, mammals.

The oldest frog-like fossil is *Triadobatrachus* found in Madagascar in deposits of the Triassic period, more than 200 million years ago. Early frog fossils are also found in deposits laid down early in the Jurassic period, almost 200 million years ago, in Argentina. By the beginning of the Tertiary period 65 million years ago, there were frogs we can classify in genera that are still represented today. So frogs that differed in no essential respect from those alive today were contemporaries of the dinosaurs, survived the great extinctions at the close of the Mesozoic, and persisted largely unchanged while mammals underwent their great evolutionary expansion.

ARCHAEOBATRACHIAN FROGS

There are four suborders of modern frogs. Some are considered more "primitive" (in evolutionary terms) than others.

The suborder Archaeobatrachia includes two families, Leiopelmatidae and Discoglossidae, whose species are thought to be the most primitive living frogs. A clearly primitive character they share is the presence of free ribs (all other frogs have their ribs fused to the vertebrae).

Leiopelmatid frogs

There are only four species of Leiopelmatidae alive today. The tailed frog *Ascaphus truei* lives in cold, fast-flowing mountain streams of the northwestern United States and southwestern

Canada. The so-called "tail" is present only in the male and is an organ adapted for insertion, so that the eggs are fertilized internally before being released from the female's body (in this type of habitat, external fertilization might be unsuccessful) and attached to the undersides of rocks in the water. The sucker-mouthed tadpoles take three years to develop and are well adapted to life in fast-running waters.

The other three species are restricted to New Zealand and are the only native frogs there. One of these, Hamilton's frog *Leiopelma hamiltoni*, occurs on just two small islands and is at risk because of its tiny area of distribution. They are all small frogs, no bigger than 50 millimeters (2 inches) body length. *L. hamiltoni* and Archey's frog *L. archeyi* lay their eggs on moist ground under rocks, logs, or vegetation; the hatchlings are in a relatively late state of larval development and climb upon the back of the attending adult male. Hochstetter's frog *L. hochstetteri* lays eggs in shallow hollows with seepage; the tadpoles are more adapted for swimming and remain in the water near to where they hatched until metamorphosis, although not feeding.

Subfossil remains, which date back to the period before humans colonized New Zealand, reveal that there were additional species of *Leiopelma*, one twice the size of the living species. Fossils from early and late in the Jurassic period (up to almost 200 million years ago) found in southern Argentina, are considered to belong to the family Leiopelmatidae.

Discoglossids

The discoglossid frogs (family Discoglossidae) are only slightly more numerous but are less geographically restricted than the leiopelmatid frogs. All but two of the 15 species (in four genera) live in the Northern Hemisphere north of the tropics. Species of the genus *Discoglossus* are semi-aquatic frogs found in Europe, North Africa, and the Middle East. Different species of fire-bellied toads (genus *Bombina*) are small to moderate-sized frogs—up to about 70 millimeters (almost 3 inches) body length in a large Asian species—that live in the Far East and Europe, where they have a mostly aquatic existence in shallow water habitats. When disturbed they arch their back and throw up their arms and legs,

▼ *The tailed frog Ascaphus truei (below left) belongs to the family Leiopelmatidae, whose members are found in New Zealand and western North America only. Although they possess primitive tail-wagging muscles, none has a true tail. The European painted frog Discoglossus pictus (bottom left) belongs to the family Discoglossidae and possesses a number of primitive features such as a non-projectile, disk-shaped tongue and round pupils. Another member of this family, the oriental fire-bellied toad Bombina orientalis (below right), displays its brightly colored underside when threatened, warning of a mildly toxic, unpalatable skin secretion.*

▲ *Mute, tongueless, and built something like a squared-off pancake with a limb at each corner, the Surinam toad Pipa pipa is one of the more bizarre members of the tropical South American frog fauna. It is almost entirely aquatic.*

showing the bright red or yellow colors beneath. Because the frog's skin secretions are distasteful, this is considered a form of warning behavior. The midwife toads (genus *Alytes*) are largely terrestrial frogs of Europe and North Africa, known for the male's care of the eggs; these adhere to his back and thighs and are carried about in and out of water until hatching time, when he enters the water, allowing the tadpoles to swim away. The two species of *Barbourula* are exceptional among discoglossids, not only in their geography—one in the southern Philippines, the other south of the Equator in Borneo—but also in being wholly aquatic. Their adaptation to life in streams includes having broadly webbed fingers and toes. The fossil record shows discoglossids to have existed in Europe as far back as the Jurassic, about 150 million years ago, and the family was present in North America in the late Cretaceous period, more than 65 million years ago.

PIPOID FROGS

Members of the suborder Pipoidea, with two families, the Pipidae and the Rhinophrynidae, are among the most peculiar of frogs. The pipids (27 species, 4 genera) have flattened bodies, are tongueless (all other frogs have tongues), and except for frogs in the genus *Pipa* have claw-like structures on three of the toes. All species are highly aquatic, with large, fully webbed feet. They don't have vocal cords but produce clicking sounds under water by using bony rods in the larynx. The size range is from about 40 to 170 millimeters (1½ to 4 inches) body length. Frogs in the genus *Pipa* live in South America east of the Andes and in eastern Panama. The Surinam toad *Pipa pipa*, which is almost 180 millimeters (7 inches), is bizarre in both appearance and breeding habits. The eggs become embedded in pockets in the skin of the mother's back, where they develop to emerge as tiny frogs. In another *Pipa* species, tadpoles emerge from the chambers and complete their development on their own.

In Africa south of the Sahara Desert there are three genera. The African clawed frogs, genus *Xenopus*, were for a time medically important as animals used in testing for human pregnancy: urine from a pregnant woman injected into a female frog causes it to extrude eggs. Development of chemical tests has spared platannas (as they

are called in South Africa) from this indignity, although they continue to be important animals for laboratory research because of the ease with which they may be kept and bred in captivity.

Fossils of pipid frogs from the early Cretaceous period, perhaps 145 million years ago, have been found in Israel, and from the late Cretaceous, more than 65 million years ago, in Africa and South America. Clearly, their distribution dates from a time when the southern continents were part of Gondwana, before Africa and South America became separate landmasses.

Some scientists classify the Mexican burrowing toad *Rhinophrynus dorsalis* with the pipid frogs; others place it in its own suborder (Rhinophrynoidea). It is the only living representative of the family Rhinophrynidae.

This rather globular, small-headed, burrowing frog has a prominent spade-like digging tubercle on each hind foot, grows to about 75 millimeters (3 inches), and ranges from southern Texas to Costa Rica. It emerges after heavy rains to call and breed in temporary ponds. Fossils show that in the late Paleocene to the early Oligocene epochs (about 37 million years ago) related species lived as far north as southern Canada.

PELOBATOID FROGS
The suborder Pelobatoidea comprises two families, Pelobatidae and Pelodytidae, although some authorities treat them as one family. These are considered to be rather primitive, but in some respects are transitional to more advanced frogs, and have in common the supposedly primitive

▲ The Mexican burrowing toad Rhinophrynus dorsalis (left) is not a true toad, but has evolved a number of toad-like features. It is usually seen above ground only following heavy rain, when it emerges to breed. Rarest of the clawed frogs of Africa, the Cape clawed frog Xenopus gilli (right) belongs to an ancient family of aquatic frogs, the Pipidae. The hind feet have black horny caps on three of the toes.

Chris Mattison

◄ The Surinam toad's breeding habits are nearly as bizarre as its appearance. In a complicated underwater ritual, the male fertilizes his mate's eggs as they are laid, and distributes them over her back, where they sink into the spongy tissue over several days. The eggs hatch into tadpoles in the pockets formed, then continue their development until they leave as small froglets.

▲ *Spadefoots, like this Couch's spadefoot* Scaphiopus couchii *(left) possess a spade-like tubercle on the bottom of their hind feet which they use to burrow rapidly backwards, circling as they descend. Although in the same family as spadefoots, the Asian horned toad* Megophrys montana nasuta *(right) is adapted for life on the rainforest floor, where its shape and "dead-leaf" coloration render it almost invisible among the leaf-litter.*

habit of the mating male clasping the female around the waist or groin, rather than higher on the back, just behind the arms. The eyes of all species have vertical pupils, and all have free-living, aquatic tadpoles.

Pelobatids and pelodytids

The family Pelobatidae has fewer than 90 species recognized in nine genera. Seven genera are found in Asia, from Pakistan and western China to the Philippines and through Indonesia to the Greater Sunda Islands in the Malay Archipelago. Their habitats vary from forested tropical lowlands and uplands to Himalayan peaks. *Megophrys* species inhabit the forest floor; they have funnel-mouth tadpoles which develop in still waters. *Megophrys montana* is notable for its cryptic coloration and form; not only does it resemble dead leaves in color and pattern, but one subspecies has pointed projections from the eyelids and nose that disrupt the frog-like outline. Also inhabiting the forest floor are *Leptobrachium* species; these frogs breed in small streams, and the tadpoles live amid rocks and gravel. Mountains as high as 5,200 meters (17,000 feet) are the home of *Scutiger* species, which lead a largely terrestrial existence but breed in cold mountain streams. An odd feature of the *Scutiger* male is that his mating grasp of the female is enhanced by two spiny patches on the chest in addition to the more usual spiny areas on the first two fingers.

"Spadefoot toads" is the common name for the other two genera in the family Pelobatidae: *Pelobates*, found in Europe, some areas of Mediterranean northwestern Africa, and western Asia; and *Scaphiopus*, found in North America from southern Canada to southern Mexico. They are toad-like burrowing frogs of moderate size, up to 100 millimeters (4 inches) body length, and their common name comes from the prominent

digging tubercle on each hind foot. The four species of *Pelobates* are partial to areas of sandy soil, where they burrow and also breed, in pools (often temporary ones) and ditches. *Scaphiopus* too spend their inactive periods in self-made burrows; of the six species, one lives in humid regions of the eastern United States, the others in arid areas, even deserts. Conditions in humid areas may be adequate for surface activity during the warm period of the year, but where rainfall is scarce the frogs are forced to spend most of the year underground. When the soil temperature is warm enough, the sound of rain falling will prompt them to emerge, and they migrate to temporary pools of water to breed. Embryonic and larval development is rapid—less than two weeks may elapse between egg-laying and metamorphosis. But even this may not be rapid enough if additional rain doesn't fall to offset evaporation in the fierce desert heat. Some spadefoot tadpoles develop a different morphology—enlarged jaw muscles and a large beak—from others of their species, and they cannibalize their pondmates, thus making the most of the available protein. The rains that call out the spadefoots also bring forth hordes of termites and other insects, and the frogs can get enough food in a few nights to survive many more months underground.

Pelobatid fossils occur as early as the late Cretaceous of North America and Asia (more than 65 million years ago), and the modern genera *Scaphiopus* and *Pelobates* appear in the Tertiary of North America and Europe, respectively, 35 to 47 million years ago.

The family Pelodytidae has two living species in the single genus, *Pelodytes*: one in Europe; and one further east, in the Caucasus region (between the Black Sea and the Caspian Sea). These terrestrial frogs look like small (50 millimeters, or 2 inches) more-typical frogs, being long-legged and lacking the digging spade. The earliest fossils are from the middle Eocene epoch in Europe (47 million years ago). From North America, where the family is extinct, there is a middle Miocene fossil some 15 million years old.

THE LAST SUBORDER: NEOBATRACHIA
The suborder Neobatrachia, thought to represent the more advanced frogs, includes more than 3,600 species, or about 96 percent of all living species of frogs. There are 18 families.

Leptodactylid frogs of the Americas
The largest family in this suborder is the Leptodactylidae, a diverse assortment of almost 800 species classified in 52 genera, inhabiting South America, the Caribbean islands, Central America, and Mexico; only five species cross into the southern United States. The world's southernmost frog is a leptodactylid, *Pleurodema bufonina*, which reaches the Strait of Magellan.

Within the vast array of leptodactylids are virtually all modes of life open to frogs. In the genus *Eleutherodactylus*, which has over 450 species, many spend their active hours in trees or shrubs, but others forage on the leaf-litter of the forest floor, and all but one *Eleutherodactylus* have direct-developing eggs (the exception is viviparous). Species of several other genera have rather cryptic lives on or within forest leaf-litter; others that live in arid regions remain underground, encased in a protective "cocoon", until rainfall permits a brief emergence for feeding and breeding; and many species have more familiar habits in and around streams and ponds. Wholly aquatic species inhabit lakes in the Andes.

Morphology and lifestyle are related, so the variations in their body form are a reflection of their ecological diversity. For example, tree-dwelling *Eleutherodactylus* are rather small, slender frogs with features found in other treefrogs, such as enlarged disks on fingers and toes, and large forward-directed eyes. Burrowing species of other genera are short-legged, with a prominent spade on each hind foot for digging.

Leptodactylid frogs of several genera lay their eggs in foam "nests"; according to the species involved, such a nest may float on open water, be in a cavity adjacent to water, or be in a burrow or other sheltered site on land. The nest is constructed while the frogs mate, the male using his feet to whip the mixture of eggs and seminal

Michael and Patricia Fogden

▲ *Eleutherodactylus is a very large genus of frogs inhabiting South America. In all but one of the more than 450 species, metamorphosis takes place entirely in the egg, the young finally emerging as fully formed small froglets.*

◄▼ *The western barking frog* Hylactophryne augusti cactorum *(left) inhabits limestone caves and crevices in southwestern North America and adjacent Mexico. The ornate horned toad or escuerzo* Ceratophrys ornata *(below left) is a robust, aggressive predator of the rainforest floor. Widely eaten by humans, the South American bullfrog* Leptodactylus pentadactylus *(bottom right) is one of the largest members of the family Leptodactylidae.*

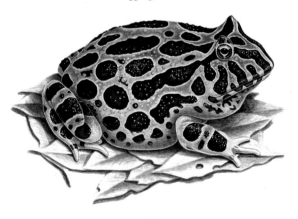

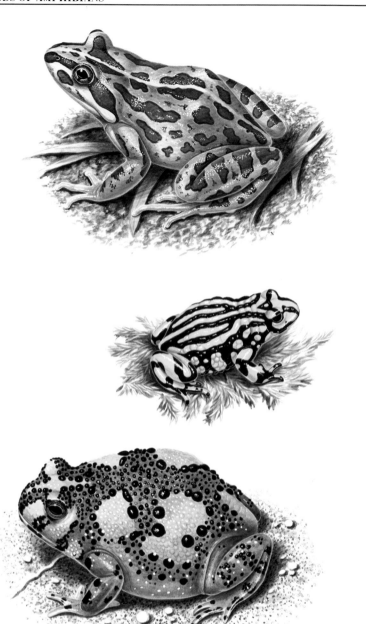

the female *Leptodactylus ocellatus* remains with the nest, and when the tadpoles hatch they stay together in a school and the female protects them against predators.

The presence of leptodactylids in South America throughout the Cenozoic (beginning 65 million years ago) is documented in the fossil record. A most remarkable fossil from the Dominican Republic on the Caribbean Island of Hispaniola is an *Eleutherodactylus* preserved whole in amber, at 37 million years old.

Myobatrachid frogs of Australasia
The myobatrachid frogs of Australia and New Guinea, numbering about 100 species in 20 genera, are the eastern hemisphere counterparts of the leptodactylids of North and South America. Indeed, some people classify them as members of the Leptodactylidae. Australia has a large proportion of arid land, where rainfall is sparse and seasonal, so it is not astonishing that many of the myobatrachids are burrowers and are seldom seen or heard except when they emerge to mate and deposit eggs in temporary pools. The more conventional burrowers such as members of the genera *Heleioporus*, *Notaden*, and *Neobatrachus* are squat, short-legged frogs with a digging spade on each hind foot. The turtle frog *Myobatrachus gouldii* and the sandhill frog *Arenophryne rotunda*, both of southwestern Australia, are exceptional in that they deposit their eggs deep in moist sand, as deep as 1 meter (3¼ feet), where they undergo direct development without a free-living tadpole stage. Both are also unusual in that they burrow head-first (rather than rump-first as most burrowing species do) into the sand or soil.

In contrast to the Leptodactylidae, the Myobatrachidae family does not include any treefrogs (this niche in Australia and New Guinea is reserved for members of the Hylidae and Microhylidae), but still there is much diversity in addition to the burrowers, especially in the tropical and temperate rainforests of eastern Australia. Examples include small frogs of the genus *Taudactylus*, at home in torrential creeks of the northeastern rainforests; and tiny *Ranidella*, about 25 millimeters (1 inch) in length, which are abundant in most of Australia's moist habitats. The latter pose a problem for herpetologists because of their morphological similarity—some species can be distinguished only by differences in their calls.

Two species of *Rheobatrachus*, the gastric brooding frogs, are highly aquatic stream-dwellers that scarcely became known to science before they disappeared for no obvious reason, possibly having become extinct. Another species with a less strange but nevertheless odd form of parental care is the pouched frog *Assa darlingtoni*, a ground-dweller of forests on the Queensland–New South Wales border. The male of this secretive

▲ *Three members of the Australasian family Myobatrachidae: the spotted grass frog* Limnodynastes tasmaniensis *(top) inhabits semi-aquatic niches; the corroboree frog* Pseudophryne corroboree *(center) lives in the Australian Alps and is so named because its vivid coloration is similar to the striped body decorations adopted by the Aborigines for ceremonial dances, or corroborees; and the crucifix toad* Notaden bennettii *(bottom), which has a specialized diet of ants and termites.*

fluid (and water, if present) into a froth. The nest protects the fertilized eggs and then the tadpoles from enemies, from overheating (if exposed to the sun), and from desiccation. A nest in a cavity can shelter the tadpoles until rains flood the site and the tadpoles escape to continue their development, as occurs in some species of *Leptodactylus*. In species of *Adenomera*, which create their nests in moist cavities on land, the tadpoles remain in the nest without feeding until they metamorphose. Hatchlings from nests on open water may merely swim off, but some surprising exceptions are known. For example, species has a brood pouch on each side of the body. When the eggs (which are laid on land) hatch, the male straddles the mass and the tadpoles make their way into the pouches. Further development and metamorphosis take place there, and the froglets emerge some seven to ten weeks later.

Frogs in several Australian genera make a foam nest for the eggs, but unlike the American leptodactylids—where the male's feet do the job—in Australia the mating female beats the foam with her hands, assisted by flanges on some of the fingers. Some create their foam nests on open water, others in water-filled burrows, and several create them on land. At least four recent genera of myobatrachid frogs (*Lechriodus*, *Limnodynastes*, *Crinia*, and *Kyarranus*) were present in Australia in the mid-Tertiary, 15 to 25 million years ago.

Toads and harlequin frogs

Toads (family Bufonidae) are native to temperate and tropical zones, deserts and rainforests, mountains and prairies, everywhere except the Australian region, Madagascar, and oceanic islands. Of about 360 species in the family, more than half belong to one of the 31 genera, *Bufo*. Toads of this genus range in size from 25 millimeters (1 inch) body length for some African

H. Ehmann

◄ *Newly hatched larvae of the pouched frog* Assa darlingtoni *of eastern Australia in the act of entering the male's pouches. The male assists the young by gentle scooping movements of his arms and legs and usually ends up with four or five in each pouch; any left over die within a few hours.*

species to 25 centimeters (10 inches) for giant toads *Bufo blombergi* and *B. marinus* of the wet forests of South America—large enough to fill a dinner plate.

No matter what their size, all *Bufo* species conform to a standard appearance: heavy-set, short-legged, with numerous wart-like glands on the body and legs, and a prominent, rounded or elongate parotoid gland behind the eye. Also, they are toothless, although this is evident only on close examination. Habits tend to be rather

◄▼ *The strikingly marked leopard toad* Bufo pardalis *(below) of southern Africa is a typical true toad, with short limbs, dry warty skin, and large parotoid glands. The Asiatic climbing toad* Pedostibes hosii *(left) shows characteristics of a tree-dwelling existence, such as long slender limbs and broad adhesive pads on the digits. Looking more like the unrelated poison frogs than its relative the toads, the variable harlequin frog* Atelopus varius *(below left) of South America is brilliantly colored, a warning of its toxicity.*

Michael and Patricia Fogden

▲ *The golden toad Bufo periglenes is known only from the rainforests of Costa Rica's Monteverde Cloud Forest Reserve. Except for the few days in each year on which they emerge and congregate to mate, they spend their lives in cavities amid the root systems of forest trees. The last emergence was in 1987, and subsequent searches have failed to find it; the species may now be extinct.*

similar among the species, too. They are ground-dwellers, typically hiding in holes during the day and emerging at night to hop about looking for invertebrates, which they snap up with their long sticky tongue.

Breeding is remarkable only for the immense numbers of eggs that many species produce. The eggs typically are laid in paired strings (one from each ovary) and may number 20,000 or more from a toad only 70 or 80 millimeters (about 3 inches) long. All frogs have a variety of skin glands, and in many species they produce defensive secretions, bad-tasting or even deadly poisonous. Among the most effective are toad poisons, concentrated especially in the parotoid glands; a dog that mouths a giant Colorado River toad *Bufo alvarius*, for example, may be fatally poisoned. This does not mean that you'll be poisoned if you handle a toad (or get warts, for that matter!), but it is a good idea not to rub your eyes after handling any frog, until you've washed your hands.

Atelopus is a Central and South American genus (more than 40 species) of small, often brightly colored, slow-moving toads, whose sucker-mouthed tadpoles live in rapidly flowing water. These diurnal toads produce potent toxins, so the conspicuous color patterns may serve to warn potential predators. Some of the remaining genera—all of which have fewer than ten species—are much like *Bufo* in structure and habits, but others diverge. For example, *Nectophrynoides* of Africa is noted for giving birth

to metamorphosed froglets; *Pedostibes* of Southeast Asia is a tree-dweller; and *Ansonia*, also of Southeast Asia, lives on the ground but breeds in riffles or cascades of water, and its tadpoles are adapted to torrents.

Bufonid fossils occur in the upper Paleocene of South America (more than 57 million years ago) and in the later Tertiary of North America, Europe, Asia, and Africa.

Hylid treefrogs

The treefrog family, Hylidae, has more than 680 species in 40 genera. More than 500 of these species live in the Americas, especially the tropics. There are more than 140 species in the Australia–New Guinea area; and a mere 12 closely related species of the genus *Hyla* are distributed across temperate regions of Eurasia, scattered from Spain to Japan, one of which also occurs in northwestern Africa.

Most species in the family are arboreal or at least climbing forms, which show the characteristic adaptations to this way of life: fingers and toes with expanded tips having adhesive properties, and eyes placed somewhat laterally and forward-directed, enhancing vision downward and binocular perspective. (Terrestrial and aquatic frogs generally have the eyes more atop the head and aimed less forward, thus giving them a larger horizontal visual field and permitting aquatic species to rest with only their eyes above water.) Treefrogs of the genus *Phyllomedusa* in tropical America have the first finger opposable to the other three, like a thumb, permitting them to grasp twigs and stems. Like other large families, the hylids have undergone adaptive radiation into several major ecological niches. There are no thoroughly aquatic species, but a number of terrestrial species are typically associated with ponds and marshes, and in wet weather they may range more widely. These are long-legged frogs, powerful jumpers with webbed toes; the tips of their fingers and toes are pointed or only slightly expanded, in contrast to those of their climbing relatives. Some of these occur in the same genus as that of tree-dwelling species—the large Australasian genus *Litoria*, for example, which also includes frogs adapted to habitats other than solely arboreal or solely aquatic. There are burrowing "treefrogs" too. Members of the Australian genus *Cyclorana* spend much of their lives underground, and one species is noted for the large amount of water it can store to tide it over periods of drought. These, and the North American *Pternohyla*, form a "cocoon" while underground for protection against desiccation.

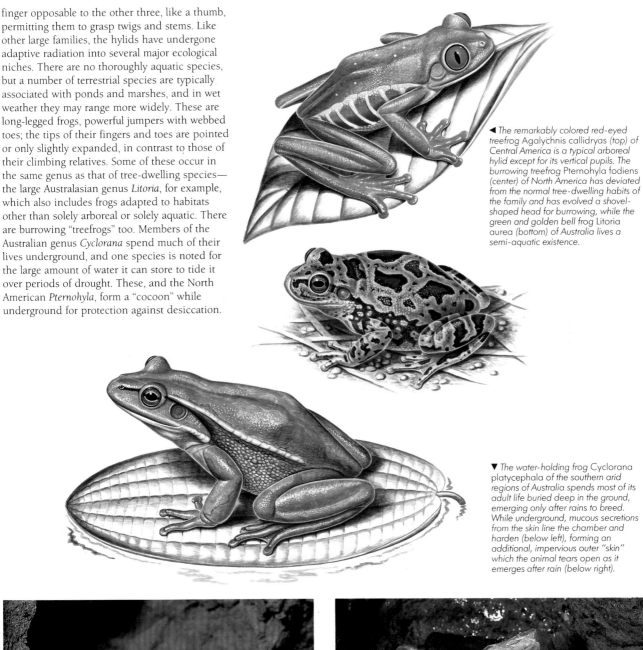

◄ The remarkably colored red-eyed treefrog Agalychnis callidryas (top) of Central America is a typical arboreal hylid except for its vertical pupils. The burrowing treefrog Pternohyla fodiens (center) of North America has deviated from the normal tree-dwelling habits of the family and has evolved a shovel-shaped head for burrowing, while the green and golden bell frog Litoria aurea (bottom) of Australia lives a semi-aquatic existence.

▼ The water-holding frog Cyclorana platycephala of the southern arid regions of Australia spends most of its adult life buried deep in the ground, emerging only after rains to breed. While underground, mucous secretions from the skin line the chamber and harden (below left), forming an additional, impervious outer "skin" which the animal tears open as it emerges after rain (below right).

▲ *Adults of many poisonous frogs have conspicuous warning colors, but the phenomenon is less widespread among tadpoles. But there are exceptions, including perhaps this colorful tadpole of the Central American treefrog* Hyla ebraccata.

The breeding habits of hylids are, with some notable exceptions, fairly conservative, and most species have aquatic tadpoles. Eggs of torrent-dwelling species are fixed firmly to rocks. Those in still water may be attached to aquatic vegetation, or spread in a thin film on the surface, thus ensuring adequate oxygen in warm waters suffering from oxygen depletion. A common tendency in regions of high humidity is to place the eggs on vegetation emerging from the water or even on leaves high in trees overhanging the water, so that when the tadpoles hatch they drop into the water. Some forest species breed in water-filled tree holes, and others in water-holding epiphytic plants such as bromeliads.

Tadpoles of many hylid species are midwater pond-dwellers, with laterally placed eyes (the better to see below as well as above) and a broad-finned tail tapering to a filamentous tip. These tadpoles may be seen hanging at an angle in the water with the tail tip vibrating while they filter microscopic food particles. Stream- and torrent-dwelling tadpoles, such as those of *Nyctimystes* in New Guinea, are quite different: depressed body, low tail fins, dorsal eyes, and mouthparts formed into an oval sucker.

Several tropical American genera (all classified in the subfamily Hemiphractinae) have the peculiar habit of brooding the eggs on the female's back: in the genus *Hemiphractus*, the eggs

are exposed on her back; in the genus *Gastrotheca*, they are completely enclosed within a pouch; and there's a whole range of body–behavior adaptations in between. One species of this subfamily deserves special mention apart from its breeding habits: *Amphignathodon guentheri* is the only living species of frog with teeth in its lower jaw. Other frogs may be toothless or possess teeth on the upper jaw, or on the upper jaw and the roof of the mouth, and some have large fang-like bony projections at the anterior ends of the lower jaws, but this species is unique in having true teeth.

Fossil hylid frogs are known from the Paleocene of South America (more than 57 million years ago) and the middle to later Tertiary of North America, Europe, and Australia.

Glass frogs

The glass frogs, family Centrolenidae, are a group of about 75 species, most of them tree-dwellers, inhabiting moist forests from southern Mexico to Bolivia, plus southeastern Brazil–northeastern Argentina. The "glass frogs" name derives from a scarcity of pigment in the skin of the abdomen, which makes the internal organs visible. Most of the species are small (maximum size about 30 millimeters, or 1¼ inches body length) and green, and are classified in the genus *Centrolenella*. The genus *Centrolene* includes just two larger species,

◀ Glass frogs, such as Centrolenella fleischmanni, are small tree-dwellers, inhabiting moist forests from southern Mexico to Bolivia, and southeastern Brazil and Argentina.

with a length up to 75 millimeters (3 inches).

Male *Centrolenella* call from leaves overhanging streams and then remain near the eggs laid on leaves at these calling sites. Egg masses placed in such situations are safe from many predators but are parasitized by flies which lay their eggs on the mass so that the maggots consume the frog eggs. Frog larvae that survive the parasites fall to the stream, where they live in gravel or debris. These tadpoles are elongate with muscular tails and very low fins (broad fins are useful only to tadpoles that swim in open water). *Centrolene* differs from *Centrolenella* not only in its much larger size, but

in living and breeding in rocky waterfalls, where the egg masses are stuck to rock surfaces. No fossil centrolenid is known.

Dendrobatids

The family Dendrobatidae has some of the most colorful and interesting frogs. Most are rather small, the smallest less than 15 millimeters (½ inch) body length, although one species reaches 62 millimeters (almost 2½ inches). There are more than 130 species in six genera. They inhabit moist tropical regions in Central and South America, from Nicaragua to southeastern Brazil

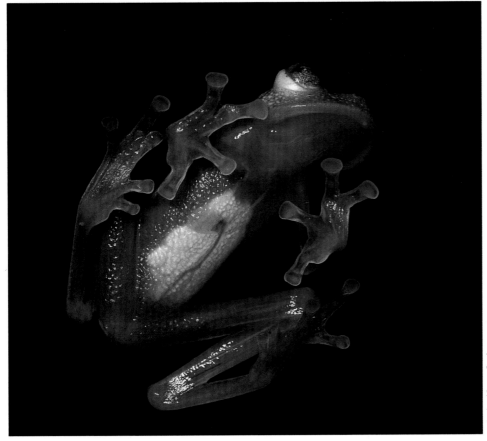

◀ The bare-hearted glass-frog Centrolenella colymbiphyllum and its relatives are so-called from the translucent skin of the underparts, through which the heart and other internal organs can easily be seen.

Michael and Patricia Fogden

▲ Many poison frogs of the genus
Dendrobates of Central and South
America display the kind of brilliant
colors and bold patterns normally
associated only with butterflies,
hummingbirds, and coral-reef fishes.
The skin secretions of these small frogs
are among the most toxic substances
known, and are used by forest
Amerindians to poison their blow-gun
darts. This species is Dendrobates
tinctorius.

and Bolivia. Unlike the majority of frogs, almost
all dendrobatids are diurnal (active during the
day). They lay small numbers of eggs in moist
sheltered places, and a parent (usually the male)
guards the eggs. When the tadpoles hatch they
wriggle onto the parent's back, are carried to
water, and released to complete their
development. Some species have a peculiar
variant of this behavior: the female releases the
tadpoles into a water-holding bromeliad plant,
and she returns occasionally to deposit an
unfertilized egg as food for the tadpoles.

The dendrobatids fall into two main groups.
One includes a large number of mostly dull-
colored species of the genus Colostethus which

live alongside streams or on the forest floor and,
with one known exception, are non-toxic. The
second group consists of the poison frogs (genera
Dendrobates, Phyllobates, Epipedobates, and
Minyobates) which are very colorful and whose
skin glands excrete alkaloid poisons that act on
the nervous system. In addition there is the genus
Aromobates, with a single species A. nocturnus,
unique on several counts: it is the largest
dendrobatid, it gives off a foul, presumably
protective odor but is not poisonous, and it is a
nocturnal stream-dweller.

The toxicity of the poisonous dendrobatids
varies greatly from species to species. Presumably
the poisons are a defense against predators, and

the bright color patterns act as a warning. One species deserves special mention: *Phyllobates terribilis* is so poisonous that it is unsafe even to handle. The toxins in one frog's skin would be sufficient to kill more than 20,000 laboratory mice, and less than 200 micrograms introduced into a human's bloodstream could be fatal. The toxicity of this species and others less-poisonous has long been known to certain groups of Indians of western Colombia, who use the frogs to poison their blow-gun darts (not arrows, as the popular literature often states). The toxin of *P. terribilis* is so abundant and potent that merely rubbing the point of a dart across a living frog's back is sufficient to make it deadly in hunting.

No fossils of dendrobatid frogs are known.

Ranids or "true" frogs

The "true" frogs, family Ranidae, have the widest distribution of any frog family: North America (even in Alaska), Central America, and northern South America; Europe and across Asia south of the Arctic Circle, through the East Indies to New Guinea, the extreme north of Australia, and the Fiji islands; and most of Africa, and Madagascar.

Jany Sauvanet/AUSCAPE International

▼ The poison frogs of the family Dendrobatidae are masters of aposematic, or warning, coloration. Pictured are the funereal poison frog *Phyllobates lugubris (below), the strawberry poison frog Dendrobates pumilio (bottom left) and the orange and black poison frog D. leucomelas (bottom right).*

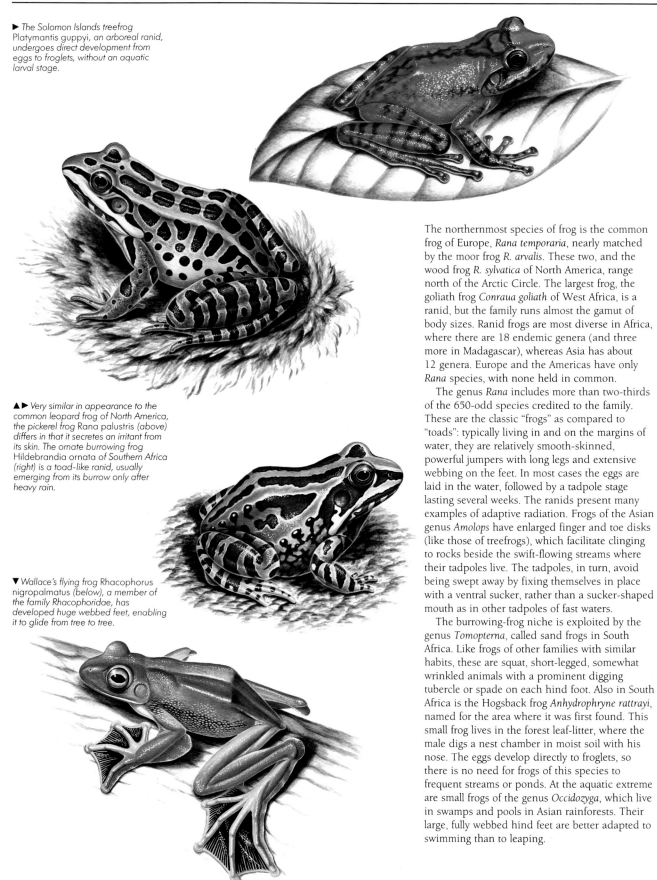

▶ The Solomon Islands treefrog Platymantis guppyi, an arboreal ranid, undergoes direct development from eggs to froglets, without an aquatic larval stage.

▲ ▶ Very similar in appearance to the common leopard frog of North America, the pickerel frog Rana palustris (above) differs in that it secretes an irritant from its skin. The ornate burrowing frog Hildebrandia ornata of Southern Africa (right) is a toad-like ranid, usually emerging from its burrow only after heavy rain.

▼ Wallace's flying frog Rhacophorus nigropalmatus (below), a member of the family Rhacophoridae, has developed huge webbed feet, enabling it to glide from tree to tree.

The northernmost species of frog is the common frog of Europe, *Rana temporaria*, nearly matched by the moor frog *R. arvalis*. These two, and the wood frog *R. sylvatica* of North America, range north of the Arctic Circle. The largest frog, the goliath frog *Conraua goliath* of West Africa, is a ranid, but the family runs almost the gamut of body sizes. Ranid frogs are most diverse in Africa, where there are 18 endemic genera (and three more in Madagascar), whereas Asia has about 12 genera. Europe and the Americas have only *Rana* species, with none held in common.

The genus *Rana* includes more than two-thirds of the 650-odd species credited to the family. These are the classic "frogs" as compared to "toads": typically living in and on the margins of water, they are relatively smooth-skinned, powerful jumpers with long legs and extensive webbing on the feet. In most cases the eggs are laid in the water, followed by a tadpole stage lasting several weeks. The ranids present many examples of adaptive radiation. Frogs of the Asian genus *Amolops* have enlarged finger and toe disks (like those of treefrogs), which facilitate clinging to rocks beside the swift-flowing streams where their tadpoles live. The tadpoles, in turn, avoid being swept away by fixing themselves in place with a ventral sucker, rather than a sucker-shaped mouth as in other tadpoles of fast waters.

The burrowing-frog niche is exploited by the genus *Tomopterna*, called sand frogs in South Africa. Like frogs of other families with similar habits, these are squat, short-legged, somewhat wrinkled animals with a prominent digging tubercle or spade on each hind foot. Also in South Africa is the Hogsback frog *Anhydrophryne rattrayi*, named for the area where it was first found. This small frog lives in the forest leaf-litter, where the male digs a nest chamber in moist soil with his nose. The eggs develop directly to froglets, so there is no need for frogs of this species to frequent streams or ponds. At the aquatic extreme are small frogs of the genus *Occidozyga*, which live in swamps and pools in Asian rainforests. Their large, fully webbed hind feet are better adapted to swimming than to leaping.

Because oceanic islands are typically barren of native frogs, the presence of two ranid species of the genus *Platymantis*—one a tree-dweller, the other terrestrial—in the Fiji islands is noteworthy. Frogs of this genus all live on islands, ranging from the southern Philippines to New Guinea, and eastward through the Solomon Islands before making the great jump to Fiji. Four closely related but morphologically diverse genera also inhabit the Solomons and share with *Platymantis* the direct mode of embryonic development. The history of how these frogs attained their present distribution can never be known for sure, but it undoubtedly involves passive distribution on islands moving very slowly over millions of years of tectonic activity. Numerous fossils from North America and Europe, none older than Oligocene (37 million years), are referred to the genus *Rana*.

Rhacophorid treefrogs

The rhacophorids, most of them treefrogs, are relatives of the largely aquatic and terrestrial ranids, and inhabit temperate and tropical parts of Africa and Asia, including Madagascar and Japan. This is a family of modest extent, with nearly 200 species in ten genera, ranging in size from 15 to 120 millimeters (½ to 4¾ inches) body length. The flying frog *Rhacophorus nigromaculatus* of Southeast Asia is a member of this family.

The African genus *Chiromantis* has some interesting adaptations. In addition to the digital disks common to all arboreal frogs, *Chiromantis* has the inner two fingers opposable to the outer two, providing a firm grip on twigs. The frogs have unusual resistance to desiccation and can spend dry periods fully exposed. They lay their eggs in a tree above water. As the eggs are

▲ *Like many frogs of the temperate zone, the European brown or common frog Rana temporaria congregates at ponds in early spring to lay large communal masses of eggs. It has been suggested that one advantage of this behavior is that losses during cold weather are reduced: the black embryos readily absorb heat while the gelatinous envelopes provide insulation.*

▲ *The gray treefrog* Chiromantis xerampelina *of arid southern Africa congregates in trees to mate in groups of up to 30 or so individuals, beating the eggs and seminal fluids with their feet to form a foam nest.*

▼ *Two members of the family Hyperoliidae: the painted reed frog* Hyperolius marmoratus *(left) and the Senegal running frog* Kassina senegalensis *(right). As its name suggests, the latter species walks or runs rather than hops.*

produced, the mating frogs use their feet to beat the eggs and accompanying liquid into a froth which hardens, protecting the developing eggs. Several pairs of frogs may work together building a communal nest. The larvae remain for a time in the nest before dropping into the water to complete development. Foam nests feature in the life history of most rhacophorid frogs.

Small tree-dwelling animals are unlikely candidates for fossilization, and there are no paleontological records for the Rhacophoridae—as indeed there are none for many other families.

Reed and lily frogs

The hyperoliids (Hyperoliidae) are another family of modest size—just over 200 species, in 16 genera—related to the ranids. They are mostly small species, about 15 to 80 millimeters (½ to 3 inches) body length, living in Africa and Madagascar, with one endemic species on the Seychelles islands in the Indian Ocean. A modest adaptive radiation has produced tree-dwelling frogs and terrestrial frogs as well as a majority that climb but tend to remain for the best part of the time in low vegetation near water.

The genus *Hyperolius* includes half the species in the family. These small frogs live mostly in marshy or swampy areas, resting on reeds and sedges, on which many of the species deposit their egg masses. They are often boldly colored and patterned, with considerable individual variation in markings. This, and a general similarity of body form and structure, make it difficult to distinguish between the species, but a knowledge of the mating calls helps biologists to differentiate between them.

The genus *Leptopelis* includes both relatively large tree-dwelling frogs and burrowing frogs that rarely climb. Species of the genus *Afrixalus* lay their eggs on a leaf and then fold the edges of the leaf together, cementing them over the egg mass with secretions from the oviduct. Curiously, this may be done either in or out of water; in the latter case, the hatchling tadpoles must fall into the water to survive.

Hyperoliids are unknown as fossils.

Squeakers

The small family Arthroleptidae (about 70 species in eight genera), distributed in Africa south of the Sahara, is placed within the Ranidae by some authors. The voice of *Arthroleptis* species is the reason why people have named them "squeakers". These are small frogs that live on and within leaf-litter of the forest floor. The eggs, deposited in cavities or burrows in moist earth, undergo direct development; in some species the froglets are completely metamorphosed when they hatch, whereas in others the tail remains to be absorbed.

One arthroleptid is unique among all frogs. This is the so-called "hairy frog" *Trichobatrachus robustus* of Cameroon and Equatorial Guinea. In the breeding season, males develop vascularized hair-like structures on the flanks and thighs. They are reported to sit under water on egg masses in streams, and apparently the "hairs" serve to augment respiration through the skin, increasing the time the frogs can remain submerged.

No fossil arthroleptids are known.

Shovel-nosed frogs

The shovel-nosed frogs, family Hemiotidae, of Africa south of the Sahara, are eight moderate-sized species (up to 80 millimeters, or 3 inches body length) in the single genus *Hemisus*. They are odd-looking frogs, round-bodied with short legs and a small pointed head with the tip of the snout hardened. Shovel-nosed frogs are burrowers, generally living in open country near

▼ Common in the coastal lowlands of southern Africa, the waterlily frog Hyperolius pusillus *lays its eggs in the cavities between overlapping waterlily leaves.*

Michael and Patricia Fogden

Michael Fogden/Oxford Scientific Films

▲ *The microhylids include a small genus of frogs, confined to Africa, with the unusual habit of laying their eggs in underground chambers. The embryo remains in the egg until metamorphosis is complete. These frogs inhabit arid regions and appear above ground only after rain. This is the common rain frog* Breviceps mossambicus.

▼ ▶ *The Asian painted frog* Kaloula pulchra *(below) is a large microhylid that is now almost exclusively found associated with human settlements. The red-banded crevice creeper* Phrynomerus bifasciatus *(right, above) of southern Africa is a rock-dweller whose skin is rubbery in texture, giving this species its other common name of rubber frog. The eastern narrow-mouthed toad* Gastrophryne carolinensis *(right, below) is a much smaller species from the southeastern United States.*

pools, and are seldom seen above ground. The eggs are laid in an underground cavity, and the female remains with them until they hatch. Using her snout, she digs a burrow leading to water nearby, and the larvae then swim out and assume a more normal tadpole existence. Their bizarre appearance notwithstanding, these frogs are thought to be related to the ranids and are treated as a subfamily of Ranidae by some authors.

Microhylid frogs

The family Microhylidae occurs in the Americas from the southern United States to Argentina, equatorial and southern Africa, and eastern India and Sri Lanka, through Southeast Asia to New Guinea and northern Australia. More than 300 species are recognized in 65 genera, the largest number of genera of any family of frogs. Microhylids comprise almost half the species of frogs in New Guinea and a sizable proportion in Madagascar, but are less significant elsewhere. Most are small frogs—several species are less than 15 millimeters (½ inch) body length—but others reach 80 to 90 millimeters (3 to 3½ inches). Morphology varies greatly, from rotund burrowers to typical treefrogs. A majority of the species live in moist tropical regions, but the evolutionary radiation of the group has placed species in arid habitats as well, and in a variety of terrestrial and arboreal niches. Some species are streamside frogs, but there do not seem to be any primarily aquatic microhylids.

Ground-dwelling microhylids live in both arid and humid tropical habitats. Among the most peculiar in arid regions are the rain frogs of the African genus *Breviceps*. (In arid regions they are likely to be seen only when it rains.) These frogs have small heads, short limbs, and round bodies, a shape accentuated by their habit of puffing up with air when disturbed. The arms are so short that the male cannot clasp the female around the body when mating (the usual method of maintaining contact between mating frogs), so instead the male and female become stuck together by secretions from skin glands on the male's ventral surface, giving an effect not unlike that of two golf balls glued together. The eggs are laid in an underground chamber prepared or

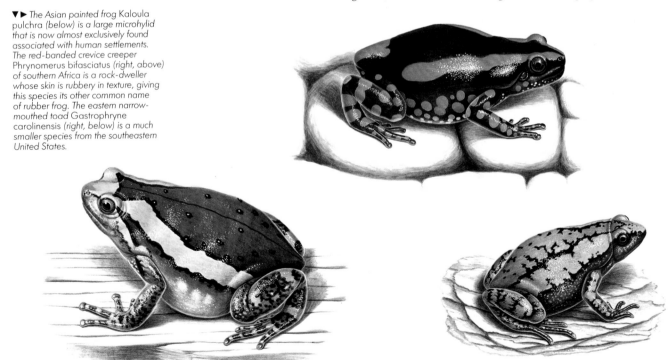

enlarged by the female. The tadpoles do not feed but live on yolk provided in the egg, and the female remains with the nest until the tadpoles metamorphose and leave.

The moist leaf-litter of tropical forests is prime habitat for many microhylid species. Some burrow in the deep litter or soil and rarely emerge on the surface. Others come to the surface at night to wait for or actively seek food. The litter also serves as a daytime retreat for small climbing species that ascend into low vegetation at night to feed and advertise for mates. Microhylids with wide bodies but narrow, pointed heads commonly feed on termites and ants. Other species with more normal frog proportions have the catholic tastes of other frogs; one New Guinean species even eats other frogs. Many microhylids, notably in Madagascar and New Guinea, are tree-dwellers.

All the 100 or so species of microhylids found in New Guinea and Australia have direct development, skipping the tadpole stage. The tree-dwellers therefore do not need to leave the trees to seek pools or streams in which to breed, but may find appropriate arboreal sites—such as an epiphytic plant called the ant plant, which has a chamber that holds moisture.

Microhylid tadpoles differ from those of other families in features of their anatomy. With rare exceptions, they lack the horny beak and denticles ("teeth") of other frog larvae. In some species the mouth is formed into a funnel shape and is used, from below, in ingesting food from the water surface. (This adaptation occurs also in tadpoles of other families.)

The only pre-Pleistocene fossils of this family are from the early Miocene of Florida, about 24 million years ago. They are classified in the living genus *Gastrophryne* which inhabits this region near the present-day limits of the distribution of microhylids.

Four South American families

Some small families of frogs (lacking fossil records) are recognized not so much because of their distinctiveness but because their species cannot be fitted unambiguously into any of the larger groups. Four South American families— Allophrynidae, Brachycephalidae, Rhinodermatidae, and Pseudidae—are examples. *Allophryne ruthveni*, the only member of its family, is a small arboreal frog of northern South America variously considered most closely related to bufonids, leptodactylids, or hylids.

The three species of brachycephalid frogs live in the Atlantic forests of southeastern Brazil. Their features include small size—*Psyllophryne didactyla* grows to less than 10 millimeters (less than ½ inch) body length, so not only is it the smallest frog, but it is also the smallest four-legged animal. The two species of the genus *Brachycephalus* are only slightly larger. Tiny frogs tend to have fewer

digits than usual, and brachycephalids, for example, have only three functional toes on each foot. At least one (and probably all three) species of brachycephalids have direct embryonic development, hatching as tiny frogs. In the family Rhinodermatidae the two species of the genus *Rhinoderma* are noted for their habit of oral brooding. They are small ground-dwelling frogs, about 30 millimeters (1¼ inches) long, found in the cool temperate forests of southern Chile and adjacent Argentina.

▲ Three South American oddities: Darwin's frog *Rhinoderma darwinii* (top) is known for its unusual means of rearing young (see page 520); the gold frog *Brachycephalus ephippium* (center) is a diminutive burrower with a number of unique specializations for subterranean excavation; and the paradox frog *Pseudis paradoxa* (bottom), so named because the tadpoles may be up to four times the length of the adults.

▶ *Restricted to the Seychelles in the Indian Ocean, the sooglossids are a small family whose relationship to other frogs is poorly understood. The eggs of these species are laid on land and undergo direct development into froglets; sometimes they hatch as non-feeding tadpoles that are carried on the back of the adult, unable to eat because they have no mouthparts.*

Jany Sauvanet/NHPA

The paradox frog *Pseudis paradoxa* gained its name because the tadpoles can reach remarkably large size, up to 250 millimeters (10 inches) in length, yet after metamorphosis the largest the frogs get is about 70 millimeters (2¾ inches). The three species in the family Pseudidae—two *Pseudis* and one *Limellus*—are almost totally aquatic, although an ability to survive dry periods

▼ *The Cape ghost frog Heleophryne purcelli belongs to a little known family confined to southern Africa. It is adapted for life in and around mountain streams, where the adhesive pads on its digits facilitate climbing on slippery rocks in fast-flowing water.*

buried in mud has been reported for the paradox frog. The family ranges through tropical lowlands over much of northern and eastern South America from Colombia to Argentina.

Ghost frogs and Seychelles frogs

There are another couple of small families whose evolutionary affinities have been disputed. The so-called "ghost frogs", family Heleophrynidae, comprise the genus *Heleophryne* with four species confined to the Cape and Transvaal regions of extreme southern Africa. The common name may have been coined because one of the species is found in a place called Skeleton Gorge; certainly the frogs are not vaporous or otherwise ghostly.

Heleophrynid frogs are up to about 60 millimeters (2⅓ inches) long and are rather flattened, with prominently enlarged tips to the fingers and toes. They are therefore well adapted to fit into crevices and cling to rock surfaces along the cool, shaded mountain streams that are their habitat and where their tadpoles live. Like other tadpoles adapted for life in swift-flowing water, the tadpoles of ghost frogs have their mouthparts modified into a large sucking disk which allows them to cling to slippery rocks while feeding.

The suggestion that the heleophrynid frogs should possibly be classified within the Australian family Myobatrachidae implies a relationship going back many millions of years to when Africa and Australia were part of the Gondwanan supercontinent.

Evidence from chromosomes and behavior suggests that the family Sooglossidae, of the Seychelles, may also be related to the Myobatrachidae of Australia. There are three species—two in the genus *Sooglossus* and one in *Nesomantis*. They are small terrestrial frogs, up to 40 millimeters (1½ inches) body length, and they deviate from typical frog behavior in their method of breeding. Eggs are laid on the ground rather than in water and follow two modes of development: direct to small frogs in one species of *Sooglossus*; in the other species, tadpoles are carried on the back until they metamorphose. In

S. seychellensis, the tadpoles are carried not by the male, as is usual, but by the female.

CONSERVATION

A few species of frogs are listed as endangered by one agency or another, but it is not individual species so much as endangered habitats that need to be conserved. Destruction of rainforests in tropical regions around the world has undoubtedly eliminated many species of frogs before they even became known to scientists. Wetlands in temperate areas and isolated sources of water in arid regions also merit special attention. Island faunas, too, are especially vulnerable to habitat destruction. Even where no specific cause can be identified, there are many instances of species apparently having disappeared or virtually so—for example, gastric brooding frogs in Australia and frogs of the family Ranidae native to southern California.

DISASTROUS INTRODUCTIONS

People have an unfortunate propensity for moving animals from their native area to exotic locations. Most attempted introductions probably fail, but success may create ecological disaster, or at least a lot of disturbance—consider the gypsy moth in North America and the rabbit in Australia.

Relatively few frog species have become established in places foreign to them, and most such introductions are probably benign. For example, formerly frogless Hawaii now has poison frogs from Central America among other species; some Australian frogs are established in New Zealand, which has only three native species; and a clawed frog of Africa *Xenopus laevis*, is spreading in southern California. But two widely introduced American frogs stand out from the rest: the bullfrog *Rana catesbeiana* and the marine or cane toad *Bufo marinus*.

The bullfrog is a large semi-aquatic species native to eastern North America. It is adaptable, prolific, voracious, and tasty. The last quality has resulted in its widespread introduction into western North America, where it competes with native frog species and may be a factor in their local extermination. It also occurs now in Puerto Rico, Italy (introduced in the 1930s), and Japan, and probably other regions as well.

The marine toad's native range is from extreme southern Texas to northern South America. Like the bullfrog, it is large, prolific, and adaptable, but being poisonous (and having relatively small hind legs), it is not eaten by humans. The excuse for its widespread introduction is control of insects that are agricultural pests. It occurs now on many Caribbean islands, Taiwan, the Philippines, New Guinea, and numerous islands in the Pacific, and

Jean-Paul Ferrero/AUSCAPE International

is spreading over northeastern Australia where it was introduced in 1935.

The marine toad's role in pest control is questioned, as a toad eats pest insects and beneficial insects indiscriminately. It has no natural enemies, and breeds all year round. Other aspects of the introduction are also clearly negative: dogs, cats, native mammals, birds, reptiles, and other amphibians can die from attempting to eat toads; and native frogs not only are preyed upon, but may be displaced ecologically.

▲ One of the factors in the cane toad's success is that it will eat almost anything that moves. It can handle surprisingly large prey, and has a voracious appetite.

Order Testudinata
2 suborders, 12 families,
c. 87 genera, c. 245 species

SIZE

Smallest Speckled Cape
tortoise *Homopus signatus*,
shell length 95 millimeters
(3¾ inches); weight
140 grams (5 ounces).
Largest Leathery turtle
Dermochelys coriacea, shell
length up to 2.4 meters
(95 inches); weight up
to 860 kilograms
(1,890 pounds).

CONSERVATION WATCH

There are 79 species of
turtles and tortoises listed in
the IUCN Red Data Book of
threatened animals. Among
the most threatened species
are: western swamp turtle
Pseudemydura umbrina, green
turtle *Chelonia mydas*,
hawksbill turtle *Eretmochelys
imbricata*, Kemp's ridley
Lepidochelys kempii, olive
ridley *Lepidochelys olivacea*,
leathery turtle *Dermochelys
coriacea*, river terrapin
Batagur baska, painted
terrapin *Callagur borneoensis*,
South American river turtle
Podocnemis expansa,
angonoka *Asterochelys
yniphora*, and Bolson tortoise
Gopherus flavomarginatus.

TURTLES AND TORTOISES

F.J. OBST

About 245 species of turtles, tortoises, and terrapins are distributed worldwide in tropical and temperate zones. They are the only reptiles that have a shell built into the skeleton, allowing them to more or less conceal themselves entirely within the shell. Of all the reptiles alive today, turtles and tortoises are not only the oldest forms but they have also changed very little in their 200-million-year history.

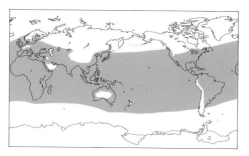

TWO MAIN LINEAGES

The oldest group of fossil turtles is known from the Triassic, about 230 million years ago. The turtles soon evolved into two main lineages which have survived until today: the cryptodirans or hidden-necked turtles (suborder Cryptodira) and the pleurodirans or side-necked turtles (suborder Pleurodira).

Cryptodirans can, by a vertical cobra-like bending of the vertebral column of the neck, draw the head directly into the shell (although in some modern species, only partial withdrawal is possible). In contrast, pleurodirans merely fold the head under the front edge of the upper shell by a sideways movement, either to the right or to the left. As well as these external characteristics, there are important differences in the structure of the skull and the skeleton.

Cryptodirans include sea turtles, which live in tropical and temperate oceans around the world, and also the majority of species that live on land or in rivers and lakes; they are found on all continents, although only one species reaches northern Australia. Pleurodirans are found only in Australasia, South America, and central and southern Africa.

In ecological terms, turtles have adapted to a wide range of habitats. They have successfully established themselves in dry landscapes—deserts, savanna, and plains—as well as grasslands, woodland, and mountains. Diverse species of freshwater turtles occur in still waters, such as ponds, and running waters, such as tropical rivers. A few species of freshwater turtles, like the true sea turtles, leave their aquatic habitats only to lay eggs, whereas other species are amphibious and regularly move about on land. There are a few species of land tortoises living in

▶ A green turtle *Chelonia mydas* at
sea, Galapagos Archipelago. Marine
turtles live at sea all their lives but come
ashore on beaches to lay their eggs.
Most species are endangered, having
been heavily exploited by humans
for food.

G.M. Wellington

such dry landscapes that they are unlikely to encounter open bodies of water at any time throughout their lives.

CHARACTERISTICS IN COMMON

A turtle's shell consists of two parts: an upper part, called the carapace, and a lower part, called the plastron. Each part typically has an inner bony layer and an outer layer of horny plates. The visible layer is made up of large horny plates, but these actually cover a thicker layer of bony segments which makes up the true protective shell. Where the carapace meets the plastron, there are openings for the head, legs, and tail. The number and arrangement of horny plates tends to differ from species to species, and these arrangements are often useful in identifying species, although there are also individual differences within species.

Only three families lack horny plates on the shell: the softshell turtles, the Papuan soft-shelled turtle, and the leathery turtle. In these there is a thick leathery covering instead of the horny plates; in the softshell turtles this covering is flexible, at least at the edges.

A universal feature in modern turtles is the absence of teeth on the jaws. The oldest fossils from the Triassic period did have very small teeth but these were on the palate, and the jaws themselves were toothless. Replacing the teeth in modern turtles and tortoises are horny ridges which cover the upper and lower jaws. In meat-eating turtles these horny ridges are knife-sharp and work like shears. In plant-eating species the outer edge of each horny jaw ridge is serrated, making it easier to bite off sections of hard woody plants, and there is often a serrated outer edge to the jaw in order to grip more readily.

All turtles possess strong limbs. Even the heaviest land turtle can lift its body off the ground when walking. In land turtles the fingers and toes have more or less grown together to form solid "clump-feet", whereas in aquatic freshwater turtles the individual digits are distinct and clearly recognizable, and many species have webbing between fingers and toes. The sea turtles are exceptional, in that their digits have again become fused during evolution to form paddle-shaped limbs with which they propel themselves through water, and in contrast to other turtles, the forelimbs are more strongly developed than the hind limbs. Sea turtles are also the only group that have to drag their body across the ground when they come ashore on sandy coasts to lay eggs.

▲ The giant tortoise Chelonoidis elephantopus of the Galapagos Islands is among the largest of the land tortoises, second only to the Aldabra tortoise in size. Some individuals may exceed 1 meter (3¼ feet) in length, and approach 200 kilograms (about 450 pounds) in weight. But Galapagos tortoises remain vulnerable despite their size: rats prey on their eggs and hatchlings, and feral goats compete for scarce fodder.

▶ *Great African tortoises Geochelone sulcata mating. Sniffing and butting are two common preludes to copulation among most tortoises: the sense of smell seems important in establishing age, sex, and readiness to mate, and a male repeatedly butts the female's shell until she responds by becoming quiescent, enabling him to mount.*

Ashod Papazian/NHPA

▼ *Marine turtles, like this loggerhead Caretta caretta, come ashore by night to lay their eggs, often in huge numbers. The female digs a hole in the sand above the high tide-line, deposits her clutch of 100 eggs or more, buries them, then returns to the sea. In due course the eggs hatch and the hatchlings go to sea without further assistance from the female.*

Jean-Paul Ferrero/AUSCAPE International

REPRODUCTION

All turtles lay eggs in a nest chamber, and the young develop in the eggs at a temperature corresponding with the surrounding sand or soil, without any further parental interest after the eggs are covered over. From the day they hatch, the young turtles must fend for themselves, their lives largely determined by the inherited instinctive behavior pattern of their species.

The incubation period varies to some extent with the microclimate within the egg chamber, but it is also genetically programed. The shortest incubation time seems to be about a month (for a softshell turtle); the average time is two to three months. The longest incubation time has been recorded in some land tortoises, which require one and a half years to incubate, although such times are reliable only when based on direct observation of the eggs or, less reliably, on breeding in captivity. In many species the hatchlings spend the winter in the egg chamber and appear for the first time the following spring, giving a false impression of long incubation. With the exception of a few larger species of sea turtles and freshwater turtles, the reproductive rate in turtles and tortoises is low. Clutches of the smaller species generally comprise only a few eggs (one to six), although many of the larger species may lay up to 150 eggs or more.

SIDE-NECKED TURTLES
Snake-necked turtles
Snake-necked turtles (family Chelidae) of South America, and Australia and New Guinea are well adapted to life in fresh water. The long neck of most species allows them to draw breath at the surface without exposing the rest of the body to potential predators, and they can stay underwater for lengthy periods while searching for food, particularly insects and their larvae, crayfish, tadpoles, and also small frogs and fish. Some species also feed on water plants, as well as fruits that fall into the water from surrounding trees.

The eight species in the genus *Chelodina*, known as Australian snake-necked turtles, live in Australia and New Guinea. Largest is the giant snake-necked turtle *C. expansa* of southeastern Australia, with a shell length of up to 42 centimeters (16½ inches); the total length of the head and neck is a further 31 centimeters (12¼ inches).

In seasonally dry areas some species will burrow deep into the mud at the bottom of lagoons and swamps to aestivate (remain dormant) until the next rain. When they need to cross dry areas between lagoons they generally wander during the night. They are helped by their ability to hold water in the anal sac, a structure that is also found in many land tortoises inhabiting desert regions.

The five species of short-necked turtles (genus *Emydura*) and three species of Australian snapping turtles (genus *Elseya*) have shorter necks, more or less normal in length. They live in Australia and New Guinea, mostly in flowing water, and can swim extremely well. Like many other stream-dwellers, the short-necked turtles are often seen sunbathing in the early morning, lying beside or on top of one another in favorable positions on the river bank, and if disturbed dive rapidly into the water.

As recently as 1980 a new genus and species was discovered in eastern Australia, the Fitzroy turtle *Rheodytes leukops*, found only in the Fitzroy River drainage in Queensland. It was given the name "*leukops*" because of the remarkably white iris of the eye.

In southwestern Australia, near the city of Perth, lives the rarest and most threatened turtle species in the world, the western swamp tortoise *Pseudemydura umbrina*. Probably fewer than 20 of

these little creatures survive in the wild (in a small reserve), but a captive breeding program in Perth Zoo has recently had significant success. If the original habitat can be made secure from predators and human interference, the wild population can be supplemented from the captive breeding program.

South American members of the family Chelidae include both snake-necked and short-necked groups, and the two species of American snake-necked turtles (genus *Hydromedusa*) are superficially very similar to the Australian *Chelodina*. Among the stream-dwellers there are at least eight species in the toad-headed genus *Phrynops*, the best known being Geoffroy's side-necked turtle *P. geoffroyanus* of Brazil and Paraguay. The largest is the spotted-bellied side-necked turtle *P. hillarii*, with a shell length of up to 44 centimeters (17¼ inches) and weighing up to 1.2 kilograms (2½ pounds); it lives in the Rio Paraná and Rio Paraguay and their tributaries in eastern South America. New *Phrynops* species have been found in recent years, and it seems likely that others are still to be discovered.

In the streams of the Chaco region in south-central South America, which can be dry for months at a time, live smaller turtles in the genus *Acanthochelys* whose ecology is similar to that of the pond turtles (family Emydidae), described later. One species, the Chaco side-necked turtle *A. pallidipectoris*, is notable for the long horny spurs on its upper thighs. It measures only

▼ *The eastern snake-necked turtle* Chelodina longicollis *is common in swamps and wetlands of southeastern Australia. It is a member of the family Chelidae, a group of aquatic and semi-aquatic turtles with representatives in Australasia and South America. The head is retracted into the shell by one or more horizontal (rather than vertical) folds of the neck.*

Jean-Paul Ferrero/AUSCAPE International

head and neck, with flaps of skin on the sides, look remarkably like a fallen leaf. Thus camouflaged, the matamata lies motionless in shallow pools in the forest or in slowly flowing streams, occasionally lifting its snout to the surface to breathe. Fish do not recognize it as a predator and swim carelessly close to its mouth.

Helmeted side-necked turtles

Side-necked turtles of the family Pelomedusidae occur only in South America, Africa, and Madagascar. Despite the small number of species in South America, they are so abundant that they are regarded as characteristic animals of South

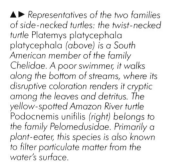

▲▶ *Representatives of the two families of side-necked turtles: the twist-necked turtle* Platemys platycephala platycephala *(above) is a South American member of the family Chelidae. A poor swimmer, it walks along the bottom of streams, where its disruptive coloration renders it cryptic among the leaves and detritus. The yellow-spotted Amazon River turtle* Podocnemis unifilis *(right) belongs to the family Pelomedusidae. Primarily a plant-eater, this species is also known to filter particulate matter from the water's surface.*

18 centimeters (7 inches) in shell length, and when the streams dry up it buries itself deep in the mud until the rainy season begins. The remaining two or three species in this genus live in eastern South America.

In northern South America, in the river systems of the Amazon and the Orinoco, are two members of the family Chelidae that have a number of unusual features. The twist-necked turtle *Platemys platycephala*, measuring only 18 centimeters (7 inches) in shell length, lives in shallow, slow-flowing streams in the rainforest. Its bright yellow/orange/brown coloring is reminiscent of turtles in the forests of Southeast Asia and probably serves a similar purpose: as a disruptive coloration that camouflages the turtle on the leaf-strewn forest floor or stream bed.

An equally effective camouflage is achieved by the matamata *Chelus fimbriatus* but using a different strategy. Its shell, which is up to 45 centimeters (17½ inches) long, is flattened and ridged, so that it looks like a piece of bark, and its

American wildlife. The giant South American river turtle *Podocnemis expansa* and the yellow-spotted Amazon River turtle *P. unifilis* are not only conspicuous because of their size—the shell length exceeding 100 centimeters (3¼ feet) and almost 70 centimeters (2¼ feet) respectively—but also because of the enormous number that seek out nesting spots on the sand banks of the Amazon and Orinoco rivers.

Unlike many other large freshwater species, *Podocnemis* turtles have a predominantly vegetarian diet.

The Madagascar big-headed side-necked turtle *Erymnochelys madagascariensis* differs very little from its South American relatives, and its presence on this island off the southeast coast of Africa is dramatic evidence of Madagascar's origin as a part of a large southern continent, Gondwana, which included South America, Africa, Australia, and Antarctica. The family is represented in Africa by widespread species of the genus *Pelusios*, all of which have a movable front section of the belly shields. Ranging in size from 12 to 45 centimeters (5 to 17½ inches) long, these roundish turtles have strong-smelling musk glands whose secretions deter potential predators. They live in both flowing and still waters, preferring to hide in the mud, where they find their food—mollusks, worms, and insects. As the water dries up, they bury themselves in the mud and aestivate there until the rainy season.

The African helmeted turtle *Pelomedusa subrufa* is similarly widely distributed over eastern and southern Africa. It also occurs in the outermost southwest tip of Asia in the Yemen, on the Arabian Peninsula, and so is the only side-necked turtle occurring in Asia. In habits and form it differs little from the *Pelusios* turtles.

HIDDEN-NECKED TURTLES
Alligator turtles (snapping turtles)
The family Chelydridae today contains only two species living in North and Central America. They are remarkable for their long tail, which has large shield-like scales similar in appearance to those of crocodiles. They also have a remarkably large head which cannot be completely withdrawn into the shell. If an alligator turtle is turned on its back it can be seen that the plastron (the shell covering the belly) is reduced in size and cross-shaped. The limbs cannot be withdrawn into the shell but

can be drawn tightly up against it.

The American snapping turtle *Chelydra serpentina* is found from southern Canada to southern Ecuador and is the smaller of the two species, with a shell length up to 47 centimeters (18½ inches). It consumes a predominantly animal diet, including salamanders, fish, frogs, and smaller birds and mammals. Much of the time is spent in the water, although these turtles also like to sunbathe in the mornings on the banks of streams and swamps.

The alligator snapping turtle *Macroclemys temmincki* of the southeastern United States, from the Mississippi Valley southwards to Texas and Florida, is the largest freshwater turtle in North America. The record size is a shell length of 66 centimeters (26 inches), and a weight of 114 kilograms (250 pounds); the head of this particular individual was 24 centimeters (9½ inches) long and about as broad.

As well as a large powerful head, the species has a hooked beak and a circle of large keeled scales around each eye, which makes the eyes appear larger. There is also a series of long soft scales on the neck, giving it a prickly appearance. The shields of the carapace are strongly overlapping like roof tiles, while the flanks have an additional row of horny shields, considered to be a primitive characteristic.

◄ *The helmeted turtle* Pelomedusa subrufa *is widespread across eastern and southern Africa, occurring also in Madagascar and even the extreme southwestern corner of the Arabian Peninsula. It favors temporary floodwaters as well as ponds and streams, and frequently moves from one area to another as its shallow ponds dry up, or alternatively aestivates in the mud until the rains return.*

Anthony Bannister/NHPA

▶ *The alligator snapping turtle Macroclemys temmincki of the eastern United States takes its name from the pronounced bony ridges on the carapace as well as from its swift and savage bite. It has a pink, fleshy appendage on its tongue, which it moves in its wide-open mouth to lure fish and other unwary prey to within reach of its snapping jaws.*

▼ *Female olive ridleys Lepidochelys olivacea massing to lay their eggs on Nancite Beach, Costa Rica.*

The alligator snapping turtle is more aquatic than its smaller relative, the American snapping turtle. Large older animals rarely leave the water, and then only the females for egg-laying. They eat virtually everything they can capture—even large snails and mussels are unable to resist their strong jaws—and are especially well adapted to catching fish using a mechanism found in no other turtles. The tongue has a reddish-colored worm-like appendage which, when the mouth is held wide open underwater, moves about in a life-like manner. Any fish that tries to eat the decoy worm will itself become a victim of lightning-fast jaws.

Hardback sea turtles

The seven species of sea turtles (family Cheloniidae) are survivors of a much larger group which reached its greatest diversity during the Jurassic and Cretaceous periods, 200 to 65 million years ago.

All recent sea turtles of the family Cheloniidae are fairly uniform in structure, form, and life history. However, one additional species is so distinctive as to warrant its recognition in a separate family—the leathery turtle is the sole survivor of an otherwise extinct family and is discussed separately.

The Cheloniidae all have rather flat shells with a complete covering of large horny plates. The forelimbs are more strongly developed than the hind limbs, a feature distinguishing them from freshwater turtles. The bony shell is less substantial than that of freshwater turtles; between the bony plates of both the carapace and the plastron are broad gaps, filled in by fibrous skin, and these in turn are covered by the strong horny plates. In this way the very heavy shell of land turtles, which is unnecessary for a turtle spending most of its life at sea, is much reduced without loss of structural support and stability.

Sea turtles leave the water only to lay their eggs, so after leaving the beach as hatchlings, the males spend their entire life in the sea. They often sunbathe at the surface, drifting or resting on floating fields of seaweed, or in shallow water left by the receding tide on coral reefs. There is some evidence that females may return to lay their eggs on the same beaches as those from which they themselves hatched; certainly females return year after year to the same beaches to nest. Unfortunately, their homing instinct has allowed humans to predict their arrival for breeding, with the consequence that sea turtles have been almost exterminated from many of their breeding grounds and their existence is threatened globally.

The green turtles (genus *Chelonia*) with one species in the Atlantic, Indian, and western Pacific oceans (*C. mydas*) and *C. agassizii* in the eastern Pacific, together with the flatback turtle *Natator depressus*, are mostly sought for their flesh, but to some extent green turtles are also hunted for the horny plates of their shells.

Ridley sea turtles are also hunted for their meat. The olive ridley *Lepidochelys olivacea* is found throughout much of the Atlantic and Pacific, while Kemp's ridley *L. kempi* is found only in the Gulf of Mexico and warm waters of the Atlantic. The latter species seems to use only a single nesting beach on the eastern coast of Mexico, making it especially vulnerable to overexploitation and extinction. Whereas green turtles attain a shell length of up to 150 centimeters (5 feet), ridley sea turtles rarely have a shell length of more than 70 centimeters (2$^{1}/_{3}$ feet).

Somewhat larger (up to 90 centimeters or 3 feet shell length) is the hawksbill turtle *Eretmochelys imbricata*, which seems to be restricted to warm tropical seas. The horny shields of the upper shell are beautifully marbled or flamed, and are much sought after for ornaments, putting the species at great risk from overexploitation.

The largest member of the family is the loggerhead turtle *Caretta caretta*. It reaches a shell length of up to 213 centimeters (7 feet), although the average is about 150 centimeters (5 feet). Its

▼ Green turtles Chelonia mydas mating. These marine turtles migrate sometimes thousands of miles to breed. The sexes rendezvous to mate at sea near the nesting beaches, but no pair bond is formed, and both males and females may mate with others several times during the brief mating period.

▲ Sea turtles display some of the most specialized aquatic adaptations among the chelonians, such as a lightened, hydrodynamically-shaped shell and large paddle-like forelimbs with reduced claws. The attractively marked Pacific hawksbill Eretmochelys imbricata bissa has been the source of commercial "tortoise shell", used in the manufacture of items from eyeglass frames to haircombs, with devastating consequences for wild populations.

normal weight is about 150 kilograms (330 pounds), but some individuals have been up to 450 to 500 kilograms (about 1,000 pounds). It has a worldwide distribution extending into the cooler waters of higher latitudes, and is the only sea turtle that still breeds on the Mediterranean coast. Fortunately its flesh is not eaten nor is its shell commercially useful. A major threat is tourism, as people and boats disturb the nesting beaches and discourage breeding, while large numbers are drowned in fishing nets. Consequently, it is not surprising that despite its large clutch size (females usually lay more than 100 eggs in a clutch and may lay several clutches in a season) loggerheads are as endangered as other sea turtles.

Leathery turtle
The largest turtle alive today is the leathery turtle *Dermochelys coriacea*, also known as the luth or leatherback, the only living representative of the family Dermochelyidae. One exceptional individual has been reported with a shell length of almost 2.5 meters (8¼ feet) and weighing about 860 kilograms (1,900 pounds), but specimens

---

SAFETY IN NUMBERS

The most hazardous period in the life of a sea turtle is when it leaves the nest and crosses the open beach on its way to the sea, running the gauntlet of predatory birds and crabs. But there is safety in numbers—the more small turtles there are, the better the chance they will not be picked off one-by-one. When the turtles hatch, they dig simultaneously toward the surface, gradually moving their chamber upward as the sand is deposited beneath them. Finally, and preferably at night, they burst forth and scuttle for the waves, where they are safe from their land-based enemies, if not from ocean-dwelling predators.

▶ Flat-back turtle hatchlings Natator depressus.

Jean-Paul Ferrero/AUSCAPE International

with shell lengths of more than about 1.5 meters (5 feet) are uncommon. The powerful forelimbs project as paddles with a greater span than the shell length itself, so the sight of a fast-moving leathery turtle in the open sea is very impressive. The large head is characterized by big eyes and a conspicuous hooked beak. Unlike the other sea turtles, whose shells are covered by horny plates, the shell of this species is covered only by a leathery skin. There are seven tubercular longitudinal ridges on the carapace (upper shell).

The leathery turtle feeds largely on jellyfish, but its diet may also include mollusks, echinoderms, and crustaceans. Apparently fish are rarely eaten. It is a cosmopolitan species, occurring in tropical and temperate seas throughout the world and extending into the colder waters of higher latitudes. However, like other sea turtles the females return to the same nesting beaches over and over again to lay their eggs, the same beaches at which they themselves hatched. Although specimens are often washed up exhausted and die on the cool coasts of Europe or North America, it seems certain that the leathery turtle can either maintain its body temperature above that of the surroundings for longer periods or that it is more tolerant of cool conditions.

Softshell turtles

The softshell turtles of the family Trionychidae include species that display a wide range of adaptations to an aquatic existence in rivers and lakes, but they are surprisingly uniform in form and habits.

Softshell turtles are characterized by not having the usual horny shields of the epidermis, but instead a leathery skin. While the central part of the carapace has a bony layer, the outer ring of solid bones has been lost during evolution. A subsequent strengthening of the plastron (lower shell) has evolved in many species through the development of dermal bones which are not attached to the bony shell and are visible as coarse, hard spots in the plastron. The leathery skin, with its smooth surface, flexible edge and especially the very elastic hind third of the shell, is well adapted to fast and energy-efficient swimming in open water, as well as movement in the muddy bottom of streams and lakes.

Softshell turtles "settle" themselves on the bottom with an undulatory movement, where they lie hidden from predators, although they are generally able to defend themselves effectively with knife-sharp horny jaws, usually concealed under swollen lips. They have a strongly vascularized throat, which is able to extract oxygen from the water and allows them to avoid having to come to the surface to breathe. The leathery skin can also exchange oxygen with the surrounding water, so softshell turtles tend to be able to stay underwater for longer periods than other turtles. Most softshell turtles are strictly carnivorous and feed on mollusks, crustaceans, aquatic insects, worms, frogs, and fish. Even smooth flatfish are successfully grasped and held fast before they are consumed. A few species also eat fruit and aquatic plants. Despite their many adaptations to aquatic life, softshell turtles in northern latitudes regularly come onto land to sunbathe.

Softshell turtles are now classified in about 15 genera and are found in North America, Africa, and Asia but not in South America or Australia, although they occur as fossils in Australia. The North American genus *Apalone* has three species, one of which, the spiny softshell turtle *A. spinifera*,

Jane Burton/Bruce Coleman Limited

▲ The softshell turtles of the family Trionychidae have a characteristic habit of lying partly buried in the muddy bed of rivers and ponds, using their ability to obtain oxygen through their skins to reduce the need to come to the surface to breathe.

▼ Like the sea turtles, softshell turtles have independently evolved numerous adaptations for an aquatic existence, including a reduction in shell armor, large webbed feet with few claws, and a snorkel-shaped snout for breathing while remaining beneath the surface. The eastern spiny softshell Apalone spinifera spinifera is a colorful species of southeastern North America.

▲ *The southern loggerhead musk turtle* Sternotherus minor minor *is a small freshwater turtle of southern North America. As individuals of this species mature and graduate from a juvenile diet of insects, the head and jaws grow disproportionately large to accommodate the adult's main diet of mollusks, hence its common name. The "musk" refers to a smelly fluid expelled by these turtles when disturbed.*

has a distribution extending from southern Canada to northern Mexico.

The largest number of softshell turtles are found in Asia. In Asia Minor is the Euphrates softshell turtle *Rafetus euphraticus*, whose range overlaps with the African softshell turtle *Trionyx triunguis*, a large species (up to 95 centimeters, or 37 inches shell length) from southern Turkey and along the Mediterranean coast as far as Egypt and Somalia, as well as in west Africa from Mauritania to northern Namibia. In some areas it inhabits the brackish estuaries of larger rivers.

The Asiatic softshell turtle *Amyda cartilaginea* occurs throughout much of the East Indies, including Java, Sumatra, and Borneo. With a shell length of 70 centimeters (27½ inches) it is quite large, but it is exceeded by two other species in this region: the narrow-headed softshell turtle *Chitra indica*, which reaches 115 centimeters (45 inches) in shell length, and the Asian giant softshell turtle *Pelochelys bibroni*, with a shell length of 129 centimeters (51 inches).

In Africa and India live several softshell species whose plastron has large flaps of skin that conceal the turtle's feet. Two African species (in genera *Cyclanorbis* and *Cycloderma*) inhabit the rivers and some lakes in tropical and central Africa. The genus *Lissemys* is found in India but also has two species in Ceylon and Indo-China. The Indian flapshell turtle *L. punctata*, like many other turtles, can apparently survive the dry season by burying itself deep in a riverbed or the bank of a drying stream or swamp.

Papuan softshell turtle
The Papuan softshell turtle or pitted-shell turtle (family Carettochelyidae) is externally similar and most closely related to other softshell turtles, in that instead of having its shell covered with horny plates, there is a layer of thick leathery skin, but unlike softshells in the family Trionychidae, the shell of the Papuan softshell turtle is not flat and plate-like but is domed with a ridged keel along

its midline—a shape found in many other swamp turtles. The bony skeleton in the shell is completely supported and has a strong bony margin which provides a solid support and a bridge with the plastron. While these are smaller than in most "hardshell" turtles, all of the normal bony structures are present.

The family Carettochelyidae is represented today by a single species, *Carettochelys insculpta*, which is found only in New Guinea and tropical northern Australia. It has a long trunk-like snout with tubular nostrils which allow the turtle to breathe at the surface without putting the rest of its head out of the water. It is largely vegetarian, preferring the fruits of bog plants and trees that grow beside its aquatic habitats, such as pandanus and figs. It will, however, take animal food such as mollusks, crustaceans, and worms. Like the sea turtles, it uses its forelimbs rather than its feet to propel itself through the water.

Mud and musk turtles
The family Kinosternidae includes small, cryptic turtles confined to North and South America. The average size is 15 to 20 centimeters (6 to 8 inches) shell length. All species have a solid carapace, which is often characterized by three long keels and is covered by strong, occasionally overlapping, horny shields. The plastron in most species is large, and two distinct hinges allow the front- and hind-most portions to move in such a way that the turtle can completely close its shell front and back. In some species only the front section of the plastron is hinged, allowing only partial closure of the shell, while in a few others the plastron is reduced to an immovable cross-shaped structure which offers little protection to the limbs and other soft parts of the body.

Probably the best known genus is that of the mud turtles (genus *Kinosternon*), with about 15 species occurring from the United States, through Central America to northern South America. Most of these are inconspicuous, brown-colored turtles, which spend the greater part of their day on the bottom of streams and lakes feeding on mollusks. In the mornings they leave the water to bask in the sun and achieve their preferred body temperature. Many species, especially the smaller ones, climb shrubs and even trees beside the water.

A second genus, the musk turtles (*Sternotherus*), has only four species, confined to the central and southern United States. All have a reduced plastron. They are often referred to as "stinkpot" turtles because of the extraordinarily strong musky smell they exude when captured. This odor is not confined to *Sternotherus* but occurs in all members of the family and many other aquatic turtles. The habits of *Sternotherus* are similar to those of other kinosternids. Associated with their small size they produce fewer eggs: one to five is the normal clutch size.

In Central America there are members of two further genera which, because of differences in anatomy and cell structure are sometimes classified in their own family, the Staurotypidae. Firstly, there are the cross-breasted musk turtles (genus *Staurotypus*): the large Mexican giant musk turtle *S. triporcatus*, almost 40 centimeters (16 inches) in shell length, and the Chiapas giant musk turtle *S. salvinii*, to 25 centimeters (10 inches) shell length. As the common name implies, their plastron is reduced to a strong, bony cross, but with a large head and a sharp horny beak they can nevertheless defend themselves effectively. Both species range from Mexico to Honduras or El Salvador.

In Mexico and in Guatemala there is another member of this group, the narrow-bridged musk turtle *Claudius angustatus*, whose shell length is only 15 centimeters (6 inches). It has a conspicuously large head with an impressive beak, and is unable to conceal it within the shell. When it bites, this turtle holds on ferociously and can cause a painful wound, while at the same time releasing copious amounts of a smelly secretion from the cloaca.

Central American river turtle

The sole living representative of family Dermatemydidae is the Central American river turtle *Dermatemys mawi* of Mexico and northern Central America. It looks like a typical large freshwater turtle, although its stronger, flatter shell has a conspicuous row of additional shields on the bridge, where the carapace meets the plastron. This impressive turtle, whose shell length is 65 centimeters (25½ inches), inhabits both fresh and brackish water in rivers, lagoons, and estuaries. It tends to seek out the warmer upper levels of deeper water, letting itself drift while raising its body temperature. When disturbed it dives quickly into deeper water.

Its food consists mainly of aquatic plants, and the horny edge of each jaw is serrated to assist it in cutting hard woody plants. However, it will also take any available animal food. They leave the water only to lay eggs, up to 20 in a clutch, twice a year, and usually close to water.

New World pond turtles

In their conquest of freshwater habitats, apparently two similar groups evolved in parallel. The pond turtles (many of which are called terrapins) of the New World (family Emydidae) and the pond turtles of the Old World (family Bataguridae) apparently shared a common ancestor, and most researchers today believe that that common ancestor was also shared by land tortoises (family Testudinidae).

The Emydidae, with eight genera, are most diverse in southern North America. A few species

extend north to Canada, others occur in the Caribbean and through Central and South America. A single species extends into South America, and one species, the European pond turtle *Emys orbicularis*, is the only representative of this family in Europe, North Africa, and the Middle East.

Pond turtles are distinguished by having a full bony shell covered with horny plates and, in some genera, well-developed hinges on the plastron which can completely close the shell. They also have well-developed limbs with webbed feet. Most are semi-aquatic and occur in swamps, rivers, and even coastal lagoons. Some are more terrestrial and live in woodlands far from water. In fact, no other turtle family lives in such a wide variety of habitats.

The best known types are the three genera of ornamented turtles—the painted turtles (*Chrysemys*), the sliders (*Trachemys*), and the cooters (*Pseudemys*)—so-called because of their brightly colored shells, heads, and limbs. One of the most beautiful is the painted turtle *Chrysemys picta*, which occurs as a number of subspecies from southern Canada to the far south of the United States. It has a bright red or yellow design on the brown carapace and the head. The shell length is about 25 centimeters (10 inches).

Scarcely less spectacular are the different geographic forms of the slider *Trachemys scripta*, especially those that have large ocellate patterning on the sides of the carapace—giving rise to another common name, the peacock-eyed turtle. The species is found from the northern United States, throughout Central America, to South America. Other *Trachemys* species are found on islands in the West Indies, and members of the related genus *Pseudemys* are distributed throughout the eastern United States. Male ornamented turtles are generally smaller than the females. They also have noticeably longer claws

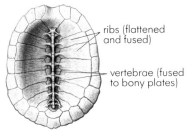

TURTLE'S CARAPACE
(bridge removed)

ribs (flattened and fused)

vertebrae (fused to bony plates)

VIEW FROM BELOW

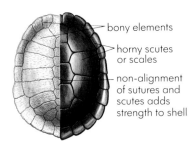

bony elements

horny scutes or scales

non-alignment of sutures and scutes adds strength to shell

VIEW FROM ABOVE

▲ The shell of chelonians has two main parts: the upper shell, or carapace, and the lower shell, or plastron. The shell is constructed of interconnected bony plates which, in the carapace (shown above), include the expanded and fused ribs, with the backbone fixed permanently in place. Both the carapace and plastron have a second covering layer of large horny plates, or scutes. The suture lines between scutes do not align with those of the bony plates, adding to the shell's strength.

▼ A male painted turtle Chrysemys picta belli (family Emydidae) of North America.

▶ The eastern box turtle Terrapene carolina carolina is a terrestrial member of the family Emydidae, and has evolved a tortoise-like appearance. The plastron in this species is hinged, allowing full closure of the shell with head, limbs, and tail withdrawn, giving the turtle a box-like appearance. This individual is a male, as indicated by its red-colored eyes.

▼ The sawback or false map turtle Graptemys pseudogeographica of North America. The sexes differ so greatly in size in this and related genera that some separation in habitat and diet is often evident, with the strongly vegetarian females tending to live in deeper waters, while the much smaller males favor a more carnivorous diet in the shallows.

Joe E. Blossom/NHPA

on the forelimbs which they use in a complex dance ritual for the females prior to mating; they approach the female from the front and rhythmically stroke both sides of her head with these long claws, stimulating her to mate.

Closely related to the ornamented turtles and occurring in the same areas of the eastern and southern United States are ten species of map turtles (genus *Graptemys*). In size they range from 15 to 30 centimeters (6 to 12 inches) shell length, and males are again much smaller than females. A characteristic of this genus is the presence of humps or ridges on the central shields of the carapace. Map turtles are not as colorful as ornamented turtles, although their heads have decorative patterns, and in many species the outer shields of the carapace are beautifully decorated with eye-shaped markings. All these turtles like to bask in the sun, especially in the early morning when they congregate on the banks of ponds, ditches, and rivers to raise their body temperature, although always prepared to plunge into deeper water when alarmed. Adults tend toward a vegetarian diet, whereas the young tend to be more carnivorous, but there is considerable variation in diet between species and even between the sexes and individuals.

Allied to the ornamented turtles is the chicken turtle *Deirochelys reticularia*, with a conspicuously long neck. Like the diamondback terrapin *Malaclemmys terrapin*, it is highly sought after for its flesh. The diamondback terrapin prefers brackish water in bogs, lagoons, and estuaries along the eastern and southern coast of the United States, where there is a broad range of food including crustaceans, mollusks, and aquatic plants. It has been exterminated in many areas as a result of being collected for food.

The genus of box turtles (genus *Terrapene*), of eastern and southern United States and Mexico, have a plastral hinge which allows the shell to close completely. Most box turtles live on land, the Carolina box turtle *T. carolina* in moist deciduous forests and grasslands, whereas the ornate box turtle *T. ornata* prefers the drier, sandy landscapes of the prairies. Only the Coahuilan box turtle *T. coahuila* is strictly aquatic; it occurs only in the Cuatro Ciénegas basin in northern Mexico. Blanding's turtle *Emydoidea blandingi*, with its plastral hinge, resembles the box turtles but with its long neck also resembles *Deirochelys* species. Like true pond turtles it prefers larger bodies of

standing water in the central regions of the
United States, and it was for many years
erroneously included in the genus *Emys*.

The European pond turtle *Emys orbicularis* is
geographically and evolutionarily distinct. This
medium-sized pond turtle (up to 28 centimeters,
or 11 inches, shell length) is semi-aquatic and is
found in North Africa and in most of Europe,
eastwards to the Aral Sea. Although it eats both
plant and animal material, the vast majority of its
diet is meat.

The last genus of New World pond turtles,
Clemmys, contains four species. The best known
and one of the smallest freshwater species is the
spotted turtle *C. guttata*, which lives in small pools
and ditches. The largest of the four is the wood
turtle *C. insculpta*, which has a shell length of
23 centimeters (9 inches) and lives in deciduous
forests of eastern North America, where it is
largely independent of water.

Land tortoises
Strictly speaking, the term "land tortoises" applies
only to members of the family Testudinidae. Land
tortoises are found in Europe, Africa, Asia, and all
of the Americas. The best known are the
European species in the genus *Testudo*, especially
the Greek or Hermann's tortoise *T. hermanni* and
the spur-thighed tortoise *T. graeca*, which occur
almost continuously throughout the countries
bordering the Mediterranean Sea. Two additional
species have more limited distributions: the
marginated turtle *T. marginata* is found only in
Greece and Sardinia, while the Egyptian tortoise
T. kleinmanni, the smallest species (up to

15 centimeters, or 6 inches, shell length) occurs
from Libya to Israel. Depending on the latitude at
which they occur, these tortoises have a short or
long hibernation period during which they bury
themselves in the ground. They depend heavily
on seasonal supplies of fresh herby plants, for all
land tortoises are predominantly vegetarian,
although they will eat insects, worms, and
mollusks, and even carrion and the dung of
hoofed animals. Greek and spur-thighed land

▲ *Ornate box turtles* Terrapene ornata
*mating. Copulation presents special
difficulties for land turtles with high,
strongly domed carapaces. Several
turtles have evolved spurs and other
aids for enabling the male to stay in
position; the box turtles of North
America have movable shells, often
used by the female as a clamp to hold
her mate's hind feet in position as he
falls on his back to mate.*

Stan Osolinski/Oxford Scientific Films

Hellio & Van Ingen/NHPA

◄ *A European pond turtle* Emys
orbicularis *feeds on an introduced
sunfish. An omnivorous and semi-
aquatic species, this turtle feeds mainly
on frogs, worms, mollusks, fishes, and
even rodents and small birds. In the
northern parts of its range it buries
itself in mud to hibernate for much
of the winter.*

tortoises have been valued as pets for many years, but this trade has threatened the survival of numerous southern European and North African populations.

The richest place on Earth for land tortoises is Africa. The genus *Geochelone* predominates, with the African spurred tortoise *G. sulcata* of the Saharan region reaching 75 centimeters (30 inches) shell length and 80 kilograms (175 pounds) in weight. Like the gopher tortoises of America this species survives dry periods by burrowing deeply into the soil. In hot, dry regions all land tortoises are active only in the morning and late afternoon. During the heat of the day they spend their time resting in the shade of shrubs and trees or in burrows in the earth. If exposed to the heat of the midday sun the bulky body, covered with horny plates and a massive bony shell, would quickly overheat and the tortoise perish.

The leopard tortoise *G. pardalis* is widely distributed in the southeast of the continent and is regarded as one of the typical animals of the savanna. This species has the longest incubation time for its eggs: up to 460 days. The eggs of all land tortoises have a calcified shell which is resistant to damage and rapid dehydration. To enable them to dig a nesting chamber in the often-hard ground, female land tortoises may release urine and water (stored in their anal sac) to saturate the ground and make digging easier.

Also widely distributed in the dry regions of Africa is Bell's hinge-back tortoise *Kinixys belliana*. This and two other species in the genus have a transverse hinge on the back of the carapace which allows the rear part of the shell opening to be completely closed as protection from many predators—although it is not safe from the bite of the hyena, which is able to completely penetrate the shell. The two other species live in rainforests of Central and West Africa, feeding mostly on worms, snails, and insects on the moist forest floor.

Southern Africa has at least two genera of land tortoises with restricted ranges. The South African land tortoises, *Psammobates*, with shell lengths of 14 to 24 centimeters (5½ to 9½ inches), depend on specific food plants. The smallest land tortoises are five *Homopus* species (9 to 15 centimeters or 3½ to 6 inches, shell length) found in dry regions of southernmost Africa. One of these, *H. bergeri*, was thought to be extinct until recently rediscovered in a very small area.

Another species recently rediscovered after almost 90 years is the most remarkable land tortoise of East Africa, the African pancake tortoise *Malacochersus tornieri*. Hatchlings have a closed and arched shell, but the adult tortoise is not only very flat but also very soft. Between the bony plates in the shell there are open spaces, which increase in proportion as the tortoise grows larger (to 17 centimeters, or 6½ inches, shell length). The horny plates covering the shell develop normally and reveal little of the reduced bony layer beneath. This tortoise lives in the mountains of East Africa and is an able climber. With its flexible shell and by inflating its body with air, it is able to wedge itself in rock crevices to avoid being pulled out by predators. Females

▼ *One of the most beautifully marked tortoises, the radiated tortoise* Asterochelys radiata *(below) belongs to a genus with only two species, both confined to Madagascar, and both endangered. The appropriately named pancake tortoise* Malacochersus tornieri *(below, right) of east Africa is an unusual species, well adapted to its rocky habitat. The shell is not only remarkably flat, but is flexible as well, allowing the tortoise to squeeze into narrow crevices, and wedge itself in by inflating its lungs and expanding its shell.*

lay one or two eggs several times each year, and with this small reproductive rate the species is especially vulnerable to collection for the pet trade.

There are four land tortoises endemic to Madagascar. Best known is the radiated tortoise *Asterochelys radiata* found in dry regions of the island's southwest. Unfortunately the local people regard it as a delicacy. In the mid-west of Madagascar lives the Angonoka or northern Madagascar spur tortoise *A. yniphora*, whose males have a long spur on the plastron which is used to drive off rivals during the mating season, rather as antlers are used in many mammals. Similar spurs are found in males of the South African bowsprit tortoise *Chersina angulata* and the American gopher tortoises (genus *Gopherus*).

Two small species are found in the south of Madagascar: the Malagasy spider tortoise *Pyxis arachnoides*, so-named because of the spider-web design on its yellow carapace, and the Madagascar flat-shelled spider tortoise *P. planicauda*, which grow to about 12 centimeters (4¾ inches) shell length. For most of the year they are inactive, living in burrows underground, and when the rainy season begins they have only a few weeks to feed, mate, and lay eggs.

On the Seychelles, the Mascarene Islands (Mauritius and Réunion), and Aldabra live the largest land tortoises on Earth. The Aldabra tortoise *Aldabrachelys elephantina*, with a shell length of up to 130 centimeters (51 inches), is larger than its distant relative the Indefatigable Island tortoise *Chelonoidis elephantopus nigrita* from the Galapagos. Although these two giant species belong to different evolutionary lineages, their shells are structured in the same way: the bones of the shell have a honeycomb structure, which encloses many small air chambers. If the bony shell were solid it would be difficult for these giant tortoises to carry around such weight.

The nearest relatives of the Galapagos giant tortoise live on the South American mainland. The South American red-footed tortoise *Chelonoidis carbonaria* and the South American yellow-footed tortoise *C. denticulata* inhabit forests of tropical regions. With shell lengths of up to 80 centimeters (31½ inches) they are not significantly smaller than the smaller races of Galapagos Island tortoises, but because the shell is slender and elongated they never achieve the weight of Galapagos tortoises of similar shell length. A third species, the Chaco tortoise *C. chilensis*, lives in the grasslands of Argentina and Paraguay. Despite its scientific name, it does not occur in Chile.

North America is home to the gopher tortoises (genus *Gopherus*), from Florida to California and southwards to northern Mexico. The fifth species, *G. lepidocephalus*, was only recently discovered in the south of Baja California but may already be nearing extinction. The other four species, which are also endangered, fortunately receive statutory

Tui de Roy/Oxford Scientific Films

protection. Found from desert and semi-arid regions to moister woodlands, all gopher tortoises are burrowers which tend to spend the heat of the day below ground. They are especially active in the early morning and evening, when they feed on various plants, like all land tortoises finding food largely through a combination of smell and sight.

The Asian brown tortoise *Manouria emys* is something of a giant among the land tortoises of Southeast Asia, reaching 60 centimeters (23½ inches) shell length. This flat, plain brown tortoise occurs in rainforests, especially with open bodies of water, in which it likes to bathe and where it feeds on aquatic plants and animals. While other land tortoises bury their eggs in the ground, this species scratches together a mound of soil and leaves in which it buries its eggs. The nest may hold up to 50 eggs, and because of its elevation above the forest floor it is protected from flooding. A more colorful relative is *M. impressa*, of northern Indo-China. It has a much flatter shell, in brown or black patterned with bright yellow or orange, effectively camouflaging the tortoise on the leaf-strewn floor of the tropical forest. Three species of *Indotestudo* and various species of *Geochelone* are found in India. The Indian star tortoise *G. elegans* is one of the most beautiful of all land tortoises, with its dark striped patterning on each shield of the carapace. It lives in savanna and dune habitats on the Indian mainland and Sri Lanka.

▲ *Darwin's finches (genus Geospiza) picking ticks from a Galapagos giant tortoise. Despite the massive appearance of the enormous carapace, the shells of these large tortoises are internally honeycombed—presumably to reduce weight—and surprisingly fragile and prone to injury.*

▲ *Two colorful Southeast Asians (family Bataguridae): the spined or cogwheel turtle* Heosemys spinosa *(top) has remarkable spines, most prominent in young individuals. In addition to the physical deterrent they provide, the spines, combined with the turtle's coloration, help to conceal the animal in leaf-litter. The Malayan snail-eating turtle* Malayemys subtrijuga *(above) is an aquatic species of slow or still waters that feeds primarily on mollusks.*

Old World pond turtles

The Old World pond turtles (family Bataguridae) form the largest family of turtles, with approximately 21 genera. Apart from a single genus (*Rhinoclemmys*), in Central and South America, all are found in Europe, North Africa, or Asia. Not surprisingly there are many parallels with the family Emydidae (described earlier) in form, structure, and biology. Indeed species of the two families can be so similar that only finer anatomical differences can be used to distinguish between them, but fortunately, the families have separate geographic distributions (apart from the exceptions mentioned here).

River turtles have a distinctive solid shell and strong, fully webbed feet. As a general rule they are the largest of the pond turtles, the Malaysian giant turtle *Orlitia borneensis*, for example, having a shell length of 80 centimeters (31½ inches). These turtles are characteristic animals of the larger rivers of India, Indo-China, and Indo-Malaysia, including Borneo. Several of them, including *Callagur*, *Hieremys* and some *Batagur* species, are also found in brackish waters. They tend to be omnivorous when young but strictly vegetarian when adult.

The painted terrapin *Callagur borneoensis*, of southern Indo-China, Sumatra, and Borneo, often lays its eggs on the same coastal beaches where sea turtles make their nests. On hatching, the young first enter the sea and live there briefly until they make their way into estuaries and rivers. The painted terrapin is also remarkable for the distinctive color pattern adopted by males during the mating season: a red crown on the normally gray-olive head, almost silvery-white on the nape, and a distinctive brightening of colors on the carapace. This kind of seasonal color change is common in lizards but rare in turtles.

Apart from large river turtles the family Bataguridae also includes a number of small species, which are also mostly aquatic, leaving the water only to sunbathe or lay their eggs. Among these are the eyed turtles (genus *Morenia*) with two species in Bangladesh and Burma, three species of the Indian roofed turtles (genus *Kachuga*), the Chinese stripe-necked turtle *Ocadia sinensis*, the spotted pond turtle *Geoclemys hamiltoni*, which ranges through the whole of northern India, from Pakistan to Bangladesh, and the black marsh turtle *Siebenrockiella crassicollis* from Indo-China and the Greater Sunda Islands. Most of these aquatic turtles are largely vegetarian in diet, although there are some notable exceptions—for example, the Malayan snail-eating turtle *Malayemys subtrijuga*.

In contrast, the four species in the genus *Mauremys* are decidedly amphibious. The Caspian turtle *M. caspica* and Mediterranean turtle *M. leprosa* have an omnivorous diet and live in streams in arid regions and mountainous areas around the Mediterranean Sea. The streams tend to dry up completely during the summer, so the turtles migrate overland, often long distances, to find new sources of water. In extreme cases they are forced to bury themselves to avoid desiccation and wait until the next rain. Similar habits are found in the Japanese turtle *M. japonica*, which lives in the southern highlands of Japan.

Another amphibious group are the hinged tortoises of India and southwards through the Indo-Malayan archipelago to the Greater Sunda Islands. Some specialists classify them in two distinct genera, *Cuora* and *Cistoclemmys*; other people place them in a single genus, *Cuora*. All have a plastral hinge which allows the turtle to withdraw its head and limbs into a completely enclosed shell. Some species are truly amphibious and spend much of their time in water. Others, like the Indochinese box turtle *Cuora galbinifrons*, in the mountain forests of southern China and Vietnam, are essentially terrestrial and can survive long periods without water. This species has a brilliantly colored shell whose warm yellow, orange, and brown pattern provides excellent camouflage on the leafy forest floor.

Other turtles in the family are even more terrestrial. In southern China and Indo-Malaysia, in the same habitat as *C. galbinifrons*, lives the keeled box turtle *Pyxidea mohouti* and the black-breasted leaf turtle *Geoemyda spengleri*—the latter species extending to Sumatra, Borneo, and the islands of Japan. Both of these species have a serrated edge to the keels that extend back along the carapace, so as well as being camouflaged by the light and dark brown color, the serrated edges of the shell give the turtle a leaf-like appearance when still.

These various forest dwellers, which ecologically are "land turtles", have many other representatives. Probably the best known is the bizarre spiny turtle *Heosemys spinosa*, which when young has a circular shell edged with regular spines. In adults the spines are less conspicuous, except on the hind edges of the shell.

The only genus in this family that is found in the Americas is *Rhinoclemmys*, with seven to nine species in Central and South America. Some are terrestrial, others amphibious. All of them are more or less brilliantly colored, especially the head and sometimes the shell itself, which may have beautiful eye-shaped markings, as in the painted wood turtle *R. pulcherrima*, or spots, as in the Mexican spotted wood turtle *R. rubida*. All land turtles, both the Asiatic and the American, are omnivorous, although predominantly herbivores. They feed on fallen fruits, but generally avoid mushrooms and other fungi.

Big-headed turtle

The East Asian big-headed turtle *Platysternon megacephalum* may be closely related to the land tortoises (family Testudinidae) or the Old World pond turtles (Bataguridae), but sometimes is classified in a separate family, Platysternidae.

It lives in southern China and northern and central Indo-China. The big-headed turtle is rarely more than 20 centimeters (7¾ inches) in shell length. Its conspicuous, large and powerful head cannot be fully retracted under the shell, nor can the long tail with its large horny scales. The flattened shell, although of normal proportions and development, is too small to fully enclose the fleshy parts of the turtle. On either side of the bridge that links the carapace and plastron is an additional row of large shields, a primitive characteristic that also occurs in alligator turtles (snapping turtles) of North and Central America and the Central American river turtle.

The big-headed turtle lives in cool, fast-flowing mountain streams and, even though it is not a powerful swimmer, is well adapted to grasp and climb among the large boulders on the stream bottom. Specimens kept in captivity have drowned if the water in their tank is too deep, and others that have escaped from their aquaria have been found not on the floor but near the ceiling, having climbed up the curtains!

By day the big-headed turtle lies hidden among stones, but at night it emerges to hunt snails, crabs, and fish. With its beak-like jaws it can grasp its prey tightly, and with very strong jaw muscles it can even bite through the thick shells of its prey.

Its reproductive rate is low, and only one to two eggs per clutch have been recorded. However, in its natural environment the big-headed turtle has few predators, offsetting its low reproductive rate. The greatest risk to this species is from collectors for the pet trade.

◄ *The big-headed turtle, here represented by the North Vietnamese race* Platysternon megacephalum shiui, *is a unique Asian turtle. Found in cool mountain streams it is primarily aquatic but a poor swimmer, preferring to walk along the bottom. It is a good climber, however, and is occasionally seen basking on the lower branches of streamside bushes or trees. Its huge head cannot be withdrawn into the shell and is consequently heavily armored, as is the long tail.*

LIZARDS

AARON M. BAUER

Order Squamata
Suborder Sauria
17 families, 407 genera,
3,865 species

SIZE
Smallest Monito gecko
Sphaerodactylus parthenopion
of Virgin Gorda Island,
head–body length 17 millimeters (²/₃ inch), total length
34 millimeters (1¹/₃ inches);
weight 0.12 grams (⁴/₁₀₀₀
ounce).
Largest
Komodo monitor *Varanus
komodoensis*, head–body
length 75 to 150 centimeters
(2½ to 5 feet), total length
170 to 310 centimeters (5½
to 10¼ feet); weight 35 to
165 kilograms (77 to 364
pounds).

CONSERVATION WATCH
There are 42 species of
lizards listed in the IUCN
Red List of threatened
animals. They include the
following: Cape Verde giant
skink *Macroscincus coctaei*,
Gunther's gecko *Phelsuma
guentheri*, Culebra Island
giant anole *Anolis roosevelti*,
Anegada ground iguana
Cyclura pinguis, San Joaquin
leopard lizard *Gambelia silus*,
Hierro giant lizard *Gallotia
simonyi*, and St Croix ground
lizard *Ameiva polops*. The
Rodrigues day gecko
Phelsuma edwardnewtonii is
presumed extinct.

Lizards today occupy almost all landmasses except Antarctica and some Arctic regions of North America, Europe, and Asia. During the extinctions that occurred at the end of the Cretaceous period, 65 million years ago, lizards survived but dinosaurs and other large reptiles did not. Other surviving reptiles—turtles, crocodilians, and tuatara—have not evolved into as many different forms. Indeed, since the only other large group of living reptiles, the snakes, evolved from lizards, it may be said that more than 90 percent of living reptiles are the descendants of the early lizards. Probably fewer than 800 species of dinosaurs existed during the entire span of the group's existence (about 140 million years), compared with more than 3,800 lizard species existing today, regardless of the number of fossil species.

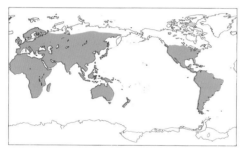

SMALL SIZE AND SUCCESS

Part of the reason for the much greater diversity of lizards is their small size; few living lizards exceed 30 centimeters (1 foot) in total length, and only a handful exceed 1 meter (3¼ feet). A given geographic region can support a greater diversity of smaller animals than it can of large animals. This is related both to the lower demand for food and other resources per individual and the diversity of microhabitats available to small animals. In a forest, for example, many habitats may be available to a typical lizard (the soil, leaf-litter, holes in tree trunks, treetops) and each might be occupied by a different lizard species. On the scale of a large dinosaur, however, the forest as a whole might represent just a single habitat. The partitioning of a habitat and its resources is well demonstrated in groups of lizards that have many species.

Also related (in part) to small size, most lizards have limited ability to spread geographically. Mountain ranges and expanses of water, such as rivers, lakes, and seas, are significant barriers for lizards and have promoted speciation (the evolution of new species), resulting in many forms that occur only in a small geographical area.

ORIGIN AND EARLY EVOLUTION

Lizards and their evolutionary offshoots—the snakes and amphisbaenians—constitute a group known as the Squamata, or "scale reptiles", and are representatives of diapsid reptiles. The lineage leading to lizards (the Lepidosauromorpha)

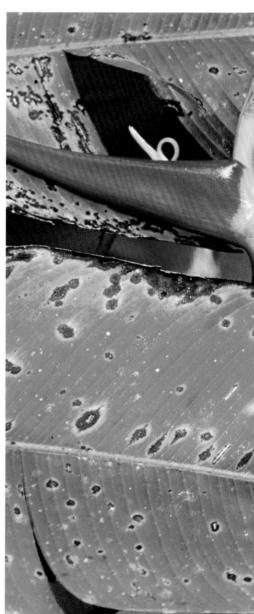

diverged from the other major lineage of diapsids (the Archosauromorpha including crocodilians, dinosaurs, and birds) during the Permian period, 285 to 245 million years ago. Lizards probably first appeared in the Triassic, 245 to 200 million years ago, but fossils definitely assignable to living lizard families are not known until the middle to late Cretaceous, about 120 to 100 million years ago. However, the presence of fossil lizards closely resembling living groups from the Jurassic, 200 to 145 million years ago, suggest that most of the body plans typical of the living lizard families were established almost 200 million years ago.

Lizards as a whole are closely related to the beak-heads or rhynchocephalians, represented today by two species of tuatara in New Zealand. Both orders share such features as the presence of a hooked fifth metatarsal bone, a part of the foot that acts as a heel in locomotion. Lizards, however, differ in skull structure, the skull often being highly mobile; when the quadrate bone (the point of attachment of the jaw) rotates forward, it pushes against the palate, which in turn lifts the lizard's muzzle. This modification of the feeding apparatus has also contributed to the success of lizards as a lineage. Lizards also differ from tuatara in that they have hemipenes, another feature they share with snakes (discussed later).

▼ Dwarfed by the heliconia leaf on which it sits, a male green basilisk Basiliscus plumifrons basks in a Costa Rican rainforest. The huge variety of different lizard species is largely a function of their small size: the smaller an animal is, the greater the range of microhabitats potentially available to it.

Michael and Patricia Fogden

◄ *Day-active reptiles inhabiting deserts need behavioral mechanisms to shed heat as well as absorb it. In the sand dunes of the Namibian desert in Africa, a sand-diving lizard Meroles anchietae lifts two feet alternately while balancing on the remaining two in order to reduce the transfer of heat from the hot sand— an activity that might well be termed "thermal dancing".*

Most lizards have eyelids and external ear openings (snakes do not), but these features reflect the retention of primitive characteristics, rather than the evolution of conditions peculiar to lizards. Thus, lizards can be defined as those squamates that lack the derived and often highly specialized features that define snakes and amphisbaenians.

THERMOREGULATION

Lizards, like other reptiles, do not have the ability to regulate their body temperatures physiologically. The metabolic heat they produce is minimal and is quickly dissipated to the environment through the skin. Under normal conditions, the body temperature of an exposed lizard quickly approaches that of the surrounding environment.

At low temperatures lizards face a dilemma. It is necessary to expose the body to sunlight in order to warm up. Yet at such temperatures the abilities of lizards to move effectively are reduced, and they are vulnerable to predation by mammals and birds, which are not as constrained by external temperatures. Because they cannot run well at low temperatures, some lizards show behavioral compensation of a sort. At very low temperatures lizards are unable to respond to most stimuli, but at slightly higher though still not ideal temperatures, lizards are often highly aggressive, relying on less energetically expensive bluffs or bites than on escape. Lizards that can endure tail loss (autotomy) are usually more prone to do so at low temperatures in order to distract predators and gain the additional time needed to escape.

It is common for lizards to begin their morning basking by exposing the head only or by basking in the shelter of a crack or crevice. Heat uptake through solar radiation may be enhanced by the darkening of the skin. This is accomplished by the hormonally controlled dispersal of melanin (the same substance found in human skin) within the cells of the skin. Once the animal has heated to an acceptable level it may then emerge to search for food or engage in other activities. Nocturnal lizards may either bask (usually at protected sites) or obtain heat through thigmothermy, the transfer of heat by contact with a warm surface. Many nocturnal geckos spend the daylight hours under bark or thin flakes of stone, absorbing heat by conduction from their surroundings.

Lizards must also cope with the stress induced by high temperatures. Although the preferred body temperatures for many lizards are high (up to 42°C, or 107°F, for sustained periods), temperatures only slightly higher may be fatal. Most lizards limit their activities to certain periods of the day. Those living in open tropical environments or in deserts may be active only in the morning and afternoon so that they avoid the heat of midday. Others remain active throughout the day but shuttle back and forth between sun and shade, regulating their body temperature to

◄ *(Opposite) Despite the tropical situation of the Galapagos islands, the seas surrounding them are cold. Insofar as thermoregulation is concerned, marine iguanas accordingly have the best of both worlds—living along the shoreline they can bask on rocks in the tropical sun, warming themselves quickly after foraging in the cool water.*

Gunther Deichmann/AUSCAPE International

▲ *A number of species of desert-inhabiting lizards evade the torrid extremes of heat at the surface by burrowing in the sand, like this burrowing skink* Lerista labialis *in Australia's Simpson Desert.*

▼ *Among lizards, two strategies have evolved to deal with sand: some species, like sandswimmers, burrow in loose sand and tend to have reduced limbs, while the inhabitants of the firmer, windward faces of sand dunes often excavate their burrows with large webbed feet, like this gecko* Palmatogecko rangei *of the Namibian desert munching its cricket prey.*

Anthony Bannister/NHPA

picking up heat radiated from the surface. *Meroles anchietae*, a lacertid lizard which inhabits the sand sea of the Namib Desert in Africa, is especially noteworthy in this regard. It raises a foreleg and the opposite hind leg simultaneously, balancing on the other two legs for a few seconds to allow the skin of the feet to cool. The lizard then alternates the lifting and lowering of the diagonal pairs for as long as it is exposed to uncomfortably hot sand.

For many lizards, both extreme heat and extreme cold may be encountered at different seasons. In temperate regions, winter is generally a period of inactivity, and metabolic rates may drop to minimal sustainable levels as temperatures remain so low that basking and foraging activity cannot take place. In warmer zones such as temperate to subtropical deserts, lizard activity in winter may be restricted to the middle of the day, when the temperature is suitable. Typically animals in such areas remain inactive during midday in the summer, when a lethally high temperature is likely to be encountered.

The length of the basking period is controlled hormonally by the pineal gland, a brain structure that lies beneath the roof of the skull and is often known as the "third eye". In many lizards, especially those that are primarily day-active, the pineal gland is not covered by bone but lies beneath a foramen (or window) in the skull, covered by a translucent scale.

within a degree or two of their preferred value. Many desert lizards avoid the excessive heat by burrowing, because at a depth of only a few centimeters the temperature may be drastically cooler than at the surface; and at a depth of 30 centimeters (11¾ inches) it may be as much as 35°C (95°F) cooler. Other lizards, such as geckos, remain in burrows or other protected sites during the day and only venture out at night when desert temperatures may be more equitable. Lizards may also orient the broadest part of their bodies into the wind in order to lose heat by convection. Still others move their bodies off the ground to avoid

DESERT ADAPTATIONS
Although lizards are found in almost all habitats, they are often thought of as desert animals. In countries such as Namibia in southwest Africa, and in Australia, where arid habitats predominate, lizards are especially numerous.

Deserts impose a number of challenges to lizards; high daytime temperatures must be endured, and low humidity and a lack of free water tend to result in dehydration. Water may be obtained entirely through the food consumed or may come from one of several other sources. In the coastal deserts such as the Namib in Africa and the Atacama in South America, fog originating over the ocean provides the water used by lizards and other animals and plants. A number of lizards living in arid zones are able to effectively use rain as a source of drinking water. Normally, light rains quickly soak into the dry desert soil and become unavailable to surface-dwelling reptiles, but at least two types of agamid lizards have developed a mechanism for salvaging such a resource: the Australian thorny devil *Moloch horridus*, and the Asiatic toad-headed agamid *Phrynocephalus mystaceus* have their scales arranged in such a way that water is channeled to the mouth by capillary action. A similar mechanism to collect fog moisture or dew may also function in these or other lizards.

Dehydration is combated by features that slow down the rate of water loss from the body. All lizards have a covering of keratin (a substance similar to that of human fingernails) which acts as an effective barrier to water. Lizards are also able to reduce water loss through excretion, because they are able to limit the amount of filtrate (and thus water) passed out of the kidney. The nitrogenous wastes produced by lizards consist mainly of uric acid, which is relatively insoluble and can be concentrated and stored with little need for dilution by water.

In arid regions with shifting sands, special demands are placed on lizards. While many lizards inhabit deserts, most occupy specialized microhabitats such as rock islands or clumps of vegetation. The open, shifting sands of dunes are especially challenging and have been exploited by a limited number of iguanids, geckos, lacertids, skinks, and girdle-tailed lizards. Two general strategies for living and moving in the dunes are employed. Most dune-dwellers, such as the sandfish *Scincus scincus* of Asia and North Africa and the sand lizard *Uma notata* of North America, utilize loose sand, which does not contain enough moisture to allow the formation of tunnels or burrows. The sand lizard generally buries itself for protection from predators and high temperatures, whereas the sandfish actually moves freely through the substrate in an action referred to as "sand swimming". Both animals have countersunk lower jaws to prevent sand from entering the mouth, and both have valved nostrils and modifications of the ear openings, also to exclude sand. In the case of the sandfish, smooth scales reduce friction as it moves through the sand, while in the sand lizard, small granular scales may perform a similar function. Enlarged scales form fringes at the borders of the toes in these and other ecologically similar lizards; the fringes increase the surface area of the feet for progression through sand or for initial burial by shimmying or lateral undulations of the body.

A much smaller number of lizards are specialists inhabiting the more compact, windward dune faces. These lizards build open tunnels and so lack features such as valves over their nostrils needed for sand burial. Instead their modifications are for sand excavation. The gecko *Palmatogecko rangei* has very large, webbed feet for exactly this purpose.

REPRODUCTION AND LONGEVITY

All bisexual lizard species (those with both males and females) exhibit internal fertilization. Lizards have a horizontally aligned vent, which is the exit of the cloaca, a common vestibule for the digestive and uro-genital tracts. Male lizards possess a pair of intromittent organs, the hemipenes. When not in use the hemipenes lie adjacent to the cloaca within the base of the tail.

Ken Griffiths/NHPA

▲ *Lizards exhibit the full spectrum of reproductive possibilities, from egg-laying to live birth, but egg-laying tends to be the rule. In some, eggs are laid in chambers excavated in the soil then covered over. Here an Australian eastern water dragon Physignathus lesueurii puts the finishing touches to her clutch.*

Austin James Steven/Bruce Coleman Limited

◄ *A hatching Seychelles green gecko Phelsuma abbotti patiently works to tear itself free of the leathery prison of its eggshell.*

During sexual activity one hemipenis is everted by the action of muscles and fills with blood. The fully everted structure may be relatively simple or highly complex, as in many chameleons, which exhibit folds and spines that differ from species to species. In copulation, which follows courtship behavior, only a single hemipenis is inserted into the female's cloaca, and the sperm travel along a groove in the hemipenis. Retraction of the hemipenis is accomplished by drainage of the blood sinuses and activation of retractor muscles that invert the structure as it is withdrawn. In at least several species of anoles, males tend to alternate which hemipenis is used when mated repeatedly.

Oviparity (egg-laying) is the predominant mode of reproduction among lizards, but viviparity (live-bearing) has evolved on many different

▲ *Kicking and biting, two rival male frill-necked lizards* Chlamydosaurus kingii *battle it out. The enormous frills are used to bluff the opponent into believing the owner is bigger than he really is.*

occasions (at least 45) and is seen in nine families. The Brazilian skink *Mabuya heathi* even has a mammalian-like placenta; the fertilized egg is minute and through gestation grows 38,000-fold, entirely as a result of maternal nourishment. In fact, most lizards fall on a continuum between live birth and oviparity, as about half of the developmental period, on average, occurs within eggs retained in the female's body. In most instances, embryos of viviparous reptiles obtain their nutrition from yolk, and live birth is an extension of egg retention. The placental condition of *M. heathi* is one of only a few instances of extensive maternally provided nutrition in lizards.

Live-bearing has both advantages and disadvantages. On the one hand, viviparous lizards can protect their young throughout the prenatal development period. They are also able to regulate the temperature of the developing young at a level greater than that provided by a nest, decreasing the time required for development. The evolution of viviparity in lizards is probably related to the occupation of cold climates, which would strongly favor these advantages. On the other hand, the added weight and bulk of embryos must be carried by the female for an extended period of time. This may substantially reduce the speed and agility of the adult and make her more prone to predation. Increased basking and thus exposure in gravid females, as seen in some Australian skinks, may also increase the risk of predation. Likewise, the need to carry the young for the full term may

prevent the production of multiple clutches in a given season. In geckos of the genus *Naultinus*, for example, gestation may last as much as eight months, limiting an individual to only one litter (of two young) every one or two years. Lifetime reproductive output may thus be quite low for live-bearers, but survival rates are higher.

Although lizards do not live as long as some other reptiles, some species have impressive longevity. A slow worm, *Anguis fragilis*, lived in captivity for more than 50 years, and some wild populations of geckos are known to include individuals over 20 years old. On the other hand, a few, mostly very small species may live only one or two years. Life-spans of more than ten years are probably exceptional for most species in the wild, although large varanids and helodermatids may commonly exceed this—the Komodo monitor *Varanus komodoensis* has a life-span of as much as 50 years. Most reptiles continue to grow long after reaching sexual maturity. Nonetheless, studies of very old individuals suggest that growth ceases at some point, and growth rates typically slow down long before this point is reached.

COMMUNICATION

Many lizards spend much of their life as solitary individuals, yet at times they communicate with one another through a set of highly stereotyped behaviors. These include aggressive behaviors directed by males at other rival males, and courting behaviors between the sexes. Other types of social interactions are also probably important in some species, but our knowledge of these is rudimentary. Males of virtually all iguanids and agamids engage in patterns of head-bobbing and/or push-ups as a means of communication. Comparable information is conveyed by movements of the tail in other lizards. The day-active semaphore geckos (genus *Pristurus*) have a laterally compressed tail that is used to signal members of its own species across significant distances in their relatively open habitat. Many reptiles are brightly colored, and such patterns are frequently important cues in mate recognition and selection. In some male lizards, particularly iguanians, color displays may be enhanced by the erection of crests or dewlaps. Members of the genus *Anolis* show species-specific color patterns of the dewlap that are exposed when the animal displays. Females may exhibit bright colors as well. In the keeled earless lizard *Holbrookia propinqua*, females exhibit hormonally controlled color changes associated with their reproductive cycle. Such changes may signal that they are receptive to males and are usually accompanied by changes in female behavior.

Autarchoglossan lizards rely on color as well. The orange head color of breeding males of the *Eumeces* skinks elicits aggressive behavior from males of the same species. Skinks and other autarchoglossans, however, also obtain important information about their environment on the basis of chemical cues. Skinks are able to recognize potential mates and rivals, as well as food items

and predators, from pheromones or other substances produced by these animals. Chemical information is obtained through tongue-flicking; the autarchoglossan tongue is especially well designed to collect chemical cues and to deliver them to the vomeronasal organ, a chemical-sensing structure in the roof of the mouth.

Many lizards, especially males, bear glandular structures on scales of the thighs and/or around the vent; such glands reach the surface by way of prominent pores. These are found on representatives of all families except the skinks, chameleons, and anguimorph lizards. While the function of these glands remains poorly understood, they appear to play a role in sexual behavior. Although much lizard communication is visual or chemical, a few lizard groups rely, in part, on vocalizations. The most spectacular voices are possessed by geckos, but representatives of nine other families have been known to vocalize. At least in geckos and some lacertids, calls seem to have particular value in night-time communication, when vision is poor. Whereas some gecko species are mute, or produce only small squeaks, others are highly vocal. The large tokay *Gekko gecko* barks loudly when disturbed and also produces a low warning growl. Other geckos actually call to attract mates. The bell geckos of southern Africa (genus *Ptenopus*) are the most accomplished at this. The mechanism of vocalization involves the passage of

▼ *Two flying lizards face-off on a rainforest log in Borneo. The prominent dewlap on the throat of these lizards (genus* Draco*) is used, like a semaphore flag, in social interactions between mates and territorial rivals, but the code has not yet been cracked.*

Jean-Paul Ferrero/AUSCAPE International

Leo Meier/Australian Picture Library

▲ *Bedecked with horns, knobs, warts, and armored scales, the thorny devil or moloch is in appearance one of the most bizarre of the world's lizards. It inhabits arid regions of Australia, feeds mainly on ants, and has no close relatives. The various protuberances are arranged in such a way that dew or rain flows gradually into the crevices, and is then channeled to the corners of the mouth.*

▶ *With enormous extended frills, wide open mouth, and loud angry hisses, a frill-necked lizard warns off an attacker. This spectacular lizard inhabits tropical woodlands across northern Australia.*

air through the larynx, which may be equipped with vocal cords. Like frogs, lizards produce recognizable calls specific to their own species.

THREE MAJOR LINEAGES

Modern lizards are representatives of three major lineages, the Iguania, Gekkota, and Autarchoglossa. Although each has a variety of morphological features common to its component families, the diversity in body form and biological characteristics within any one of the lineages is staggering.

THE IGUANIANS

The iguanians include the agamids, chameleons, and iguanids, and probably diverged from other groups early in the history of lizards, although no fossils have been found from before the late Cretaceous. The iguanians are fully limbed and are visually oriented. Their tongues are large and often used in capturing prey and other food gathering. Prey are most often ambushed rather than pursued. Crests, fans, and dewlaps are common in the group.

Agamids

The family Agamidae is a group of about 325 species in 40 genera living throughout the warmer regions of the Old World (except Madagascar and most oceanic islands). In form they closely resemble the American iguanids, to which they are closely related. One of the distinguishing characteristics of this family is the presence of acrodont teeth—these are unsocketed teeth borne on the rim of the jaws. Chameleons also have this type of dentition. Most other lizards have pleurodont teeth—also unsocketed, but on the inner face of the jaw bones. Agamids are diurnal (day-active), visually oriented lizards, and all but a few species are oviparous (the female produces eggs that hatch outside her body). Some species live on the ground, others are tree-dwellers, and many others make their home among rocks. All are fully limbed and lack fracture planes in the tail, so the tail is never shed.

The largest radiations of agamids occur in

Michael and Patricia Fogden

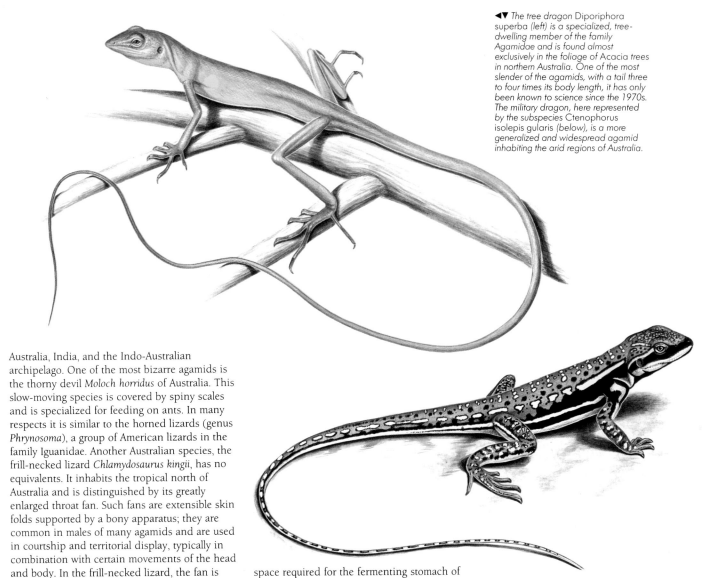

◄▼ The tree dragon Diporiphora superba (left) is a specialized, tree-dwelling member of the family Agamidae and is found almost exclusively in the foliage of Acacia trees in northern Australia. One of the most slender of the agamids, with a tail three to four times its body length, it has only been known to science since the 1970s. The military dragon, here represented by the subspecies Ctenophorus isolepis gularis (below), is a more generalized and widespread agamid inhabiting the arid regions of Australia.

Australia, India, and the Indo-Australian archipelago. One of the most bizarre agamids is the thorny devil Moloch horridus of Australia. This slow-moving species is covered by spiny scales and is specialized for feeding on ants. In many respects it is similar to the horned lizards (genus Phrynosoma), a group of American lizards in the family Iguanidae. Another Australian species, the frill-necked lizard Chlamydosaurus kingii, has no equivalents. It inhabits the tropical north of Australia and is distinguished by its greatly enlarged throat fan. Such fans are extensible skin folds supported by a bony apparatus; they are common in males of many agamids and are used in courtship and territorial display, typically in combination with certain movements of the head and body. In the frill-necked lizard, the fan is disproportionately large and conspicuous. It is used in social interactions between individuals and in displays to deter predators. The frill-necked lizard is one of several agamids whose powerful hind legs and long tail facilitate running bipedally—usually over short distances as part of a rapid escape behavior—but the frill-necked lizard may also stand on its hind legs and spread its magnificent frill in a defensive bluff attack.

Several agamids, such as the mastigures (genus Uromastyx), which are desert-dwelling lizards of western Asia and North Africa, reach large size. Mastigures are stocky, terrestrial lizards that use their short powerful legs to excavate burrows. They are strikingly similar in appearance to the chuckwalla Sauromalus obesus, an American desert iguanid. Mastigures and the chuckwalla are vegetarians. At least in iguanian lizards, large size seems often to be associated with the additional

space required for the fermenting stomach of plant-eaters. Mastigures show a further specialization for plant-eating in having the blade-like cutting surfaces instead of more-typical teeth in the front of the jaw, whereas the back teeth are situated for crushing.

Another desert-dwelling group are the toad-headed agamids (genus Phrynocephalus), small lizards with fringed toes used for moving over sand, and for burial and excavation. They are insect-eaters and live in central Asia and the Middle East. In social interactions with one another, toad-heads coil the tail upward over the head. Although most toad-headed agamids are oviparous, a few species in mountainous regions retain eggs in the body until hatching, so that they give birth to live young.

The agamids of Asia include many tree-dwelling species of the forests. Most have relatively compressed bodies. One of these, the earless agama *Cophotis ceylanica*, is the only tropical agamid known to give live birth. Only about five young are born, in keeping with the relatively low number of eggs (1 to 27) typifying oviparous members of the family. This is a slow-moving, chameleon-like agamid with a strongly prehensile tail and a fleshy or scaly nasal ornament, which is believed to help it recognize individuals of its own species. Another Asian species, the water dragon *Hydrosaurus amboinensis*, is also semi-arboreal, but lives in close association with streams. It has a strongly compressed tail and toe-fringes that increase the surface area for swimming efficiency. The water dragon is the largest agamid, with a total length of about 1 meter (3¼ feet). The most specialized of the arboreal agamids inhabit the forests of Southeast Asia; the "flying dragons" (genus *Draco*) not only climb but glide through the air as well.

Relatively few species of agamids occur in Africa, but they include several tree-living species and numerous rock-living forms. The Namibian rock agama *Agama planiceps* is a specialist of boulder piles or rocky outcrops. Like many other agamids it is a sun-lover and spends much time basking and displaying to other members of its species. As in many agamids, there is marked color difference between the sexes (known as sexual dichromatism). Males are deep blue or purple with an orange or red head and tail. Females and juveniles have brownish or grayish bodies with yellow marks on the head and orange shoulder spots. Color change is common in some species; for example, males of the Indian bloodsucker *Calotes versicolor* turn bright red following victory in battles with rivals.

Chameleons

The family Chamaeleonidae is the most distinctive of all lizard families. Nonetheless, chameleons have several anatomical and behavioral features that clearly link them to the Agamidae, from which they probably evolved; for example, like agamid lizards they have acrodont dentition—unsocketed teeth borne on the rim of the jaws. Chameleons are distributed throughout much of Africa (except the Sahara) and extend eastward to India and north to Spain. There are 128 species in six genera, all strikingly similar in body form. Ranging in size from tiny species of *Brookesia* less than 25 millimeters (1 inch) in total length, to giants of more than 550 millimeters (21½ inches) in total length like *Chamaeleo oustaleti*. All have laterally compressed bodies, prehensile tails, prominent, independently mobile eyes, and partially fused toes.

Another characteristic feature of chameleons is their projectile tongue, used to capture small prey which form the bulk of their diet. Insects and other arthropods are taken by most species, although small birds and mammals may be eaten by larger types. Each turret-like eye can be moved independently of the other, so a chameleon has the excellent depth perception necessary for

▼ The sail-tailed water lizard, or soasoa *Hydrosaurus ambionensis* of Sulawesi is a large semi-aquatic agamid that basks on rocks and branches at the water's edge, retreating to the water when pursued. Aided by fringes on the toes of its hind feet, it can run on the surface of the water for some distance before sinking and swimming away.

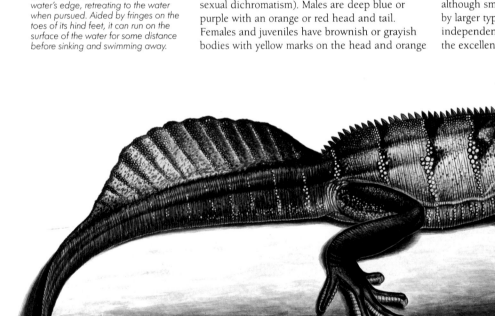

◀ *A common or flap-necked chameleon* Chamaeleo dilepis. *The chameleons are predominantly an African group; their three most distinctive external characteristics are their prehensile tail, partly-fused opposable toes, and their extraordinary turreted, independently rotating eyes.*

▼ *Two Johnston's chameleons* Chamaeleo johnstonii *mating. The males of several species of chameleons have conspicuous horns, used in species recognition and perhaps in battles over mates or territories.*

aiming its extremely long tongue and for judging distance in dense vegetation.

Most chameleons live in humid forest areas, and they are especially numerous in the rainforest belt of eastern Madagascar and highland forests of east Africa and the Cameroons. But they are also successful in Mediterranean climates and even in deserts. Their feet are equipped with opposable sets of partially fused toes, a condition called zygodactyly, which is an adaptation for grasping and is used by the tree-dwelling species for climbing. The limb girdles are also highly modified to permit the slender limbs to be held close to the body, as required by the small diameter of perches occupied by the lizard. Their prehensile tail acts as a fifth limb and unlike the tails of many other lizards, they are never shed. Although most chameleons are arboreal and move clumsily on the ground, arid-zone species such as the Namaqua chameleon *Chamaeleo namaquensis* spend most of their time on the ground, as do many of the stump-tailed chameleons (genus *Brookesia*) of Madagascar. The latter are tiny and drab in color, resembling dead leaves, and the tail is reduced to a small protuberance. Despite the apparent difficulty in walking on flat surfaces, both types of terrestrial chameleons retain the zygodactylous condition.

Males of virtually all species are territorial. Jackson's chameleon *Chamaeleo jacksonii* is one of many species in which males possess horns or other ornamentation on the head—devices that play a role in species recognition between the sexes and may be used in intense combat among the males. Breeding males of the South African dwarf chameleons (genus *Bradypodion*) may take on incredible colors when presented with another

► *A tongue-lashing, chameleon-style. A Mediterranean chameleon* Chamaeleo chameleon *scores another hit. The tip of the chameleon's remarkable tongue often exceeds 5 meters (about 16½ feet) per second as it speeds toward the unsuspecting fly.*

CHAMELEON TONGUES

Lizard tongues vary greatly in their structure, enabling scientists to use them as evidence of relationships between different groups. Those of autarchoglossan lizards (and snakes) are generally forked and often very slender, and are highly efficient as chemical-detection organs, collecting molecular cues about the environment. In geckos and flap-foot lizards the tongue is fleshier and serves in lapping up water or nectar; it is also used for "spectacle-wiping", cleaning the surface of the transparent scale over each eye. Some lizards, such as the Australian blue-tongued skink *Tiliqua scincoides*, have a strikingly colored tongue which

is used in defensive displays. Only in one lineage of lizards, the Iguania, does the tongue play an active role in grasping and bringing food to the mouth; in iguanids and agamids this ability is moderately well developed, but in chameleons it is developed to an amazing degree.

The tongue of all lizards is supported by the hyoid skeleton; the tip of the tongue is supported by one rod of the hyoid skeleton, the lingual process. Complex musculature moves the tongue as a whole, or certain parts of the tongue, as needed for swallowing, seizing prey, or other functions. In chameleons the tongue musculature

Stephen Dalton/NHPA

is exceedingly complex. During feeding, chameleons first judge the distance to their prey, then orient themselves to aim the tongue correctly. The mouth is opened and the hyoid apparatus is pulled forward, causing the tongue to protrude. An accelerator muscle, which extends the tongue out of the mouth, lies near the tip of the tongue. When activated it contracts, exerting a force around the lingual process of the hyoid. This serves to project the tongue tip out of the mouth (much as squeezing a wet bar of soap will cause it to shoot from your hand), bringing with it the retractor muscle, which at rest lies coiled around the more basal portion of the lingual process. The length of the retractor, usually equal to or greater than the head–body length of the chameleon, determines the distance that the tongue may be extended, but active firing of this muscle and/or the accelerator may allow the chameleon to extend the tongue a lesser distance. The outer surface of the tongue tip bears a glandular pad which contacts and grasps the prey by a combination of wet adhesion and muscular activity. Once the prey has been contacted the retractor muscle fires, pulling the entire apparatus back into the mouth. Finally, the entire hyoid is retracted, and the mouth closes. The tongue tip may exceed speeds of 5 meters (16½ feet) per second during the projection phase, requiring less than one-hundredth of a second for the tongue to reach the prey.

▲ *Chameleons are famed for their highly developed camouflage, but not all color changes are in response to the environment. The Knysna dwarf chameleon Bradypodion damaranum (top) is predominantly green at rest, but males develop the bright color pattern illustrated as a threat display to other males. The Malagasy chameleon Chamaeleo lateralis (above, right) is also usually dull green in color when resting among the foliage but takes on the striking disruptive pattern shown here when disturbed.*

male and engage in hissing, posturing, and biting. These species may breed several times in a year, giving birth to live young, a trait shared with several *Chamaeleo* species.

Most species lay eggs, however, with the number per clutch largely dependent on body size. During mating males often bite females. The hemipenes of many chameleons are highly complex and are species-specific in their architecture. Females are capable of storing viable sperm for long periods. Meller's chameleon *Chamaeleo melleri*, one of the largest species, lays up to 70 eggs, and clutches of 30 to 40 are the rule for many species. In most cases, females descend to bury the eggs in the ground or in rotting logs or other moist protected spots.

One well-known feature of chameleons is their ability to change color. This trait is shared by a variety of other lizard groups but is especially well developed in representatives of the subfamily Chamaeleonidae. The ability to blend into their surroundings, which is enhanced by their leaf-like shape, is especially useful in slowly stalking intended prey. Color change is also associated with temperature and behavior. At night when temperatures are cool, chameleons blanch and show up like white leaves on the ends of twigs. By sleeping in such spots the lizards are protected from predators such as snakes, whose weight could not be supported by the slender branches. An alternative strategy, employed by small and mid-sized chameleons, is to drop to the ground and remain motionless when disturbed. Among the leaf-litter they are virtually impossible to see.

Iguanids

Iguanid lizards (family Iguanidae) are New World relatives of the agamids and chameleons. They are distinguished from these other iguanian groups by their pleurodont dentition—unsocketed teeth on the inner face of the jaw bones. More than 550 species in 54 genera are distributed from southwestern Canada to the southern tip of South America; another five species occur on the oceanic islands of Fiji and Tonga in the Pacific, and seven inhabit Madagascar, where agamids are absent. Iguanids are highly diverse in body form and structure and have one of the greatest size ranges among all lizards. Most of them lay eggs, but several groups have evolved viviparity (giving birth to live young) independently of one another. All species are day-active. Recently some zoologists have provided evidence that instead of being only one family, the Iguanidae should be broken up into eight different families.

Some of the most familiar iguanids are the iguanines. These include the green iguana *Iguana iguana* of Central and South America, which reaches a total length of more than 2 meters (6½ feet). A female can produce up to 71 eggs in a clutch. The green iguana is essentially arboreal, but egg-laying occurs on the ground and hatchlings emerge en masse from buried nests. The genus *Iguana* has close relatives, the terrestrial rhinoceros iguanas (genus *Cyclura*) throughout the islands of the Caribbean, but some of these have become extinct because of the actions of humans. The lizards in the iguanine group share the feature of being vegetarian. (As described later, most other iguanids, which are considerably smaller in size, are insect-eaters although many will prey on other animals if available.) Closely related to the mainland and Caribbean iguanas are the iguanas of the Galapagos islands, genera *Conolophis* and *Amblyrhynchus*. These giant lizards must have arrived on the Galapagos as a result of over-water rafting from South America, because the volcanic Galapagos have never been attached to another landmass. Charles Darwin and countless other visitors to the islands were impressed by the marine iguana *Amblyrhynchus cristatus*, a black-skinned lizard that feeds almost exclusively on marine algae, for which it dives. It is the only living lizard dependent on the marine environment. In order to rid itself of excess salt ingested as a result of its feeding, the marine iguana must excrete concentrated salt in the form of crystals from a nasal salt gland. Another problem faced by this lizard is one of temperature regulation, for although the Galapagos islands straddle the Equator, the equatorial undercurrent brings cold waters to the shore. When not

foraging, the marine iguana must spend much of the time basking on the rocks near the sea in order to raise their body temperature to the point where digestion can occur and further foraging can take place. The black coloration of the lizard facilitates rapid heat absorption. Thousands of individuals may bask at a given site.

Banded iguanas (genus *Brachylophus*) occur in Fiji and Tonga. These too must have arrived over

▼ Most lizards are carnivorous, feeding especially on insects. But in its diet of marine algae, the marine iguana Amblyrhynchus cristatus of the Galapagos islands presents a conspicuous exception. Here a male grazes in a tide pool.

Mark Jones/AUSCAPE International

◀▼ The marine iguana (left) is the only marine lizard. Normally black in color, for rapid re-warming after emergence from the cool water, this male is in full breeding dress. Like the marine iguana, the Fijian crested iguana Brachylophus vitiensis (below) probably evolved from mainland South American iguanas that accidentally drifted across the Pacific on floating vegetation.

Michael and Patricia Fogden

▲ A male ground anole Anolis humilis in a Central American rainforest. Mostly arboreal, male anoles have a conspicuous sail-like fin, called a dewlap, on the throat, which is extended in display.

water from the Americas, a journey of more than 9,000 kilometers (5,600 miles).

Anoles and their relatives are primarily a South American and West Indian group, although one species, the green anole *Anolis carolinensis*, reaches well into the southeastern United States. Anoles are a huge group, accounting for 50 percent of all known iguanids. Their radiation has been particularly spectacular in the West Indies, where their ecology has been well studied. Anoles bear pads on their toes, similar to but slightly less well-developed than those of geckos. Most are effective climbers, but species in any one region tend to segregate according to their preferred perch heights, ensuring that food and other resources are available to each species. For example, in Puerto Rico, *Anolis cuvieri* occupies the crowns of trees, *A. evermanni* prefers tree trunks, *A. pulchellus* is active on bushes and in grass, and *A. cooki* may be found on the ground. Size, diet, shade tolerance, and other features further segregate the

anole species in any given area. Male anoles display using their brightly colored dewlap, and successful matings result in the production of a single egg. Many species have more than one clutch a year.

The sand lizards, horned lizards, fence lizards, and their allies constitute a particularly diverse group of iguanids, widely distributed from extreme southern Canada to Panama. The most obviously distinctive genus is *Phrynosoma*, the horned lizards. These are flattened, with round bodies and spiny dorsal scales. They bear prominent backward pointing "horns" on the head, which are used in defense. Horned lizards are ant-feeding specialists, reminiscent of the thorny devil of Australia. Horned lizards dig burrows or cover themselves with sand. Four species give birth to live young, whereas others lay eggs. The short-horned lizard *P. douglassii* occurs in shortgrass prairie, sagebrush country, and even open forests at moderate to high

◄ The color and shape of the dewlap in anoles varies from species to species. This is a displaying male Cuban brown anole Anolis sagrei sagrei.

elevations in the United States and in the more arid regions of southwestern Canada; it is viviparous, giving birth to 5 to 48 young.

Certain horned lizards are capable of employing a rather specialized defense, in which blood is squirted from the eyes. This is accomplished by the lizard restricting the blood flow out of its head until mounting pressure ruptures delicate capillaries in the eyes. The thin stream of blood may travel up to 1.2 meters (4 feet) and may act as a deterrent if it contacts the eyes or mouth of a mammal predator.

Among several groups of sand lizards, Uma is the most specialized genus. These animals are active on the surface but bury themselves in sand dunes to escape predators and to sleep. Some of the sand lizards are capable of withstanding body temperatures as high as 45° to 47°C (112° to 116°F) and have highly complex posturing behaviors to regulate their body temperature. The zebra-tailed lizard Callisaurus draconoides shows sexual dichromatism (the males exhibiting bright blues patches on the belly and flanks that are absent in the females), and in most sand lizards pregnant females take on a color pattern indicative of their reproductive state. The zebra-tailed lizard is an especially fast runner and is often seen to run on its hind legs. It uses the black and white banded underside of the tail as a signaling device. The genus with the greatest number of species is Sceloporus, the fence lizards and their allies. These are a widely distributed

John Cancalosi/AUSCAPE International

◄ The regal horned lizard Phrynosoma solare. Members of this North American group have evolved an exceptionally bizarre defense against predators: when under threat they can restrict blood flow from the head until mounting pressure ruptures small blood vessels in and around the eyes, resulting in a spurt of blood that may leap a meter (3¼ feet) or more.

group with broad ecological preferences. Some, like the large *S. magister*, occur in rocky desert areas, while others, like the eastern fence lizard *S. undulatus*, favor forest edges and other sunny but not necessarily dry habitats. Typical for the family, eastern fence lizards are sexually dichromatic, with males having blue pigmentation on the sides and underside. These are accentuated during head-bobbing and other sexual and aggressive behaviors. Of about 70 species of fence lizards, 40 percent give birth to live young; this trait seems to have arisen several times within genus *Sceloporus* and mostly characterizes those species that occur at high elevations.

Basiliscines are a small group of moderate to large forest-dwelling iguanids. Their range centers on Central America, extending north to Mexico and south to Ecuador. The best-known lizards in this group are the basilisks (genus *Basiliscus*).

These are large green or brown lizards with prominent crests and dewlaps and long, powerful legs. The common basilisk *B. basiliscus* bears not only head ornamentation but also a sail-like crest on its back and tail. Basilisks are capable of running on their hind legs, and may even run (for brief periods) on water. Their toes bear a fringe of scales on their lateral surface. At rest, this fringe folds over the axis of the toes, but in water the fringe is forced upward and serves to increase the surface area of the foot. When threatened or in search of food, a basilisk may run bipedally toward the water, where the combined effect of the fringes and its momentary high speed allows the animal to run on the surface of the water for a short distance before sinking. The tail is essential as a counterbalance in bipedalism, and basilisks are not able to shed their tail when caught by a predator, although most other iguanid lizards

▶▼ *The South American equivalent of the Asian sail-tailed lizards, the basilisk or Jesus lizard* Basiliscus basiliscus *(right), also sports a large fin-like crest, and can run bipedally for a short distance across the surface of the water aided by fringes on its toes. The collared lizard* Crotaphytus collaris *(below, right) of arid southwestern North America can run for a distance on its hind legs at top speeds.*

retain this ability. Basilisks are oviparous, but their close relatives, the tree-dwelling helmeted iguanids (genus *Corytophanes*), are another of the many iguanid types that give birth to live young. Helmeted lizards are laterally compressed and have a large casque and crest on the head, which increases the apparent size of the head, thus serving as a deterrent to potential predators.

Crotaphytines are desert and plains-dwelling lizards of western North America. Two very closely related groups are included, the leopard lizards (genus *Gambelia*) and the collared lizards (*Crotaphytus*). They all have large heads and long limbs and tails and are exceptionally fast runners. The larger leopard lizards feed mainly on other reptiles, consuming a large number of smaller iguanid lizards.

Hoplocercines are lizards of tropical South America. The Brazilian spiny-tailed iguanid *Hoplocercus spinosus* uses its tail as a defensive weapon when attacked. The spiny-tailed lizard is not truly a burrowing animal, but it digs shallow retreats in the soil and may use its tail to block the entrance and prevent a predator from reaching it. Spiny tails occur in several other groups of iguanids as well, such as the tropidurines.

Tropidurines are a South American group, although some species live in the West Indies and the Galapagos. They include *Liolaemus magellanicus*, the southernmost lizard in the world, reaching Tierra del Fuego at the southern tip of South America. *Liolaemus* is a large genus including many high-altitude species in the Andes and its foothills. Like most other tropidurines, they are generally terrestrial lizards. Most members of this genus give birth to live young, in keeping with the harsh climate of their mountain home. Lava lizards (genus *Tropidurus*) are found on the mainland of South America and on the Galapagos, where their large numbers and bright colors make them highly conspicuous. Males engage in head-bobbing and push-ups; and confrontations, often involving displays rather than physical contact, are common. Size and color differences between the sexes are quite obvious in members of this genus.

On the island of Madagascar there are seven species of iguanids. These are lizards of moderate size, generally sun-loving, and all are egg-laying. Six species belong to genus *Oplurus* (two are tree-dwellers, and four live among rocks). The other iguanid lizard, *Chalarodon madagascariensis*, inhabits sandy areas where it makes burrows. Males in combat may be especially violent, and combat may involve acrobatic twists in the air while engaged in head biting. The presence of iguanids in Madagascar remains a biogeographic mystery to scientists but certainly seems to date from an early period in the history of the Iguania as a whole.

Michael and Patricia Fogden

THE GEKKOTANS

Only two living families, the Gekkonidae and Pygopodidae, constitute the Gekkotan lineage. The group is an ancient one, as evidenced by *Eichstaettisaurus*, a fossil gekkotan from the Jurassic, 200 to 145 million years ago, which bears some striking similarities to living forms. Although differing greatly in body form, the fully limbed geckos (Gekkonidae) and the reduced-limbed flap-footed lizards (Pygopodidae) share a number of important features. The absence of temporal arches lends a lightly-built appearance to the skull and renders it especially mobile. Males of most species have cloacal bones—paired structures embedded below the skin on either side of the vent. Both males and females have cloacal sacs, whose function we do not yet know. Females generally lay two eggs, although one or very rarely three eggs may be laid. Gekkotans are predominantly nocturnal, and the replacement of moveable eyelids by a fixed transparent spectacle characterizes most of the group. The eyes are large, and the pupils of nocturnal geckos dilate greatly at night to let in sufficient light to allow activity. In bright light the pupils close to a vertical slit or a series of pinholes.

▲ *A high-casqued lizard* Corytophanes cristatus *in threat display, Costa Rica. Inhabiting mainly rainforests, they have a characteristic habit of perching head-downward on tree trunks.*

David Hughes/Bruce Coleman Limited

▲ *The Namib gecko* Palmatogecko rangei *of Africa. With large eyes, geckos as a group are characterized mainly by their adaptations for night-time hunting by sight.*

▼ *Recent studies of the leopard gecko* Eublepharus macularius *have shown that the sex of their offspring is determined by the temperature at which the eggs are incubated.*

Geckos

Next to the skinks, geckos (family Gekkonidae) are the most numerically diverse group of lizards. More than 900 species in 90 genera are known, and they occupy all continents except Antarctica. Although they are especially numerous in the tropics and subtropics, some species live in cooler areas from northern Italy to southern New Zealand, and a few have even adapted to harsh alpine conditions. Most geckos are small lizards, reaching a total length of no more than 15 centimeters (less than 6 inches), but a few exceed 30 centimeters (1 foot), and one species, the giant New Zealand gecko *Hoplodactylus delcourti*, which until recently inhabited New Zealand, was over 60 centimeters (2 feet) long. The largest living gecko, Leach's giant gecko *Rhacodactylus leachianus*, lives in the rainforests of New Caledonia and reaches 40 centimeters (15¾ inches) total length and may weigh 600 grams (1½ pounds).

One possible reason for the success of the geckos has been their nocturnal habits. The majority of geckos are active at night or are crepuscular (active at twilight and before sunrise). No other large group of lizards has specialized to such a degree in night-time activity. Geckos are therefore able to take advantage of the lack of competition from other lizards during the hours when a huge number of insects and spiders are available as prey.

Four subfamilies of geckos have been recognized. Members of one subfamily, the Eublepharinae, retain many primitive characteristics. Most obviously, they have moveable eyelids and their feet do not have the complex climbing system typical of many other geckos (see page 589). All species lay two eggs with a leathery shell, and all are primarily nocturnal. In at least one species, the leopard gecko *Eublepharis macularius* of southwest Asia, the sex of an individual is determined by the temperature experienced by the embryo in its egg, a condition known chiefly in turtles and crocodilians. At present the subfamily comprises only 22 species occurring in Africa, Asia, and North and Central America. Some authors regard them as constituting a separate family of lizards.

One species, the banded gecko *Coleonyx variegatus*, is a small, thin-skinned resident of the arid southwest of the United States and adjacent areas of Mexico, in a wide variety of microhabitat types, and it is an insect-eater. The banded gecko

is inactive for as much as half of the year, when night temperatures fall below about 23°C (73°F). The male uses visual and chemical signals to locate and identify a female. Courtship involves tail flicks by the male, and he ultimately mounts the female, biting her nape, and pinning her body down, while maneuvering his tail under hers to reach her cloaca with his hemipenis.

All other geckos and their close relatives, the flap-footed lizards, lack eyelids. Instead the eye is covered by a transparent scale, called the spectacle, which is cleaned by periodic licking. (Similar transparent windows in the eyelids have evolved in night lizards as well as some skinks and lacertids.)

Many geckos are terrestrial and live in deserts or other arid habitats. However, perhaps the feature most closely associated with geckos is their ability to climb, even on smooth surfaces such as glass. This feature has certainly played a major role in the diversification of geckos and their success throughout the world. Climbing ability is dependent upon claws and/or scansors, the adhesive pads on the toes. In many instances the type of toe pad is revealed by the name of the genus, and each type seems to have particular advantages on certain types of surfaces.

The Diplodactylinae is the second subfamily of geckos. The group is restricted to New Zealand, where they are the only gecko group represented, and Australia and New Caledonia where they form the bulk of the gecko fauna. It is probable that the flap-footed lizards, which are also restricted to the Australian region, evolved from within this group of geckos. Most diplodactylines lay eggs, but the geckos of New Zealand (*Naultinus* and *Hoplodactylus*) and a single New Caledonian species, the rough-snouted giant gecko *Rhacodactylus trachyrhynchus*, give birth to live young. The evolution of this feature has not been fully explained, but in the New Zealand geckos it was associated with cool temperatures experienced by the animals and the need to

regulate the temperature of the embryos. Gestation may take almost a year, and like other geckos, only two young are produced in each litter. The most successful genus in the subfamily is *Diplodactylus*: 38 species, generally only 45 to 85 millimeters (1¾ to 3¼ inches) head–body length, are distributed across mainland Australia. The name of the genus comes from the divided scansors or toe pads that characterize the group. Although many *Diplodactylus* species are terrestrial, most are capable of climbing as well. Certain members of this genus, such as the western spiny-tailed gecko *Diplodactylus spinigerus*, have large glandular structures in their tails which produce a sticky substance that is squirted from the tail up to a meter (and with some accuracy) by muscular contraction. Although the secretion is not toxic, it may be distasteful to predators or may foul the feeding apparatus of large insects and spiders that prey on these geckos. One of the oddest

Hans and Judy Beste/AUSCAPE International

▲ Virtually all geckos lack eyelids, the eye being instead protected with a fixed transparent scale. Licking the eye to clean it is a gesture common to most geckos, as in this Cogger's velvet gecko *Oedura coggeri* of Australia.

▼ Velvet geckos possess tiny, even scales which give their skin a velvety texture, hence the common name. The genus *Oedura* is endemic to Australia; the southern spotted velvet gecko *O. tryoni* shown here is found on granite outcrops on the east coast.

Australian geckos is the spiny knob-tailed gecko *Nephrurus asper*, whose tail is short and ends in a small knob, like the other members of its genus. The function of the knob is unclear, but it may play a sensory role. These geckos spend the daylight hours in burrows and emerge at night to feed on insects, spiders, and other lizards. By far the most numerous and widely distributed geckos are those of the subfamily Gekkoninae. The species are classified in almost 60 genera and, with the exception of New Zealand and parts of the United States, they reach to the limits of the family's distribution. Some of the largest genera are each characterized by a different type of toe structure, which has apparently contributed to their success in adapting to different habitats.

In Africa, south of the Sahara, members of the genus *Pachydactylus* are well represented. These are mostly terrestrial forms that attain their greatest diversity in the rocky arid regions of Namibia and South Africa. As many as eight species may live in the same area, partitioning the available resources. From this genus several others have evolved. Most spectacular of these is the Namib web-footed gecko *Palmatogecko rangei*, which has translucent skin, extremely thin legs,

and large webbed feet used for digging burrows in the sands of the coastal Namib Desert. The webs are fleshy, but contain small cartilages that support a delicate system of muscles which coordinate the fine sand-scooping motions of the feet. Adults frequently feed on small spiders, but larger white lady spiders in turn regularly capture and eat the smaller geckos.

In North Africa *Tarentola* is the dominant gecko genus. It also extends to southern Europe, where *T. mauritanica* is a common house gecko. Like a number of geckos throughout the world, this species has adapted well to the presence of humans and thrives on the walls of buildings where lights attract suitable insect prey.

A few species of geckos have been successful in colonizing vast geographic areas, being transported accidentally with humans and their possessions. The house gecko *Hemidactylus frenatus*, for example, ranges from India across Southeast Asia and far into the Pacific. It continues to spread to islands in the central Pacific, sometimes causing local extinctions of other gecko species. One of the features that makes these animals successful in this regard is their hard-shelled egg. Most reptiles have egg

▶ Several geckos have unusually shaped tails, but the knob-tailed gecko Nephrurus asper has one of the oddest: as its name suggests, the tail is short and ends in a small knob of uncertain function. One of Australia's largest geckos, this species lives in arid, rocky environments across the northern half of the continent.

Jean-Paul Ferrero/AUSCAPE International

shells that are leathery, but geckos of the subfamilies Gekkoninae and Sphaerodactylinae have hard-shelled eggs which are highly resistant to desiccation and can even survive prolonged immersion in salt water. The genus *Hemidactylus* is one of the most successful of all gecko genera, with 75 species ranging throughout Africa (a great diversity in Ethiopia and Somalia), Asia (many Indian forms), islands of the Pacific, and South America. A similarly successful group (85 species) are the bent-toed geckos, *Cyrtodactylus*, *Cyrtopodion*, and their allies. Although formerly regarded as a single genus, several lineages of bent-toed geckos are now recognized. These are rock-, ground-, or tree-dwelling species with reduced climbing pads on their toes. The claws are large and prominent, however, and most are efficient climbers. Bent-toed geckos are most diverse in Asia, extending from the Mediterranean, through the arid and mountainous regions of western Asia to tropical Southeast Asia, the Indo-Australian archipelago, and out into the islands of the Pacific. The mourning gecko *Lepidodactylus lugubris* is a particularly widespread species in the Pacific. In addition to having a calcareous eggshell, its spread has probably been enhanced by parthenogenesis, the ability of a female to reproduce without the need for fertilization of the eggs and hence without the need of a male.

The vast majority of geckos are insectivorous, attacking any arthropods small enough to ingest. Some of the larger geckos, however, may take vertebrate prey such as small mammals, birds, and especially other lizards. The tokay *Gekko gecko* is a

good example. It has even been known to attack snakes. Tokays are also especially vocal lizards and growl or bark loudly in social and defensive behaviors. The genus *Gekko* is one of several highly successful genera with their center of distribution in tropical Asia. Gecko colors are usually drab browns or grays, but a few species, such as tokays which are bluish with orange spots, deviate from this pattern. By far the most colorful species are the day geckos (genus *Phelsuma*) of Madagascar and the Indian Ocean islands. In addition to taking arthropod prey such as insects and spiders, these bright green, red, and blue lizards seek out flowers and eat nectar and pollen. This habit may, in fact, be widespread among geckos. (It has been reported in the New Zealand diplodactyline geckos of the genera *Hoplodactylus* and *Naultinus*.) Day geckos have greatly reduced claws and rely entirely on their toe pads when climbing.

The last group of geckos, the subfamily Sphaerodactylinae, is an American offshoot of the subfamily Gekkoninae. The ancestors of these lizards probably came from Africa, either traveling on the South American continent as it drifted

▲ *The blue-tailed day gecko* Phelsuma cepediana *(top) is a colorful gecko from the islands of Réunion and Mauritius. Being active during daytime and a tree-dweller, it has round pupils and greatly reduced claws on its digits, relying solely on its toe pads when climbing. The ring-tailed gecko* Cyrtodactylus louisadensis *(above) is a widespread species of the Australasian region and is the largest gecko on the Australian continent. For an obvious reason the common name of this genus is bent-toed geckos.*

G.I. Bernard/NHPA

▲ Many nocturnal geckos rely heavily on crypsis, with subtle and intricate patterns of sober shades of gray, brown, and green to protect them by day. But day-active species, like this multicolored gecko Gonatodes ceciliae of Trinidad, are often brilliantly colored and boldly patterned.

▼ The yellow-headed gecko Gonatodes albogularis fuscus is a member of the subfamily containing the smallest geckos, the Sphaerodactylinae. One of the few sexually dimorphic geckos, only the males display the yellow color on the head.

significant predators. Some species have very fragile skin which may be an adaptation to escape such attackers. Similar weak skin is also found in some gekkonines, especially island species. Although many species of sphaerodactylines are diurnal (day-active), they are secretive and active in protected microhabitats. The white-throated gecko *Gonatodes albogularis* and its allies are generally the most conspicuously day-active sphaerodactylines. Sexual dichromatism is seen in this species, with males exhibiting a bright yellow head, whereas females are more subtly patterned; sexual dimorphism, such as this, is rare in geckos and is found in only a few groups, most of them diurnal (day-active).

Pygopodids (flap-foots)

The flap-footed lizards of the family Pygopodidae have extremely long bodies and reduced limbs, so that they look remarkably like snakes. During evolution, forelimbs have been lost entirely, whereas hind limbs are invariably just small, flattened flaps lying close to the cloaca. Though small, the limbs may be used in some types of locomotion (through vegetation) and in courtship and defensive behaviors. Most of the 36 species of pygopodids are endemic to Australia—one species is restricted to New Guinea, and another occurs on both landmasses. Despite their striking dissimilarity in overall appearance, flap-footed lizards are closely related to geckos, and may in fact be an offshoot of a particular group of geckos endemic to Australia and the southwest Pacific, the diplodactylines. Like most geckos they do not have eyelids. Also, in common with the majority of geckos, all pygopodids lay two eggs per clutch.

Some pygopodids, such as the species of *Aprasia*, are chiefly burrowers that feed on insects, whereas others have become specialized for life above ground and they retreat to the burrows of spiders or other animals or into the dense clumps of spinifex grass found in arid Australia. Members of the genus *Delma* are slender, surface lizards and employ a unique jumping mechanism in their repertoire of defenses. When agitated they use their long tail to generate an upward thrust and

westward in the Cretaceous period 145 to 65 million years ago or rafting across the narrow and newly formed Atlantic Ocean shortly thereafter. Their closest relatives are probably among the semaphore geckos (*Pristurus*) of the Middle East, or *Quedenfeldtia*, a small North African form. These are mostly tiny geckos, including the smallest living lizard, Monito gecko *Sphaerodactylus parthenopion* of the Virgin Islands, which reaches only 17 millimeters (²/₃ inch) head–body length. Most of the 84 species of the genus *Sphaerodactylus* are island-dwellers, living among fallen palm fronds or other litter in forests or open areas in the West Indies. A few, such as the Cuban dwarf gecko *S. bromeliarum*, occupy bromeliads and leaf axils in trees. The Americas are generally poor in geckos, and members of this genus account for the majority of all American gekkonids. These lizards are so small that arthropods such as spiders may be their most

STROLLING ON THE CEILING

Many lizards climb, but only a few have evolved specialized structures that enable them to scale smooth surfaces without the aid of claws. Such climbing devices are found in many representatives of the family Gekkonidae, a few skinks, and the iguanid lizards of the genus *Anolis*. The toes of climbing geckos bear enlarged overlapping plates called lamellae on their undersurfaces. Each lamella is covered by a field of setae, which are microscopic projections of the skin; these tiny spatula-shaped prongs vary in length from species to species, but in most geckos they measure 10 to 100 micrometers. When the gecko climbs, the expanded tips of the setae are brought into very close contact with the surface, and weak forces between the molecules form a temporary bond between the two surfaces. Although the forces acting on any one seta are minute, there may be more than one million setae per toe, so the additive adhesive force may be quite impressive. In order to ensure that as many of the setae as possible actually come into contact with the surface, geckos also have a complex internal mechanism within the toe.

A lamella is the most external portion of a structure called a scansor. Inside the scansor there is an extensive network of blood vessels. These in turn are linked to a sinus, a small blood reservoir beneath the bones of the toe. The gecko can shut off this part of the circulatory system from the rest of the body by a series of valves; and when the animal pushes onto the bones of the toe above the sinus, it pressurizes the network of vessels, causing them to expand and push onto adjacent lamellae, forcing them onto the substrate. In this way the expanded toe pad is made to conform closely to any irregularities in the substrate, maximizing the number of setae that can be involved in adhesion. Tendons in the scansors also permit fine control so that each scansor can be moved independently of the others.

The mechanism of adhesion poses a special problem when the gecko is walking. In order to lift the foot a gecko must depressurize its blood sinus and the network of blood vessels and break the weak bonds that hold the setae to the surface. This is accomplished by rolling the toes up from tip toward the base, thus forcing blood back toward the foot and peeling the setae away from the surface. All of this happens with every single step the gecko takes!

The origin of the whole adhesive complex is obscure, but the setae themselves appear to be elaborations of surface features that are present on the skins of all lizards. The climbing ability of geckos has certainly contributed to their success in habitats in trees and rocks. This form of adhesion is perhaps most useful on hard, smooth surfaces to which lizards with claws cannot cling. Paradoxically, surfaces such as glass, which

Mike Birkhead/Oxford Scientific Films

▲ *Geckos are a familiar part of the normal household fauna in homes throughout the tropics and subtropics of the world. Mainly nocturnal, they scuttle freely over walls and ceilings in their hunts for moths and other insects. This is the bronze house-gecko Aeluronyx seychellensis of the Seychelles.*

David Corke/ZEFA

◄ *A gecko's foot, showing the lamellae on the underside of the toes.*

intuitively seem highly challenging for a climbing animal, are quite easily negotiated by geckos. The basis for the molecular bonds between the substrate and the setae is the amount of adhesive force of the surface. This is known as surface energy. It can be measured in relative terms by putting a drop of water on to a substrate; if the drop spreads out, the surface has a high degree of attraction for the water molecules and thus high surface energy; if the drop remains beaded-up, the attractive force is small and surface energy is low. Thus glass, with reasonably high surface energy is no challenge for a gecko, but the waxy surface of a plant leaf may be.

The adhesive mechanisms of the other climbing lizards have not been investigated in such detail but do not seem to be so complex. Nor are these lizards as accomplished climbers as geckos.

lift the entire body from the ground. Several jumps may be completed in succession, with the animal changing direction erratically each time. *Delma* species, like other pygopodids, are capable of vocalization, and this may also be used as part of the defensive strategy of the lizard. Burton's snake-lizard *Lialis burtonis*, with its characteristic elongate snout, is the largest Australian pygopodid at 59 centimeters (23¼ inches) total length, and feeds exclusively on lizards and snakes. It has pointed teeth that are hinged at their bases, which aid in capturing its prey; the smooth-scaled skinks that constitute much of the diet easily slide past the "folded teeth", but if they attempt to move backward, out of the snake-lizard's mouth, the teeth are pushed into their erect posture preventing escape. This species exhibits amazing variability in color and pattern. It is also the most widespread pygopodid, ranging across all of Australia except the extreme southwest and southeast, and occurs also in New Guinea. Another widespread flap-footed lizard, the black-headed scaly-foot *Pygopus nigriceps*, is characterized by a black crown and nape and closely resembles a venomous elapid snake that occurs in the same area. On this basis it has been regarded as a potential mimic of the snake.

THE AUTARCHOGLOSSANS
The third major lineage of lizards is the Autarchoglossa, which contains the twelve remaining lizard families. These lizards are mostly terrestrial, burrowing, or living among rocks. Few generalities can be applied to the group as a whole, but many species have osteoderms (bony plates) in the skin, and most rely heavily on chemical cues in their environment.

Autarchoglossa is further divided into two subgroups: the Anguimorpha (five families) and the Scincomorpha (seven families).

GROUP ONE: THE ANGUIMORPHS
The five types of anguimorph lizards are the anguids, beaded lizards (helodermatids), the Bornean earless monitor, true monitors (including goannas), and knob-scaled lizards. Although relationships are unclear, snakes share a number of features with some of the anguimorph lizards and some scientists believe it may be this group from which they evolved.

Anguids
The family Anguidae is a moderately diverse group of about 90 species, most of which inhabit the Northern Hemisphere. The family includes fully limbed and limbless species. Their scales contain bony osteoderms that give the lizards a hard, rigid feel. In two of the four subfamilies, Gerrhonotinae and Anguinae, the large rectangular scales of the back give way to small granular scales on the underside of the body. These cover an osteoderm-free fold of skin that runs most of the length of the body.

Alligator lizards and their allies are members of the subfamily Gerrhonotinae. This group extends from southwestern Canada to the tropics of Central America and is especially diverse in Guatemala and Mexico. Nearly 40 species are known, and almost half of them belong to the tree-dwelling genus *Abronia*. Most species in this genus occur in forests in mountainous regions. Other alligator lizards are terrestrial or semi-burrowing and occupy a wide range of habitats from desert mountains to tropical lowlands.

▶ *The legless lizard Delma inornata. Restricted to Australasia, members of this genus are mainly nocturnal, and chiefly remarkable in their unique habit of leaping erratically when threatened, thrust wriggling into the air by the muscles in their slender but powerful tails.*

Esther Beaton/AUSCAPE International

George Bernard/NHPA

◄ *Many alligator lizards (genus Abronia) inhabit rainforests of Central and South America, some species reaching 30 centimeters (11 inches) or more in length. Members of this genus have prehensile tails.*

The southern alligator lizard *Elgaria multicarinata*, and the northern alligator lizard *E. coerulea* occur together in much of western North America. The two species are similar in appearance, but differ in reproductive mode: the southern alligator lizard is oviparous, laying 6 to 17 eggs in burrows or other protected sites in the oak-grasslands it inhabits; the northern alligator lizard, which is somewhat smaller, gives birth to 2 to 15 live young about three months after mating. The northern alligator lizard is especially adapted to colder conditions and tends to occupy the margins of coniferous forests or other slightly cooler and moister sites than the southern

alligator lizard. Both species are typical of terrestrial members of the subfamily, in that they are generally secretive and feed on insects and spiders, small mammals, birds, and reptiles. *Elgaria* species can bite fiercely in defense but do not appear to be territorial. Although chiefly ground-dwelling, they also climb with the aid of a partially prehensile tail. The largest member of the subfamily is the Texas alligator lizard *Gerrhonotus liocephalus*, which reaches 20 centimeters (8 inches) head–body length. Like most members of the group, it is widely but incorrectly believed by many local people to be venomous.

Most members of the second anguid subfamily,

Aldo Brando Leon/Oxford Scientific Films

▲ The subfamily Diploglossinae, or galliwasps, has a distribution centered mainly on the islands of the Caribbean. This is the white-spotted, or Malpelo galliwasp Diploglossus millepunctatus, restricted to Malpelo island off the coast of Colombia.

the birds themselves are eaten (the latter as carrion), as are the feces of the birds. Stranger still, galliwasps mob adult boobies returning to the nest with food intended for their chicks, causing the birds to regurgitate prematurely and thereby provide the lizards with food.

There is a tendency for the limbs to be reduced in some galliwasps. This is most accentuated in the South American worm lizards, genus *Ophiodes*, a small group of diploglossines retaining only rudimentary hind limbs. They are endemic to mainland South America and are strikingly similar to the anguines.

The anguines—the slow worm and glass lizards—are the most geographically widespread of the anguid subfamilies and the only group to occur in the Old World. They are characterized by the absence of limbs. The slow worm *Anguis fragilis*, found throughout much of Europe and Britain, is a highly adaptable lizard and survives today even within the confines of the City of London, occupying the railway right-of-ways that cross the city. The slow worm gives birth to 3 to 26 live young after about three months of gestation, although this is variable and if conditions are unfavorable the young may remain in the uterus of the mother.

The largest genus in the subfamily is *Ophisaurus*, with 12 species. Its representatives occur in North America, southern Europe, across Asia to Borneo, and marginally in North Africa. Some recent fossil and phylogenetic studies seem to support the hypothesis that the Old World and New World species had separate origins and evolved limblessness independently. All members of the genus are oviparous and the female remains with the eggs during incubation, although she does not actively defend the nest. The scheltopusik *Ophisaurus apodus*, of southeastern Europe and southwestern Asia, is the largest member of the genus and of the entire anguid family; large individuals may reach 1.4 meters (4½ feet) in total length. This species feeds on small mammals, snails, insects, and other reptiles. Smaller species in the genus are primarily insectivorous but may take nestling mammals and small amphibians and reptiles too. The tails of these lizards are extremely long—two and a half times the body length in one subspecies of *O. attenuatus*, the slender glass lizard of North America. The "glass lizard" name has been given to these animals because of their tendency to "break" or autotomize the tail. The slender glass lizard occurs in tallgrass prairie habitats, where it "swims" through the grass in search of prey or hides under vegetation or in burrows co-opted from other animals. These burrows are used for hibernation, as nests, and as refuges from predators and inclement weather. In cool weather the slender glass lizard is active during the day, but as the weather becomes warmer it prefers the

Diploglossinae, are known as galliwasps. There are 37 species recognized at present. They are skink-like animals with smooth overlapping scales and relatively short legs. Although there are some mainland species, most galliwasps are found on the islands of the Caribbean. Like the alligator lizards this group also contains both egg-laying and live-bearing forms. As few as one or as many as 34 offspring may result from galliwasp clutches or litters. Most galliwasps are primarily terrestrial or semi-burrowing, although many species often climb trees and other vegetation. Both nocturnal and diurnal species exist; a couple of the largest species, *Diploglossus anelpistus* and *D. warreni*, at about 28 centimeters (11 inches) head–body length, are night-active. Although galliwasps live in many types of habitats at all elevations up to 2,500 meters (760 feet), most prefer somewhat moister microhabitats. The Malpelo galliwasp *D. millepunctatus*, endemic to rocky Malpelo Island off the Pacific coast of Colombia, is a relatively large galliwasp, growing to at least 25 centimeters (10 inches) head–body length. It has some of the strangest feeding habits of any lizard. Malpelo is a small island, and resources are limited; few potential prey items of an appropriate size for the galliwasp can be found. Crabs and seabirds (mostly boobies) are plentiful, however, and provide most of the lizard's diet. The crabs and twilight and before dawn or, in especially hot periods, the night-time. During winter months the lizards hibernate in deep tunnels of rodents or moles. The slender glass lizard mates in late spring, and 5 to 16 eggs are laid in early summer, hatching by late summer or early fall. Females breed only every second year, as they lose significant weight while egg tending and must renew energy reserves the year after laying. As might be predicted from the low reproductive output, glass lizards are relatively long-lived, reaching at least 10 years of age for American species and perhaps 20 for the scheltopusik.

There are only two species in the anguid subfamily Anniellinae. They are the California limbless lizards (genus *Anniella*). Both are small (10 to 16 centimeters, or 4 to 6¼ inches, head–body length), limbless, and burrowing, and they inhabit coastal and inland regions of

California and Baja California. They are regarded by some herpetologists as representing a separate family, the Anniellidae, closely related to anguids. Anniellids lack external ear openings but have unreduced eyes with moveable lids. The California limbless lizard *Anniella nigra* occurs in California, where it lives in sandy soil around the bases of bushes; coastal sand dunes are favored sites, as are regions of alluvial soils in the inland where the substrate is suitable for the lateral undulations used by the lizard in sand-swimming. Like most anguids it is primarily insectivorous, catching adult and larval insects near the surface during the late afternoon or evening. California legless lizards are viviparous, giving birth to small litters (one to four young) which are relatively large when born in the fall. Relatively cool temperatures and access to soil moisture are favored by anniellids.

▼ *The slow worm is one of the hardiest and most adaptable of European lizards. Long, slender, and snake-like— a resemblance extending even to a forked tongue—it is often common even at the heart of major cities. It gives birth to living young after an extremely variable gestation period. Remarkably long-lived for such a small creature, some captive individuals have been known to live for 20 years or more.*

Jane Burton/Bruce Coleman Limited

Beaded lizards

The family Helodermatidae contains only two living species: the Mexican beaded lizard *Heloderma horridum*, which occurs along the Pacific coast of Mexico and Guatemala; and the gila monster *H. suspectum*, from the southwestern United States to northwestern Mexico. These are the only venomous lizards in the world. They are most closely related to the true monitor lizards and the Bornean earless monitor, although Helodermatidae has existed as a recognizable group for at least 65 million years, since the late Cretaceous, and a number of species occurred in the Tertiary of North America. Both species are large—the head-body length of the Mexican beaded lizard can be 52 centimeters (20½ inches), or total length 1 meter (3¼ feet), and the gila monster's head-body length can be 33 centimeters (13 inches)—and slow moving. The scales are small, bead-like and do not overlap, and all but some of those on the underside are underlaid by bony osteoderms. The two species are generally similar in appearance. They are stocky with broad heads and are typically black with varying amounts of yellow or pinkish spotting or reticulations. The tails are short in both species, especially so in the gila monster, and are used for fat storage.

Beaded lizards spend much of their time in burrows and generally avoid extremely high temperatures. Only in the cooler spring months and overcast days are they active during the day. They forage on the ground but may climb trees in search of prey as well, relying largely on chemical cues, such as smell. This is especially true of the more slender-bodied Mexican beaded lizard, which occupies not only arid regions but also tropical deciduous forests. Gila monsters also inhabit a variety of habitats but are most often associated with rocky slopes in areas of desert scrub, grassland, or oak woods. Both species feed on a variety of prey but rely heavily on the nest young of rodents and other small mammals, and bird and reptile eggs. Mammalian prey, detected through smell and sound, are usually dug out of the ground with the powerful forelimbs. The gila monster mates in late spring, often after intense combat between rival males. As many as 12 eggs are laid by the female and require about ten months before hatching.

Earless monitor lizard

The family Lanthanotidae contains only one species, the Bornean earless monitor *Lanthanotus borneensis*. It is closely related to the true monitors (family Varanidae) and is also allied to the beaded

VENOMOUS LIZARDS

Beaded lizards (family Helodermatidae) are the only lizards to possess venom. Evidence for the existence of a venom-delivery system is present even in an early member of the family, *Paraderma bogerti*, which lived more than 65 million years ago. Unlike venomous snakes, the lizards have venom glands in the lower jaw, which bulges conspicuously. Individual ducts lead from the glands to each of the lower teeth, which have grooves in the front for carrying venom. Venom is delivered with every bite, but there is no forceful ejection from the glands, and the lizard must chew the venom into its victim. The venom is generally not needed to subdue prey, as these lizards often feed on eggs or relatively defenseless prey (young mammals), so although venom may have evolved in association with feeding it is now used mainly as defense for these slow-moving lizards, which may be exposed for long periods while foraging in open habitat. Beaded lizards are immune to their own venom, which may be injected during territorial combat between males. The striking black and pink coloration of the gila monster *Heloderma suspectum* may serve either as camouflage or as a warning to potential predators.

The bite of the gila monster results in localized swelling and severe pain and may cause vomiting or faintness but is usually not fatal for humans. The mechanical damage done by the teeth and powerful jaws may also be significant. A gila monster may maintain a defensive bite for many minutes, during which the venom will continue to be released.

John Cancalosi/AUSCAPE International

▲ *The gila monster inhabits arid regions of Mexico and the American southwest.*

lizards, to which it bears some resemblance. The earless monitor is dark brown in color and grows to about 20 centimeters (8 inches) head-body length. The body scales are small, but several rows of enlarged tubercles run down its back. It is earless only in that it lacks a tympanum, or external eardrum, but it is quite capable of detecting sounds. This poorly-known nocturnal lizard lives in the northern part of the island of Borneo, where it burrows and swims in search of prey, including earthworms. It may be able to climb as well, but is slow-moving on land. Its limbs are small, and burrowing is accomplished mainly by movements of the head; in water it moves by lateral undulations of the body. Like its closest relatives, the Bornean earless monitor is oviparous (egg-laying). Details of its physiology, behavior, and ecology remain unknown.

Monitor lizards

The family Varanidae includes the monitors or goannas. All 34 species are classified in the genus *Varanus*, although some researchers recognize more species or divide the genus into subgenera, or species groups. All monitors have a relatively

similar body form, with long necks, well-developed limbs, strong claws, and powerful tails. Varanids are a strictly Old World group, occurring throughout Australia, Asia, and Africa. By far the majority of species are Australian, and this region is regarded as the main center of monitor evolution. Like beaded lizards, the monitors are strictly oviparous (egg-laying).

The largest living lizard species, the Komodo monitor (or Komodo dragon) *Varanus komodoensis*, is a member of this group. It may grow to about 3 meters (10 feet) in total length and weigh as much as 165 kilograms (364 pounds). The Komodo dragon, which was not described scientifically until 1912, occurs only on Komodo and neighboring islands in the Lesser Sunda Chain of Indonesia, where it may be locally abundant. Like all monitors, it is active during the day. Juveniles take a variety of relatively small prey items, while adults feed on carrion and fresh prey. Eggs, lizards, and small mammals are eaten, but larger mammals, especially deer and even water buffalo weighing 500 kilograms (1,100 pounds) are very important sources of food for larger individuals. The large recurved teeth of the

▲ The largest of all lizards, the Komodo dragon is restricted to Komodo and a few neighboring islands in central Indonesia. Although they feed largely on carrion, these huge monitors also prey on animals as large as deer and water buffalo, and there are several well-documented cases of fatal attacks on humans.

▶ About 25 of the 34 species of monitors occur in Australia, where they are generally known as goannas. Most inhabit arid or semi-arid country, are active by day, and are mainly terrestrial, although they are accomplished tree-climbers. This is Varanus panoptes, a little-known species of the north and west of Australia.

David Curl

dragon are serrated and effectively slice through prey tissues. Powerful forelimbs and claws, used chiefly in digging, may also be used to disembowel large mammals. Attacks on humans are rare but do occur and sometimes result in death.

Other *Varanus* species are also carnivorous, taking mammalian or reptilian prey, but most varanids are largely insectivorous. Gray's monitor *V. olivaceus* of the Philippines, however, is a specialist, subsisting on fruit and mollusks.

Even larger monitors lived in the Pleistocene (2 million to 10,000 years ago) of Australia. *Megalania prisca* was a giant goanna that reached 7 meters (23 feet) in length and may have preyed upon some of the giant marsupials that are now extinct. The largest lizards ever to have lived were also relatives of the monitors, but these were marine giants known as mosasaurs.

Not all varanids are large, however. Some Australian species, such as the short-tailed monitor *V. brevicauda*, reach only 12 centimeters (4¾ inches) head–body length. Most monitors are terrestrial, but the emerald tree monitor *V. prasinus*, a bright green species inhabiting New Guinea and the northern tip of Australia, is an arboreal specialist with a long prehensile tail. Another Australian species, Mertens' water monitor *V. mertensi*, is one of several highly aquatic species. In such water-loving species the tail is greatly compressed and is used to generate powerful lateral undulations in swimming. It may also be used to corral fish or other aquatic prey.

During the breeding season male monitors engage in combat for access to females. In actions reminiscent of those seen in some snakes, they raise up on their hind legs and tail base, and "wrestle". The contest is over when the victor has pushed over his opponent. Monitors rely heavily on chemical senses and frequently tongue-flick to test for airborne cues; the tongue is slender and deeply forked, closely resembling that of snakes. Although sight and hearing are also keen in monitors, courtship and mating behavior is primarily based on chemical and tactile cues. Varanids lay 7 to 37 eggs, typically in the soil or in tree stumps or hollows. The Nile monitor *V.*

niloticus, which ranges throughout most of Africa, may lay its eggs inside termite mounds; the architecture of the mound provides ideal humidity and ventilation, while the hardness of the mound surface protects the eggs from foraging predators. Although the adult does not guard the eggs, it may return to release the hatchlings from the termite mound. Similar behavior is known to occur in a number of arid-zone monitors.

Knob-scaled lizards

The knob-scaled lizards (family Xenosauridae) include two genera, widely separated geographically and perhaps not closely related to one another. Lizards in both genera have bumpy scales and bony osteoderms in the dermis of the skin. All four species are moderate in size, ranging from 10 to 15 centimeters (4 to 6 inches) head–body length.

Xenosaurus is a genus of three species distributed from Mexico to Guatemala. They occur in a variety of habitats from semi-arid scrub to high-elevation cloud-forest, although they prefer moist or even wet habitats. All are believed to be nocturnal or crepuscular (dawn- and twilight-active), terrestrial, and secretive, seeking out holes and crevices. The common Mexican knob-scaled lizard *Xenosaurus grandis*, the best-known species, occupies crevices in rocky outcrops or cavities in trees, often in forested regions; it feeds on termites and other insects, and the female gives birth to two to six live young. Xenosaurs have a

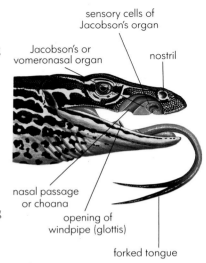

▲ *In snakes and some lizards the sense of smell is greatly enhanced by the Jacobson's or vomeronasal organ, located on the roof of the mouth. Although present in most reptiles, this organ is particularly well developed in species with bifid, or forked tongues. With the mouth held closed, the tongue is extended and flicked repeatedly, picking up chemical scents which are then detected by the highly sensitive sensory cells lining the Jacobson's organ.*

◄ *The largest of the Australian goannas is the perentie Varanus giganteus. The specimen illustrated has inflated its neck as part of its threat display. Highly agitated varanids will also arch their tail and, unlike crocodilians which are falsely reputed to do so, will often wield it as a weapon.*

Anthony Bannister/NHPA

▲ *The armadillo girdle-tailed lizard relies on its heavily armored tail for protection, either when wedged in rock crevices or, when threatened in the open, by clamping it in its mouth to shield its more vulnerable underparts.*

gestation lasts seven or eight months, and as many as eight young may be born.

GROUP TWO: THE SCINCOMORPHS

Relationships among the scincomorph lineage have not been fully studied. The families included are the cordylids, dibamids (tentatively), lacertids, skinks, teiids, and microteiids. They account for nearly half the species of all lizards. Scincomorphs have pleurodont dentition— unsocketed teeth on the inner face of the jaw bones—and rarely show crests or other ornamentation. Limbless forms are especially numerous in the group. Both visual and chemical signals are used during interactions between individuals of the same species and other species.

Girdle-tailed lizards

The family Cordylidae contains 60 species comprising two rather distinctive types of lizards: girdle-tailed lizards and their allies, and plated lizards. The former are found in Africa, and the latter in Africa and Madagascar. In the Eocene epoch, 57 to 37 million years ago, lizards in this family also occurred in Europe. All species are heavily armored, with bony osteoderms (bony plates), and all are day-active sun-lovers. The scales are usually large and rectangular and are arranged in regular rows around the body.

The true girdle-tailed lizards are mainly rock-dwellers and live in semi-arid to highly arid areas. Their bodies and heads are flattened, and most are fully limbed. Species in the genus *Cordylus* typically have whorls of spines on the tail, which aid in active defense and in wedging the animal into rock crevices. Most species are drab, with browns, blacks, and straw colors dominating. These lizards are live-bearing, and although only one to six young are born in a season, the chance of survival is generally good. *Cordylus* species are late to reach maturity and may live for more than ten years (25 years in captivity). The armadillo girdle-tailed lizard *C. cataphractus* is a large, insect-eating species that responds to attack by curling into a ball and grasping its own tail in its mouth, thereby forming an armoured ring in which only the osteoderm-reinforced dorsal surface is visible to the potential predator. Groups of related individuals often live together in deep rock cracks in the South African arid zone. Another large member of the genus, the sungazer or giant cordylid *C. giganteus*, which reaches 40 centimeters (15¾ inches) in total length, is highly unusual because it lives in burrows rather than rock crevices and may form large colonies in the highveld savanna of Orange Free State and adjacent areas. The burrows are as much as 2 meters (6½ feet) long and give protection from predators and the weather. Like other girdle-tailed lizards in temperate areas with cold winters, sungazers become inactive for part of the year.

somewhat flattened body form in association with their rock-dwelling habits.

The crocodile lizard *Shinisaurus crocodilurus* is a semi-aquatic lizard known only from a series of isolated localities in Guanxi province, southern China. It has a rather high, laterally compressed head. Its name comes from the large tubercles on its back, arranged in rows and extending onto the tail, giving the animal a crocodile-like appearance. Until recently almost nothing was known about this species, and it was regarded as one of the rarest lizards in the world. Its food consists of fish, tadpoles, and insect larvae captured in flowing water, although terrestrial prey may be an important dietary component too. The crocodile lizard is active during the day or at twilight and spends the night perched on branches overhanging streams. If threatened it may retreat to the water. Like the Mexican knob-scaled lizards, the crocodile lizard is viviparous;

One or two large young (up to 15 centimeters, or 6 inches, total length) are born at the end of summer in February–March. Because it inhabits land capable of supporting agriculture, the sungazer has had to compete with humans for space. It is threatened but is legally protected.

Not as spiny as the girdle-tailed lizards, but better adapted for life in narrow rock crevices, are the flat lizards (genus *Platysaurus*). These are covered with small granular scales and are fully limbed. The entire body and particularly the head is amazingly depressed. Flat lizards occupy crevices and spaces beneath rock flakes on isolated rock outcrops or boulders. They are especially diverse in Zimbabwe and the northern Transvaal. Large numbers of individuals are frequently found together in retreats. In many species there is a striking sexual dichromatism, with males displaying vibrant colors (usually reds and blues or greens) that are species-specific. Females and juveniles of all species lack bright colors but have a striped back. Flat lizards are the only egg-laying members of the family Cordylidae. Two elongate eggs are laid in the moister microclimate of deep rock cracks, and a single site may be shared by ten or more females.

The grass lizards (genus *Chamaesaura*) have only tiny, spike-like hind limbs, used for stability when at rest. They have a very long tail and are "grass-swimmers", moving by pushing the body against vegetation in the high grassland habitats, and their extreme length allows for their weight to be distributed across many individual plants.

Plated lizards (also known as gerrhosaurines) constitute the remainder of the family, although some herpetologists classify them in their own family. Four genera occur in mainland Africa and another two in Madagascar. Their bodies are generally not spiny, as in the girdle-tailed lizards, and they have a prominent lateral fold similar to some of the anguid lizards. All of them are

Anthony Bannister/NHPA

▲ Living on rocky outcrops in arid regions of southern Africa, the flat lizards have an extraordinarily flattened shape to facilitate hiding in crevices. The male dwarf flat lizard Platysaurus guttatus illustrated is in breeding colors. Surprisingly, the underside is the most colorful feature of these lizards, as males display to each other by elevating their forequarters.

◄ Early Afrikaans folklore held that the sungazer lizard faces always into the sun. Terrestrial and day-active, members of this group (genus Cordylus) are sometimes known as zonures.

probably oviparous (egg-laying), although little is known about many species. The giant plated lizard *Gerrhosaurus validus* is the largest species, reaching a total length of 70 centimeters (27½ inches). It occupies rock cracks in boulder piles of the African savanna, and feeds not only on insects and other arthropods, but also small vertebrates and plant material, especially fruit. Two to five eggs are laid in midsummer, and the large hatchlings emerge about 11 weeks later. The desert plated lizard *Angolosaurus skoogii* differs from all other cordylids in its very specialized habitat preferences. It is an inhabitant of the sand dunes of the northern Namib Desert, where it lives in small colonies and forages on the slipfaces of the dunes for insects, plant debris, and small desert melons. The body is cylindrical and the snout is wedge-shaped, with the rim of the lower jaw hidden when the mouth is closed. The toes bear broad rectangular fringes. When disturbed the lizard dives into the dune and disappears from sight.

Species in the genus *Tetradactylus* have a range of body forms, from fully limbed to nearly limbless (that is, those with small hind limbs only). Unlike certain other reduced-limbed lizards, however, these gerrhosaurs all retain large fully-functional eyes, indicating that they are surface-active rather than burrowing. Like the grass lizards described above, they have long tails and appear to "swim" across the grass of their habitats. One species, *T. eastwoodae*, has apparently become extinct because its habitat has been destroyed by humans.

The Madagascan gerrhosaurs (*Zonosaurus* and *Trachyleptychus*) are poorly known, but some species may live near water. They are somewhat similar to the mainland African plated-lizards in appearance and probably in ecology.

Dibamids

The 11 species in the family Dibamidae are among the least well-known lizards. At various times they have been regarded as closely related to anguid lizards, geckos and pygopodids, skinks, amphisbaenians, and snakes. Dibamids are small (less than 25 centimeters, or 10 inches, total length), long-bodied burrowers, with highly reduced limbs. Females lack limbs all together, while males have tiny flap-like hind limbs. The scales of the body are overlapping, and those of the snout and mandibles are enlarged and plate-like. The eyes are small and covered by a scale, and there are no external ears. The genus *Dibamus*, which is distributed through Southeast Asia, the Philippines, and the islands of the Indo-Australian archipelago (including the western part of New Guinea), accounts for all but one of the species in the family. Most species have been found in the soil or under debris in humid forests, but at least one from Vietnam has been collected in a tree! *Anelytropsis papillosus*, a similar animal, lives in rather more arid habitats in northeastern Mexico. Dibamids are insectivorous and oviparous. The eggs have a calcified shell, a feature that apparently evolved for similar ecological reasons in the Gekkota.

Lacertids

The family Lacertidae, sometimes referred to as the "true lizards", are an Old World group. Of more than 200 species, most inhabit the Mediterranean region, although others extend southward to the tip of Africa, eastward to Japan,

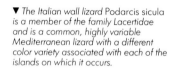

▼ *The Italian wall lizard* Podarcis sicula *is a member of the family Lacertidae and is a common, highly variable Mediterranean lizard with a different color variety associated with each of the islands on which it occurs.*

LIZARDS WITHOUT LIMBS

The reduction or loss of limbs and the elongation of the body are common and recurrent trends throughout lizard evolution. Of the 17 families of lizards, six include species that have greatly reduced limbs and two (the Dibamidae and Pygopodidae) include only reduced-limbed animals. In the skinks alone, loss of skeletal elements in the limbs has occurred about 30 times. The origin of limblessness may be associated with two different types of habitats. In the case of dibamids, teiids, skinks and some pygopodids, limb loss appears to correlate with the occupation of underground habitats. Anguids, cordylids, and some other pygopodids are surface-active, and their reduced limbs may have evolved during their association with dense grassland vegetation or other low thick plants, or similarly complex landscapes.

Limb loss almost certainly evolves by stages. Skeletal elements in the limbs may be lost gradually—initially only a single toe or finger may be reduced. Or it may occur more drastically, with the loss of numerous elements through the disturbance of developmental pathways. The same environmental demands that give advantages to smaller limbs, also favor a more elongate, snake-like body, and in general the greater the reduction of limb elements, the greater the number of vertebrae present.

Although these "stages" can only be inferred, there are several genera that exhibit a variety of limb reduction in different species; *Bachia* among the tropical American microteiids, *Tetradactylus* among the cordylids of Africa, and the Australian *Lerista* among the skinks are three of the best examples. The genus *Lerista* includes more than 50 species and gives a very complete view of the stages in the evolution of limb reduction. In nearly all reduced-limbed lizards it is the forelimbs that disappear first. Only in certain species of the microteiid *Bachia* are the forelimbs retained and hindlimbs lost.

Even in species that have no limbs at all, at least some remnant of the limb girdles remains internally.

Mark Newton/AUSCAPE International

▲ *Burton's legless lizard* Lialis burtonis *occurs in a range of habitats virtually throughout Australia and New Guinea. Like all pygopodids it lays a clutch of two eggs.*

J.A.L. Cooke/Oxford Scientific Films

◀ *Pygopus lepidopodus is one of the largest and commonest of Australian "legless" lizards. This close-up shows its rudimentary hind limbs.*

Pygopodids, for example, retain vestiges of the pelvic girdle, although the connection of the girdle to the vertebral column is lost. Remnants of the pelvis and femur are retained even in some snakes.

and north to the Arctic Circle. With a few exceptions the family is reasonably uniform in body form and structure. Head scales are large and often contain bony osteoderms; body scales are usually small and granular on the back but large and rectangular on the underside. Most species are terrestrial, and all but a few are exclusively diurnal—lacertids prefer high body temperatures, and many are inactive when weather conditions are unfavorable.

Wall lizards (genus *Podarcis*) are familiar to most Europeans, as they are conspicuous in both natural habitats and those influenced by humans. The common wall lizard *P. muralis* is the most widely distributed species in the genus. It is typical of many lacertids in preferring somewhat open country, where it may be found climbing on rocks, trees, or buildings. Typical of the family, males are generally patterned differently from the females. In the more ground-living lizards of the genus *Lacerta*, males are especially distinctively colored, often featuring greens in contrast to the browns of females. Many species of lacertids display highly variable color and/or scalation in different locations, especially island populations of some species. Five species of *Lacerta* consist of females only, which reproduce partheno-genetically (without a male). The species common in Europe are mainly insect-eaters, although larger species such as the ocellated lizard *L. lepida* also eat vertebrates and fruit.

The most widespread lacertid, and perhaps the most widespread of all lizards, is the viviparous lizard *L. vivipara*, which grows to only 65 millimeters (2½ inches) head–body length. As its name suggests, this lizard gives birth to live young, a trait shared with only two other lizards in the family. This feature has allowed the lizard to be successful in harsh climates that could not be successfully colonized by egg-laying species. By retaining the developing embryos within the body, the female is able to regulate the temperature and moisture experienced by the embryos more closely than if they were left to hatch in some protected site. Interestingly, in the warmer south of its range in Spain and adjacent France, the viviparous lizard is reputed to lay eggs. The occurrence of two reproductive modes in one species highlights the relative ease with which viviparity may evolve from egg retention. Litter size in the viviparous lizard is generally 4 to 11. Clutch size in other lacertids is roughly related to size, with smaller species producing one or a few eggs, and larger forms laying 10 or more. Mating occurs in spring, and eggs are laid by early

summer, hatching in time for young to feed before winter hibernation or inactivity.

The largest living lacertid, the Canary Island lacertid *Gallotia stehlini*, at 46 centimeters (18 inches) is considerably smaller than its now-extinct relatives, also on the Canary Islands. These giants reached sizes of more than 1 meter (39½ inches) total length. Apart from their large size, *Gallotia* species are the most vocal of lacertids and one of the few groups active during twilight or at night.

Members of the genus *Takydromus* range through eastern Asia. They are fully limbed, but like *Tetradactylus* and *Chamaesaura* in the family Cordylidae they have extremely long tails and are able to "swim" through dense groundcover.

Very few lacertids have adapted to life in tropical forests, but several major groups of lacertids have become specialized for a sand-dwelling existence. The most extreme case is that of the shovel-snouted lizard *Meroles anchietae*, endemic to the coastal Namib Desert. Like many dune lizards, its wedge-shaped head features a countersunk lower jaw, and its feet are strongly fringed. This species is an evolutionary extreme in a genus that, in general, shows some degree of modification for sand-dwelling. *Acanthodactylus*, a genus from North Africa and the Middle East, shows many of the same characteristics and occupies comparable habitats in the Sahara and its surroundings, although these lizards are often

▶ *The viviparous lizard* Lacerta vivipara *forms eggs, but these are generally retained in the body until the time of hatching, resulting in the birth of miniature replicas of itself, a habit otherwise unusual in the lacertid lizards. This hardy and widespread species occurs in Europe and northern Asia.*

L. Campbell/NHPA

Anthony Bannister/NHPA

◄ *With its spade-shaped snout and underslung jaw, the sand-diving lizard* Meroles cunierostris *is one of a number of closely related lacertid lizards adapted to loose sand conditions in the deserts of southern Africa.*

more generalized in their ability to use a variety of microhabitats. Another African species, the bushveld lizard *Heliobolus lugubris*, is largely a termite-eater in grassy savannas and open country. Although the adults are not especially strikingly patterned, the juveniles are black with conspicuous white markings. They may be seen walking in a scuttling manner with the back arched high. This odd behavior is a defensive mechanism, as the juvenile in this posture resembles the oogpister, a noxious beetle occurring in the same habitat.

Skinks

The family Scincidae contains more than 1,300 species in about 85 genera, making it the largest of all lizard families. Skinks are cosmopolitan in their distribution, although there are relatively few in Europe, northern Asia, South America or much of North America. Many skinks are secretive and inhabit leaf-litter, rock crevices, or rotting logs. Limb reduction, a common trend in many lizard

families, has occurred in numerous groups of skinks and a variety of limbed and limbless forms are burrowers. Nearly all skinks are covered by relatively smooth, overlapping scales, giving them a fish-like appearance, as reflected in such common names as "sandfish" (*Scincus*) and "land mullet" (*Egernia*). Most skinks are primarily gray or brown in color, although bright colors are sometimes seen in males and juveniles. Many of the more secretive forms are nocturnal or crepuscular (twilight-active), but many others are diurnal (day-active).

Four subfamilies of skinks are recognized: feylinines, acontines, scincines, and lygosomines. Feylinines are a small group of limbless burrowing skinks found only in tropical central and west Africa. The six species are apparently all viviparous (giving birth to live young).

The acontine skinks, 17 species of limbless African lizards, are more widespread and occupy a wider range of habitats. Many occur in arid regions of southern Africa, where they live in

young of the five-lined skink *E. fasciatus* have bright blue tails, although the coloration is lost as the animals grow. Tail loss is common among skinks, and in many populations most adults have regenerated this appendage. These skinks also show sexual dimorphism: males are larger than females and lose the patterning seen in juveniles, whereas this is retained by females. In the breeding season males of the five-lined skink develop orange color on the head, and the width of the head increases as the cells of the jaw muscles increase in size (hypertrophy). Males recognize each other visually and chemically and engage in combat for territories and access to females. There is little courtship display in most skinks, and tactile and chemical information is exchanged before copulation in most cases.

Many of the remaining genera of scincine skinks are small groups living in Africa, Madagascar, the Philippines, and the Middle East.

▲▶ *Like many other lizards, most skinks can discard their tail as a defense mechanism. As a consequence, many species have evolved brilliantly colored tails to deliberately draw the predator's attention away from the vulnerable head and trunk. This adaptation is often accompanied by a striped body pattern, a confusing target for a predator. Compare the unrelated red-tailed skink* Morethia ruficanda exquisita *(above), an Australian lygosomine, and a juvenile five-lined skink* Eumeces fasciatus *(right), a scincine from North America.*

sandy soils. The Cape legless skink *Acontias meleagris* of South Africa may be found beneath rocks or surface debris; if exposed it rapidly dives downward into the soil. In addition to having lost their limbs, these skinks have large headshields that are used in burrowing and, as is typical of many burrowing skinks in all subfamilies, the eyes are small and there is no external ear or exposed tympanum. The giant legless skink *A. plumbeus* may reach a length of 55 centimeters (21½ inches) and gives birth to as many as 14 young. Smaller viviparous species usually have only one to four young in a litter.

The scincines may be the most primitive living skinks. They are widely distributed, except for Australia and the Pacific islands, and members of the genus *Eumeces* are the most numerous skinks in North America. Some, such as the Great Plains skink *E. obsoletus*, show a high degree of parental care, and females remain with the eggs during the incubation period; they may protect the eggs from predation and/or from mold or spoilage. The

Among the reduced-limb forms are the 15 species of the African genus *Scelotes*, which show a variety of stages in limb loss. The sandfish (genus *Scincus*) are also members of this group. These are sand-swimming lizards of North Africa and southwest Asia. Their modifications for life within the sand, where they spend most of their time, include a countersunk lower jaw, exceptionally smooth scales, and protected ears. Unlike *Scelotes*, the sandfish are egg-layers.

The lygosomines are the largest group of skinks and account for all of the species occurring in Australia and islands of the Pacific Ocean, as well as many Asian and African and a few Central and South American species. Members of the *Mabuya* genus occur on several continents, but it is probable that not all species are closely related. Most species are placed in the genus because they lack the more distinctive features characteristic of other groups; therefore, most are rather generalized skinks, either terrestrial or rock-dwelling and usually insect-eating. Many of the

African skinks are classified in this genus. They include a large number of largely rock-dwelling forms, as well as some specialized species. The wedge-snouted skink *Mabuya acutilabris*, for example, occurs in southern Africa where it occupies burrows at the bases of grass tussocks and bushes in areas of sandy soil. It is more active than many other skinks and behaves somewhat like the lacertid lizards, which occur in the same area. Reproductive mode varies in the genus, and in the Cape skink *M. capensis* both live birth and

egg-laying have been documented. In the South American Brazilian skink *M. heathi*, the fertilized egg is very small at first, and the developing embryo receives nutrition not from yolk but from the mother via a mammalian-like placenta. This is the most highly developed example of live birth in all reptiles.

Among the lygosomines of the Australian region are a group of moderate to very large skinks, allied to land mullets (genus *Egernia*) which are among the most bizarre lizards in the family. One of the

▲ Feeding mainly on worms, the silver sand lizard or two-legged skink Scelotes bipes *is common across much of southern Africa.*

▼ *A relative of the blue-tongued skinks, the Australian pink-tongued skink* Cyclodomorphus gerrardii *has a blue tongue as a juvenile, but adults display a bright pink tongue when threatened.*

Jean-Paul Ferrero/AUSCAPE International

▲ *So-called from its spiny scales, the crocodile skink* Tribolonotus gracilis *inhabits New Guinea.*

▼ *Big and burly, the eastern blue-tongued skink of Australia is likely, when threatened, to stand its ground, puff itself up, hiss, and stick out its brightly colored tongue.*

C.B. & D.W. Frith/Bruce Coleman Limited

strangest of these is the Solomon Islands prehensile-tailed skink *Corucia zebrata*. This large tree-dwelling lizard feeds on plants at night and gives birth to a single large young. Related to the prehensile-tailed skinks is the Australian blue-tongued skink *Tiliqua scincoides*, one of the largest skinks, reaching 33 centimeters (13 inches) head–body length. Like other very large skinks it is partially herbivorous, eating plant material as part of a broad diet. Live birth has evolved independently in at least 22 lineages of skinks, most frequently in lygosomines. In the blue-tongued skink about ten young are born, whereas in the shingleback *T. rugosa* only two, very large twins comprise the litter. The shingleback is also huge in comparison to most skinks and is distinguished by the large rough scales that cover its body. In this species the tail is especially short and resembles the head; it may serve to disorient predators and misdirect attacks. Another genus in this group is *Tribolonotus*, the spiny skinks of New Guinea and the Solomon Islands. These are the least "skink-like" members of the family, in that the scales are uniquely covered with small bumpy tubercules.

Another group of lygosomines, including *Emoia* (72 species) and *Cryptoblepharus*, is also chiefly distributed in the Australian region. These are among the only lizards other than geckos to have successfully colonized the islands of the central and eastern Pacific. Also included in this group are the genus *Carlia* (21 species in Australia and more elsewhere) and all of the skinks of New Zealand (about 30 species).

The third group of lygosomines includes many sand-swimmers or burrowing forms from Australia, New Guinea, and Asia. Among these are *Ctenotus* (about 70 species); and *Lerista*, a genus of 51 Australian species of skinks with small or reduced limbs, most of them are sand-swimmers inhabiting the arid regions of the continent. Although the small limbs (if present) may be used when on the surface, movement in the sandy soil is by lateral (side-to-side) undulations of the body. *L. bougainvillii* is one of the most fully limbed forms, while *L. apoda* has lost all external traces of both fore- and hind limbs.

Tropical Asia is home to a number of tree-dwelling lygosomine skinks. One of these, the green-blooded skink *Prasinohaema virens*, which also occurs in the New Guinea region, is particularly noteworthy because it has scansors on its toes (a development parallel to geckos) and green pigment in the blood. The mucous

---

AUTOTOMY: SHEDDING THE TAIL

Most lizards are relatively small and cannot repel attackers by using force, although many will bite, lash with the tail, cry out, or defecate when attacked. Another mechanism of escape is autotomy, the loss of body parts. Although autotomy is common in many invertebrates, it occurs in vertebrates only in some salamanders, a few mammals (several rodents) and the majority of lizards. At least some lizards of all families except the Agamidae, Chamaeleonidae, Helodermatidae, Lanthanotidae, Xenosauridae, and Varanidae, may drop the tail or part of it when grasped by a predator. In many cases the tail continues to wriggle long after it is detached, perhaps serving to distract the predator and thus buying time for the lizard to escape.

Tail loss may carry a high cost. The tail is a common site of fat deposition in lizards, and loss of these important stores may decrease survival rates during winter and/or reproductive output. In addition, the lizards must cope with the temporary loss of any specialized tail functions, such as those in locomotion, grasping, or social behavior. Males of the side-blotched lizard *Uta stansburiana* (an iguanid) that are deprived of their tails may even suffer a decrease in social status and thus opportunity to breed.

In most lizards capable of autotomy, the rupture of the tail occurs not between adjacent vertebrae but within a single vertebral unit. A zone of weakness, developed prior to birth or hatching, runs through each of the autotomic vertebrae. This division corresponds to a boundary between two muscle segments and is continued to the surface, where in this region the skin is somewhat weaker. Blood loss is minimal in autotomy, and a lizard that escapes will regrow the tail over a period of months, although the lost vertebrae are replaced by a cartilaginous rod and the muscles and scales that regrow are generally irregular. The dwarf day geckos of Africa (genus *Lygodactylus*) and some other species have specialized tail forms, such as adhesive tail pads similar to those on the toes. Even these, with only minor modifications, are faithfully reproduced by the regeneration process.

Autotomy of a different type characterizes some geckos. In the bronze gecko *Ailuronyx seychellensis* of the Seychelles and the Madagascan genus *Geckolepis*, as well as some others, the skin on the animal's back is weakened by gaps in its fibrous structure. When predators grasp the lizard, the skin (or rather, most of the skin, for a thin layer remains to protect the underlying tissues) rips away. As in tail autotomy the loss of part of the body may distract a predator long enough to allow the lizard to escape.

Michael and Patricia Fogden

▲ *A tokay gecko Gecko gecko with a regenerated tail.*

membranes, bones, and eggs of this skink are also green. The pigment is similar to that found in bile (a digestive fluid) which is also green. The function of this pigment, if any, is unknown.

Macroteiids

Macroteiid lizards (family Teiidae) are remarkably similar in appearance to the lacertids, and they are geographically complementary—lacertids occur in the Old World, and macroteiids in the New World (the Americas). There are about 100 species in nine genera. All macroteiids have well-developed limbs and are primarily terrestrial and day-active. Head scales are usually large, dorsal scales are tiny, and ventral scales are enlarged and rectangular. Egg-laying is the universal mode of reproduction in the group.

Whiptail lizards (genus *Cnemidophorus*) are the only macroteiids in the United States and are the smallest members of the family. Whiptails are active foragers, at times moving almost constantly in search of prey. Insects and other arthropods are the principle components of the diet. Prey such as termites are often excavated by digging. The western whiptail *C. tigris*, from western North America, is typical in its preference for high temperatures. This lizard is active at body temperatures of about 40°C (104°F). As a result its metabolic rate is high, as is the demand for more energy (which means more food). During periods of inclement weather and the colder months of the year the lizard retreats to a burrow, usually near the base of a shrub or bush. The female lays small clutches of eggs, but may produce two clutches per year. Low clutch size (one to six eggs) is the rule for whiptails as a whole. While

▼ *The checkered whiptail* Cnemidophorus tesselatus *(below) is the largest teiid native to North America. A virtually all-female species, these lizards are able to lay fertile eggs in the absence of males; in fact few males have ever been found. Essentially a large version of the whiptails, the jungle runner* Ameiva ameiva *(bottom) is a wide-ranging species consisting of several races spread throughout much of Latin America.*

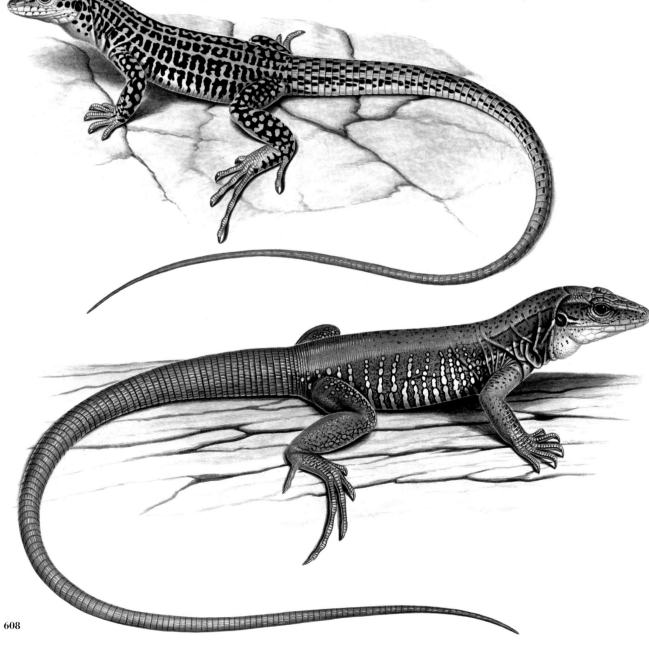

the western whiptail is a bisexual species, parthenogenesis (reproduction without fertilization) characterizes at least 12 species. *Cnemidophorus* species are widely distributed in North America and extend south to Argentina. They are joined in Mexico and further south by the genus *Ameiva*, the jungle-runners. Both genera occur in a variety of habitats from desert to tropical forests. Some species in both groups are sexually dichromatic, with the males having brightly colored markings on the back and sides, used in displays to rival males and in courtship.

Larger macroteiids have a broad diet and will eat insects, vertebrates, and even some plant material. *Callopistes* inhabits valleys in the Andean region and along the coast of Chile and Argentina and is a specialist feeder on other lizards, mostly iguanids. One of the largest species, the tegu *Tupinambis teguixin*, occurs throughout much of tropical and subtropical South America and is primarily a predator on small mammals. Some tegus grow almost as big as the more massive monitor lizards, with head–body lengths of up to 45 centimeters (17¾ inches). As many as 32 eggs, a record for the family, may be laid by females of this species. The caiman lizard *Dracaena guianensis* rivals the tegu as the largest macroteiid, with a total length of up to 1.3 meters (4¼ feet). It has unusally large, flattened teeth, which it uses to crush the shells of mollusks, its primary food. It is one of a few semi-aquatic macroteiids and has a compressed tail and large tubercles on the back.

Many of the larger macroteiids are captured for food by rural people in South America. Tegus are especially valued as meat and for their skins, which are sold commercially to be made into fashionable lizard-skin leathergoods.

Microteiids

Microteiid lizards (family Gymnophthalmidae) are far more structurally diverse than macroteiids, with which they are grouped by some scientists. Like macroteiids, all are oviparous (egg-laying) and there are some parthenogenetic (unisexual) forms. There are 130 species in 30 genera, ranging from Mexico southwards through South America, achieving their greatest diversity in tropical South America. Most are small, leaf-litter-dwelling species that are active at night or intermittently active in their sheltered microhabitat throughout much of the day. Like many other groups, the microteiids include reduced-limb forms; these occur in eight different genera, suggesting that reduction has been a recurrent theme in microteiid evolution. Members of the genus *Bachia* show a wide range of stages in reduction. Unlike other lizards (but in common with the amphisbaenian *Bipes*), the hind limbs are reduced more than the forelimbs. *Bachia trinasale* possesses only two small forelegs, the hind limb rudiments being entirely internal. Members of this

Jany Sauvanet/NHPA

genus are burrowers, digging tunnels in the forest soil beneath leaf-litter. If threatened on the surface, however, they may leap erratically. In contrast, the litter-dwelling microteiids of the genus *Echinosaura* "freeze" when threatened and rely on camouflage to avoid predation. Lizards in genus *Neusticurus* are water-loving and prefer swampy areas within the tropical forest. When disturbed they retreat to water. Most species of microteiids prefer shaded conditions and are most active in the rainforest or on overcast days. *Cercosaura ocellata*, however, seeks out sunny spots on the forest floor, as do the very skink-like genera *Tretioscincus* and *Gymnophthalmus*.

Night lizards

The night lizards (family Xantusiidae) are a group of 19 species of New World lizards. The name may be somewhat misleading, since it has been found that some species are active by day, but usually in hidden microhabitats. The species that have been studied generally have a low preferred body temperature, averaging 23°C (73°F), nearly 10 degrees lower than most lizards. All species are viviparous, giving birth to one to eight live young after a gestation of three months. Night lizards have been considered as relatives of geckos, with which they share certain features, such as the absence of eyelids and the presence of a protective spectacle, but they are probably most closely related to skinks, lacertids, and teiids.

The largest genus is *Lepidophyma*, with 14 species distributed from Mexico to Panama, inhabiting both tropical lowland forests and rocky semi-arid zones at high elevation. Most species have rows of enlarged tubercles on their back. Little is known about the biology of these animals, but most seem to be nocturnal or active at

▲ *The common tegu* Tupinambis teguixin *is one of the largest members of the family Teiidae, a group of energetic, terrestrial lizards confined to the Americas.*

▶ *The tree-dwelling flying lizards (genus* Draco*) of the Indonesian region have five to seven pairs of ribs, much lengthened, hinged, and connected by a membrane in such a way that when pulled forward they form a parachute. Such a surface allows glides of up to 60 meters (about 65 yards) or more, with a considerable degree of control.*

Jean-Paul Ferrero/AUSCAPE International

Stephen Dalton/NHPA

▲ *In the flying geckos (genus* Ptychozoon*) of Southeast Asia folds of skin extend along the sides of the body and tail, increasing the total surface area presented to the air and thus allowing some degree of control over the rate and direction of fall.*

GLIDING THROUGH THE TREES

Several lizards have a body built for gliding. None of them can use flapping or powered flight, but they are able to slow their rate of descent and move through the air in a controlled manner. When the rate of descent is less than 45° from the horizontal this is referred to as gliding; when the angle is steeper it is parachuting. Only the flying dragons (genus *Draco*, in the family Agamidae) can glide effectively, but parachuting occurs in the "gliding" geckos (*Ptychozoon*), and the "gliding" lacertid (*Holaspis*). Each of these lizards has a way of enlarging their surface area to increase drag and thus slow the descent. In *Holaspis* the increased surface consists of fringed scales on the toes and along the sides of the tail. In *Ptychozoon* fleshy fat-containing flaps extend along the sides of the body and the tail, and webs span the spaces between the toes. In *Draco* there is an altogether different type of mechanism: five to seven ribs are lengthened and support a membrane that can be extended or folded back against the body by

muscular action. In "flight" the ribs are pulled forward, stretching out the membrane to provide the maximum surface area for lift.

Some lizards, such as the green anole *Anolis carolinensis* and the butterfly agamid *Leiolepis belliana*, have no structural modifications for parachuting but are behaviorally equipped to assume a posture that maximizes air resistance when they fall (or are dropped) from a significant height.

Gliding or parachuting may be a mode of escape or a means of locomotion. The distance covered by a gliding gecko when it parachutes is relatively short (less than 10 meters, or 33 feet), but it allows the animal to rapidly leave an area of potential danger or move between trees in the forest. Aerial movements of *Draco* may be more impressive, with horizontal distances of up to 60 meters (65 yards) being covered during only a small vertical descent, and fine movements of the membrane and tail give precise control of direction, speed, distance, and angle of descent.

twilight, and some occupy rotting logs or even caves. Most species are insect-eaters, but Smith's night lizard *L. smithii*, one of the cave-dwelling species from Mexico, feeds on fallen fruit from fig trees near its rocky retreats. Some southern populations of *L. flavimaculatum* consist only of females. The Cuban night lizard *Cricosaura typica* is the sole representative of a second genus of the family.

A third genus, *Xantusia*, occurs in northern Mexico and the southwestern United States. They are characterized by small granular dorsal scales contrasting with large smooth head shields and large rectangular ventral scales. These lizards have small litters but are reasonably long-lived, regularly reaching the age of five years, and as many as 12 to 15 years in the largest species. The desert night lizard *X. vigilis* was considered to be rare until it was discovered that individuals of some populations spend much of their life among fallen debris of Joshua trees and other plants, where their density may be very high. Like others in its genus, this species matures slowly and is long-lived. The granite night lizard *X. henshawi* lives in southern California, but unlike the desert night lizard, it is distributed only in regions with granite boulders, where it occupies crevices. Its litter size of one or two per year is one of the lowest for any lizard. The island night lizard *X. riversiana* occurs only on the Channel Islands off the coast of southern California, where it reaches densities of several thousands per hectare in prime habitat. The lack of competitors and the protected habitat allows enormous concentrations to build up in spite of the low reproductive rate, but the island night lizard is considered threatened because it occupies only a few small islands—the species as a whole would be susceptible to predation and habitat destruction if mammals were introduced to the islands. The island night lizard is much larger than mainland species: 90 millimeters (3½ inches) head–body length, compared to 35 to 70 millimeters (1⅓ to 2¾ inches). And unlike

most other *Xantusia* species, it eats plants as well as insects.

CONSERVATION

While very few lizards are dangerous to humans, humans are most decidedly dangerous to lizards. Many of the larger lizards, especially monitors and large teiids such as the tegu, continue to be exploited for their skins, which are fashioned into leathergoods.

Accidental introduction of rats and intentional introduction of cats and other predators has greatly diminished populations of some lizards, especially on islands. In New Zealand skinks and geckos that once occupied the main islands now occur only on small offshore islands which are less disturbed. On Guam, the brown tree snake *Boiga irregularis* has caused the extirpation of both birds and lizards. Even other lizards introduced into new ecosystems may have negative effects on native forms. In several areas of the Pacific, for example, the house gecko *Hemidactylus frenatus*, introduced to many islands during and after the Second World War, seems to be causing a dramatic decline in numbers of the Indo-Pacific gecko *H. garnotii*, which had previously been numerically dominant.

Our greatest impact on lizards has certainly been the destruction of their habitats. Alteration of habitats in agriculture, recreation, and urban expansion effectively destroy most resident lizard populations. This is especially true because most lizards cannot flee to distant undisturbed areas or even to adjacent plots of land. Only in rare cases, such as that of the Coachella Valley fringe-toed lizard *Uma inornata*, in California, have conservationists been successful in setting aside land specifically for lizard preservation. In all too many cases, lizard species are affected long before biologists understand their ecology and habitat requirements, and in some cases the lizards become extinct before there is time for them to be described scientifically.

▼ *The granite night lizard* Xantusia henshawi *is an inhabitant of granite outcrops in arid and semi-arid regions of North and Central America; its somewhat flattened body allows it to slide in and out of crevices easily.*

SNAKES

Order Squamata
Suborder Serpentes
c. 10 families, *c.* 420 genera,
c. 2,500 species

Size

Smallest Many threadsnakes
(family Leptotyphlopidae)
and blind wormsnakes
(family Anomalepidae) are
less than 15 centimeters
(6 inches); maximum weight
less than 2 grams
(¹⁄₁₀ ounce).
Largest Anaconda *Eunectes
murinus,* length up to
10 meters (33 feet); weight
250 kilograms (550 pounds).

Conservation Watch

The following species are
listed as endangered in the
IUCN Red Data Book of
Threatened Animals: Round
Island boa *Bolyeria
multicarinata,* Round Island
keel-scaled boa *Casarea
dussumieri,* Puerto Rican boa
Epicrates inornatus, Central
Asian cobra *Naja oxiana,*
Lebetine viper *Vipera
schweizeri,* and Latifi's viper
Vipera latifii. Many others are
listed as vulnerable or of
indeterminate status.

RICHARD SHINE

Many reptiles show evolutionary tendencies toward lengthening the body and reducing the size of the limbs, but it is the snakes that have developed most successfully in this way. They have diversified dramatically during recent geological history and now inhabit most parts of the planet outside the polar regions—from alpine meadows to the open ocean. They are all carnivorous but use a wide variety of methods to find and overpower their prey. Their diets are equally diverse: some species feed on the tiny eggs and larvae of ants, whereas others can eat animals as large as antelopes, tapirs, and wallabies.

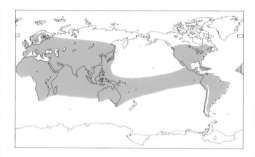

WHAT IS A SNAKE?

Many people assume that a snake is simply a reptile without legs. However, many kinds of lizards have legs so small that they are difficult or impossible to see without close inspection, making it hard to tell whether the animal is a snake or a lizard. Unfortunately there is no simple diagnostic character that is easy to use in all situations. Most lizards (although not all) have at least a vestige of the hind limbs, even if it is just a small bud or a flap of skin, whereas snakes generally do not. Most lizards have external eardrums, which snakes do not. Most lizards have relatively long tails, whereas snakes usually have short ones. Most lizards have movable eyelids, whereas snakes have a fixed transparent scale over each eye. The differences that are absolutely consistent and reliable in distinguishing between snakes and lizards are all fairly subtle, mostly involving the structure of bones in the head. This won't be of much use if you want to identify the long, thin object that you've just seen disappearing under your house, but it does emphasize the great similarity between snakes and lizards—which is why they are placed in the same order, Squamata, within the class Reptilia.

HOW SNAKES MOVE

The most obvious distinguishing feature of snakes is their shape. Lengthening of the body and reduction or loss of limbs has occurred many times in the evolution of the vertebrates—for example, in eels, salamanders, caecilians, and lizards. This evolutionary change in body shape has profound consequences for an animal's

biology, most obviously in the way the animal moves around. Snakes have several features in their vertebral column that are related to limbless locomotion. Firstly, the number of vertebrae are greatly increased, providing a much more flexible backbone. Humans have only 32 vertebrae, whereas some snakes have more than 400. Secondly, snakes have extra projections from each vertebral element, so that adjoining vertebrae are connected more tightly, helping to provide stability to this extremely long backbone.

Even without legs, several different methods of locomotion are possible. The most familiar technique used by snakes is lateral undulation. All snakes seem to be capable of using this method when they are swimming or when they are moving over solid surfaces that have enough irregularities for them to obtain sufficient grip. The snake moves forward by pushing its body against these irregularities and can often travel quite rapidly. Speeds of up to 10 kilometers (6 miles) per hour have been reliably measured. Much faster speeds have been claimed, but none of these are anywhere near the kinds of speeds

▼ *The brilliant coloration of the green tree python* Chondropython viridis *of Papua New Guinea and northern Australia camouflages it among epiphytic plants in its rainforest habitat.*

Jean-Paul Ferrero/AUSCAPE International

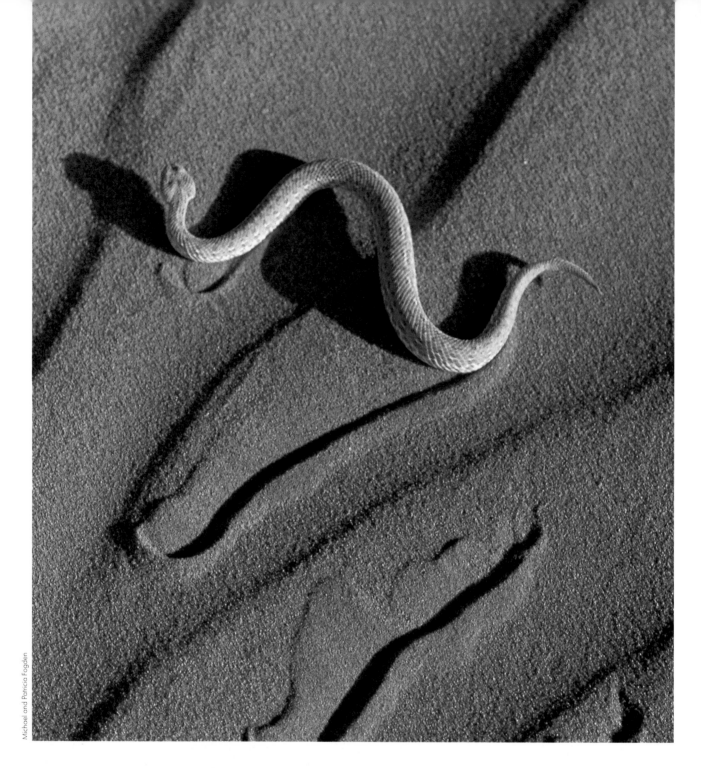

▲ *The most efficient mode of locomotion across loose sand, sidewinding has developed independently in several species of desert-dwelling vipers. Here a Peringuey's desert adder Bitis peringueyi sidewinds its way across a dune in the Namib Desert.*

often attributed to snakes in comic books. No snake in the world is capable of overtaking a galloping horse, as in one common myth. If the surface of the ground is very smooth, snakes can move by other techniques such as rectilinear locomotion, concertina locomotion, slide-pushing, and sidewinding. In rectilinear locomotion (the "caterpillar crawl") the snake stretches out in a straight line and depends on movements of the ventral skin (the underside) relative to the rest of

the body. The snake pulls itself forward by muscular contraction while anchoring its belly-scales using friction on the ground. It then pulls the ventral scales forward to a new friction point and repeats the process. This ventral anchoring and forward movement take place simultaneously at independent segments of the body. Large pythons, boas, and vipers use rectilinear locomotion frequently, especially when creeping up to prey across open ground. The snake's movements are subtle and difficult to detect when it is moving in this way.

In concertina locomotion and sidewinding, the snake uses a point of contact with the ground as purchase, then lifts its trunk clear of the ground to establish another point of contact. Sidewinding is particularly well suited to soft substrates such as sand and soft mud, where it would be difficult to find firm irregularities allowing lateral undulation. Small desert-dwelling vipers and mudflat-inhabiting colubrid snakes rely heavily on sidewinding to move around.

A few colubrid snakes of the Indo-Pacific jungles are known as "flying snakes" because of their unusual method of moving from tree to tree. They launch themselves from high trees and by flattening their bodies can glide for considerable distances without being injured when landing.

A NARROW ADVANTAGE

A snake is really just a long tube. Unfortunately, elongation means that the mouth is very small relative to the size of the body, and therefore to the amount of food required. Elongate limbless vertebrates have adopted several ways to overcome this problem. Some eat large numbers of very small prey items, which can easily be ingested even by an animal with a small head. This is the most common solution used by lizards, and a few snakes, such as wormsnakes. Other elongate vertebrates catch larger prey and tear off pieces small enough to swallow. Amphisbaenians use this approach. Most snakes, however, have a third solution: drastic modifications to the skull, to enable the snake to ingest prey items that are very large relative to its own size.

As a result the head of a snake is very different from that of other reptiles. There has been a general loosening of attachments to permit greater flexibility, so that the snake's skull contains several points at which adjacent bones can move relative to each other. Most importantly, the two halves of the lower jaw are not rigidly fused together, but instead are joined at the front by an elastic ligament that allows them to stretch far apart. The opening to the windpipe can be extruded to one side so that the snake can keep breathing while it is engaged in subduing and swallowing a large prey item, a process that may take many hours. Most elements of the snake skull are reduced, permitting greater flexibility, although the floor of the braincase is thickened and provides protection for the brain against injury from struggling prey.

This complex reorganization of skull structure allows snakes to swallow truly prodigious meals. Many snakes routinely feed on prey much larger in diameter than their own heads, and some species have been observed eating prey weighing considerably more than themselves. Popular attention has been focused on large snakes and

Michael and Patricia Fogden

▲ *A spider accidentally stumbles across a blunt-headed tree snake subduing an anole.*

Stan Osolinski/Oxford Scientific Films

▲ *In a spectacular demonstration of skull flexibility, an African rock python Python sebae distends its mouth to swallow an impala. Although many snakes habitually swallow animals larger than their heads, prey items the size of this antelope are uncommon.*

large meals, such as pythons swallowing pigs, antelopes and wallabies, but the achievements of smaller snakes are just as impressive. Comparative studies suggest that vipers can swallow large prey items more quickly and efficiently than can other types of snakes.

Because they can eat such huge meals relative to their own size, snakes can survive for long periods without feeding. This ability is increased by the low rate at which snakes use energy for their own bodily processes. As ectothermic ("cold-blooded") animals, they don't need to expend large amounts of energy just to maintain a high body temperature (as do birds and mammals). Thus, many snakes probably eat very infrequently, perhaps only a few times a year. This means that snakes (and other ectotherms, such as lizards and salamanders) may be able to survive in areas where the food supply is low or erratic. The ability to survive under these conditions is probably one of the reasons that snakes and lizards have been more successful in many desert areas than the supposedly "advanced" birds and mammals.

Ectothermy confers other advantages as well. Because mammals and birds must maintain high stable body temperatures, they face enormous problems in losing heat to the environment. To reduce that heat loss, such animals must be covered in insulation (fur, feathers, or blubber) and have to be fairly round in shape. Elongate bodies have a much greater relative surface area, and so a greater area over which heat can be lost to the surrounding air. Ectotherms don't face this problem, and in terms of energy loss and gain they can "afford" to be any shape at all. An elongate shape is often an advantage—for example, it allows the animal to enter narrow crevices for food or shelter—and this is probably the reason why elongate shapes are so common in ectotherms (fishes, some amphibians, reptiles) but not endotherms (birds, mammals). For the same reason—that is, the ratio of surface area to volume—warm-blooded animals must be relatively large. Ectotherms don't have to be, and so they can operate with much smaller body sizes.

The elongate shape of snakes also has advantages in terms of heat gain and loss. Because snakes, especially small ones, have such a high surface area relative to their body volume, they are able to gain heat rapidly when they bask in the sun or press against a sun-warmed rock. They also have the ability to slow down the rate at which they heat or cool, either by changing shape (a tightly coiled snake has a much lower surface area through which heat can be lost or gained) or by physiological changes (especially, redirection of blood flow between surface vessels and deeper ones). The overall result is that an elongate shape gives snakes a great degree of control over their body temperatures.

FOOD AND FEEDING

Because snakes cannot bite or tear their prey to pieces (unlike most lizards), small snakes simply cannot swallow very large prey. Perhaps as a consequence of this gape-limitation, the body size of a snake has a major influence on its feeding habits. Smaller snakes eat smaller prey. In some cases this just means that juvenile and adult snakes eat the same species of prey but that the young snakes eat smaller (younger) prey individuals. This is true, for example, of many snake species with specialized diets, like some of the smaller Australian venomous snakes; they eat scincid lizards throughout their lives, starting out by eating newly hatched skinks and graduating to adult skinks as they themselves grow larger. In snakes that attain larger sizes, the increase in body size often means a change in the type as well as size of prey; for example, many pythons and vipers feed on lizards when they are young, graduating to larger mammals as they grow larger.

Although many snakes fall into this pattern of having relatively broad diets, which change from place to place and with the size of the snake, other snakes feed exclusively on a single type of prey throughout their lives. Some are very highly specialized indeed. The Australian bandy-bandy *Vermicella annulata*, a brightly black-and-white-banded burrowing snake, feeds only on blindsnakes of the genus *Ramphotyphlops*; it tracks its prey by scent and often eats blindsnakes as large as itself. Several other snake species specialize on lizard eggs; the lizard nests are presumably located by scent, and the snakes' teeth are modified so that each egg shell is slit as it is swallowed. Similar tooth modifications have developed in a variety of unrelated elapid and colubrid snakes that eat the eggs of reptiles as their primary source of food.

Other snakes take the much larger eggs of birds, and the African egg-eating snakes of the genus *Dasypeltis* are specialized colubrids with almost no teeth—only a few tiny ones in the very rear of the lower jaw. Where other snakes have teeth, these egg-eaters have a series of thick folds of gum tissue arranged in accordion-like folds. These folds act as suction cups on the smooth surface of the egg. After the snake swallows the egg—a remarkable feat in itself—it bends its neck sharply so that the egg is pushed against a series of sharp, downward-projecting spines that pierce the shell. These spines are formed by elongated projections from the snake's backbone. The egg's contents then flow down into the snake's stomach, owing to a set of special muscles that close the throat forward of the egg. As the egg

empties and travels further down the snake's throat, the blunt ends of some vertebrae crush the egg, forming a compact bundle of empty shell that can be easily regurgitated when the forward throat muscles have relaxed.

Many sea snakes are also highly specialized feeders. For example, some sea kraits (*Laticauda* species) feed primarily on eels. Other sea snakes (for example, some *Hydrophis* species) have remarkably slender forebodies and tiny heads that enable them to reach deep into crevices to obtain their prey. The turtle-headed sea snakes (*Emydocephalus* species) have vestigial teeth and feed only on the eggs of bottom-spawning marine gobies (a type of fish).

▲ Steadying a bird's egg within its coils, an egg-eating snake *Dasypeltis scabra* begins the arduous task of swallowing its food. Possessing few teeth, this snake uses folds of gum tissue to grip the egg as it is swallowed.

▼ Almost completely engulfed, the egg will soon be pierced by projections on the snake's backbone and the shell regurgitated.

▲ *By flicking its tongue a South American colubrid snake (genus Tachimenys) detects the scent of a tree frog and stalks to within striking distance.*

SENSE ORGANS

Snakes rely on a variety of sense organs to find their prey. Scent is probably the most generally important, and the forked tongue of snakes has been beautifully fashioned to gather information about chemicals in the environment. The two tongue-tips are widely separated. The tongue is in constant motion, regularly extruded from the mouth to sample particles in the air, the water, or on the ground, then withdrawing into the mouth to bring these particles to Jacobson's organ in the roof of the mouth. There the chemicals are analyzed, giving the snake accurate information about the presence of predators or prey in its local environment. In this way, Jacobson's organ has a similar function to the taste and smell organs of humans.

Vision is also important. The eyes of snakes differ considerably from those of other vertebrates and even from those of other reptiles such as lizards. For example, lizards focus their eyes by distorting the lens to change its radius of curvature, whereas snakes focus by moving the lens in relation to the retina. There are also several distinctive features in the eyes of snakes suggesting that their original ancestors may have had greatly reduced eyes—perhaps because they were burrowing creatures. When snakes later adopted life above ground again, larger eyes and better vision re-evolved but not in exactly the same way as before. The eyes of snakes are covered by transparent caps rather than eyelids (this also occurs in some types of lizards), giving them the "unblinking stare" so often interpreted as a sign of malevolent intentions.

Snakes have traditionally been thought to be totally deaf, because they have no external ear openings, eardrums, tympanic cavities, or eustachian tubes. However, they are capable of detecting even faint vibrations through the ground or water, and recent research suggests that some snakes may actually be able to hear airborne sounds as well. The pit organs of boid snakes (pythons and boas) and pit vipers allow them to detect warm-blooded prey because of the slight temperature difference between the prey and its surroundings. In practice, snakes use a combination of all these different senses to find their preferred food. We still have much to learn about the way in which information from these diverse sensory inputs is combined and interpreted in the brains of snakes.

WORMSNAKES

Three separate families of snakes—Anomalepidae, Typhlopidae, and Leptotyphlopidae—are often called "wormsnakes" because they resemble worms in size and general appearance. For example, their eyes are reduced to small, darkly pigmented spots which can tell the difference between light and dark, but probably little else, and their bodies are smoothly cylindrical. The blunt head merges smoothly with the rest of the body, and the tail is short and tipped with a small spine used to anchor the snake so that it can move forward more easily as it burrows through the soil. All are burrowers. In size they range from less than 10 centimeters (4 inches) long as adults, up to heavy-bodied snakes almost 80 centimeters (32 inches) in length. The three families are anatomically more primitive (that is, more like ancestral snakes known from fossils) than other living snakes. For example, most of them retain traces of a pelvic girdle, suggesting the presence of limbs at one time in their history. Their small size means that they do not leave a good fossil record, and this group probably arose much earlier than the oldest wormsnake fossils yet found, which are from the Eocene, about 50 million years ago. Some wormsnakes are very slender, like the 64 species of "threadsnakes" (family Leptotyphlopidae) of the southern United States, the West Indies, Central America, Africa, Arabia, and Pakistan. The 20 species of "blind wormsnakes" (family Anomalepidae) are found in continental Central and South America. The third family (Typhlopidae), known as "blindsnakes", is more diverse and contains about 150 species; most of these are found in Africa, Asia, and Australia but some species also occur in Central America, and one species (discussed later) is found almost worldwide. Although they differ considerably in some anatomical features—for example, threadsnakes have teeth only on their lower jaws, and blindsnakes only on their upper jaws—the general similarity between members of these three families suggests that they are closely related. All are non-venomous, feeding on soft-bodied invertebrates such as worms, or the eggs and larvae of ants and termites.

Wormsnakes rely on scent, rather than their rudimentary eyes, to locate their food. They are adept trail-followers, flicking their tongues in and out to pick up any faint chemical traces left by foraging ants, and analyzing these chemicals with the Jacobson's organ in the roof of the mouth. They can then follow these trails back to their source and find the ant brood. But how can a tiny snake enter an ant colony and defend itself against the bites and stings of the worker ants trying to protect the brood? The Australian

blindsnakes (*Ramphotyphlops* species) manage this simply by being reasonably large and having thick, smooth scales, so that the ants can't get a good enough grip to bite them. The Central American threadsnakes (*Leptotyphlops* species) are smaller, and they apparently repel the attacking ants by secreting repellant chemicals, writhing around to smear this secretion all over their bodies. It is such an effective technique that these tiny snakes can actually join hordes of marauding "army ants" as they travel through the forest, the snakes feeding on eggs and larvae from ant nests taken over by the army ants.

Most wormsnakes reproduce by egg-laying, and

▲ The head of this giant blindsnake Typhlops schlegelii of South Africa shows several adaptations for burrowing. Enlarged scales form shields to protect the shovel-shaped snout and to cover the greatly reduced eyes, which are virtually useless underground.

◄ With their blunt heads, small cylindrical bodies, and rudimentary eyes it is easy to see how blindsnakes, such as Ramphotyphlops nigrescens, came to be known as wormsnakes.

Michael and Patricia Fogden

Esther Beaton/AUSCAPE International

EGGS OR BABIES?

Most snakes reproduce by laying eggs, but some species have evolved a different system. The developing eggs are retained within the mother's body instead of being laid in a nest, so that the young snake does not have to face the world until it is fully formed and ready for an independent life. This evolutionary transition from egg-laying (oviparity) to live-bearing (viviparity) has occurred at least 30 times in the ancestors of living snakes.

In some cases an entire group is live-bearing (like the filesnakes, family Acrochordidae), whereas in others a single group contains both egg-laying and live-bearing members. For example, within the family Boidae all of the boas are live-bearers, whereas all of the pythons are egg-layers. In a few cases, both types of reproduction occur within closely related species. The European smooth snakes, genus *Coronella* (family Colubridae) offer a good example. The two species in this genus are similar in most respects, and their geographic ranges overlap considerably. Nonetheless, the southern smooth snake *C. girondica* lays eggs, whereas the more northern species *C. austriaca* bears live young.

Why has live-bearing evolved in so many types of snakes? The geographic distributions of live-bearers give us a clue. Live-bearers are mostly found in climates cooler than their egg-laying relatives, and they are the only species to penetrate into severely cold areas. It seems that soils in these areas are too cold to allow successful incubation of eggs laid in the ground, whereas eggs retained within the mother's body can be kept much warmer because she can bask in the sun and select warmer shelters. Pregnant females of many live-bearing species spend most of their time basking, and this seems to accelerate the development of the eggs so that birth can occur while temperatures

Anthony Bannister/NHPA

are still favorable for activity and the young snakes can find safety for the winter.

If viviparity confers such advantages, why aren't all snakes live-bearers? The answer is that female snakes that retain developing young suffer *disadvantages* as well: they are physically slowed down by the volume of the litter they carry around, and they usually do not feed during most of their pregnancy. Females of egg-laying species do not suffer these "costs" for as long, and therefore may be less vulnerable to predators and be able to begin feeding again (and perhaps, laying another clutch of eggs) much sooner than the live-bearers.

▲ *Splitting the shell membranes, baby sand racers* Psammophis sibilans *hatch from their leathery eggs.*

Michael and Patricia Fogden

◄ *A female hog-nosed viper* Porthidium nasutus *gives birth, its offspring already on the alert as it emerges from the birth membrane.*

females of one small American threadsnake, *Leptotyphlops dulcis*, may coil around their eggs and stay with them until hatching. It is difficult to imagine what benefits maternal attendance could offer these tiny snakes, but perhaps the female discourages small invertebrates that might otherwise prey on her eggs. Production of live young (viviparity) has been reported in some African blindsnakes (genus *Typhlops*).

Perhaps the most remarkable aspect of the reproductive biology of blindsnakes comes from the tiny flowerpot snake *Ramphotyphlops braminus*. This species is one of the smallest of all snakes (less than 15 centimeters, or 6 inches, as an adult) and has the broadest geographic range of any snake, including many small isolated islands in the Pacific Ocean. This huge range almost certainly results from the snake being accidentally spread around the globe by humans, who unwittingly transfer it in small containers of soil such as flowerpots (hence its common name). Like all wormsnakes, the flowerpot snake is completely harmless, although it has been reported to cause some problems in India by crawling into the ears of people sleeping on the ground. Its reproductive biology is quite bizarre, because all flowerpot snakes are females. They are triploid (that is, they have three sets of each chromosome in each cell, rather than two as in most animals) and reproduce parthenogenetically, with females giving birth to daughters who produce daughters who produce daughters, and so on, without any genetic contribution from a male. The same phenomenon occurs in several groups of lizards and some salamanders. Ultimately this reproductive system is likely to be an evolutionary dead end, because a parthenogenetic species lacks the genetic variation brought about by sexual reproduction. In the shorter term, however, it has obviously been a very sucessful strategy for these miniature snakes.

PIPE SNAKES AND SHIELD-TAILED SNAKES

This group is a confusing mixture of medium-sized fossorial (burrowing) and semi-fossorial species, whose evolutionary relationships are obscure. They have a number of primitive features, such as a large left lung and traces of a pelvic girdle, which are lost in the so-called "advanced" snakes.

The pipe snakes, or aniliids (family Aniliidae), include one species of *Anilius* in northern South America, but there is considerable disagreement on what other snakes should also be included in the family. Some scientists would include *Anomochilus weberi* from Sumatra in the family, whereas others believe that it belongs in a separate group. Another unusual snake, *Loxocemus bicolor* of southern Mexico and Central America, may also be related to the pipe snakes; it burrows in rotting foliage and loose earth and digs shallow

tunnels in the ground, but is not exclusively burrowing. Seven *Cylindrophis* species in Southeast Asia are traditionally regarded as aniliids but may actually be more closely related to another group of Asian burrowing snakes, the shield-tailed snakes (family Uropeltidae). The genus *Cylindrophis* includes both egg-laying and live-bearing species, whereas all of the shield-tailed snakes are probably live-bearers.

Burrowing pipe snakes and shield-tailed snakes use their head to force their way through the soil, and the bones of their skull are solidly united, unlike those of most other snakes. A second unusual feature in pipe snakes, but one shared with boid snakes (described later), is the presence of rudimentary hind limbs, in the form of small spurs on either side of the vent. The shield-tailed snakes are burrowers and take their common name from the greatly enlarged scale near the tip of the tail. The scale may be compressed from side to side (in a few species it resembles a

▼ *Two pipe snakes (family Aniliidae): the blotched pipe snake Cylindrophus maculatus (below) of Sri Lanka, has a defense display which involves hiding its head, then flattening and raising its brightly patterned tail in mimicry of a cobra's head and hood. The relationship of the South American coral pipe snake Anilius scytale (bottom) to other pipe snakes is often debated. Its bright color pattern is in apparent mimicry of venomous coral snakes. However, its main defense tactic is to hide its head and present the blunt tail as an alternative "head".*

diagonally-cut end of a salami sausage) and bears several small ridges or small spines, which hold a clump of soil. The soil blocks the tunnel behind the snake, protecting it from predators (including pipe snakes) which might otherwise capture and eat the shield-tail. The spectacular iridescent colors of the shield-tails (and many other types of snakes that burrow in wet soil) are not due to pigments in the skin, but to the microscopic structure of the scales. These bear small ridges that reduce friction against the surrounding soil while the snake is burrowing, and incidentally diffract light, producing an attractive iridescence with all the colors of the rainbow.

SUNBEAM SNAKE

The sunbeam snake *Xenopeltis unicolor*, a medium-sized ground-dwelling snake of India and Southeast Asia, seems to be only distantly related to other living snakes and is usually placed in a family by itself, the Xenopeltidae. Vestiges of a pelvic girdle and rudimentary hind limbs are primitive features it shares with the pipe snakes, although this may not indicate close relationship. It grows to about 1 meter (3¼ feet) in length and has a round body and glossy scales like those of the shield-tails, presumably for the same reasons. One very unusual feature is that the teeth of the lower jaw are set on a loosely hinged dentary bone. Surprisingly, the sunbeam snake doesn't seem to show any corresponding peculiarities in its diet, because it feeds on a wide variety of vertebrates including other snakes, frogs, and rodents (mice and rats).

THE LARGEST SNAKES: PYTHONS AND BOAS

The 60 species of boid snakes (family Boidae) include the largest of all living snakes, but also many smaller species—for example, the pygmy python *Liasis perthensis* of the western Australian deserts grows to only 60 centimeters (24 inches),

and feeds mostly on geckos and small mice. Nonetheless, it is the larger species that attract the most public interest. The record for the "longest snake in the world" goes to either the reticulated python *Python reticulatus* of Asia, reliably measured up to 10 meters (almost 33 feet), or the semi-aquatic anaconda *Eunectes murinus* of South America, also measured up to at least 10 meters and much more heavy-bodied than the reticulated python. Other giants include the Indian python *Python molurus*, to 9 meters (29½ feet), the African python *P. sebae*, 8 meters (26½ feet) and the Australian scrub python *Morelia amethistina*, more than 7 meters (23 feet).

Many authorities believe that the snakes usually grouped together as "boids" are actually a combination of different types that are only distantly related to each other. The wood snakes (often considered a separate family, the Tropidophiidae) are found from Mexico to northern South America, and many offshore islands as well. These 20 species of small ground-dwelling boa-like snakes seem to be the living representatives of a very ancient lineage, with fossils known from the Paleocene, about 60 million years ago. Their closest relatives may be the two small species of bolyerine snakes *Bolyeria multicarinata* and *Casarea dussumieri*, found only on tiny Round Island in the Indian Ocean, although fossils have also been found on nearby Mauritius. There has been massive environmental degradation on Round Island, mostly because of grazing by introduced pigs, which has destroyed much of the habitat available to the boas on the island. Their extinction seems almost inevitable.

The two other groups of boids both contain species that attain very large body sizes. Pythons (subfamily Pythoninae) are found mostly in the Old World and consist entirely of egg-laying species. In contrast, boas (subfamily Boinae) are mostly in the New World and are viviparous

◀ Best known of the large snakes, the boa constrictor Boa constrictor is by no means the largest, and is dwarfed by several other boids. Commonly thought of as a jungle snake, the numerous subspecies are found in a variety of habitats from semi-desert to rainforest.

Michael and Patricia Fogden

species—that is, the female retains the developing eggs within her oviduct, instead of laying them, and gives birth to fully formed offspring.

Many people have a mental picture of a python as a large, heavy-bodied snake lying on a branch in a tropical forest, waiting to drop onto some unlucky explorer. The truth of the matter is quite different. Many pythons are quite small and do not live in tropical forests at all. Indeed, some are burrowers living in desert sands. In fact, although they have representatives in Africa and Asia, true pythons are most diverse in Australia. While it is true that all pythons are non-venomous, their feeding habits vary from species to species. Some

▼ The blood python Python curtus is a short, stout python of Southeast Asia, where it inhabits swamps, marshes, and rainforest streams. Its common name is derived from the deep red coloring of some individuals.

feed on frogs, some on lizards, some on mammals, some on birds. They all rely on constriction to subdue their prey, throwing a series of coils around the animal as soon as it is seized. These coils suffocate the prey by tightening every time it exhales and preventing it from drawing another breath—although stories of prey animals being crushed to jelly, and of all their bones being broken by constriction, are simply untrue. One desert species, the woma python *Aspidites ramsayi* of central Australia, uses an interesting variation to conventional constriction. This species catches many of its prey down burrows, where there isn't enough room for it to throw coils around the prey. Instead, the woma just pushes a loop of its body against the unlucky mammal so that it is squeezed (and soon suffocated) between the snake and the side of the burrow. Unfortunately for the woma, this technique doesn't immobilize the prey as quickly as would "normal" constriction, so many adult woma pythons are covered in scars from retaliating rodents.

All pythons are egg-layers, like most other reptiles, but pythons are unusual among reptiles in the care they afford their developing eggs. The female may build a "nest" of vegetation, by coiling under loose leaf-litter, for example, or simply selecting an appropriate well-insulated burrow. She coils around the clutch after laying and remains with her eggs until they are ready to hatch. She may leave to bask in the sun or to drink but will not feed until after her maternal duties are complete. Although pythons are generally ectothermic like other reptiles—that is, they rely on heat from the environment to keep themselves warm—brooding female pythons actually generate heat by shivering, and this keeps the eggs at a high and stable temperature throughout development. It is very expensive in terms of the female's own energy reserves: she may lose up to half her own body weight between egg-laying and the end of incubation, and it may take her two or three years to regain enough energy reserves to breed again. However, it means that the eggs develop rapidly to hatching, and are safe when the air temperature is low. Perhaps for this reason, some pythons can reproduce successfully even in relatively cold areas. Like all other snakes, pythons do not take care of their offspring after hatching; their maternal responsibilities finish when the young snakes emerge from the eggs.

Boas are similar in many ways to the pythons,

▼ *Loosening its coils after constricting a rat, a Burmese python* Python molurus bivittatus *repositions the prey to begin swallowing. One of the largest snakes, the rapid growth rate and attractive patterning of this python have resulted in its popularity in both the leather and live animal trades.*

Michael Leach/Oxford Scientific Films

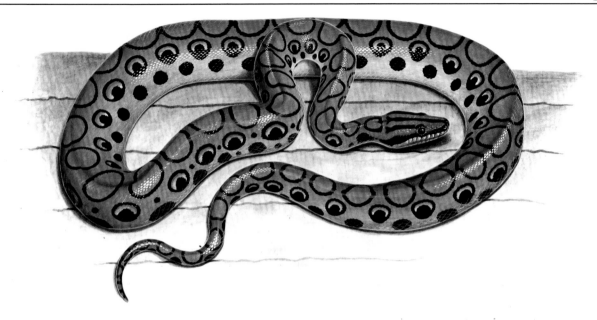

except that they bear live young instead of laying eggs. The best-known boas may be large species, but there are also smaller types. They are mostly found in Central and South America, although two genera occur in North America, the rubber boa *Charina bottae* and the rosy boa *Lichanura trivirgata*, both relatively small and secretive. Small groups have also made their way somehow to Madagascar (*Sanzinia* species) and to New Guinea and nearby Pacific islands (*Candoia* species); their ancestors may have rafted across from Central America, like the ancestors of the Pacific iguana lizards—or perhaps their descendants rafted the other way. The anaconda *Eunectes murinus* is a giant semi-aquatic snake of South American rivers, where it spends much of its time lying in wait at the water's edge for unwary mammals and caiman, a type of crocodilian. Probably the best-known species is the *Boa constrictor*, a medium-sized boa at 4.5 meters (14¾ feet), distributed from northern Mexico to Paraguay and Argentina, and often kept as a pet. The danger posed by large boid snakes is undoubtedly grossly exaggerated, but there are reliable reports of predation on humans (usually children) by unusually large African pythons, Indian pythons, reticulated pythons, and anacondas.

FILESNAKES

In many respects, the three totally aquatic species in this family (the Acrochordidae) are among the most unusual of snakes. Their skin is loose and baggy and looks as if it is one or two sizes too large for the snake. The skin is covered with small granular scales like the surface of a file, giving these snakes their common name. They have lost the enlarged ventral scales, the belly-shields that characterize most terrestrial snakes, and have trouble moving around on land. Another name

that describes them well is "elephant's trunk snake", applied to a freshwater Asian species, *Acrochordus javanicus*. The Arafura filesnake *A. arafurae* lives in rivers and freshwater billabongs in Australia and New Guinea, and the third species, the little filesnake *A. granulatus*, inhabits estuarine and coastal areas of the Indo-Pacific.

When you see a filesnake underwater, its strange skin begins to make sense. The loose skin flattens out ventrally as the snake swims, giving it a flattened profile like a sea snake and thus enabling it to swim more efficiently. Its rough skin is used to capture prey in a rather surprising way. Filesnakes feed on fishes, and constrict them in the same way that pythons constrict their prey on land. Squeezing a slippery fish would be impossible without small roughened scales that

▲ Named for the iridescent sheen of its scales, the Brazilian rainbow boa Epicrates cenchria cenchria (top) feeds on the ground and is often found near village outskirts, where there is a steady supply of rodents. Sand boas are unusual boids in that they are exclusively Old World in distribution and adapted for a burrowing lifestyle. Illustrated is the Kenyan sand boa Eryx colubrinus loveridgei (above), the southern-most representative of this group.

▲ *An underwater close-up of an Arafura filesnake* Acrochordus arafurae *reveals the coarse textured skin with which it holds its slippery fish prey.*

▼ *Gaping in threat, a toad-eater* Xenodon rabdocephalus *displays the enlarged rear teeth with which it punctures toads and frogs.*

can penetrate the slime covering the fish's scales. Filesnakes are unusual in behavior, ecology, and physiology as well as body form and structure. They seem to be specialists in coping with low rates of energy availability: they eat rarely, have little capacity for sustained exercise, and females may reproduce only once every few years. Still, they are able to acquire energy fairly rapidly if food does become available, and can occur in remarkably high numbers in suitable areas. In tropical Australia, Aboriginal people harvest many of these snakes just before the beginning of the annual monsoonal rains, when water levels are at their lowest. Aborigines catch the snakes by groping around blindly in the muddy water, recognizing the snakes by the distinctive feel of their rough skin. They are cooked by the simple technique of throwing them on hot campfire coals, and then eaten in their entirety.

Marine filesnakes are also harvested at sea by humans, but this is a commercial industry based on the value of their skins for clothing and fashion apparel.

COLUBRID SNAKES

Most living snakes belong to a single family, the Colubridae, often called the "harmless snakes", even though a few of them have very toxic venom. About 1,600 species are recognized, and they occur on all continents except Antarctica (which has no snakes at all). This remarkable assemblage is extraordinarily diverse, but apparently fairly recent in geological terms. The earliest definite colubrid fossils come from the Oligocene, about 30 million years ago, and the colubrids are one of the most spectacular success stories in the world's evolutionary history since that time. They are the most common snakes almost everywhere snakes occur, with the notable exception of Australia. They have lost all trace of the pelvic girdle, and in most species the skull is highly modified so that it is very flexible, allowing large prey items to be swallowed.

Venom has evolved independently in several different groups of snakes. Colubrid venom is really just modified salivary secretion from Duvernoy's gland, in the upper jaws. Because saliva contains components that break down tissue and begin the process of digestion, it's easy to imagine the evolutionary pathway leading to the appearance of venom in snakes. All that is required in the early stages is that some toxic saliva trickles down into the prey item as it is held in the snake's mouth. Snakes with more toxic saliva, or with larger teeth so that the saliva could penetrate the prey more easily, had an advantage because their prey were killed more quickly. Over many generations, natural selection favored snakes that had more and more toxic venom, and larger and more elaborate teeth with which to deliver the venom to the prey.

The earliest venomous snakes may have been "rear-fanged" (opisthoglyphous)—that is, their enlarged teeth (fangs) were at the rear of their mouth—and this is the type of system seen in the venomous colubrid snakes today. In bio-mechanical terms, the rear of the mouth is where the tooth can exert the greatest force on a prey item; many non-venomous snakes have enlarged teeth at the rear of their mouths, which they use for slitting relatively hard objects (like eggs) or puncturing prey animals that inflate themselves with air (like toads) and hence may be difficult to ingest. Most snakes subdue and swallow their prey without venom, but several groups have evolved toxic secretions to kill prey more rapidly and perhaps to begin the process of digestion before the prey item reaches the stomach. Among the colubrids, potent venoms have arisen in a number of species, perhaps the best known being three arboreal African snakes; the boomslang *Dispholidus typus* and the two species of vine snakes, genus *Thelotornis*. Snakes of both these genera have caused human fatalities, including famous herpetologists who underrated their danger.

◄ A rear-fanged colubrid, the long-nosed tree snake Ahaetulla nasuta has grooves in front of its horizontally-pupiled eyes to allow unobstructed forward vision for hunting.

HOW OFTEN DO SNAKES REPRODUCE?

In tropical areas where conditions are suitable for reproduction year-round, females of some egg-laying species may produce more than one clutch of eggs in a single year. However, highly seasonal reproductive cycles are common even in tropical snakes. This is probably because most tropical areas actually show significant seasonal variation in characteristics such as rainfall, which may influence the food supply for hatchlings. In temperate-zone habitats, snakes generally produce only a single clutch or litter per year, because winter temperatures are too low for reproduction (and often, for any activity at all).

In many kinds of snakes, adult females don't reproduce every year: instead, a female is likely to skip opportunities for reproduction, perhaps producing offspring only every second or third year. These low reproductive frequencies seem to result from the female's difficulty in accumulating the amount of energy needed to produce a full clutch every year. Particularly in live-bearing species living in cold areas (where the activity season may be only a few months long), a female may have little time to feed between the time she gives birth and the time she must enter her winter retreat. For this reason, populations of geographically widespread species may vary in the frequency at which the females reproduce: females in colder areas may be able to reproduce only once every few years, whereas females of the same species living in warmer areas can repro-duce every year. This pattern is seen in several species, including both the European viperids and the Australian elapids.

Females may also skip a year between successive reproductions if food supply is limited. In the Saharan desert viper *Echis colorata*, some females living near oases (where food is plentiful) reproduce every year, whereas those in drier areas do not. In the European adder and the Australian water python *Liasis fuscus*, few females reproduce during years when rodent populations are low and food is therefore hard to find. Female Arafura filesnakes have very low reproductive frequencies, with an average of only 10 percent of females reproducing each year. The only years of significant reproduction in this species come after exceptionally prolonged flooding, when the snakes can capture many fishes in shallowly flooded areas. In all of these species, however, males tend to mate every year, because they do not need to gather enough energy to produce a large clutch of eggs or young.

▼ The considerable energetic cost of producing relatively large eggs, such as those being laid by this yellow-faced whip snake Demansia psammophis, is a major factor determining the frequency of reproduction in snakes.

Some authorities believe that the family Colubridae is such a huge group that it should be divided into many smaller lineages, but there are still difficulties in identifying relationships. Recent studies using a combination of biochemical and morphological (body structure) techniques have identified some groupings within the Colubridae, but have left many uncertainties; for example, the "burrowing vipers" of the genus *Atractaspis* found in Africa and the Middle East are very distinctive, with no clear relationships to other kinds of snakes. So let us begin with the groups within the Colubridae that do seem to represent natural evolutionary lineages.

Subfamily Homalopsinae

The homalopsines are an array of about 40 rear-fanged aquatic snakes of Asian and Australian waters. One land-dwelling genus, *Brachyorrhos*, probably belongs here too. All are live-bearers, and many are mangrove-dwellers. Some species are very distinctive, like the tentacled snake *Erpeton tentaculum*, so-called because of the strange protuberances on its head. Most of the homalopsines are fish-eaters, but the white-bellied mangrove snake *Fordonia leucobalia* is a specialist on crustaceans, especially crabs. It stalks them at night on mangrove mudflats left by the receding tide. The hard shell of a crab makes it difficult to seize, so the snake uses a special technique. When the snake is close enough, it launches a strike, not *at* the crab, but *above* it. By striking above the crab, the snake's forebody pushes the crab down into the soft mud, and the snake can then turn and bite it more carefully to introduce venom. This seems to stop the crab's struggles quickly, and the snake then proceeds to eat the unfortunate crustacean—or, if the crab is too big, to remove and eat the legs only. This is probably the only snake that can actually tear pieces off a prey item (because the crab sheds its legs quite readily); all others must eat the prey entire.

Subfamily Xenodermatinae

The xenodermatine colubrids of India and Southeast Asia are unusual in having upturned edges on the scales bordering the lip, and expanded bony plates on the spines of the vertebrae. *Xenodermus javanicus* is a frog-eating species found in moist soft earth in wet cultivated areas, such as dykes between rice fields, in parts of Asia.

Subfamily Calamariinae

The calamariines or "dwarf snakes" of East Asia form another clearly differentiated colubrid group, possibly related to the xenodermatines. There are about 80 species, all of them small snakes that seem to feed mostly on earthworms and insects. These small secretive snakes are slow-moving and relatively defenseless, and hence are often eaten by other snakes.

Subfamily Pareatinae

One Southeast Asian group, the pareatines, are particularly interesting because of their specialized diet. Like a distantly related genus of colubrids from tropical Central and South America, *Dipsas*, they feed on snails and have evolved some remarkable modifications of their body structure and behavior to suit them to this unusual diet. The lower jaw is strengthened by the fusion of adjacent scales and can be inserted into the opening of a snail's shell; the long front teeth then hook the snail's body and pull it out with twisting movements. The snake does not consume the snail's shell, which it probably could not digest anyway.

▶ A puzzle to taxonomists, the spotted harlequin snake Homoroselaps lacteus of southern Africa is venomous and has fixed front fangs like an elapid and yet has been classified with both the vipers and the colubrids. Often found in termite mounds, this species feeds on blind snakes and legless lizards.

Subfamily Boodontinae

The boodontine colubrids of Africa and Madagascar are a very large group, including both harmless species and rear-fanged species. Some are aquatic fish-eaters, some are terrestrial with broad diets, some feed mostly on mammals, some on lizards, some specialize on eating other snakes, and one genus (*Duberria*) specializes on slugs. Both egg-laying and live-bearing occur within the group, and even within a single genus (*Aparallactus*). Several of the species within this large and diverse group are known as "house snakes" because they often enter houses to feed. The western keeled snake *Pythonodipsas carinata* is an unusual-looking species that closely resembles horned vipers found in the same area; this mimicry may confuse predators and hence reduce the keeled snake's vulnerability to them.

Stephen Dalton/NHPA

Subfamily Natricinae

One of the most successful colubrid groups in North America, Asia, and Europe is the subfamily Natricinae: garter snakes and their relatives. This group probably had its origin somewhere in the Old World, possibly in Asia, but has spread widely through the New World as well. The Old World forms are mostly egg-layers, like the common grass snake *Natrix natrix*, which is common over much of Europe, even at high latitudes. Egg-laying snakes are usually not able to survive severely cold climatic regions, because they require high soil temperatures for their eggs to develop. Female grass snakes overcome this problem by migrating long distances to find

▲ *The most widespread European snake, the grass snake* Natrix natrix *is associated with water throughout much of its range.*

▼ *Like its close relatives in the genus* Dipsas, *the snail-eater* Sibon annulata *of Central America has a specialized diet of snails.*

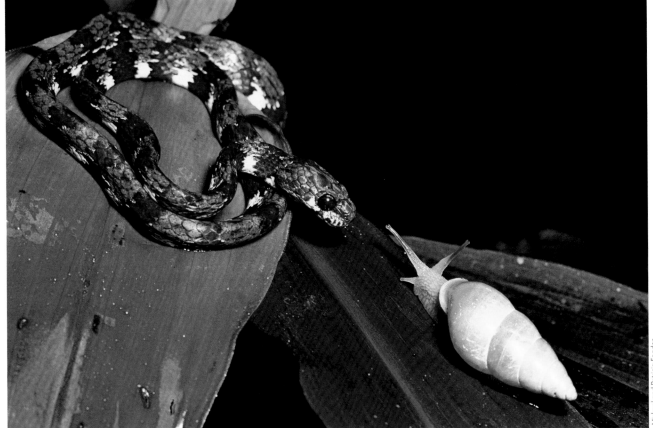

Michael and Patricia Fogden

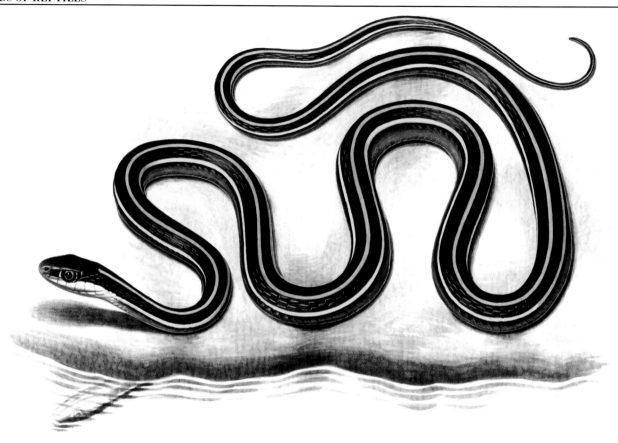

▲ *A slender version of the closely related garter snakes, the eastern ribbon snake* Thamnophis sauritus sauritus *is a typical New World natricine snake in that it is live-bearing and semi-aquatic, rarely being found far from water.*

suitable incubation sites, but even so, such sites are scarce in areas like northern Sweden. How, then, can grass snakes survive in these areas? They do so by taking advantage of manure heaps on farms. These huge piles of cow manure are heated by the action of microbes, and snakes from a wide area may converge onto a single farm and lay thousands of eggs in the same manure pile. In this way, their eggs are kept at high-enough temperatures to complete development and to hatch before the first frosts of winter. In natural conditions, before manure piles were available, the snakes probably nested under large, flat sun-warmed rocks on south-facing slopes because these would have offered the only nest sites warm enough to allow for successful embryonic development.

Not all natricines are egg-layers, however. One semi-aquatic Asian species, *Sinonatrix annularis*, is viviparous, as are all of the New World natricines such as garter snakes and watersnakes. The North American natricines are a spectacularly diverse group and remarkably common in many areas throughout their range. In some severely cold areas in south–central Canada, red-sided garter snakes *Thamnophis sirtalis parietalis* gather together in huge groups to spend winter in the few sites where underground crevices are deep enough for them to escape the winter freeze. These small

snakes may travel many kilometers from their hibernation sites to their summer feeding grounds, the frog-rich swamps and meadows. When the first warm days of spring bring the hibernating snakes out to bask at the entrance to the den, the numbers of snakes to be seen are nothing short of astounding. Males emerge first and wait by the den entrance for the females, which are larger. Reproductively receptive females secrete a special chemical substance, a pheromone, which stimulates mating activity by the males. Because of the huge numbers of snakes present, a female may be surrounded by a writhing ball of dozens of amorous suitors. It seems to be a case of "first come, first served", but some males have an advantage because they also produce the female pheromone or something very like it. This confuses the other males, who don't know who to try to mate with, and as a result these "she-males" are more successful than normal males at mating.

Because garter snakes and watersnakes are so common and so widely distributed in North America, they are among the best-known snakes in many respects. For example, we know a great deal about their feeding habits. Different species of garter snakes found in the same area tend to eat different things, although it is not clear whether this is because of competition between species. Scientists can find out the food preferences of a newborn garter snake by testing its reaction to different prey odors presented to the snake on cotton swabs. It turns out that prey preferences seem to be genetically programed and that they differ between species of garter snakes, and even between different populations of the same species.

Natural selection has also modified the snakes' tolerance to toxins in its prey. For example, garter snakes living in areas where toxic newts are common are very resistant to the poisons produced by the newts. In areas where the newts don't occur, or where they are not as toxic, the local garter snakes don't show this resistance.

Subfamilies Xenodontinae and Pseudoxenodontinae

The xenodontine colubrids of the temperate and tropical New World, including the West Indies and the Galapagos, are less well known ecologically than the other members of the colubrid family. Recent research indicates that the West Indies xenodontines feed mostly on anoline lizards ("false chameleons") and to a lesser extent on frogs, whereas boas and vipers living in the same area eat mainly birds and mammals. The North American hognose snakes (genus *Heterodon*) feed primarily on toads and use their broadened noses for digging their prey out of burrows. The unusual nose is also used in a remarkable defensive display. These heavily-built snakes look rather like vipers, and the resemblance is strengthened by the way they flatten their head and neck when harassed, and hiss and strike vigorously. If this formidable display fails to deter the attacker, the snake adopts a very different strategy: turning over on its back and feigning death.

The pseudoxenodontine colubrids are an Asian group of small to medium-sized snakes that feed mostly on frogs and toads. When attacked, they show a "death-feigning" display similar to that of the North American hognose snakes; the neck and forebody are flattened, the mouth is opened, the lips are drawn back, the tail is vibrated, and the snake rolls over onto its back.

Several different types of toad-eating snakes seem to have independently evolved this kind of dramatic display, and some researchers have suggested that physiological modifications for toad-eating may play some role in the behavior. Toad-eating snakes tend to have large adrenal glands, and perhaps are more likely to go into some kind of shock when attacked.

▼ *The western hognose snake* Heterodon nasicus *uses its shovel-shaped snout to dig for toads, its main prey. These are then dispatched with the snake's enlarged rear teeth.*

John Cancalosi/AUSCAPE International

Jack Dermid/Bruce Coleman Limited

Subfamily Colubrinae

The colubrine subfamily is very diverse and wide-ranging, found over the entire range of the Colubridae family. The evolutionary relationships among this subfamily are particularly difficult to unravel. For example, although all of the North American natricines seem to result from a single ancestral group that came from Eurasia, this isn't true of the North American colubrines; several different migrations seem to have occurred, so that several different evolutionary lineages of colubrines may be present in North America. These include the spectacularly colorful king snakes and milk snakes (genus *Lampropeltis*), which have enormous variation in color even among individuals within a single population.

Colubrines have adapted to a wide variety of ecological niches, with the Indian wolf snake *Lycodon aulicus* often found in houses, where it preys on lizards and mice. The Asian "kukri snakes" of the genus *Oligodon* got their common name from the supposed resemblance of their enlarged rear teeth to the ceremonial dagger (kukri) used by local tribes. Kukri snakes feed mostly on the eggs of reptiles, and their teeth are modified to slit the shell as the egg is swallowed.

Many colubrines have enlarged rear teeth, with or without the development of significant venom. One of the groups with venom—quite toxic in some species—is a genus of slender, tree-dwelling snakes, *Boiga*, which includes some of the most spectacular colubrids. They are nocturnal foragers, remarkably adept climbers, and prey on birds and any other small vertebrate that they encounter. One species, the brown tree snake *B. irregularis*, is widely distributed through Australasia and was accidentally introduced to the tiny Pacific island of Guam after the Second World War, probably in

▲ *The scarlet king snake* Lampropeltis triangulum elapsoides, *a harmless colubrine, is one of several snakes whose banded patterns are believed to mimic the coloration of the highly venomous coral snakes. Like the king cobra, the "king" refers to its habit of eating other snakes.*

▶ *The Mandarin ratsnake* Elaphe mandarina *is a brilliantly colored colubrine of high altitude regions in China. Like other ratsnakes it is a constrictor and feeds on warm-blooded prey, particularly rodents, giving the group its common name.*

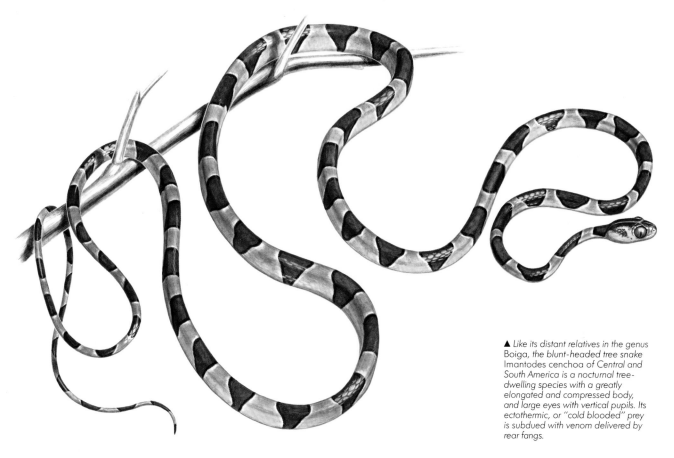

▲ Like its distant relatives in the genus Boiga, the blunt-headed tree snake Imantodes cenchoa of Central and South America is a nocturnal tree-dwelling species with a greatly elongated and compressed body, and large eyes with vertical pupils. Its ectothermic, or "cold blooded" prey is subdued with venom delivered by rear fangs.

military equipment brought back from the jungles of Guadalcanal and similar areas. Guam had no endemic snakes, but is (or was) home to an interesting array of bird species found nowhere else in the world. They had evolved on this snake-free island and so were easy prey to the depredations of the introduced tree snake. Within the past 50 years the tree snakes have increased enormously in numbers, and the native birds of Guam have been driven to the brink of extinction or beyond. Without the birds to control their numbers, the insects have increased unchecked, and the entire jungle ecosystem is under severe threat. This tragic example highlights not only the dangers of introducing "foreign" animals to new areas but also the important ecological role played by snakes. Presumably the evolution of various types of snakes and their dispersal into new areas over evolutionary history has profoundly influenced the nature of the ecosystems around us today.

FIXED-FRONT-FANGS: FAMILY ELAPIDAE

The dangerously venomous snakes belong to two different groups. Both first appear in the fossil record quite recently (about 20 million years ago, in the Miocene) and have fangs at the front of the mouth, and hence in a much better position to deliver venom with a rapid strike. One group (the proteroglyphs), which includes cobras and their relatives, consists of species with "fixed" fangs; the fangs are attached to the upper jawbone like normal teeth. This sets an upper limit to the size of the fangs, which must be small enough for the snake to be able to close its mouth without the downward-projecting fangs piercing the lower jaw and dragging along on the ground.

The other major group of highly venomous snakes, the vipers, have evolved an ingenious solution to this constraint. Their fangs are attached to a small bone that can rotate so that the fangs lie back along the length of the upper jaw when the snake's mouth is closed but can swing forward into striking position when the mouth is opened. Their dentition is known as solenoglyphous, meaning "pipe-tooth", in reference to the large hollow fangs of these snakes. Many vipers have enormously long fangs, whereas the fixed-front-fang group have relatively short fangs. For example, the fangs of the king cobra Ophiophagus hannah (the largest proteroglyph, which grows to more than 5 meters or 16½ feet) are not much larger than those of the adder, one of the smallest viperid snakes.

Most fixed-front-fang snakes, whether they live in the oceans or on the land, are long slender

▲ *Flourishing the trademark of the cobras, the most widely recognized elapids, an agitated black-necked spitting cobra* Naja nigricollis *spreads its hood in a defense display.*

THE BIGGER THE BETTER

Females grow larger than males in most types of snakes. This is true of almost all boids, so the largest snake alive today is almost certainly a female, probably an anaconda in some South American river. Sometimes the disparity between the sexes can be extreme; for example, in one mating pair of Arafura filesnakes *Acrochordus arafurae* the male weighed only a tenth as much as his mate.

It is the general rule in the animal kingdom that the female is larger than the male, and Charles Darwin in the nineteenth century suggested an explanation for this phenomenon. The number of offspring that a female produces depends upon her own body size, so larger females have more eggs or babies. The range in clutch sizes can sometimes be quite wide, even within a single species; for example, in the carpet python *Morelia spilota* young females produce only about six eggs, whereas older larger females often have more than 30. This means that genes resulting in large body sizes for females are likely to increase the number of offspring a female produces during her lifetime, and thus be favored by natural selection. Over long periods of time, genes for large size in females should become more and more common in the population, and females are likely to grow larger than males unless there is some opposing selective force favoring even-larger size in males.

If we look at the kinds of snakes in which males do equal or exceed females in size, we find evidence that the relative size of the two sexes seems to depend on the mating system. Males tend to be larger than their mates mostly in species whose males engage in physical combat with each other during the mating season. This behavior differs from species to species, but generally consists of two males intertwining their

animals that gather their food by actively searching for it. Many that live on land rely on scent to locate prey trails and follow them to the hiding-places of their prey. For example, many elapid snakes are small species that catch sleeping lizards in their nocturnal retreats. Although there are numerous exceptions, most of the fixed-front-fang snakes feed on ectothermic ("cold-blooded") prey such as fishes, frogs, lizards, and other snakes.

Traditionally fixed-front-fang snakes were assigned to two families—the terrestrial snakes such as cobras, kraits, mambas, taipans, and coral snakes in one family, and marine or sea snakes which spend all or part of their lives in the sea in the other family. A number of recent studies have shown this classification to be much too simplistic, and a number of new and conflicting classifications have been proposed. However, in all these classifications a number of distinct groups are recognized, and these are treated below (as subfamilies) under the umbrella of a single family: the Elapidae.

Terrestrial elapids are diverse and abundant on all tropical and subtropical landmasses. So far, 180 species of terrestrial elapids are known. Many elapids are famous for the danger they represent to humans. They include large and formidable species such as the black mamba *Dendroaspis polylepis*, the green mambas *D. angusticeps* and *D. viridis*, and a variety of other mambas and cobras

in Africa; cobras, king cobras and kraits in Asia; coral snakes in Central and South America; and taipans, brown snakes, death adders, and the like in Australia. However, even though it is the large species that attract popular attention, most elapids are actually small snakes that pose no threat to humans.

Australasian elapids

Australia is the only continent where most of the snakes are venomous, and it also has the snakes with the most toxic venom. Both of these attributes are due to elapid snakes. Australia, together with New Guinea and the island archipelago extending from its eastern edge, has long been isolated from the rest of the world, and during much of its geological history it was even

bodies while each attempts to push the other's head downward. It seems to be a test of physical strength and one that is almost always won by the larger of the two males. Detailed studies of the adder *Vipera berus* confirm that winning fights like this increases a male's mating opportunities, so that—according to Darwin's hypothesis of sexual selection—we would expect male–male combat to favor the evolution of large body size in males. In snakes that do not engage in male–male combat, such as garter snakes *Thamnophis* or filesnakes *Acrochordus*, larger size does not seem to enhance a male's reproductive success.

Although there are many exceptions to the rule, the overall pattern fits this prediction. Males do tend to be at least as large as their mates in snake species with male–male combat.

▲ In grass snakes Natrix natrix, as in most snakes, the male is smaller than the female. The dark-colored snake in this courting pair is the male.

◀ Two male Malagasy giant hognose snakes Lioheterodon madagascariensis intertwine as they combat over access to females.

more isolated as it floated northward after the breakup of the great southern continent, Gondwana. This separation meant that relatively few groups of plants and animals found their way to Australia. The successful ones faced little competition from other groups and thus have radiated very extensively. This is clearly true of eucalyptus trees and marsupials (kangaroos, koalas, etc.) but also true of various groups of amphibians and reptiles. It seems as though elapid snakes reached Australia just after it collided with Asia about 25 million years ago. The Australian elapids (subfamily Oxyuraninae) have diversified to fill a wide variety of ecological niches, with one genus (the death adders, *Acanthophis*) evolving to look and act remarkably like the vipers of other continents. The first

◀ A stocky ambush predator, the death adder Acanthophis antarcticus has evolved an appearance more like that of unrelated vipers than of its relatives in the family Elapidae. Like many vipers, the death adder can lure prey with its modified tail.

scientists to describe Australian snakes believed that *Acanthophis* was actually a viper. Other Australian elapids have evolved to resemble colubrid "whipsnakes" and small banded burrowing snakes in other countries. This kind of convergent evolution happens when animals of different evolutionary backgrounds are exposed to similar environments and therefore similar evolutionary pressures. In the case of death adders, the important similarities with viperid snakes seem to be that these heavily-built elapids ambush their prey rather than searching actively for it like most of the other proteroglyphous snakes.

In Australia, elapids occur in a very wide variety of habitats. Small offshore islands often have distinctive populations of brown snakes or tiger snakes, and even on adjacent islands there may be major differences between snakes. For example, some island tiger snakes (genus *Notechis*) reach more than 2 meters (6½ feet) in length, whereas snakes on other islands do not attain even 1 meter (3¼ feet). Tenfold differences in average body weights of tiger snakes have been recorded from islands just a few kilometers apart. The reason for this size variation seems to be the food supply: tiger snakes grow large only on islands where large prey are available, and this means only where there are large aggregations of nesting seabirds, especially muttonbirds. For most of the year food for snakes is scarce, and they survive by catching occasional small lizards; food is abundant for only a few weeks each summer, after the muttonbird chicks hatch and before they grow too large for the snakes to swallow. On islands without nesting colonies of muttonbirds, the tiger snakes can find only small prey and never grow very large.

Mambas

Africa has a variety of elapids ranging from small secretive creatures such as the shield-nosed snakes (genus *Aspidelaps*) to cobras and the renowned mambas (subfamily Dendroaspinae). The black mamba *Dendroaspis polylepis*, which can grow to 4 meters (13 feet) in length, is the most terrestrial of the four mamba species, and the subject of many horrifying tales. The animal's aggressiveness, speed, and venom toxicity are highly exaggerated in most accounts, but there is no doubt that these slender olive-brown elapids are among the most dangerous of all snakes. The threat display of an angry mamba—head and neck held high, mouth gaping open—is a truly terrifying sight.

▼ *With its neck thrown into a distinctive "S" curve and its mouth agape, the defense pose of the eastern brown snake* Pseudonaja textilis *is a warning wisely heeded. This fast-moving venomous snake is part of a successful species group found throughout Australia.*

Mike W. Gillam/AUSCAPE International

Dr Norman Myers/Bruce Coleman Limited

Cobras and their allies: subfamily Elapinae

Although cobras are distributed widely in Asia, they are most diverse in Africa, where there are 10 species: in the arid northeast, the Egyptian cobra *Naja haje*; in hot, humid western and central African jungles the forest cobra *N. melanoleuca*, the spitting cobra *N. nigricollis*; in the rocky fields and mountains of South Africa, the Cape cobra *N. nivea*; in the great freshwater lakes the water cobra *Boulengerina annulata* which feeds on fish, emerging from among lakeshore boulders at dusk to hunt for prey; and in southern Africa a small cobra, the rhinghals *Hemachatus haemachatus* only distantly related to the others. The rhinghals is distinctive in its reproductive habits, being the only cobra species to bear live young instead of laying eggs.

Although venom undoubtedly evolved as a means of immobilizing prey, many types of venomous snakes also use venom to deter potential attackers. Two groups of African cobras, and one Asian species, have evolved a particularly effective means of defense in this regard: spraying venom toward the eyes of an attacker. In "conventional" cobras the venom flows from a small aperture near the tip of the fang, but in the "spitters" this aperture is closer to the base of the fang and is rounded rather than elongate in

Anthony Bannister/NHPA

▲ The green mambas are the most arboreal members of the family Elapidae. Here an East African green mamba *Dendroaspis angusticeps* glides through a colony of weaver nests (*Ploceus* species), possibly in search of nestlings.

◄ By exhaling forcefully as venom drips from its fangs, a Mozambique spitting cobra *Naja mossambica* can spray venom at the eyes of an intruder up to 3 meters (9¾ feet) away.

637

► *If the bright warning coloration of the Arizona coral snake Micruroides euryxanthus euryxanthus fails to deter a predator, it defends itself by hiding its head, raising its tail, and everting its cloacal lining with a popping sound.*

▼ *Waving its curled tail to draw attention away from its concealed head, a Costa Rican coral snake Micrurus micrurus exhibits a defensive posture seen in many unrelated terrestrial snakes.*

Michael and Patricia Fogden

shape. Also, the venom canal inside the fang reaches the outlet at a right angle to the tooth. Muscular action and vigorous exhalation cause two fine sprays that can travel several meters, and are aimed at reflective surfaces such as eyes. Venom delivered in this way will not kill a person, but can cause temporary blindness.

Asia contains many interesting species of elapids, including kraits and long-glanded coral snakes (in which the venom glands extend for the entire front third of the body), but the most famous Asian elapid undoubtedly is the king cobra. This magnificent animal is the largest venomous snake in the world, reaching more than 5 meters (16½ feet) in length. It is found in India and eastwards to southern China and the Philippines. The head of a king cobra can be as large as a man's hand, and these snakes are truly formidable when aroused. There are reliable reports of elephants dying within a few hours after being bitten by this species. Fortunately the king cobra is not an aggressive animal, even during the nesting season when the female deposits her eggs in twigs and foliage and remains to guard them. The king cobra feeds primarily on other snakes, including venomous species.

Coral snakes

The coral snakes belong to the subfamily Micrurinae. There are about 50 species (genera *Micrurus* and *Micruroides*) in the American tropics, mainly in South America. Most are brightly banded, and all are slender. Some species are aquatic, but most seem to be terrestrial foragers of the forest floor. Lizards and small snakes are probably their main prey, although some species also take mammals, birds, frogs, and invertebrates. The startling bands of color warn potential predators to stay away from these snakes, and some predatory birds have an innate fear of their color pattern. On other continents many semi-fossorial snakes are similarly marked with bright bands, even when the snake itself is harmless, so the bands may function to confuse predators encountering the snakes in dim light—as the snake thrashes around, the bands seem to flicker and fuse together and make it difficult for the predator to determine the exact position of the snake as it tries to escape. In the case of coral snakes, this coloration has been developed further as a warning symbol. Many unrelated harmless or mildly venomous snakes living in the same areas as the coral snakes have evolved color patterns that match those of the local coral snakes very closely, perhaps confusing predators and giving the snakes more chance to escape. Some coral snakes also employ an unusual posture to deflect the predator's attention from their vulnerable head: they hide their head among the coils of the body and the tail is waved around in the air in the way that you would expect the head to move. Any predator seizing the snake's tail by mistake is likely to receive an unpleasant surprise when the head suddenly appears.

Serpents of the sea

Two main evolutionary lines of fixed-front-fang snakes have taken to the oceans. The sea kraits, or laticaudine snakes (subfamily Laticaudinae) are brightly banded species of the Indo-Pacific region. Their common name comes from their resemblance to land kraits, a banded terrestrial Asian species of the genus *Bungarus*. Like the land krait, sea kraits have highly toxic venom but are extremely reluctant to bite, even in self-defense. One "sea krait" *Laticauda crockeri* is actually restricted to a landlocked lake in the Solomon Islands, but the other five species are truly marine. They feed on eels from the coral reefs but spend much of their time ashore on small coral islands.

Although they have flattened paddle-like tails that help them to swim rapidly, they retain broad belly scales and so are capable of moving around proficiently on land. They mate and lay their eggs ashore and can be found in large numbers on some small tropical islands.

The other group of fixed-front-fang snakes have become even more fully aquatic than the sea kraits, because they no longer need to return to land to breed. The "true" sea snakes or hydrophiines (subfamily Hydrophiinae), the latter name meaning "water-lovers", retain developing embryos within the female's body and produce live young, rather than laying eggs like the sea kraits. The tail is flattened laterally to act as an oar, the belly scales are reduced in size, and the nostrils are located on the top of the snout so that the snake can breathe even when most of its head is under water. The nostrils are sealed by valves that exclude water when the snake dives. The lung is much longer than in terrestrial snakes, extending almost the entire length of the snake, from just behind the head to the posterior end of the body cavity. It may play a role in adjusting the snake's buoyancy during diving, as well as storing oxygen-rich air for respiration. Some species can also take up oxygen directly from the surrounding water, through their skin, so that they can remain

◀▼ *Laticaudines, like the yellow-lipped sea krait* Laticauda colubrina *(left), appear to have evolved their marine habits independently of the true sea snakes. They have retained their cylindrical shape and enlarged belly scales for crawling on land, which they frequently do to seek shelter, bask and mate. The pelagic sea snake* Pelamis platurus *(below), has a laterally compressed body for swimming and narrow belly scales, making movement on land extremely clumsy and difficult. A live-bearer, it does not leave the water, even to give birth.*

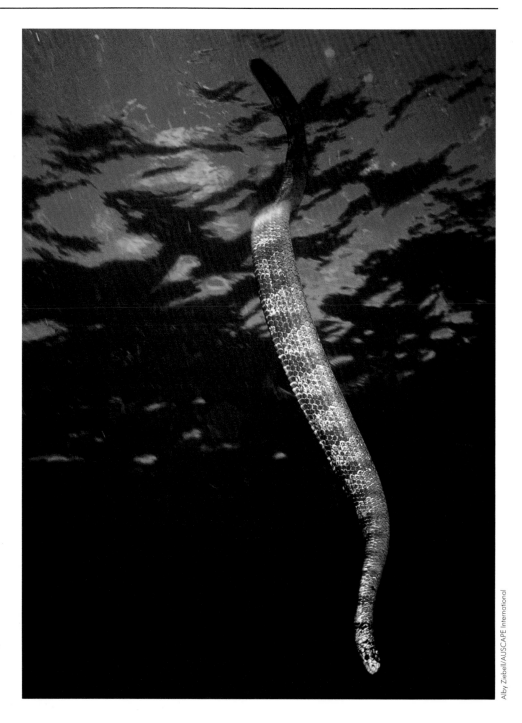

► *Named for a beak-like scale on the snout of mature males, the turtle-headed sea snake* Emydocephalus annulatus *is an Australian sea snake with a highly specialized diet of fish eggs.*

Alby Ziebell/AUSCAPE International

submerged for long periods of time. Their adaptation to saltwater life includes specially modified glands at the base of the tongue, which function to concentrate and excrete excess salt from the snake's bloodstream. Most of the hydrophiids are so specialized for movement in water that they are almost helpless on land.

Biochemical evidence indicates that the hydrophiids might have evolved from the terrestrial Australian elapids, probably quite recently in evolutionary terms. They have been very successful, and the 50 species are widespread through the Indian and Pacific oceans. Most species are restricted to relatively shallow water and feed on fishes and their eggs around coral reefs. However, one species, the yellow-bellied sea snake *Pelamis platurus*, has adopted a pelagic lifestyle, drifting across the open oceans at the apparent mercy of the winds and currents. Its brightly spotted black and yellow tail serves as a warning to predatory fish not to attack the snake, which has very toxic venom. How can a snake drifting along on the surface of the ocean find and catch fish to eat? The yellow-bellied sea snake relies on the tendency of small fish to gather under any floating object, so the snake is soon "adopted" by a school of fish that swarm around its tail. But how can the snake seize the fish, when they are gathered around its tail and not its head? The answer is simple and elegant: the snake begins to swim backward, so that the fish now gather around its head. One rapid sideway strike, and the snake has its meal.

VIPERS

The highly venomous vipers tend to feed in a very different way from the fixed-front-fang snakes; they lie and wait to ambush unwary prey, especially mammals. Most vipers (and their close relatives, the pit vipers) are relatively heavy-bodied snakes, often beautifully camouflaged in their natural environments. They coil beside a mammal trail, or in the branches of a fruiting tree where birds are likely to gather, or beside a desert shrub where lizards will come for shade. There they wait for prey to wander within range. With their superb camouflage many of these snakes are almost invisible. Some actually lure prey within striking range by wriggling the tip of their tail, modified into an insect-like shape, to imitate the movements of a small invertebrate. Because they

do not have to move long distances actually searching for prey but must strike rapidly and accurately when a prey animal approaches, ambush-hunters tend to be relatively muscular and thus heavy-bodied. Because they eat large prey items, they need large heads. Because many of their prey items are not only large but also covered in fur or feathers, they need long fangs to penetrate deep into the victim's body and deliver the venom effectively. These snakes have thus evolved a very efficient means of killing large and potentially dangerous prey items. Unlike constricting snakes, which remain in contact with the prey while it struggles, vipers only need to inject venom with a quick strike and then wait. Even if the prey runs away to die, the snake can follow its scent trail.

Although the scientific name of the family, Viperidae, apparently comes from the Latin words *vivus* and *paro*, meaning "giving birth to live young", in fact there are many egg-laying (oviparous) vipers as well, both in the "true" vipers (subfamily Viperinae) and the pit vipers (subfamily Crotalinae).

"True" vipers

The 60 species of "true" vipers are widely distributed in Africa, Europe, and Asia. One species, the adder *Vipera berus*, has a remarkable geographic range from Britain, across Europe and Asia, to Sakhalin Island north of Japan. This species even lives in severely cold areas, extending into the Arctic Circle at 67°N and in

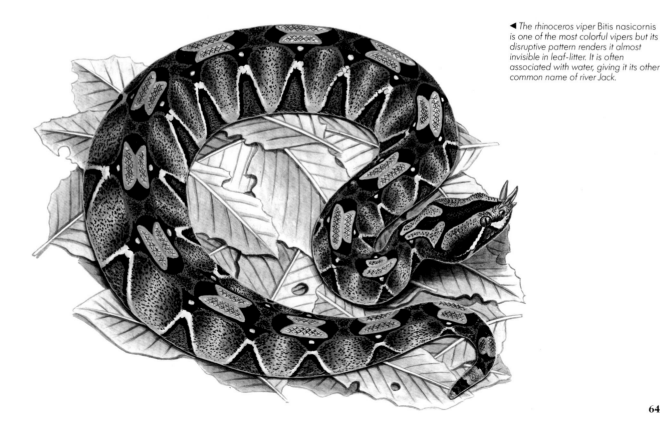

◄ The rhinoceros viper Bitis nasicornis is one of the most colorful vipers but its disruptive pattern renders it almost invisible in leaf-litter. It is often associated with water, giving it its other common name of river Jack.

parts of southern Siberia. In such cold areas, the adder can be active for only a few months each year and must spend the rest of the year deep within soil crevices to avoid the lethally low temperatures. The males emerge first in spring. After a few weeks of basking to ready themselves for reproduction, they shed their old skins and become far more brightly colored and active, roaming around in search of recently emerged females. If two males encounter each other near a female, they judge each other's size. The smaller male will usually flee, but if the two are evenly matched a battle may ensue. Instead of biting each other, they wrestle, with the two males intertwined until one is overpowered and gives up the fight. The winner then has the chance to court the female, and perhaps mate with her. Pregnancy takes about two months, and females spend most of their time basking so that the embryos develop at high temperatures and therefore more rapidly.

Females don't feed while they are pregnant, so they are often emaciated by the time they give birth. Many die at this time, and even those that survive are unable to gather enough energy in the autumn for reproduction the following year. Thus,

a female may be able to reproduce only once every two or three years, or even less often if food supplies are low or if the activity season is too short.

Vipers are found in tropical areas as well— for example, the large African puff adder *Bitis arietans*, Gaboon viper *B. gabonica*, and rhinoceros viper *B. nasicornis*. The saw-scaled viper *Echis carinatus* and its relatives are abundant in deserts of the Middle East. In these hot areas the reproductive cycles are likely to be quite different from those of the adder, but we know very little about tropical vipers in this regard. We do know that males of at least some species fight with each other, and that females of some species are egg-layers whereas others are live-bearers, but their detailed biology and behavior remains a mystery to researchers.

Pit vipers

The subfamily Crotalinae is the second major group of viperid snakes. Their common name, "pit vipers", comes from the deep pit between the eye and the nostril on each side of the head. Here there are sensory organs that detect heat. They are incredibly sensitive, detecting temperature differences of as little as 0.003°C. The ability to

▼ A puff adder *Bitis arietans* rears its head in a threat posture. Normally undetected because of its cryptic coloration, this widespread African species puts on an impressive display when threatened, inflating its body and hissing loudly.

Anthony Bannister/NHPA

recognize such tiny differences means that pit vipers can accurately locate and strike warm-blooded prey even on pitch-black nights. Most of the 140 or so species are ground-dwellers, but some are tree-dwellers and a few are semi-aquatic. They are widely distributed throughout the Americas, Europe, Asia, and Africa, but have not reached Australia or invaded the oceans.

▲ Beautifully camouflaged among fallen leaves, a West African gaboon viper Bitis gabonica rhinoceros awaits its next meal. This large species possesses the longest fangs of any snake, with reported lengths exceeding 5 centimeters (2 inches).

◀ This prey-eye view of an eyelash viper Bothriechis schlegeli shows clearly the heat-sensitive pits near the eyes, the chief diagnostic feature of the pit vipers.

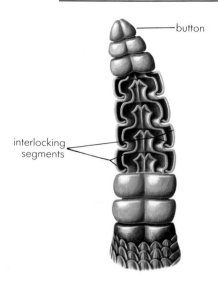

button

interlocking
segments

▲ *Trademark of the rattlesnakes, the rattle is developed from enlarged and thickened scale covers that are retained after molting. These form interlocking segments that hit against each other when they move, producing the characteristic buzzing sound. A new segment is added after each molt, the oldest being the end segment, or button.*

▼ *The urutu* Bothrops alternatus *is one of a large group of closely related South American crotaline pit vipers, some of which are commonly referred to as fer-de-lance, a name originally given to the species found on the island of Martinique in the Caribbean.*

The most famous of the pit vipers are probably the 30 species of rattlesnakes (genera *Crotalus* and *Sistrurus*) of North America, although three species (all of them belonging to the genus *Crotalus*) occur in Central and South America. These snakes have the tip of the tail modified into a remarkable warning device, the rattle. It is formed of specially shaped dry scales, one of which is added every time the snake sheds its skin (which doesn't happen on a regular annual basis, so the number of rattle segments cannot be used to determine the snake's age). When the tail is vibrated, the rattle segments move against each other to create the characteristic "buzz" that can often be heard from many meters away. It seems likely that the rattle evolved as a warning to large grazing mammals, such as bison, which share the prairie habitats of many rattlesnake species. Interestingly, one island population of rattlesnakes in the Gulf of California, living in an area without such hazards, has lost the rattle during its evolutionary history.

The other main group of pit vipers in North America, extending to Central America and Asia, is the genus *Agkistrodon* and its relatives. The North American representatives are the copperhead *A. contortrix* and the cottonmouth *A. piscivorus*. The cottonmouth (or water moccasin) derives its common name from the white coloration inside its mouth, exposed when the snake gapes in its dramatic defensive posture. Cottonmouths are swamp-dwellers and often

occur in large numbers in suitable habitats in the southern United States. On offshore islands they may live under heron rookeries, surviving almost entirely on fish dropped by clumsy herons.

The pit vipers have radiated extensively in Central and South America. A single egg-laying species, presumably the descendant of a separate pit viper invasion of the Americas, is the bushmaster *Lachesis muta* of lowland and lower montane rainforest. This huge snake, almost 4 meters (13 feet) in length, is a classic ambush predator, selecting a suitable ambush site beside a mammal trail and waiting, sometimes for weeks, until prey wanders within range. The scientific name *Lachesis* comes from one of the three Fates of Grecian mythology—the one who determines the length of the thread of life—because of the great danger to human life posed by a bite from this species.

Living in the same habitats as the bushmaster are a wide variety of pit vipers that give birth to live young, the most famous species being the fer-de-lance *Bothrops atrox*. In Central America this snake is also known by the Spanish name *barba amarilla* ("yellow beard") in reference to its yellow chin. Like the bushmaster it is a terrestrial ambush predator.

Pit vipers are diverse in Asia and include a range of terrestrial and arboreal forms. Some of the terrestrial species, like the habu *Trimeresurus flavoviridis*, are large brown snakes that strike readily when alarmed and cause many human

◀ Displaying a pattern unique among rattlesnakes, the lance-headed rattlesnake Crotalus polystictus is a small pit viper of the Mexican Plateau.

▼ Adopting the characteristic defense posture of rattlesnakes, a western diamondback rattlesnake Crotalus atrox vibrates its rattle to warn an intruder. Rattlesnakes probably evolved in North America when large grazing mammals like the bison more commonly posed a threat of trampling.

▶ *Wagler's palm viper* Trimerisurus wagleri *(opposite) is a tree-dwelling pit viper of Southeast Asia. A tractable species, it is also known as the temple pit viper as it is frequently kept in "snake temples" where it is freely handled by the priests.*

fatalities in Japan every year. Others, like the palm viper *T. wagleri* found in the Sunda Islands and Malaysia, are spectacularly colored (bright green) tree-dwelling species that are very reluctant to bite humans; they are often kept as good-luck charms in "snake temples" and in trees near houses.

SNAKEBITE

Snake venom is a complex mixture of various chemical substances, mostly enzymes. It is a somewhat cloudy liquid, manufactured and stored in venom glands in the upper jaw and delivered to the prey either by injection through a hollow fang or by seeping into wounds caused by enlarged teeth. Some animals have very high resistance to particular snake venoms, whereas others are extremely sensitive to even small amounts of venom. The main constituents of snake venoms are enzymes such as proteinases which destroy tissue, hyaluronidase which increases tissue permeability (so that the venom can spread through the body more rapidly), phospholipases which attack cell membranes, and phosphatases which attack high-energy chemical compounds. The venoms of many colubrid snakes also contain L-amino acid oxidase, a substance that causes great tissue destruction. The venom of fixed-front-fang snakes (elapids and sea snakes) contains basic polypeptides to block nerve transmission and thus cause rapid death of the prey animal (or an unlucky human) by paralysing the diaphragm so that the victim stops breathing. In contrast, the venom of vipers has a high level of proteinase, which results in more severe tissue damage around the site of the bite and thus more profuse bleeding.

Which type of snake is the most deadly? The answer depends on several factors. The most potent venom is usually that of the fixed-front-fang snakes, the inland taipan *Oxyuranus microlepidotus* of central Australia being the

record-holder in this regard; a single bite from this species can contain enough venom to kill almost 250,000 mice. However, fixed-front-fang snakes have relatively short fangs and often do not produce as much venom as some of the large vipers. Also, some species are much more likely to bite than others, or are more likely to be encountered by humans.

Venomous snakes kill many thousands of people every year, especially in tropical or subtropical areas where people live in huge numbers and are often barefoot and where medical facilities are limited. For example, some authorities estimate that the saw-scaled vipers *Echis carinatus* in the Middle East may be responsible for about 20,000 fatalities each year. In Asia and Africa venomous snakes take a significant toll of human lives. In contrast, an average of fewer than five people per year die of snakebite in the whole of Australia, despite the fact that most of the Australian snakes are venomous, and some have the most toxic venoms in the world. Low population densities, adequate footwear, and excellent medical facilities are responsible for this low death toll.

CONSERVATION

Like most other living organisms, snakes have suffered at the hands of humans. The biggest threat to snakes, as to other animals, is the continued destruction of natural habitats. Rapid increases in human populations, and exploitation of the natural environment for logging, agriculture, and grazing, have decimated or eliminated snakes from many areas. The key to conserving snakes will be to conserve the places where they live. The snakes most at risk are those with a restricted distribution (like the Round Island boas) and those that depend on specific types of easily damaged habitats. For example, one small Australian elapid, the broad-headed snake

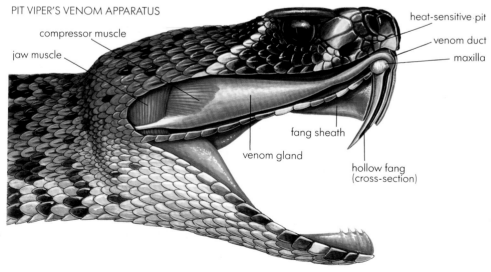

PIT VIPER'S VENOM APPARATUS

compressor muscle

jaw muscle

heat-sensitive pit

venom duct

maxilla

fang sheath

venom gland

hollow fang (cross-section)

▶ *In most venomous snakes, venom is delivered by a highly evolved injection system, acting much like a syringe and hypodermic needle. The prey is first stabbed with the elongated fangs, then the venom glands are compressed by muscular contraction, forcing the venom through the venom ducts and the hollow fangs into the wound.*

DEFENSIVE STRATEGIES

Snakes defend themselves with a remarkable variety of behaviors. One of the most common, though least spectacular, is crypsis; the snake remains absolutely motionless despite the close approach of a potential predator. Viperid and boid snakes rely heavily on this behavior, and many of these snakes are beautifully marked with complex patterns that blend almost perfectly with their natural backgound. A large American copperhead coiled in sun-dappled foliage, or a puff adder in the leaf-litter, can be impossible to see unless it moves. Other snakes, especially more slender-bodied species, depend on speed for escape.

"Whipsnakes" from a variety of colubrid and elapid groups, although not closely related, share a common set of features such as large eyes, slender body, long tail, high selected body temperature, and a diurnal (day-active) life, which enable them to locate and capture fast-moving prey, usually lizards. These snakes move considerable distances and thus (unlike the "ambush" species) often encounter potential predators when they are in the open, far from cover. Camouflage is unlikely to be effective, and so they flee. These snakes are often unicolored or striped, unlike the "cryptic" species which more usually have a blotched pattern. Interestingly, this correlation between color pattern and mode of escape from predators even occurs within a single species, and sometimes within a single litter. Recent research on the garter snake *Thamnophis ordinoides* shows that among the newborn, blotched individuals are more likely to "freeze" and rely on camouflage when chased, whereas their striped litter-mates are more likely to rely on speed of escape.

If neither of these methods are effective, snakes still have many strategies to protect them from predators. Some hide their heads and wave their tails in the air, apparently to attract the predator's bites away from the head and neck to the less-vulnerable tail. The tail is often brightly colored in these species, and in pipe snakes is greatly flattened so that it resembles the hood of an angry cobra. A few American colubrids (*Gyalopion*) and coral snakes (*Micruroides*) take this even further by "cloacal popping", forcing air out of the cloaca to make a distinct popping noise. Why this noise should deter a predator is unclear. Nor do we know the significance of the bizarre defense display of the Australian bandy-bandy *Vermicella annulata*, which slowly raises body loops above the ground when threatened. Species such as the hognose snake (genus *Heterodon*) feign death by rolling onto their back, opening their mouth and extruding the tongue. Presumably, the predator loses interest in such an obviously "dead" snake.

Other snakes defend themselves more vigorously, striking at the attacker and often inflating part of the forebody to make themselves look larger and more formidable. Cobras (*Naja*) and hognose

Anthony Bannister/NHPA

snakes flatten their heads and/or necks horizontally, whereas Australian tree snakes (*Dendrelaphis*) and boomslangs inflate their necks dramatically. Sometimes brightly colored skin is visible between the expanded scales. Mambas and cottonmouths gape to display their fangs, while some large Australian elapids such as the king brown snake *Pseudechis australis* display chewing movements of their jaws.

The display can be auditory as well as visual. Most angry snakes hiss loudly in defense, and some, like the American bullsnake *Pituophis melanoleucus*, have a specially modified epiglottis to enhance the volume of the hiss. The rough-scaled desert viper *Echis carinatus* can produce a similar hissing sound by rubbing its body coils against each other. Many types of snakes twitch their tail-tips rapidly when alarmed, and this can produce considerable noise if the snake is lying in dry grass. The ultimate development of the tail as a sound-producing organ, however, is undoubtedly in the rattlesnakes. The "buzz" of a large diamondback *Crotalus atrox* is clearly audible for many meters, and is usually enough to convince a predator to look elsewhere for its meal.

▲ Buried to its eyes in sand, a Peringuey's desert adder Bitis peringueyi avoids detection by both predators and prey.

▼ A bandy bandy Vermicella annulata slowly raises loops of its body when threatened, possibly to confuse predators.

Pavel German

Hoplocephalus bungaroides, has become endangered because it relies on weathered sandstone boulders for shelter, the type of boulders that are rapidly being removed from natural habitats because they are popular as garden decorations. There is little point in "protecting" such a species with legislation, even international legislation, unless its habitat can somehow be preserved. This is a difficult task in many countries, where the immediate need to feed hungry people takes precedence over the needs of other species.

Some types of snakes are also threatened by commercial exploitation. The ones most at risk are large, brightly colored species whose skins are of value to the fashion industry. For example, pythons and boas are ruthlessly killed in many parts of their natural range because they are relatively slow-moving (easy to kill) and large enough to provide a valuable skin as well as a useful meal for local people. In some parts of Asia the killing of snakes is so intensive that the numbers of these animals have been considerably reduced. As a consequence, rats and mice previously kept in check by predatory snakes are now so numerous that they have become a major agricultural problem. Some sea snakes are also harvested for their skins and meat, and huge numbers of sea kraits are taken every year from small coral islands where they come ashore to rest and reproduce. The ecological impact of this harvest has never been thoroughly investigated.

Many snakes are killed because humans hate and fear them for the supposed danger they represent. This fear is entirely legitimate in some cases, but in most areas snakes do far more good (by controlling agricultural pests) than harm. A high proportion of snakebites occur when people try to kill snakes, often in remote areas where the snake poses no threat to human safety. Most people cannot reliably distinguish between harmless snakes and venomous snakes—even in countries like the United States, where the two types are very different in appearance—and kill harmless species such as American watersnakes (genus *Nerodia*) in mistake for venomous snakes such as the cottonmouth in areas hundreds of kilometers away from where the venomous species actually lives. The infamous "rattlesnake roundups" of the southern United States represent one of the best-publicized examples of the absurd enmity that many people feel for these magnificent but much-maligned creatures.

▼ *Suffering the plight of most harmless snakes, the banded water snake* Nerodia fasciata fasciata *is commonly mistaken for a venomous species, in this case the cottonmouth* Agkistrodon piscivorus, *and killed on sight.*

Order Squamata
Suborder Amphisbaenia
4 families, 18 genera,
c.143 species

SIZE
Smallest Tanzania thread
amphisbaenian *Chirindia
rondoense*, body length 9 to
12 centimeters (3½ to
4¾ inches); diameter 2 to
3 millimeters (⅛ inch).
Largest South American red
worm-lizard *Amphisbaena
alba*, body length up to
72 centimeters (28½ inches);
diameter up to 3 centimeters
(1⅕ inches).

CONSERVATION WATCH
Almost a third of the species
are known from only a single
specimen, so we have no
information about the size of
the population. One species
(from Puerto Rico) is listed as
endangered.

▼ *Amphisbaena fuliginosa of
South America is typical of the most
widespread family of amphisbaenians
or worm-lizards, the Amphisbaenidae.*

AMPHISBAENIANS

CARL GANS

Worm-lizards and other amphisbaenians have long represented an evolutionary conundrum. Many specialists have at different times in their career reported that they were lizards, only later (or earlier, in some cases) to consider them as an independent offshoot within the order Squamata (to which lizards and snakes belong). Certainly they are among the very few completely subterranean and self-tunneling reptiles. They are so well matched to underground environments that they seem very different from the lizards and snakes alive today.

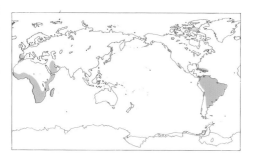

TRUE BURROWERS

All amphisbaenians seem capable of creating their own tunnels. Burrowing is always done by the head, and so the skull has been deformed during their evolution to match the needs of whichever burrowing method is used. The deformed skulls also reflect the special needs of the feeding and

sensory systems of tunnel-dwelling animals. The skulls are heavily ossified; and the small scales are arranged in rings around the body. In all but one species, the mouth is recessed into the lower surface of the head.

Amphisbaenians generally have a long slender body and a scaly skin (they shed the entire skin in one piece); the males have paired hemipenes, which are not penises but rather pouches, one of which is turned outward during copulation. All these features suggest that they are members of the order Squamata. However, they differ from lizards and snakes in at least a half dozen derived characteristics, such as a right- rather than left-lung reduction, a uniquely shaped egg tooth (used to pierce the egg when hatching), special skull bones, and the ring-like segments of the skin. All living and fossil species discovered so far have been classified in one of four families:

Chris Mattison

Carl Gans

◄ *Almost entirely subterranean in habits, worm-lizards are characterised by rather loose, baggy skins; blunt, heavily ossified skulls; underslung jaws; tiny eyes; and lack of evident ear-openings. They live in warm to tropical areas around the world. This is* Bipes biporus *of Mexico.*

Trogonophidae (desert ringed lizards), Bipedidae (ajolotes), Rhineuridae, and Amphisbaenidae (worm-lizards).

Members of the family Bipedidae retain forelimbs, but external limbs are otherwise lost. Remnants of a shoulder girdle can be seen in the skeletons of the Trogonophidae, but not in the other families.

Most amphisbaenian species seem to be oviparous (egg-layers), but at least three cases of viviparity (giving birth to live young) have been documented.

THE SKIN

An amphisbaenian's head is always covered with shields larger than the scales of its body, and in some of the most advanced species these are fused together on the anterior part of the head. The scales of the trunk of the body are rectangular, or at least parallelograms, and are arranged on annuli (rings) that circle the body. Normally the segments on the underside are larger than those above. The cloacal slit is surrounded by a patch of enlarged segments, with the last annulus (ring) of the trunk sometimes having a row of precloacal pores, which in males may be involved during mating, although no observations as yet confirm this.

The skin of the trunk is loose around the circumference of the body, and muscles connect pairs of segments so that these can be narrowed or widened. There are three sets of muscles on each side of each segment: two reach backward and one reaches forward. These allow the animal to propel itself by rectilinear progression. A portion of the body is fixed against the soil, while the backward-reaching muscles contract and pull up the rest of the body. Each forward-directed muscle then pulls the local portion of the skin out of contact with the soil and to a new, more forward position. It is quite different from the way snakes move; snakes fix the skin tightly near the middle of the back, whereas amphisbaenians have the skin free around the circumference.

Amphisbaenians are also capable of moving by concertina and lateral undulatory movements. Like many advanced snakes and limbless lizards, an amphisbaenian will use whatever movement is best suited to the soil and the shape of the tunnel—different propulsive patterns with different parts of their body.

The external appearance of different species gives us clues about the way they create their tunnels. For example, perhaps to reduce friction during penetration of the soil, adjacent segments of skin are fused not only on the head shields but also in the pectoral region of the spade-snouted species, for this part of the body pulls forward, rubbing against the ground when the head is lifted to compress the soil.

PREY AND PREDATORS

Worm-lizards are effective predators, and they can bite pieces out of larger animals. Even the smaller species have massive jaw-closing muscles. The upper jaw has five to nine teeth on each side, with a single central tooth that fits between the enlarged teeth of the lower jaw. Amphisbaenians can therefore exert powerful bites, and their

▶ A white-bellied worm-lizard *Amphisbaena alba* from Brazil gapes in threat. Worm-lizards are formidable predators, with sharp teeth and powerful jaws. Members of the genus *Amphisbaena* are locally (but erroneously) considered poisonous, and are called "ant-kings" from the widespread belief that they are raised by ants or termites. Recent observations confirm the connection with these insects, but the exact relationship remains a mystery.

Martin Wendler/NHPA

HEARING UNDERGROUND

Terrestrial reptiles generally have external eardrums on the sides of the head or at the base of an external ear canal. From these a stapedial link crosses the air-filled middle ear to an oval window that opens into the liquid-filled tube of the inner ear. Here, rows of hair cells transduce the sound vibrations into electrical signals which pass to the brain, where they are perceived as sounds. As signals reach the two ears at slightly different times and at different magnitudes (the head serving as a sound shadow), the animal not only hears the sound but may be able to orient to it.

Reptiles living underground tend to lose the external eardrums and sometimes even their middle ears. This suggests not only that their ear openings are protected from predator or parasite attack, but also that they rely more on other senses. Members of three of the four families of amphisbaenians, however, have developed a unique hearing system that lets them sample sounds emanating from the tunnel ahead of them. Bones of the throat now connect with the stapedial plate on either side of the face, and scales of the skin on the face act as a substitute forward-facing eardrum. Experiments have shown that the system indeed detects airborne sounds,

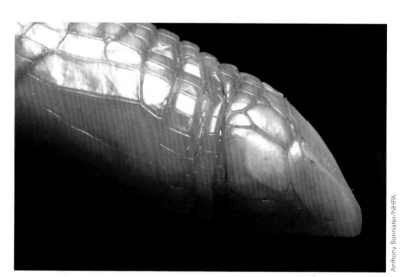

Anthony Bannister/NHPA

and is as sensitive to these vibrations as to those applied directly to the skull. This indicates that they hear and orient to the sounds of prey, and possibly to others of their own kind.

▲ Specialized scales on the face function as substitute eardrums in an auditory system unique to worm-lizards.

Chris Mattison

interlocking teeth can cut and tear out pieces from larger prey; worm-lizards have been observed in their tunnels biting and then spinning their body to free the portion of flesh.

Most reports suggest that amphisbaenians are omnivorous, attacking and subduing any small animals encountered in their environment; but many of these are observations on captive animals, and we still know relatively little about their feeding habits. Seasonal changes of diet may well be important. Recently we learned that some centuries-old reports of the association of the giant *Amphisbaena alba* with the nests of South American leafcutter ants are indeed correct, although we don't yet know what may be the mutual benefit to the worm-lizards and the ants.

Amphisbaenians are in turn threatened by many predators. There are those that forage on the surface, such as birds and small mammals; indeed, one seemingly rare worm-lizard was first described by a scientist dissecting the crop of a small eagle. They are also in danger from pigs that dig in the soil and root out any "protein" encountered, and monkey troops that forage over wide regions, lifting pieces of wood and rocks under which amphisbaenians may shelter. Finally, there are tunnel predators, such as some African snakes, each of which is claimed to specialize on a particular local amphisbaenian.

Perhaps 30 percent of amphisbaenians show a darkened color pattern suggesting camouflage,

which suggests that they may make occasional visits to the surface. A few species have been observed basking on the surface. And when the skin is darkly pigmented it coincidentally seems to be less permeable to water than the skin of deeply burrowing species, which normally has substantial water loss. Perhaps some of them do make voluntary visits to the surface. Alternatively they may move to the surface in response to flooding after heavy rains and to attacks by various predatory ants.

All but the African amphisbaenians have short tails, the end of which is commonly reinforced with dense connective tissues that may be supported by fused terminal vertebrae. In spite of this, many species show "caudal autotomy", or voluntary tail-shedding—they break off their tail when a predator gets hold of it. The broken tip may move and distract predators, whereas the stump may heal, so perfectly that some people have been fooled into thinking that the individual belongs to a "new", "short-tailed" species. Other species use their tail differently for defense: the tip has a cone-covered surface which attracts dirt and thus supplies a block against any predator following the worm-lizard down its tunnel.

THE AMPHISBAENIAN FAMILIES

The family Trogonophidae is the most distinct group of amphisbaenians. Species occur in North Africa and also from western Iran across the

▲ *Except for their obvious scales, amphisbaenians might well be mistaken for caecilians in their size, shape, behavior and general appearance. A scarred tail stump on this* Amphisbaena fuliginosa *confirms that worm-lizards, like many skinks and lizards, exhibit caudal autotomy — that is, they can voluntarily shed their tail if grabbed by a predator.*

Arabian Peninsula to Aden, Socotra Island, and eastern Somalia. They tunnel using an oscillating movement, and their body structure has special modifications for this such as a downward-pointing tail, which the animal digs in to give extra force for burrowing. The teeth are permanently fused to the crest of the bones, rather than being in a groove and attached to the sides of the bones. The species that live in Arabia and Somalia show the most advanced characteristics and have a general tendency to become shorter and stouter.

The family Bipedidae consists of three species (genus *Bipes*) in northwestern Mexico. All of them have hands, which lie so far forward and are so large that the animals bear the popular name of "little lizard with big ears". The fingers are used during tunnel-building to scrape away soil and widen the tunnel.

Florida is the home of the only remnant species of the once-widespread family Rhineuridae. Fossils have been found across the western United States, dating back in time to the Paleocene, about 60 million years ago. *Rhineura*

floridana has a spade-shaped head, but with a different arrangement of bones from those in the other amphisbaenian families.

The true amphisbaenids (family Amphisbaenidae) are more than 120 very diverse species with a variety of tunnel-forming specializations. The most primitive occur in the West Indies and northern South America. Other species range south to Patagonia and cross the Andes in Peru. In Africa, they live south of the Sahara Desert, but not in the Central African highlands or at the southernmost tip of the continent. Some have a keel or crest on the head, others have a spade-shaped head.

Species in the genus *Blanus* may represent a fifth family. They live in Spain and North Africa and also in Lebanon, northern Israel, and northern Iraq. Fossils have been found in north–central Europe. Some of the species alive today have several primitive features such as a mouth that isn't recessed under the snout and residual hind limbs, but we really need to know more about them before classifying them in one family or another.

▶ *The only amphisbaenian inhabiting Europe, Blanus cinereus is a member of a small group of uncertain relationships, distributed around the Mediterranean region. Characterized by an unspecialized jaw and residual hindlimbs (not visible here), they may constitute a distinct family.*

◄ *Known locally by a name that translates as "little lizard with big ears" from the size and far-forward position of its forelimbs, Bipes biporus and two other species together constitute the family Bipedidae, restricted to northwestern Mexico. All other worm-lizards are limbless.*

Carl Gans

BURROWING STYLES

An amphisbaenian makes a tunnel by forcing its head into the soil using either compression of the soil in front of it, or by oscillation, in which the end surface of the tunnel is regularly scraped away, later to be forced into the walls. Most species use a compression method (perhaps one of several variants described below) to make permanent tunnels in many different types of soils. Oscillation methods of tunnel-building are used by members of the Trogonophidae family; they mostly live in sandy soils, which are less compressible than other soils, but the grains may be packed in different patterns.

Soil-compressing amphisbaenians force their heads directly into the soil, and some use slight sideways movements to enhance penetration. In the simplest pattern, the amphisbaenian's snout is somewhat rounded, and a series of pushes drives it forward. Between the drives, some species pull up the back of the body, curve the neck, and then produce the next penetrating drive by straightening these curves.

Intermittent penetration by soil-compressing species—who drive the head in part-way then rotate it, thus compressing the loose soil into the tunnel wall—has resulted in two quite different modifications to the amphisbaenians' skeletons. In the first, the skull has become spade-shaped; the head drives into the tunnel end at a fairly low level, and is then rotated upward, the blade of the spade compacting the loose soil into the roof of the tunnel. In the second, the skull has formed a spectacular crest, or keel, and the amphisbaenian widens its tunnel by packing the soil into either side. In both types the rotation of the skull involves the axial muscles of the segments behind the head, so tunneling in hard soils is possible without an increase in the animal's body diameter, as the force-generating muscles are distributed along the trunk. The spade-snouted animals do

Carl Gans

have an advantage in that they drive the head upward only and thus need only a single set of axial muscles. In contrast, the keel-heads tend to alternate compression to left and right, and only half of the available musculature is involved in either drive.

The oscillation method of tunnel-making may require a continuous application of force against the end of the tunnel where the surface is being scraped. Some species have a keratinized edge on each side of the head, which serves as a scraper. The scraping movement is complex, as the head twists about two different axes. A short and stout body helps the amphisbaenian to apply force from its tail; a pointed tail can be dug in easily to apply this force. A body shape that is triangular or rectangular in cross-section prevents the amphisbaenian from spinning during oscillation.

▲ *Agamodon angeliceps is a member of the family Trogonophidae, a group that tends to live in sandy soils and uses oscillating head movements in tunneling. The somewhat rectangular cross-section of the body resists twisting, while the rim on each side of the spade-shaped face-plate shaves soil toward the roof, from one side, then the other.*

Order Rhynchocephalia
1 family, 1 genus, 2 species

SIZE
Males total length up to 60
centimeters (23¾ inches);
weight up to 1,300 grams
(2¾ pounds).
Females total length up to
50 centimeters (19¾ inches);
weight 550 grams (1¼
pounds).

CONSERVATION WATCH
Eight of about 30 populations
are considered vulnerable or
endangered, mainly through
predation by introduced rats.
Sphenodon punctatus is listed
as rare in the IUCN Red List
of threatened animals.

TUATARA

DONALD G. NEWMAN

Despite appearances, tuatara are not lizards. They are the sole living members of the order Rhynchocephalia ("beak-headed"). Fossil rhynchocephalians were small to medium-sized reptiles that were common throughout the world between about 225 and 120 million years ago, long before the first dinosaurs appeared. Later their numbers declined, and about 60 million years ago they apparently became extinct everywhere except in New Zealand.

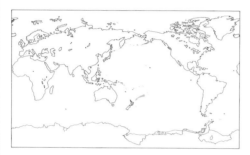

DISTINGUISHING FEATURES

Tuatara are of extraordinary zoological interest and are often referred to as "living fossils". However, recent research suggests that tuatara display many advanced features when compared with their nearest extinct relatives. They cannot, as is often claimed, have persisted unchanged from their ancestors for 100 to 200 million years.

Tuatara differ from lizards in many ways. For example, they lack external ears; they have hook-like extensions on some of their ribs (a bird-like feature); they have two large apertures on each side of the skull immediately behind and above the cavity housing the eyeball (lizards have only one); and the males do not have a penis.

Tuatara have a legendary "third eye", which is part of a complex organ situated on the top of the brain. This has a lens, retina, and nerve connection to the brain, but early in the growth of the tuatara the organ is covered by opaque scales. Many lizards also have a "third eye" of similar complexity. While in lizards this organ is involved in regulation of body temperature, experimental tests have failed to reveal the function, if any, of the "third eye" in tuatara.

Tuatara may be gray, olive, or occasionally brickish red in color, and are spotted to varying degrees. In the nineteenth century several species and subspecies were named on the basis of differences of color and body proportion, but most differences are now known to occur within populations. Thus the tuatara became almost

▶ *Tuatara were once widespread throughout New Zealand, but were exterminated on the main islands within the last few centuries, apparently through a combination of hunting, changes in land use, and introduced predators such as rats. Their populations are now restricted to about 30 small offshore islands, where, even on many of these last refuges, their status remains precarious.*

Michael Schneider/New Zealand Geographic

John Cancalosi/AUSCAPE International

universally regarded as a single species, *Sphenodon punctatus*. However, work still in progress on genetic variation among tuatara populations has supported the early scientists' contention that at least two species of tuatara exist. The population living on North Brother Island in Cook Strait, is now considered by researchers genetically different enough to be regarded as a distinct species, *Sphenodon guntheri*.

Male and female tuatara can be easily distinguished. The male is substantially larger than the female and is generally twice as heavy. The adult male has a striking crest along the back of the neck and another down the middle of the back, both of which bear conspicuous soft spines. The female also has neck and body crests, but neither crests nor spines are well developed. The

male has a proportionately narrower abdomen and a larger head than the female.

Like most lizards, tuatara are capable of caudal autotomy—shedding their tail as a means of escaping from enemies. The lost portion regrows but differs in color and pattern from the original and is shorter.

WHERE TUATARA LIVE
Tuatara or their ancestors may have reached New Zealand over 140 million years ago, when it was part of the Gondwana landmass, but no fossils older than 25,000 years have been found. Those located show that tuatara once ranged throughout the two main islands of New Zealand. Now, however, they are restricted to 30 small islands off the northeast coast of the North Island and in

▲ *In tuatara the sexes differ considerably in size. A large male may extend 60 centimeters (2 feet) from snout to tail-tip, and weigh over 1 kilogram (2 pounds), but females seldom exceed 500 grams (17½ ounces) in weight. Males also usually have more conspicuous spines on the nape and back.*

Cook Strait. Half of these islands have an area of 10 hectares (25 acres) or less.

The total number of tuatara surviving today probably exceeds 100,000. Stephens Island (150 hectares, or 370 acres), in Cook Strait, supports at least 30,000 and possibly more than 50,000. There, density equivalents of over 2,000 tuatara per hectare (2½ acres) have been recorded, although average densities are considerably lower.

Most tuatara islands have difficult access, being wholly or partially cliff-bound. They are frequently exposed to strong winds and the natural vegetation is generally stunted. Many have been cleared by Maori or European people for

AND THEN THERE WERE TWO . . .

Alison Cree

In the nineteenth century two living and one extinct species of tuatara were named. Later it was contended that differences in features such as color and body proportion, upon which the proposed species were distinguished, were no greater than could be observed within individual colonies. Thus the tuatara became universally regarded as a single species, *Sphenodon punctatus*.

During the southern summer of 1988–89 zoologists from Victoria University of Wellington, New Zealand, began a comprehensive survey of the genetic and morphological variation in tuatara. Expeditions have been made to 24 of the 30 islands on which tuatara populations are thought to survive. Six islands have not been visited because of the extreme rarity of tuatara or

difficulties of access. From blood samples taken from animals, three distinct groups of tuatara have been identified. Most importantly, the population living on North Brother Island in Cook Strait, was distinguished as a separate species, *S. guntheri*, a name first proposed in 1877. Among other populations, a northern group and a western Cook Strait group were distinguished. Further, analysis of seven body characters (for example, body length, length of longest neck spine) showed that the same three groups could be differentiated morphologically, although considerable overlap occurred. Western Cook Strait and northern populations were thought sufficiently distinctive from each other to warrant assigning them subspecific status within *S. punctatus*.

▲ An adult male of the North Brother Island tuatara population, Sphenodon guntheri. Tuatara have long been treated as a single species, but recent research has demonstrated the existence of two species, very similar in appearance but differing substantially in genetic structure.

cultivation and grazing. Lighthouse stations (now automated) exist on some. The major factor common to all is the presence of breeding populations of burrow-nesting seabirds (petrels and shearwaters) which, in part through the influence they exert over soils and vegetation, influence the tuatara.

Many tuatara islands support several species of petrels and shearwaters, but islands where tuatara still maintain their numbers are rodent-free and have very large breeding populations of the fairy prion *Pachyptila turtur* and/or the diving petrel *Pelecanoides urinatrix*. These small petrels provide tuatara with many benefits, principally housing and food. They excavate burrows in which many tuatara live (although the reptiles also dig their own), and by turning over the soil and incorporating their mineral-rich guano they create conditions that may increase production of ground-dwelling invertebrates which form the bulk of the tuatara diet. Tuatara also eat lizards, the adults, chicks, and eggs of small petrels, and even their own young.

FEEDING AND BREEDING

Tuatara are active mostly at night and spend the day in burrows or, if it is sunny, basking at burrow entrances. Only one tuatara at a time has been found in each burrow. As with reptiles generally, the number of tuatara abroad at night is largely dependent upon environmental temperature. Maximum activity occurs between 17° and 20°C (63° and 68°F)—low for a reptile— then declines steadily as temperature falls, becoming negligible below 7°C (45°F). Other weather variables can also affect activity. For instance, tuatara slowly dehydrate unless they have access to water. There are no ponds or streams on many islands where they live, so after prolonged dry spells the reptiles are particularly active on rainy nights, when they may frequently be found soaking in puddles.

When hunting, tuatara usually adopt a sit-and-wait strategy. Prey moving in their vicinity is seized—smaller items with the tongue, and larger items by being impaled on the incisor-like teeth. Small, hard items such as beetles are crushed and dismembered, and larger prey are literally carved up. On each side of the mouth the tuatara has two rows of teeth on the upper jaw and one row on the lower jaw; the lower row fits between the two upper rows, and when chewing, the lower jaw is moved back and forth in a sawing motion, akin to the action of a carving knife.

Courtship and mating take place in late summer–autumn (January–March). During January males establish territories, each including the territories of several females. They display at conspicuous locations within their territories, presumably to attract females and to advertise their presence to other males. This involves

Michael Schneider/New Zealand Geographic

▲ *Tuatara are essentially nocturnal, but often sunbathe outside their burrows by day.*

increasing the size of the trunk and throat by inflating the body and elevating the crest and the spines.

Territorial encounters between males take the form of "face-off" displays and fights. In face-off displays, two males line up side by side about a meter (3 feet) apart, often with their heads facing in opposite directions, then slowly gape open their mouths and rapidly snap them shut. One male may be chased away, or a fight may ensue. During fights, one male locks his jaws around the head or neck of the other male, and the two scuffle on the ground, often emitting croaking calls at the same time. Facial wounds, broken jaws, and lost tails are not uncommon.

If a female is attracted, a displaying male will circle her using a slow, intermittent, stiff-legged walk. This may continue for about 20 minutes and ends when she either disappears down a burrow or allows the male to mount her. The male mounts the female from behind and moves to lie on top of her flattened body, using his hind legs to lift her tail so that his vent can come into contact with hers for transfer of sperm.

Eggs are laid in the following spring or early summer (October–December). Females gather at sunny, unforested sites to nest (the soil in forests being too cold for successful incubation). A female may first dig chambers or blind tunnels as trial nests, but eventually lays her clutch of up to

19 soft-shelled eggs (generally seven to ten) in one nest during a single night. Several more nights are spent filling the nest entrance with soil. She may return to her nest for eight consecutive nights, even remaining alongside it by day. This is because other females often dig up nests already containing eggs, then lay their own eggs at the same site. When at their nests, females may defend them vigorously from other females, such encounters being every bit as intense as those between territorial males. On average, females produce eggs just once every four years; males have an annual reproductive cycle.

Tuatara have one of the longest incubation periods of any reptile—12 to 15 months, the actual time depending on temperatures during this period. Unlike eggs of most reptiles, those of the tuatara temporarily cease development during winter, when conditions can be quite cool.

As in crocodilians and turtles, young tuatara are born with a horny "tooth" on the tip of their snout, which they use to cut their way out of the parchment-like egg shell. Upon emergence they average 5 grams (¼ ounce) in weight and 100 millimeters (4 inches) in length. They retain a remnant of the allantoic sac, which shrivels and falls off within a few days; the horny tooth falls off after about two weeks. Young are normally a cryptic fawn and brown color. They can move very quickly and are not often seen. They are

▼ Mating tuatara Sphenodon punctatus. They mate annually, in February and March, but females lay eggs only every four years or so. Unlike most reptiles, the male tuatara has no penis, and copulation depends on direct contact between the partners' cloacae. No pair bond is formed; the two remain together for about an hour, then go their separate ways.

Michael Schneider/New Zealand Geographic

PRESERVATION OF A "LIVING FOSSIL"

Tuatara live only in New Zealand. They were once widespread over the two main islands, but during the past 150 years have become extinct there, as well as on at least 10 offshore islands. Habitat destruction and predation by cats, rats, pigs, and other mammals introduced by Maori and European settlers are probably the main causes of extinction.

While about 30 populations are thought to survive, eight are vulnerable or endangered because of the presence of an introduced rat (the kiore, *Rattus exulans*). During recent surveys, no juvenile tuatara were found on seven of the eight rat-inhabited islands. Although we don't yet have experimental evidence, kiore probably prey on tuatara eggs and juveniles, and compete for food with juveniles and adults.

Ten years ago eradication of introduced rodents from offshore islands seemed an impossibility, but recent success with islands of up to 200 hectares (500 acres) has changed that view. In a typical rodent eradication program, baits containing an anticoagulant poison are placed systematically over an island. Poisoning is carried out in spring, when the number of rats is low and there is little food available. Monitoring continues for at least a year after the poisoning before the island is considered rat-free. Tuatara are removed before poison is laid, and any progeny raised in captivity can be released on the island once it has been made free of rats.

Few attempts to breed tuatara in captivity have been successful, although recently, significant advances have been made. Enhanced knowledge of the extensive behavioral repertoire of tuatara suggests that social interactions should be encouraged in captivity (previously adult tuatara

Brett Robertson

were often maintained in single pairs). We now know how to incubate eggs artificially, and that eggs can be induced from females ready to lay by administering a hormone called oxytocin. Using these techniques, eggs have been collected from wild females of the North Brother Island tuatara. This species occurs on only one small, 4-hectare (10-acre) island, which supports fewer than 300 adults. The eggs were successfully incubated in the laboratory, and the hatchlings will be used to establish a second population of this vulnerable species.

▲ There is no parental care in tuatara. Females lay a clutch of (usually) 7 to 10 eggs in underground nests that are then abandoned. After 12 to 15 months of incubation, the young tuatara hatches, using a special egg-tooth on the snout, visible on this youngster, Sphenodon guntheri, to tear open the eggshell. It then burrows its way to the surface. Tuatara eggs have been incubated successfully in research laboratories.

most active during the day—possibly a strategy made necessary by the cannibalistic tendencies of the larger adults.

As with the duration of incubation, the rate of growth of tuatara is largely dependent on temperature. The warmer the conditions, the faster the growth. At the southern part of their range animals mature at between 11 and 13 years, and to the north about two years earlier. Sexually mature tuatara may continue to grow for a further 20 years or more; on Stephens Island most animals reach full size at somewhere between 25 and 35 years of age. A number of Stephens Island adults recaptured after 30 years showed no appreciable growth, so the life-span of the species must be at least 60 years. One animal, collected as an adult, is known to have survived in captivity for more than 70 years.

Michael Schneider/New Zealand Geographic

◀ A tuatara confronts a snack. Tuatara will eat almost anything that moves, including the eggs, nestlings, and adults of many seabirds, but their normal diet consists mainly of large terrestrial invertebrates such this weta, one of a group of ground crickets found commonly in New Zealand.

SIZE
Smallest Cuvier's dwarf caiman *Paleosuchus palpebrosus*, males about 1.5 meters (5 feet) total length; females about 1.2 meters (4 feet).
Largest Indo-Pacific or saltwater crocodile *Crocodylus porosus*, males about 7 meters (23 feet) total length, weight more than 1,000 kilograms (1 ton); females rarely reach 4 meters (13 feet) total length.

CONSERVATION WATCH
The following species are listed in the IUCN Red List of threatened animals: Chinese alligator *Alligator sinensis*, spectacled caiman *Caiman crocodilus*, broad-nosed caiman *Caiman latirostris*, black caiman *Melanosuchus niger*, American crocodile *Crocodylus acutus*, African slender-snouted crocodile *C. cataphractus*, Orinoco crocodile *C. intermedius*, Johnston's crocodile *C. johnsoni*, Philippines crocodile *C. mindorensis*, Morelet's crocodile *C. moreletii*, Nile crocodile *C. niloticus*, New Guinea crocodile *C. novaeguineae*, mugger *C. palustris*, Indo-Pacific or saltwater crocodile *C. porosus*, Cuban crocodile *C. rhombifer*, Siamese crocodile *C. siamensis*, dwarf crocodile *Osteolaemus tetraspis*, false gharial *Tomistoma schlegelii* and gharial *Gavialis gangeticus*.

CROCODILES AND ALLIGATORS

WILLIAM E. MAGNUSSON

The order Crocodilia includes the largest of the living reptiles, and they are among the largest of the vertebrates that still venture onto land. During the Mesozoic era 245 to 65 million years ago, the Archosauria ("ruling reptiles") dominated the land, but the only giant archosaurs that have survived to modern times are the crocodilians. The dinosaurs left other descendants—the birds—but it is only the crocodilians (crocodiles, alligators, caimans, and gharials) and a few turtles that today reflect the majesty of the time when giant reptiles ruled the Earth.

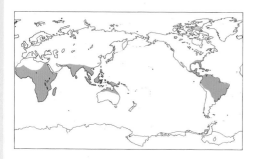

PREDATORS IN TROPICAL RIVERS
Most of the 23 or so species of crocodilians occur in tropical parts of the world, but a few, notably the American and Chinese alligators, extend into temperate regions. All species are semi-aquatic and do not venture far from estuaries, swamps, lakes, streams, or rivers. Their generally large size and short legs mean that they are agile on land only over very short distances. In the water, however, they are superbly adapted predators.

The external characteristics that differentiate crocodilians from other reptiles are mainly adaptations to a semi-aquatic lifestyle. For example, the elongate head and numerous peg-like teeth are characteristic of many aquatic animals with backbones, from fish to dolphins. Such heads are probably too ungainly for purely land-dwelling creatures. The eyes, with their transparent protective third eyelids (nictitating membrane), and the nostrils, with their watertight valves, are set high on the head so that these remain at the surface while the rest of the animal is completely submerged. Most swimming is done by strokes of the powerful blade-like tail.

On land, crocodilians can lift their bodies high off the ground and walk with their legs positioned almost directly below the body, as does a mammal. However, they cannot maintain this position at speed, and most species revert to a typically reptilian splayed-leg belly slide if hard-pressed. One species, Johnston's crocodile *Crocodylus johnsoni*, gallops over short distances to

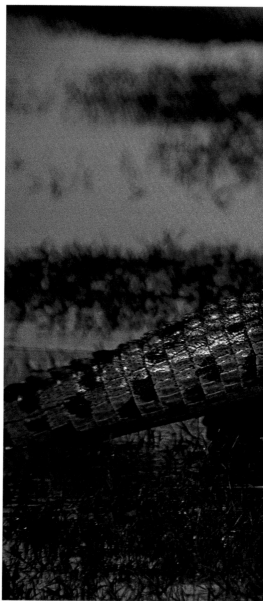

escape back to the water, sometimes with all four limbs off the ground at the same time.

Internally, crocodilians have some features that are more similar to those of mammals than those of other reptiles. Crocodilians have a bony secondary palate. This, combined with a fleshy valve at the base of the tongue, completely isolates the respiratory system from the mouth, allowing them to open their mouths underwater without risk of drowning. The intestinal cavity is separated from the pectoral cavity by a muscular septum, analagous to the diaphragm of mammals, presumably increasing the efficiency of ventilation of the lungs. The heart has four separate chambers, but crocodilians have retained an

aperture, the foramen of Panizza, between the left and right ventricles and are able to shunt blood to the regions of the body that most require it while warming in the sun or diving.

Three families (or subfamilies) of living crocodilians are recognized by taxonomists. Although their lineages have been distinct for tens of millions of years, differences between the families are few.

THE ALLIGATOR FAMILY

Alligators and caimans are classified in the family Alligatoridae and are therefore known collectively as alligatorids. In all alligatorids the teeth of the lower jaw fit into pits in the upper jaw and cannot

▼ *Lord of the waterways: the Nile crocodile* Crocodylus niloticus. *Although agile only for short distances on land, there are few more dangerous or formidable predators than the crocodile, even to creatures much bigger than humans. Adult male Nile crocodiles sometimes exceed 1 tonne (1 ton) in weight and 5 meters (over 16 feet) in length when fully grown.*

Jen and Des Bartlett/Bruce Coleman Limited

▲ *Restricted to the southeastern United States, the American alligator Alligator mississippiensis is one of only a few crocodilians whose range extends well beyond the tropics. Its blunt, broad jaws help it catch prey in thick vegetation.*

be seen when the jaws are closed. The scales on the underside of the body have no sensory pits.

The two species of true alligators occur in widely separated regions of the world. The American alligator *Alligator mississippiensis*, which grows to 6 meters (20 feet) in length, occurs in the southeastern United States. The Chinese alligator *A. sinensis* inhabits the lower Yangtze River and its tributaries in China; it rarely exceeds 2 meters (6½ feet). Both species have broad heavy heads, which are presumably adaptations to living in heavily vegetated swamps. In open air or in water, thin snouts would be more maneuverable, but a heavy head has more momentum to catch prey by smashing through intervening vegetation. When the ponds and swamps inhabited by the American alligator freeze over, as they do occasionally, most of the larger animals survive by lying in shallow water and maintaining a breathing hole through the ice. Occasionally their snouts are frozen into the ice, but provided their nostrils are not covered, they usually survive. Most other species of crocodilians would be killed by such low temperatures.

Alligatorids in South and Central America are all called caimans. The largest is the black caiman *Melanosuchus niger*, which can grow to more than 6 meters (20 feet). Biochemical evidence indicates that it is more closely related to other caimans

than to alligators, but externally it is similar to the American alligator. Unlike most crocodilians, which may be vividly marked as hatchlings but soon assume the drab colors of adults, the black caiman remains distinctively marked throughout life. Hatchlings have light gray heads and black bodies patterned with lines of white dots. As they grow, the gray on the head becomes brown and the lines of white dots fade, but even adults longer than 5 meters (16½ feet) are more colorful than the hatchlings of most other species. Black caimans inhabit flooded forests and grassy lakes around slow-flowing rivers throughout the Amazon basin in Brazil and to the north in the coastal rivers of the Brazilian state of Amapa, in French Guiana and in Guyana. They were once the most commonly seen caimans throughout their range, but hunting (for their skins) has drastically reduced their numbers. Part of the problem seems to be that the black caiman occurs in the same areas as the common caiman, and that this species sustains hide hunting in areas long after hunting for the black caiman alone would have become uneconomic.

The genus *Caiman* has two species: the common caiman *C. crocodilus*, which occurs from southern Mexico to northern Argentina; and the broad-snouted caiman *C. latirostris*, which occurs in streams and marshes along the coast of Brazil from the state of Rio Grande do Norte south to Uruguay, as well as in the inland basins of the São Francisco, Doce, Paraíba, Paraná, and Paraguay rivers in Brazil, Paraguay, and Argentina. Both species are extremely adaptable and can be found in a variety of habitats, including large rivers in tropical rainforest, seasonally flooded savannas, permanent swamps, and mangroves. The common caiman is by far the most hunted crocodilian in the world, accounting for 60 to 80 percent of the skins in trade. However, its bony skin is of comparatively low value, and a hide of this species typically attains only about a tenth of the price paid for *Alligator* or *Crocodylus* skins.

Dwarf caimans of the genus *Paleosuchus* are among the least studied of the crocodilians. Schneider's dwarf caiman *P. trigonatus* occurs throughout the forested areas of the Amazon and Orinoco basins and the coastal rivers between them; it is rarely found away from thick tropical rainforest. Cuvier's dwarf caiman *P. palpebrosus* occurs throughout the same region but generally lives in more open habitats such as gallery forest in savannas, or the margins of large lakes and rivers. It also extends across the high plains of the Brazilian shield, south to the state of São Paulo.

Dwarf caimans are sometimes called smooth-fronted caimans because they lack the bony ridge between the eyes that is typical of *Caiman* and *Melanosuchus*. Their skins are heavily ossified—a build-up of calcium carbonate and calcium phosphate in the connective tissue—and are much less flexible than those of most other crocodilians. Even their eyelids are heavily ossified, and they have been compared to turtles because of this dermal armor. Whereas Schneider's dwarf caiman has a fairly typical head shape, the skull of Cuvier's dwarf caiman is high, smooth, and dog-like. There are as yet no convincing explanations as to why it should have

evolved this unique skull shape. Both species are small. The total length of Schneider's dwarf caiman is about 1.7 meters (5½ feet) for an adult male and 1.3 meters (4¼ feet) for a female. Cuvier's dwarf caiman is probably the world's smallest crocodilian, at about 1.5 meters (5 feet) for an adult male and 1.2 meters (4 feet) for a female. Both species have rich brown eyes, unlike the yellowish eyes of most other crocodilians.

THE CROCODILE FAMILY
Members of the family Crocodylidae have a notch in the upper jaw on each side which accommodates the fourth tooth in the lower jaw when the mouth is closed; this tooth is therefore

Michael and Patricia Fogden

▲ Lacking extensive osteoderms (bony buttons) in the belly scales, the black caiman Melanosuchus niger of Amazonia yields a so-called "classic" skin, much in demand for the manufacture of luxury items like handbags. Intense hunting during the 1950s and 1960s quickly reduced the species to critically low population levels.

▼ Cuvier's dwarf caiman Paleosuchus palpebrosus is the smallest and most heavily armored crocodilian, with even its eyelids protected by bony plates.

▲ A group of Nile crocodiles basking on a river bank in Kenya.

Crocodylids possess salt glands on the tongue, which can be used to excrete excess salt in saline environments. Even species that live in totally fresh water have functional salt glands, and it is thought that the tolerance of crocodylids for salt water may explain their wide distribution throughout the tropics.

The genus *Crocodylus* includes 13 fairly similar species. The last was named in 1990, and in the near future another species of the Indo-Pacific region will probably be reclassified as two separate species. All are moderate to very large, with narrow snouts. None has a broad snout like an alligator or a snout as thin as that of a gharial.

The Indo-Pacific crocodile *C. porosus* ranges from India to northern Australia and the Solomon Islands. The presence of this very large species in estuaries and mouths of large rivers may serve to prevent regular contact between the six or more freshwater species that occur on the major land-masses in the region. This species is probably the largest of the crocodilians, and lengths of more than 7 meters (23 feet) and weights of more than 1,000 kilograms (almost 1 ton) have been recorded.

In the New World the American crocodile *C. acutus* seems to have a similar role; it occurs in estuaries from the southern tip of Florida in the United States to Colombia and Venezuela. Within its range, three other crocodiles occur in freshwater

always visible. The scales on the underside of the body have sensory pits, seen as a small dot in the middle of each scale, even in tanned skins.

Stan Osolinski/Oxford Scientific Films

SALT SOLUTIONS

Like many vertebrates living in salt water, crocodiles must deal with extremes: too much salt and too little fresh water in their environment. In order to maintain the correct salt balance in their body fluids, they must excrete the excess salt they absorb directly through their skin and that they ingest while eating. However, reptiles cannot excrete sufficient amounts of salt unless their kidneys are flushed by large amounts of fresh water—a major hurdle for an animal living in an estuarine environment.

Crocodiles and other marine reptiles have overcome this problem by having glands capable of excreting high concentrations of salt without the loss of valuable water. The salt glands of marine turtles are modified tear glands, while the Galapagos Island's marine iguana expels a salty solution from glands in its nostrils. Crocodylids have salt glands on their tongue. These glands, which are actually modified salivary glands, secrete drops of salty solution. The crocodile's mouth is sealed off from its internal organs by a fleshy cartilaginous valve at the back of the throat, effectively making the tongue part of the external body surface.

This special method of excretion affords

crocodylids a high tolerance to salt water, and may explain their extensive distribution throughout the tropics. The lack of salt glands in the American alligator and in caimans (both members of the family Alligatoridae) may be the reason why these species have never successfully invaded salt water.

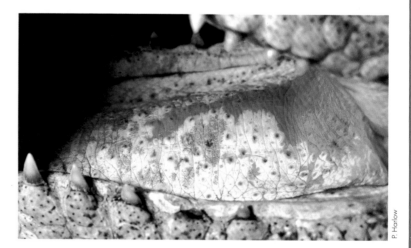

P. Harlow

▲ A close-up of a crocodile's tongue shows the scattered, pore-like openings of the salt glands. These modified salivary glands excrete excess salt as a highly saline fluid to maintain an appropriate salt/water balance in the crocodile's body.

habitats: Morelet's crocodile *C. moreletii* in Mexico, Belize, and Guatemala; the Cuban crocodile *C. rhombifer* in Cuba; and the Orinoco crocodile *C. intermedius* in Venezuela and Colombia.

The wide-ranging Nile crocodile *C. niloticus* occurs throughout Africa (except for the waterless areas of the Sahara) and in Madagascar; this species occurs in estuaries, and in the interior of the continent it also inhabits rivers, swamps, and lakes. The only other member of the genus that occurs on the continent, the African slender-snouted crocodile *C. cataphractus*, is restricted to heavily forested areas of West and Central Africa. The ecological separation of this species and the Nile crocodile has not been studied.

The only crocodylid that does not belong to the genus *Crocodylus* is the dwarf crocodile *Osteolaemus tetraspis*. This tiny species is superficially similar to the dwarf caimans, and possibly its ecology in the heavily forested areas of West Africa is similar to that of the dwarf caimans in South America's Amazon basin. Virtually nothing is known of its life history or population dynamics.

THE GHARIAL FAMILY

The family Gavialidae contains two species of extremely thin-snouted crocodilians. Both are large. The false gharial *Tomistoma schlegelii* grows to lengths in excess of 4 meters (13 feet) and for the gharial *Gavialis gangeticus* lengths of of 6.5 meters (21 feet) have been recorded. The false gharial is classified in this family mainly from biochemical evidence; some authorities place it in the family Crocodylidae.

The extremely thin snouts of gharials are generally considered to be adaptations to a diet of fish, and it is unlikely that the snout of the true gharial could support the stresses involved in the capture of mammals as large as those taken by other crocodilians of similar size. Very thin snouts are also probably ineffective in habitats with thick obstructive vegetation.

The false gharial occurs on the Malay Peninsula of Thailand and Malaysia, and on the islands of Sumatra, Borneo, Java, and possibly Sulawesi (Celebes). It is distinctly marked with dark blotches, and adults are almost as colorful as juveniles. It is found in freshwater swamps, lakes, and rivers, but no detailed studies have been made of its ecology. The true gharial is one of the most distinctive crocodilians. The head with its thin snout and the weak legs appear disproportionately small in relation to the large body. Adult males develop a large fleshy knob on the tip of the snout that apparently serves to modify the sounds they make during social interactions. It spends more time in the water than most other crocodilians and frequents relatively fast-flowing reaches of the Indus, Bhima, Mahanadi, Ganges, Brahmaputra, Kaladan, and Irrawaddy rivers in Pakistan, India, Nepal, Bhutan,

▲ *Long thought to resemble the Indian gharial through convergent evolution only, recent biochemical studies have placed the Malayan or false gharial* Tomistoma schlegelii *in the same family (Gavialidae).*

▼ *The gharial* Gavialis gangeticus *of northern India is primarily a fish-eater, as suggested by its long slender snout. As males are usually characterized by a conspicuous knob at the tip of the snout, this is probably a female or subadult male. This species may grow to more than 6 meters (19 ¾ feet) in length.*

Jean-Paul Ferrero/AUSCAPE International

667

Bangladesh, and Burma. Its habitat has a high human population density, and net-fishing, dams, and egg collecting have seriously depleted populations. Fortunately, an intensive management and restocking program by the Indian government seems to be bringing the species back from the edge of extinction.

FOOD

Crocodilians eat a wide range of foods and generally adapt to what is most available in their habitat. Hatchlings eat mainly insects and other small invertebrates, then as they grow they eat more shrimp, snails, or crabs, depending on availability. Most species generally change over to fish as they approach maturity, but they will also take mammals, birds, and other vertebrates that live around the water's edge. Large adults may specialize on mammals, and Nile crocodiles have been seen to kill full-grown Cape buffaloes. Despite popular opinion, cannibalism is relatively rare among crocodilians and can usually be traced to social rather than nutritional causes.

Although they are master hunters in the aquatic arena, their large bodies are not very maneuverable on land, and potential prey are generally safe if more than about 4 meters (13 feet) from the water's edge. Humans are particularly clumsy mammals which spend a lot of time around the water, but fortunately for our species, most crocodilians do not usually regard us as suitable food items. Large crocodilians such as the American alligator, the black caiman, and the Indian mugger occasionally kill humans in defense of nests or territories, or for food, but such attacks are very rare considering the frequency of associations between these species

and high densities of people. In contrast, two species, the Indo-Pacific crocodile and the Nile crocodile, habitually hunt humans, and extreme care is necessary when near the water within the ranges of these species.

REPRODUCTION

All crocodilians lay eggs. The clutches of the smallest species average 10 to 15 eggs, while the Indo-Pacific crocodile may lay more than 50. All of the alligatorids, the false gharial and most crocodiles make nests of vegetation, soil, and debris, scraped up into a mound. The gharial and five species of crocodiles bury their eggs in holes in sandy beaches or other friable soil. The American crocodile sometimes makes mounds of sand in which it buries its eggs instead of burying them directly in the beach.

Most species do not simply abandon their eggs after laying. The mother, and sometimes the father, remain in the vicinity of the nest to defend it against predators. If the predator is a human or a bear, this nest defense can result in the death of the mother. Incubation takes 60 to 100 days, depending on the species and the incubation temperature, then the young crocodiles pip through the eggs and begin to call within the nest. In response to this, the adult will scrape a hole in the nest to release the babies. Often, roots or termite workings will have completely encased the eggs, so help from a parent is necessary for the babies to escape from the nest. Once the hatchlings are free of the nest, the adult will gently pick them up in its mouth and carry them to water. This incredibly delicate feat by a mother that weighs thousands of times more than her offspring was not generally believed possible until

M.P. Price/Bruce Coleman Limited

▲ *Baby gharials hatching in Nepal. A solicitous parent, the female gharial buries her clutch of 30 to 50 eggs in a cavity dug in sand, then remains in the vicinity until they hatch 60 to 80 days later, digging out the nest and helping the young to break free of the eggshells.*

▼ *An Orinoco crocodile Crocodylus intermedius swallows a capybara, a pig-sized South American rodent. Rare and little-known, the Orinoco crocodile has a restricted distribution in Venezuela and Colombia.*

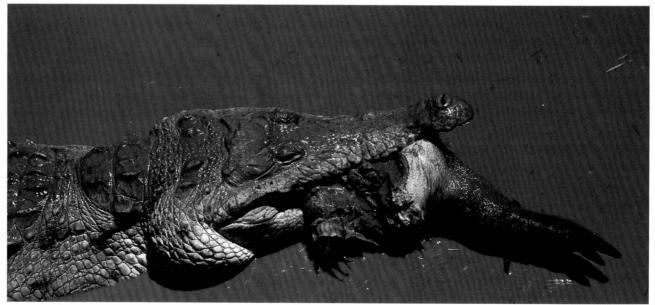

David MacDonald/Oxford Scientific Films

filmed in the 1970s. A Nile crocodile has even been seen to pick up an unhatched egg from which the baby was calling and then roll it between the tongue and the roof of her mouth until the shell broke. The mother gently lowered the baby into the water before swallowing the empty egg shell.

Whether mounds of vegetation or holes in beaches, nests of crocodilians function as precise incubators. In the case of hole nests, the overburden of soil acts as a heat reservoir which dampens out the extreme temperature fluctuations at the surface. In mound nests, the temperature is maintained not only by the sun's heat but also from decaying vegetable material in the mound. In all nests, metabolic heat generated by the embryos raises the temperature by several degrees in the latter part of incubation. Whether they live in Himalayan mountain streams, or tropical jungles, or temperate swamps, all species of crocodilians incubate their eggs at about 30°C (86°F). Prolonged exposure to temperatures below about 27°C (81°F) or above about 34°C (93°F) kills embryos of most species.

The temperature at which the eggs are incubated is not only important for the survival of the embryos; it also determines the sex of the offspring. Unlike birds and mammals, the sex of whose embryos are determined at the moment of fertilization, the embryo in a newly laid crocodilian egg has no gender. The temperature at which the egg is incubated during the first few weeks determines whether the embryo will develop into a female or a male.

In the American alligator, Schneider's dwarf caiman, and probably the common caiman, temperatures less than about 31°C (88°F) produce females, and temperatures higher than about 32°C (90°F) produce males; intermediate temperatures produce both sexes. In some crocodylids, low temperatures produce all females, and intermediate temperatures, 31° to 33°C (88° to 91°F) produce males and females; however, higher temperatures again produce mainly females. The temperature of incubation has also been shown to affect color patterns and preferred temperatures of the offspring.

It has been suggested that the link between sex and temperature may be adaptive because of differences between the sexes in mating systems. Males of most species of crocodilians have strong social hierarchies, and it is probably only the largest that get to breed. It is therefore wasteful for a female to produce male offspring at less than optimal temperatures because this will result in low rates of growth after hatching. In contrast, all females will probably have a chance to breed, irrespective of their final size. It therefore makes sense to produce males at optimal temperatures for growth and females at other temperatures. While this theory is plausible, much more

Edward Robinson/Oxford Scientific Films

research needs to be done on the implications of temperature-dependent sex determination.

SOCIAL BEHAVIOR

Crocodilian social behavior is more complex than that reported for any other reptiles. Several researchers have seen what may be cooperative hunting by crocodilians. However, the interpretation of these observations is difficult because the motivation for the behavior may have been merely a tendency to group together at a food resource while maintaining a minimum personal distance (the studies were made in the wild). In contrast, social behavior during reproduction can easily be studied in captivity, and the motivation for the behavior is more obvious. Each species varies in its degree of sociality. Some, such as Schneider's dwarf caiman, live in territories from which they generally exclude adult members of their own sex. Others, such as some populations of the Nile crocodile, undertake seasonal migrations to nesting beaches where each male sets up a temporary territory for his harem of breeding females.

A crocodilian's jaws are terrible weapons, and adults are quite capable of killing each other. However, this rarely happens. The loser in a stand-off can indicate submission by vocalizations if it is a juvenile, or by lifting its head vertically to expose its throat. It is usually then allowed to slink off without further aggression. Bites, when they occur, tend to be on the base of the tail, and not on the relatively vulnerable chest region.

▲ Throat bulging, an American alligator bellows at its neighbors in Everglades National Park, Florida. Perhaps the noisiest of the crocodilians, both sexes of the American alligator have an extraordinarily varied vocabulary. Territorial males utter loud bellowing notes, easily audible at 150 meters (nearly 170 yards), and repeated about every ten seconds; neighbors respond in choruses that may last half an hour or more. Adults also utter a range of grunting and hissing sounds, and courting couples exchange quiet cough-like notes.

David Curl

▲ A crèche of Indo-Pacific crocodiles. For the first few weeks after hatching, baby crocodiles remain in groups close to the water's edge, guarded by the mother and sometimes other adults, feeding on insects and other small aquatic life, and diving instantly at any sign of danger.

Social signals include body postures such as the submission signal, head slaps on the water, grunts and roars that are audible to human listeners, and subaudible low-frequency sounds for communication under water. Most of these have been studied only in relation to reproduction, but they may also have more subtle everyday uses. Adults and subadults will respond to the distress calls of juveniles and defend them against predators whether they are related or not. In one instance, three researchers had their inflatable boat sunk by an adult Schneider's dwarf caiman when they imitated the distress calls of a hatchling common caiman. Hatchlings will often stay together in a group, accompanied by the mother for periods ranging from weeks to years, depending on the habitat and the species. During this period the hatchlings communicate among themselves and with the mother by means of audible grunts.

ECONOMIC EXPLOITATION

Crocodilian leather is fashionable, and products made from crocodilian skins fetch high prices. The most sought-after leather is from "classic skins" such as the Indo-Pacific crocodile, the American alligator, and the Nile crocodile. Belly skins of these species have few osteoderms (bony buttons below each scale), and the surface polishes to a glossy sheen. However, in recent years the increasing scarcity of classic skins has led to the use of skins of the common caiman. Wild specimens of this species have thick osteoderms, and the leather cannot be given a glossy sheen unless coated artificially.

Unrestricted hunting during the 1950s and 60s reduced most species with classic skins to close to the point of economic extinction. National and international legislation was enacted during the

late 1960s and early 70s by many producer countries, and the focus moved to conservation. The success of those conservation actions is reflected in the fact that controlled economic exploitation is now allowed in most of the countries that initially enacted complete protection for their crocodilians. The countries that led the way to controlled economic exploitation were Papua New Guinea for the Indo-Pacific crocodile, the United States for the American alligator, Zimbabwe for the Nile crocodile and Venezuela for the common caiman. The exact form of management varies from country to country but most have a mix of the following three strategies.

Hunting

Adult and subadult crocodilians are hunted in the wild, but there is usually some legal limit on the the methods that can be used and the size that can be taken. In Papua New Guinea, only crocodiles above a certain size limit are protected. The smallest individuals are not hunted because the skins are too small for the trade. Thus the wild breeding stock of large, old crocodiles is protected and only an intermediate size group exploited. Presumably enough of these escape to grow and replenish the older age group. Populations of crocodilians that have been subjected to size-selective harvesting in Papua New Guinea, the United States, and Venezuela seem to be doing quite well. Although it might appear that hunting is a bad thing for a species, this is not necessarily so; crocodilians have shown themselves resistant to intense persecution if their habitats are maintained intact. In developing countries one of the most compelling reasons to maintain habitats is that the local people depend on these habitats economically. Hunting maintains the economic incentive and, paradoxically, in many areas is seen as an important conservation tool.

Ranching

Crocodilian ranching involves the collection of eggs or hatchlings (occasionally juveniles) in the wild for rearing in captivity. Growth rates in captivity, with the appropriate temperature and feeding regimes, are far higher than those attained in the wild, and the animals are usually slaughtered at between one and three years of age. Raising the animals in captivity usually results in better-quality, uniform-grade skins that achieve higher prices. It is also easier to control the conditions of slaughter for the production of the meat, which is increasingly becoming a restaurant delicacy. In many cases it is also easier for the official wildlife agency to monitor the exploitation because all killing takes place at the rearing stations. The collection of eggs or hatchlings in the wild also gives people an economic incentive to maintain natural habitats.

Farming

The breeding and raising of crocodilians on closed-cycle farms is the least-used method of exploiting crocodilians, but it is likely to become more important in the future as natural habitats are destroyed. It is potentially possible to select individuals for domesticated races of crocodilians that produce more eggs in captivity, grow faster, and have more fashionable skins. Research along these lines is being undertaken in the United States and Zimbabwe. Such activities are essentially irrelevant to the conservation of natural species and their habitats. However, farming has recently shown that it is not always neutral in terms of conservation. The shifting of species of crocodilians around the world for farming operations poses a grave threat to many aquatic systems. Should the exotic species escape, the local crocodilians may be wiped out by exotic diseases or through competition and predation by the introduced species. Many aquatic animals that have evolved to coexist with one species of crocodilian may not be able to survive if another is suddenly thrust upon them. It would be ironic if captive breeding, which was instrumental in the recovery of some species of crocodilians, should change the status of these species from threatened to threatening.

LOCOMOTION

Crocodiles swim with lateral strokes of the tail which propel the body slowly but apparently effortlessly through the water. This type of locomotion must be very efficient, because some individuals have been seen at sea hundreds of kilometers from land. Crocodiles swim fast only to escape from danger or to pursue prey. The tail is capable of propelling the whole body vertically out of the water to snap prey from overhanging branches or to power the thrust as a large crocodile explodes out of the water to take a mammal that has ventured too close to the bank.

On land a crocodilian's tail is useless, and it must rely on its short legs for locomotion. Although superficially lizard-like, the body is too heavy to be pushed along the ground in a normal splayed-leg lizard-run for any distance. Crocodilians generally "belly slide" only if they are on mud or some other slippery surface, or if hard-pressed by a predator.

For long-distance or leisurely movements over dry land, crocodilians use a "high walk" similar to that of a walking mammal, with the legs positioned almost under the body and the belly held well off the ground. Some species, such as the common caiman *Caiman crocodilus*, regularly cover long distances over land. However, movement is slow, and although they may eat carrion it is unlikely that they can catch prey away from the water.

The most spectacular crocodilian gait is that of Johnston's crocodile *Crocodylus johnsoni*. To escape back to water over short distances, it gallops using the same stride sequence as a horse at high speed. The hind limbs propel the body forward and high off the ground. In fact, all four limbs may be off the ground at the same time. The forelimbs take the weight of the body as it lands, the body is flexed, and the hind limbs are swung forward to provide the next forward thrust. This method of escape can be used only by small crocodilians, and the belly slide is probably more effective unless the ground is firm or is covered by obstacles such as fallen logs.

John Carnemolla/Australian Picture Library

▲ As predators, crocodiles rely heavily on their terrifying ability to explode into sudden violent activity, completely unexpectedly—to humans at least—in such otherwise lethargic-seeming creatures. The sudden lunge of a crocodile can lift it almost entirely clear of the water.

◀ Except for brief charges, crocodiles are generally slow and relatively awkward on land. A conspicuous exception is Johnston's crocodile *Crocodylus johnsoni* of Australia, which is capable of a full gallop in startled sprints back to the safety of the water.

Gunther Deichmann/AUSCAPE International

Page numbers in *italics* refer to illustrations.

682